# THEORY OF GROUND WATER MOVEMENT

# THEORY OF GROUND WATER MOVEMENT

BY

P. YA. POLUBARINOVA-KOCHINA

TRANSLATED FROM THE RUSSIAN BY

J. M. ROGER DE WIEST

1962

PRINCETON UNIVERSITY PRESS

PRINCETON, NEW JERSEY

L. C. Card 62-12616

Princeton Legacy Library edition 2015
Paperback ISBN: 978-0-691-62538-6
Hardcover ISBN: 978-0-691-65183-5

Printed in the United States of America

## AUTHOR'S PREFACE TO THE RUSSIAN EDITION

The realization, in the Stalin period, of plans to construct gigantic hydraulic structures, created an upsurge of the interest in the theory of ground water flow among the scientific and engineering workers of the Soviet Union.

At the basis of the present book are lectures, which I gave several times at the State University of Moscow, named after M. V. Lomonosov, for the students of hydromechanics in the mechanical engineering-mathematical sciences department. The content of these lectures is considerably broadened in the book. The first part of the book is devoted to the steady movements of ground waters, for which extensive and widely varied mathematical investigations are available.

In the second part, questions of unsteady flows are treated, less developed at the present time, but having an important significance in forecasting the regimen of ground waters in the erection of hydraulic structures, the irrigation of plains, etc.

In the first chapter, an attempt is made to give a general presentation of the contemporary concepts of the physical properties of the system "water-soil-air," on which basis the theory of seepage must be built, but which are still little developed and almost not taken into account for the hydromechanic investigation into the theory of seepage.

The material of the book is essentially arranged after the methods of mathematical investigation and, in part, in view of lecturing. I wanted to give such an exposition of the mathematical methods of research into the hydromechanics of ground water flow, that the features, prevailing in this book, would make it possible to assimilate these methods and to be in a position to apply them in the solution of new seepage problems. However, as a result of the magnitude of the available material in this branch of hydromechanics, I had to restrict myself in the selection of problems. Since the author is not a specialist-practical worker, this selection is perhaps dictated by some subjective ideas, but also by the desire to give illustrative material for the mathematical methods.

The basic content of the book consists of an exposition of the theory of the flow of ground water, considered as an incompressible fluid in this case, when one may consider the law of seepage to be linear and may neglect the presence of air in the soil. Therefore, the paragraphs that touch upon other cases, are preceded by an asterisk. Also in such paragraphs are grouped the solutions to some problems, possessing an interest of principle, but still little used for application. Whenever it is possible, the problems are brought to numerical or graphical results, and it is indicated where data of calculations can be found, if, for reason of space economy, the latter are omitted.

At present, in the theory of seepage, there exist a number of questions of basic significance, for which a common viewpoint has not yet been agreed upon. So is there some question about the most general expressions of the equations of the theory of seepage. I do not touch upon them at all, even more so because these equations did not have time yet to find practical applications. The issues about the seepage forces, which operate on soil particles, and therefore on the base of hydraulic structures, remain questionable and subject to discussion. Due to the great importance of this matter, I present it and moreover, as mathematicians use to do, I consider that, although the exposed ideas undergo changes, the method of computation remains basically the same.

G. N. Polozjii added, to my instruction, his own analysis on problems along the calculus of variations and other matters, C. N. Numerov added extra research and solutions of his own problems, mentioned here in paragraph 4 and 7 of Chapter VI, and therefore I convey to both of them deep gratitude. G. K. Mikhailov undertook a vast task in the preparation of the book for printing; he carried out several extra calculations; moreover I took advantage of his advice, in particular in the first chapter; besides this, he wrote paragraph 12 of Chapter VII. Help of various kinds has been given to me by M. M. Semchinova and other collaboratrices of the Institute of Mechanics of the Academy of Sciences of the Soviet Union, and also by N. N. Kochina. I convey to all of them my sincere thanks.

The present book, in which for the first time extensive material about the theory of the flow of ground waters is brought together, has, without question, many shortcomings. In particular, the list of literature is incomplete. A rather detailed enumeration of works up to 1947 is added to the list [7] of literature in Chapter VII.

For all instructions and remarks about the book, I shall be grateful. Please send them to the address: Moscow, Orlikov per., 3, Gostekhizdat.

P. Polubarinova-Kochina.

## INTRODUCTION

The main content of the present book consists of an exposition of the mathematical theories of the flow of ground waters, the fundamentals of which are layed out in the works of N. E. Zjoukovsky and N. N. Pavlovsky.

We note that the fundamental investigations about the equations of motion of gaseous oil and gases in porous medium belong to L. C. Leibenzon. We do not touch upon this matter, essentially considering water as an incompressible liquid (what in special cases also is applied to oil).

The origin of the study of seepage phenomena in porous medium (sand) goes back to Darcy (1856), who carried out experiments about the flow of water in pipes filled with sand and established the steady law of this movement. Similar experiments later on were made by many investigators on a wide scale. Dupuit and Boussinesq gave the hydraulic theory of the flow of ground waters. F. Forchheimer developed the hydraulic theory of wells. The first investigations of N. E. Zjoukovsky about the flow of underground waters, theoretical and experimental, date back to 1888.

We indicate here the names of the first Russian investigators. Thus K. E. Lembke (1886) gave an approximate solution of several problems of unsteady flow of ground water, using for the first time a method, which now is called the method of successive change of steady states.

N. E. Zjoukovsky took benefit of the results of Lembke's observation for the first time in his own works related to the wells of Kostromsky province. A small amount of research in the field of underground water motion belongs to I. A. Evnevitsch (1890), famous for his own works in hydraulics. A significant number of Russian hydrotechnicians worked in the field of irrigation, closely related to seepage problems. The book by A. A. Krasnopolsky [5] about unconfined and artesian wells, played an important role for its time (1912). In this book were available the theoretical results valid at that time, but also the investigations of the same author about seepage in fissured rocks (nonlinear law of seepage).

We note that the problem of underground waters — their origin,their tie with sea waters - already attracted the attention of M. V. Lomonosov.

The theory of seepage reached its real and significant development only in Soviet times. The rigourous mathematical theory of the theory of the flow of ground waters under hydraulic structures was founded by N. N. Pavlovsky (1922). His investigations were continued and developed in the numerous works of his students and followers. It also belongs to Pavlovsky to have conducted the first investigations about the application of the method of inverses to the theory of unconfined flows. Simultaneously with them, research in this field was initiated by other authors (V. V. Vedernikov and others), continued and widely developed. Of great merit is Pavlovsky's application of the experimental method of the electro-hydromechanical analogs (Egda), which at present are widely used by the scientific research and design institutes. One ought also to mention the great organizational activities of N. N. Pavlovsky in the field of the development of scientific research applied to hydraulics.

The works of N. N. Pavlovsky attracted the attention of N. E. Zjoukovsky, who afresh began to devote himself to the theory of ground waterflow and who in 1920 gave his own method for the solution of problems in confined and unconfined flows. The latter method is at present widely used for the solution of many problems in the theory of seepage.

New mathematical methods, based upon the presently known results of the theory of functions and the equations of mathematical physics, have been developed in the Soviet Union in the past decade. The leading role played by Soviet scholars in the development of the mathematical theory of ground waterflow becomes evident from the content of the present book.

## Translator's Remarks

Mrs. P. Ya. Polubarinova- Kochina is best known to Western Scientists through her contribution "Theory of Filtration of Liquids in Porous Media," to the Advances in Applied Mechanics, Part II, pp. 153-225, edited by Richard von Mises and Theodore von Karman (Academic Press, Inc., New York, 1951).

In 1952, when she completed the present work, Theory of Ground Water Movement, she was a corresponding member of the Academy of Sciences of the U.S.S.R. and a lecturer in hydromechanics at the State University of Moscow. In this book, reference is made to over thirty of her original and significant contributions on the hydromechanics of porous media (groundwater and oil flow). Most of these are published in the Doklady of the Russian Academy of Sciences or in the Prikladnaya Matematiki i Mekhaniki (P.M.M.; Journal of Applied Mathematics and Mechanics).

At present, the author is attached to the Institute of Hydrodynamics, Novosibirsk and it was from there that, while apologizing for the "numerous errata," she kindly authorized me to translate her book. Unfortunately, we did not receive a list of errata to the original text, although we indeed corrected many – not all – mistakes in the process of translation.

One could hardly say that a ten-year-old book completely fills the existing need for an advanced work in flow through porous media; nor could one say that this book covers the entire field. But it is a classic, much like Morris Muskat's Flow of Homogeneous Fluids Through Porous Media, originally published in 1937 and translated into the Russian language some thirteen years later. The Russians have also translated the works of K. Terzaghi, D. W. Taylor, and many others, and have made considerable use of all of them. While reviewing papers submitted for publication in leading U.S. Journals, we became aware of the fact that authors at times unknowingly duplicated efforts in trying to solve problems already solved in this book or became involved in research problems for which the present book would have been a valuable tool. Before setting out on this translation we mailed out five sample chapters of the book to more than one hundred experts in the field. We received much encouragement from these people to go ahead with the project. These were reasons enough for making the work available in English.

We would like to emphasize that this translation is primarily intended as a partial answer to the real need for such works in this field. It is not a literary masterpiece: without question, certain aspects of the book could be more elegant, but at the cost of considerable additional time and production expense. It suffices to mention that, since copies of the

Russian edition are not available, we had only a photocopy of the original to work from and had to re-draw and hand-letter all the figures. In Chapters I, II, IX, XI, XII, and XIII, we tried to communicate as literally as possible the words of the original author. In the remaining chapters of the book the style is more relaxed, but not free of foreign undertones.

I gratefully acknowledge the support given by the School of Engineering of Princeton University, and the excellent work of the Princeton University Press.

Princeton, New Jersey
July 25, 1962.

R. De Wiest

CONTENTS

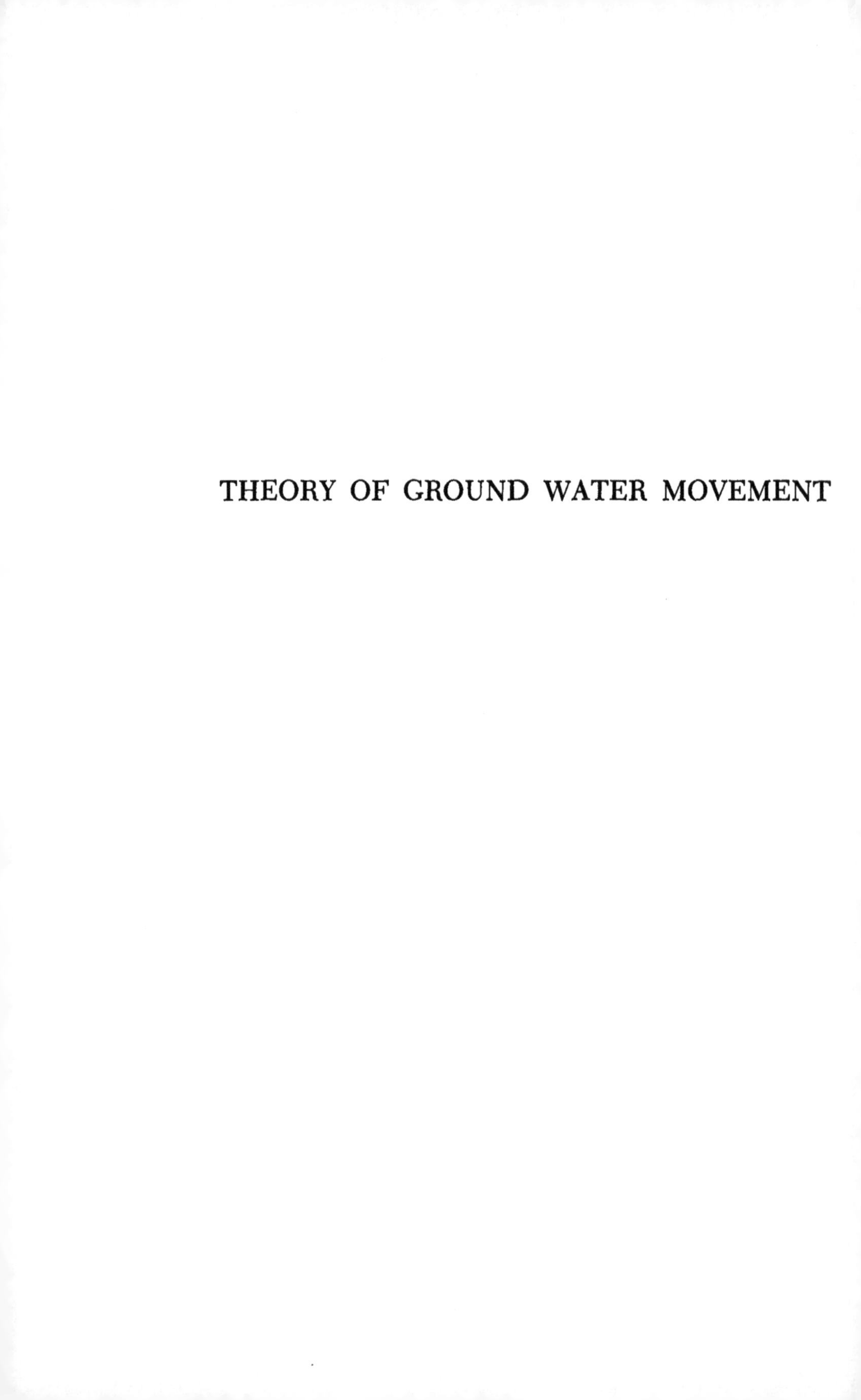

# THEORY OF GROUND WATER MOVEMENT

# PART I.

## Steady Flow of Ground Waters

## CHAPTER I.

## PHYSICAL AND MATHEMATICAL FUNDAMENTALS OF THE THEORY OF GROUND WATER FLOW

### §1. Soil Composition.

Before going into the exposition of the mathematical theories of ground water flow, we shall briefly outline, without indulging in details, the fundamental knowledge about the properties of soil.

Soil must be considered as a single system, consisting of mineral particles, colloidal particles, surrounded by waters containing dissolved salts and elements in gaseous phase (air, water vapors). From this viewpoint, what is customarily called soil, makes up strictly speaking the solid phase of soil.

Below, the definition of the term "ground waters" will be given (Par. 4). We note here already, that N. E. Zjoukovsky called these waters "underground" [1] and talked about their percolation under dams [3]. These expressions give at once a clear representation about how the phenomenon is considered. Deep-rooted with us is the term "seepage", equivalent with the concept of percolation, i.e., the slow movement in porous media.

The movement of the ground waters occurs in the upper layer of the earth crust. From the viewpoint of study, the regimen of underground waters usually presents a basic interest when the upper layers are permeable to water at a depth of several meters or several decameters. However, the complete study of the bedding of the ground waters and of the interaction between waters belonging to separate layers, calls also for a knowledge of what happens at great depths. The present methods of drilling permit us to obtain boreholes at a depth of several kilometers.

Soils may be divided into "consolidated" and "unconsolidated" or "loose" conglommerates. Unconsolidated soils are made by the crushing of dense rocks. They form soils of two basic kinds: soils of the sandy type (cohesionless) and of the clayey type (cohesive) [36]. A typical sand in the dry state consists of a loose substance, which through moistening temporarily passes into a bound state. The moisture of a sand fluctuates between small limits, and for every degree of moistening, a sand is devoid of plasticity, i.e., the property to retain its given shape. Sand is easily permeable to water, does not swell, possesses no significant capillary water rise and does not show shrinkage under drought.

Clay can exist in three states: liquid, plastic and solid. The bond of clay may be so high, that sometimes blasting operations are used for the exploitation of clay. The moisture content of clay may fluctuate between very wide limits. Clay is poorly permeable to water, and in the plastic state practically impervious, swells strongly, possesses a high capillary rise, shows considerable shrinkage under drying conditions, and therefore its decrease in volume is accompanied by the formation of cracks. From the point of view of shape of the particles, the following distinction between sands and clays exist. Sands have the form of grains, approaching a cubic or rounded shape, the clayey particles have the form of scales or plates. The specific surface of the clay particles is significantly larger than that of the sandy ones, this constitutes one of the causes of the large molecular forces acting between the clay and water particles.

The properties of the commonly occuring soils are intermediate between the properties of sands and clays (sandy loams). Soils whose content of clayey particles exceeds 60% (in volume) of the total amount of particles, are called "heavy clays"; also, those whose quantity of clayey particles varies between 30 and 60%, are called clays. "Loams" (divided into "heavy", "medium" and "light") have from 10 to 30% clayey particles, "sandy loams" (heavy and light ones) - from 3 to 10%, and, finally, sands may have up to 3% clayey particles.

For construction practice, a classification of soil particles, according to their dimensions, [30], exists and is given in Table 1.

There exist also other subdivisions of the particles according to their magnitude. Thus, for example, if the soil particles have a diameter larger than 3mm., then soil scientists say [12], that the particles form a stony part of soil; particles from 3 to 0.01 mm. are called by them "physical" clay; particles smaller than 0.001 mm. are sometimes called silt, smaller than 0.0001 mm. - colloidal particles. Below (Par. 4) will be shown the important role of the colloidal particles in the interaction between water and soil particles and their influence upon the physical properties of the soil [10]. A cross-section of the upper layer of the earth crust at a depth of the order of several decameters shows that the soil usually has a stratified structure (represented, for example, in Fig. 1). At the border of each layer, the soil consists of particles of different shape and magnitude.

To obtain an idea about the composition of a given soil, one has to carry out an analysis of the magnitude of its parts and fractions, for example, by screening the soil through a series of sieves with apertures of various sizes. The result of the analysis gives the so-called "curve of the mechanical or granulometric analysis" of the soil, which one has adopted to construct as follows: along the abscissa axis, the diameters of the particle in mm. are laid out, along the ordinate axis - the ratio, expressed in percentage, of the summed weight of the particles which are smaller than a given diameter, to the weight of all the particles of the analyzed soil sample. In

Fig. 2 and 3 are given the curve of the mechanical analysis of medium-sized soil [11], constructed on linear and semi-logarithmic scales.

We note, that under "subsurface" are understood the shallow layers of the ground, cultivated and changed through combined action of the climate, vegetable and living organisms and human activities, and being fertile.[13] The seepage of water in the subsurface region, in comparison with the seepage in the soil, has its own special features which we sometimes will mention. In the study of the phenomenon that takes place as well in the subsurface region as in the deeper soil, also the name "subsurface-soil" is used.

TABLE 1.

Diameters of various soil components.

| Name of Components | | Diameters in mm. |
|---|---|---|
| Clayey | | < 0.005 |
| Silty | small | 0.005 - 0.01 |
| | large | 0.01 - 0.05 |
| Sandy | very small | 0.05 - 0.25 |
| | small | 0.25 - 0.5 |
| | medium | 0.5 - 2.0 |
| | large | 1.0 - 2.0 |
| Gravellous | small | 2.0 - 4.0 |
| | medium | 4.0 -10 |
| | large | 10 -20 |
| Pebble | | 20 -60 |
| Boulder | | >60 |

TABLE 2.

Granulometric composition of sand from borrow pit.

| Diameters of particles in mm. | 1.0-0.25 | 0.25-0.10 | 0.10-0.05 | 0.05-0.01 | 0.01-0.005 | 0.005-0.0025 | <0.0025 |
|---|---|---|---|---|---|---|---|
| Percentages in weight of dry soil. | 40.70 | 51.32 | 3.08 | 2.27 | 1.32 | 0.66 | 0.66 |

The desire to characterize the size of the soil by one single parameter leads to the concept of "granulometric index," which by different authors is defined in a different way - by several, for example, as the mean diameter of the particles. The so-called "effective diameter of the particles $d_{10}$" is the index that received the widest diffusion. It is equal to the diameter of the opening of the sieve, through which is screened 10% by weight out of the soil sample (and on which 90% by weight remains).

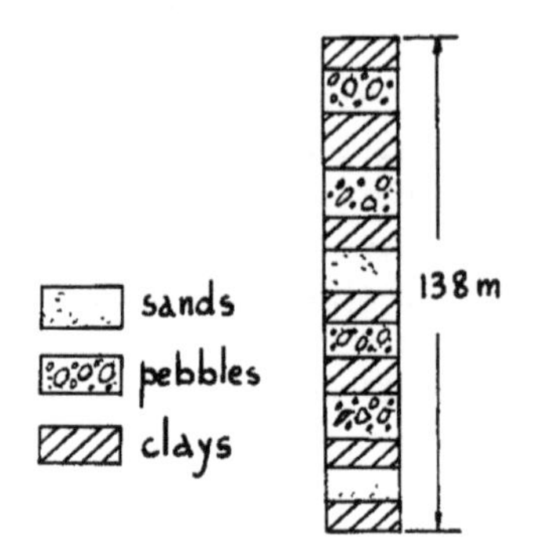


Fig. 1

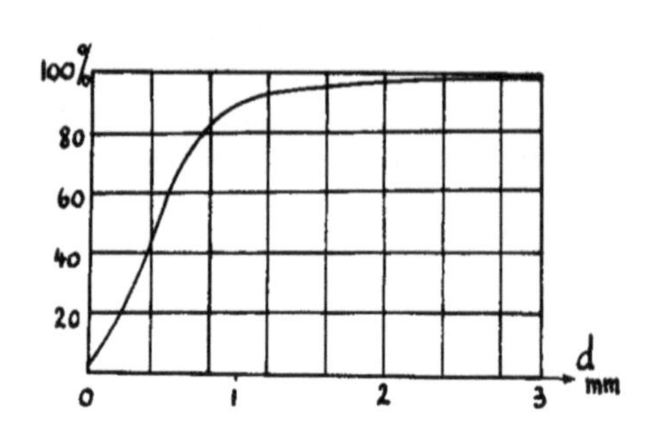


Fig. 2

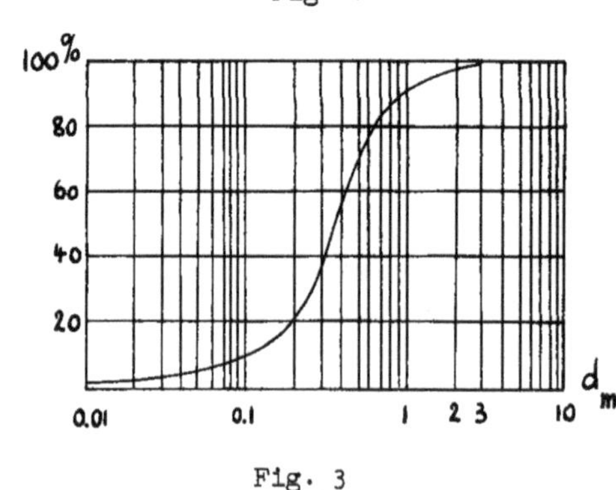


Fig. 3

As example in Table 2 the granulometric composition of a sample of ravine sand is given [15].

In the composition of sand cited above (Table 2), five small fractions give for sum: 0.66 + 0.66 + 1.32 + 2.27 + 3.08 = 7.99% i.e., 2.01% less than 10 %. This means that $d_{10}$ lies in the interval between 0.10 and 0.25 mm. Interpolation gives $d_{10}$ = 0.11 mm.

§2. Soil Porosity.

Let us take a sample of soil of volume V. Let the volume of all the pores in this sample be $V_1$. The ratio $V_1$ over V is called "porosity" of the soil. We designate this quantity with the letter $\sigma$: $\sigma = V_1/V$. In other words, the porosity is the sum of all the pores, present in the unit volume of soil.

The porosity depends upon the nature of the soil, its geologic age, the pressure at which the soil is tested, the cultivation of the subsurface soil, and so on; the porosity depends upon the compaction of the soil - what is often observed in laboratory conditions, - and may change in time [17]. The cross-sectional dimension of a separate pore varies from 2-3 centimeters to small fractions of a micron; in practical problems of seepage, the cross-sectional dimension of the pores is of the order of fractions of a millimeter.

* If the soil were composed of small spheres of the same diameter, then it would be possible to find the porosity for different dispositions of the particles. So, in the case of a cubical disposition of the spheres (when the spheres may be considered to be inscribed in a cubical lattice), we note th t $V_1 = d^3 - \pi/6\ d^3$ (difference between volumes of cube and

sphere). We obtain for the porosity $\sigma = 1 - \pi/6 = 0.476$ .

In books about the theory of seepage, the cubical disposition is considered as the least tight of all the regular dispositions [6], just like the tightest disposition is shown to be the "rhombohedral" one, which is obtained when one lays the tightest row of spheres on a flat surface (Fig. 4) and then superimposes such rows one on top of the other so that the spheres of the second row fill the gaps of the first. For such disposition the porosity will be equal to $\sigma = 1 - \pi\sqrt{2}/6 = 0.260...$

The latter meaning of $\sigma$ is really the smallest, because it depends upon a cubical packing, and because such packing is least tight only among truly homogeneous packings. In order to explain this, let us note that in the rhombohedral packing every sphere comes into contact with twelve spheres: so, in Fig. 4 the striped sphere is contiguous with six spheres, whose centers lie in the same flat plane, and, moreover, with the three spheres of the upper and lower layer which are not presented in the sketch. In the cubical packing, every sphere also is in contact with six spheres.

Packings have been examined, in which every sphere touches four spheres. One of these, the "tetrahedral" in which the four mentioned spheres are seated at the vertices of a tetraheder (inside which the first sphere is located), gives a porosity 0.660 (Fig. 5,a). Another packing, still more spread out, gives a porosity equal to 0.876 (Fig. 5,b)[27]. This is, evidently, the largest porosity for a truly firm lattice of single spheres, in which the spheres hold each other (the latter circumstance meets the requirements that each sphere touches with four spheres, whose centers do not lie in a single flat plane).

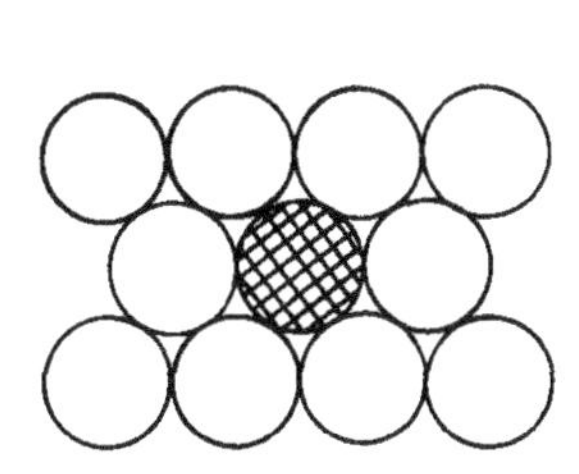

Fig. 4

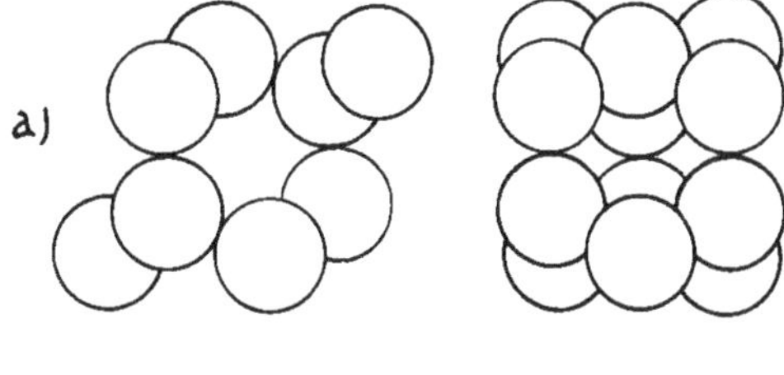


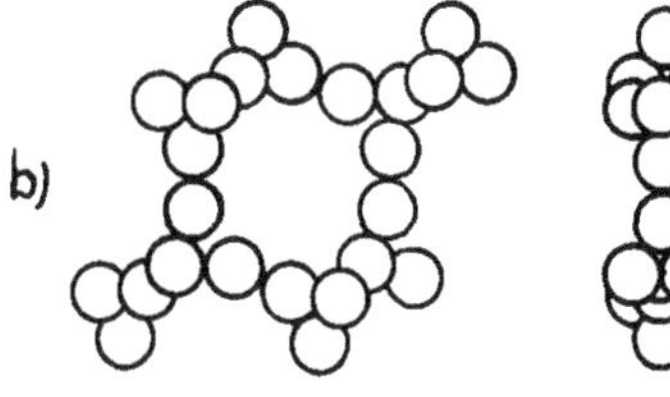


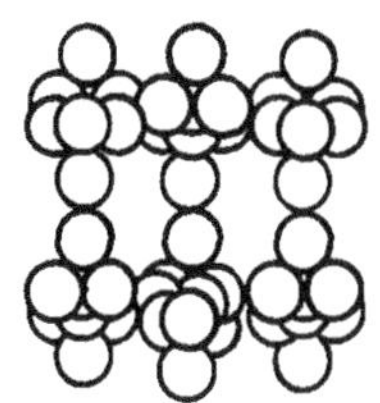

Fig. 5

In natural soils, the porosity must vary between wide limits, just like the soil structure may be complicated and diverse. Thus, subsurface soils have a complicated, crumpled structure. [13]. In Fig. 6 is given an example of heterogeneous clayey soil species, composed of large sandy grains (1), small crummy grains

(2), unpacked clayey particles in the pores between the large fractions (3), and packed clayey particles in contact with each other between the large parts (4).

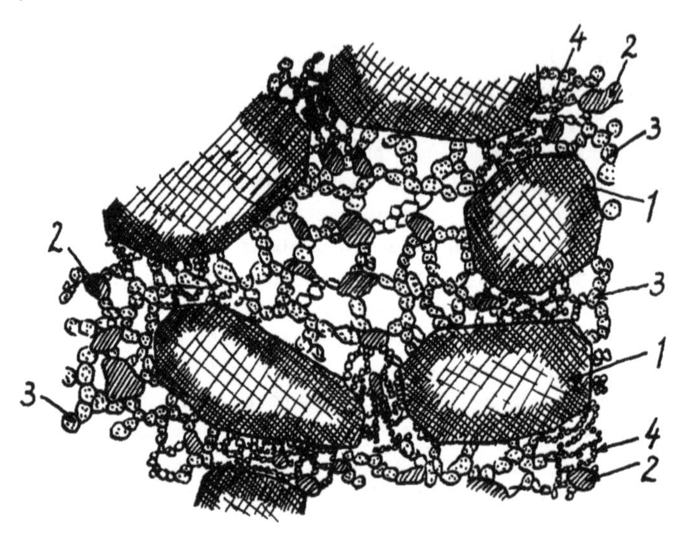


Fig. 6

In Table 3, porosity values for a series of soils are given [18].

TABLE 3.

Values of porosity for various soils.

| Designation of soil | Porosity |
|---|---|
| Gravel (with particle diameter from 2 to 20 mm.) | 0.30 - 0.40 |
| Sands (with particle diameter from 0.05 to 2 mm.) | 0.30 - 0.45 |
| Sandy loam | 0.35 - 0.45 |
| Loam | 0.35 - 0.50 |
| Clayey soil | 0.40 - 0.55 |
| Peaty soil | 0.60 - 0.80 |

In A. A. Rode's [35] book, limits for the porosity of subsurface soils are given as 25 - 90% (and wider).

For so-called cracked mountain soils, the concept of "crackedness" has been introduced, i.e., the ratio of the volume of cracks of the sample of given soil to the total volume of the sample [19].

After a rainstorm, it happens that the water table, usually lying at some depth below the shallow earthlayers, strongly rises Thus, if the precipitation was of the magnitude of 2 - 3 cm., then the ground waters rose 20 - 30 cm. If immediately above the free surface of the water table the soil were dry, then, taking into account the values of porosity, given in Table 3, it would be possible to expect a rise of the ground waters, exceeding the amount of precipitation 3 to 4 times. The 10-to-12-fold stronger rises that prevail are explained by the fact that near the free surface of the ground waters the soil is moistened, and this decreases the volume of gaps between the soil particles that could be replenished with water. One still has to note, that for the rise of the ground waters to take place, not all of the volume of these gaps must be replenished with water, because a certain amount of packed air remains in the pores. During the drop of the ground water level, not all of the volume of the porous space releases its water, as long as the soil remains moist.

In connection with this the concept of "active porosity" of the soil has been introduced, taken out of the drainage zone or zone closely adjacent to the free surface of the ground waters: this is the ratio of the volume of of pores, not occupied by water that is firmly linked with the soil, to the total volume of sample. The active porosity may be of the order of 0.1 and smaller.

## §3. Electromolecular Forces in Soils.

Between solid soil particles and surrounding water, electromolecular forces of interaction exist [14]. Water consists of polar molecules with hydrogen ions, loaded positively, and oxygen ions, loaded negatively. Under influence of an external electric field, polarization of molecules and orientation of water dipoles in the electric field take place.

As the dielectric constant of the solid particles strongly differs from the dielectric constant of water, under the contact of the soil particles with water, an electric field with surplus energy at the surface of the soil particles is generated, and the water dipoles are attracted towards the surface of the mineral parts.

The attractive forces operate at very short distances from the surface of the particles, not exceeding 0.25 - 0.5$\mu$ ($\mu$ - micron = 0.001 mm.). The magnitude of the molecular forces is high near the surface of the particles (of the order of tens of thousands kilograms per square centimeter) and decreases quickly with the distance. The region of strong action is measured by the thickness of a layer of the order of a few molecular coatings.

In Fig. 7, a scheme is shown of the molecular attraction forces at the border separating water and a solid particle. Fig. 7,a represents at an enlarged scale part of Fig. 7,b. On this figure is shown the orientation of the water dipoles near a clayey particle. The electrical charge of clayey particles has an important significance for the drainage of soils by the electrodrainage method.

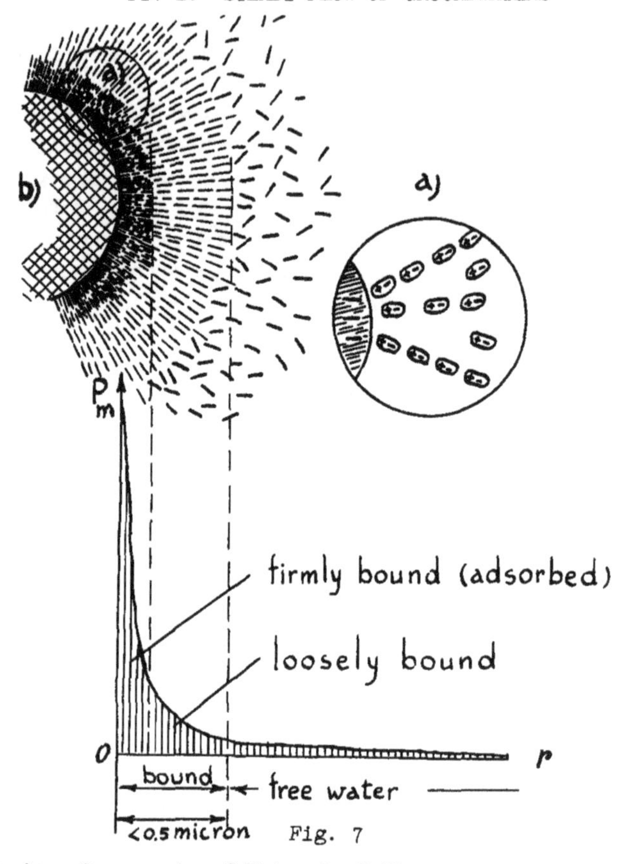


Fig. 7

§4. Various Components of Water in Soil.

The water molecules, that surround the soil particle immediately, are subject to such large attraction forces (see Par. 3), that they are found to be firmly attached to the surface of the particle. The water particles, which cannot be separated even by centrifugal action, develop forces several tens of thousand times superior than the force of gravity. These molecules form the so-called "firmly bound water." The layer of firmly bound water, having a thickness of the order of a few tens of molecular coats, in its turn links and orientates the water molecules by which it is surrounded. The latter form the "loosely bound water." It is difficult to draw a boundary-line between firmly bound and loosely bound water. The fundamental classifications of the kinds of moisture in shallow and deep soils were given by A. F. Lebedev. [5] He distinguishes the following sorts of water in soils: (1) water vapor, (2) hygroscopic water, (3) pellicular water, (4) gravitational water, (5) water in solid state, (6) crystalline water and (7) chemically bound water. A. F. Lebedev did not study the last three categories. Let us examine the first four.

(1) "Water vapor" fills completely the gaps of the soil and shifts from regions of higher pressure to regions of much lower pressure.

A. F. Lebedev considered the condensation of in the ground moving water vapors to be one of the reasons of the formation of ground waters, especially in deep layers.

(2) "Hygroscopic water" is water that condenses at the surface of the particles. When dry soil gets in touch with humid air, the soil particles absorb moisture, and the whole soil volume increases until some magnitude is reached, corresponding to the maximum hygroscopic effect. The maximum hygroscopy may have these values for various soils: for sands about 1%; for silt - up to 7%; for clay - up to 17% of the total dry matter.

After A. F. Lebedev, hygroscopic water can move in the ground, only while transforming into the vaporous phase.

(3) "Pellicular water" is formed on the particles under influence of the molecular forces of adhesion. This water holds together with a huge force and cannot be separated by centrifugal action with an acceleration 70 thousand times exceeding the acceleration of gravity. Pellicular water is able to move as a liquid from much denser films to much thinner. The gravitational force does not show any influence on the movement of pellicular moisture, and this moisture freezes only at minus 1.5°C. The moisture of the soil, corresponding to the maximum thickness of film, is called by Lebedev "maximum molecular moisture content." Pellicular water disappears under drying soil conditions.

(4) "Gravitational water" is the free water, that is not subject to the action of the attraction forces towards the surface of the solid particles. This water moves under influence of the gravity force, and is subject to hydrodynamical pressure. We will call gravitational water "ground water" and subsequently will study the laws of motion of precisely this ground water. Among ground waters, usually also capillary water is numbered - that water which partially or completely fills the pores of the soil and has a surface that exhibits concave meniscs. (Par.9). We note that some alkaline waters may give convex meniscs and a negative height of capillary rise, i.e., capillary pull down. Recent investigations have shown, that the capillary rise of water takes place thanks to the hydration energy of ions and molecules at the border surface of the solid and liquid phase, i.e., that capillarity has an electrochemical nature.

The concepts of A. F. Lebedev introduced more clarity and complemented the past investigations of a series of scientists. At present, an important significance is attached to the role of colloidal particles in soil. Colloidal matter is deposited on the surface of the mineral grains under the form of thin coatings. [9]. Concentrated water vapors are absorbed by these coatings, when they enter into chemical interaction with them Electrically loaded colloidal particles, called mycelium, collect water vapors up to the well known limit of maximum hygroscopy, about which we talked before. This limit is characterized by the cessation of the outflow of condensation heat. The colloidal cover can further swell, by absorbing capillar-fluid moisture without outflow of heat.

Thus, particles of humid soil are covered with an hygroscopic layer, and under further wetting of the soil, after release of all the condensation heat, they acquire pellicular water. Actually soil particles are so grouped together amongst each other that the pellicular layer does not develop till its maximum possible limit. In the corners of the pores between soil particles, water is formed that exhibits a menisc under influence of surface tension. For further wetting of the soil, the accumulations of water in the corners of the pores increase and their menisci come in touch with each other. In the pore, a clear space is left along which water may move around.

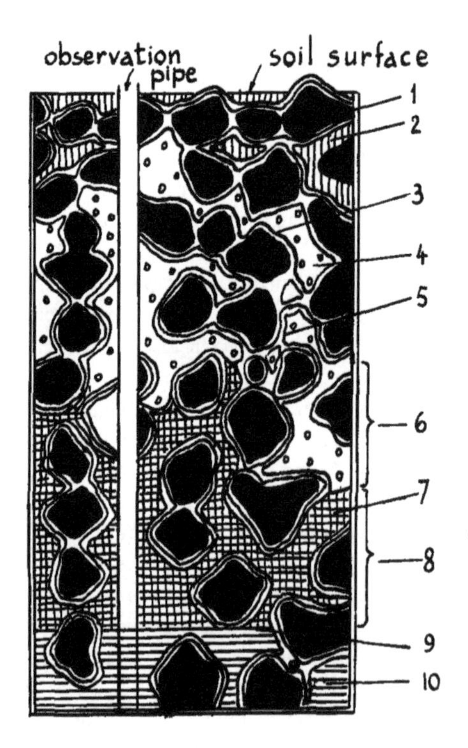


Fig. 8

When water fills all the pores (a certain amount of air may remain in them), then it obtains the ability to move (with the exception of pellicular water) under influence of gravity. Such water, as we already said, is called gravitational or ground water. The movement of such water in porous medium is called "seepage."

Thus, the physical picture of the flow of water in the soil is complicated. In any soil, completely or incompletely saturated with water, water is available that does not participate in the flow, linked by absorption forces with the soil skeleton. On account of the coating of the soil particles with firmly bound water, the volume of the solid phase increases and the pore volume of the ground mass decreases. The amount of bound water differs from a few percents of the porosity for sands to the full porosity for some clays. In the latter, movement of water is only possible under the application of forces exceeding the absorption forces.

In the clayey-colloidal fractions the phenomena of coagulation and peptisation [11] take place, leading to the formation of aggregates or to their destruction, changing, when superimposed, the granulometric composition of, and with this also the specific surface of the solid phase of the soil.

In Fig. 8 is represented a sketch of various forms of water in soil, after N. A. Katchinsky [12]. In this sketch, the number 1 designates a soil particle, 2 - water of precipitated rain, 3 - a film of hygroscopic water, 4 - soil entrapped air with water vapors, 5 - pellicular water under which N. A. Katchinsky understands water loosely linked with soil, 6 - zone of open capillary water, where water and air alternately fill the pores, 7 - capillary water, 10 - ground water. If we look at a separate clear path between the grains of soil, and this when the water did not yet become gravitational, then we may have the picture, represented on Fig. 9: capillary menisci are in the corners between the grains and the interior region is filled with air.

As for ground water, it usually does not, as we already noted before, fill completely the entire region between the soil particles, so that in this region small air bubbles remain. It is possible to draw a sketch of the movement of ground water such as represented in Fig. 10, where the interior circle corresponds to a small air bubble.

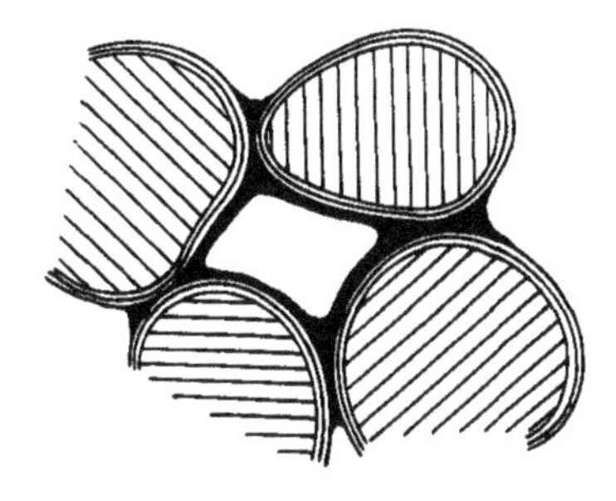

Fig. 9

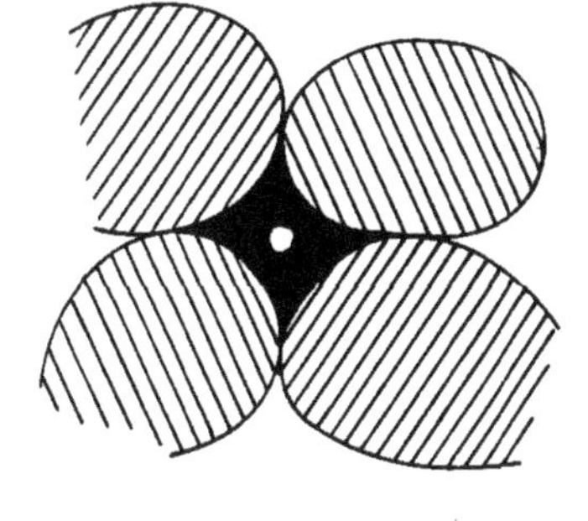

Fig. 10

§5. Seepage Velocity.

Under seepage velocity is understood the flux of fluid, i.e., volume of fluid, flowing per unit time across a unit area, isolated in porous medium.

Let us recall, how the flux of fluid is defined in hydrodynamics. Let us take an element of area S (Fig. 11), through which the fluid travels with a velocity, designated by the vector $\vec{u}$. The amount of fluid, flowing through S per unit time, is equal to the volume of the cylinder, built on S and $\vec{u}$, and since the height of this cylinder is equal to $u_n$ - component of the velocity normal to the elemental area S, the flow rate Q across S is equal to $Q = Su_n$.

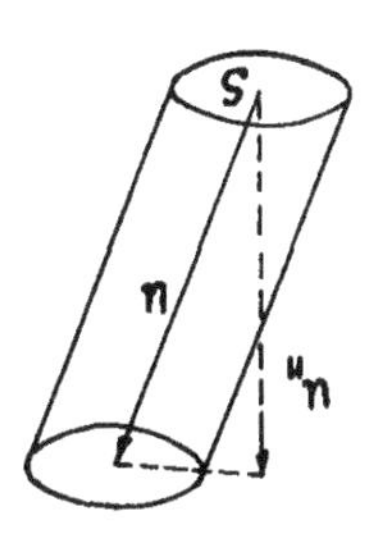


Fig. 11

If, in a given point, we rotate an elemental area and erect the normal to it, then the direction n, for which the flowrate is maximum, will be the direction of the velocity vector. Now, let us apply this to the flow of fluid in porous medium. Consider an elemental area in the soil, composed of sections of soil grains and gaps between these sections. (Fig. 12). The flow of fluid between the grains of soil bears a complicated character, and therefore it has been adopted not to consider the velocity of the fluid in the different points, but the average value of these velocities.

Let the vector of the average velocity of the fluid particles in the region of the area $S$ be $\vec{u}$. The area of the gaps located in $S$, will be designated by $S_1$. Let us put $m = S_1/S$ and let us call $m$ "the superficial porosity." The flowrate across the area $S$ will be

Fig. 12

$$Q = S_1 u_n = mSu_n .$$

The flowrate through an elemental surface, the area of which is equal to one unit, is expressed as $mu_n$ and is called the "seepage velocity."

The vector of the seepage velocity has the magnitude $m\vec{u}$, equal to the maximum value of $mu_n$ for various positions of the elemental surface $S$, and directed along the normal to that position of $S$, through which the maximum flowrate passes.

If the vector $\vec{u}$ of the velocity of particles has the components

$$\frac{dx}{dt}, \frac{dy}{dt}, \frac{dz}{dt},$$

then the components of the vector of the seepage velocity $\vec{v}$ will be:

$$u = m\frac{dx}{dt}, \quad v = m\frac{dy}{dt}, \quad w = m\frac{dz}{dt} . \tag{5.1}$$

In order to define the superficial velocity of a certain sample, it would be possible to proceed as is done in some special laboratories: the sample, extracted from the ground with the help of a cylindrical pipe with sharp edges (such a sample is called "core"), is impregnated with matter that glues it together and is then exposed to a series of thin slices. When the slices are put under the microscope, it is possible to measure the area of the gaps and to take their ratio to the area of the cross-section of the sample. The average of these values for all the slices gives the average porosity of the sample under consideration. However, this method is complicated. Moreover it is obvious that the described method of the determination of the superficial porosity, for our cylindrical sample gives the value of the average porosity. Indeed, let us designate by $S_1(z)$ the area of the pores at a cross-section of the cylinder at a distance $z$ above its base. (Fig. 13). Let $m(z)$ be the porosity of this cross-section, i.e.,

$$m(z) = \frac{S_1(z)}{S} .$$

Where $S$ is the area of the base of the cylinder. Then the average value of the porosity $m$ will be equal to

$$m = \frac{1}{h}\int_0^h m(z)dz .$$

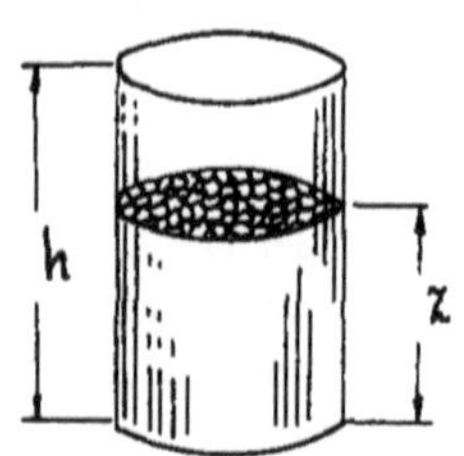


Fig. 13

Here $hS = V$ is the volume of the cylinder under consideration, but the integral is equal to the volume $V_1$ of all the pores in the sample. Therefore the average superficial porosity $m$ is equal to the average volumetric porosity $\sigma$; $m = V_1/V = \sigma$. In the future, we will not distinguish between the superficial and the volumetric porosity, and we will designate one and the other value by the letter $m$ or the letter $\sigma$.

## §6. Experimental Seepage Laws.

Before giving some results of experimental investigations of water in pipes filled with soil, let us recall some information from hydrodynamics. For an incompressible inviscid fluid flowing in a pipe - horizontal or inclined - with smooth walls, the steady state Bernoulli equation is

$$\frac{p}{\rho g} + z + \frac{v^2}{2g} = \text{const.} \tag{6.1}$$

Here $\rho$ is the density of the fluid, $g$ - the acceleration of gravity, $p$ - the pressure, $v$ - the velocity; $z$ is the "geometric height," $p/\rho g$ is called the piezometric height; the quantity $v^2/2g$ bears the name "velocity head."

Bernoulli's equation states that for all points of the pipe the sum of the three quantities remains a constant value. The sum of the first two terms of equation (6.1) is called the "head" or "piezometric head." We will designate it by $h$.

$$h = \frac{p}{\rho g} + z \quad . \tag{6.2}$$

Now we can re-write equation (6.1) in the form

$$h + \frac{v^2}{2g} = \text{const.} \tag{6.3}$$

From this it is evident that, if the fluid were flowing in the pipe "without friction" and with constant velocity, the head in all points of the pipe would have one and only one value. The friction along the rough walls of the pipe is taken into account in hydraulics by the introduction of a correction term in equation (6.3).

Let us examine, how the flow of fluid in porous medium develops. Numerous experiments, carried out in many laboratories, related to the steady flow of water (and other fluids, for example oil) have led to the following results.

Let us take two points along the axis of the pipe (Fig. 14), a distance $\Delta s$ from each other, and let us insert piezometers at these points. In the piezometers the water rises correspondingly to heights $h_1$ and $h_2$, measured from an arbitrary horizontal plane (Fig. 14), and to which $h$ in formula (6.2) is referred.

In hydraulics, one considers the quantity $J$, called "piezometric slope" or "gradient of the piezometric head," defined as the derivative of $h$

along the path s, taken with the minus sign:

(6.4) $$J = -\frac{dh}{ds} = -\lim_{\Delta s \to 0} \frac{\Delta h}{\Delta s}$$

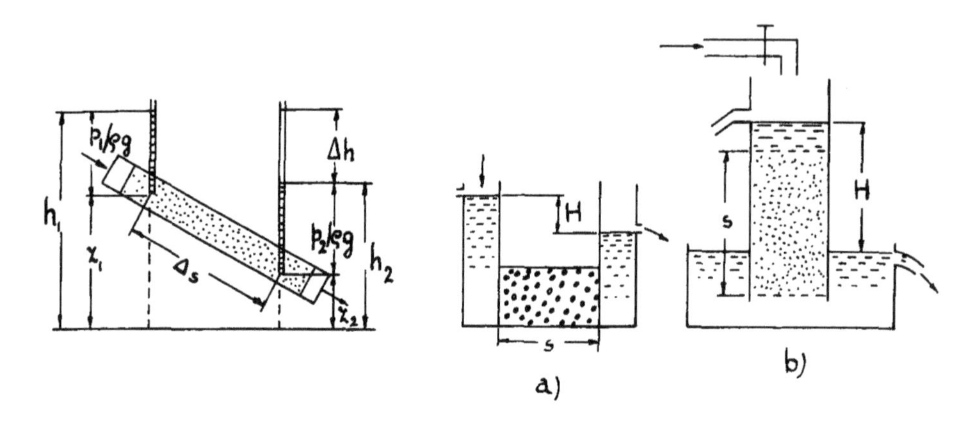


Fig. 14 Fig. 15

Experiments have shown, that the seepage velocity is a function of the piezometric slope, or that the slope is a function of the velocity:

(6.5) $$v = f(J) , \quad J = \Phi(v) .$$

Such is the character of the flow under consideration, that for seepage in porous medium the fluid experiences a considerable resistance of friction of fluid particles on the surface of the soil particles.

For many soils - sands, clays, peaty soils, thinly fissured rocky soils, and so on - there is a linear dependence of the seepage velocity upon the piezometric slope

(6.6) $$v = kJ = -k \frac{dh}{ds}$$

where the proportionality coefficient $k$ is called the "seepage coefficient." We note that the seepage coefficient has the dimensionality of a velocity; it is equal to the seepage velocity for a piezometric slope equal to one.

The equality (6.6) was first established by Darcy [22] and is sometimes called "Darcy's Law."

Fig. 15, gives two schemes of experiments to check the law of seepage. The sand in the pipes is prevented from wash-out by means of a screen.

To examine these schemes, the piezometric slope $J$ will be taken equal to the ratio of the difference in head $H$ to the length of the seepage path $s$, so that the seepage velocity will be

(6.7) $$v = k \frac{H}{s} .$$

In dense clays and heavy loams, in which the water is of a molecularly bound nature, seepage starts only when the gradient exceeds a certain value $i_0$, called the "initial gradient" (7). In this case equation (6.6) is replaced by

$$v = -k\left(\frac{dh}{ds} - i_0\right). \tag{6.8}$$

For very dense clays $i_0$ may attain values equal to 20 - 30.

§7. Seepage Coefficient.

The seepage coefficient of a given soil sample may be determined by one of the devices represented in Fig. 15, a, b. Fig. 16 gives a more detailed scheme of such a device, supplied with piezometric tubes [24].

If it is necessary to determine the seepage coefficient in natural conditions, then one must take samples of soil (possibly of undisturbed structure) and test them in the apparatus of Fig. 16. However this method is not sufficiently reliable, because it cannot give the characteristics of the entire region as a whole. One has recourse to field methods for the determination of the seepage coefficient through well pumping (see Chapter IX), which give a more reliable value of this quantity. By this method one obtains some average value of the seepage coefficient for the flow region under consideration. In Table 4, the order of magnitude of the seepage coefficient for various soils is given [15]. The seepage coefficient differs because of soil composition, magnitude and form of its grains, and also because of the fluid in the soil, partially for its viscosity and, consequently, its temperature. Usually, the seepage coefficient, determined in one or the other way, is reduced to a temperature of 0°C. or 10°C. Then the coefficient of seepage for a temperature t°C. is computed by a formula of the form

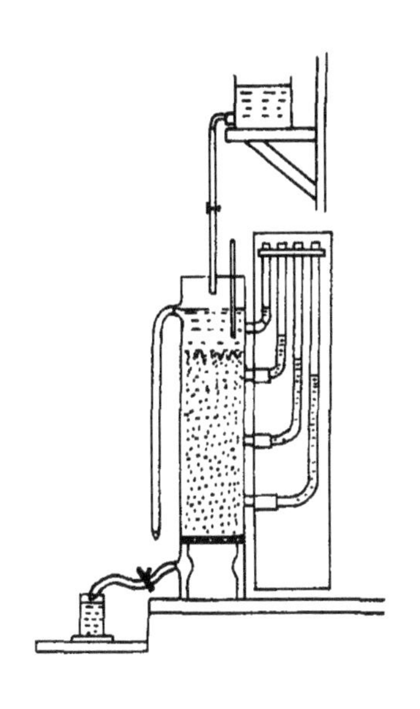

Fig. 16

$$k = k_0(1 + 0.0337t + 0.000221t^2) \quad \text{(Poiseuille's formula)}$$

$$k = k_{10}(0.7 + 0.03t) \quad \text{(Hazen's formula)} \quad .$$

N. E. Zjoukovsky [2], in 1888, paid attention to the dependence of the seepage coefficient upon the atmospheric pressure. Observations have shown, that the height of ground water in wells decreases with increase of atmospheric pressure and increases with a decrease of atmospheric pressure. After Zjoukovsky, small elastic air bubbles (may be introduced by rainwater) expand in the water-bearing layers under the drop in pressure and consequently push out the ground waters. The water level rises in a well and make it flow faster. For an increase of pressure, the picture is the reverse.

But even in the case of steady flow, the presence of air bubbles in ground water changes with the degree of water permeability of the soil, which does not remain constant during the completion of a more or less prolonged test. The change in the coefficient of seepage in time may also depend upon physical and chemical processes, - for example the leaching of dissoluble salts, - and even biological processes in the soil.

TABLE 4.

Average values of the seepage coefficient for various soils.

| Designation of soil | Average value of the seepage coefficient $k$, in cm/sec. |
|---|---|
| Clean sand | 1.0 - 0.01 |
| Clayey sand | 0.01 - 0.005 |
| Sandy loam | 0.005 - 0.003 |
| Carbonated loam | 0.001 - 0.00005 |
| Clay | 0.0005 - 0.000005 |
| Saline clay | 0.000001 - 0.0000003 |
| Carbonated loess | 0.0005 - 0.0001 |
| Noncarbonated loess | 0.00005 - 0.00001 |
| Salt-marsh | 0.001 - 0.0001 |
| Peat, little decomposed | 0.006 - 0.002 |
| Peat, moderately decomposed | 0.0008 - 0.0002 |
| Peat, young sphagnum | 0.002 - 0.0002 |
| Peat, old sphagnum | 0.0002 - 0.0001 |

In the theory of oil seepage, the so-called "permeability of the soil" [6], is considered instead of the seepage coefficient.. And indeed, in the study of the flow of oil in a horizontal pipe, the seepage velocity was found to be equal to

$$v = -\frac{k_0}{\mu}\frac{dp}{ds}, \tag{7.1}$$

where the coefficient $k_0$ differs with the composition of the soil and is called permeability of the soil; $\mu$ is the viscosity of oil, and $p$ the pressure.

When the flow in a vertical plane or space is considered, then one has to change $p$ into $p + \rho gz$ in this formula.

As it is easy to see, the seepage coefficient $k$ and the permeability $k_0$ are correlated

$$k = \frac{k_0\rho g}{\mu} = \frac{k_0 g}{\nu}$$

where $\nu = \mu/\rho$ is the kinematic viscosity of the fluid.

There are formulas that express the dependence of the seepage coefficient upon porosity and given mechanical constitution of the soil. We reproduce here Kozeny's formula [6]

$$k = \beta \frac{d^2}{\mu} \frac{\sigma^3}{(1 - \sigma)^2}$$

where $\mu$ is the viscosity of the fluid, $d$ the effective diameter of the particles (see Par. 1), $\sigma$ the porosity, $\beta$ a coefficient, which Kozeny considers to be constant (for water). E. A. Zamarin [16] proposed for $\beta$ the expression $\beta = 8.4(1.275 - 1.5\sigma)^2$ .

§8. Limits of Applicability of the Linear Seepage Law. Non Linear Laws.

Numerous experiments have shown that the linear law of Darcy (66) has definite limits of applicability. So, for homogeneous soils - sands, gravels- Darcy's law holds only under the fulfillment of the inequality [4] for the Reynold's number

$$R = \frac{vd}{\nu} \leq A \tag{8.1}$$

(designation of symbols same as at end of Par. 7). Here $A$ is a number for which various authors give different values, between limits from 3 to 10.

If we take $\nu = 0.018$ cm$^2$/sec. (for water), then

$$vd \leq 0.070 - 0.180 \tag{8.2}$$

where $v$ is expressed in cm/sec., $d$ - in cm.[18].

The presence of the upper limit of applicability of Darcy's law is usually linked with a deviation from the laminar motion in porous medium towards turbulent [4]. However, already for laminar motion, digressions from the linear law [26] are possible.

There must also be a lower limit of applicability of Darcy's law, when the action of the molecular forces, pointed out by N. N. Pavlovsky [4], becomes important.

Beyond the limits of the linear seepage law, for example for large grained soils or for very fast movements, other relations between $v$ and $J$ have been established on the basis of experiments. So, a polynomial of the second degree has been considered: $J = Av + Bv^2$ or a polynomial of the third degree or a monomial relation of the form $v = CJ^n$ .

The constants $A$, $B$, $C$, $n$ are determined experimentally. S. A. Khristianovich [25] gave a method to study the movement of ground waters for non linear seepage laws. The problem of the flow of water towards a well has been examined by a series of authors, starting with N. E. Zjoukovsky (1888) [1]. N. K. Gjirinsky [29] came to the conclusion that for the majority of soils (loose), deviations from Darcy's law may take place only in the immediate vicinity of the borehole. Therefore he has shown that usually the zone of violation of Darcy's law also is the zone of disturbance of the natural composition of the soil under influence of the boring and developing (i.e., linked with the ejection of particles) processes occurring during the development of a well.

In fissured rock, evidently, more significant deviations from the linear seepage law are possible. We note that along the concrete foundation of a dam, so-called contact seepage may take place, in which a layer of water is formed between the soil and the weir [18]. The movement of liquid in such a layer is not subject to the usual laws of seepage.

§9. Capillarity.

If we put the end of a pipe filled with sand in water, then we see that the water at first rapidly rises in the pipe, then moderately and finally its motion becomes invisible for the naked eye. We designate the height of the maximum rise of the water (really, its average value over the contour of the pipe) by $h_k$ and call it the "static height of capillary rise".

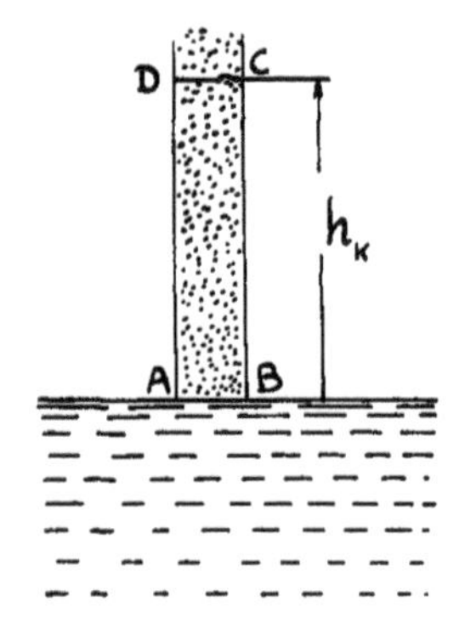

Fig. 17

We suppose that the basis of the pipe AB (Fig. 17) is in touch with the surface of the water in the vessel. Then along AB there will be atmospheric pressure. Molecualr forces, compelling the liquid to rise between the grains of sand, operate so that in the wetted part of the pipe a pressure (vacuum) lower than atmospheric will be established.

If we put the atmospheric pressure equal to zero, then the pressure in the region ABCD is to be considered negative and at the plane CD equal to $-\rho g h_k$. Or, if we designate the atmospheric pressure by $p_a$, then at the plane CD $p = p_a - \rho g h_k$. This condition, which must be satisfied at the free surface of ground waters, has been indicated by N. E. Zjoukovsky [3] (1920), more precisely defined by V. V. Vedernikov and applied by him, and later on by other authors in the solution of a series of problems (see Chapter V).

We note that the upper layer of the ground - the subsurface soil - is the "structural" part [13], i.e., in a sense that in this layer elementary particles are combined into an aggregate of mineral particles, and cemented with humus. Between the crumples, large pores of 1 - 5 mm. diameter are formed. In the structural part of the soil, capillarity is almost not observed. This phenomenon is also weak in coarse-grained soils. However, there are soils in which the capillary rise attains heights from 3 to 5 meters.

Table 5 gives the values of heights of capillary rise in some soils [15].

As is well known, the height of the rise in a capillary tube, wetted with water, is different for various curvatures of the meniscus and for a circular pipe is equal to $2\sigma/\gamma r$, where $\sigma$ - the surface tension of the water, $r$ - the radius of the pipe, $\gamma$ - the unit weight of water.

TABLE 5

Values of height of capillary rise in some soils

| Soil | $h_k$(cm) | Soil | $h_k$(cm) |
|---|---|---|---|
| Clay | 400 - 200 | Turkestanish loess | 350 - 250 |
| Sandy loam | 300 - 150 | Podsol | 35 - 40 |
| Sandy soil | 150 - 100 | Peat . | 120 - 150 |
| Sandy uppersoil | 100 - 50 | Saline soil, salt-marsh | 12 |

The pores of the upper soil layers constitute a complicated "labyrinth of voids" and therefore, for the height of the capillary rise a more complicated relation holds. For sands the approximate formula [15]

$$h_k = 0.45 \frac{1 - \sigma}{\sigma} \frac{1}{d_{10}} \text{ cm} ,$$

has been proposed, where $\sigma$ - porosity, $d_{10}$ - effective diameter of the soil particles in cm.

*§10. Permeability to Water of Incompletely Saturated Soils.

This problem has little been investigated theoretically and its physical nature is not yet fully understood. We communicate here briefly the results of the investigations of S. F. Averjanov [20], being the first attempt of a theoretical interpretation of the problem of the dependence of the seepage coefficient upon the moisture content of soil.

We designate the porosity of the soil by $\sigma$, the relative amount of bound water in the voids of the volume of soil by $w_o$, the full moisture content under natural conditions, i.e., taking into account the presence of a certain quantity of included air, (in the voids of the soil volume) by $w_1$. The real degree of moisture or moisture content is designated by $w$. Then the inequalities $\sigma \geq w_1 > w > w_o$ hold. If there is no included air, $w_1 = \sigma$.

*We suppose that the soil is not fully saturated, and has a moisture content $w$. We designate by $k_w$ the seepage coefficient for unsaturated soil. We will have for $w = w_o$, $k_w = 0$; for $w = \sigma$ we must have $k_w = k$, where $k$ - seepage coefficient for completely saturated soil. In the general case we put $k_w = \alpha k$

S. F. Averjanov examines the following scheme of the movement of a viscous fluid in a cylindrical pipe (Fig. 18). In the central part of the cylindrical pipe is a region with radius $R_1$ filled with air; the annular region, the borders of which are formed by the circles with radii $R_o$ and $R_1$, is filled with moving viscous fluid.

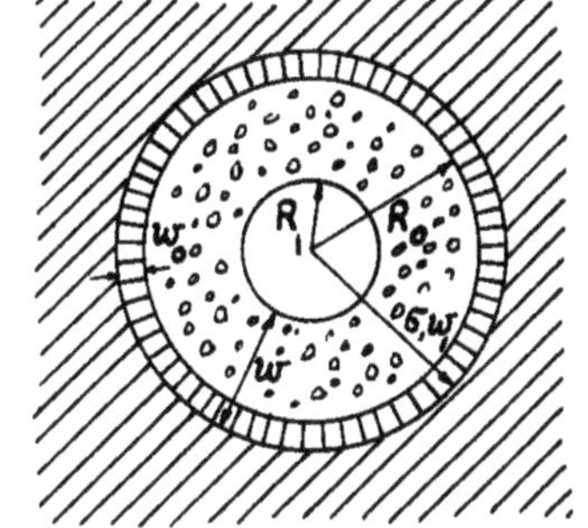


Fig. 18

Also the circle with radius $R_0$ is the borderline between gravitational and bound water. S. F. Averjanov supposes that the velocity along the pipe wall is equal to zero for $r = R_0$, but for $r = R_1$, attains the largest value.

Then, as is well-known from the course in hydrodynamics, the velocity of the flow of a viscous fluid in a cylindrical pipe [32] is given:

$$v = \frac{\Delta p}{4\eta\Delta\ell} r^2 + a \ln r + b , \tag{10.1}$$

where $v$ - velocity of flow at a distance $r$ from the longitudinal axis of the pipe for a pressure difference $\Delta p$, occurring over a distance $\Delta\ell$; $\eta$ - kinematic viscosity coefficient. In our case we obtain

$$v = \frac{\Delta p}{4\eta\Delta\ell}\left[R_0^2 - r^2 + 2R_1^2 \ln \frac{r}{R_0}\right] .$$

We determine the flowrate $Q$ by the formula ($\rho$ - density of fluid)

$$Q = 2\pi\rho \int_{R_1}^{R_0} vr\, dr .$$

This gives:

$$Q = \frac{\pi\rho\Delta p}{8\eta\Delta\ell}\left[\left(R_0^2 - R_1^2\right)\left(R_0^2 - 3R_1^2\right) + 4R_1^4 \ln \frac{R_0}{R_1}\right] . \tag{10.2}$$

We introduce the relative moisture content $\tilde{w}$, taking into account the bound water

$$\tilde{w} = \frac{w - w_0}{\sigma - w_0} = 1 - \frac{\sigma - w}{\sigma - w_0} .$$

But, as is evident from Fig. 18, we have:

$$\frac{\sigma - w}{\sigma - w_0} = \frac{\pi R_1^2}{\pi R_0^2} , \quad \frac{R_1}{R_0} = \sqrt{\frac{\sigma - w}{\sigma - w_0}} = \sqrt{1 - \tilde{w}} .$$

We see that the ratio $R_1/R_0$ can be expressed in terms of $\tilde{w}$. In the same way, S. F. Averjanov arrives at the dependence of $\alpha$ on $\tilde{w}$, which for simplification he puts in the form

$$\alpha = \tilde{w}(3\tilde{w} - 2) - 2(1 - \tilde{w})^2 \ln (1 - \tilde{w}) \approx \tilde{w}^n ,$$

in which $n = 3.5$ is taken.

A comparison of available experimental results, collected from different authors, with the theoretical curve gave a good result (Fig. 19). We note that B. P. Gorbunov made an attempt to establish a relation between seepage properties of soil and the quantity of bound water, and also the magnitude of the specific surface of the soil and that he considered methods to determine these quantities [21].

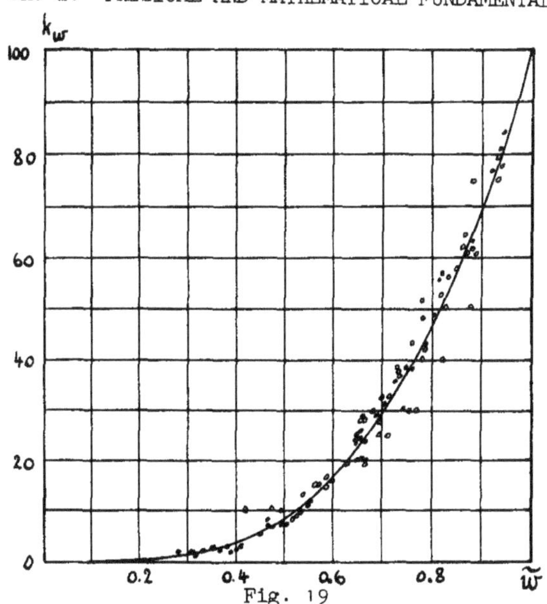


Fig. 19

§11. Equations of Motion of Ground Waters.

Previously we have mentioned that water in porous media may be found in various compositions. To a small degree, it is absorbed by soil grains (hygroscopic water); with increasing moisture, it begins to cover the soil grains under the form of a film; under further increase of moisture, it fills the narrow pores in the form of menisci; finally, water may fill all the pores (with the exclusion, perhaps, of isolated parts, filled with air bubbles). In the latter state, water acquires the property to move under influence of the gravity force. It is called gravitational or ground water. Laws are applicable to this kind of water, that are related to the laws of hydrodynamics. We will be occupied with problems of the flow of so-called gravitational water.

Everywhere out of the soil pores, a complicated movement issues, with velocity and acceleration changing in magnitude and direction from point to point. Pores have different direction and different wall shape, but in every volumetric soil compartment, the same diversified values of the velocity, in magnitude and direction, must exist. Therefore it is impossible to examine the velocities of separate particles.

It should be correct to consider the average value of the velocities in a certain volume. In the theory of seepage, it is conventional to consider not the velocity, but the discharge through an isolated surface. We will designate by $\vec{v}$ the vector of the flowrate through a unit surface. If the average velocity of the particles of a certain volume is equal to $\vec{V}$, then we may refer it to the center of gravity of this volume, the coordinates of which we designate by (x, y, z). Let the vector $\vec{V}$ have the components dx/dt, dy/dt, dz/dt . Then, with the porosity of the soil designated by $\sigma$, we may write:

(11.1) $$\vec{v} = \sigma \vec{V}$$

i.e., if the components of vector $\vec{v}$ are designated by $u, v, w$ then

(11.1[1]) $$u = \sigma \frac{dx}{dt}, \quad v = \sigma \frac{dy}{dt}, \quad w = \sigma \frac{dz}{dt} .$$

In hydrodynamics [32], for the derivation of the equation of motion, the acceleration of the fluid particles is examined, and is expressed in terms of the velocity as:

$$\frac{dv_x}{dt} = \frac{\partial v_x}{\partial t} + \frac{\partial v_x}{\partial x} v_x + \frac{\partial v_x}{\partial y} v_y + \frac{\partial v_x}{\partial z} v_z , \quad \text{and so on} .$$

Considering that the velocities $v_x$, $v_y$, $v_z$ and their derivatives along the coordinates $\partial v_x/\partial x, \ldots$ are small, we will neglect their product, leaving in the expressions of the components of the acceleration only the terms

$$\frac{\partial v_x}{\partial t}, \quad \frac{\partial v_y}{\partial t}, \quad \frac{\partial v_z}{\partial t} .$$

Now let us write the equations of motion of separate fluid particles in the pores, in this form:

(11.2) $$\begin{cases} \dfrac{\partial v_x}{\partial t} = -\dfrac{1}{\rho}\dfrac{\partial p}{\partial x} + f_x , \\ \dfrac{\partial v_y}{\partial t} = -\dfrac{1}{\rho}\dfrac{\partial p}{\partial y} + f_y , \\ \dfrac{\partial v_z}{\partial t} = -\dfrac{1}{\rho}\dfrac{\partial p}{\partial z} + f_z - g . \end{cases}$$

Here $p$ is the pressure in the fluid, $\rho$ the density, $g$ the gravitational acceleration. The $z$ axis is directed upward. By $f_x$, $f_y$, $f_z$ are indicated the components of the resistance forces, which are applied to the fluid particles in the pore. These resistance forces depend upon the friction of fluid particles with soil particles. The forces of internal friction - fluid particles with fluid particles - are negligibly small in comparison with the forces of external friction.

It is possible to apply equations (11.2) to some volume, sufficiently small, to take into account the variation of motion from one place to the other, but sufficiently large in comparison with the dimensions of the pore. Then it is possible to understand under $v_x$, $v_y$, $v_z$, $p$, $f_x$, $f_y$, $f_z$, their average values over the volume. Similarly, under $f_x$, $f_y$, $f_z$ we will understand the components of the friction forces, as obtained from experiment.

In this way, Darcy's law (Chapter I, Par. 4), can be interpreted as a linear law of friction, i.e., such a law, in which the friction is proportional in the first degree to the seepage velocity. Let us write this dependence in the form

(11.3) $$f_x = -\frac{gu}{k}, \quad f_y = -\frac{gv}{k}, \quad f_z = -\frac{gw}{k} .$$

Considering that the coefficient $k$ retains the same value for the unsteady movement as for the steady, we write equations (11.2) with the help of (11.1) in the form

$$(11.4)\qquad \begin{cases} \frac{1}{\sigma}\frac{\partial u}{\partial t} = -\frac{1}{\rho}\frac{\partial p}{\partial x} - \frac{gu}{k}\ , \\ \frac{1}{\sigma}\frac{\partial v}{\partial t} = -\frac{1}{\rho}\frac{\partial p}{\partial y} - \frac{gv}{k}\ , \\ \frac{1}{\sigma}\frac{\partial w}{\partial t} = -\frac{1}{\rho}\frac{\partial p}{\partial z} - \frac{gw}{k} - g\ . \end{cases}$$

It is possible to show that for actual values of the seepage coefficient ($k$ = 1 to 100 meters per 24 hours), the terms containing differentiation in time, may be neglected.

Indeed, each of the equations (11.4) takes on the form [26]

$$\frac{1}{\sigma}\frac{\partial U}{\partial t} = f - \frac{1}{\lambda}U\ .$$

We introduce the difference $V = U - \lambda f$, expressing the digression of $U$ from $\lambda f$. We obtain

$$\frac{1}{\sigma}\frac{\partial V}{\partial t} = -\frac{\lambda}{\sigma}\frac{\partial f}{\partial t} - \frac{V}{\lambda}\ .$$

If $\partial f/\partial t$ remains limited in all times, then

$$\frac{\lambda}{\sigma}\frac{\partial f}{\partial t}$$

will be very small. Indeed with $k = \frac{100\ \text{m}}{24\ \text{hrs}} = \frac{1}{864}\ \frac{\text{m}}{\text{sec}}$, $g = \frac{10\ \text{m}}{\text{sec}^2}$ we obtain

$$\lambda = \frac{k}{g} = \frac{1}{8640}\ \text{sec}.$$

and the term $\lambda/\sigma\ \partial f/\partial t$ may really be neglected, if $\partial f/\partial t$ is not very large (see Chapter XII). The remaining equation

$$\frac{\partial V}{\partial t} = -\frac{\sigma}{\lambda}V$$

integrated, gives $V = V_0 e^{-\frac{\sigma t}{\lambda}}$. The right side of this equation tends so fast to zero, that already after a fraction of a second one may consider $V = 0$; $U = \lambda f$, i.e., one may neglect the terms

$$\frac{1}{\sigma}\frac{\partial u}{\partial t}\ ,\quad \frac{1}{\sigma}\frac{\partial v}{\partial t}\ ,\quad \frac{1}{\sigma}\frac{\partial w}{\partial t}\ .$$

Then the equations (11.4) may be written, by introduction of the piezometric head $h$, in the form

$$(11.5)\qquad \vec{v} = -k\ \text{grad}\ h\ ,\quad h = \frac{p}{\rho g} + z\ .$$

Linear inertia terms remain in special cases.

Movements, in which the inertia forces may be neglected in comparison with the friction forces, are called "creeping" by L. Prandtl [33]. In these movements, the resistance is usually proportional in the first degree to the velocity.

The equations (11.4) contain four unknown functions: $u$, $v$, $w$ and $p$. In addition to these equations, there is the equation of continuity. We consider flowing water as incompressible. The final equation of continuity is arrived at in the same way as in the hydrodynamics of incompressible fluids. We isolate an elemental volume in the form of a parellelipipedum with edges $dx$, $dy$, $dz$, parallel to the coordiante axes. The flowrate through the surface perpendicular to the x-axis and at a distance $x$ from the origin, is equal to $\sigma u\, dy\, dz$. The flowrate through the surface, parallel with the first and having an abscissa $x + dx$, is equal to

$$\left[\sigma u + \frac{\partial}{\partial x}(\sigma u)\, dx\right] dy\, dz\ .$$

The difference between outflowing and inflowing amount of fluid in the direction of the x-axis is equal to

$$\frac{\partial}{\partial x}(\sigma u)\, dx\, dy\, dz\ .$$

In the direction of the $y$ and $z$ axes, similarly we obtain expressions for the changes in flowrate

$$\frac{\partial(\sigma v)}{\partial y}\, dx\, dy\, dz\ ,\quad \frac{\partial(\sigma w)}{\partial z}\, dx\, dy\, dz\ .$$

Adding the three expressions thus obtained, we have the total excess of the amount of fluid that leaves the volume $dx\, dy\, dz$ over the amount of fluid that enters the volume. The sum must be equal to zero in the case of an incompressible fluid. After dividing by $dx\, dy\, dz$ we obtain the equation of continuity.

$$\frac{\partial(\sigma u)}{\partial x} + \frac{\partial(\sigma v)}{\partial y} + \frac{\partial(\sigma w)}{\partial z} = 0 \tag{11.6}$$

If the porosity $\sigma$ of the soil does not depend upon the coordinates, then we have

$$\frac{\partial u}{\partial x} + \frac{\partial v}{\partial y} + \frac{\partial w}{\partial z} = 0 \tag{11.7}$$

i.e., the equation of continuity has the same form as in the hydrodynamics of imcompressible fluids. From the equations (11.5) it follows that the seepage velocity has a potential (for constant $k$)

$$\varphi(x,\ y,\ z) = -\ k\left(\frac{p}{\rho g} + z\right) + C = -kh + C \tag{11.8}$$

($C$ - arbitrary constant), so that

$$\vec{V} = \text{grad}\ \varphi = -k\ \text{grad}\ h\ ,\quad h = \frac{p}{\rho g} + z\ . \tag{11.9}$$

From the equation of continuity it follows that the velocity potential satisfies Laplace's equation $\Delta\varphi = 0$ or $\nabla^2\varphi = 0$, where

$$\Delta\varphi \equiv \nabla^2\varphi \equiv \frac{\partial^2\varphi}{\partial x^2} + \frac{\partial^2\varphi}{\partial y^2} + \frac{\partial^2\varphi}{\partial z^2} \quad . \tag{11.10}$$

## *§12. Equations of Motion for Nonlinear Seepage Laws.

This problem is treated in the work of S. A Khristianovich [25]. As was stated in Par. 6 of this chapter, the movement of ground waters in the general case is subordinated to a relation of the form

$$I = \Phi(W) \tag{12.1}$$

where the hydraulic slope $I = -\partial H/\partial s$ is linked with the seepage velocity $W$. For Darcy's law $I = W/k$; in other cases experimental data lead to a law of the form $I = aW + bW^2 + cW^3$; $I = 1/\kappa\, W^\gamma$ and so on.

S. A. Khristianovich considers the equation $\text{grad } H = -\Phi(W)\, \vec{W}/W$, or

$$\frac{\partial H}{\partial x} = -\Phi(W)\frac{u}{W}, \qquad \frac{\partial H}{\partial y} = -\Phi(W)\frac{v}{W} \qquad (W = \sqrt{u^2 + v^2}) \tag{12.2}$$

Considering that the liquid is incompressible, we add to these equations the equation of continuity (we limit ourselves to the two-dimensional case)

$$\frac{\partial u}{\partial x} + \frac{\partial v}{\partial y} = 0 \tag{12.3}$$

We introduce the notation

$$K = \frac{W}{\Phi(W)} \quad . \tag{12.4}$$

One may conditionally consider $K$ as a seepage coefficient, depending upon the velocity. Elimination of $H$ from (12.2) gives:

$$\frac{\partial\left(\frac{u}{K}\right)}{\partial y} - \frac{\partial\left(\frac{v}{K}\right)}{\partial x} = 0 \; . \tag{12.5}$$

Instead of $u, v$ one may take as new functions the magnitude of the velocity vector $W$ and its angle with the abscissa axis $\theta$: $u = W\cos\theta$, $v = W\sin\theta$. Then, instead of (12.3) and (12.5), we obtain this system

$$\begin{gathered} \cos\theta\frac{\partial W}{\partial x} + \sin\theta\frac{\partial W}{\partial y} - W\sin\theta\frac{\partial\theta}{\partial x} + W\cos\theta\frac{\partial\theta}{\partial y} = 0 \\ \Phi'(W)\left[\sin\theta\frac{\partial W}{\partial x} - \cos\theta\frac{\partial W}{\partial y}\right] + \Phi(W)\left[\cos\theta\frac{\partial\theta}{\partial x} + \sin\theta\frac{\partial\theta}{\partial y}\right] = 0 \; . \end{gathered} \tag{12.6}$$

From the continuity equation is derived the important function $\psi$, so that

$$\frac{\partial\psi}{\partial x} = -v = -W\sin\theta \; , \quad \frac{\partial\psi}{\partial y} = u = W\cos\theta \quad . \tag{12.7}$$

Let us take a function $F(x, y)$. If we introduce new independent variables $H$ and $\psi$, then we obtain on the basis of (12.2) these equations:

$$
(12.8) \qquad \begin{cases} \dfrac{\partial F}{\partial x} = -\,\Phi(W)\cos\theta\,\dfrac{\partial F}{\partial H} - W\sin\theta\,\dfrac{\partial F}{\partial \psi} \,, \\[2ex] \dfrac{\partial F}{\partial y} = -\,\Phi(W)\sin\theta\,\dfrac{\partial F}{\partial H} + W\cos\theta\,\dfrac{\partial F}{\partial \psi} \,. \end{cases}
$$

With the help of these formulas, the system (12.6) is put into the form

$$
(12.9) \qquad \frac{\partial \theta}{\partial \psi} - \frac{\Phi(W)}{W^2}\,\frac{\partial W}{\partial H} = 0 \,, \quad \frac{\partial \theta}{\partial H} + \frac{W\Phi'(W)}{[\Phi(W)]^2}\,\frac{\partial W}{\partial \psi} = 0 \,.
$$

Before going to further transformations, let us examine the problem of the reduction of a system of equations of the elliptical type into its canonic form.

Given the system

$$
(12.10) \qquad \begin{cases} a_{11}\dfrac{\partial u}{\partial x} + a_{12}\dfrac{\partial v}{\partial x} + b_{11}\dfrac{\partial u}{\partial y} + b_{12}\dfrac{\partial v}{\partial y} = e_1 \,, \\[2ex] a_{21}\dfrac{\partial u}{\partial x} + a_{22}\dfrac{\partial v}{\partial x} + b_{21}\dfrac{\partial u}{\partial y} + b_{22}\dfrac{\partial v}{\partial y} = e_2 \,. \end{cases}
$$

The coefficients $a_{ik}$, $b_{ik}$, and $e_k$ of which are essentially functions of $x$, $y$, $u$ and $v$.

To find the equations of the characteristics, we add to equations (12.10) the equations

$$
(12.11) \qquad du = \frac{\partial u}{\partial x}dx + \frac{\partial u}{\partial y}dy \,, \quad dv = \frac{\partial v}{\partial x}dx + \frac{\partial v}{\partial y}dy \,.
$$

We shall solve the system of equations (12.10) - (12.11) with respect to the derivatives $\partial u/\partial x$, $\partial u/\partial y$, $\partial v/\partial x$, $\partial v/\partial y$. Therefore we write (12.11) as

$$
(12.12) \qquad \frac{\partial u}{\partial x} = \frac{du}{dx} - \frac{\partial u}{\partial y}\frac{dy}{dx} \,, \quad \frac{\partial v}{\partial x} = \frac{dv}{dx} - \frac{\partial v}{\partial y}\frac{dy}{dx} \,.
$$

We designate $dy/dx = \lambda$ and insert the values of (12.12) in (12.10). We obtain a system of two equations, linear with respect to $\partial u/\partial y$ and $\partial v/\partial y$:

$$
(12.13) \qquad \begin{cases} (b_{11} - a_{11}\lambda)\dfrac{\partial u}{\partial y} + (b_{12} - a_{12}\lambda)\dfrac{\partial v}{\partial y} = e_1 - a_{11}\dfrac{du}{dx} - a_{12}\dfrac{dv}{dx} \,, \\[2ex] (b_{21} - a_{21}\lambda)\dfrac{\partial u}{\partial y} + (b_{22} - a_{22}\lambda)\dfrac{\partial v}{\partial y} = e_2 - a_{21}\dfrac{du}{dx} - a_{22}\dfrac{dv}{dx} \,. \end{cases}
$$

From this we find $\partial u/\partial y$, $\partial v/\partial y$:

$$
\frac{\partial u}{\partial y} = \frac{\Delta_u}{\Delta} \,, \quad \frac{\partial v}{\partial y} = \frac{\Delta_v}{\Delta} \,,
$$

where

$$
\Delta = \begin{vmatrix} b_{11} - a_{11}\lambda & b_{12} - a_{12}\lambda \\ b_{21} - a_{21}\lambda & b_{22} - a_{22}\lambda \end{vmatrix} \,,
$$

$$
\Delta_u = \begin{vmatrix} e_1 - a_{11}\dfrac{du}{dx} - a_{12}\dfrac{dv}{dx} & b_{12} - a_{12}\lambda \\[2ex] e_2 - a_{21}\dfrac{du}{dx} - a_{22}\dfrac{dv}{dx} & b_{22} - a_{22}\lambda \end{vmatrix} \,,
$$

$$\Delta_v = \begin{vmatrix} b_{11} - a_{11}\lambda & e_1 - a_{11}\dfrac{du}{dx} - a_{12}\dfrac{dv}{dx} \\ b_{21} - a_{21}\lambda & e_2 - a_{21}\dfrac{du}{dx} - a_{22}\dfrac{dv}{dx} \end{vmatrix}$$

The equations of the characteristics are obtained from the conditions that along their direction no unique solution exists for $\partial u/\partial y$, $\partial v/\partial y$, i.e., from the equations $\Delta = 0$, $\Delta_u = 0$, $\Delta_v = 0$ . One of these equations is a consequence of the two others. Developing the determinants $\Delta$ and $\Delta_u$ and introducing the notation

$$\begin{vmatrix} a_{11} & a_{12} \\ a_{21} & a_{22} \end{vmatrix} = A , \quad \begin{vmatrix} b_{11} & b_{12} \\ b_{21} & b_{22} \end{vmatrix} = C , \quad \begin{vmatrix} b_{12} & a_{12} \\ b_{22} & a_{22} \end{vmatrix} = D ,$$

$$\begin{vmatrix} b_{12} & a_{11} \\ b_{22} & a_{21} \end{vmatrix} = E , \quad \begin{vmatrix} a_{12} & b_{11} \\ a_{22} & b_{21} \end{vmatrix} = F , \quad \begin{vmatrix} e_1 & b_{12} \\ e_2 & b_{22} \end{vmatrix} = M , \quad \begin{vmatrix} a_{12} & e_1 \\ a_{22} & e_2 \end{vmatrix} = N ,$$

$$2B = F + E ,$$

we obtain these equations:

$$A\lambda^2 + 2B\lambda + C = 0 , \quad (E + A\lambda)\frac{du}{dx} + D\frac{dv}{dx} + M + N\lambda = 0$$

Solving the first equation of $\lambda$, we have:

$$\lambda = -\frac{B}{A} \pm \frac{\sqrt{B^2 - AC}}{A} .$$

If

(12.14) $$B^2 - AC < 0$$

it is stated that the given system is of the elliptic type. Supposing that this condition is fulfilled, we put $\lambda_1 = P + iQ$ where

$$P = -\frac{B}{A} , \quad Q = \frac{\sqrt{AC - B^2}}{A} .$$

We may write the equations of the characteristics in the form

(12.15) $$dy = \lambda_1 dx ,$$

(12.16) $$(E + \lambda_1 A)\, du + D\, dv + (M + N\lambda_1)\, dx = 0 .$$

Let us assume that we have found a solution of the system (12.10) $u = u(x, y)$, $v = v(x, y)$. To determine the characteristics, corresponding to this solution, let us put $P$ and $Q$ instead of $u$ and $v$ for their values $u(x, y)$, $v(x,y)$. Let the general integral of (12.15) by $\mu(x, y) + i\nu(x, y) = \text{const}$. Differentiating this expression and making use of (12.15), we find

$$\frac{\partial}{\partial x}(\mu + i\nu) + (P + iQ)\frac{\partial}{\partial y}(\mu + i\nu) = 0$$

Separation of the real and imaginary parts gives

$$(12.17)\qquad \frac{\partial\mu}{\partial x} + P\frac{\partial\mu}{\partial y} - Q\frac{\partial\nu}{\partial y} = 0\,, \quad \frac{\partial\nu}{\partial x} + Q\frac{\partial\mu}{\partial y} + P\frac{\partial\nu}{\partial y} = 0\ .$$

We examine the system of equations (12.10) and (12.17), taking for independent variables $\mu, \nu$ but for unknown functions $x, y, u, v$. The transformation formulas are:

$$(12.18)\qquad \frac{\partial\mu}{\partial x} = \frac{1}{\Delta}\frac{\partial y}{\partial\nu}\,, \quad \frac{\partial\mu}{\partial y} = -\frac{1}{\Delta}\frac{\partial x}{\partial\nu}\,, \quad \frac{\partial\nu}{\partial x} = -\frac{1}{\Delta}\frac{\partial y}{\partial\mu}\,, \quad \frac{\partial\nu}{\partial y} = \frac{1}{\Delta}\frac{\partial x}{\partial\mu}\,,$$

where

$$\Delta = \frac{\partial x}{\partial\mu}\frac{\partial y}{\partial\nu} - \frac{\partial y}{\partial\mu}\frac{\partial x}{\partial\nu} = Q\left\{\left(\frac{\partial x}{\partial\mu}\right)^2 + \left(\frac{\partial x}{\partial\nu}\right)^2\right\}.$$

We carry out the substitution of the variables in (12.17) and (12.10), and in the last equations we replace the derivatives $\partial u/\partial x$, $\partial u/\partial y$ ... by the derivatives with respect to $\mu, \nu$ using the expressions (12.17) and (12.18). We obtain the system

$$(12.19)\qquad \begin{cases} \dfrac{\partial y}{\partial\nu} - P\dfrac{\partial x}{\partial\nu} - Q\dfrac{\partial x}{\partial\mu} = 0\,, \quad \dfrac{\partial y}{\partial\mu} + Q\dfrac{\partial x}{\partial\nu} - P\dfrac{\partial x}{\partial\mu} = 0\,, \\[2mm] \dfrac{F - E}{2}\dfrac{\partial u}{\partial\mu} + \sqrt{AC - B^2}\,\dfrac{\partial u}{\partial\nu} - D\dfrac{\partial v}{\partial\mu} = M\dfrac{\partial x}{\partial\mu} + N\dfrac{\partial y}{\partial\mu}\,, \\[2mm] \dfrac{F - E}{2}\dfrac{\partial u}{\partial\nu} - \sqrt{AC - B^2}\,\dfrac{\partial u}{\partial\mu} - D\dfrac{\partial v}{\partial\nu} = M\dfrac{\partial x}{\partial\nu} + N\dfrac{\partial y}{\partial\nu}\,, \end{cases}$$

which we may call the "canonic system." In it, $x, y, u, v$ are functions of the parameters $\mu, \nu$. The solution of this system for a determinant $\Delta$ which differs from zero will also be a solution of the initial system.

The obtained system does not change if we replace $\mu, \nu$ by the variables $\mu^*, \nu^*$ where $\mu^* + i\nu^* = f(\mu + i\nu)$, an analytic function. If the equations of the characteristics allow an integrated combination, i.e., if there exist an expression $p(x, y, u, v) + iq(x, y, u, v) = \text{const.}$, satisfying equations (12.15), (12.16), then, if we take $p$ and $q$ as new unknowns instead of $x, y$ or $u, v$, two equations of the system will be the Cauchy-Riemann equations, or $p + iq$ will be an analytic function of $\mu + i\nu$.

Let us now go back to system (12.9). Instead of $x, y, u, v$ in this system, there are $\psi, H, W, \theta$. The canonic system with the parameters $\mu, \nu$ for the system (12.9) will have this form

$$(12.20)\qquad \begin{gathered} \frac{\partial\psi}{\partial\nu} + \frac{W}{\Phi(W)}\sqrt{\frac{W\Phi'(W)}{\Phi(W)}}\,\frac{\partial H}{\partial\mu} = 0\,, \quad \frac{\partial\psi}{\partial\mu} - \frac{W}{\Phi(W)}\sqrt{\frac{W\Phi'(W)}{\Phi(W)}}\,\frac{\partial H}{\partial\nu} = 0\,, \\ \sqrt{\frac{\Phi'(W)}{W\Phi(W)}}\,\frac{\partial W}{\partial\nu} - \frac{\partial\theta}{\partial\mu} = 0\,, \qquad \sqrt{\frac{\Phi'(W)}{W\Phi(W)}}\,\frac{\partial W}{\partial\mu} + \frac{\partial\theta}{\partial\nu} = 0\ . \end{gathered}$$

Instead of the function $W$ one may introduce a fictitious seepage velocity $\widetilde{W}$ and the variable $S$ by means of the equality

$$(12.21)\qquad S = \ln\widetilde{W} = \int\sqrt{\frac{\Phi'(W)}{W\Phi(W)}}\,dW\ .$$

For Darcy's law $W = c\widetilde{W}$, for the power law $\widetilde{W} = cW^{\sqrt{\gamma}}$. Now system (12.20) can be rewritten:

(12.22)
$$\frac{\partial \ln \widetilde{W}}{\partial \nu} - \frac{\partial \theta}{\partial \mu} = 0 , \qquad \frac{\partial \ln \widetilde{W}}{\partial \mu} + \frac{\partial \theta}{\partial \nu} = 0 ,$$
$$\frac{\partial H}{\partial \mu} = - L \frac{\partial \psi}{\partial \nu} , \qquad \frac{\partial H}{\partial \nu} = L \frac{\partial \psi}{\partial \mu} ,$$

where

(12.23)
$$L = \frac{\Phi(W)}{W} \sqrt{\frac{\Phi(W)}{W\Phi'(W)}} = L(\widetilde{W}) .$$

For the power law
$$L = \frac{1}{\sqrt{\gamma}} \frac{W^{\gamma-1}}{\kappa} .$$

By means of formula (12.8), it is possible to find the derivatives of $x$ and $y$ with respect to $H$, $\psi$.

$$\frac{\partial x}{\partial H} = - \frac{\cos \theta}{\Phi(W)} , \quad \frac{\partial x}{\partial \psi} = - \frac{\sin \theta}{W} , \quad \frac{\partial y}{\partial H} = - \frac{\sin \theta}{\Phi(W)} , \quad \frac{\partial y}{\partial \psi} = \frac{\cos \theta}{W} .$$

From here, we find for $x, y$ expressions in the form of curvilinear integrals.

$$x = - \int \frac{\cos \theta}{\Phi(W)} dH + \frac{\sin \theta}{W} d\theta , \quad y = - \int \frac{\sin \theta}{\Phi(W)} dH - \frac{\cos \theta}{W} d\theta .$$

We examine the conditions at the boundaries of the flow region. Along rectilinear impervious walls, $\psi = \text{const.}$ and $\theta = \text{const.}$ Along rectilinear boundaries of a reservoir, $H = \text{const.}$, $\theta = \text{const.}$ Along a free surface, $\psi = \text{const.}$ and the pressure is constant, i.e., $p = \rho g(H - y) = \text{const.}$ By differentiation of this equation along the free surface, we obtain:

$$\frac{\partial H}{\partial x} + \frac{\partial H}{\partial y} \frac{dy}{dx} - \frac{dy}{dx} = - u \frac{\Phi(W)}{W} - \frac{v^2}{u} \frac{\Phi(W)}{W} - \frac{v}{u} = 0$$

or

$$\Phi(W) + \sin \theta = 0 .$$

We obtain a closed curve in the $u, v$ plane. Along a seepage surface - a rectilinear segment at an angle $\theta_0$ with the axis - we have constant pressure. By differentiation along this segment of constant pressure $p$, we obtain:

$$\frac{\partial H}{\partial x} + \frac{\partial H}{\partial y} \operatorname{tg} \theta_0 - \operatorname{tg} \theta_0 = 0$$

or

$$u \frac{\Phi(W)}{W} + v \frac{\Phi(W)}{W} \operatorname{tg} \theta_0 + \operatorname{tg} \theta_0 = 0 .$$

which can be written in the form

$$\Phi(W) \cos (\theta - \theta_0) + \sin \theta_0 = 0 .$$

For the solution of concrete problems, it is recommended to look in the $\mu, \nu$ plane for the region which exhibits a flow that satisfies Darcy's law and then to look for a correction of the digression of this law.

V. V. Sokolovsky [34] examined the particular case where

$$\Phi(W) = \frac{W}{k\sqrt{1 - \left(\frac{W}{m}\right)^2}} .$$

This expression, selected from the consideration of integrability of the equations of the characteristics, is artificial; however it may, in some interval of the values of $W$, give an estimate of the solution in the case of digression from the linear seepage law.

CHAPTER II.

# TWO-DIMENSIONAL FLOWS IN A VERTICAL PLANE

## A. General Considerations.

### §1. Equations of Motions in a Plane.

The equations of two-dimensional flow of ground waters may be written in the form [see equations (11.9) of Chapter I]

$$(1.1)\qquad u = \frac{\partial\varphi}{\partial x} = -k\frac{\partial h}{\partial x}, \quad v = \frac{\partial\varphi}{\partial y} = -k\frac{\partial h}{\partial y},$$

Taking the y-axis vertical, oriented upward, and the x-axis horizontal, we have:

$$(1.2)\qquad h = \frac{p}{\rho g} + y + C, \quad \varphi = -k\left(\frac{p}{\rho g} + y\right) + C_1.$$

If the y-axis is oriented downward, then

$$(1.2')\qquad h = \frac{p}{\rho g} - y + C, \quad \varphi = -k\left(\frac{p}{\rho g} - y\right) + C_1.$$

According to formulas (11.10) of Chapter I,

$$(1.3)\qquad \Delta\varphi = \frac{\partial^2\varphi}{\partial x^2} + \frac{\partial^2\varphi}{\partial y^2} = 0.$$

From the equation of continuity

$$(1.4)\qquad \frac{\partial u}{\partial x} + \frac{\partial v}{\partial y} = 0$$

it follows that there exist a streamfunction $\Psi(x,y)$ such that

$$(1.5)\qquad u = \frac{\partial\Psi}{\partial y}, \quad v = -\frac{\partial\Psi}{\partial x}.$$

Comparing (1.1) and (1.5) we obtain

$$\frac{\partial\varphi}{\partial x} = \frac{\partial\Psi}{\partial y}, \quad \frac{\partial\varphi}{\partial y} = -\frac{\partial\Psi}{\partial x}.$$

These are the d'Alembert-Euler (Cauchy-Riemann) conditions. If they are fulfilled, then as is well known, the linear combination of the functions $\varphi$ and $\Psi$

$$(1.6)\qquad \omega = \varphi + i\Psi$$

is a function of the complex variable $z = x + iy$. The function $\omega$ is called the "complex potential".

§2. Boundary Conditions in Two-dimensional Steady Motion.

In Fig 20, is represented the vertical cross-section of an overflow weir, normal to the axis of the weir [8]. Its underground base, called "wetted contour", consists of the "upstream apron"1, the function of which it is to lengthen the seepage path under the weir, the "water breaker" 2, absorbing the impact of the falling water and also lengthening the seepage path, the "water passing part" or filter blanket 3, protecting the riverbed from wash-out and assuring a safe exit of the seeping waters into the tailwater, and, finally, the "cut-off walls" 4, i.e., vertical walls extending deeper from the base, serving also to lengthen the seepage path. The upstream apron of the weir base may also be absent. The filter blanket is established perviously and, for seepage computations, its foundation is not included in the underground contour of the hydraulic structure. In a retaining dam, which does not have any water overflow, filter-blanket and waterbreaker are not used [9].

When dealing with problems about seepage of water under hydraulic structures, in the body of earth dams, and also in the percolation of water from channels into the subsoil and so on, we encounter four kinds of boundaries of the flow region.

1. "Impervious boundaries". These are the underground contours of hydraulic structures (Fig. 20), and also the boundaries of the flow region with impervious soil.

Impervious boundaries are flowlines, and along them the streamfunction must have a constant value

$$\Psi = \text{constant}. \tag{2.1}$$

Usually, impervious boundaries consist of rectilinear segments. Let the equation of such a segment (for example, BC in Fig. 21) be:

$$ax + by + c = 0. \tag{2.2}$$

Equations (2.1) and (2.2) may be considered as two conditions that must be fulfilled along impervious boundaries.

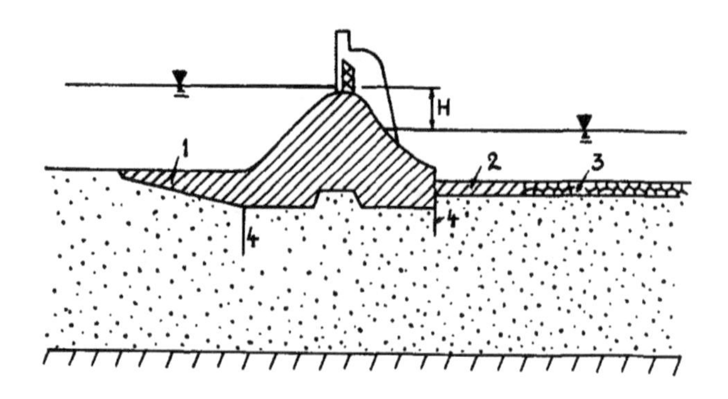


Fig. 20

2. "Boundaries of water reservoirs". For large dimensions of a water reservoir, one may consider that the water pressure is distributed according to the hydrostatic law. Therefore, in an arbitrary point $M$, located on the boundary $AB$ between soil and water body (Fig. 21), at a height $y$ measured from an arbitrary horizontal $x$-axis, we have for the pressure the expression

$$p = p_a + \rho g(H_1 - y) \tag{2.3}$$

where $p_a$ - the pressure at the surface of the reservoir, equal to the atmospheric pressure. The second term of formula (2.3) expresses the weight of the column of liquid that is applied to the unit area in point $M$.

Inserting (2.3) in formula (1.2) for $\varphi$, we obtain:

$$\varphi = - k\left(\frac{p_a}{\rho g} + H_1\right) + C, \tag{2.4}$$

Thus,

$$\varphi = \text{const.} \tag{2.5}$$

along the boundary of the water body; in other words, the boundary of the water body is a line of constant potential.

Assuming that the boundary of the water body consists of a rectilinear segment,

$$a_1 x + b_1 y + c_1 = 0, \tag{2.6}$$

we will consider in the future (2.5) and (2.6) as two conditions along the boundary of a water reservoir.

3. "Line of free surface", consists of a boundary line between wet and dry soil. Two forms of conditions for the pressure at the free surface must be considered:

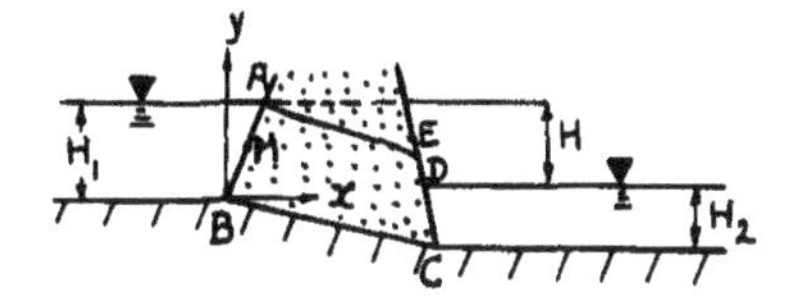


Fig. 21

(a) This pressure is equal to the atmospheric pressure, which is there because the soil pores communicate mutually with the atmospheric air. We may take the atmospheric pressure equal to zero ($p_a = 0$). Along the free surface line, which is also called "depressed curve", we obtain the condition:

$$\varphi + ky = \text{constant.} \tag{2.7}$$

When the $y$-axis is directed downwards, then (2.7) is changed into the equation

$$\varphi - ky = \text{constant.} \tag{2.7$^1$}$$

(b) Because of the presence of soil capillarity, it has been assumed [1,3], that along the free surface the pressure has a constant value, smaller than atmospheric in magnitude, keeping in mind the height of capillary rise in the soil:

$$p = p_a - \rho g h_k . \tag{2.8}$$

Observations have shown, that usually for the flow of ground waters, one has to take a value for $h_k$ in formula (2.8), smaller than that obtained for water in the pipe with soil, about which we wrote in Par. 6, Ch. I (see further in Ch. V, Par. 4). Substituting for $p$ its value (2.8) in formula (1.2), we obtain again condition (2.7), but only with another value of the constant.

One of the most important problems in the theory of the flow of g ground waters is the search for the form of the free surface, the equation of which is not known beforehand. But on the free surface, one must also satisfy another condition: the free surface must be a streamfunction, i.e.,

$$\psi = \text{constant} \tag{2.9}$$

along the free surface. So, along the free surface two equations must be satisfied, (2.7) and (2.9).

Interest is presented by the case, where precipitation or percolated water reaches the free surface (due to irrigation, melting snow and so on). In this case, " infiltration" is said to take place through the earth's surface to the free surface of the ground waters.

The following law is applied to flow of moisture to the free surface: the flux of water through some part of the free surface is proportional to the horizontal projection of the arc of this surface, or, otherwise, proportional to the difference of the abscissas of the endpoints of this arc:

$$\psi - \psi_0 = \varepsilon(x - x_0) \tag{2.10}$$

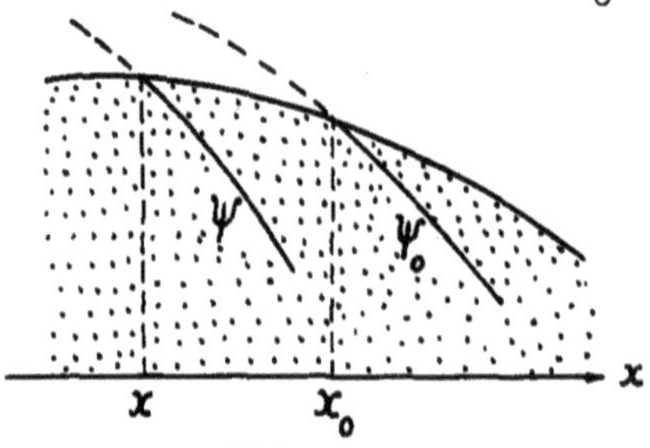

Fig. 22

Here $\psi$ and $\psi_0$ are the values of the streamfunction corresponding to points of the free surface with abscissas $x$ and $x_0$ (Fig. 22) and $\varepsilon$ - the amount of water, reaching the unit length of horizontal projection of the arc of the free surface in the unit of time. For the case under consideration — infiltration — $\varepsilon$ is positive.

The same law, expressing equality (2.10) holds also in case of evaporation from the free surface with the only difference that $\varepsilon$ is negative.

Thus, we may consider that on the free surface conditions (2.7) and (2.10) are satisfied, in which $\varepsilon$ may be positive, negative or zero.

4. "Seepage surface". In an earthdam, there may exist strips, where water flows out of the dam not into a water reservoir but directly into the atmosphere (Fig. 21, strip ED). Such strips are called "seepage surfaces". They also exist along walls of wells, drainage canals, and others. Along these, the pressure must be atmospheric i.e., satisfy eq. (2.7). If the seepage surface is rectilinear, then to this equation must be joined the equation of a straight line

$$a_2x + b_2y + c_2 = 0. \tag{2.11}$$

§3. Conditions on the Boundary of Two Soils.

Let us assume that ground water flows through two soils with different seepage coefficients $k_1$ and $k_2$, separated from each other by the line KL (Fig. 23).

Designating the complex potential of the first and second medium respectively by $\omega_1$ and $\omega_2$

$$\omega_1 = \varphi_1 + i\psi_1, \quad \omega_2 = \varphi_2 + i\psi_2 ,$$

we have for the velocity potential expressions

$$\varphi_1 = - k_1\left(\frac{p_1}{\rho g} + y\right) + C_1, \quad \varphi_2 = - k_2\left(\frac{p_2}{\rho g} + y\right) + C_2, \tag{3.1}$$

where $p_1$ and $p_2$ - the corresponding pressures in the first and second medium. Since at the boundary of two media the pressure must remain continuous, we must have $p_1 = p_2$ in each point of the line KL; from (3.1) we obtain

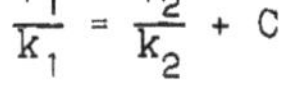

$$\frac{\varphi_1}{k_1} = \frac{\varphi_2}{k_2} + C$$

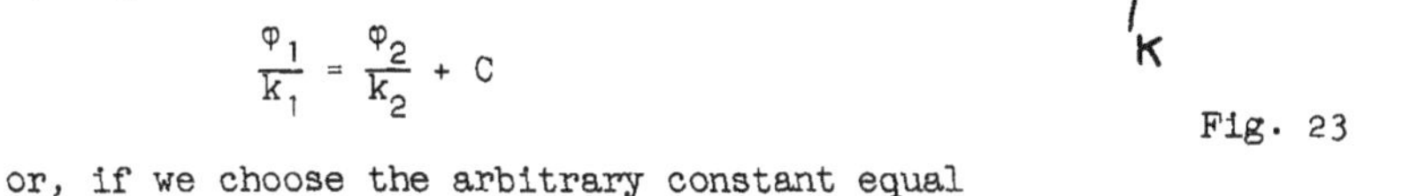



Fig. 23

or, if we choose the arbitrary constant equal to zero,

$$\frac{\varphi_1}{k_1} = \frac{\varphi_2}{k_2}$$

at the linear boundary of the two soils.

We obtain other conditions, by expressing that the normal component of the velocity to the linear boundary must be continuous. If we designate by $v_{1n}$ and $v_{2n}$ the normal components of the velocity approaching the linear boundary from the first and second medium, then we have along KL

$$v_{1n} = v_{2n} .$$

If we introduce the streamfunctions $\psi_1$ and $\psi_2$ for the media, then the last equality may be written in the form

$$\frac{\partial\psi_1}{\partial s} = \frac{\partial\psi_2}{\partial s}$$

Integrating the obtained expression along s (s-arc length of linear boundary) and taking the integration constant equal to zero, we obtain, along the linear boundary KL

$$\psi_1 = \psi_2 . \tag{3.3}$$

Equations (3.2) and (3.3) represent conditions, that must be satisfied at the boundary of the two media. We now differentiate (3.2) along the arc length of the linear boundary. We obtain

$$\frac{1}{k_1}\frac{\partial\varphi_1}{\partial s} = \frac{1}{k_2}\frac{\partial\varphi_2}{\partial s},$$

or, introducing tangential components of the velodity $v_{1s}$ and $v_{2s}$, we have:

$$\frac{v_{1s}}{k_1} = \frac{v_{2s}}{k_2}.$$

As can be seen from Fig. 23

$$\frac{v_{1s}}{v_{1n}} = \operatorname{tg}\alpha_1, \quad \frac{v_{2s}}{v_{2n}} = \operatorname{tg}\alpha_2,$$

where $\alpha_1$ and $\alpha_2$ are the angles between the normal to the linear boundary and the velocity vectors. On the basis of the dependence between the normal and tangential components of the velocity, it is possible to obtain the "law of refraction" of a two-layered medium:

$$\frac{\operatorname{tg}\alpha_1}{k_1} = \frac{\operatorname{tg}\alpha_2}{k_2} \tag{3.4}$$

§4. The Velocity Hodograph.

The complex potential $\omega$ represents a function of the variable z - complex coordinate of the points in the flow region:

$$\omega = f(z).$$

The derivative is called "complex velocity"

$$\frac{d\omega}{dz} = u - iv. \tag{4.1}$$

This velocity is, as well as $\omega$, a function of the complex variable $z$

$$u - iv = w(z). \tag{4.2}$$

To the region of the variable $z$, i.e., to the region occupied with moving fluid, must correspond a certain region in the $w$ plane, which is called the "velocity hodograph".

If the boundary of the flow region consists of rectilinear segments and a free surface, then, as we now will show, the corresponding region in the $u, v$ plane will be bounded by straight lines and a circular arc [7].

For the proof, we consider different forms of the boundary. For more clearness, instead of the $u - iv$ plane, we will consider the $u + iv$ plane. The magnitude of the angles will interest us, and they are the same as in the $u - iv$ plane (the direction of their reading will be opposite).

1. At an impervious boundary, the velocity vector is directed along this boundary. Therefore, if the impervious wall makes an angle $\alpha$ with the abscissa acis, then the projections of the velocity are linked by the equation

$$\frac{v}{u} = \operatorname{tg}\alpha \tag{4.3}$$

i.e., in the $u, v$ plane we have a straight line passing through the origin of coordinates and parallel with the boundary.

2. The boundary of a water reservoir is an equipotential line, and therefore the velocity vector is perpendicular to this boundary. If the equation of the boundary is

$$y = x \operatorname{tg} \alpha + b ,$$

then the endpoint of the velocity vector must lie on the straight line

$$\frac{v}{u} = -\operatorname{cotg} \alpha \tag{4.4}$$

passing through the origin of coordinates in the $u, v$ plane and perpendicular to the considered boundary.

3. Along the free surface, we have the equation (2.7)

$$\varphi + ky = \text{constant}.$$

Differentiating this equation along $s$, where $s$ -arc length of the free surface, we obtain

$$\frac{\partial \varphi}{\partial s} + k \frac{dy}{ds} = 0. \tag{4.5}$$

From this, after multiplying with $\frac{\partial \varphi}{\partial s}$, we obtain

$$\left(\frac{\partial \varphi}{\partial s}\right)^2 + k \frac{\partial \varphi}{\partial s} \frac{dy}{ds} = 0.$$

But $\frac{\partial \varphi}{\partial s}$ is the magnitude of the velocity vector, therefore

$$\left(\frac{\partial \varphi}{\partial s}\right)^2 = u^2 + v^2, \quad \frac{\partial \varphi}{\partial s} \frac{dy}{ds} = v ,$$

and the condition at the free surface can be written as

$$u^2 + v^2 + kv = 0. \tag{4.6}$$

This is the equation of a circle with radius $\frac{k}{2}$ and center in the point $(0, -\frac{k}{2})$, tangent to the abscissa axis in the coordinate origin and cutting off the ordinate axis a segment, with magnitude equal to the seepage coefficient $k$.

In the presence of infiltration or evaporation at the free surface, we have the equations

$$\varphi + ky = \text{constant}, \quad \psi - \varepsilon x = \text{constant}. \tag{4.7}$$

Equation (4.5) holds here also, however $\frac{\partial \varphi}{\partial s}$ no longer will be the magnitude of the total velocity, but equal to the projection of the velocity in the direction of a tangent to the free surface.

Designating by $\alpha$ the angle of the tangent with the $u$ axis, we have:

$$\frac{\partial \varphi}{\partial s} = u \cos \alpha + v \sin \alpha, \quad \frac{dy}{ds} = \sin \alpha, \tag{4.8}$$

on which basis equation (4.5) takes on the form

$$u \cos \alpha + v \sin \alpha + k \sin \alpha = 0. \tag{4.9}$$

Differentiation of the second of equation (4.7) along $s$ leads to the equation

$$u \sin \alpha - v \cos \alpha - \varepsilon \cos \alpha = 0 \tag{4.10}$$

since we have the ratio:

$$\frac{\partial\varphi}{\partial s} = - v_n = u \sin\alpha - v \cos\alpha .$$

Eliminating the variable $\alpha$ from equations (4.9) and (4.10), we obtain:

(4.11) $$u^2 + v^2 + (k + \varepsilon) v + k\varepsilon = 0.$$

This is the equation of a circle, passing through the points $(0, -\varepsilon)$ and $(0, -k)$, having its center at the point $(0, -\frac{\varepsilon+k}{2})$ and its radius equal to $\frac{k-\varepsilon}{2}$ (Fig. 24). It is evident that for $\varepsilon = 0$ we obtain equation (4.6).

For $\varepsilon < 0$, we have evaporation from the free surface (Fig. 24,b).

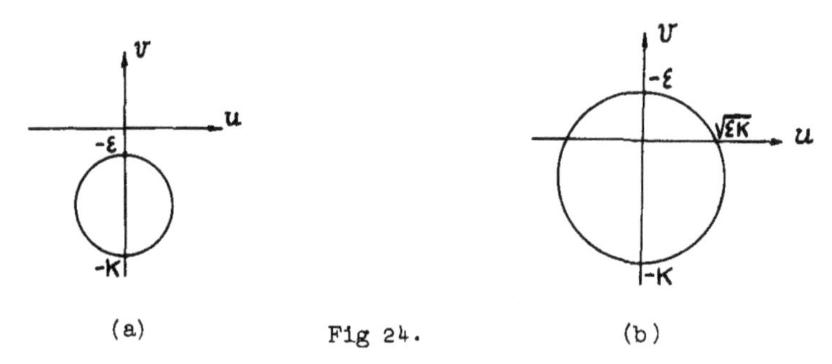


(a) Fig 24. (b)

4. Along the seepage surface, where the pressure is constant, equation (4.5) holds and so does equation (4.9), in which now $\alpha$ designates the constant angle, made by the seepage surface with the abscissa axis. Therefore, for the seepage surface one has the equation of the straight line

(4.12) $$u \cos\alpha + v \sin\alpha + k \sin\alpha = 0,$$

passing through the point $(0, -k)$ and perpendicular to the straight seepage line.

So we see that the examination of the boundary conditions encountered in a wide circle of seepage problems, leads us to the conclusion that in the plane $u + iv$, and consequently also in the $u - iv$ plane, the borders of the region are straight lines or circles. If we interpret the straight line as a circle with infinitely long radius, then we may state that the $w$ region is bounded by arcs of circles.

Before proceeding to examples of the construction of the velocity hodograph, we analyze the problem of the behaviour of mapping functions in the vicinity of the vertex of an angle. This allows us to form an idea of the values of the velocity in corner points of the flow region and facilitates the construction of the velocity hodograph.

## §5. Behaviour of the Velocity in Corner Points of the Flow Region.

Let us assume that the region $z$, which we want to map on a half plane, has boundaries consisting of segments of two intersecting straight lines AB and AC. Let us examine the behaviour of the mapping function in the vicinity of the vertex of the corner of the seepage region (Fig. 25).

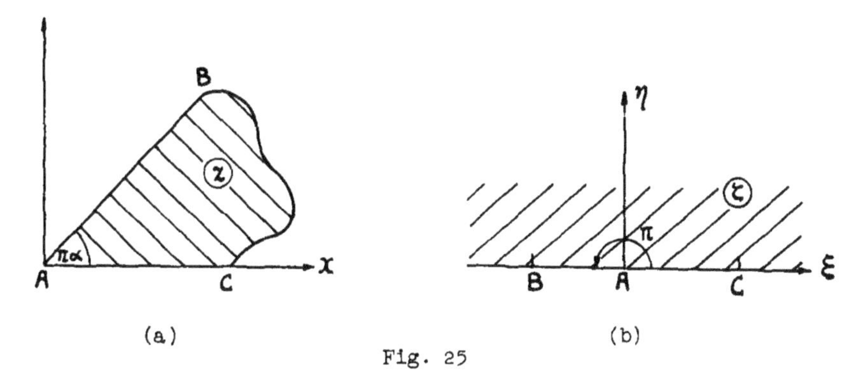


Fig. 25

In the mapping of the angle BAC on to the upper half plane $\zeta = \xi + i\eta$, the angle $\pi\alpha$ is transformed into the angle $\pi$. We shall consider, instead of $z$, the function

$$U = z^{1/\alpha}.$$

We shall put its argument equal to zero for $\zeta = \xi > 0$. Then, when rotating about the point $\zeta = 0$ over the angle $\pi$ in the positive direction the argument of $z$ obtains the value $\alpha\pi, z$ acquires the increase $e^{\alpha\pi i}$, but $z^{1/\alpha}$- the increase $e^{\pi i}$. Therefore, along $AB, U$ will assume the value

$$U = e^{\pi i}|z|^{1/\alpha} = -|z|^{1/\alpha}$$

So, $U$ takes on real values along the segment BAC. In the neighborhood of the point $\zeta = 0$, the function $U$ is analytic, singlevalued and continuous right up to the real axis. According to the principle of symmetry [14], it is possible to continue analytically such function in the lower half plane. Consequently, $U$ will be analytical in the point $\zeta = 0$ and will be decomposed in a power series of $\zeta$. Since $U$ reduces to zero for $\zeta = 0$, its series development takes on the form

$$U = z^{1/\alpha} = a_1\zeta + a_2\zeta^2 + \dots = \zeta(a_1 + a_2\zeta + \dots),$$

from which

$$z = \zeta^{\alpha}(a_1 + a_2\zeta + a_3\zeta^2 + \dots)^{\alpha}$$

Since $(a_1 + a_2\zeta + \dots)^{\alpha}$ may be developed in appower series of $\zeta$, we obtain:

$$z = \zeta^{\alpha}(b_0 + b_1\zeta + b_2\zeta^2 + \dots). \tag{5.1}$$

If an analytic function in the neighborhood of a point $\zeta = \zeta_0$ is represented in the form of a general power series

$$f(\zeta) = (\zeta - \zeta_0)^{\alpha}[b_0 + b_1(\zeta - \zeta_0) + b_2(\zeta - \zeta_0)^2 + \dots],$$

convergent inside an arbitrary circle around the point $\zeta_0$, then it is said that $\zeta_0$ is a "regular singular point" of the function, which accomplishes a conformal mapping of a given region onto a half plane.

Let us examine different cases of corner points.

1. Corner point at impervious boundary. Along this boundary $\psi$ = constant; consequently, in so far as its complex potential is concerned, point $\zeta = 0$, onto which we map the corner point under consideration, will become an ordinary point in the mapping onto the halfplane $\zeta$, in which point we will have

$$\omega = c_0 + c_1\zeta + c_2\zeta^2 + \ldots \tag{5.2}$$

With the help of (5.1) and (5.2) we find the expression of the complex velocity in the form

$$\frac{d\omega}{dz} = \zeta^{1-\alpha}(e_0 + e_1\zeta + \ldots). \tag{5.3}$$

From this expression it is evident that if the angle $\alpha\pi < \pi$, the velocity will be equal to zero for $\zeta = 0$, and if this angle exceeds $\pi$, then the velocity will be infinitely high.

2. Corner point at boundary with water body. Here along the separation line $\varphi$ = constant, and therefore as for $\omega$, the point $\zeta = 0$ will again be an ordinary point and the velocity will have the expression (5.3).

3. Corner point at the intersection of rectilinear boundary of water body and rectilinear impervious boundary. Then in point A the complex potential changes the condition Re $\omega$ = constant into the condition Im $\omega$ = constant, and its expansion about $\zeta = 0$ has the form

$$\omega = \zeta^{\frac{1}{2}}(c_0 + c_1\zeta + \ldots). \tag{5.4}$$

From (5.1) and (5.4) we obtain

$$\frac{d\omega}{dz} = \zeta^{\frac{1}{2}-\alpha}(e_0 + e_1\zeta + \ldots). \tag{5.5}$$

From this equality it is evident that if $\alpha < \frac{1}{2}$, i.e., if the angle in point A is acute, then the velocity will be zero; if this angle is obtuse, then the velocity will be infinite. In the case of the 90° angle, we will have a finite value of the velocity.

## §6. Examples of Construction of the Velocity Hodograph.

When we have unconfined flow, i.e., movement with a free surface, then it is useful to construct the velocity hodograph. In this paragraph we will examine examples of the velocity hodograph construction for earth dams. Therefore we give here some drawings, representing characteristics of the shape of several existing earth dams, constructed at different times in different places.

In Fig. 26 are represented the profiles of several earth dams of medium height [13]. (For higher dams, usually a more gentle slope is adopted. In the figure the dimensions in meters and the inclinations of the slopes are indicated.) Amongst them we note the first — the Zmejnogorsky dam, constructed near Alta in 1780 by the Russian hydrotechnician K. D. Frolov, and still existing at present. For this dam, the steep slopes are characteristic, and the upstream slope has a bulging – with respect to the dam – inflow surface. According to the investigations of the preceding paragraph, the velocity at the lower vertex of this inflow line is zero. Later we will present the velocity hodograph of such a dam (Fig. 37).

Let us consider for example the dam with inclined up-river slope CD (Fig. 27), standing on an impervious foundation DE (EA – drainage line).

In an earth dam, the water depth H of the headwater is usually small in comparison with the length of the base of the dam L. We assume that the ratio is so small that the perpendicular CN to the line CD in the point C intersects the base DE of the dam (Fig. 27). In this case the free surface must have an inflection point. Indeed, in the opposite case, the free surface must entirely lie under the straight line CN, which evidently, is impossible, as the water that is stored in the body of the dam must have an exit to the underdrain.

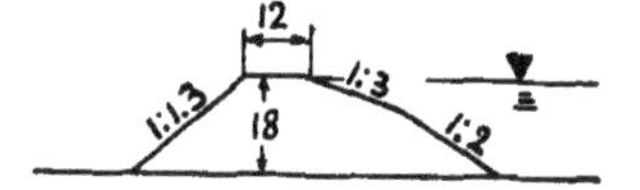


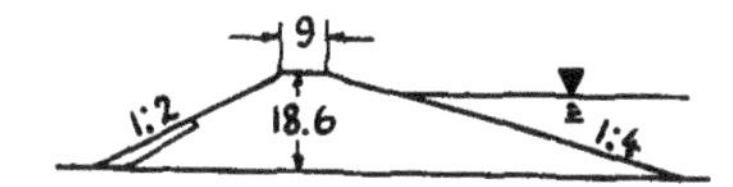


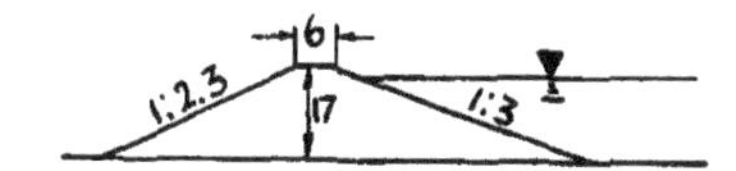


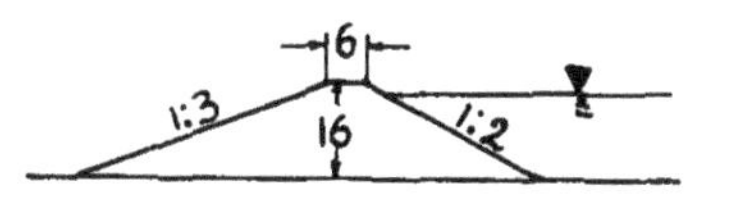


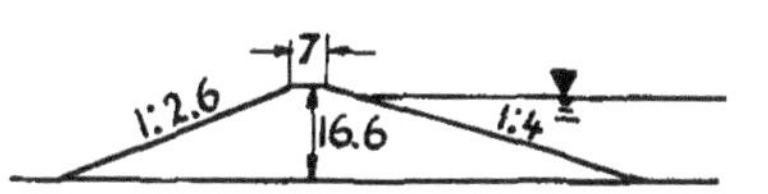


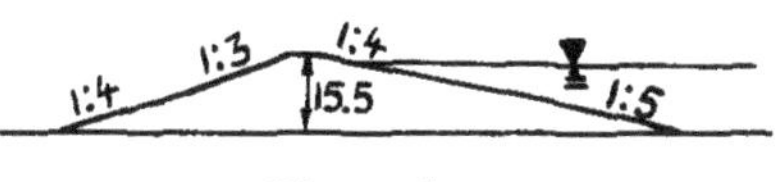


Fig. 26

The question arises: will the inflection point remain on the free surface when the height H of the water in the reservoir increases or may it disappear, and how may this occur?

To obtain the answer, let us look at the velocity hodograph (Fig. 28). It is defined in agreement with (4.6) by the circle $u^2 + v^2 + kv = 0$, representing the free surface, a straight line, passing through the origin of coordinates and perpendicular to the up-river slope of the dam, and finally, the v-axis, corresponding to the drainage line. Having examined a finite number of regions, limited by the enumerated lines, we must choose the uniquely possible shaded region of Fig. 28. In order to take account of the presence of the inflection point at the free surface, it is necessary to make a cut along the arc of the circle. We will now increase the $\frac{H}{L}$ ratio. For sufficiently large values of this ratio, the free surface will not have an inflection point: With increasing $\frac{H}{L}$, the cut CB decreases and vanishes for some definite value of $\frac{H}{L}$; for further increase

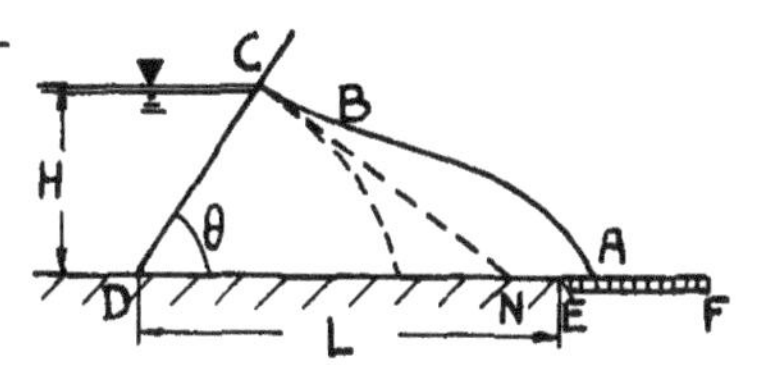


Fig. 27

of $\frac{H}{L}$, the cut CB appears along the straight line DC. This means, that the inflection point C, in which the magnitude of the velocity has a minimum value, after reaching the slope, vanishes and in its place along the slope appears a point with maximum velocity. With increasing $\frac{H}{L}$, this point will move down along the slope and the maximum value of the velocity will increase.

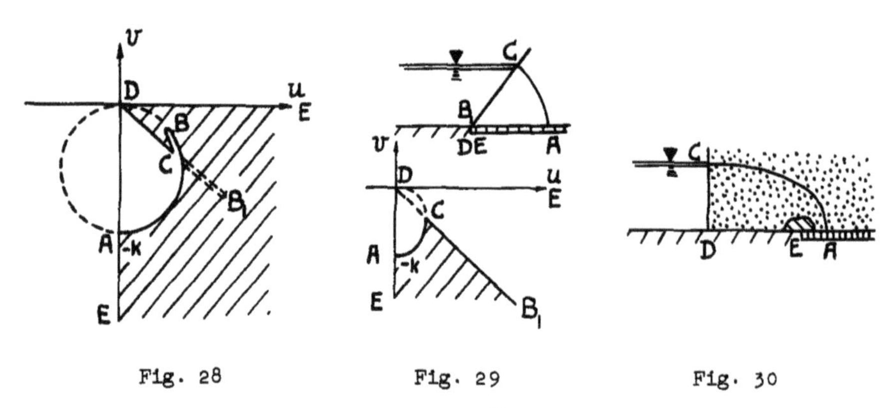


Fig. 28 Fig. 29 Fig. 30

When we consider the growth of $\frac{H}{L}$ as being such that H remains constant but that L decreases and tends to zero, then in the limit we obtain the scheme that is represented in Fig. 29, with the velocity hodograph $AEB_1CA$. We see that, in the limit, the part $EDB_1$, of the original region (Fig. 28) disappears.

Computations for two values of the angle $\theta$ have shown [5] that the inflection point disappears for the values

$$\frac{H}{L} = 1.16 \sin\theta = 0.820 \quad \text{for} \quad \theta = \frac{\pi}{4},$$

$$\frac{H}{L} = 1.19 \sin\theta = 0.232 \quad \text{for} \quad \theta = \frac{\pi}{16}.$$

Usually in dams $\frac{H}{L}$ has a smaller value than reported here and therefore, as a rule, the free surface has an inflection point.

Let us now pass on to the special case of the underdrained dam, when the angle $\theta = \frac{\pi}{2}$ (Fig. 30). In this case the velocity hodograph has the form represented on Fig. 31. There is no inflection point.

If one compares the velocity hodograph of Fig. 28 with the velocity hodograph of Fig. 31, then at first sight it may seem incomprehensible, how the point D may wander from the coordinate origin to some point on the u-axis. But here matters stand so that, for an increase of the angle $\theta$ of the dam (Fig. 27), with given H and L, always such a value of this angle can be found, for which the inflection point vanishes; and on the velocity hodograph the cut $CB_1$ (Fig.32) appears, which will increase and draw nearer to the u-axis (Fig.28). In the limit a part of the original hodograph region disappears.

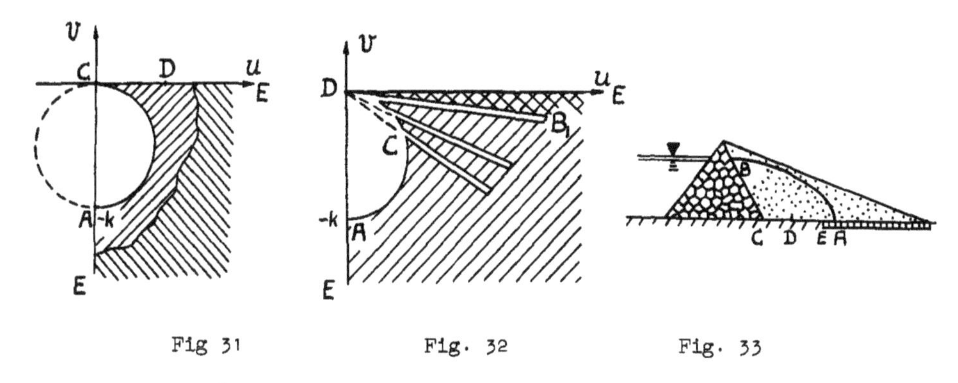


Fig 31 Fig. 32 Fig. 33

In case the angle θ of the up-river slope with the impervious basis is obtuse – such cases occur in dams with rock fill up-river part (Fig. 33) – a cut, reaching till infinity, appears along the u-axis of the velocity hodograph. (Fig. 34.) Let us now go back to the scheme of the dam with vertical slope. The only parameter, limiting the configuration of this dam is the $\frac{H}{L}$ ratio (the length of the EA segment is sought for). In the velocity hodograph (Fig. 31), the only variable vertex remains point D. Consequently, under changing $\frac{H}{L}$, this point must wander. In the limit case, when $L \to \infty$, it is shown that the free surface becomes a parabola with focus in point E, and the abscissa axis is the symmetry axis of this parabola (see Par. 10). From this it follows that for $L \to \infty$, unless $H \to \infty$, the limit $\lim_{L \to \infty} \frac{H}{L} = 0$. Both the points C and D move to infinity and consequently, in the velocity hodograph point D coincides with C in the limit. Oppositely, if $L \to 0$, then point D of the velocity hodograph must go to infinity, and, consequently, point D tends towards point E.

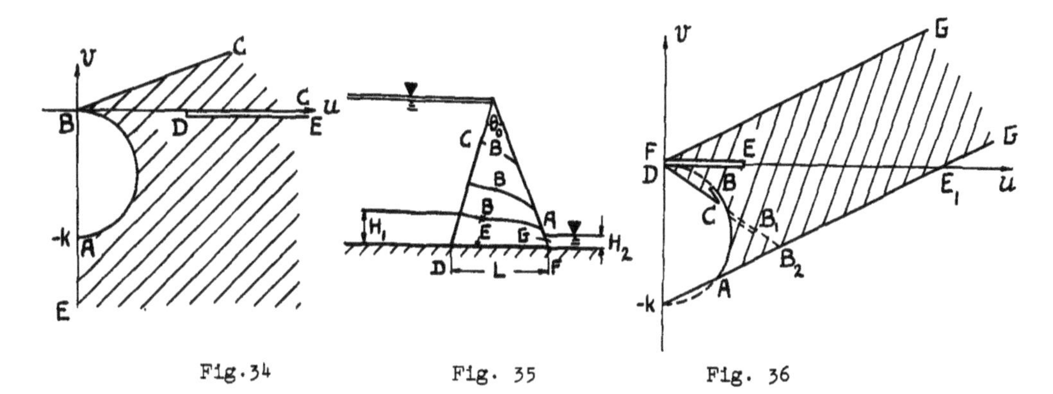


Fig.34 Fig. 35 Fig. 36

Let us assume now that the wall CD remains in place but that the drainage strip EA moves off to infinity, and, consequently, $\frac{H}{L} \to 0$. Then again on the velocity hodograph, point D must tend towards point C, and in this limit case only one point C corresponds to the segment CD in the flow region, and the point D coincides with C in the u,v plane. But in this case the whole region of the w-plane must exist of one single point — the point of zero velocity. In other words, the movement in this limit case will cease. From physical considerations it is also clear, that if the segment EA (Fig. 30) goes to infinity, the free surface in the dam will be an horizontal straight line, and flow is discontinued. We still note that in the considered limit case, the region of the complex velocity by no means shrinks to a point in a continuous way, — the region has always the shape represented in Fig. 31, where only the location of point D changes. In other words, to a small region in the plane z in the vicinity of point E (see shaded part in Fig. 30) corresponds a large region in the w-plane; for a given H and $L \to \infty$, the velocity $u - iv$ tends to zero in an irregular way [6].

Let us still consider another example of change of the velocity hodograph. In Fig. 35 is represented a dam with inclined slopes, having an acute angle $\theta_0$. Let $H_1$ and $H_2$ be the depths of head and tail water, L the length of the impervious basis. We designate $H = H_1 - H_2$.

When the value of the ratio $\frac{H}{L}$ is small, the inflection point of the free surface is represented by point B in the velocity hodograph. (Fig. 36). For an increase of the $\frac{H}{L}$ ratio, point B tends towards point C; then, instead of the cut CB the cut $CB_1$ appears. For $\theta_0 = \frac{\pi}{2}$ the points C and A of the velocity hodograph coincide, the cut $CB_1$ becomes impossible. So, for $\theta_0 > \frac{\pi}{2}$ the inflection point always exists, except in the limit case of a full head behind the dam.

When $\theta_0 > \frac{\pi}{2}$, even in the case of full head behind the dam, there will always exist a free surface at which an inflection point is available. The velocity hodograph will always have a cut AB along the circle (as point C of the velocity hodograph in this case will be below point A).

For $H_2 = 0$, i.e., when there is no tailwater, point E of the velocity hodograph assumes position $E_1$ and the upper part of the hodograph ceases to exist.

When we move the basis DF of the dam down to infinity, then we obtain:

$$H_1 \to \infty, \quad H_2 \to \infty, \quad \frac{H_1}{H_2} \to 1.$$

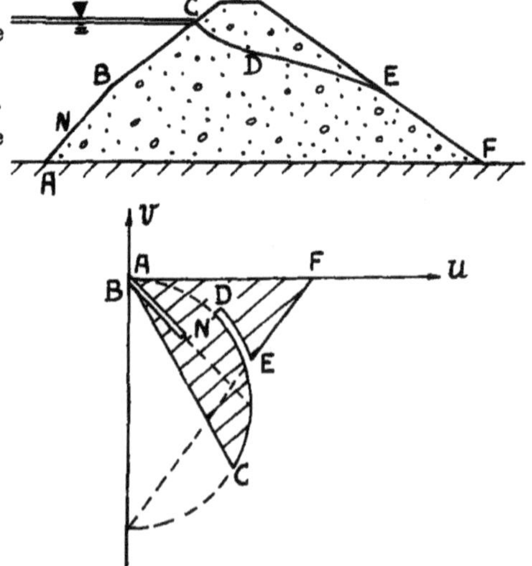


Fig. 37

In this case, the cut FED in the velocity hodograph disappears. It disappears also if, considering L and $H_1$

(or $H_2$) invariable, $H_2$ tends to $H_1$ (or $H_1$ to $H_2$). But in this case, the flow tends to rest, for which the velocity hodograph reduces to a point. In this case, as in the previous one, in the transition to rest, the continuity in the change of velocity is not preserved. Fig. 37 gives the velocity hodograph of the Frolov dam when there is no tail water. If the slope ABC consisted of a single straight line, then the cut ANB would not exist. If the angle ABC would be larger than $\pi$, then the velocity in point B would be equal to $\infty$.

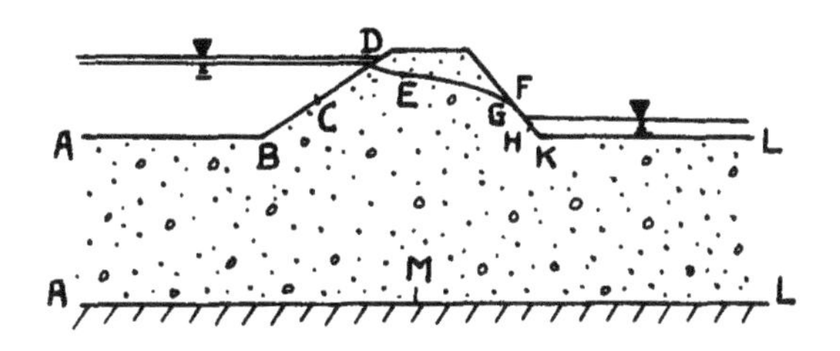


Fig. 38

In Fig. 38 a dam on a pervious foundation of finite depth is sketched; the velocity hodograph for this dam is given in Fig. 39. In Fig. 40 a similar dam is sketched, as in the previous case, but with an horizontal underdrain. For this case the hodograph is given in Fig. 41. We shall give some explanation to this figure. The points at infinity A and K have zero velocity, therefore points $A_1$ and $K_1$ will be in the origin of coordinates. Further, along AB we have $u = 0$, $v$ oriented downward and increasing with the movement from A towards B. Since in point B we have a discontinuity in the direction of the velocity, according to a property of the function of a complex variable, $u + iv$ in this point may be zero or infinite. In the present case, the angle in point B is obtuse, and consequently $u + iv = \infty$. Further, the velocity vector changes its direction, becomes perpendicular to the border BD — in the hodograph we have the segment $B_1C_1D_1$. $C_1$ is the point where the velocity has the smallest value, $D_1$ the point of intersection with the circle which corresponds to the free surface. The return point $E_1$ in the velocity hodograph corresponds to the inflection point. Along the arc of the circle we must go till the point $F_1$ and then wander along $F_1G_1$, since along the drainage strip FG the velocity is directed perpendicular to the strip. In point G, the velocity turns to infinity, changing its direction in the transition to the strip GK. In the point K at infinity, as we already mentioned, the velocity is zero. Along the impervious bottom the velocity is oriented to the right, assuming in some point L a maximum value, represented by the segment $K_1L_1$. The lines AK exhibit on the velocity hodograph the strip $A_1L_1K_1$, going in the right and reversed directions.

In Fig. 41 we have a double-valued region, as part of the u,v plane is twice covered by the change of the variable $u + iv$. When we go in the z-plane along the contour ABD, then the flow region remains on the right side, but on the velocity hodograph the corresponding region is found to the left side of the contour (since $u + iv$ is a function of $\bar{z}$ and not of $z$), what is indicated by stripes below the contour. The analyzed examples have shown, that the construction of the hodograph, generally speaking, is complicated. Therefore in complicated cases, it is desirable to go about without the velocity hodograph in the solution of problems. Such methods will be given in Chapter VI.

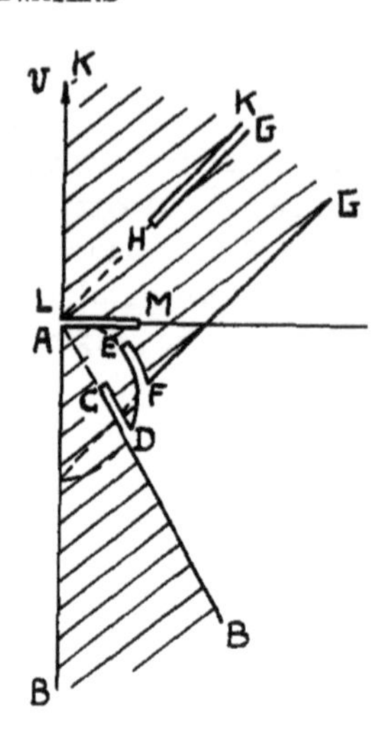


Fig. 39

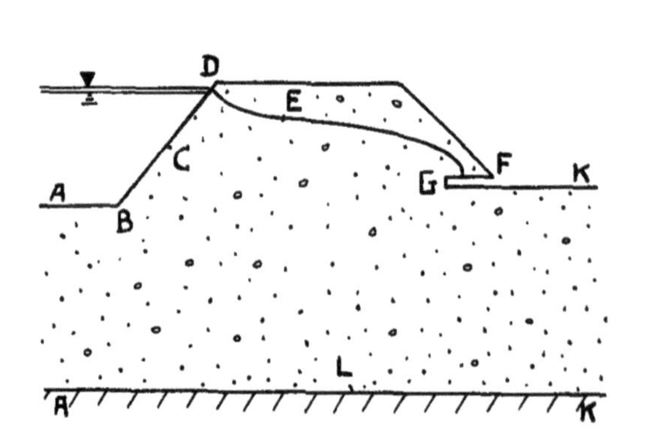


Fig 40

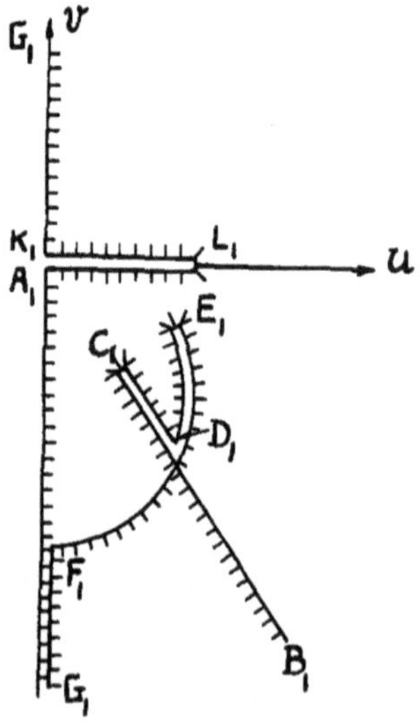

Fig. 41

§7. The Seepage Triangle [4].

Under "seepage triangle" we will understand the triangle that is constructed by vectors, which are obtained if one takes the gradient of both parts of the equation

$$\frac{p-p_a}{\gamma} = h - y.$$

This equation follows from (1.2) if we put

$$\rho g = \gamma, \quad C = -\frac{p_a}{\gamma}.$$

Where $\gamma$ - specific weight of the fluid, $p_a$ - atmospheric pressure. We obtain:

$$\frac{1}{\gamma}\,\text{grad}\ p = \text{grad}\ h - \vec{j} \tag{7.1}$$

where $\vec{j}$ - the unit vector oriented upward along the vertical. Such a triangle is sketched in Fig. 42.

Equation (7.1) may be written in another form, after multiplying by $k$ ($k$ - seepage coefficient) and adopting for notation that

$$-\,k\ \text{grad}\ h = \vec{V}, \tag{7.2}$$

where $\vec{V}$ - seepage velocity vector. We obtain

$$\vec{V} = -\,\frac{k}{\gamma}\,\text{grad}\ p - k\vec{j}\ .$$

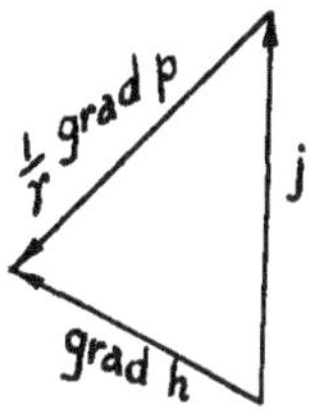

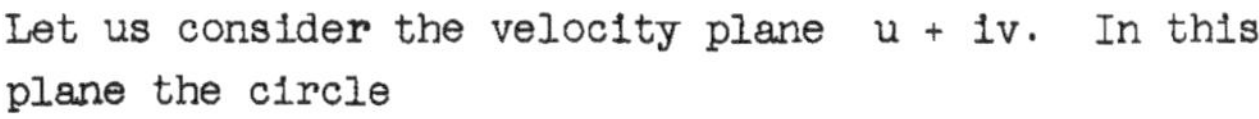

Let us consider the velocity plane $u + iv$. In this plane the circle

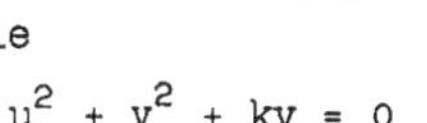

$$u^2 + v^2 + kv = 0$$

Fig. 42

is constructed in dotted line and represents the free surface of groundwater flow. From consideration of Fig. 43 it is apparent that on the free surface the velocity vector and $\frac{k}{\gamma}$ grad p are orthogonal.

## §8. Forces Acting on Soil Particles.

In problems of strength and stability of hydraulic structures, the question of the study of the forces exerted by water flowing under these structures has an important meaning (see further Par. 12 and 13).

This problem has been treated by many scientists: N. P. Pouzirevsky, N. N. Pavlovsky, K. Terzaghi, N. M. Gersevanov, and others. Moreover, the question of the physical nature of the seepage forces caused discussion [12]. We represent here an outline as given in the work of B. K Rysenkampf [4]. We choose a unit volume of porous medium and find the resultant $\vec{S}$ of the forces that act on the solid part of this volume. This force may be considered as the geometric sum of the following forces:

1. The weight $\vec{F}_1$ of the dry particles of the chosen volume. This force is oriented vertically downward and equal in magnitude to the product $n\gamma_1$, where $\gamma_1$ specific weight (weight of unit volume) of the dry particles and $n = 1 - m$, volume of the solid particles in the unit volume of porous medium. In this way it is possible to write the equation

$$\vec{F}_1 = -\,n\gamma_1\vec{j}\ ,$$

when one designates by $\vec{j}$ the unit vector, oriented upward along the vertical.

2. The force $\vec{F}_2$ of the water pressure on the solid part in the unit volume of porous medium. Since the action of the pressure in the unit volume of soil is equal to $-$ grad p, we have for the force $\vec{F}_2$ :

$$\vec{F}_2 = -\,n\ \text{grad}\ p\ .$$

3. As a third force, B. K Rysenkampf considers the force $\vec{F}_3$ - the action carried by the seeping water onto the solid part, contained in a unit volume of porous medium. This action is equal and opposite in sign to the resistance of the unit volume of porous medium. The resistance, reduced to that of the unit mass of water, as we saw in formula (11.3), Ch. I, can be defined by the equality

$$\vec{R} = -\frac{g}{k}\vec{V}.$$

Since the mass of water in the unit volume of water is equal to $m\rho$, where $\rho$ density of the water, we obtain for $\vec{F}_3$

$$\vec{F}_3 = -m\rho\vec{R} = \frac{m\gamma}{k}\vec{V}$$

$\vec{F}_3$ has the name of seepage force.

The geometric sum of $\vec{F}_1$, $\vec{F}_2$, $\vec{F}_3$ gives the resultant $\vec{S}$

$$\vec{S} = -n\gamma_1\vec{j} - n\,\text{grad}\,p + \frac{m\gamma}{k}\vec{V}.$$

We transform the obtained expression in two ways, using the equations

$$\vec{V} = -k\,\text{grad}\,h, \quad \frac{1}{\gamma}\,\text{grad}\,p = \text{grad}\,h - \vec{j}$$

and eliminate from them once $p$, the other time $\vec{V}$.

$$\vec{S} = \frac{\gamma}{k}\vec{V} - n(\gamma_1 - \gamma)\vec{j},$$

$$\vec{S} = -\text{grad}\,p - \gamma_0\vec{j}.$$

Here we introduce the notation

$$\gamma_0 = m\gamma + n\gamma_1.$$

It is evident that the quantity $\gamma_0$ represents the weight of the unit volume of porous medium, water included.

One can construct the force $\vec{S}$ with the help of diagram Fig. 44, in

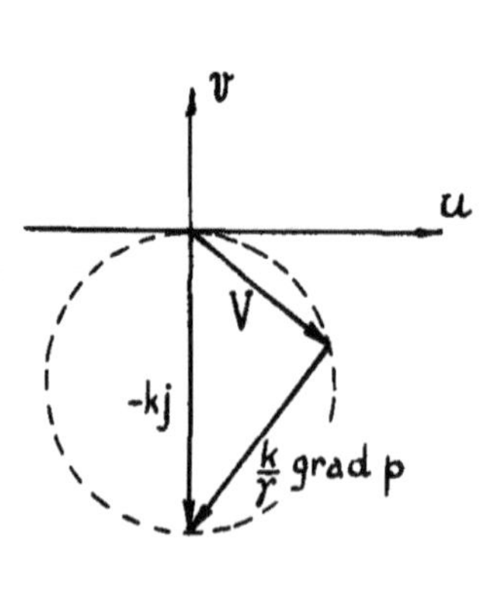


Fig. 43

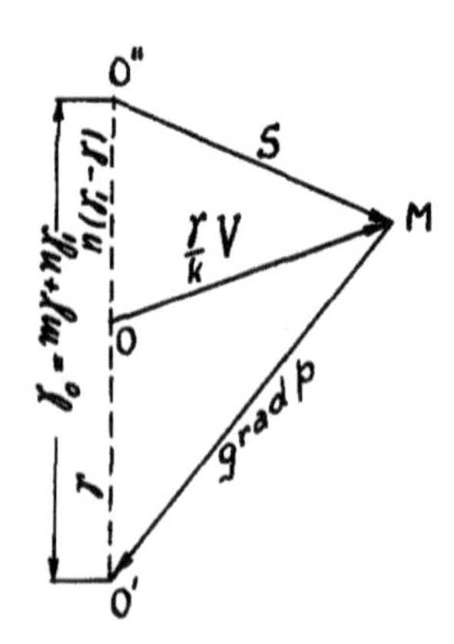


Fig. 44

which both representations of $\vec{S}$ are shown. The vector $\vec{S}$ may be represented through the force function

$$V = -\gamma h - n(\gamma_1 - \gamma)y = -p - \gamma_0 y + \text{constant}$$

in the form

$$\vec{S} = \text{grad } V .$$

It is not difficult to be convinced that $V$ satisfies Laplace's equation.

Up to this point, the whole reasoning holds both for the plane and for space.

In the case of plane movement, besides the function $V$, we may also introduce the function $U$, harmonic conjugate to $V$, so that

$$U + iV = \frac{1}{k}\gamma\omega - n(\gamma_1 - \gamma)z .$$

The lines $V$ = constant are essentially equipotential lines of $\vec{S}$, the equation $U$ = constant defines force lines of the force field $\vec{S}$. When the liquid is in rest, $\vec{V} = 0$ and

$$|\vec{S}| = n(\gamma_1 - \gamma) ,$$

which expresses Archimedes' law: the force that operates on the soil skeleton, is equal to the weight of the skeleton, diminished with the weight of fluid displaced by the skeleton.

When the pressure in the fluid is constant, then the fluid moves only under the action of the gravity force. In this case

$$|\vec{S}| = \gamma_0 = m\gamma + n\gamma_1 ,$$

i.e., to the weight of the soil frame is added the weight of the pore water.

B. K. Rysenkampf noticed that if one takes $m = 0.4$, $\gamma = 1$, $\gamma_1 = 2\frac{2}{3}$, one finds that the segments $00'$ and $00''$ (Fig. 44) have a length equal to one.

B K. Rysenkampf has demonstrated the possibility to compute the stress in soil through which water flows.

If we designate by $\sigma_x$ and $\sigma_y$ the normal stresses on elementary planes in porous medium, perpendicular to the axes $x$ and $y$, and by $\tau_{xy}$ - the tangential stress to these planes, then on the basis of the equation of equilibrium of a continuous medium, we will have

$$\frac{\partial\sigma_x}{\partial x} + \frac{\partial\tau_{xy}}{\partial y} + S_x = 0, \quad \frac{\partial\tau_{xy}}{\partial x} + \frac{\partial\sigma_y}{\partial y} + S_y = 0,$$

where

$$S_x = \frac{\partial v}{\partial x} , \quad S_y = \frac{\partial v}{\partial y} .$$

If the porous medium follows Hooke's law, then, introducing the stress function $\varphi(x,y)$ with the help of the equations

$$\sigma_x = \frac{\partial^2\varphi}{\partial y^2} - v, \quad \tau_{xy} = -\frac{\partial^2\varphi}{\partial x\partial y} , \quad \sigma_y = \frac{\partial^2\varphi}{\partial x^2} - v ,$$

we find that $\varphi$ satisfies the biharmonic equation:

$$\frac{\partial^4\varphi}{\partial x^4} + 2\frac{\partial^4\varphi}{\partial x^2\partial y^2} + \frac{\partial^4\varphi}{\partial y^4} = 0.$$

## B. Simplest Examples of Plane Flows.

### §9. Preliminary Remarks.

Now we consider some examples of ground water flow. Here we only pause at such cases for which an exact solution is obtained in a comparatively simple way. More complicated problems will be examined further with the help of special methods of mathematical analysis, worked out by various authors and applied to different groups of problems in the seepage theory.

Ground water movements are divided into confined, unconfined and semi-confined flows.

That movement in which there is no free surface is called "confined" in the theory of seepage. Such is the movement of ground water under hydraulic structures, flowing under influence of the pressure difference in head and tailwater, or the flow of water towards the borehole in a waterbearing layer, confined by impervious surfaces. In confined flow there is always a solid wall at the upper level of the flow region, the boundaries of this region can only be of two types: streamlines or lines of equipotential. "Unconfined" are called such flows in which the flow region is bounded upward by a free surface (possible cases of degeneration; see, for example, end of Par. 12 of Chapter VII).

Finally, "semi-confined" seepage is characterized by the fact that the seepage flow at first is contiguous with the underground contour of the structure, and then breaks away from it, forming a free surface. Semi-confined flow takes place under structures, below which a significant drop of ground water level occurs, for example under the storage reservoirs of hydropower plants. Semi-confined flows will be treated in Chapter X. Paragraphs 10 and 11 of this chapter are devoted to unconfined flows, the rest to confined movements.

### §10. Drainage Slit on Impervious Basis.

We search for $z$ as a function of $\omega = \varphi + i\psi$, for which we adopt the simplest – after the linear – form of functional dependence

$$z = A\omega^2 \tag{10.1}$$

where $A$ – for the time being, an undetermined constant. Let us examine if the considered flow can have a free surface. For the latter, the condition of zero pressure must be satisfied, equivalent to equation (2.7) with the constant = 0:

$$\varphi + ky = 0 ,$$

and also the condition that the free surface line is a streamline:

$$\psi = \psi_0 .$$

Inserting into the right part of (10.1) the values

$$\varphi = - ky, \quad \psi = \psi_0 ,$$

we obtain:

$$x + iy = A(-ky + i\psi_0)^2 .$$

Separation of the real and imaginary part gives for a real value of A

$$x = A(k^2y^2 - \psi_0^2), \quad y = -2Ak\psi_0 y .$$

In order to satisfy the latter of the two equalities, it is necessary that

$$A = -\frac{1}{2k\psi_0} . \tag{10.2}$$

Then the following parabola will correspond to the free surface.

$$x = -\frac{ky^2}{2\psi_0} + \frac{\psi_0}{2k} ,$$

or in other writing the parabola

$$y^2 = -\frac{2\psi_0}{k}\left(x - \frac{\psi_0}{2k}\right) \tag{10.3}$$

Now we insert the obtained value of A in equation (10.1) and separate in this equation the real and imaginary parts. We find the dependence of the coordinates $x$, $y$ on the functions $\varphi$ and $\psi$:

$$x = -\frac{1}{2k\psi_0}(\varphi^2 - \psi^2), \quad y = -\frac{\varphi\psi}{k\psi_0} . \tag{10.4}$$

For $\psi = 0$ the streamline is the negative part of the x-axis, since for this value we have, considering $\psi_0 > 0$:

$$y = 0, \quad x = -\frac{\varphi^2}{2k\psi_0} < 0 .$$

We obtain other streamlines by inserting constant values of $\psi$ in (10.4). They are confocal parabolas with focus in the origin of coordinates (Fig. 45a)

Finally, putting $\varphi = 0$ in (10.4), we obtain the equipotential

$$y = 0, \quad x = +\frac{\psi^2}{2k\psi_0}$$

This is part of the positive abscissa axis. We determine it to be the segment

$$0 < x < \frac{\psi_0}{2k} .$$

Then three boundary lines — two streamlines, $\psi = 0$ and $\psi = \psi_0$, and a line of equipotential $\varphi = 0$ — form the flow region, corresponding to the flow of groundwater from infinity towards a drainage slit OA (Fig. 45a). The equipotential lines will be parabolas, orthogonal to the parabolas of the flow lines.

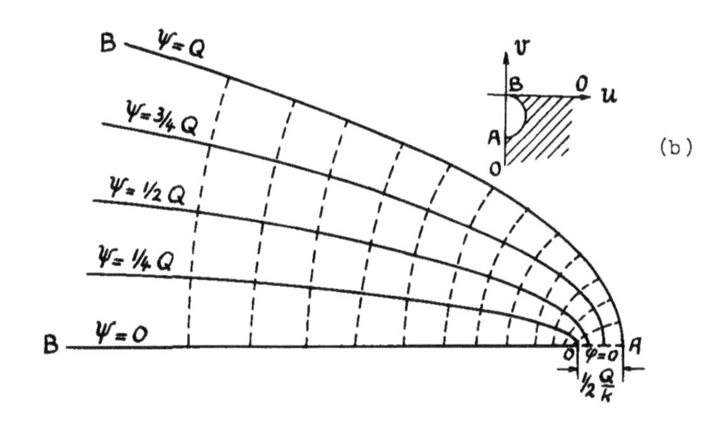


(a)

Fig. 45

We put

$$\psi_0 = Q \quad ,$$

where $Q$ — the flowrate through the drainage slit. The length of the drainage segment and the flowrate $Q$ are related as

$$d = \frac{Q}{2k} \; .$$

Consequently, the flowrate is proportional to the width of the slit:

$$Q = \psi_0 = 2kd \; . \tag{10.5}$$

The above represented problem was examined by Forchheimer and Kozeny.

Subsequently N. N. Pavlovsky [2] obtained its solution in a direct way, assigning boundary conditions to three parts of the contour which border the flow region, and constructing the velocity hodograph, which, as is easy to see, has the form of Fig. 45b.

Returning to the equation of the free surface (10.3), we take on it two points with coordinates $(x_1, y_1)$ and $(x_2, y_2)$ and insert them in equation (10.3). From the obtained identities, we find for the equation of the flow rate

$$Q = -k\,\frac{y_2^2 - y_1^2}{2(x_2 - x_1)} \quad .$$

Quite true, the same formula for the flow rate is obtained in the hydraulic theory of steady flows (see Chapter X), where it is called Dupuit's formula.

## §11. Horizontal Drain in the Absence of an Impervious Basis. Lines of Equal Ground Water Flow.

This case is obtained from the one we have considered, if to the flow region, chosen in Par. 10, we join the lower half plane (Fig. 46). All flowlines are confocal parabolas. For their construction one has to give to $\psi$ values from $\psi_0$ to $-\infty$.

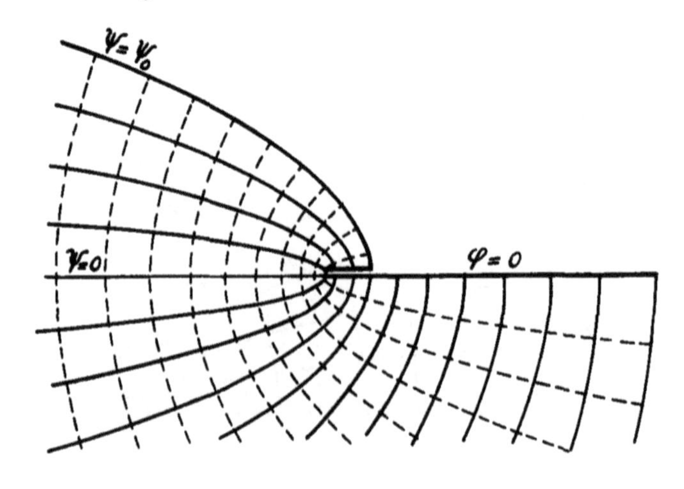


Fig. 46

Besides flowlines and lines of equal potential, lines of equal pressure or isobars of ground water flow present interest. In order to construct the family of isobars, we are reminded of the relationship between velocity potential and pressure

$$\varphi = -k\left(\frac{p}{\rho g} + y\right).$$

We can rewrite it as

$$\frac{\varphi}{k} + y = -\frac{p}{\rho g}$$

from which it becomes clear that graphical addition (see Chapter XI, Par. 6) of equipotentials and lines $y$ = constant gives the isobars.

For the problem under consideration, it is not difficult to find the equations of the isobars. It is sufficient therefore to eliminate $\psi$ from equations (10.4) and to solve for $\varphi$ the obtained equation of the fourth order. For the quantity $\varphi + ky$, proportional to the pressure, we have

$$\varphi + ky = ky \pm \sqrt{-kq(x \mp \sqrt{x^2 + y^2})} \qquad (q = \psi_0).$$

Equating $\varphi + ky$ to a constant $C$ and eliminating the irrational operation, we obtain for the isobar the equation of a curve of the fourth degree

$$k^2q^2y^2 = (C - ky)^2[(C - ky)^2 + 2kqx].$$

In Fig. 47 are constructed a family of isobars (continuous lines) by means of graphical addition of lines $h$ = constant (dotted lines) and horizontal lines $y$ = constant. (About graphical addition see Chapter XI.)

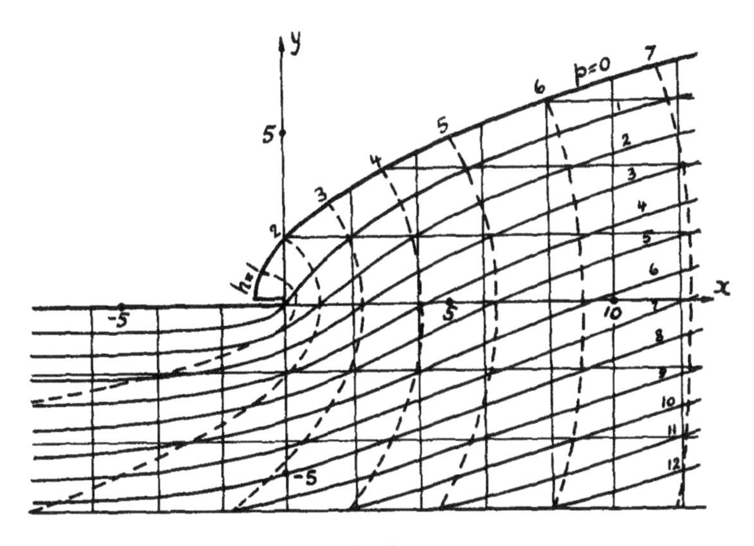


Fig 47

Besides the orthogonal grid, the family of streamlines and constant potential lines, it is also customary to examine the families of "isoveles" or "isotaches," i.e., lines along which the velocity vector has one and the same value. The isotache grids, as demonstrated by N. N. Pavlovsky, are useful for computation of the "piping" effect, i.e., the phenomenon of the washout of small soil particles, which may lead to the failure of the dam, if repeated.

We take the logarithm of the complex velocity and separate it into real and imaginary parts.

$$\ell n\, w = \ell n|w| + i \arg w = \ell n\, V - i\theta$$

V magnitude of the velocity, $\theta$ - angle between the velocity vector and the abscissa axis. Since the real and imaginary parts of $\ell n\, w$ are harmonic functions, the lines

$$\ell n\, V = \text{constant}, \quad \theta = \text{constant}$$

represent orthogonal families.

For the simplest example that we have considered, it is not difficult to find the equations of the isoclines and isotaches. Indeed, based on (10.1) and (10.2) we have:

$$\omega = \sqrt{-2k\psi_0 z}, \qquad w = \frac{d\omega}{dz} = \sqrt{\frac{k\psi_0}{-2z}} .$$

Therefore the logarithm of the complex velocity is equal to

$$\ell n\, w = \ell n \sqrt{-\frac{k\psi_0}{2}} - \frac{1}{2}\left[\ell n|z| + i \text{ arc tg } \frac{y}{x}\right] .$$

From this it is clear that the isotaches essentially are concentric circles $r$ = constant with center in the focus of the parabolas, and the isoclines - rays originating from the focus (Fig. 48).

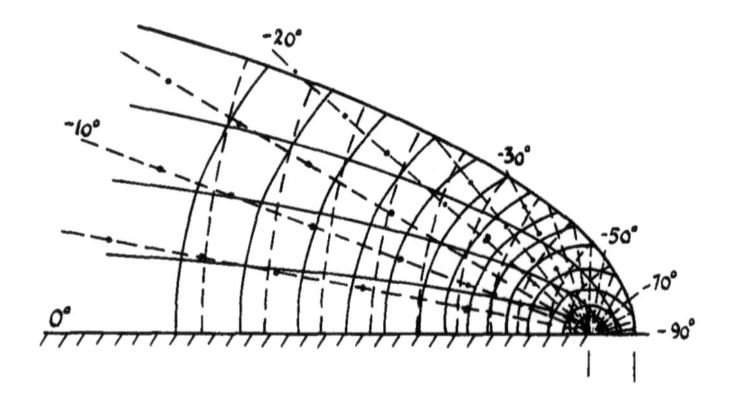


Fig. 48

The magnitude of the velocity is equal to the coefficient of seepage along the circle, which touches the free surface (at the intersection of this surface with the drainage strip).

§12. Flat Bottom Weir on a Layer of Infinite Depth.

We now consider the simplest example of confined flow of ground waters under hydraulic structures — the flow about a "flat bottom weir" [9] on soil of infinite depth (Fig. 49).

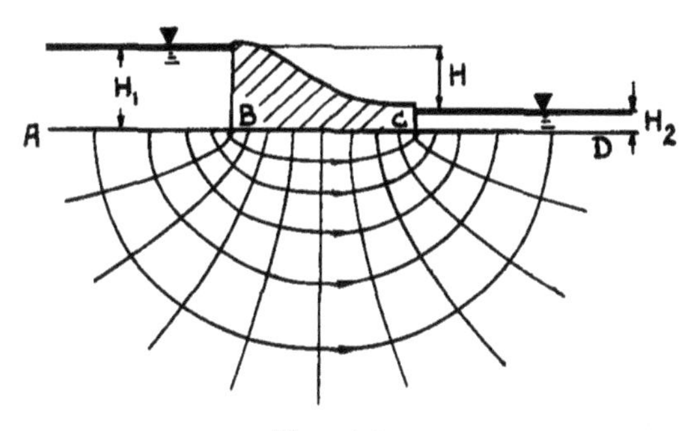


Fig. 49

In this case it is possible to construct the solution using peculiarities of the complex velocity $w = u - iv$, conceived as a function of $z$. Along the segments AB and CD (Fig. 49) we have $u = 0$; along the wetted contour BC we have $v = 0$; in other words, $w$ takes on a real value on the strip BC, i.e., for $-b < x < b$ and a purely imaginary value for $|x| > b$.

Let us examine the function $\sqrt{b - z}$. It is real for $z < b$ and has a purely imaginary value for $z = x > b$. The function $\sqrt{b + z}$ is real for $z = x > -b$ and purely imaginary for $z = x < -b$. Therefore the function $\sqrt{b^2 - z^2}$, taking on real values on the segment $(-b, b)$, remains purely imaginary outside this strip on the real axis. We must assume a natural condition in the given problem — the convergence of the velocity to zero at infinity. Then we may take for $w$ the function

$$w = \frac{M}{\sqrt{(b - z)(b + z)}} = \frac{M}{\sqrt{b^2 - z^2}} \tag{12.1}$$

where $M$ — some constant.

The expression for $w$, as is easy to see, satisfies the conditions: $v = 0$ for $|x| < b$ and $u = 0$ for $|x| > b$. But it is not the only function that satisfies the given conditions. Thus, it would be possible to multiply expression (12.1) with a rational function with real coefficients, and in which the degree of the numerator does not exceed the degree of the denominator. For example, it is possible to take for $w$ the function

$$\frac{M(z - c)^n}{(z - a)^m \sqrt{b^2 - z^2}} \tag{12.2}$$

where $m$ and $n$ integer numbers for which $n \leq m$. However, the function (12.2) will have additional features in the points $a$ and $c$; if, for example, $a = b$, then for $z = b$ a more complicated feature is obtained, than the one that corresponds to our problem: when the velocity vector in this point (Fig. 49) must be turned over a right angle, then according to formula

(12.2) it turns over the angle $\pi(m + \frac{1}{2})$.

Thus, in order that the problem be determined from the mathematical viewpoint, it is necessary to define the order of infiniteness of the function in a singular point.

If we take $-b < a < b$, then formula (12.2) gives a function which has a singular point for $z = a$ on the wetted contour. For $n = 0$, $m = 1$, this singular point is a source or a sink. We will use this remark below, in the solution of the problem about the drainage pipe under the wetted contour (Par. 14 of the present Chapter).

Keeping for our flow formula (12.1), we must define the value of the constant M. For this we return to the complex potential. By integrating the equation

$$w = \frac{d\omega}{dz} = \frac{M}{\sqrt{b^2 - z^2}} ,$$

we find

$$\omega = \varphi + i\psi = M \sin^{-1} \frac{z}{b} + N. \tag{12.3}$$

We take $\psi = 0$ along the wetted contour. We rewrite formula (1.2), expressing the dependence between the velocity potential and the pressure, in this form:

$$\varphi = - k\left[\frac{p}{\rho g} + (y - H_2)\right] . \tag{12.4}$$

The arbitrary constant $C_1$ is chosen so that the value of $\varphi$ along the tailwater is equal to zero. The atmospheric pressure is considered to be zero. Along the boundary of the upriver waterbody, we have

$$p = \rho g H_1 .$$

Therefore, along AB, since $y = 0$ along the headwater boundary, we obtain

$$\varphi = - k(H_1 - H_2) .$$

For the difference $H_1 - H_2$, the so-called "true head," the designation H is adopted:

$$H = H_1 - H_2 .$$

By means of this quantity the conditon on the headwater boundary can finally be written in the form

$$\varphi = -kH .$$

After inserting in (12.3) $z = \pm b$ and noticing that in these points $\psi = 0$, and that $\varphi$ is respectively equal to 0 and $-kH$, we find

$$\omega = \frac{kH}{\pi} \sin^{-1} \frac{z}{b} - \frac{kH}{2} = - \frac{kH}{\pi} \cos^{-1} \frac{z}{b} . \tag{12.5}$$

Therefore, we have for z:

$$z = b \cos \frac{\pi\omega}{kH} . \tag{12.6}$$

Separating here real and imaginary parts, we obtain:

$$x = b \cos \tilde{\varphi} \operatorname{ch} \tilde{\psi}, \quad y = - b \sin \tilde{\varphi} \operatorname{sh} \tilde{\psi} \tag{12.7}$$

where

$$\tilde{\varphi} = \frac{\pi\varphi}{kH} , \quad \tilde{\psi} = \frac{\pi\psi}{kH} . \tag{12.8}$$

Assigning to $\tilde{\psi}$ the constant value $\tilde{\psi} = \tilde{\psi}_1$, we obtain the streamlines – ellipsi with foci in the points $\pm$ b:

$$\frac{x^2}{b^2 \operatorname{ch}^2 \tilde{\psi}_1} + \frac{y^2}{b^2 \operatorname{sh}^2 \tilde{\psi}_1} = 1. \tag{12.9}$$

For $\tilde{\varphi} = \tilde{\varphi}_1$ we obtain equipotential lines – confocal hyperboli

$$\frac{x^2}{b^2 \cos^2 \tilde{\varphi}_1} - \frac{y^2}{b^2 \sin^2 \tilde{\varphi}_1} = 1 . \tag{12.10}$$

In Fig. 49 are constructed streamlines and equipotential lines. It is possible to estimate the magnitude of the average velocities from the density of these lines.

For the calculation of the local value of the velocity in any point $z = x + iy$, the following formula serves

$$w = u - iv = \frac{kH}{\pi\sqrt{b^2 - z^2}} \quad . \tag{12.11}$$

In particular, we find the distribution of the velocity along the wetted contour (Fig. 50),

$$u = \frac{kH}{\pi\sqrt{b^2 - x^2}}$$

and along the boundaries of head – and tailwater [(-) sign for headwater, (+) sign for tailwater]:

$$v = \mp \frac{kH}{\pi\sqrt{x^2 - b^2}} \quad .$$

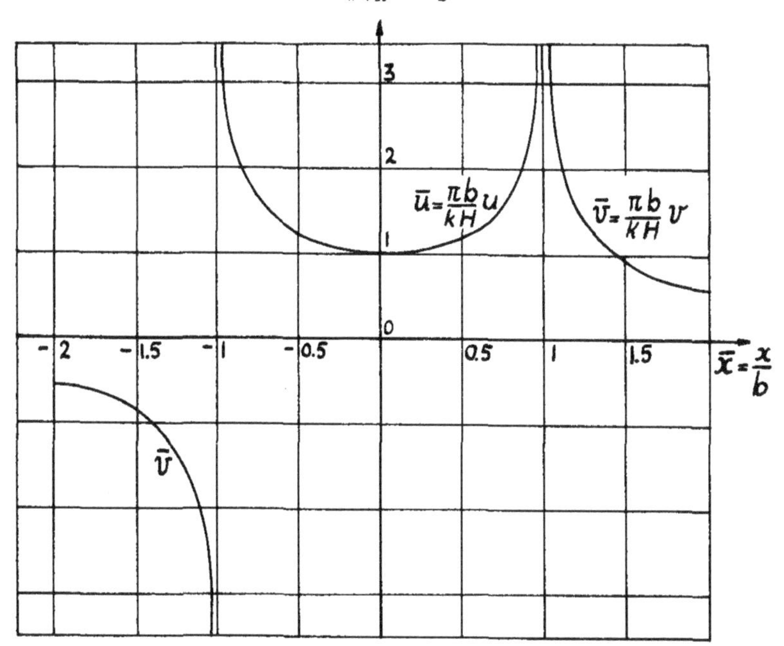


Fig. 50

Separation of the real and imaginary parts in the formula for $w$ gives the magnitudes of the velocity components $u, v$ in any point $(x,y)$ of the flow region:

$$u = \frac{kH}{\pi\sqrt{2}} \frac{\sqrt{\sqrt{(b^2 - x^2 + y^2)^2 + 4x^2y^2} + b^2 - x^2 + y^2}}{\sqrt{(b^2 - x^2 + y^2)^2 + 4x^2y^2}} ,$$

$$v = \mp \frac{kH}{\pi\sqrt{2}} \frac{\sqrt{\sqrt{(b^2 - x^2 + y^2)^2 + 4x^2y^2} - b^2 + x^2 - y^2}}{\sqrt{(b^2 - x^2 + y^2)^2 + 4x^2y^2}} ,$$

in which the upper sign takes place for $x < 0$, the lower for $x > 0$. The pressure along the foundation of the hydrotechnical structure is an important element in the motion of ground water. In our example, the pressure (within a constant) coincides with the head along the fluid bed. Since along the latter $\psi = 0$, we obtain the relationship between $x$ and $\varphi$ [from (12.6)] in the form

$$x = b \cos \frac{\pi\varphi}{kH} = b \cos \frac{\pi h}{H} ,$$

from where

$$h = \frac{H}{\pi} \cos^{-1} \frac{x}{b} .$$

Since along the wetted contour $y = 0$, the pressure is proportional to $h$:

$$p = \rho g h = \frac{\rho g H}{\pi} \cos^{-1} \frac{x}{b} .$$

In Fig. 51 the distribution of the head along the wetted contour is represented. The calculation of the discharge $\psi_0$ through the strip of tailwater from the point $x = b$ till the point $x = x_0$ presents some interest. For its computation, one may avail oneself first from equation (12.7), inserting in it $\varphi = 0$, $x = x_0$. Then we obtain:

$$\psi_0 = \frac{kH}{\pi} \operatorname{ch}^{-1} \frac{x_0}{b} = \frac{kH}{\pi} \ln \frac{x_0 + \sqrt{x_0^2 - b^2}}{b} . \qquad (12.12)$$

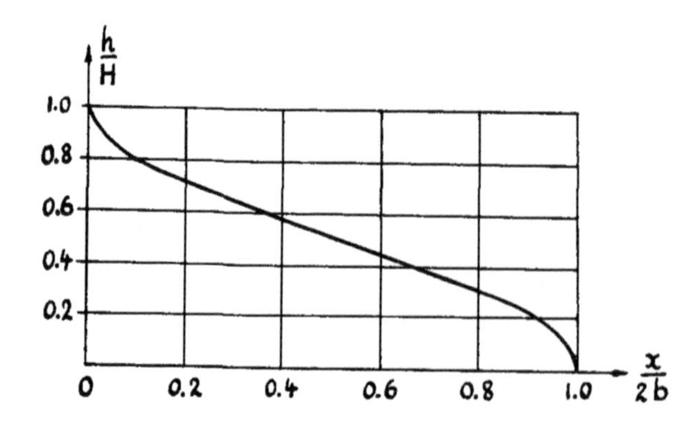


Fig. 51

When $x_0$ goes to infinity, the discharge $\psi_0$ becomes infinitely large. We will see further that for a pervious layer of finite depth the full discharge that prevails is finite.

In Par. 9 of the present chapter we introduced the force function V, for the vector $\vec{S}$ of the resultant forces, operating on the soil particles:

$$V = -\gamma(h + \alpha y), \quad \alpha = n\left(\frac{\gamma_1}{\gamma} - 1\right).$$

In Fig. 52 the construction of the lines V = constant is given for a plane wetted contour. We notice that the isobars in this example, should we take $\alpha = 1$, would be lines symmetrical with the lines V = C, about the vertical axis, passing through the middle of the wetted contour (or, otherwise, the isobars would be coinciding with the lines V = C, if we would change the position of head and tailwater); for $\alpha \neq 1$, the isobars would be somewhat deformed in comparison with the level lines.

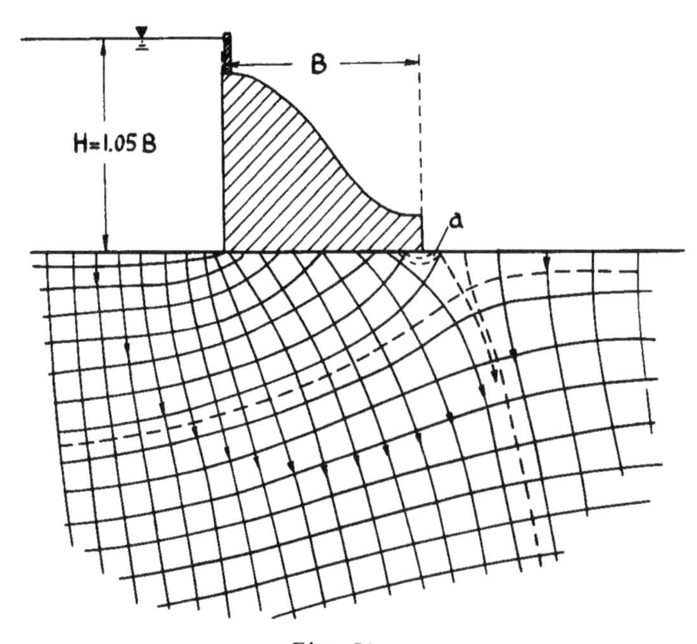


Fig. 52

In the same Fig. 52 are constructed the lines U = constant, along which the vector $\vec{S}$ is directed. In the superposition of these families, an orthogonal grid is obtained.

From the examination of Fig. 52 it is clear that vector $\vec{S}$ everywhere is oriented downward, with the exception of the region behind the tailwater section where the vertical component of this vector is pointed upward. There, boiling of the soil and washout of particles upward are possible.

Remark No. 1: One may choose the constant $C_1$ in the first of formulas (1.2) otherwise; one may, for example, put forward the condition that the values of the velocity potential along headwater and tailwater basins be equal in magnitude but opposite in sign. Then we obtain:

$$C_1 = k \frac{H_1 + H_2}{2}, \quad \varphi = -k\left(\frac{p}{\rho g} + y - \frac{H_1 + H_2}{2}\right).$$

The value of $\varphi$ along head and tailwater basins consequently will be

$$\varphi_1 = -k \frac{H_1 - H_2}{2} = -\frac{kH}{2}, \quad \varphi_2 = k \frac{H_1 - H_2}{2} = \frac{kH}{2}.$$

Complex potential and $z$ are related as

$$\omega = \varphi + i\psi = \frac{kH}{\pi} \sin^{-1} \frac{z}{b}.$$

Remark No. 2: The streamlines directly under the hydraulic structure depend upon the form of the subterranean contour of this structure. But far away in the depth of the layer and on both sides of the bases of the structure, this influence will be very small. The streamlines become close to half-circles.

In other words, if one moves away over a sufficiently large distance from the structure to the depth of the pervious layer, then one may consider the dimensions of the structure as vanishingly small (Fig. 53). Therefore the conditions at infinity for the actual contour will be the same as for the "point wetted contour," i.e., the wetted contour for which $b = 0$. In this case formula (12.11) gives:

$$w = u - iv = \frac{kH}{\pi i z} \tag{12.14}$$

(we take $i$ with the positive sign, in order to have a positive value for the velocity component $u$ in the lower half plane). It is evident that the complex potential of the "point wetted contour" will be

$$\omega = \varphi + i\psi = \frac{kH}{\pi i} \ln z + C. \tag{12.15}$$

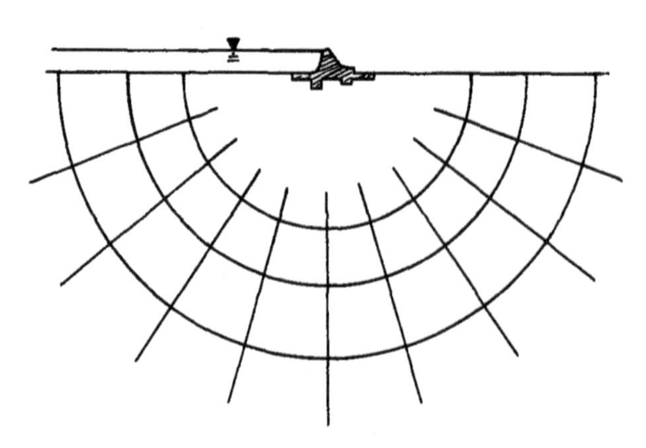

Fig. 53

In other words, a point vortex with intensity $\Gamma = 2kH$ gives the scheme of flow at long distances from the structure, and namely, where the flowlines practically coincide with half-circles (this happens for $r > 2L$, where $L$ — width of the structure.)

Remark No. 3: In hydrodynamics the problem of circulation flow about a rectilinear segment (Fig. 54) is treated. If the length of the segment is $2b$, the intensity of circulation $\Gamma$, then complex velocity and complex potential are given by the expressions:

$$w = u - iv = \frac{\Gamma}{2\pi} \frac{1}{\sqrt{b^2 - z^2}} ,$$

$$\omega = \varphi + i\psi = \frac{\Gamma}{2\pi} \sin^{-1} \frac{z}{b} + C .$$

Equating these expressions with (12.11) and (12.13), obtained for the plane fluid bed, we observe that one flow goes over into the other, if we take

$$\frac{\Gamma}{2} = kH .$$

## §13. Sheetpile in Pervious Layer of Infinite Depth.

In order to decrease the pressure under the dam and to decrease the exit velocity of the flow under hydraulic structures, a series of cut-offs is established, usually consisting of metal plates. In order to prevent the occurrence of slits between elements, they are given a special form (in Fig. 55 an example of the profile of two elements is given).

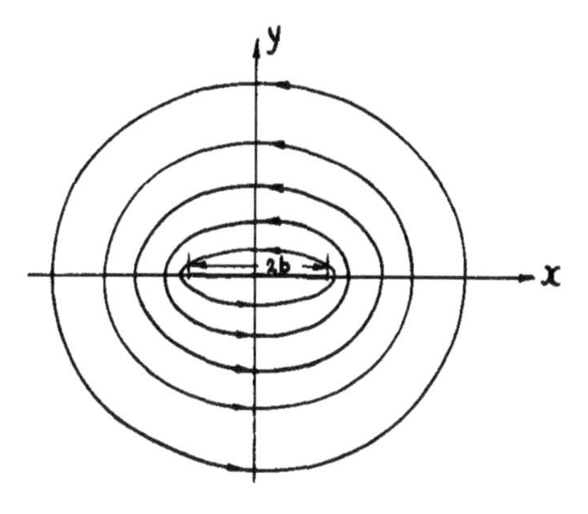


Fig. 54

If, in the problem shown at the end of the preceding paragraph, one changes the roles of $x$ and $y$, or, in other words, one turns the sketch in Fig. 54 over 90°, then another flow is obtained, represented in Fig. 56. A vertical segment of length $\ell$, around which ground water flows under influence of a head difference is sketched in this figure.

Fig. 55

Here complex velocity and complex potential have the form

$$w = \pm \frac{kHi}{\pi} \frac{1}{\sqrt{\ell^2 + z^2}} ,$$

$$\omega = \pm \frac{kHi}{\pi} \ell n \left(z + \sqrt{z^2 + \ell^2}\right) + C .$$

The plus sign is taken for the right side of the sketch. The streamlines in essence are half ellipses, the equipotential lines are hyperbolas orthogonal to the ellipses. A flownet is constructed in Fig. 56. In Fig. 57 the lines $V$ = constant and $U$ = constant are constructed, as this was done in paragraph 12 for the case of the two-dimensional wetted contour. From

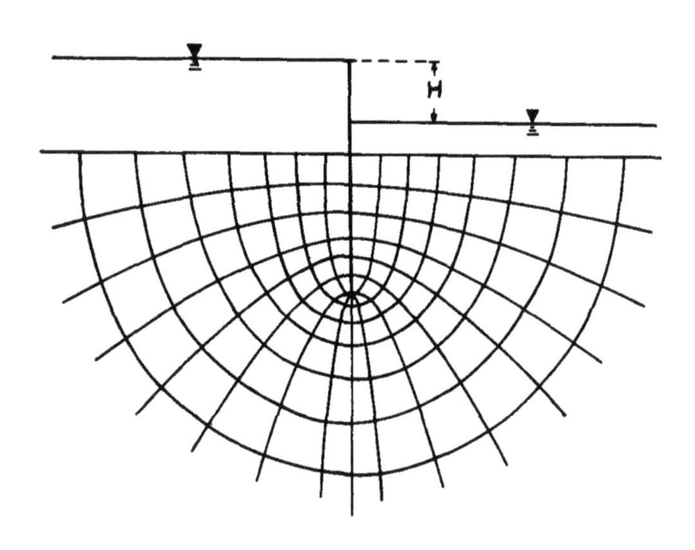


Fig. 56

Fig. 57

this figure it is apparent that, behind the cut-off, there is a region, designated by the letter a, inside which the force $\vec{S}$ has a vertical component, directed upward. In this region, the soil particles may come to the surface [4]. In Fig. 58 three cases are shown, corresponding to three values of the ratio H over $\ell$. In the last one of these, for $H = \ell$, the region "a" is very small, so that it was difficult to locate it on the drawing — its position is indicated by an arrow. However, such a region must always exist behind an extremely slender sheetpile.

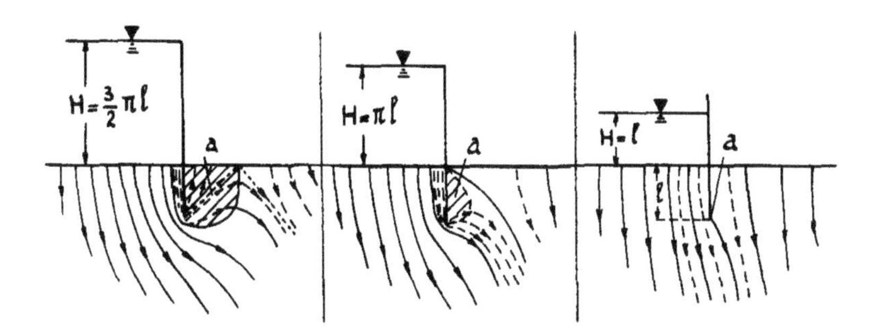


Fig 58

One may pose the problem: to find a depth $\ell$ of the driven sheetpile, so that along the boundary of the tailwater basin vertical component $S_y$ not be negative. To solve this problem, one must have in mind that along the boundary of the tailwater basin

$$u = 0, \qquad v = \frac{kH}{\pi\sqrt{\ell^2 + x^2}} .$$

The smallest value of the vertical component of the velocity is equal to $\frac{kH}{\pi\ell}$, therefore at the point of exit

$$S_y = \frac{\gamma H}{\pi\ell} - n(\gamma_1 - \gamma) .$$

If we want to satisfy the following inequality in this point,

$$S_y \leq 0$$

then we must take a length of sheetpile not smaller than

$$\ell_1 = \frac{\gamma H}{\pi n(\gamma_1 - \gamma)}$$

Such a length is sufficient to have $S_y$ directed inwardly to the soil along the boundary of the tailwater basin, but there will always remain a region behind the sheetpile, in which the force $\vec{S}$ is oriented upward (Fig. 58).

§14. Wetted Contour with Drainage Hole.

Sometimes, to decrease the pressure, a hole is established under the base of the weir. We suppose that a pipe, of semi-cylindrical shape, is placed immediately under the base. (Fig. 59.) Let the center of the pipe be at the point $x = a$. According to the remark, made in paragraph 3, we may look for a solution in the form

$$w = u - iv = \frac{kH}{\pi\sqrt{b^2 - z^2}} + \frac{M}{(z - a)\sqrt{b^2 - z^2}} , \tag{14.1}$$

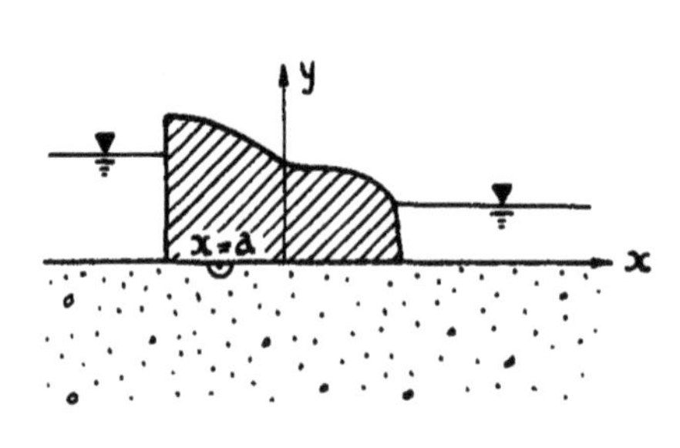


Fig. 59

for each of the terms of the right side satisfies the boundary conditions for $w$, but the second term gives the particular feature of the drainage type in the point $z = a$.

If, in the point $z = a$, drainage at the intensity $Q'$ takes place, then $w$ must have the form

$$w = -\frac{Q'}{2\pi(z - a)} + F(z) ,$$

where $F(z)$ — an holomorphic function in the point $z = a$. Therefore $Q' = 2Q$ where $Q$ — the discharge of the pipe, per unit length of the pipe. The expression

$$-\frac{Q'}{2\pi} = -\frac{Q}{\pi}$$

is the residue of $w(z)$, which can be computed as the limit

$$\lim_{z \to a} [(z - a) w(z)] = \frac{M}{\sqrt{b^2 - a^2}} .$$

From this,

$$M = -\frac{Q}{\pi}\sqrt{b^2 - a^2}$$

and

$$w = \frac{kH}{\pi\sqrt{b^2 - z^2}} \left[ 1 - \frac{Q\sqrt{b^2 - a^2}}{kH(z - a)} \right] .$$

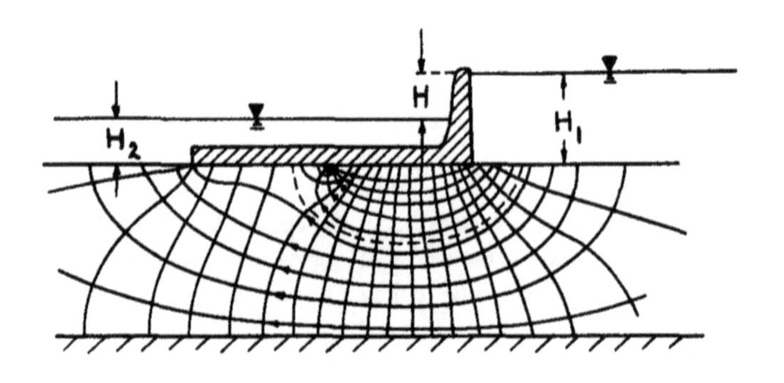


Fig. 60

In Fig. 60, the hydrodynamic flownet is given for $a = 0$ [9]. We must emphasize that the streamlines in this figure must be orthogonal to the rectilinear boundaries of the water bodies, and the equipotential lines orthogonal to the impervious boundaries.

More complicated schemes are analyzed in the work of N. T. Melechenko [10] and in the book by V. I. Aravin and S. N. Numerov (see literature to Chapter I, [18]).

## CHAPTER III.

## CONFINED SEEPAGE UNDER HYDRAULIC STRUCTURES

### A. Polygonal Regions in Problems of Confined Seepage. Uniqueness Theorem.

§1. Statement of Problem. (N. N. Pavlovsky)

In 1922, a book by N. N. Pavlovsky [1] was published in which the author presented the complete mathematical theory of confined flow under hydraulic structures such as overflow weirs, locks, spillways, etc. N. N. Pavlovsky examined a large number of problems, for many of which he gave quantitative formulas and related graphs.

N. N. Pavlovsky considered the general problem: the contour of the foundation of the hydraulic structure and the boundary of the impervious bedrock are composed of straight line segments, the boundaries between soil and head and tailwater reaches are also straight lines (Fig. 61). In other

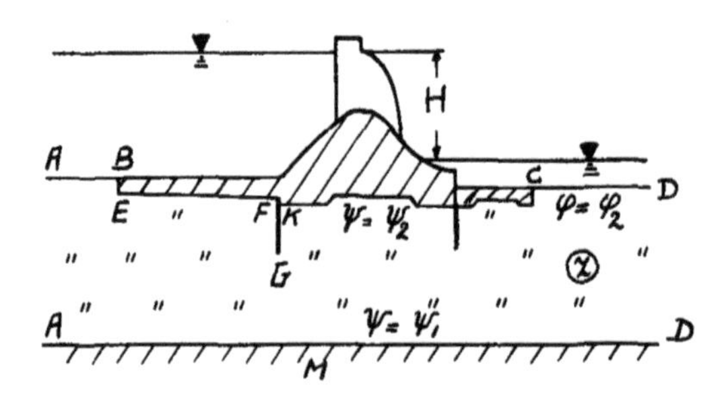


Fig. 61

words, the flow region $z$ is represented by a polygon. In the plane of the complex potential $\omega$, one also obtains a polygon and more particularly a rectangle (Fig. 62; the vertices of the rectangle are designated by the same letters as the vertices of the polygon in the flow region). We introduce the auxiliary complex variable $\zeta$ – sometimes called the "parametric variable", – in the upper or lower half plane onto which we shall map two regions: the region of the function $z$ and the region of the function $\omega = \varphi + i\psi$. Having $z = F_1(\zeta)$, $\omega = F_2(\zeta)$, we can find all the elements of the flow; for example, we find the velocity by means of the formula

$$\frac{d\omega}{dz} = \frac{F_2'(\zeta)}{F_1'(\zeta)} \quad . \tag{1.1}$$

The conformal mapping of the polygon onto the half-plane is carried out by means of the Christoffel-Schwarz formula. We examine this formula now.

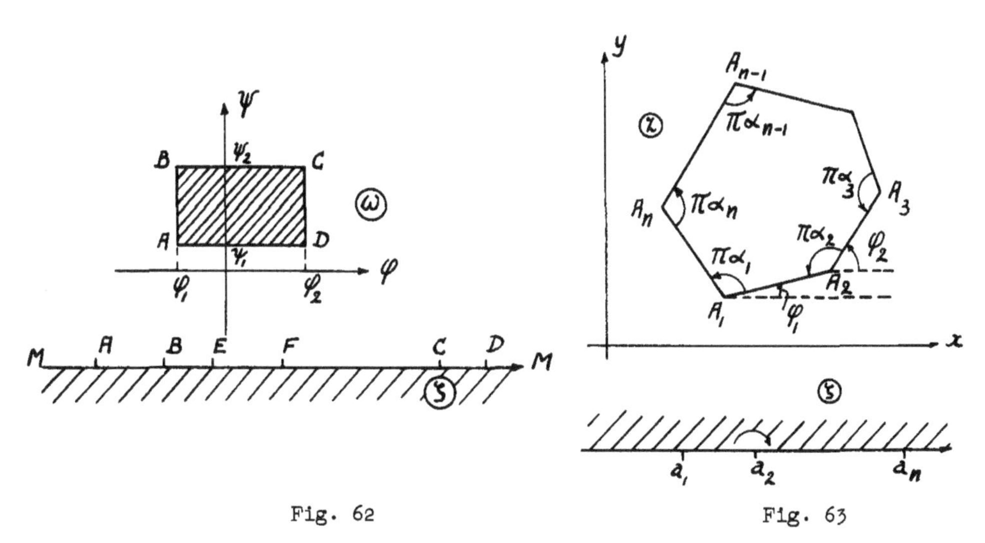


Fig. 62 Fig. 63

§2. Conformal Mapping of a Polygon onto a Half Plane.

Let us consider a polygon with vertices $A_1$, $A_2$, ... $A_{n-1}$, $A_n$ in the z plane (Fig 63.). We map it onto the upper half plane $\zeta$. We assume that the vertices of the polygon map onto points of the real axis with abscissae $\zeta = a_1, a_2, \ldots a_{n-1}, a_n$. The Christoffel-Schwartz [26] formula has the form

$$z = M \int (\zeta - a_1)^{\alpha_1 - 1}(\zeta - a_2)^{\alpha_2 - 1} \ldots (\zeta - a_n)^{\alpha_n - 1}\, d\zeta + N. \tag{2.1}$$

Here $\alpha_1$, $\alpha_2$, ... $\alpha_n$ are multiplied by $\pi$ to represent the magnitudes of the interior angles of the polygon.

We check the correctness of equation (2.1) and show the method to determine the constants M, N, $a_1$, ... $a_n$.

We start with the examination of the segment $A_1A_2$ and rewrite the function under the integral (2.1) such that it is real and positive in the $\zeta$ plane for the segment $A_1A_2$. It suffices therefore to change the differences $\zeta - a_2, \ldots \zeta - a_n$ into the differences $a_2-\zeta, \ldots, a_n-\zeta$, and to substitute a new unknown constant $M_1$ for the constant M. We change the indefinite integral (2.1) into a definite one between limits $a_1$ and $\zeta$:

$$z = M_1 \int_{a_1}^{\zeta} (\zeta - a_1)^{\alpha_1 - 1} (a_2 - \zeta)^{\alpha_2 - 1} \ldots (a_n - \zeta)^{\alpha_n - 1}\, d\zeta + z_1 \quad . \tag{2.2}$$

Here $z_1$ is the complex coordinate of the vertex $A_1$ of the polygon. We designate the function under the integral (2.2) by $\Phi_1(\zeta)$:

$$\Phi_1(\zeta) = (\zeta - a_1)^{\alpha_1 - 1}(a_2 - \zeta)^{\alpha_2 - 1} \dots (a_n - \zeta)^{\alpha_n - 1}$$

and examine the differential $dz$ along the segment $A_1A_2$:

$$dz = M_1\Phi_1(\zeta)d\zeta \quad .$$

Assume that the segment $A_1A_2$ makes the angle $\varphi_1$ with the abscissa axis. Let

$$M_1 = \rho e^{i\varphi_1} , \tag{2.3}$$

where $\rho$ is a positive constant, to obtain

$$z = \rho e^{i\varphi_1}\int_{a_1}^{\zeta} \Phi_1(\zeta)d\zeta + z_1, \qquad dz = \rho e^{i\varphi_1}\Phi_1(\zeta)d\zeta \ .$$

We observe that as long as $\Phi_1(\zeta) > 0$ along $A_1A_2$, $dz$ really corresponds to a rectilinear segment, making the angle $\varphi_1$ with the abscissa axis.

Arriving at the point $A_2$, we must put $\zeta = a_2$, $z = z_2$, where $z_2$ is the coordinate of the vertex $A_2$. We obtain from (2.2), after insertion of the value (2.3):

$$z_2 = \rho e^{i\varphi_1}\int_{a_1}^{a_2} \Phi_1(\zeta)d\zeta + z_1 \quad . \tag{2.4}$$

We designate the length of $A_1A_2$ of the polygon by $\ell_{12}$. Then we may write:

$$z_2 - z_1 = \ell_{12}e^{i\varphi_1} ,$$

and the equation (2.4), after dividing by $e^{i\varphi_1}$, gives:

$$\ell_{12} = \rho \int_{a_1}^{a_2} \Phi_1(\zeta)d\zeta \ . \tag{2.5}$$

From this it would be possible to determine $\rho$, if the parameters $a_1$, $a_2$, ... $a_n$ were known.

We turn now to the segment $A_2A_3$. Therefore, in the $\zeta$ plane, we must go around the point $\zeta = a_2$ along a semi-circle in clockwise sense (Fig. 63). The value $a_2 - \zeta$ now becomes negative and if we have chosen $\arg(a_2 - \zeta) = 0$ for $\zeta = \xi < a_2$, then the argument $(a_2 - \zeta)$ will be $(-\pi)$ after turning around $\zeta = a_2$. We may write that for the segment $A_2A_3$

$$a_2 - \zeta = e^{-\pi i}|a_2 - \zeta| = (\zeta - a_2)e^{-\pi i} \ .$$

The factor $(a_2 - \zeta)^{\alpha_2 - 1}$ becomes

$$(\zeta - a_2)^{\alpha_2 - 1}e^{-\pi i(\alpha_2 - 1)} = (\zeta - a_2)^{\alpha_2 - 1}e^{(\pi - \pi\alpha_2)i} \ .$$

But the angle $\pi - \pi\alpha_2$ is the angle made by the vector $\overline{A_1A_2}$ if one turns it into the position of the vector $\overline{A_2A_3}$.

We introduce the notations

$$\varphi_1 + \pi - \pi\alpha_2 = \varphi_2 ,$$

$$(\zeta - a_1)^{\alpha_1 - 1}(\zeta - a_2)^{\alpha_2 - 1}(a_3 - \zeta)^{\alpha_3 - 1} \dots (a_n - \zeta)^{\alpha_n - 1} = \Phi_2(\zeta) ,$$

and obtain instead of (2.4):

$$z = \rho e^{i\varphi_2} \int_{a_1}^{a_2} \Phi_1(\zeta)d\zeta + \rho e^{i\varphi_1} \int_{a_2}^{\zeta} \Phi_1(\zeta)d\zeta + z_1 =$$

$$= z_2 - z_1 + \rho e^{i\varphi_1} e^{(\pi - \pi\alpha_2)i} \int_{a_2}^{\zeta} \Phi_2(\zeta)d\zeta + z_1$$

or

$$z = z_2 + \rho e^{i\varphi_2} \int_{a_2}^{\zeta} \Phi_2(\zeta)d\zeta . \tag{2.6}$$

$\Phi_2(\zeta)$ assumes positive values along $A_2A_3$. We designate by $\ell_{23}$ the length of the segment $A_2A_3$ of the polygon. Then we obtain from the last formula, after insertion of $\zeta = a_3$ for the upper limit of integration:

$$z_3 - z_2 = \rho e^{i\varphi_2} \int_{a_2}^{a_3} \Phi_2(\zeta)d\zeta .$$

But $z_3 - z_2 = \ell_{23} e^{i\varphi_2}$ and therefore we have for $\ell_{23}$ the equation

$$\ell_{23} = \rho \int_{a_2}^{a_3} \Phi_2(\zeta)d\zeta . \tag{2.7}$$

We continue in the same way and obtain equations, analogous to (2.5) and (2.7), for other sides of the polygon. However, the equations for the last two sides, $A_{n-1}A_n$ and $A_nA_1$, will not be independent from the preceeding ones. Indeed, for given vertices $A_1$ and $A_{n-1}$ and given angles $\pi\alpha_1$ and $\pi\alpha_{n-1}$ in these vertices, the position of the vertex $A_n$ is completely determined, and the equations applying to the sides $A_{n-1}A_n$ and $A_nA_1$ do not give us any new correlations. We have for the last equation of type (2.5), (2.7)

$$\ell_{n-2,\ n-1} = \rho \int_{a_{n-2}}^{a_{n-1}} \Phi_{n-2}(\zeta)d\zeta , \tag{2.8}$$

where

$$\Phi_{n-2}(\zeta) = (\zeta - a_1)^{\alpha_1 - 1} \dots (\zeta - a_{n-2})^{\alpha_{n-2} - 1}(a_{n-1} - \zeta)^{\alpha_{n-1} - 1}(a_n - z)^{\alpha_n - 1} .$$

The formulas (2.5), (2.7)...(2.8) give n-2 equations, containing n+1 parameters: $\rho, a_1, a_2, \dots a_n$. Consequently, three of these parameters remain arbitrary. We may arbitrarily choose three of the quantities $a_1, a_2, \dots a_n$.

It is found that by means of the linear fractional transformation with real coefficients

$$\zeta = \frac{\alpha\zeta_1 + \beta}{\gamma\zeta_1 + \delta}, \tag{2.9}$$

we map the half plane $\zeta$ onto the half plane $\zeta_1$ with corresponding real axes; the parameters $\alpha, \beta, \gamma, \delta,$ or more precisely the ratio of three of them to the fourth, may be chosen so that three arbitrary points of the real $\zeta$-axis are mapped onto three given points of the real axis in the $\zeta_1$ plane.

It is easy to prove that the expression under the integral of formula (2.1) preserves its form, if we go from the variable $\zeta$ to the variable $\zeta_1$ by means of formula (2.9). Indeed, we obtain a correlation of the form

$$M_1(\zeta - a_1)^{\alpha_1 - 1}(\zeta - a_2)^{\alpha_2 - 1} \dots (\zeta - a_n)^{\alpha_n - 1} d\zeta = \\ = M_2 \frac{(\zeta_1 - b_1)^{\alpha_1 - 1} \dots (\zeta_1 - b_n)^{\alpha_n - 1} d\zeta_1}{(\gamma\zeta_1 + \delta)^{\alpha_1 + \dots + \alpha_n - n + 2}}.$$

But according to a property of the angles of a polygon

$$\alpha_1 + \alpha_2 + \dots + \alpha_n = n - 2$$

and, consequently, the denominator of the right side reduces to one. Often, one selects three special values for $a_k$, namely $0, 1, \infty$. We observe that in formula (2.1) the limit for $a_n = \infty$ can be taken, after the formula is written in the form

$$z = M_3 \int (\zeta - a_1)^{\alpha_1 - 1} \dots \left(1 - \frac{\zeta}{a_n}\right)^{\alpha_n - 1} d\zeta + N;$$

where $M_3 = M(-a_n)^{\alpha_n - 1}$.

Because $\frac{\zeta}{a_n} \to 0$ for $a_n \to \infty$, the Christoffel-Schwartz formula preserves its form when one of the vertices of the polygon is mapped onto the point of infinity of the $\zeta$-plane, but the factor corresponding to this point in formula (2.1) becomes one.

If a vertex, for example $A_2$, tends to infinity in such a way that the adjacent sides become parallel (Fig. 64a), then one must take $\alpha_2 = 0$. For further turning of the sides $A_1A_2$, when they cease to be parallel, but when the vertex $A_2$ remains at infinity (Fig. 64b), the angle $\pi\alpha_2$ must be considered negative,

a)

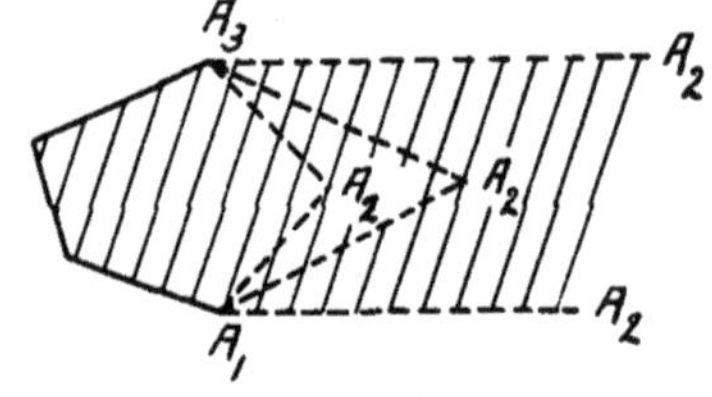


b)

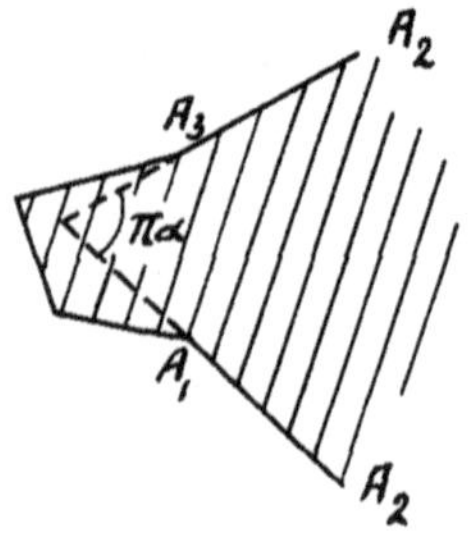


Fig. 64

and namely

$$\pi\alpha_2 = -\pi\alpha ,$$

where $\alpha$ is the magnitude of the angle formed by the prolongation of the sides $A_1A_2$ and $A_3A_2$.

Formula (2.1) is also correct for polygons that are located on several sheets of the Riemann surface [22, 23].

§3. Mapping of a Rectangle onto the Half Plane.

Because of the symmetry of the mapped figure (the rectangle A B C D in Fig. 65), it is convenient to map two vertices — we choose A and B — onto the points $\zeta = \pm 1$. The vertices C and D then transform into points with symmetrical position in the $\zeta$-plane, designated by $\pm \frac{1}{k}$, where k — some number, smaller than one, that must be determined further. Instead of the values $\zeta - \frac{1}{k}$, $\zeta + \frac{1}{k}$ in formula (2.1), one may take $1 - k\zeta$, $1 + k\zeta$; instead of $\zeta - 1$, $\zeta + 1$, one takes $1 - \zeta$, $1 + \zeta$. Since the vertices have right angles, all the exponents $\alpha_s$ are equal to $\frac{1}{2}$ and $\alpha_s - 1 = -\frac{1}{2}$ ($s = 1, 2, 3, 4$). The Christoffel-Schwartz formula takes the form:

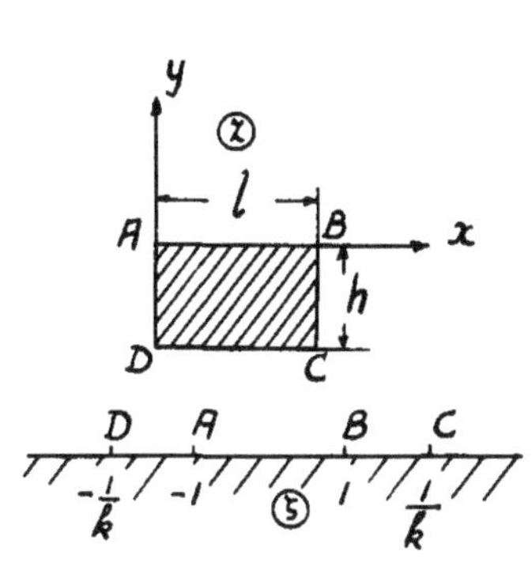

Fig. 65

(3.1)
$$z = \frac{l}{2K}\int_{-1}^{\zeta}\frac{d\zeta}{\sqrt{(1-\zeta^2)(1-k^2\zeta^2)}} ,$$

where

(3.2)
$$K = \int_0^1 \frac{d\zeta}{\sqrt{(1-\zeta^2)(1-k^2\zeta^2)}} = \int_0^{\frac{\pi}{2}} \frac{d\varphi}{\sqrt{1-k^2\sin^2\varphi}} .$$

Introducing the notation of the elliptical integral of the first kind,

$$F(\varphi, k) = \int_0^{\varphi}\frac{d\varphi}{\sqrt{1-k^2\sin^2\varphi}}$$

or, since

$$\sin\varphi = \zeta ,$$

(3.3)
$$F(\text{arc}\sin\zeta, k) = \int_0^{\zeta}\frac{d\zeta}{\sqrt{(1-\zeta^2)(1-k^2\zeta^2)}} ,$$

we may write:

(3.4)
$$z = \frac{l}{2K}[K + F(\text{arc}\sin\zeta, k)] .$$

If we designate, as is generally accepted, the elliptical integral of the first kind for the modulus $k' = \sqrt{1-k^2}$ by $K'$, then we obtain the ratio

(3.5) $$h = \frac{\ell K'}{2K} .$$

If the lengths $h$ and $\ell$ are given, then we find from them

(3.6) $$\frac{K'}{2K} = \frac{h}{\ell} .$$

In special tables one can find the modulus $k$, when the ratio $\frac{K'}{K}$ is known.

§4. Basic Rectangle of Confined Seepage Problems.

Turning to a typical case of flow about the foundation of an hydraulic structure, represented in Fig. 66, we find the boundary conditions of the flow region for the functions $\varphi$ and $\psi$ or the head $h$ and function $q$ proportional with them: (1)

$$h = -\frac{\varphi}{\kappa}, \quad q = -\frac{\psi}{\kappa}$$

We call $q$ the "reduced seepage rate". We choose the value of $\varphi$ as was done in Par. 12 of Chapter II for the two dimensional flow about a dam foundation, i.e., $\varphi = -\kappa H$ along the bottom of the headwater basin, $\varphi = 0$ along the bottom of the tailwater basin (Fig. 66). Along the foundation of

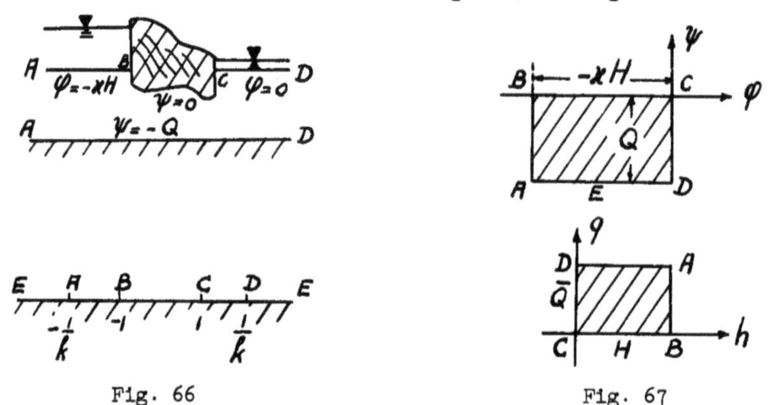


Fig. 66 Fig. 67

the structure, we take $\psi = 0$, along the boundary of the bedrock, let

$$\psi = -Q \qquad (Q > 0) .$$

Then for. $h$ and $q$ we have

$h = H$ along the boundary of the headwater basin

$h = 0$ along the boundary of the tailwater basin

$q = 0$ along the foundation of the structure

$q = \bar{Q} = \frac{Q}{\kappa}$ along the bedrock .

We understand by $\bar{Q}$ the "reduced seepage rate" i.e., the rate for $\kappa = 1$,

(1): (Kappa) $\kappa$ is used in this paragraph instead of $k$ to avoid confusion.

calculated for a dam of unit width. The real flow rate is $Q$.

In the mapping of the rectangle A B C D onto the lower half plane (Fig. 66) we obtain:

$$(4.2)\quad \begin{cases} \omega = \varphi + i\psi = -\dfrac{\kappa H}{2K}\,[K - F(\text{arc sin}\,\zeta,\,k)] \;, \\ h + iq = \dfrac{H}{2K}\,[K - F(\text{arc sin}\,\zeta,\,k)] \;. \end{cases}$$

The inverse of the elliptic integral

$$u = \int_0^\zeta \frac{d\zeta}{\sqrt{(1-\zeta^2)(1-k^2\zeta^2)}}$$

is the elliptic sine

$$\zeta = \text{sn}\,u \;.$$

After formula (4.2) we obtain for $\zeta$

$$(4.3)\quad \zeta = \text{sn}\left(K + \frac{2K\omega}{\kappa H}\right) = \frac{\text{cn}\,u}{\text{dn}\,u}$$

where

$$(4.4)\quad u = \frac{2K\omega}{\kappa H}$$

$$\text{cn}\,u = \sqrt{1 - \text{sn}^2\,u}, \qquad \text{dn}\,u = \sqrt{1 - k^2\,\text{sn}^2\,u} \;.$$

Substituting $\overline{Q}$ and $H$ for $h$ and $\ell$ in (3.5), we obtain (Fig. 67)

$$(4.5)\quad \overline{Q} = \frac{HK'}{2K}\;, \qquad Q = \frac{\kappa HK'}{2K} \;.$$

If we knew $H$ and $Q$, then we would be able to find $k$ from the ratio

$$\frac{K'}{2K} = \frac{Q}{\kappa H} \;.$$

But usually the seepage rate $Q$ is unknown, and the modulus $k$ is determined from the conditions that some lengths in the $z$-plane are known.

If there is no bedrock, i.e., if it is found sufficiently deep, then one may consider the soil as reaching infinitely deep, and the rectangle in the $\omega$-plane becomes a half strip. We may obtain the mapping of the half strip onto the half plane immediately, but we may also put $k = 0$ in the previous formulas and account for the fact that for $k = 0$,

$$K = \frac{\pi}{2} \;.$$

The elliptic integral of the first kind gives arc sin $\zeta$ for $k = 0$. From equation (4.2) we obtain:

$$(4.6)\quad \omega = \varphi + i\psi = \frac{\kappa H}{\pi}\left(\text{arc sin}\,\zeta - \frac{\pi}{2}\right) = -\frac{\kappa H}{\pi}\,\text{arc cos}\,\zeta \;,$$

$$(4.7)\quad h + iq = \frac{H}{\pi}\,\text{arc cos}\,\zeta \;,$$

from which

$$(4.8)\quad \zeta = \cos\frac{\pi\omega}{\kappa H} \;.$$

§5. Uniqueness Theorem.

The problem of confined seepage under an hydraulic structure belongs to the kind of so called "mixed boundary value problems of potential theory," i.e., problems in which along different sections of the contour alternately $\varphi$ and $\frac{\partial\varphi}{\partial n}$ are given. ($\varphi$ = velocity potential, constant along boundaries of reservoirs, $\frac{\partial\varphi}{\partial n} = 0$ along impervious boundaries.) As is known, when along the entire boundary of the region the function $\varphi$ is given and one has to find the value of $\varphi$ inside of the region, the problem is a so called Dirichlet problem; if the value of the normal derivative $\frac{\partial\varphi}{\partial n}$ is given along the contour of the region, the problem of finding the function $\varphi$ inside of the region is called a Neumann problem. In courses of analysis, the uniqueness of solutions of the Dirichlet and Neumann problems for well defined boundary conditions is shown.

N. N. Pavlovsky [1] investigated the uniqueness of solutions of problems in confined seepage under hydraulic structures.

We assume, at first, that we have a flow region as represented in Fig. 68. This region is bounded by lines along which $\varphi$ = constant or $\psi$ = constant; along the latter, $\frac{\partial\varphi}{\partial n} = 0$. To prove the uniqueness of the solution of such a problem, it is not necessary to assume rectilinear boundaries – they may be curvilinear, with satisfactory conditions for a properly posed Dirichlet or Neumann problem [26].

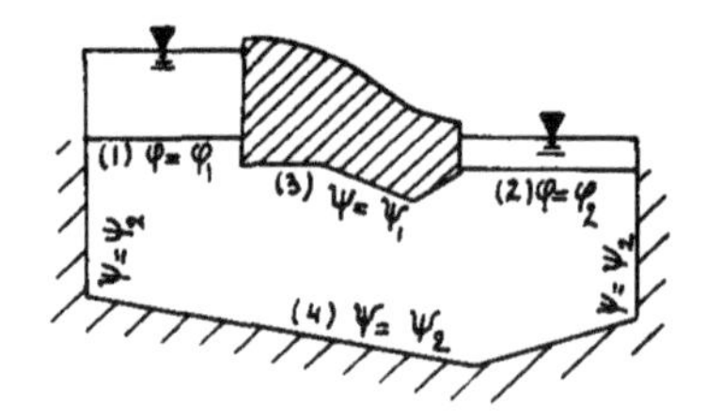

Fig. 68

We examine first the case where seepage takes place in a finite region D, bounded by streamlines and equipotential lines (Fig. 68).

Let us assume that there exist two harmonic functions, $\varphi(x, y)$ and $\varphi_1(x, y)$, satisfying the boundary conditions: $\varphi$ = constant at the boundaries (1) and (2), $\frac{\partial\varphi}{\partial n} = 0$ at the boundaries (3) and (4). We show that $\varphi(x,y) \equiv \varphi_1(x, y)$.

Therefore we form the difference

$$w = \varphi(x, y) - \varphi_1(x, y) .$$

We write Green's formula for two functions $u(x, y)$, and $v(x, y)$, having continuous first and second order derivatives in D:

$$\iint_D \left(\frac{\partial u}{\partial x}\frac{\partial v}{\partial x} + \frac{\partial u}{\partial y}\frac{\partial v}{\partial y}\right)dxdy = \int_C u\frac{\partial v}{\partial n}\,ds - \iint_D u\Delta dxdy .$$

Here $\frac{\partial v}{\partial n}$ is the derivative along the exterior normal to the contour C of the region.

Taking $u = v = w$, we obtain:

$$\iint_D \left[\left(\frac{\partial w}{\partial x}\right)^2 + \left(\frac{\partial w}{\partial y}\right)^2\right] dxdy = \int_0 w \frac{\partial w}{\partial n} ds - \iint_D w\Delta w \, dx\, dy \; .$$

Since $\Delta\varphi = \Delta\varphi_1 = \Delta w = 0$, the right side of the equation reduces to a line integral. But this line integral is also zero, since at the boundaries (1) and (2) the function $w = 0$, and at the boundaries (3) and (4) its normal derivative is zero.

Therefore the equation becomes

$$\iint_D \left[\left(\frac{\partial w}{\partial x}\right)^2 + \left(\frac{\partial w}{\partial y}\right)^2\right] dx\, dy = 0 \quad .$$

The integral of a sum of squares can only be zero if each term is zero:

$$\frac{\partial w}{\partial x} = 0, \qquad \frac{\partial w}{\partial y} = 0 \; .$$

From this it follows that $w = c$, but since at some parts of the contour $w = 0$, we must have the identity

$$w \equiv 0.$$

Thus the uniqueness of the solution of the mixed boundary value problem for a finite region of seepage is proved.

If seepage takes place in a pervious stratum of infinite depth, then one may bound the region with a half circle of large radius $R$, along which $\frac{\partial \varphi}{\partial n}$ has the order of magnitude of the velocity (see Chapter II, Par. 12), i.e.,‡ the order of $\frac{1}{R^2}$. Therefore, since $ds = R\, d\theta$ and since $\varphi$ is of order $\ell n\, R$, we have in first approximation

$$\int_{(R)} w \frac{\partial w}{\partial n} ds \approx \int_0^{\pi} \frac{\ell n\, R}{R^2} R\, d\theta = \pi \frac{\ell n\, R}{R} \to 0 \quad \text{for} \quad R \to \infty \quad .$$

Consequently, the previous conclusions remain also valid here.

For the region that is bounded by bedrock but at one side reaches till infinity, the uniqueness theorem may also be derived as a limit case from the finite region.

## B. Seepage Under Flat-bottom Foundations.

### §6. Flat-bottom Foundation on a Layer of Finite Depth.

Now we examine several categories of seepage problems under hydraulic structures and begin with the above mentioned problem. We examined a problem of this kind, but for soil of infinite depth, in Par. 12 of Chapter II. Here we investigate more complex cases in Fig. 69, the flowregion is a strip and must be mapped onto the half plane $\zeta$ as shown. By means of formula (2.1) we obtain

$$z = Mk^2 \int_0^{\zeta} \frac{d\zeta}{1 - k^2\zeta^2} = \frac{M}{2k} \ell n \frac{1 + k\zeta}{1 - k\zeta} \; . \tag{6.1}$$

‡ i.e., $\frac{\partial w}{\partial n}$ is of order $\frac{1}{R^2}$ .

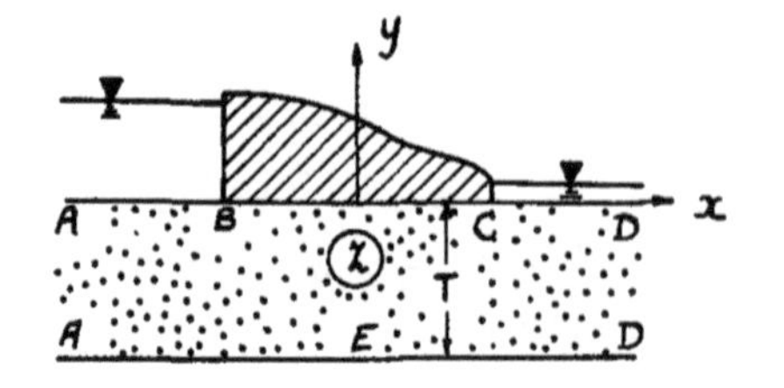


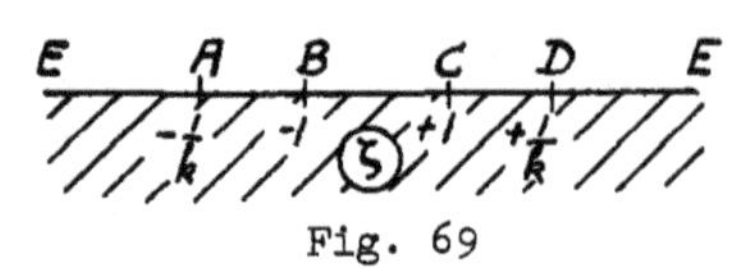


Fig. 69

Let the length of the base BC be $2\ell$. Then, for $\zeta = \pm 1$, we have $z = \pm \ell$ and

(6.2) $$\ell = \frac{M}{2k} \ell n \frac{1 + k}{1 - k} .$$

Furthermore, walking around the point $\zeta = \frac{1}{k}$ in the positive sense in the lower half plane corresponds to jumping from CD to segment DE in the z-plane, and thus gives for $\ell n(1 - k\zeta)$ an increase of $\pi i$; consequently, the increase of $z$, or $\Delta z$, will be

$$\Delta z = \frac{M}{2k} (- \pi i) = - \frac{M\pi i}{2k} .$$

M, as is evident from formula (6.2) is a positive number; consequently, we obtained for $\Delta z$ an imaginary number. This corresponds to the change of the imaginary part of $z$, i.e., the ordinate $y$, from the value $y = 0$ to $y = - T$ (T = thickness of layer). In other words, $\Delta z = - Ti$, or

$$- Ti = - \frac{M\pi i}{2k} ,$$

and

$$M = \frac{2kT}{\pi} .$$

According to formula (6.2)

$$\ell = \frac{T}{\pi} \ell n \frac{1 + k}{1 - k} .$$

From this

(6.3) $$\frac{1 + k}{1 - k} = e^{\frac{\ell\pi}{T}}, \qquad k = \text{th} \frac{\pi\ell}{2T} .$$

In equation (4.3) we had

$$\zeta = \text{sn } u, \qquad u = \frac{2K\omega}{\kappa H} + K;$$

so that

(6.4) $$z = \frac{T}{\pi} \ell n \frac{1 + k\zeta}{1 - k\zeta} = \frac{T}{\pi} \ell n \frac{1 + k \text{ sn } u}{1 - k \text{ sn } u} .$$

The seepage rate is computed by formula (4.5)

(6.5) $$Q = \frac{\kappa HK'}{2K} , \qquad k = \text{th} \frac{\pi\ell}{2T} .$$

Formula (6.4) gives the dependence of $z$ on the complex potential. From (6.4) we obtain

(6.6) $$k\zeta = \text{th} \frac{\pi z}{2T} .$$

We differentiate (6.4) and integrate again, noticing that

$$\left[\ell n \frac{1 + k \text{ sn } u}{1 - k \text{ sn } u}\right]' = 2k \frac{\text{cn } u}{\text{dn } u} \qquad (\text{dn } u = \sqrt{1 - k^2\text{sn}^2 u}) ,$$

and that the ratio $\frac{\text{cn } u}{\text{dn } u}$ can be developed into the trigonometric series [24]

$$\frac{\text{cn } u}{\text{dn } u} = \frac{\pi}{kK}\left\{\frac{\cos\frac{\pi u}{2K}}{\text{sh}\frac{\pi K'}{2K}} - \frac{\cos\frac{3\pi u}{2K}}{\text{sh}\frac{3\pi K'}{2K}} + \frac{\cos\frac{5\pi u}{2K}}{\text{sh}\frac{5\pi K'}{2K}} - \cdots\right\}$$

After integration along $u$, we obtain for $z$ the series

$$z = \frac{4T}{\pi}\left\{\frac{\sin\frac{\pi u}{2K}}{\text{sh}\frac{\pi K'}{2K}} - \frac{\sin\frac{3\pi u}{2K}}{3\,\text{sh}\frac{3\pi K'}{2K}} + \frac{\sin\frac{5\pi u}{2K}}{5\,\text{sh}\frac{5\pi K'}{2K}} - \cdots\right\} .$$

Here $u = \frac{2K\omega}{\kappa H} + k$; consequently

$$\frac{m\pi u}{2K} = \frac{m\pi\omega}{\kappa H} + \frac{m\pi}{2} .$$

Since $m$ is odd,

$$\sin\frac{m\pi u}{2K} = (-1)^{\frac{m-1}{2}} \cos\frac{m\pi\omega}{\kappa H}$$

and

$$z = x + iy = \frac{4T}{\pi}\left\{\frac{\cos\frac{\pi\omega}{\kappa H}}{\text{sh}\frac{\pi K'}{2K}} + \frac{\cos\frac{3\pi\omega}{\kappa H}}{3\,\text{sh}\frac{3\pi K'}{2K}} + \frac{\cos\frac{5\pi\omega}{\kappa H}}{5\,\text{sh}\frac{5\pi K'}{2K}} + \cdots\right\}, \tag{6.7}$$

We put

$$\frac{\pi\omega}{\kappa H} = -\frac{\pi(h + iq)}{H} ,$$

$$\cos\frac{\pi}{H}(h + iq) = \cos\frac{\pi h}{H}\,\text{ch}\frac{\pi q}{H} - i\sin\frac{\pi h}{H}\,\text{sh}\frac{\pi q}{H} .$$

We separate in (6.7) the real part from the imaginary and find:

$$\left\{\begin{aligned} x &= \frac{4T}{\pi}\left\{\frac{\cos\pi\bar{h}\,\text{ch}\,\pi\bar{q}}{\text{sh}\,\alpha} + \frac{\cos 3\pi\bar{h}\,\text{ch}\,3\pi\bar{q}}{3\,\text{sh}\,3\alpha} + \cdots\right\}, \\ y &= -\frac{4T}{\pi}\left\{\frac{\sin\pi\bar{h}\,\text{sh}\,\pi\bar{q}}{\text{sh}\,\alpha} + \frac{\sin 3\pi\bar{h}\,\text{sh}\,3\pi\bar{q}}{3\,\text{sh}\,3\alpha} + \cdots\right\}, \end{aligned}\right. \tag{6.8}$$

where

$$\bar{h} = \frac{h}{H}, \quad \bar{q} = \frac{q}{H}, \quad \alpha = \frac{\pi K'}{2K} .$$

With these formulas, one can construct the families of lines $q$ = constant and $h$ = constant

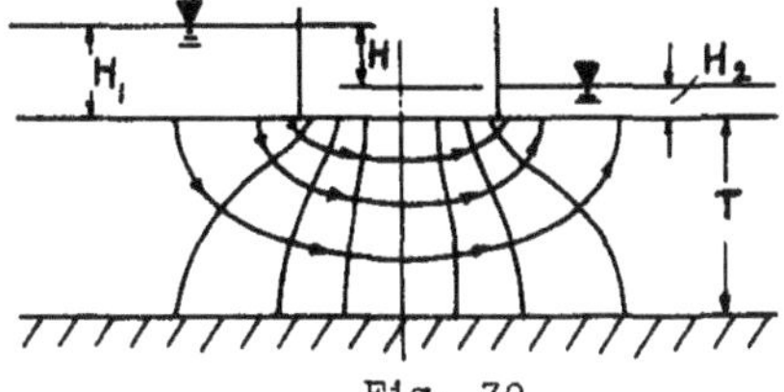


Fig. 70

We obtain the head along the foundation if we insert $\zeta$ from (6.6) in (4.2) for $z = x$

$$h = \frac{H}{2}\left\{1 - \frac{1}{K}F\left(\text{arc sin}\frac{\text{th}\frac{\pi x}{2T}}{\text{th}\frac{\pi \ell}{2T}}, k\right)\right\}.$$

To find the velocity, we differentiate (4.2) and (6.4):

$$\frac{d\omega}{dz} = \frac{d\omega}{d\zeta} : \frac{dz}{d\zeta} = \frac{\kappa H\pi}{4kKT}\sqrt{\frac{1 - k^2\zeta^2}{1 - \zeta^2}},$$

By means of (6.4) we find for the right half of Fig. 70:

$$w = u - iv = \frac{\kappa H\pi}{4KT}\frac{\text{ch}\frac{\pi\ell}{2T}}{\sqrt{\text{sh}\frac{\pi(\ell + z)}{2T}\text{sh}\frac{\pi(\ell - z)}{2T}}}$$

In particular, along the boundary of the lower reservoir, where $z = x > \ell$

$$v = \frac{\kappa H\pi}{4KT}\frac{\text{ch}\frac{\pi\ell}{2T}}{\sqrt{\text{sh}\frac{\pi(\ell + x)}{2T}\text{sh}\frac{\pi(x - \ell)}{2T}}}$$

and along the impervious bedrock, where $z = x - Ti$,

$$u = \frac{\kappa H\pi}{4KT}\frac{\text{ch}\frac{\pi\ell}{2T}}{\sqrt{\text{ch}\frac{\pi(x + \ell)}{2T}\text{ch}\frac{\pi(x - \ell)}{2T}}}.$$

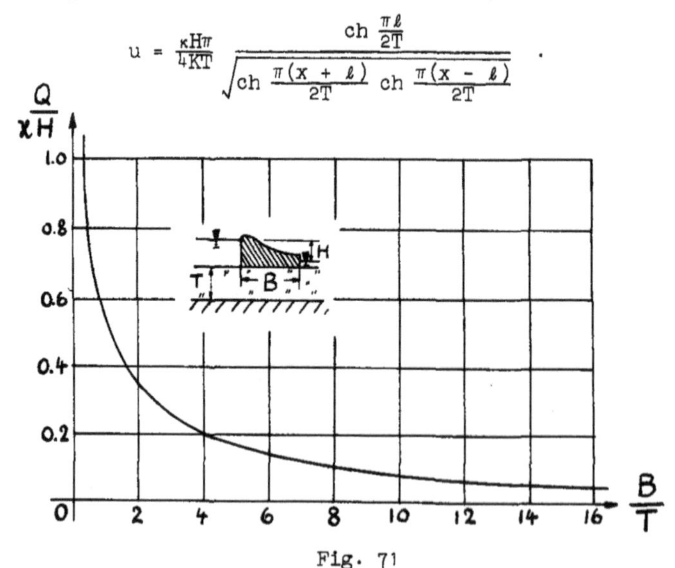


Fig. 71

Streamlines and equipotential lines are sketched in Fig. 70. The dependence of the reduced discharge on the dimensions of the flowregion is given by the curve of Fig. 71. Graphs for the distribution of the head (pressure) along the foundation of the dam are given in Fig. 72.

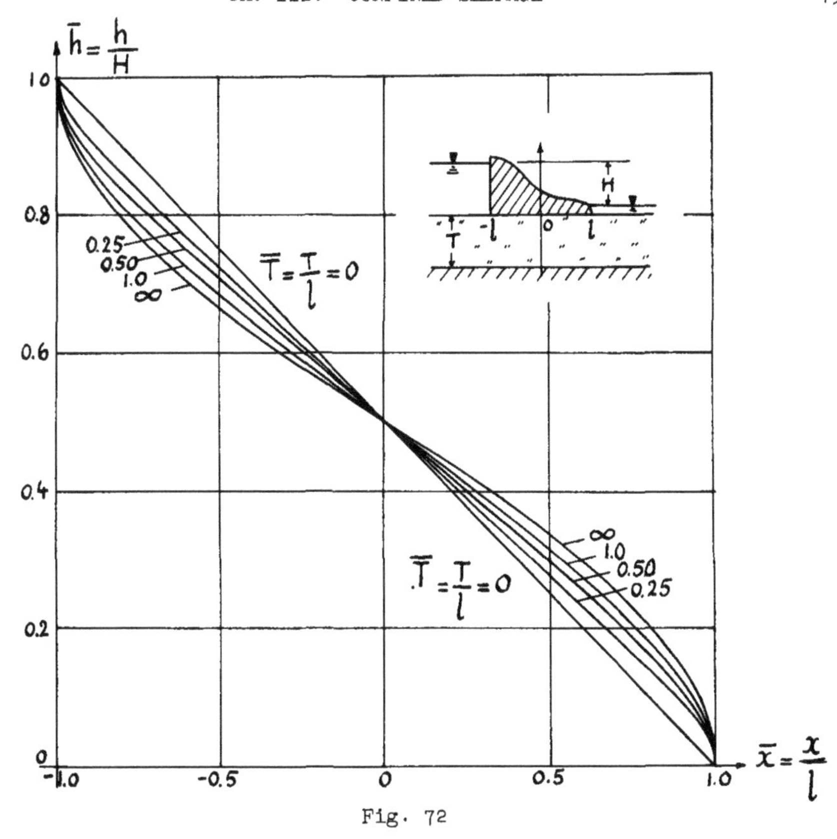


Fig. 72

§7. Flat Bottom Foundation on a Drained Stratum.

In nature one encounters the case where a more pervious stratum underlies the soil on which the dam foundation rests. The seepage coefficient of the former stratum is considerably larger than that of the latter. One may assume that the head has a constant value $H_0$ at the boundary of the two strata. This problem may be considered as a special case of the flow about a foundation on two-layered soil. (Chapter VIII, Par. 1)

The flow-region here is the same as in the previous case (Fig. 73). Its mapping onto the half-plane (Fig. 74) gives

(7.1) $$\zeta = \frac{1}{k} \operatorname{th} \frac{\pi z}{2T} \quad ,$$

where

(7.2) $$k = \operatorname{th} \frac{\pi \ell}{2T} \quad .$$

We assume that $H_1 > H_0 > H_2$. Then we obtain a semi strip with cut in the plane of the complex potential. (Fig. 74.)

Let point F of the $\omega$-plane transform into the point

$$\zeta = a .$$

The other points are indicated in Fig. 74.

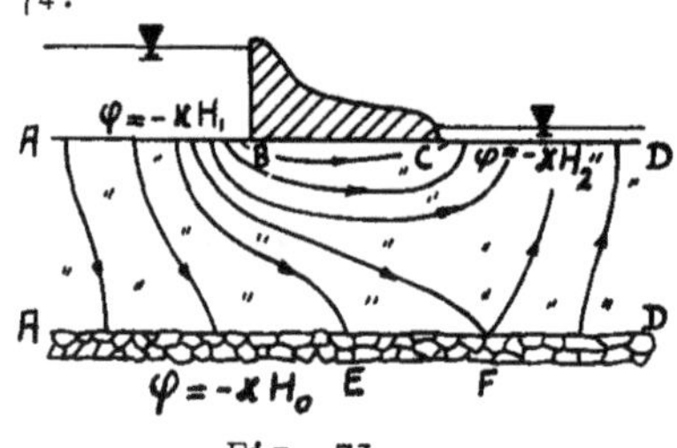

Fig. 73

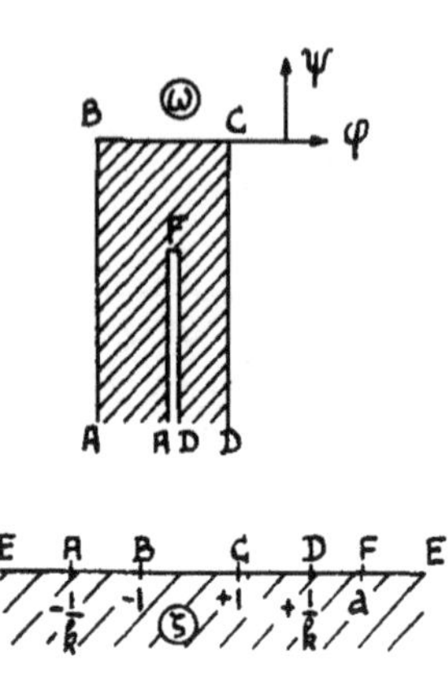

Fig. 74

We obtain:

$$\omega = M\int_1^\zeta \frac{(\zeta - a)d\zeta}{(1 - k^2\zeta^2)\sqrt{1 - \zeta^2}} - \kappa H_2 = \tag{7.3}$$
$$= \left[-\frac{1}{kk'}\text{ arc tg }\frac{k}{k'}\sqrt{1 - \zeta^2} + \frac{a}{k'}\text{ arc tg }\frac{\sqrt{1 - \zeta^2}}{k'\zeta}\right] - \kappa H_2 .$$

Walking around the points C and D in the positive direction and around B and A in the negative direction along half circles in the lower half plane gives for $\omega$ values that correspond to the jumps in the passage from segment CD to FD and from AB to AF. We obtain the equalities:

$$-M\pi\frac{1 - ka}{2kk'} = \kappa(H_2 - H_0), \qquad M\pi\frac{1 + ka}{2kk'} = \kappa(H_0 - H_1) .$$

From this system of equations we find:

$$ka = \frac{H_1 - H_2}{H_1 + H_2 - 2H_0}, \qquad M = -\frac{\kappa}{\pi}(H_1 + H_2 - 2H_0)kk' .$$

Equation (7.3) gives for $\omega$:

$$\omega = \frac{\kappa(H_1 + H_2 - 2H_0)}{\pi}\text{ arc tg }\frac{k}{k'}\sqrt{1 - \zeta^2} - \tag{7.4}$$
$$- \frac{\kappa(H_1 - H_2)}{\pi}\text{ arc tg }\frac{\sqrt{1 - \zeta^2}}{k'\zeta} - \kappa H_2 .$$

Inserting the values of (7.1) and (7.2), we finally obtain:

$$\omega = \frac{\kappa(H_1 + H_2 - 2H_0)}{\pi}\text{ arc cos }\left(\frac{\text{ch }\frac{\pi z}{2T}}{\text{ch }\frac{\pi\ell}{2T}}\right) - \tag{7.5}$$
$$- \frac{\kappa(H_1 - H_2)}{\pi}\text{ arc cos }\left(\frac{\text{sh }\frac{\pi z}{2T}}{\text{sh }\frac{\pi\ell}{2T}}\right) - \kappa H_2 .$$

From this equation one easily finds the head along the dam foundation, for $z = x$ and $h = -\frac{\varphi}{\kappa} - H_2$:

(7.6)
$$h = \frac{2H_0 - H_1 - H_2}{\pi} \text{ arc cos } \left( \frac{\text{ch } \frac{\pi x}{2T}}{\text{ch } \frac{\pi \ell}{2T}} \right) + \frac{H_1 - H_2}{\pi} \text{ arc cos } \left( \frac{\text{sh } \frac{\pi x}{2T}}{\text{sh } \frac{\pi \ell}{2T}} \right) .$$

To compare this case with the previous one, we compute the value of $h$ in the middle of the dam foundation for $x = 0$. We find

$$h = h_0 = \frac{H}{2} + \frac{2H_0 - H_1 - H_2}{\pi} \text{ arc cos } \frac{1}{\text{ch } \frac{\pi \ell}{2T}}$$

$$(H = H_1 - H_2) ,$$

while in the previous case $h = h_0' = \frac{H}{2}$ (see Fig. 72). Therefore, if $H_0 > \frac{H + H_2}{2}$, then $h_0 > h_0'$.

If $H_0 < \frac{H + H_2}{2}$, then $h_0 < h_0'$. We assumed that $H_1 > H_0 > H_2$. If this inequality is not satisfied, then the $\omega$ region in Fig. 74 changes, but all formulas remain unchanged.

For $H_0 = \frac{H_1 + H_2}{2}$ all computations simplify. The streamlines become symmetrical with respect to the ordinate axis, now a line of constant head $h = \frac{H}{2}$.

In the case of non homogeneous soil (Chapter 8), maximum conditions occur. The flowrate through the line CD (Fig. 73) in the case $H_2 < H_0 < H_1$ under consideration, becomes infinitely large, part of the water flowing from the upper to the lower reservoir, part of the water flowing from the upper reservoir to the lower soil stratum and part of the water flowing from this soil stratum into the lower reservoir.

We also give the formula for the velocity along the bottom of the lower reservoir

$$v = \frac{\kappa}{2T} \frac{H \text{ ch } \frac{\pi x}{2T} + (2H_0 - H_1 - H_2) \text{ sh } \frac{\pi x}{2T}}{\sqrt{\text{sh } \frac{\pi(x + \ell)}{2T} \text{ sh } \frac{\pi(x - \ell)}{2T}}} .$$

§8. Two Dams With Flat Foundations in Series on a Stratum of Finite Depth.

The sketch of the dams is given in Fig. 75. Here we reproduce some

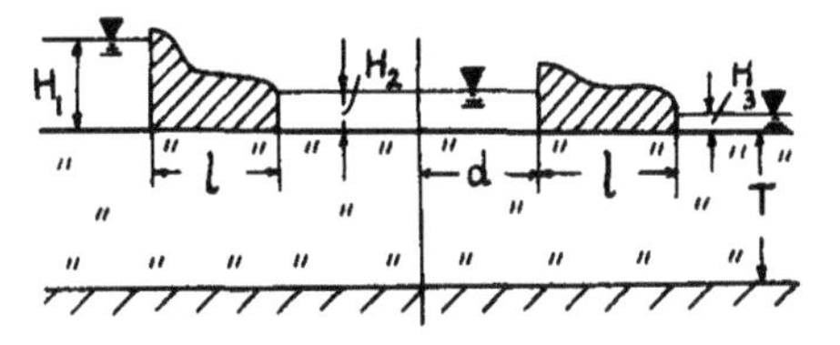


Fig. 75

basic results without comments. We designate the heads $H_1$, $H_2$, $H_3$ on the figure.

The flowrate in the region between the two dams is:

$$Q = \kappa |H_1 - 2H_2 + H_3| \frac{2K}{K'} ,$$

where

$$k = \frac{\operatorname{sh} \frac{\pi d}{2T}}{\operatorname{sh} \frac{\pi(d + \ell)}{2T}} .$$

The relation between the complex potential and the z-coordinate may be represented in the form

$$\omega = - \frac{\kappa}{\pi}(H_1 - 2H_2 + H_3) F (\operatorname{arc} \sin \lambda, k') ,$$

where

$$\lambda = \frac{\operatorname{sh} \frac{\pi(d + \ell)}{2T}}{\operatorname{sh} \frac{\pi z}{2T}} \sqrt{\frac{\operatorname{sh} \frac{\pi(z + d)}{2T} \operatorname{sh} \frac{\pi(z - d)}{2T}}{\operatorname{sh} \frac{\pi \ell}{2T} \operatorname{sh} \frac{\pi(2d + \ell)}{2T}}} .$$

We close with the remark that N. N. Pavlovsky also investigated the flow about an embedded rectangular dam foundation. We give the results later as special cases of a more general problem (Par. 14).

Results are available for the study of flow about flat bottom dam foundations on layers of finite depth and with drain. They are described in the book by V. I. Aravin and S. N. Numerov [6] and in the book by N. T. Melechenko [19]. In the latter work the problem is treated for a simple dam foundation when the bedrock has some inclination.

## C. Flow About Structures With Cut-off Walls.

### §9. Mapping onto a Half Plane of a Polygon when All the Sides of this Polygon Converge into One Point.

Let us consider a "star-shaped" polygon, i.e., one that has all its sides (or their prolongations) going through one point (Fig. 76). Let this point be the origin of coordinates of the $\omega$-plane, and let the region that must be mapped onto the half plane have the boundary ABCDEFA. We designate the angles made by the lines AB, BCD, DE,... with the abcissa axis as $\pi\alpha$, $\pi\alpha_2$, $\pi\alpha_3$,....

We construct the function

$$w = Ce^{\pi\alpha_n i} (a_1 - \zeta)^{\alpha_n - \alpha_1} (a_2 - \zeta)^{\alpha_1 - \alpha_2} \dots (a_n - \zeta)^{\alpha_{n-1} - \alpha_n} \tag{9.1}$$

and show that it carries out the conformal mapping of the star-shaped region ABCD... onto the upper half plane $\zeta$. Here C is a real constant, $\pi\alpha_n$ is the angle made with the abcissa axis by the line segment ANFE, where N maps into the point $\zeta = \infty$. We take the logarithm of (9.1):

$$\begin{aligned} \ell n\, w = \ell n|w| + i \arg w = \ell n\, C + \pi\alpha_n i + (\alpha_n - \alpha_1)\, \ell n\, (\alpha_1 - \zeta) + \\ + (\alpha_1 - \alpha_2)\, \ell n\, (\alpha_2 - \zeta) + \dots + (\alpha_{n-1} - \alpha_n)\, \ell N\, (\alpha_n - \zeta) . \end{aligned} \tag{9.2}$$

We consider first real values of $\zeta$, between $-\infty$ and $a_1$. Since for these points $\arg(a_1 - \zeta) = \dots = \arg(a_n - \zeta) = 0$, we see that all the terms of the right part of (9.2) have real values, except $\pi\alpha_n i$. Therefore

$$\arg w = \pi\alpha_n$$

along the segment AB, and this satisfies one condition of the problem.

Further, in the passage to the segment AB, we go around the point A in the negative sense along a half circle. Therefore the increase for $\ln(a_1 - \zeta)$ is $-\pi i$, and for $(\alpha_n - \alpha_1)\ln(a_1 - \zeta)$ the increase is $-\pi i(\alpha_n - \alpha_1)$. This is also the change of $\ln w$ when we pass A. Therefore

$$\arg w = \pi\alpha_n - \pi(\alpha_n - \alpha_1) = \pi\alpha_1 ,$$

and this corresponds to the value of arg w along the segment AB. Walking around the point B gives a change in $\ln w$ of

$$-\pi i(\alpha_1 - \alpha_2) .$$

Consequently, we obtain

$$\arg w = \pi\alpha_2$$

along the segment BCD.

Continuing in this way, we convince ourselves of the correctness of formula (9.1).

We notice that the mapping of a star-shaped region onto a half plane may be carried out by means of the Schwarz-Christoffel formula. Since the comformal mapping of a star-shaped region onto a half-plane is unique (for three fixed vertices), the Christoffel-Schwarz integral must coincide with the expression (9.1); consequently, in the present case this integral is calculated in finite form.

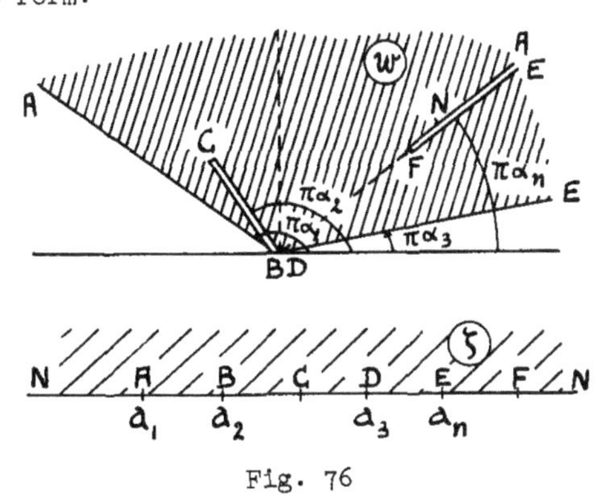


Fig. 76

One ought to notice that formula (9.1) contains only the affixes $a_1, \dots a_n$ of those vertices in which the direction of the adjacent line segments changes.

§10. Flow About an Inclined Cut-off Wall.

In Chapter II we have examined the flow about a vertical cut-off in a stratum of infinite depth. Here we examine a cut-off with inclination. Sometimes walls, to decrease seepage losses, are built with inclined cut-offs [15]. The problem of the flow about a vertical cut-off in anisotropic soil reduces to that of an inclined cut-off, as we will see further. The inclined cut-off is sketched in Fig. 77.

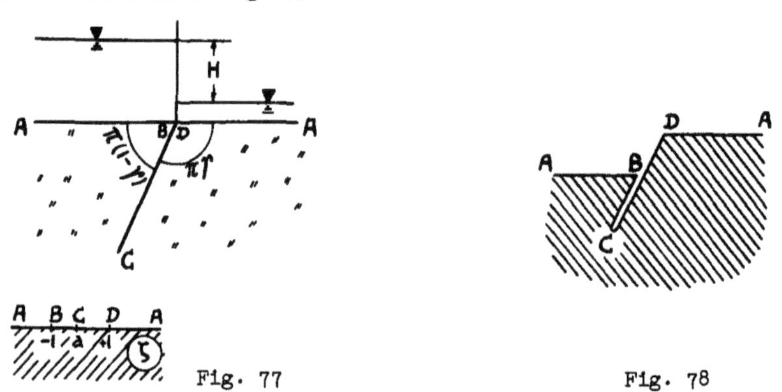


Fig. 77 Fig. 78

We map the region ABCDA (Fig. 77) onto the lower half-plane $\zeta$, in such a way that vertices B, C, D transform into points $\zeta = -1,\ a,\ 1$. According to formula (9.1) we have

$$z = C(1 + \zeta)^{1-\gamma}(1 - \zeta)^{\gamma} . \tag{10.1}$$

We compare this result with that given by the Christoffel-Schwarz formula:

$$z = A\int_{-1}^{\zeta}(1 + \zeta)^{-\gamma}(\zeta - a)(1 - \zeta)^{\gamma-1}d\zeta . \tag{10.2}$$

The integral of the right side of this equation maps the polygon of Fig. 78 and in general is not expressed in elementary functions. But in the present case, when vertices B and D of the polygon coincide, the expression in elementary functions is possible. We may equate (10.1) and (10.2):

$$A\int_{-1}^{\zeta}(1 + \zeta)^{-\gamma}(\zeta - a)(1 - \zeta)^{\gamma-1}d\zeta = C(1 + \zeta)^{1-\gamma}(1 - \zeta)^{\gamma} .$$

Differentiating the identity, we obtain:

$$A(1 + \zeta)^{-\gamma}(\zeta - a)(1 - \zeta)^{\gamma-1} = C(1 + \zeta)^{-\gamma}(1 - \zeta)^{\gamma-1}(\zeta + 2\gamma - 1) .$$

From this one obtains

$$A = - C, \qquad a = 1 - 2\gamma .$$

Thus, by comparison with the Schwarz-Christoffel formula, we were able to determine the value of the parameter a, the affix of the tip of the cut-off in the $\zeta$-plane. It might have been possible to determine this value from the condition that for $\zeta = a$, the velocity is infinitely large.

To determine the constant C, we note that the tip of the cut-off corresponds to

$$z = Se^{-\pi\gamma i}$$

where S is the length of the cut-off.

Insertion of $\zeta = a$ in formula (10.1) gives:

$$Se^{-\pi\gamma i} = C(1 + a)^{1-\gamma}(1 - a)^{\gamma} \quad .$$

From this:

$$C = \frac{Se^{-\pi\gamma i}}{(1 + a)^{1-\gamma}(1 - a)^{\gamma}} \quad ,$$

and equation (10.1) becomes

$$z = Se^{-\pi\gamma i}\left(\frac{1 + \zeta}{1 + a}\right)^{1-\gamma}\left(\frac{1 - \zeta}{1 - a}\right)^{\gamma} \quad , \tag{10.3}$$

where $a = 1 - 2\gamma$.

The region in the complex potential plane is a half strip and its conformal mapping onto the half-plane $\zeta$ gives:

$$\omega = \frac{\kappa H}{\pi} \text{ arc sin } \zeta \ ,$$

and from this

$$\zeta = \sin \frac{\pi\omega}{\kappa H} \quad . \tag{10.4}$$

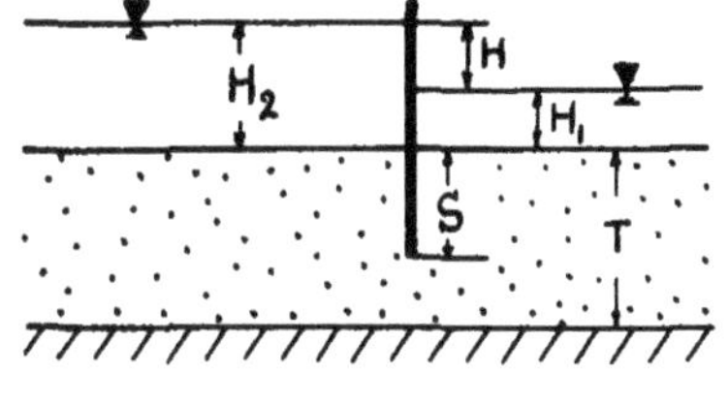


Fig. 79

In Chapter XI we give the hydrodynamical flownet for an inclined cut-off.

§11. Cut-off For Layer On Impervious Bedrock Or On Draining Substratum.

The solution for the case of Fig. 79 is obtained in the same way as for the flat bottom foundation of Par. 6. Moreover, the solution for the single cut-off may be obtained as a particular case of a more general problem, which we analyze further (Par. 13). Therefore we represent here the related results without derivation:

The flowrate beneath the cut-off is

$$Q = \kappa H \frac{K'}{2K} \ ,$$

and

$$k = \sin \frac{\pi S}{2T} \quad .$$

The complex potential has the form

$$\omega = -\frac{\kappa H}{2}\left\{1 + \frac{1}{K} F(\text{arc sin } \lambda,\ k)\right\} \ ,$$

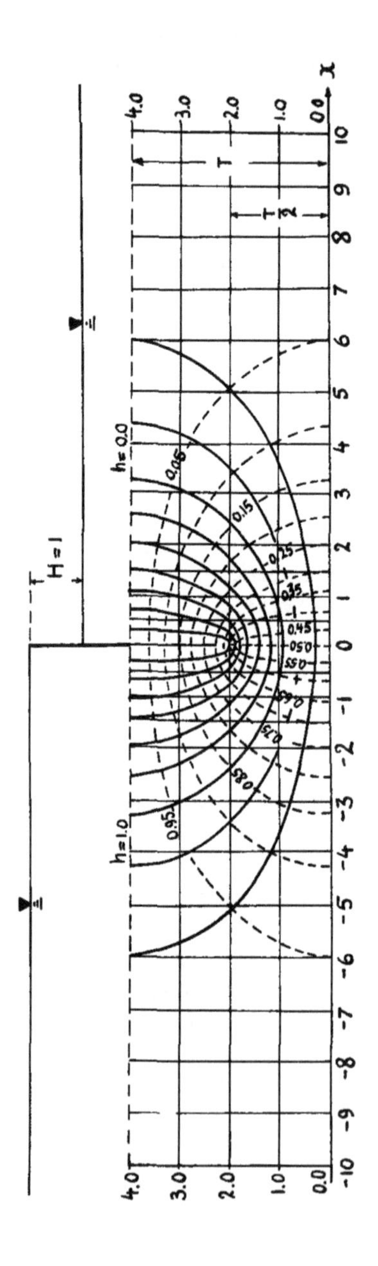


Fig. 80

$$\lambda = \frac{\sqrt{\operatorname{ch}^2 \frac{\pi z}{2T} - \cos^2 \frac{\pi S}{2T}}}{\sin \frac{\pi S}{2T} \operatorname{ch} \frac{\pi z}{2T}} .$$

The seepage velocity is equal to

$$w = u - iv = \frac{\kappa H \pi}{4KTi \sqrt{\operatorname{ch}^2 \frac{\pi z}{2T} - \cos^2 \frac{\pi S}{2T}}} .$$

The flownet for the special case $S = \frac{T}{2}$ [12] is given in Fig. 80. A graph of the dependence of the reduced flowrate

$$\overline{Q} = \frac{Q}{\kappa H}$$

and the velocity at the point $z = 0$ on the ratio $\frac{S}{T}$ is given in Fig. 81. The velocity at $z = 0$ is

$$v_0 = \frac{\kappa H \pi}{4KT \sin \frac{\pi S}{2T}} = \frac{\kappa H \pi}{4KTk} .$$

For the case of soil on a draining substratum (Fig. 82), analogous to that of Par. 7, we restrict ourselves to the final formulas.

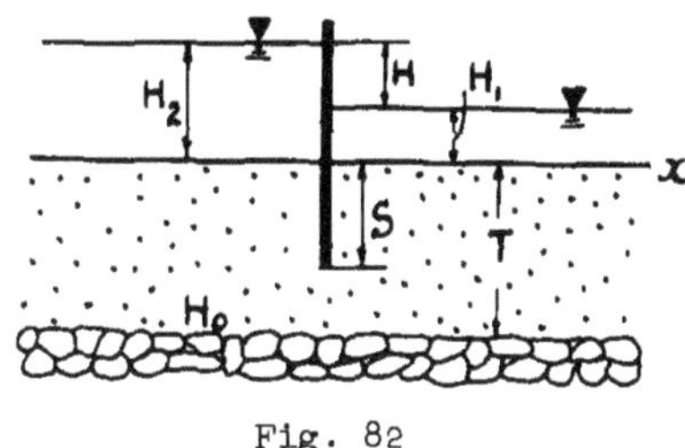


Fig. 82

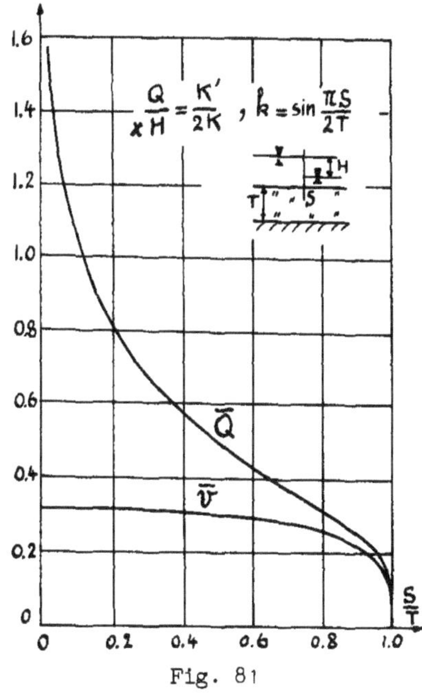


Fig. 81

The complex potential here has the form

$$\omega = -\kappa\left[\frac{1}{2T}(2H_0 - H_1 - H_2)\frac{z}{i} + \frac{H}{\pi} \operatorname{arc} \sin\left(\frac{\operatorname{sh} \frac{\pi z}{2T}}{\sin \frac{\pi S}{2T}}\right)\right]$$

$$(H = H_1 - H_2) .$$

From this we obtain the distribution of head along the tailwater side of the cut-off:

$$h = \left(H_0 - \frac{H_1 + H_2}{2}\right)\frac{y}{T} + \frac{H}{\pi} \operatorname{arc} \sin\left(\frac{\sin \frac{\pi y}{2T}}{\sin \frac{\pi S}{2T}}\right) ,$$

Along the headwater side we have

$$h = H + \left(H_0 - \frac{H_1 + H_2}{2}\right)\frac{y}{T} - \frac{H}{\pi} \operatorname{arc} \sin\left(\frac{\sin \frac{\pi y}{2T}}{\sin \frac{\pi S}{2T}}\right) .$$

The seepage velocity is given by the formula

$$w = \frac{\kappa}{2T}\left\{H_0 - \frac{H_1 + H_2}{2} + \frac{H \operatorname{ch} \frac{\pi z}{2T}}{\sqrt{\operatorname{sh}^2 \frac{\pi z}{2T} + \sin^2 \frac{\pi S}{2T}}}\right\} .$$

§12. Flat Bottom Foundation With Cut-off On a Stratum of Infinite Depth.

Let the length of the base be $B = 2\ell$, that of the cut-off $S$, and $\ell_1$ and $\ell_2$ the segments on the base from the cut-off (Fig. 83). Let the depth of the pervious stratum be $T$.

To explain the solution of the problem by N. N. Pavlovsky [1], we choose the coordinate axes as shown on Fig. 83. We map the flow region onto the half-plane $\zeta$. Let the points 1, 2, ... 7 of the z region correspond to the points $\zeta_1, \zeta_2, \dots \zeta_7$ of the $\zeta$ region, in this way

$$(12.1)\quad \begin{cases} \zeta_1 = -\sigma, & \zeta_2 = 0, \\ \zeta_3 = +\sigma, & \zeta_4 = \beta_2, \\ \zeta_5 = -\beta_1, & \zeta_6 = 1, \\ \zeta_7 = -1 . \end{cases}$$

The z region has right angles in the points $\zeta = \pm\sigma$, an angle $2\pi$ in point 2 for which $\zeta = 0$, and angles of value zero in the points $\zeta = \pm 1$. Therefore the conformal mapping of the flow region onto the lower half of the $\zeta$ plane gives:

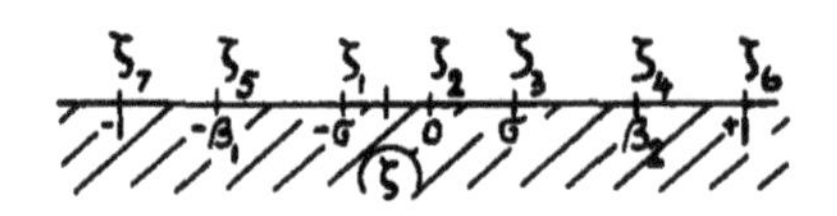


Fig. 83

$$z = A\int_0^{\zeta} \frac{\zeta d\zeta}{(1 - \zeta^2)\sqrt{\sigma^2 - \zeta^2}} + Si .$$

By means of the substitution $\sqrt{\sigma^2 - \zeta^2} = t$ we find

$$(12.2)\qquad z = -\frac{A}{\sigma'}\left[\operatorname{arc\,tg} \frac{\sqrt{\sigma^2 - \zeta^2}}{\sigma'} - \operatorname{arc\,tg}\frac{\sigma}{\sigma'}\right] + Si ,$$

where

$$\sigma' = \sqrt{1 - \sigma^2} .$$

For $\zeta = \sigma$, we have $z = 0$, and therefore:

$$(12.3)\qquad 0 = \frac{A}{\sigma'} \operatorname{arc\,tg} \frac{\sigma}{\sigma'} + Si .$$

Consequently,

$$(12.4)\qquad z = -\frac{A}{\sigma'} \operatorname{arc\,tg} \frac{\sqrt{\sigma^2 - \zeta^2}}{\sigma'} .$$

Passing to $\zeta = \xi > \sigma$ when we turn around the point $\zeta = \sigma$ over $\pi$ in the positive sense, i.e., from the $\eta$ axis towards the $\xi$ axis, we obtain that $\sqrt{\sigma - \zeta}$ goes over into $-i\sqrt{\zeta - \sigma}$ and $\sqrt{\sigma^2 - \zeta^2}$ into $-i\sqrt{\zeta^2 - \sigma^2}$ .

Then arc tg $\frac{\sqrt{\sigma^2 - \zeta^2}}{\sigma'}$ goes over into

$$- i \tanh^{-1} \frac{\sqrt{\zeta^2 - \sigma^2}}{\sigma'}$$

and for the segment $\zeta = \xi > \sigma$ we have

$$z = \frac{Ai}{\sigma'} \operatorname{Tanh}^{-1} \frac{\sqrt{\zeta^2 - \sigma^2}}{\sigma'} .$$

To find the value of the constant $A$, we rewrite the integral for $z$, by changing $\sqrt{\sigma^2 - \zeta^2}$ into $- i \sqrt{\zeta^2 - \sigma^2}$. We find

$$z = Ai \int \frac{\zeta d\zeta}{(1 - \zeta^2) \sqrt{\zeta^2 - \sigma^2}} + B .$$

In particular the transition from the segment $\xi < 1$ to the segment $\xi > 1$ and the corresponding integration along a half circle around the point $\zeta = 1$ gives an increase of $(- \pi i)$ so that

$$Ai \frac{- \pi i}{- 2 \sqrt{1 - \sigma^2}} = - \frac{A\pi}{2\sigma'} .$$

On the other side, the corresponding increase of $z$ in the transition from the boundary of the lower reservoir to the bedrock is $Ti$. Therefore

$$- \frac{A\pi}{2\sigma'} = Ti \tag{12.5}$$

and consequently, for $z$:

$$z = \frac{2T}{\pi} \tanh^{-1} \frac{\sqrt{\zeta^2 - \sigma^2}}{\sigma'} .$$

If we solve this equation for $\zeta$, we find:

$$\zeta = \pm \sqrt{\sigma^2 + \sigma'^2 \operatorname{th}^2 \frac{\pi z}{2T}} .$$

The plus sign corresponds to the right side of the half-plane, the minus sign to the left side.

Equations (12.3) and (12.5) combined give

$$\frac{2Ti}{\pi} \operatorname{arc\,tg} \frac{\sigma}{\sigma'} = Si ,$$

and therefore

$$\frac{\sigma}{\sigma'} = \operatorname{tg} \frac{\pi S}{2T} ,$$

i.e.,

$$\sigma = \sin \frac{\pi S}{2T}, \qquad \sigma' = \cos \frac{\pi S}{2T} . \tag{12.6}$$

Consequently, for $\zeta$ we have

$$\zeta = \pm \cos \frac{\pi S}{2T} \sqrt{\operatorname{th}^2 \frac{\pi z}{2T} + \operatorname{tg}^2 \frac{\pi S}{2T}} . \tag{12.7}$$

In particular, for the endpoints $z = - \ell_1$ and $z = \ell_2$ of the base, we find

$$\begin{cases} \zeta_4 = \beta_2 = \cos \frac{\pi S}{2T} \sqrt{\operatorname{th}^2 \frac{\pi \ell_2}{2T} + \operatorname{tg}^2 \frac{\pi S}{2T}} , \\ \zeta_5 = - \beta_1 = - \cos \frac{\pi S}{2T} \sqrt{\operatorname{th}^2 \frac{\pi \ell_1}{2T} + \operatorname{tg}^2 \frac{\pi S}{2T}} . \end{cases} \tag{12.8}$$

We now turn to the conformal mapping of the rectangle of the plane $\omega = \varphi + i\psi$ onto the half-plane $\zeta$. Since the vertices of the rectangle transform into the points $\zeta_4$, $\zeta_5$, $\zeta_6$, $\zeta_7$, we have

$$\omega = A\int_1^\zeta \frac{d\zeta}{\sqrt{(\zeta - \zeta_4)(\zeta - \zeta_5)(\zeta - \zeta_6)(\zeta - \zeta_7)}} \quad . \tag{12.9}$$

We introduce the function $u$ by means of the substitution

$$\operatorname{sn}^2(u, k) = \frac{(\zeta_4 - \zeta_7)(\zeta - \zeta_6)}{(\zeta_6 - \zeta_7)(\zeta - \zeta_4)}, \quad k^2 = \frac{(\zeta_4 - \zeta_5)(\zeta_6 - \zeta_7)}{(\zeta_6 - \zeta_5)(\zeta_4 - \zeta_7)} \quad . \tag{12.10}$$

We substitute $\zeta - \zeta_4$, $\zeta - \zeta_5$, ... in (12.9) and obtain

$$\frac{\omega}{A} = \int_1^\zeta \frac{d\zeta}{\sqrt{(\zeta - \zeta_4)(\zeta - \zeta_5)(\zeta - \zeta_6)(\zeta - \zeta_7)}} = \frac{2u}{\sqrt{(\zeta_4 - \zeta_7)(\zeta_6 - \zeta_5)}} \quad .$$

Then the inversion of integral (12.6) gives

$$\zeta = \left(\zeta_6 - \zeta_4 \frac{\zeta_6 - \zeta_7}{\zeta_4 - \zeta_7} \operatorname{sn}^2 \frac{\mu\omega}{A}\right) : \left(1 - \frac{\zeta_6 - \zeta_7}{\zeta_4 - \zeta_7} \operatorname{sn}^2 \frac{\mu\omega}{A}\right), \tag{12.11}$$

where

$$\mu = \frac{1}{2}\sqrt{(\zeta_4 - \zeta_7)(\zeta_6 - \zeta_5)} \quad .$$

We examine the vertices of the rectangle. For $\zeta = \zeta_6$, we have as should be, $\omega = 0$. For $\zeta = \zeta_4$, we have

$$\operatorname{sn}^2 \frac{\mu\omega}{A} = \infty \quad ,$$

i.e., [24]

$$\frac{\mu\omega}{A} = iK' \quad .$$

Since for $\zeta = \zeta_4$, we must have $\omega = -Qi$, the previous equation gives

$$-Qi = \frac{A}{\mu} K'i$$

or

$$Q = -\frac{AK'}{\mu} \quad .$$

Further, for $\zeta = \zeta_5$,

$$\operatorname{sn} \frac{\mu\omega}{A} = \frac{1}{k} \quad .$$

Consequently

$$\omega = \frac{A}{\mu}(K + iK') \quad .$$

But for the vertex $\zeta_5$,

$$\omega = -\kappa H - Qi \quad .$$

We equate these expressions and find:

$$A = -\frac{\kappa\mu}{K} H \quad ,$$

and

$$Q = \frac{\kappa HK'}{K} \quad .$$

The modulus according to (12.10) and (12.1) is expressed as

$$k = \sqrt{\frac{2(\beta_1 + \beta_2)}{(1 + \beta_1)(1 + \beta_2)}} ,$$

where $\beta_1$ and $\beta_2$ are determined by formula's (12.8). Thus we have the complete solution of the problem.

We notice that for a cut-off in the center of the base, when $\beta_1 = \beta_2 = \beta$,

$$k = \frac{2\sqrt{\beta}}{1 + \beta} \tag{12.12}$$

where

$$\beta = \cos\frac{\pi S}{2T} \sqrt{\mathrm{th}^2\frac{\pi \ell}{2T} + \mathrm{tg}^2\frac{\pi S}{2T}} .$$

Formula (12.12) corresponds to the Landen [24, 25] transformation. Between the complete elliptic integrals with moduli $k$ and $\beta$, the following relation exists:

$$\frac{K'(k)}{K(k)} = \frac{K'(\beta)}{2K(\beta)} .$$

Therefore, for a symmetrical location of the cut-off in the center of the base

$$Q = \frac{\kappa H K'(\beta)}{2K(\beta)}$$

A. M. Mkhitarian (Fig. 84, 85) [17] carried out a series of calculations of $Q$ for symmetrical and non-symmetrical disposition of the cut-off wall. N. T. Melechenko [19] made detailed nomographs for this case. The influence of the cut-off on the pressure distribution along the base is given in Fig. 86: in front of the cut-off, the pressure is increased, behind the cut-off the pressure strongly decreases.

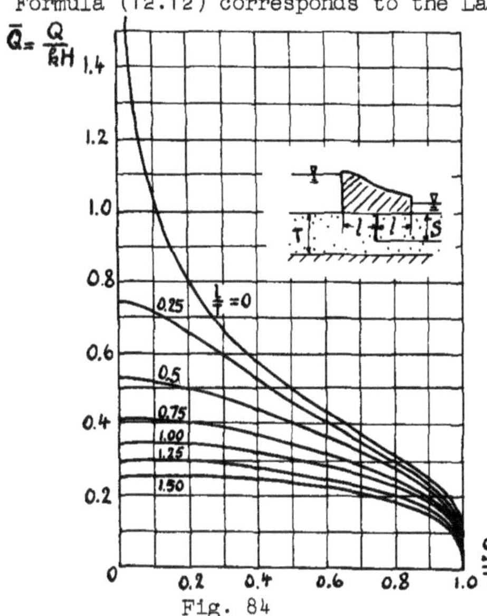

Fig. 84

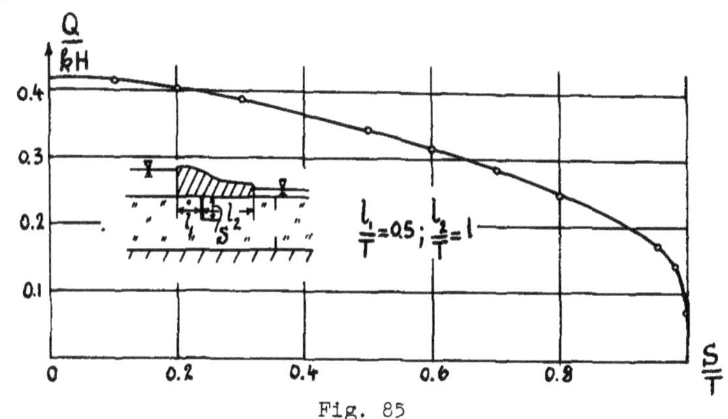

Fig. 85

decreases. Flownets corresponding to different locations of the cut-off [1, 11] are given in Figs. 87, 88 and 89.

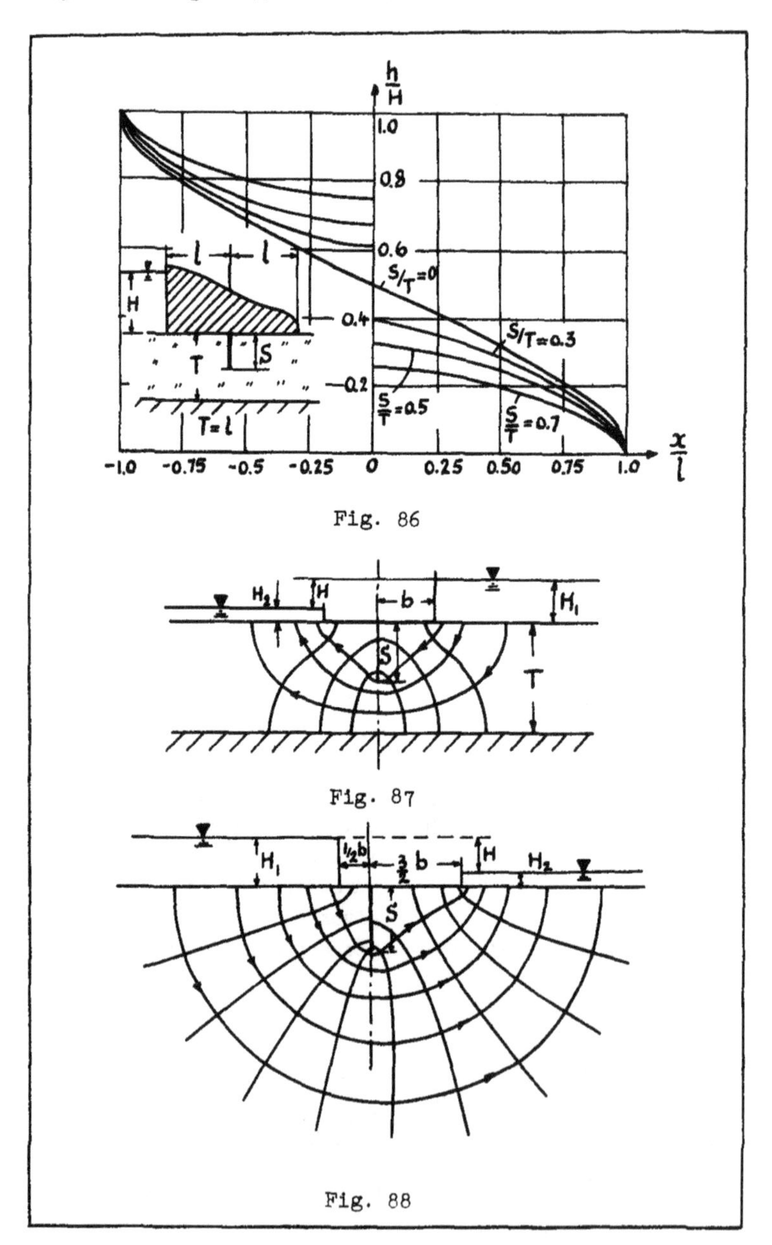


Fig. 86

Fig. 87

Fig. 88

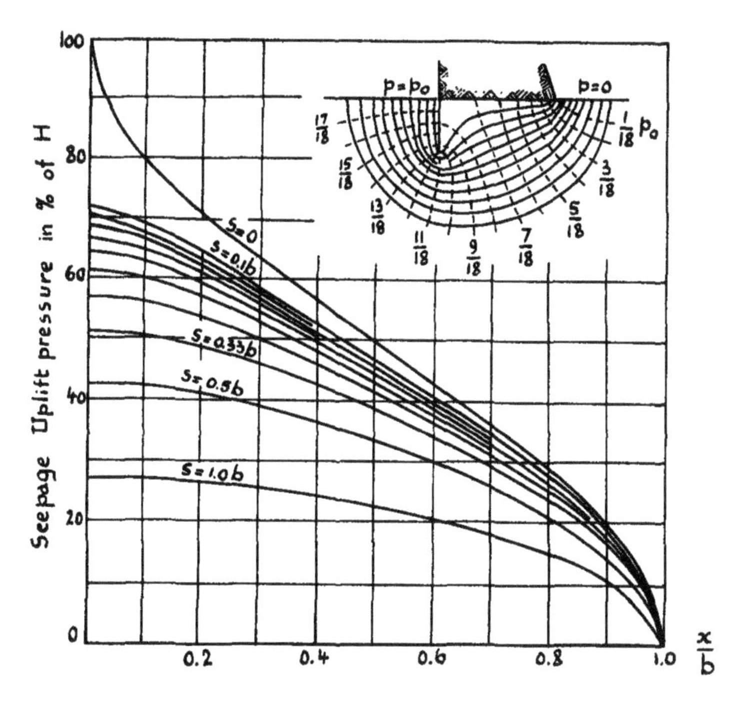


Fig. 89

*§ 13. Dam With Two Cut-offs On a Pervious Stratum of Finite Depth.

P. F. Filchakov [9] examined the case that includes all schemes of weirs with cut-offs, taken up in this chapter, and some new ones. This case is sketched in Fig. 90. Here the dam foundation is embedded, the cantilever flat.

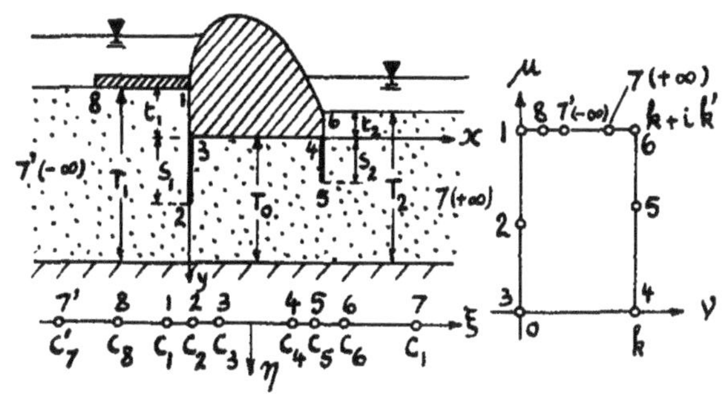


Fig. 90

The function that maps the flow region onto the half-plane has the form

$$z = A \int \frac{(\zeta - c_2)(\zeta - c_5)d\zeta}{(\zeta - c_7')(\zeta - c_7)\sqrt{(\zeta - c_1)(\zeta - c_3)(c_4 - \zeta)(c_6 - \zeta)}} + C \,. \tag{13.1}$$

Putting

$$\mathrm{sn}^2 a = -\frac{n}{k^2} \quad , \tag{13.9'}$$

we may write for $\prod(u, n, k)$ [25]:

$$\prod(u,n,k) = \int_0^u \frac{du}{1 - k^2\mathrm{sn}^2 a\ \mathrm{sn}^2 u} =$$

$$= u + \frac{\mathrm{sn}\ a}{\mathrm{cn}\ a\ \mathrm{dn}\ a}\left[\frac{\theta'(a)}{\theta(a)} u + \frac{1}{2} \ln \frac{\theta(u-a)}{\theta(u+a)}\right] \quad . \tag{13.10}$$

Here $\theta$ is the Jacobian elliptic function, for which we have the following formula [24]:

$$\frac{\theta'(a)}{\theta(a)} = \mathrm{zn} a = \mathcal{E}(\mathrm{am}\ a) - \frac{E}{K} a \quad , \tag{13.11}$$

where

$$\mathcal{E}(\mathrm{am}\ a) = \int_0^a \mathrm{dn}^2 u\ du, \quad E = \int_0^{\pi/2} \sqrt{1 - k^2\sin^2\varphi}\ d\varphi \quad . \tag{13.12}$$

Substituting expression (13.10) in (13.8) and collecting the terms with $u$, we find:

$$z = \frac{2A}{L}\{Du + [D_7'\Phi(u, a_7') - D_7\Phi(u, a_7)]\} + C \quad , \tag{13.13}$$

where we have introduced the notation

$$\Phi(u, a) = \frac{1}{2} \ln \frac{\theta(u-a)}{\theta(u+a)} \quad . \tag{13.14}$$

Furthermore $k$ and $\mathrm{sn}(u, k)$ are determined by equations (13.3) and (13.2), and the constants $a_7$ and $a_7'$ according to (13.6) and (13.9')

$$\mathrm{sn}^2(a_7, k) = \frac{(c_6 - c_3)(c_7 - c_1)}{(c_6 - c_1)(c_7 - c_3)}, \quad \mathrm{sn}^2(a_7', k) = \frac{(c_6 - c_3)(c_7' - c_1)}{(c_6 - c_1)(c_7' - c_3)} \quad . \tag{13.15}$$

The constants $D_7$, $D_7'$ and $D$ have these values:

$$D_7 = \frac{L}{(c_7' - c_7')} \frac{(c_7 - c_2)(c_7 - c_5)}{\sqrt{(c_7 - c_1)(c_7 - c_3)(c_7 - c_4)(c_7 - c_6)}} \quad , \tag{13.16}$$

$$D_7' = \frac{L}{(c_7' - c_7)} \frac{(c_7' - c_2)(c_7' - c_5)}{\sqrt{(c_7' - c_1)(c_7' - c_3)(c_7' - c_4)(c_7' - c_6)}} \quad , \tag{13.17}$$

$$D = \frac{(c_5 - c_3)(c_2 - c_3)}{(c_7 - c_3)(c_7' - c_3)} + \frac{1}{K}[D_7' \prod(a_7') - D_7' \prod(a_7)] \quad . \tag{13.18}$$

$\prod(a)$ stands for the following expression

$$\prod(a) = K\ \mathcal{E}(\mathrm{am}\ a) - aE \quad . \tag{13.19}$$

By means of formulas (13.2), (13.3), we may also find

$$k'^2 = \frac{(c_6 - c_4)(c_3 - c_1)}{(c_4 - c_1)(c_6 - c_3)} \quad ,$$

(13.20)
$$\mathrm{cn}^2(u, k) = \frac{(c_3 - c_1)(c_4 - \zeta)}{(c_4 - c_3)(\zeta - c_1)}, \quad \mathrm{dn}^2(u, k) = \frac{(c_3 - c_6)(c_6 - \zeta)}{(c_6 - c_3)(\zeta - c_1)} .$$

In order to find more relations, we examine subsequently points 3, 4, 6, 7, 1 (Fig. 90)

For point 3, $z = 0$, $\zeta = c_3$, $\mathrm{sn}^2 u = 0$, $u = 0$, $\Phi(0, a) = 0$, and therefore

(13.21)
$$C = 0$$

In point 4, $z = B$ (length of dam), $\zeta = c_4$, $\mathrm{sn}^2(u, k) = 1$, $u = k$, and therefore

$$\Phi(K, a) = \frac{1}{2} \ln \frac{\theta(K - a)}{\theta(K + a)} = 0 ,$$

and equation (13.13) gives:

(13.22)
$$B = \frac{2ADK}{L} .$$

In point 6

$$z = B - it_2, \quad \zeta = c_6, \quad \mathrm{sn}^2(u, k) = \frac{1}{k^2}, \quad u = K + iK', \quad \Phi(K + iK', a) = i\frac{\pi a}{2K},$$

and therefore, from (13.13)

$$B - it_2 = \frac{2A}{L}\left\{D(K + iK') + \frac{\pi i}{2K}(D_7'a_7' - D_7a_7)\right\} .$$

By means of (13.22)

(13.23)
$$- t_2 = \frac{2A}{L}\left\{DK' + \frac{\pi}{2K}(D_7'a_7' - D_7a_7)\right\} .$$

In point 1

$$z = - it_1, \quad \zeta = c_1, \quad \mathrm{sn}^2(u, k) = \infty, \quad u = iK', \quad \Phi(iK', a) = \frac{\pi i}{2K}(a - K)$$

and consequently,

$$- t_1 = \frac{2A}{L}\left\{DK' + \frac{\pi}{2K}[(D_7'a_7' - D_7a_7) - K(D_7' - D_7)]\right\} .$$

Subtracting this from equation (13.23), we find:

(13.24)
$$t_1 - t_2 = \frac{A\pi}{L}(D_7' - D_7) .$$

For point 7, $z = \infty$, $\zeta = c_7$.

We find from conditions at $z = \infty$, the point that is mapped into $\zeta = c_7$, a simple pole of the integrand (13.1):

$$T_2 = \frac{\pi A(c_7 - c_2)(c_7 - c_5)}{(c_7 - c_7')\sqrt{(c_7 - c_1)(c_7 - c_3)(c_7 - c_4)(c_7 - c_6)}} .$$

By means of (13.16) we find for $D_7$:

(13.25)
$$A = \frac{LT_2}{\pi D_7} .$$

We introduce the substitution

$$\mathrm{sn}^2(u, k) = \frac{(c_4 - c_1)(\zeta - c_3)}{(c_4 - c_3)(\zeta - c_1)} , \tag{13.2}$$

$$k^2 = \frac{(c_4 - c_3)(c_6 - c_1)}{(c_4 - c_1)(c_6 - c_3)} . \tag{13.3}$$

Putting

$$R = \sqrt{(\zeta - c_1)(\zeta - c_3)(c_4 - \zeta)(c_6 - \zeta)} , \tag{13.4}$$

we obtain by differentiation of (13.2) for the function $u$, only differing by a constant multiplier from the complex potential $\omega$:

$$\frac{2du}{\sqrt{(c_4 - c_1)(c_6 - c_3)}} = \frac{d\zeta}{R} , \tag{13.5}$$

$$\zeta - c_m = \frac{c_3 - c_m - (c_1 - c_m)\dfrac{c_4 - c_3}{c_4 - c_1}\,\mathrm{sn}^2 u}{1 - \dfrac{c_4 - c_3}{c_4 - c_1}\,\mathrm{sn}^2 u}$$

and

$$\frac{(\zeta - c_2)(\zeta - c_5)}{(\zeta - c_7')(\zeta - c_7)} = \frac{c_3 - c_1}{c_7 - c_7'}\left[\frac{(c_7' - c_2)(c_7' - c_5)}{(c_7' - c_1)(c_7' - c_3)}\,\frac{1}{1 + n'\mathrm{sn}^2 u} - \right.$$

$$\left. - \frac{(c_7 - c_2)(c_7 - c_5)}{(c_7 - c_1)(c_7 - c_3)}\,\frac{1}{1 + n\,\mathrm{sn}^2 u}\right] + \frac{(c_5 - c_1)(c_2 - c_1)}{(c_7 - c_1)(c_7' - c_1)} .$$

Here the following notations are introduced:

$$-n' = \frac{(c_4 - c_3)(c_7' - c_1)}{(c_4 - c_1)(c_7' - c_3)} , \qquad -n = \frac{(c_4 - c_3)(c_7 - c_1)}{(c_4 - c_1)(c_7 - c_3)} . \tag{13.6}$$

We insert these expressions in formula (13.1) and introduce the notation

$$L = \sqrt{(c_4 - c_1)(c_6 - c_3)} , \tag{13.7}$$

to obtain:

$$z = \frac{2A}{L}\left\{\frac{(c_5 - c_1)(c_2 - c_1)}{(c_7 - c_1)(c_7' - c_1)}\,u + \frac{c_3 - c_1}{c_7 - c_7'}\left[\frac{(c_7' - c_2)(c_7' - c_5)}{(c_7' - c_1)(c_7' - c_3)}\,\Pi(u, n', k)\right.\right. \tag{13.8}$$

$$\left.\left. - \frac{(c_7 - c_2)(c_7 - c_5)}{(c_7 - c_1)(c_7 - c_3)}\,\Pi(u, n, k)\right]\right\} + C .$$

Here $\Pi(u, n', k)$, $\Pi(u, n, k)$ are elliptic integrals of the third kind

$$\Pi(u,n,k) = \int_0^u \frac{du}{1 + n\,\mathrm{sn}^2 u} . \tag{13.9}$$

For point $7'$, $z = -\infty$, $\zeta = c_7'$.

As in the previous case, we find

(13.26) $$A = \frac{LT_1}{\pi D_7'} .$$

From (13.25) and (13.26) we find

(13.27) $$D_7 T_1 = D_7' T_2$$

or with the values of $D_7$, $D_7'$,

(13.28) $$\frac{(c_7 - c_2)(c_7 - c_5)}{\sqrt{(c_7 - c_1)(c_7 - c_3)(c_7 - c_4)(c_7 - c_6)}} T_1 = \frac{(c_7' - c_2)(c_7' - c_5)}{\sqrt{(c_7' - c_1)(c_7' - c_3)(c_7' - c_4)(c_7' - c_6)}} T_2 .$$

The mapping function (13.13) may be written in the form

(13.29) $$z = \frac{B}{K} u + \frac{2}{\pi}[T_2 \Phi(u, a_7') - T_1 \Phi(u, a_7)] .$$

We expand the function for different segments of the flow region. Along segment 3 — 4, with the notation

(13.30) $$u = \nu + i\mu$$

we have

$$z = x (0 \leq x \leq B), \quad c_3 \leq \zeta \leq c_4, \quad u = \nu (0 \leq \nu \leq K) .$$

For computations one may use the following series for $\Phi(u, a)$:

(13.31) $$\Phi(u, a) = \frac{1}{2} \ln \frac{\theta(u - a)}{\theta(u + a)} = - \sum_{m=1}^{\infty} \frac{\sin \frac{\pi m u}{K} \sin \frac{\pi m a}{K}}{m \operatorname{sh} \frac{\pi m K'}{K}} .$$

Along segment 4 — 5 — 6, $z = B + iy (- t_2 \leq y \leq s_2)$,

$$c_4 \leq \zeta \leq c_6, \quad u = K + i\mu \quad (0 \leq \mu \leq K') .$$

By means of formula (13.31) we find:

(13.32) $$\Phi(K + i\mu, a) = i \sum_{m=1}^{\infty} (-1)^{m+1} \frac{\operatorname{sh} \frac{\pi m \mu}{K} \sin \frac{\pi m a}{K}}{m \operatorname{sh} \frac{\pi m K'}{K}} .$$

It is possible to write the mapping function in the form

(13.33) $$z - B = i \left\{\frac{B\mu}{K} + \frac{2}{\pi}[T_2 \Phi(\mu, a_7') - T_1 \Phi(\mu, a_7)]\right\} ,$$

where, according to a property of the elliptic functions [24]

$$\operatorname{sn}^2(\mu, k') = \operatorname{sn}^2[i(K - u), k'] = \left[\frac{i}{k'} \frac{\operatorname{cn}(u, k)}{\operatorname{sn}(u, k)}\right]^2 = \frac{(c_6 - c_3)(\zeta - c_4)}{(c_6 - c_4)(\zeta - c_3)} .$$

In particular, for point 6, where $\zeta = c_6$, $\mu = k'$,

$$\Phi(K + iK', a) = \sum_{m=1}^{\infty} (-1)^{m+1} \frac{\sin \frac{\pi m a}{K}}{m} = \frac{\pi a}{2K} .$$

Consequently,

$$z - B = i\left(\frac{BK'}{K} + \frac{T_2 a_7' - T_1 a_7}{K}\right) = - it_2 ,$$

as it should be.

Along the segment $[6, +\infty); (-\infty, 1)]$, we may put:

$$z = x + it_n \quad (B \leq x \leq +\infty; \ -\infty \leq x \leq 0) ,$$

in which

$t_n = - t_2$ along the boundary of the tailwater reservoir $(c_6 \leq \zeta \leq c_7)$;

$t_n = T_0$ along the impervious bedrock $(c_7 \leq \zeta \leq \infty; \ -\infty \leq \zeta \leq c_7')$;

$t_n = - t_1$ along the boundary of the headwater reservoir $(c_7' \leq \zeta \leq c_1)$;

On the entire segment

$$u = K + iK' - \nu \qquad (K \geq \nu \geq 0) .$$

Therefore [24, 25]

$$\mathrm{sn}(\nu, k) = \mathrm{sn}(K + iK' - u, k) = \frac{\mathrm{dn}(u, k)}{k\ \mathrm{cn}(u, k)}$$

and

$$\frac{\theta(u - a)}{\theta(u + a)} = \frac{\theta[K + iK' - (\nu + a)]}{\theta[K + iK' - (\nu - a)]} = e^{\frac{i\pi a}{K}} \frac{H_1(\nu + a)}{H_1(\nu - a)} .$$

Introducing the notation

$$\Phi^*(\nu, a) = \frac{1}{2} \ln \frac{H_1(\nu + a)}{H_1(\nu - a)} ,$$

we have

$$\Phi^*(u, a) = \Phi(K + iK' - \nu, a) = \frac{\pi a i}{K} + \Phi^*(\nu, a) .$$

We may rewrite (13.29) as follows:

$$z - B + it_n = - \frac{B\nu}{K} + \frac{2}{\pi}[T_2 \Phi^*(\nu, a_7') - T_1 \Phi^*(\nu, a_7)] , \tag{13.34}$$

where

$$\mathrm{sn}^2(\nu, k) = \frac{(c_4 - c_1)(\zeta - c_6)}{(c_6 - c_1)(\zeta - c_4)} .$$

In points 7 and 7', the function $\Phi^*(\nu, a)$ has a logarithmic singularity. For the computation of $\Phi^*(\nu, a)$, one may use the expansion

$$\Phi^*(\nu, a) = \frac{1}{2} \ln \frac{H_1(\nu + a)}{H_1(\nu - a)} =$$

$$= \frac{1}{2} \ln \frac{\cos \frac{\pi(\nu + a)}{2K}}{\cos \frac{\pi(\nu - a)}{2K}} + \sum_{m=1}^{\infty} (-1)^m \frac{\sin \frac{\pi m \nu}{K} \sin \frac{\pi m a}{K}}{\mathrm{sh}\, \frac{\pi m K'}{K}} \frac{e^{-\frac{m\pi K'}{K}}}{m} . \tag{13.35}$$

Along segment 3 — 2 — 1

$$z = iy \quad (-t_1 \le y \le s_1), \quad c_1 \le \zeta \le c_3, \quad u = i\mu \quad (0 \le \mu \le K') ;$$

and consequently,

$$\mathrm{sn}(\mu, k') = -i \frac{\mathrm{sn}(u, k)}{\mathrm{cn}(u, k)} ,$$

and after formula (13.31) we obtain

$$\Phi(u, a) = \Phi(i\mu, a) = -i \sum_{m=1}^{\infty} \frac{\mathrm{sh} \frac{\pi m \mu}{K} \sin \frac{\pi m a}{K}}{m \ \mathrm{sh} \frac{\pi m K'}{K}} . \tag{13.36}$$

The expression for $z$ takes on this form:

$$z = i \frac{B\mu}{K} + \frac{2}{\pi}[T_2 \Phi(i\mu, a_7') - T_1 \Phi(i\mu, a_7)] , \tag{13.37}$$

where

$$\mathrm{sn}^2(\mu, k') = \frac{(c_4 - c_1)(c_3 - \zeta)}{(c_3 - c_1)(c_4 - \zeta)} .$$

We now examine the equations to determine constants $c_1, c_2, \ldots$

$$-c_1 = c_6 = 1, \quad -c_7' = c_7 .$$

Constants $c_2, c_3, c_4, c_5, c_7$, must be determined. We need five equations therefore. We insert $\zeta = c_5$, $\zeta = c_2$ respectively in (13.33) and (13.37), eliminate $D_7'$ from (13.24) and (13.27), eliminate $\frac{A}{L}$ from the formula for $t_1 - t_2$, $T_2$. We obtain these equations:

$$\frac{s_2}{B} = \frac{\mu_5}{K} - \frac{D_7' \Phi(\mu_5, a_7') - D_7 \Phi(\mu_5, a_7)}{DK} , \tag{13.38}$$

in which

$$\mathrm{sn}^2(\mu_5, k') = \frac{(1 - c_3)(c_5 - c_4)}{2(c_5 - c_3)} , \qquad k'^2 = \frac{(1 - c_4)(1 + c_3)}{(1 + c_4)(1 - c_3)} ,$$

$$\frac{s_1}{B} = \frac{\mu_2}{K} - i \frac{D_7' \Phi(i\mu_2, a_7') - D_1 \Phi(i\mu_2, a_7)}{DK} , \tag{13.39}$$

where

$$\mathrm{sn}^2(\mu_2, k') = \frac{(1 + c_4)(c_3 - c_2)}{(1 + c_3)(c_4 - c_2)} ,$$

$$\frac{t_2}{B} = \frac{\pi}{2K} \frac{D_7 a_7 - D_7' a_7'}{DK} - \frac{K'}{K} , \tag{13.40}$$

$$\frac{t_1 - t_2}{B} = \frac{\pi}{2} \frac{D_7' - D_7}{DK} , \tag{13.41}$$

$$\frac{T_2}{B} = \frac{\pi}{2} \frac{D_7}{DK} . \tag{13.42}$$

We now examine some particular cases.

## *§14. Weir With Two Symmetrical Cut-offs, and Foundation Pit With Symmetrical Sheetpiles.

For a weir with two cut-offs as shown in Fig. 91,

$$s_1 = s_2 = s, \quad t_1 = t_2 = t, \quad T_1 = T_2 = T$$

and to satisfy the previously chosen conditions

$$-c_1 = c_6 = 1, \quad -c_7' = c_7 = \gamma$$

one must have, because of the symmetry

$$-c_2 = c_5 = \sigma, \quad -c_3 = c_4 = \alpha .$$

The mapping function has the form

$$z = \frac{B}{K} u + \frac{2T}{\pi}[\Phi(u, a_7') - \Phi(u, a_7)] \quad ,$$

where $B$ is the length of the dam and

$$\mathrm{sn}^2(u, k) = \frac{(1+\alpha)(\zeta+\alpha)}{2\alpha(1+\zeta)}, \quad k^2 = \frac{4\alpha}{(1+\alpha)^2}, \quad \mathrm{sn}^2(a_7, k) = \frac{(1+\alpha)(1+\gamma)}{2(\gamma+\alpha)} ,$$

$$\mathrm{sn}^2(a_7', k) = \frac{(1+\alpha)(\gamma-1)}{2(\gamma-\alpha)} .$$

Parameters $\alpha, \gamma, \sigma$ are determined from the system of equations,

$$t = \frac{T(a_7 - a_7') - BK'}{K} , \quad T = \frac{\pi B}{2}\frac{D_7}{DK} , \quad \frac{s}{B} = \frac{\mu_5}{K} + \frac{D_7}{DK}[\Phi(\mu_5, a_7') - \Phi(\mu_5, a_7)] ,$$

where

$$\mathrm{sn}^2(\mu_5, k') = \frac{(1+\alpha)(\sigma-\alpha)}{2(\sigma+\alpha)} , \quad k' = \frac{1-\alpha}{1+\alpha} .$$

Constants $D_7$, $D_7'$, $D$ and $A$ are found from the equations

$$D_7' = D_7 = \frac{1+\alpha}{2\gamma} \frac{\gamma^2 - \sigma^2}{\sqrt{(\gamma^2-1)(\gamma^2-\alpha^2)}} , \quad D = \frac{\sigma^2 - \alpha^2}{\gamma^2 - \alpha^2} + \frac{D_7}{K}\left[\Pi(a_7') - \Pi(a_7)\right] ,$$

$$A = \frac{2T\gamma\sqrt{(\gamma^2-1)(\gamma^2-\alpha^2)}}{\pi(\gamma^2-\sigma^2)} .$$

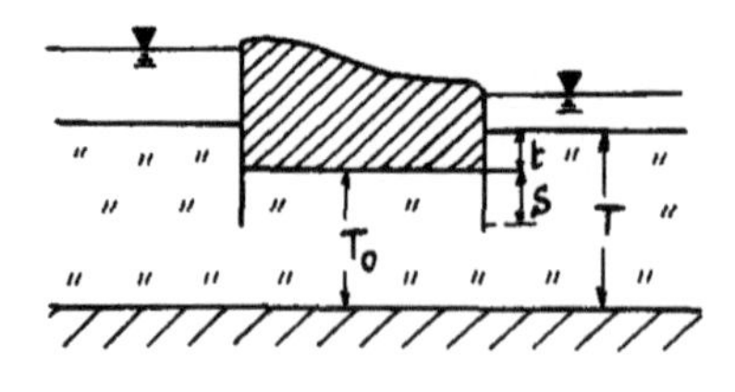


Fig. 91

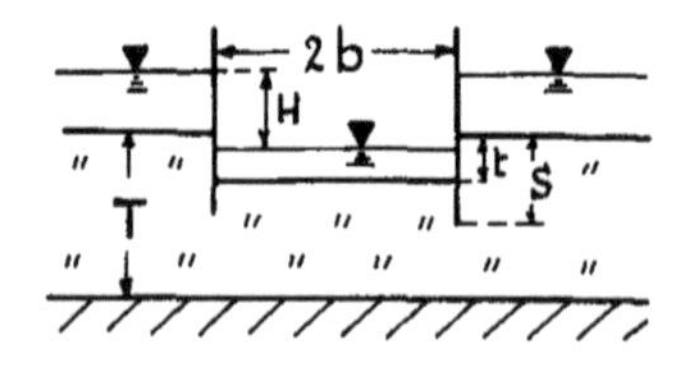


Fig. 92

The same flow region as considered above is also available in the case of water flowing towards a foundation pit, bounded by two identical rows of sheetpiles (Fig. 92; here $B = 2b$, $s$ has not the same value as in Fig. 91).

Such a problem was analyzed by V. S. Koslov [5]. He has given several auxiliary nomographs in the case where the bedrock is infinitely deep. In

Fig. 93 we reproduce a graph to compute the seepage rate in function of $t/b$ and $s/b$. Here $q_r = Q/kH$.

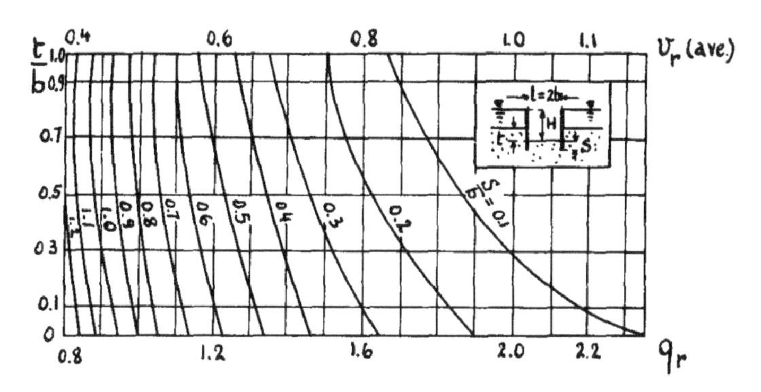


Fig. 93

In Fig. 94 and 95, graphs are given for the dependence upon $t/b$ and $s/b$ of the largest and smallest value of the velocity along the bottom

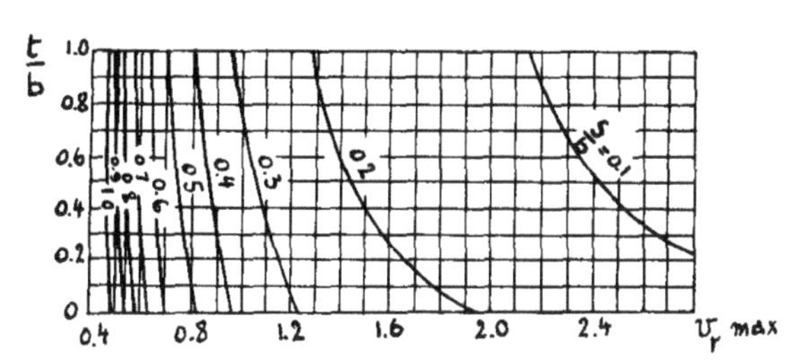


Fig. 94

of the pit. The following notations are used, $v_{r,\ max} = b/(kH)v_{max}$; $v_{r,min} = b/(kH)v_{min}$.

The distribution of the reduced velocity is given along the bottom of the foundation pit for values of $t = 0.4$, $s = 0.6$, $b = 1$. Here $v = kH/bv_r$, where $v$ is the seepage velocity.

The flownet of Fig. 97 corresponds to the particular case of Fig. 91, where the weir is not embedded.

## * §15. Embedded Weir.

For an embedded weir without cut-offs, we have (Fig. 98) $s_1 = s_2 = 0$, and consequently, to satisfy the conditions $-c_1 = c_6 = 1$, $-c_7' = c_7 = \gamma$, we have:

$$c_2 \equiv c_3, \quad c_4 \equiv c_5 \ .$$

The mapping function has the form of (13.29). Equations (13.22) – (13.27) give for the determination of the constants $\gamma$, $c_3$, $c_4$:

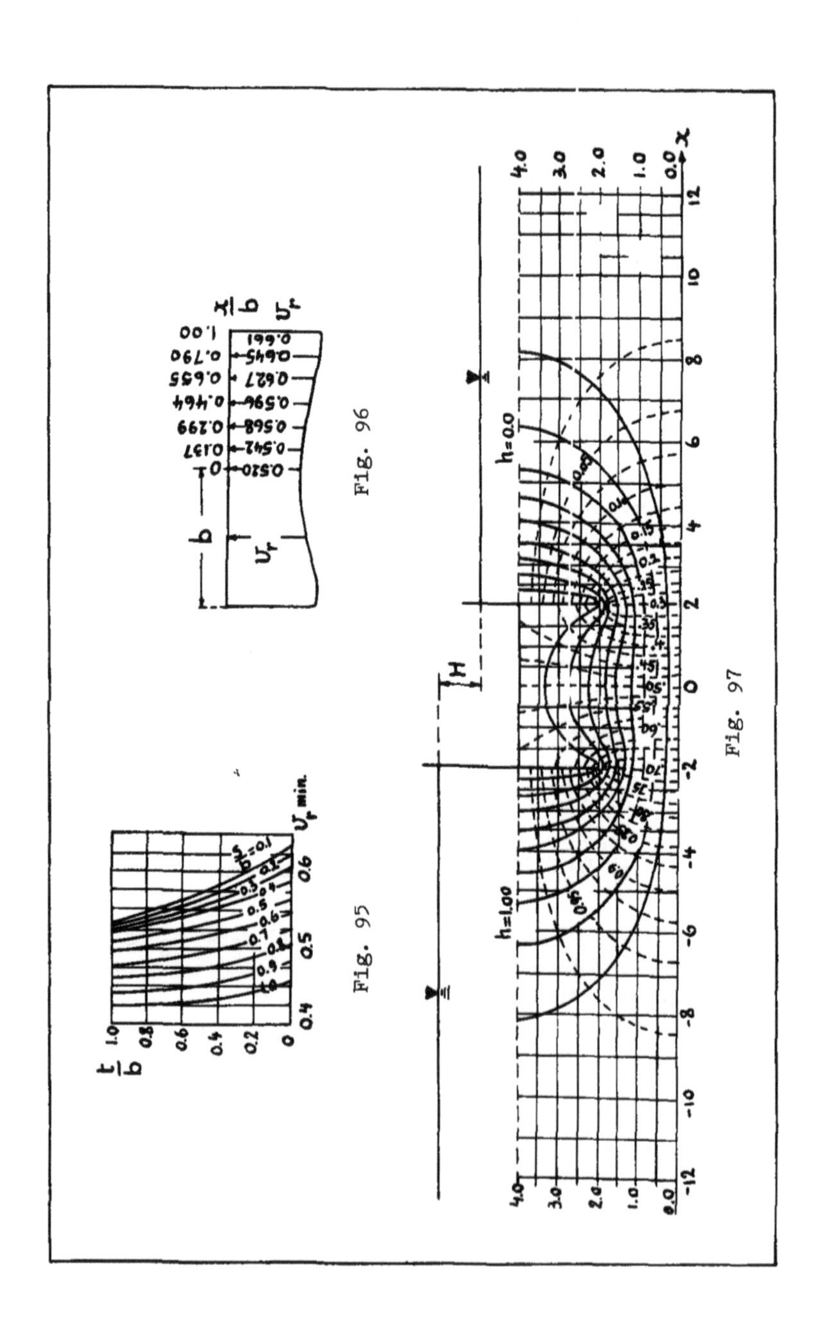

Fig. 95

Fig. 96

Fig. 97

$$\frac{t_2}{B} = \frac{\pi}{2K}\,\frac{D_7 a_7 - D_7' a_7'}{DK} - \frac{K'}{K}\ ,$$

$$\frac{t_1 - t_2}{B} = \frac{\pi}{2}\,\frac{D_7' - D_7}{DK}\ ,$$

$$\frac{T_2}{B} = \frac{\pi}{2}\,\frac{D_7}{DK}\ .$$

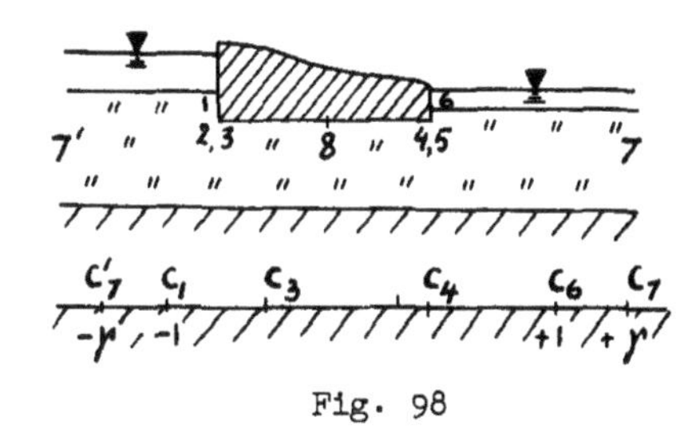


Fig. 98

In the non symmetrical case, when $t_1 \neq t_2$,

$$L^2 = (1 + c_4)(1 - c_3),\qquad k^2 = \frac{2(c_4 - c_3)}{L^2}\ ,\qquad k'^2 = \frac{(1 + c_3)(1 - c_4)}{L^2}\ ,$$

$$D_7 = \frac{L\sqrt{(\gamma - c_3)(\gamma - c_4)}}{2\gamma\sqrt{\gamma^2 - 1}}\ ,\quad D_7' = \frac{L\sqrt{(\gamma + c_3)(\gamma + c_4)}}{2\gamma\sqrt{\gamma^2 - 1}}$$

and

$$DK = D_7'[K\,\mathcal{E}\,(\text{am}\ a_7') - a_7'E] - D_7[K\,\mathcal{E}\,(\text{am}\ a_7) - a_7E]\ ,$$

$$\text{sn}^2(a_7,\ k) = \frac{(1 - c_3)(\gamma + 1)}{2(\gamma - c_3)}\ ,\quad \text{sn}^2(a_7',\ k) = \frac{(\gamma - 1)(1 - c_3)}{2(\gamma + c_3)}\ .$$

The constant $A$ is determined by the formula

$$A = \frac{2T_2\gamma\sqrt{\gamma^2 - 1}}{\pi\sqrt{(\gamma - c_3)(\gamma - c_4)}}\ .$$

For a symmetrically embedded weir, we have

$$t_1 = t_2 = t,\qquad T_1 = T_2 = T,$$

and

$$-c_3 = c_4 = \alpha,\qquad D_7' = D_7\ .$$

The equations for $B/T$ and $t/T$ are put in the form

$$\frac{\pi B}{2T} = (a_7 - a_7')E + K[\,\mathcal{E}\,(\text{am}\ a_7') - \mathcal{E}\,(\text{am}\ a_7)]\ ,$$

$$\frac{\pi t}{2T} = K'[\,\mathcal{E}\,(\text{am}\ a_7) - \mathcal{E}\,(\text{am}\ a_7')] - (K' - E')(a_7 - a_7')\ .$$

Therefore

$$k = \frac{2\sqrt{\alpha}}{1 + \alpha}\ ,\quad A = \frac{2T\gamma}{\pi}\sqrt{\frac{\gamma + 1}{\gamma - 1}}\ ,\quad \text{sn}^2 u = \frac{(\alpha + 1)(\zeta + \alpha)}{2\alpha(\zeta + 1)}$$

$$\text{sn}^2(a_7,\ k) = \frac{(1 + \alpha)(\gamma + 1)}{2(\gamma + \alpha)},\quad \text{sn}^2(a_7',\ k) = \frac{(\gamma - 1)(1 + \alpha)}{2(\gamma - \alpha)}\ .$$

The complex potential $\omega$ is linked with the variable $\zeta$ in the following way:

$$\omega = \frac{H}{2K}\,F\!\left(\zeta,\ \frac{1}{\gamma}\right)\ .$$

We observe that $\mathcal{E}$ (am a) is none other than the elliptical integral of the second kind, since $\mathcal{E}$(am a) = arc sin (sn a) and

$$\mathcal{E}\,(\mathrm{am}\ a) = \int_0^a \mathrm{dn}^2 u\, du = \int_0^{\mathrm{sn}\ a} \frac{\sqrt{1-k^2t^2}}{\sqrt{1-t^2}}\, dt \qquad (t = \mathrm{sn}\ u)\ .$$

N. N. Pavlovsky considered the case of the weir with rectangular foundation on a layer of infinite depth. We reproduce here the final results (here $B = 2\ell$)

$$z = \frac{k^2\ell}{E - k'^2K}\{\mathcal{E}\,(\arcsin\zeta, k) - k'^2F(\arcsin\zeta, k)\} - ti, \quad \zeta = \frac{1}{k}\sin\frac{\pi\omega}{xH}\ .$$

The modulus $k$ is determined by the equation

$$\frac{t}{t} = \frac{E' - k^2K'}{E - k'^2K}$$

N. N. Pavlovsky gave $k$ a series of values, found the corresponding values of $t/\ell$ and constructed a graph. In Fig. 99, the dependence of $t/\ell$ on $k^2$ for $0 < t/\ell < 1$ is given and simultaneously the dependence of the reciprocal value $\ell/t$ on $k^2$, for $0 < \ell/t < 1$. Thus the graph spans all values $0 < \ell/t < \infty$. If $0 \leq k^2 \leq 0.001$, for calculations with required accuracy of three decimals one may take

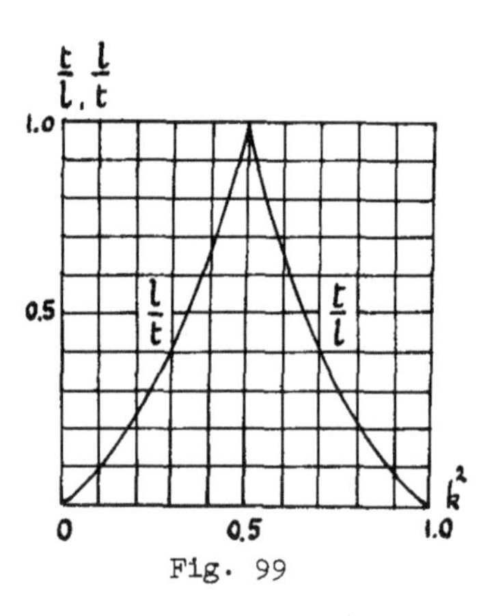

Fig. 99

$$\frac{\ell}{t} = \frac{E - k'^2K}{E' - k^2K'} \approx \frac{\pi}{4}k^2\ .$$

If $0.999 \leq k^2 \leq 1$, then

$$\frac{t}{\ell} = \frac{E' - k^2K'}{E - k'^2K} \approx \frac{\pi}{4}k'^2\ .$$

The dependence of the velocity on $\zeta$ is given by:

$$w = u - iv = \frac{\kappa H}{\pi k\ell}\,\frac{E - k'^2K}{\sqrt{1-\zeta^2}}\ .$$

The exit velocity is obtained for $\zeta = \frac{1}{k}$, and we have therefore (see Fig. 98),

$$v_6 = \frac{\kappa Hk}{\pi\ell k'}\,(E - k'^2K)\ .$$

In the middle of the base of the weir, $\zeta = 0$, and therefore the minimum velocity $u_8$ (see Fig. 98) is

$$u_8 = \frac{\kappa H}{\pi\ell k}\,(E - k'^2K)\ .$$

The ratio of the two minimum velocities appears to be

$$\frac{u_8}{v_6} = \frac{k'}{k^2} .$$

Since usually $t/\ell$ is small for weirs, $k$ will be close to one, and $k'$ close to zero. Therefore $u_8 < v_6$, and this inequality will be more emphatic the longer the weir is.

§16. Overfall Weirs.

Overfall weirs are constructed on sites with steep slope (Fig. 100). The scheme of the previous paragraph also holds as a special case for the overfall weir with cut-offs.

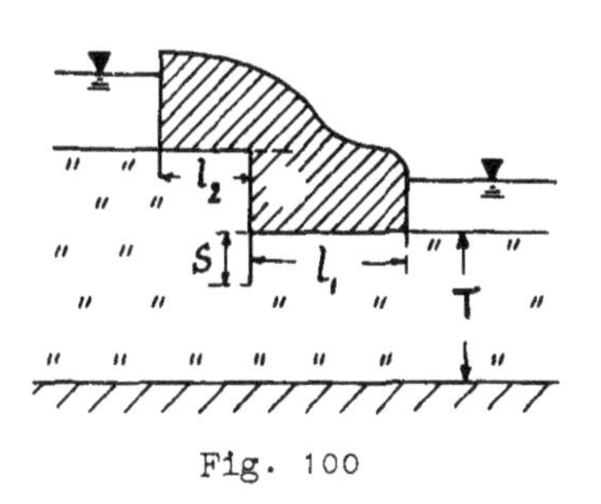


Fig. 100

M. I. Bazanov [7] applied the method of N. N. Pavlovsky to the calculation of the seepage rate under a single stepped overfall, with and without cut-off. The same schemes had been examined by N. N. Pavlovsky and E. A. Zamarin [2, 3] in the case of pervious strata of infinite depth. In the first case, the seepage rate is computed by the formula

$$Q = \kappa H \frac{K'}{K} ,$$

where the modulus $k$ follows from

$$k = \sqrt{\frac{2(\beta + \delta)}{(1 + \beta)(1 + \delta)}} .$$

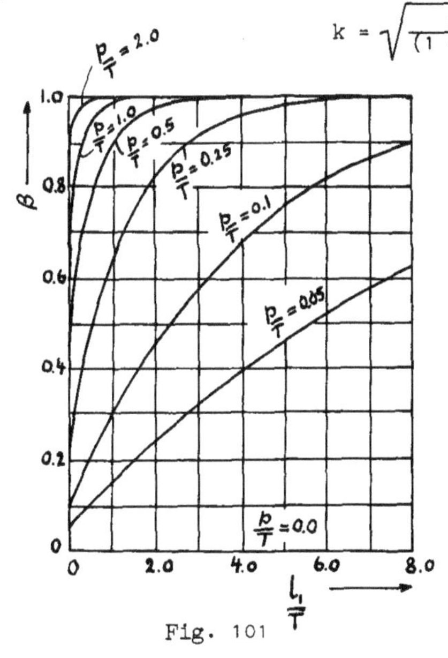


Fig. 101

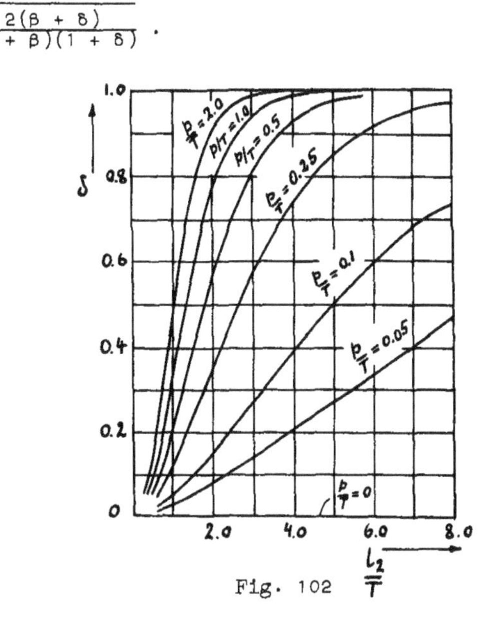


Fig. 102

M. I. Bazanov introduced graphs (Fig. 101 and 102), giving the dependence of $\beta$ and $\delta$ on $p/T$ and $\ell_1/T$ and on $p/T$ and $\ell_2/T$ (see meaning on Fig. 100). In Fig. 103 is given a graph of the exit velocity $v_r$ the reduced head $h_r$ and flowrate $\psi_r$ for a particular case.

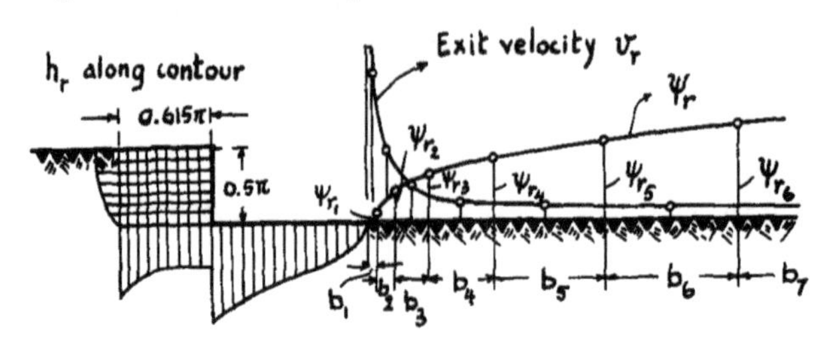


Fig. 103

For an overfall with cut-off, $\beta$ and $\delta$ are determined in a more complicated way, since they depend also on $s/p$, i.e., the ratio of the length of the cut-off to the step. The dependence of $\beta$ and $\delta$ on the ratio's $p/T$, $\ell_1/T$, $\ell_2/T$ for $s/p = 0.5$ and 1 is given by M. I. Bazanov in graphical form.

## D. Structures With Multiple Cut-offs

### §17. The Method of Fragments.

The formula to map the flow region onto the half-plane becomes complicated when the weir has several cut-offs, as is apparent from paragraph 13. The number of parameters to be determined increases, since each cut-off adds two or three new vertices to the flowregion. N. N. Pavlovsky [1], V. S. Koslov [4], P. F. Filchakov [9] and other authors have investigated a series of cases with two, three and even four cut-offs, but as we saw, the computations in these cases are complicated.

N. N. Pavlovsky proposed the "method of fragments", which we apply to to a structure with multiple cut-offs on a pervious stratum of finite depth. This method divides the flow region into fragments by means of vertical lines drawn through the ends of the cut-offs. The dividing lines (dash lines in Fig. 104) are taken for lines of constant potential. Actually they differ

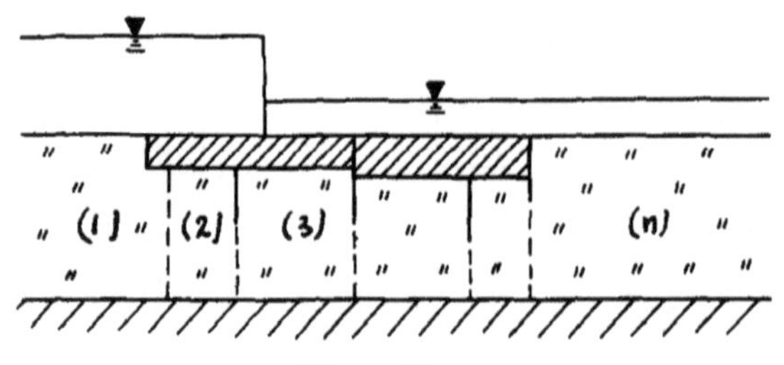


Fig. 104

from equipotential lines, but they approach the latter as the depth of the pervious stratum becomes smaller.

We assume that we may calculate the flow for each of the n fragments. In particular, we assume that the computed flow rate of each fragment is

$$Q = \frac{\kappa H_m}{\Phi_m} \qquad (m = 1, 2, \ldots, n), \tag{17.1}$$

where $H_m$ is the head loss for the seepage through the $m^{th}$ fragment, and $\Phi_m$ the "form modulus" of this fragment, i.e., a dimensionless quantity, depending upon the form of the fragment. So, for the $n^{th}$ fragment (Fig. 104), based on the formulas of Par. 11, we have $\Phi_n = \frac{2K}{K'}$ for $k = \sin\frac{\pi S}{2T}$.

Since the flowrate through all the fragments must be the same,

$$Q = \frac{\kappa H_1}{\Phi_1} = \frac{\kappa H_2}{\Phi_2} = \ldots = \frac{\kappa H_n}{\Phi_n} .$$

Adding numerators and denominators of these ratios, we obtain:

$$Q = \kappa\frac{\sum H_m}{\sum \Phi_m} .$$

But the sum of all the headlosses is equal to the total head H (difference between headwater and tailwater). Finally the discharge is calculated from

$$Q = \frac{\kappa H}{\sum_{m=1}^{n} \Phi_m} . \tag{17.2}$$

Each of the headlosses may now be computed from

$$H_m = \frac{\Phi_m}{\sum_{s=1}^{n} \Phi_s} H . \tag{17.3}$$

Knowing all the head losses $H_n$, we may compute the flow for each typical fragment #, and consequently, for an arbitrary region. It has been shown that the method of fragments, based upon the introduction of artificial equipotential lines in the region of confined flow, must overestimate the flowrate as compared with that of the accurate solution [18].

## §18. Structures With Multiple Cut-offs On a Stratum of Infinite Depth.

In case of a stratum of infinite depth, the method of fragments becomes inapplicable. N. T. Melechenko [8] proposed for this case the method

# S. V. Kozlov constructed nomographs, published in his book and also partially in V. I. Aravin's and S. N. Numerov's book.

of the "eliminated cut-offs. We show the essence of the method for the scheme of Fig. 105, with two cut-offs. After discarding the second cut-off, we map the flowregion with the first cut-off onto the lower half-plane $\zeta$. We take for mapping function

$$\zeta = \sqrt{z^2 + s^2} \quad . \tag{18.1}$$

It maps the cut-off of length $s$ onto the segment ABC, with length $2s$, of the $\xi$ axis. Separating in (18.1) the real and imaginary part, we obtain

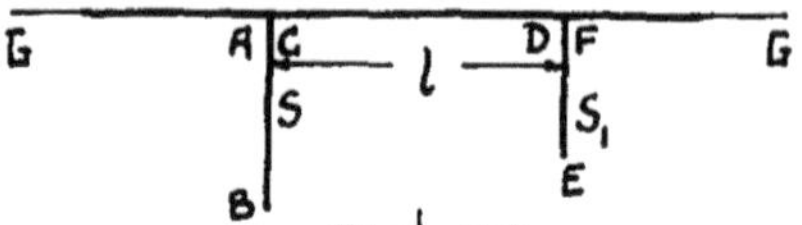

Fig. 105

$$\left\{ \begin{aligned} \xi^2 &= \left[ \sqrt{(s^2 + x^2 - y^2)^2 + 4x^2y^2} + s^2 + x^2 - y^2 \right] , \\ \eta^2 &= \left[ \sqrt{(s^2 + x^2 - y^2)^2 + 4x^2y^2} - s^2 - x^2 + y^2 \right] . \end{aligned} \right. \tag{18.2}$$

Fig. 106 represents a family of lines $x$ = constant, $y$ = constant, in the planes $x, y$ and $\xi, \eta$ for $s = 1$. The second cut-off is at a

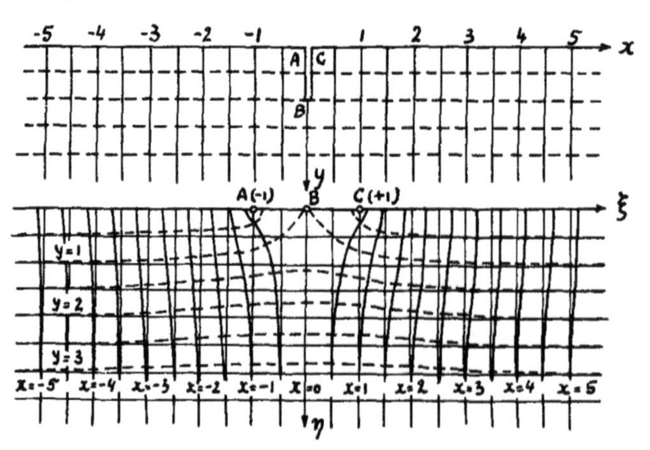


Fig. 106

distance $\ell$ from the first. Putting $x = \ell$ in (18.2) and letting $y \to \infty$, we find

$$\lim_{y \to \infty} \xi^2 = \ell^2 \ .$$

Consequently that curve into which $x = \ell$ is mapped after (18.1) has the asymptotic form

$$\xi = \ell$$

This curve cuts a segment of the $\xi$ axis

$$\xi_1 = \sqrt{\ell^2 + s^2} = \ell \sqrt{1 + \frac{s^2}{\ell^2}} = \ell_1 \quad . \tag{18.3}$$

The larger $\ell$, the less $\xi_1$ differs from $\ell$ and consequently the less the second cut-off is distorted.

Thus, in the $\zeta$-plane we have, instead of the two rectilinear cut-offs, one somewhat distorted cut-off. We neglect its deviation from the vertical, designate its vertical length by $s_2$ and have now a much simpler single cut-off scheme.

If we carry out a new mapping, onto the half-plane $\zeta'$, then the contour GABCDEFG maps into a segment with length L, representative of the base length of a weir in general. The new mapping is

$$\left\{\begin{aligned} z' + \ell_1 &= \sqrt{z^2 + s_2^2} \;, \\ z &= \sqrt{\left(\sqrt{\ell^2 + s_2^2} + \zeta'\right)^2 - s_2^2} \;. \end{aligned}\right. \tag{18.4}$$

The head along such weir base, as we saw (Chapter II, §12) is defined by

$$h = \frac{H}{\pi} \text{ arc cos} \left(\frac{2\xi'}{L} - 1\right) ,$$

where $\xi'$ is the distance from the left end of the base.

By means of inverse mappings, we carry this distribution law back to the initial contour with cut-offs, and obtain an approximate value of the head along this contour.

The computation of the exit velocity may be carried out as follows. After the first mapping, when we obtain the weir with one cut-off, we compute the velocity for that case, namely $\frac{d\omega}{d\zeta}$. The velocity in the z-plane will be

$$w = \frac{d\omega}{dz} = \frac{d\omega}{d\zeta} : \frac{dz}{d\zeta} ,$$

and this gives a simpler expression for the velocity, than in the direct solution of the initial problem.

When the number of cut-offs exceeds two, then the method of straightening the cut-offs may be extended till a simple scheme is obtained.

P. F. Filchakov drew the attention on the fact that the lines y = constant of Fig. 106 become straight with the depth and become almost horizontal at a depth of 4 to 5 times the length of the cut-off. Therefore the method of straightening of the cut-offs is applicable even for infinitely thick strata. P. F. Filchakov [10] gave tables for the computation of weirs with cut-offs on infinitely thick strata. By means of the scheme of the single step overfall, he even gave an approximate method for a weir with finite embedding.

### E. Hydrodynamic Reactions On Weirbases.

#### §19. Resultant Vector of Presure Forces.

When we examine the two-dimensional flow of groundwater under hydraulic structures, we consider the part of the structure that is confined between two vertical planes, one being the plane of Fig. 107, the other, parallel to the first at a distance of unit length from it. We designate by $\alpha$ the angle of the normal to the contour with the x-axis, by $\theta$ the angle between the tangent to the contour and the x-axis. For positive arc length we take the sense from B to C along the contour of the base, and the positive direction of the normal is to the interior of the flow region. Then the direction of $n$ will correspond with that of the x-axis, that of the tangent with that of the y-axis.

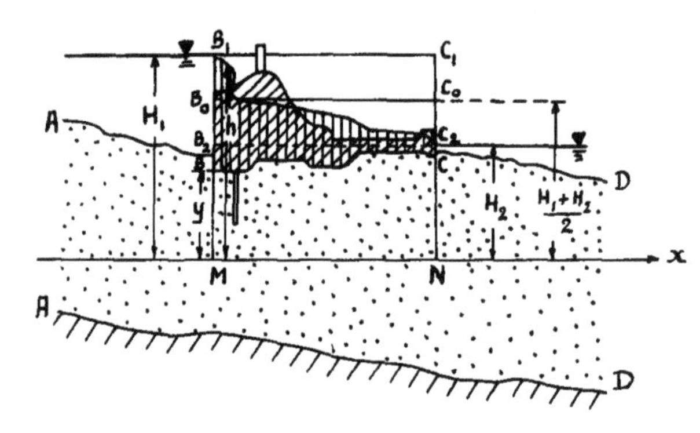


Fig. 107

We designate by $dF$ the element of pressure force that acts on a curvilinear rectangular surface composed of the arc length $ds$ and the unit length normal to the plane of the figure. Since the pressure forces are perpendicular to the contour but in opposite sense of the normal, and since

$$dF = dX + i dY , \tag{19.1}$$

we will have the following expressions for $dX$ and $dY$:

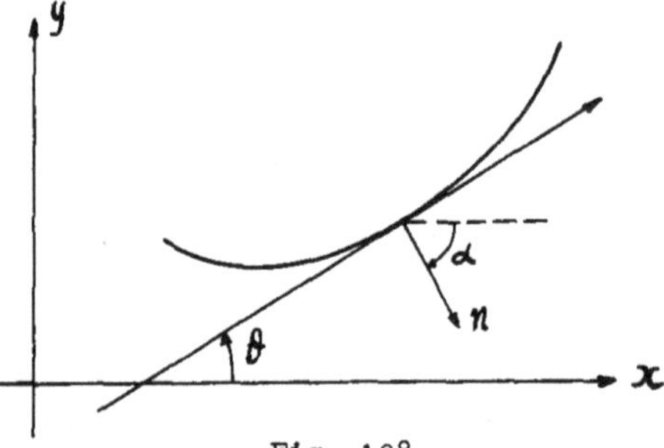


Fig. 108

$$\begin{cases} dX = - p \cos(n, x) ds = - p \cos \alpha \, ds , \\ dY = - p \cos(n, y) ds = - p \sin \alpha \, ds . \end{cases}$$

From Fig. 108 it is apparent that

$$\begin{cases} \cos \alpha = \cos\left(\theta - \frac{\pi}{2}\right) = \sin \theta , \\ \sin \alpha = \sin\left(\theta - \frac{\pi}{2}\right) = - \cos \theta . \end{cases} \tag{19.2}$$

Therefore we may write:

$$dX + i \, dY = - p(\sin \theta - i \cos \theta) ds = pi(\cos \theta + i \sin \theta) ds ,$$

But

$$\cos \theta \, ds = dx, \quad \sin \theta \, ds = dy ,$$

so that

$$dF = dX + i\, dY = pi(dx + i\, dy) = pi\, dz .$$

Carrying out the integration along the contour BC, we find the resulting vector of the pressure forces on the base of the dam

$$F = X + iY = i \int_{BC} p \, dz . \tag{19.3}$$

We had (in Chapter II) for the pressure $p$ in an arbitrary point of the flow region

$$p = \gamma(h - y) , \tag{19.4}$$

where $\gamma = \rho g$, unit weight of water. Moreover, $h$ is expressed in terms of the velocity potential $\varphi$ as

$$h = -\frac{\varphi}{k} , \tag{19.5}$$

Substituting (19.4) in (19.3), we find

$$X + iY = i\gamma \int_{BC} (h - y) dz . \tag{19.6}$$

or, by separation of the real and imaginary parts

$$X = -\gamma \int_{BC} (h - y) dy, \quad Y = \gamma \int_{BC} (h - y) dx . \tag{19.7}$$

We examine points B and C, extreme points of the base of the dam at the intersection with head and tailwater. We designate by $H_1$ and $H_2$ the levels of head and tailwater, by $y_1$ and $y_2$ the ordinates of the points B and C, and by $p_1$ and $p_2$ the pressures in those points.

$$\frac{p_1}{\gamma} = H_1 - y_1, \quad \frac{p_2}{\gamma} = H_2 - y_2 ,$$

$$h(x_1, y_1) = H_1, \quad h(x_2, y_2) = H_2 .$$

We assume that we know the distribution of the head along the base of the dam. We assume that a straight line parallel to the ordinate axis intersects the contour BC only in one point. Then the magnitude of $Y$, as determined by equation (19.7), will be equal to the weight of the shaded area of Fig. 107, if it were filled with water and per unit length of weir. Since

$$H_2 < h < H_1 ,$$

we have the inequality for $Y$

$$\gamma \int_{BC} (H_2 - y) dx < Y < \gamma \int_{BC} (H_1 - y) dx .$$

G. N. Polozjii introduced the concept of upper and lower hydrostatic pressures $\Gamma_U$ and $\Gamma_L$, as follows

$$\Gamma_U = \gamma \int_{BC} (H_1 - y)dx, \qquad \Gamma_L = \gamma \int_{BC} (H_2 - y)dx \quad . \tag{19.8}$$

It is evident that $\Gamma_L$ represents the weight of the volume per unit width of dam and area $BCC_2B_2$. Similarly, $\Gamma_U$ is the weight of the volume of water represented by the area $BCC_1B_1$ and the unit width of dam.

When the flow region is symmetrical with respect to the ordinate axis, then the function

$$h - \frac{H_1 + H_2}{2}$$

or, in other words, the head above a line averaging $y = H_1$ and $y = H_2$, will be odd, and the area between the lines $B_1C_2$ and $BC$ will be equal to that between $BC$ and $B_0C_0$ (midway between $B_1C_1$ and $B_2C_2$, see Fig. 107).

After G. N. Polozjii, we introduce the concept of average hydrostatic pressure

$$\Gamma_{ave} = \gamma \int_{BC} \left( \frac{H_1 + H_2}{2} - y \right) dx \quad . \tag{19.9}$$

In the case of a symmetrical region, the vertical component of the resultant pressure forces is equal to the average hydrostatic pressure

$$Y = \Gamma_{ave} \quad .$$

In general we may write

$$Y = \gamma \int_{BC} \left(h - \frac{H_1 + H_2}{2}\right) dx + \gamma \int_{BC} \left( \frac{H_1 + H_2}{2} - y \right) dx$$

or

$$Y = \Gamma_{ave} + \gamma \int_{BC} \left(h - \frac{H_1 + H_2}{2}\right) dx \quad . \tag{19.10}$$

Integration by parts leads to

$$\int_{BC} \left(h - \frac{H_1 + H_2}{2}\right) dx = \frac{H_2 - H_1}{2} (x_1 + x_2) - \int_{BC} x \, dh \quad .$$

When the y-axis is equidistant from $B$ and $C$, then $x_1 + x_2 = 0$ and

$$Y = \Gamma_{ave} - \gamma \int_{H_1}^{H_2} x \, dh \quad . \tag{19.11}$$

It is easy to compute $Y$ with this formula in case the family of lines of constant head is constructed.

We now turn to the horizontal component $X$ of the pressure forces. We assume that a line parallel with the x-axis intersects the contour $BC$ only in one point. (This may happen for an overfall but not for weirs.) In this case we have the inequality for $X$

$$\gamma\left(H_2 - \frac{y_1 + y_2}{2}\right)|y_1 - y_2| < |X| < \gamma\left(H_1 - \frac{y_1 + y_2}{2}\right)|y_1 - y_2| \ .$$

We designate $|y_1 - y_2|$ by $d$. If the x-axis is midway between $y_1$ and $y_2$, then $y_1 + y_2 = 0$, and

$$\gamma H_2 d < |X| < \gamma H_1 d \ .$$

One must remember that $H_1$ and $H_2$ here are counted from the average abscissa axis.

Integration of (19.7) by parts for $X$ gives:

$$X = -\gamma(H_2 y_2 - H_1 y_1) + \gamma \int_{BC} y \, dh \ . \tag{19.12}$$

Returning to the general case of contour $BC$, we designate by $\rho_1$ and $\rho_2$ the ordinates of the highest and lowest points of $BC$. Then

$$\gamma\rho_2 H < \gamma \int_{BC} y \, dh < \gamma\rho_1 H \ , \qquad (H = H_2 - H_1)$$

will be an estimate for the integral $\gamma \int y \, dh$ .

In a similar way an estimate for $Y$ is found when $B$ and $C$ are not the extreme left and right points of the base of the structure.

We notice that in formulas (19.11) and (19.12) we may substitute $\Omega = h + iq$ for $h$ under the integral; since $q$ is constant along the contour $BC$, we may take $q = 0$.

We obtain

$$X = B_{ave} - \gamma \int_{H_1}^{H_2} y \, d\Omega, \qquad Y = \Gamma_{ave} - \gamma \int_{H_1}^{H_2} x \, d\Omega$$

or

$$X + iY = B_{ave} + i\Gamma_{ave} + \gamma i \int_{H_1}^{H_2} z \, d\Omega \ ,$$

where

$$B_{ave} = \gamma(H_1 + H_2)y_2 \quad \text{for} \quad y_2 + y_1 = 0.$$

It is important to know the resultant $Y$ for the problem of uplift pressures on the dam. If we designate by $Y'$ the weight of the dam and by $Y''$ the downward resultant of the pressure forces on the embedded parts of the dam (for a dam with vertical walls, $Y'' = 0$), we obtain the necessary condition against uplift

$$Y'' + Y' > Y \ .$$

The value

$$k_u = \frac{Y'' + Y'}{Y} \ ,$$

is called the "coefficient of stability against uplift" [20]. The inequality permits to estimate $k_u$.

We notice that problems about forces exerted by the flow on the soil skeleton and on the base of the structure, have a significant practical interest, since their solution determines many details in the hydraulic project. However, although many theoretical and experimental investigations of these problems have been conducted, for the time being there is no exact understanding of the method to compute the pressure of groundwater flow on the soil skeleton and on the structure; the physical action of the flow on the structure is not clear. We refer to a special discussion devoted to this problem [21].

§20 Resulting Moment of Pressure Forces.

To elucidate the question of stability of the hydraulic structure, it is of interest to compute the resulting moment of the pressure forces with respect to a point $z_0$ in the vertical cross-section of the dam. Of particular interest is the case when $z_0$ coincides with one of the ends of the dam.

Let $z_0 = x_0 + iy_0$. The expression of the resulting moment M of the pressure forces with respect to $z_0$ has the form[27]

$$M = -\int_{BC} p[(x - x_0)\cos(n, y) - (y - y_0)\cos(n, x)]\, ds \quad . \tag{20.1}$$

According to formulas (19.2) we have

$$M = \int_{BC} p[(x - x_0)\, d(x - x_0) + (y - y_0)\, d(y - y_0)] = \frac{1}{2}\int_{BC} pd|z - z_0|^2 \quad .$$

Putting $p = \gamma(h - y)$ and integrating by parts, we have

$$M = \frac{1}{2}\gamma[H_2|z_2 - z_0|^2 - H_1|z_1 - z_0|^2] - \frac{\gamma}{2}\int_{BC} yd|z - z_0|^2 + \frac{1}{2}\gamma\int_{H_2}^{H_1} |z - z_0|^2\, dh \quad . \tag{20.2}$$

For the last term, we may obtain an inequality, if we take $z_0$ for one of the points $B(z_1)$ or $C(z_2)$. We assume $z_0 = z_1$. Then

$$M(z_1) = \frac{1}{2}\gamma H_2|z_2 - z_1|^2 - \frac{\gamma}{2}\int_{BC} yd|z - z_1|^2 + \frac{\gamma}{2}\int_{H_2}^{H_1} |z - z_1|^2 dh \quad . \tag{20.3}$$

Now we express

$$0 \le |z - z_1| \le |z_2 - z_1| \quad .$$

From this the inequality results

$$0 < \frac{\gamma}{2}\int_{H_2}^{H_1} |z - z_1|^2 dh < \frac{\gamma}{2}|z_2 - z_1|^2(H_1 - H_2) \quad .$$

The remaining terms in the formula for $M(z_1)$ are computed after the geometrical form of the dam.

For example, we compute the resultant vector and moment of the pressure forces on a dam on an infinitely deep layer, where $z_0$ is the left end of the dam, adjacent to headwater. We found the head before

$$h = -\frac{H_1 - H_2}{\pi} \text{arc sin} \frac{x}{\ell} + \frac{H_2 + H_1}{2} .$$

By means of formula (19.7) we obtain:

$$X = 0,$$

$$Y = \gamma \frac{H_1 + H_2}{2} B - \frac{\gamma(H_1 - H_2)}{\pi} \int_{-\ell}^{\ell} \text{arc sin} \frac{x}{\ell} \, dx = \gamma B \frac{H_1 + H_2}{2} = \Gamma_{ave} .$$

Here $B = 2\ell$, length of the base of the dam.

Formula (20.3) gives, for $z_1 = -\ell$, $y = 0$,

$$M(-\ell) = 2\gamma H_2 \ell^2 + \frac{\gamma}{2} \int_{-\ell}^{\ell} (x + \ell)^2 \frac{H_1 - H_2}{\pi} \frac{dx}{\sqrt{\ell^2 - x^2}} .$$

The integral of the right part is equal to

$$\int_{-\ell}^{\ell} \frac{(x^2 + 2\ell x + \ell^2) \, dx}{\sqrt{\ell^2 - x^2}} = \frac{3\pi}{2} \ell^2 .$$

Therefore, for the overturning moment with respect to point $z_0 = -\ell$

$$M(-\ell) = \frac{3H_1 + 5H_2}{16} \gamma B^2 .$$

It is evident that the moment about $z_0 = \ell$

$$M(\ell) = -\gamma \frac{3H_2 + 5H_1}{16} B^2 .$$

## §21. About the Displacement of Boundary Points in Mapped Regions.

The question about the nature of the displacement of boundary points of mapped regions is of interest in the seepage theory, since it allows us to obtain some qualitative conclusions about the change in discharge under a deformation of the contour of the flow region. Since the theorem given by G. N. Polozjii is based on the so called Schwarz' lemma, we formulate this lemma [26] here.

Schwarz' Lemma. If the function $w = f(z)$, holomorphic (or analytic) in the circle $|z| < 1$, satisfies the condition

$$f(0) = 0 \tag{21.1}$$

and if $|f(z)| < 1$ for $|z| < 1$, then

$$|f(z)| \leq |z|$$

everywhere in the circle $|z| < 1$. If the equality sign in

$$|f(z)| = |z|$$

holds in only one point of the interior [except the point $z = 0$, in which the condition holds by assumption], then the equality holds in all points of the region, and $f(z) = e^{\sigma i}z$.

In other words, Schwarz' lemma affirms that, if under condition (21.1) the modulus of the mapping function is smaller than one, it will be smaller than the modulus of $z$ in all other points inside the domain. Geometrically this means that if $w = f(z)$ maps the region of the unit circle into a region inside a curve corresponding to the unit circle, by means of the function $f(z)$ either any point approaches the origin or the mapping represents a rotation about the origin of coordinates.

We designate by $g$ (Fig. 109) the given region inside a circle of unit radius and by $g^*$ the region of the circle. Point 0 remains fixed in mapping region $g$ onto region $g^*$. When region $g$ transforms into region $g^*$, point 0 is repulsive, since it moves the point of region $g$ towards the circumference of the circle.

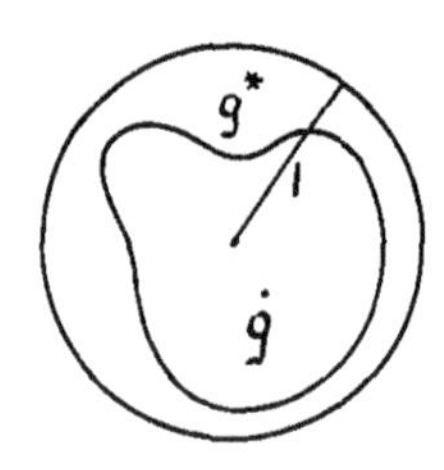

Fig. 109

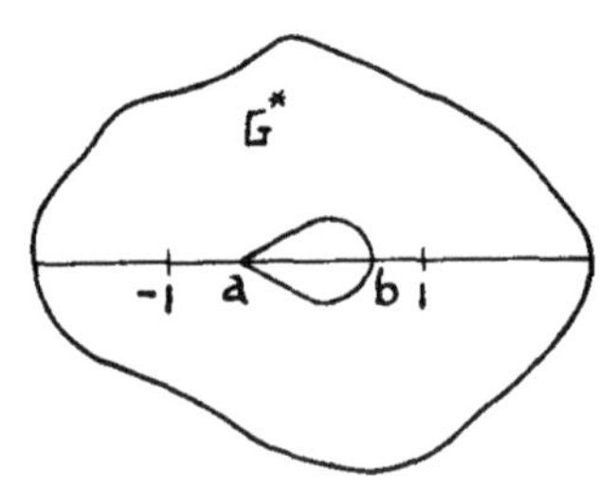

Fig. 110

If we take an arbitrary region instead of a circle, encompassing a given region $g$, Schwarz' lemma is still valid. Any point may be taken for fixed point, even the point at infinity, that may be considered as outside of the region.

We repeat now the theorems of G. N. Polozjii as given by this author [27].

Let $G$ be a simply connected domain in the $z = x + iy$ plane, $G^*$ a simply connected domain that comprises $G$ and has with $G$ a partially general boundary in the form of some Jordan Curve $\Gamma$.

<u>Theorem</u>. Given the conformal mapping of a simply connected domain $G$ onto arbitrary region $G^*$, comprising $G$ and having with $G$ a partially general boundary, being some Jordan curve $\Gamma$. On the open [i.e., without its ends] curve $\Gamma$ there may be no more than three fixed points. If there are three such points, then the two outside points are attractive, the middle one is repulsive; if there are only two such points, then one of them is attractive, the other repulsive.

Indeed, in the case of three fixed points on the open curve $\Gamma$, we may consider, without restriction of generality, that $G^*$ coincides with the upper half-plane, $G$ with a region lying in the upper half-plane, $\Gamma$ with the interval on the real axis outside the segment $[a, b]$, $-1 < a \leq b < 1$, and the fixed points with the points $z = 1$, $z = \infty$ and $z = -1$. Let H be the region, obtained by adding to the region G its symmetrical part about the real axis, H* its image in the plane of the function $w = f(z)$, mapping G onto G*. The region H, cut along the segments $[-1, a]$ and $[b, 1]$, is clearly mapped by the function $w = f(z)$ onto a plane that is cut along the segment $[-1, 1]$, and, consequently, after Schwarz' lemma, the point $z = \infty$ must be a repulsive fixed point and there will be no other fixed points in the interval $|x| > 1$. The region H, cut along the segments $[-1, a]$ and $[b, 1 - \varepsilon]$, $0 < \varepsilon < 1 - b$, is mapped by the function $w = f(z)$ onto a plane that is cut along some segment $[-1, 1 - \varepsilon']$, $\varepsilon' > 0$. The assumption that $\varepsilon' \geq \varepsilon$ leads to the contradiction that in the mapping of some region onto another region, its inside may not be bigger than one interior fixed point. Consequently, the point $z = 1$ will be a fixed attractive point and there will be no fixed points in the interval $(b, 1)$. The region H, cut along the segments $[-1 + \varepsilon, a]$ and $[b, 1]$, $0 < \varepsilon < 1+a$ is mapped by the function $w = f(z)$ onto a plane that is cut along the segment $[-1 + \varepsilon', 1]$, where $\varepsilon'$ is some positive number. The assumption $\varepsilon' \geq \varepsilon$ thus leads to a contradiction. Consequently, the point $z = -1$ will be a fixed attractive point and there will be no fixed points in the interval $(-1, a)$. In case there are only two fixed points on the open curve $\Gamma$, we map the region G so onto G*, that these two fixed points are attractive. This is possible by virtue of what is said before. We map now the region G* onto itself by means of the fractional linear transformation. For the given fixed points, we see that one of them will be attractive, the other not. The theorem is proved with this.

<u>Corollary</u> 1: When P and Q are the end points of curve $\Gamma$, and when there are three fixed points A,B,C, on the open curve $\Gamma$, then all points of the half closed loop [P, B) (i.e., comprising the point P but not the point B), except point A, and all points of the half closed loop (B, Q], except point C, receive a positive displacement in the direction of point A, and, correspondingly, in the direction of point C.

<u>Corollary</u> 2: When there are only two fixed points A and B on the open curve $\Gamma$, and when A is considered to be attractive, then all points of the half closed loop [P, B), except point A, receive a non-zero displacement towards A, and all points of the open curve (B, Q) receive a non-zero displacement in the same direction towards B.

<u>Corollary</u> 3: (About a fixed end). If the end Q of the curve $\Gamma$ is fixed and if there are two fixed points A and B on the curve $\Gamma$, then all points of the half closed curve [P, B), except A, receive a non-zero displacement towards point A, and all points of the open curve (A, Q), except B, receive a non-zero displacement away from B.

<u>Corollary</u> 4: (About fixed ends). When the ends P and Q of curve

$\Gamma$ are fixed and when there is a fixed point $A$ on curve $\Gamma$, then all points of the open curve $\Gamma$, except $A$, obtain a non-zero displacement away from $A$.

Corollary 5: Let $G^*$ be a region comprising a simply connected domain $G$ and having with $G$ a partially general boundary in the form of some Jordan curves. If in the mapping of $G$ onto $G^*$, there are three or only two fixed points on one of these curves, then on each other of these curves there can be no more than one fixed point, and if such a point exists, it is attractive.

Theorem: Let $G$ and $G^*$ be simply connected domains, partially encompassing each other and having only two Jordan curves as part of their boundaries. If there is a fixed point on one of these curves, then on the other there may not be more than two fixed points and if there are two such points, then one must be repulsive, the other attractive.

Indeed, the mapping of $G$ onto $G^*$ can be done with three fixed points by squeezing out the part of the boundary of $G$ that is inside $G^*$ and by shrinking the part of the boundary of $G$ that is outside of $G^*$. According to the previous theorem one of two fixed points lying on the same Jordan curve will be attractive, the other repulsive, and there will be no other fixed points on the given Jordan curve.

Corollary. (About the free middle point). Let the boundaries of the simply connected regions $G$ and $G^*$ be composed of a common Jordan curve $\Gamma$ and respectively of the Jordan curves $\gamma$ and $\gamma^*$, intersecting in some point $P$. In the mapping of $G$ onto $G^*$, if the ends of $\Gamma$ are fixed and one of the points of the open curve remains fixed, point $P$ obtains a non-zero displacement along the curve $\gamma^*$ in the direction of the end of this curve, approaching the boundary of region $G$ from within.

## §22. Application to the Flow of Groundwater Under Dams.

We examine some simple consequences of the theorems, [27, 28]. Consider the flow of ground water under an hydraulic structure with arbitrary foundation (Fig. 107) and bedrock with arbitrary form AD, boundaries AB and CD. Points A and D must not be at infinity. The boundaries are piece-wise smooth. We designate by $G$ the flow region ABCD. Let $G_1$ be the region with the same vertices A, B, C, D, but obtained from $G$ by means of deformation of some boundaries, which are changed into piece-wise smooth curves, lying partially or completely inside $G$. Then we have the following theorem:

Theorem 1: Under unchanged head, the seepage rate decreases, if the impervious boundaries BC or AD are changed, and increases, if the boundaries of the reservoirs are changed.

Indeed, let $g_1$ be the image of $G_1$ in the complex potential plane $\omega$. We map the rectangle $g_1$ onto the lower half-plane $\zeta$ with fixed points

$\zeta = -1, 1, \frac{1}{k}$. By virtue of the theorem about the displacement of boundary points, the point $\zeta = -\frac{1}{k}$ approaches point $\zeta = -1$ for a change of BC or DA and goes away from $\zeta = -1$ when AB or CD are changed. In the first case Q decreases, in the second case it increases.

Theorem 2: Under unchanged head H, the magnitude of the seepage velocity vector at the common part of the boundaries of G and $G_1$ decreases near the points B and C, if $G_1$ is obtained from G by a change of the impervious boundary BC, and decreases near D and A, if bedrock AD is changed. Indeed, let $g_1^*$ be the image of G in $\omega$. Let $G_1$ be obtained from G by changing BC. We map $g_1^*$ onto G* with fixed points B*, C*, D*. B* and C* will be attractive points.

Corollary. If the parts BB' and CC' become common to G and $G_1$ under a change of BC, then the pressure increases for all points of BB', decreases for all points of CC'.

In particular, it follows from the above, if the bottom of the tailwater reservoir is excavated, that the exit velocities (along the boundary of tailwater reservoir) decrease (Q increases). The exit velocities decrease also, when in the head reservoir a sill is constructed (Q decreases).

The principle of variation and its application to problems about confined seepage flow are given in another form by M. A. Lavrentiev [22, 23].

## CHAPTER IV.

## ZHUKOVSKY'S FUNCTION AND ITS APPLICATIONS:

## APPLICATION OF FUNCTIONAL ANALYSIS

### A. Direct Methods To Solve Seepage Problems.

§1. Zjoukovsky's Function.

In steady two-dimensional groundwater flow, the following relation exists between velocity potential and pressure (Chapter I, § 11)

$$\varphi = -k\left(\frac{p}{\rho g} + y\right) \quad ,$$

It can be written as

$$-\frac{k}{\rho g}\, p = \varphi + ky \quad .$$

Consequently, the expression

(1.1)
$$\theta_1 = \varphi + ky$$

differs from the pressure only by a constant multiplier.

The function $\theta_1$ is harmonic in $x, y$, since $\varphi$ and $y$ are harmonic. Conjugate with $\theta_1$ is the function

(1.2)
$$\theta_2 = \psi - kx \quad .$$

We designate $\theta_1 + i\theta_2$ by $\theta$:

(1.3)
$$\theta = \theta_1 + i\theta_2 = \omega - ikz$$
$$(\omega = \varphi + i\psi, \quad z = x + iy) \quad .$$

We call the function (1.3) and each function that differs from it only by a constant multiplier, a "Zjoukovsky function" [1].

Analogous with [1.1], we may introduce

(1.4)
$$\theta_2 = -\frac{k}{\rho g}\, p' \quad ,$$

so that

(1.5)
$$\theta = -\frac{k}{\rho g}\,(p + ip') \quad .$$

We may also call the complex pressure

(1.6)
$$P = p + ip' \quad ,$$

a Zjoukovsky function.

At the free surface, the pressure in the groundwater flow is atmospheric. In case of capillarity, the pressure is also constant (Chapter II, §2). If we take zero for atmospheric pressure, we must have at the border of the capillary fringe

$$p = -\rho g h_k , \tag{1.7}$$

where $h_k$ is the height of capillary rise in the soil.

So, at the free surface the real part of Zjoukovsky's function is a constant.

Instead of (1.1) and (1.2) we may write for the head and reduced seepage rate:

$$h = -\frac{\varphi}{k} , \quad q = -\frac{\psi}{k} .$$

Putting $\rho g = 1$, and expressing $p$ and $p'$ in meters, we have

$$p = h - y , \quad p' = q + x .$$

We show applications of Zjoukovsky's function to different problems.

§2. Zjoukovsky's Cut-off Wall.

Flow takes place about the cut-off under head $H$. Behind the cut-off, water rises over a height $CD$ (Fig. 111) and flows with a free surface $DA$.

At the free surface

$$\theta_1 = \varphi + ky = kh_k \tag{2.1}$$

Along the boundary $AB$:

$$\varphi = -kH . \tag{2.2}$$

Along the cut-off, we put

$$\psi = 0 . \tag{2.3}$$

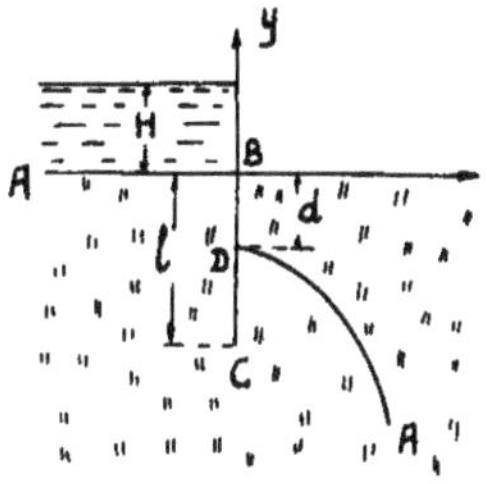

Fig. 111

Since along $AB$ $y = 0$, we have there

$$\theta_1 = \varphi + ky = -kH . \tag{2.4}$$

Furthermore, since along the cut-off $x = 0$, we have there

$$\theta_2 = \psi - kx = 0 . \tag{2.5}$$

In Zjoukovsky's plane (Fig. 112), we obtain a semi-infinite strip; its mapping onto the lower half plane gives:

$$\theta = k\frac{H + h_k}{\pi} \text{ arc sin } \zeta - k\frac{H - h_k}{2} = \omega - ikz . \tag{2.6}$$

In the complex potential plane we have a simple region (Fig. 112) with angle $\frac{\pi}{2}$, since $\varphi = -kH$ along the line $AB$ and $\psi = 0$ along the remaining boundaries. Consequently, we may put

$$\omega + kH = M\sqrt{\zeta + 1} .$$

Indeed, along $BCA$, where $\zeta = \xi > -1$, $\omega$ takes on real values (for $M$ real), i.e.,

$$\psi = 0 .$$

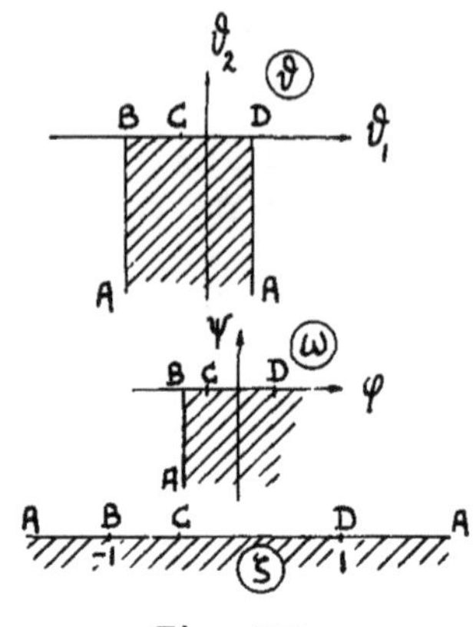

Fig. 112

Along AB, where $\zeta = \xi < -1$, we have

$$\omega + kH = - iM\sqrt{-\zeta - 1} ,$$

i.e., $\varphi + kH = 0$.

To determine M, we put $\zeta = 1$. Since in point D $\varphi = -ky + kh_k$ and since the length of BD is $y = -d$,

$$kd + k(H + h_k) = M\sqrt{2}$$

and

$$M = \frac{k(H + h_k + d)}{\sqrt{2}} ,$$

$$\omega + kH = \frac{k(H + h_k + d)\sqrt{\zeta + 1}}{\sqrt{2}} . \tag{2.7}$$

Eliminating $\omega$ from (2.6) and (2.7), we obtain for z the expression

$$iz = \frac{1}{\sqrt{2}}(H + h_k + d)\sqrt{\zeta + 1} - \frac{H + h_k}{2} - \frac{H + h_k}{\pi} \arcsin \zeta . \tag{2.8}$$

We differentiate equations (2.6) and (2.7) with respect to $\zeta$ to find the velocity in an arbitrary point of the flow region:

$$\frac{d\omega}{d\zeta} - ik\frac{dz}{d\zeta} = k\frac{H + h_k}{\pi}\frac{1}{\sqrt{1 - \zeta^2}} , \quad \frac{d\omega}{d\zeta} = \frac{k(H + h_k + d)}{2\sqrt{2(\zeta + 1)}} .$$

Dividing these equation side by side we find

$$1 - ik\frac{dz}{d\omega} = \frac{2\sqrt{2}}{\pi\sqrt{1 - \zeta}}\frac{H + h_k}{H + h_k + d} .$$

At the tip of the cut-off, in the point $z = -\ell i$, the velocity becomes infinitely high, and the magnitude of its reciprocal becomes zero. Therefore, the value for $\zeta = \zeta_0$ at the tip of the cut-off is:

$$\zeta_0 = 1 - \frac{8}{\pi^2}\left(\frac{H + h_k}{H + h_k + d}\right)^2 . \tag{2.9}$$

The substitution of this expression in (2.8) (and for $z = -\ell i$) gives the length of the cut-off

$$\ell = \sqrt{(H + h_k + d)^2 - \frac{4(H + h_k)^2}{\pi^2}} - (H + h_k) + \frac{2(H + h_k)}{\pi} \arcsin \frac{2(H + h_k)}{\pi(H + h_k + d)} . \tag{2.10}$$

The dependence between $\omega$ and $\theta$ is found by means of equations (2.6) and (2.7) as

$$\omega = - k\left[H - (H + h_k + d) \cos \frac{\pi(\theta - kh_k)}{2k(H + h_k)}\right] .$$

Along the free surface $\varphi + ky = kh_k$, $\psi = 0$, $\theta = - ikx + \kappa h_k$ and $\omega = \varphi = - ky + kh_k$. Therefore, we obtain the equation for the free surface

$$y = H + h_k - (H + h_k + d) \operatorname{ch} \frac{\pi x}{2(H + h_k)} , \tag{2.11}$$

i.e., the equation of a catenary.

This solution was given by V. V. Vedernikov [4]. N. E. Zjoukovsky constructed expressions for $\omega$ and $\theta$ as functions of an auxiliary complex variable in integral form. The integrand is found by means of the nature of the singularities. In the case of Fig. 112, the function contains a quadratic root in the denominator, since the angles of the region are $\pi/2$.

To close this paragraph, we reproduce in table 6 the dependence of $\frac{d}{H + h_k}$ on $\frac{\ell}{H + h_k}$, computed by V. V. Vedernikov. Negative values of $\frac{d}{H + h_k}$ mean that water beyond the cut-off rises higher than point B (Fig. 111).

Table 6. Dependence of $\frac{d}{H + h_k}$ on $\frac{\ell}{H + h_k}$.

| $\frac{\ell}{H + h_k}$ | 0 | 0.074 | 0.213 | 0.64 | 1.10 | 2.066 |
|---|---|---|---|---|---|---|
| $\frac{d}{H + h_k}$ | - 0.363 | - 0.2 | 0 | 0.5 | 1 | 2 |
| $\frac{\ell - d}{H + h_k}$ | 0.363 | 0.274 | 0.213 | 0.14 | 0.10 | 0.066 |

We notice that the influence of capillarity is taken into account by replacing the actual head $H$ by $H + h_k$, where $h_k$ is the height of capillary rise.

## §3. Single Drain.

This problem was also examined by N. E. Zjoukovsky [1]. We may represent a drainage ditch or drain with very small water-depth as a slit (Fig. 113).

On the lines AB and FG, representing free surfaces, the pressure is constant. We assume it is zero. We may then write the condition for the velocity potential at the free surface (see Chapter II, §2) as

$$\varphi + ky = 0 . \tag{3.1}$$

The contour BCDEF of the drain is a line of constant head $\varphi = 0$. Since for the contour $y = 0$, condition (3.1) also holds there.

Thus, for the Zjoukovsky function

$$\theta = \theta_1 + i\theta_2 = \varphi + ky + i(\psi - kx)$$

the real part is zero all along the contour of the flow region.

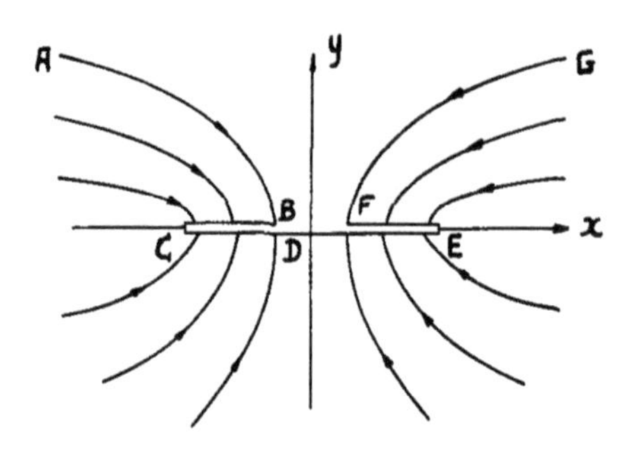


Fig. 113

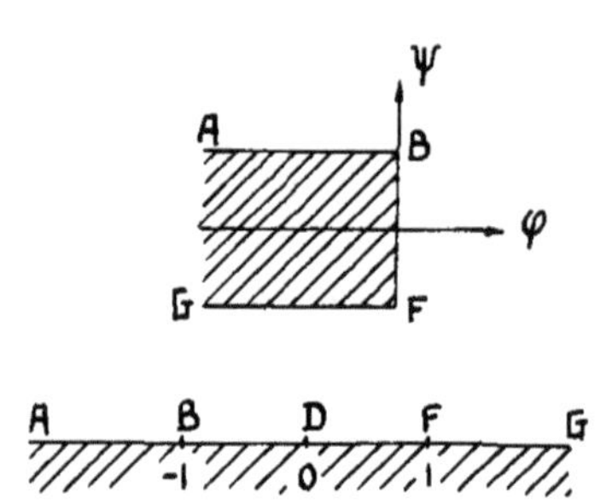


Fig. 114

The half-plane $\theta$ corresponds to the half-plane $\zeta$, and we may assume

$$\theta = a\zeta , \tag{3.2}$$

where "a" is a constant, to be determined further.

The lines AB and FG are streamlines.

We designate by $Q$ the full flowrate of the drain (per unit length). We may assume $\psi = -\frac{Q}{2}$ along FG, $\psi = +\frac{Q}{2}$ along AB.

The mapping of the region of the complex potential onto the lower half-plane $\zeta$ (Fig. 114) gives:

$$\omega = M\int_0^\zeta \frac{d\zeta}{\sqrt{1 - \zeta^2}} = M \text{ arc sin } \zeta . \tag{3.3}$$

For $\zeta = 1$, we have $\varphi = 0$, $\psi = -\frac{Q}{2}$. From equation (3.3) we obtain for $\zeta = 1$

$$-\frac{Qi}{2} = M\frac{\pi}{2} .$$

and

$$\omega = -\frac{Qi}{\pi} \text{ arc sin } \zeta . \tag{3.4}$$

Solving this equation for $\zeta$, we find

$$\zeta = \sin \frac{\pi\omega i}{Q} . \tag{3.5}$$

Putting this expression of $\zeta$ in (3.2) and substituting $\omega - ikz$ for $\theta$, we find

$$\omega - ikz = a \sin \frac{\pi\omega i}{Q} .$$

Solving this equation for $z$:

$$z = \frac{\omega}{ki} - \frac{a}{ki} \sin \frac{\pi\omega i}{Q} . \tag{3.6}$$

To determine the constant "a" we put $z = b_1/2$, where $b_1$ is the width of the segment BF. We have for the point F (Fig. 114) $\varphi = 0$, $\psi = -\frac{Q}{2}$, i.e.,

$$\omega = -\frac{Qi}{2}.$$

Equation (3.6) gives:

$$\frac{b_1}{2} = -\frac{Q}{2k} - \frac{a}{ki},$$

and therefore

$$a = -ki\left(\frac{Q}{2k} + \frac{b_1}{2}\right). \tag{3.7}$$

We find $\frac{dz}{d\omega}$, the reciprocal of the seepage velocity, by differentiating (3.6) with respect to $\omega$:

$$\frac{dz}{d\omega} = \frac{1}{w} = \frac{1}{ki} - \frac{a\pi}{kQ}\cos\frac{\pi\omega i}{Q}. \tag{3.8}$$

Point E, at the end of the drainage slit, has the property that the velocity is infinitely high there, and consequently $\frac{1}{w} = 0$. With "a" from (3.7), this gives:

$$\cos\frac{\pi\omega i}{Q} = -\frac{Qi}{a\pi} = \frac{2Q}{\pi(Q + kb_1)}. \tag{3.9}$$

In point E we have $\varphi = 0$. Let $Q'$ be the discharge through CDE, then in E

$$\psi = -\frac{Q'}{2}, \qquad \omega = -\frac{Q'i}{2}.$$

Inserting this value of $\omega$ in (3.9) and solving (3.9) for $Q'$, we find

$$Q' = 2\frac{Q}{\pi}\arccos\frac{2Q}{\pi(Q + kb_1)}. \tag{3.10}$$

Now we determine the width b of the slit along the bottom, considering CDE = b. In point E

$$z = \frac{b}{2}, \qquad \omega = -\frac{Q'i}{2}.$$

Inserting these values in (3.6) and considering (3.7) and (3.9), we find:

$$\frac{b}{2} = -\frac{Q'}{2k} + \left(\frac{Q}{2k} + \frac{b_1}{2}\right)\sqrt{1 - \frac{4Q^2}{\pi^2(Q + kb_1)^2}},$$

or after substitution of $Q'$ from (3.10)

$$b = \left(\frac{Q}{k} + b_1\right)\sqrt{1 - \frac{4Q^2}{\pi^2(Q + kb_1)^2}} - \frac{2Q}{\pi k}\arccos\frac{2Q}{\pi(Q + kb_1)}. \tag{3.11}$$

Finally, we find the equation of the branch FG of the free surface. Therefore we put in equation (3.6):

$$\varphi = -ky, \qquad \psi = -\frac{Q}{2}.$$

We obtain

$$z = -\frac{ky + \frac{Qi}{2}}{ki} - \left(\frac{Q}{2k} + \frac{b_1}{2}\right)\sin\frac{\pi i}{Q}\left(ky + \frac{Qi}{2}\right).$$

Putting $z = x + iy$, we find after some operations:

$$x = -\frac{Q}{2k} + \left(\frac{Q}{2k} + \frac{b_1}{2}\right)\operatorname{ch}\frac{\pi ky}{Q}. \tag{3.12}$$

We obtained, as in the case of Zjoukovsky's cut-off, a catenary for the free surface. It has been shown [5] that Zjoukovsky's drain problem may be obtained from Zjoukovsky's cut-off (without capillary rise in the soil), if we replace the function $z$ by $-iz$, $\omega$ by $i\theta$ and $\theta$ by $i\omega$.

Equation (3.12) may be rewritten as

$$y = \frac{Q}{k\pi} \operatorname{ch}^{-1} \frac{x + \frac{Q}{2k}}{\frac{b_1}{2} + \frac{Q}{2k}} \tag{3.13}$$

$$[\operatorname{ch}^{-1} x = \ln(x + \sqrt{x^2 - 1})] \quad .$$

If some point $(x_1, y_1)$ is known at the free surface, and also segment $b_1$, then equation (3.13) allows for the computation of the discharge $Q$ of the drain. When $x_1$, $y_1$, and $b$ are given, then we find $b_1$ and $Q$ from (3.11) and (3.13).

Any of the equipotentials (Fig. 113) may be taken for drain contour. If the given drain has a form different from these equipotentials (usually the vertical cross-section of a drainage ditch is rectangular, trapezoidal) then near the drain the free surface has a unique form, but at sufficiently large distance from the drain, the equation of the free surface is close to that of (3.13).

V. N. Aravin [14] examined examples of Zjoukovsky drains by means of a flownet for various values of $b_1$, including $b_1 = 0$. Here the branches of the free surface converge into one point, in the middle of the drain. Such a drain may be called "submerged." For $b_1 = 0$, we obtain from (3.11)

$$b = \frac{Q}{k}\left(\sqrt{1 - \frac{4}{\pi^2}} - \frac{2}{\pi} \arccos \frac{2}{\pi}\right) = 0.262 \frac{Q}{k} \quad . \tag{3.14}$$

## §4. Infiltration and Flow To a Drain.

Usually drains are designed to evacuate excess water from the soil. Therefore, drainage ditches have a slope, so that water may run down to collectors, large drains from which water flows by gravity or discharges into rivers or large canals.

Drains are fed by infiltration, i.e., the passage of water through the soil surface as result of rain, irrigation, snowmelt, etc. We examine the problem of the flow of groundwater to a system of a large number of rectilinear horizontal drains, infinitely thin slits, at unit depth, equidistant, in a pervious layer of infinite depth.

We designate by $\varepsilon$ the constant infiltration rate (Chapter I). If we look at the part of the flow (Fig. 115) between vertical streamlines, half way between adjacent drains, then we obtain a flow region with firm vertical walls and pervious horizontal boundaries (walls of the slit). In this case we apply Zjoukovsky's function in the form

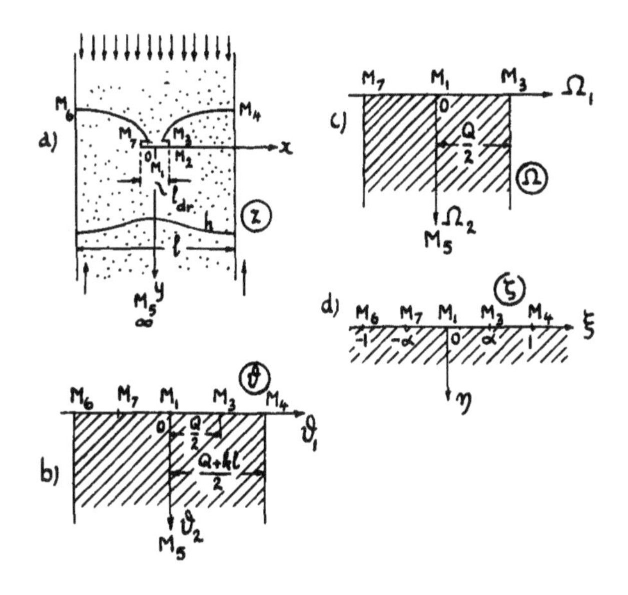


Fig. 115

(4.1)
$$\theta = -i\omega + kz = \theta_1 + i\theta_2 .$$

We introduce the function

(4.2)
$$\Omega = -i\omega + \varepsilon z = \Omega_1 + i\Omega_2 ,$$

in which the real part has a constant value at the free surface

(4.3)
$$\Omega_1 = \psi + \varepsilon x = \text{constant} .$$

If we determine $\theta$ and $\Omega$, $z$ and $\omega$ will follow from

(4.4)
$$\begin{cases} z = \dfrac{1}{k - \varepsilon} (\theta - \Omega) , \\ \omega = \dfrac{1}{k - \varepsilon} (k\Omega - \varepsilon\theta) . \end{cases}$$

We designate by $Q$ the total discharge of each drain, by $Q'$ the discharge from the ground waters. The discharge $Q''$ from infiltration is then

(4.5)
$$Q'' = Q - Q' = \varepsilon \ell ,$$

where $\ell$ is the distance between adjacent drains.

A conformal mapping of regions $\Omega$ and $\theta$ onto the half-plane $\zeta$ (Fig. 115) gives

(4.6)
$$\theta = \frac{1}{\pi} (Q' + k\ell) \text{ arc sin } \zeta$$

and

(4.7)
$$\Omega = \frac{Q}{\pi} \text{ arc sin } \frac{\zeta}{\alpha} ,$$

where, from (4.6)

$$\alpha = \sin\frac{\pi Q}{2(Q' + k\ell)} = \sin\frac{\pi Q}{2[Q + (k - \varepsilon)\ell]} \quad (4.8)$$

Inserting (4.6) and (4.7) in (4.4) and considering that $\alpha < \zeta < 1$ along the free surface, we find:

$$z = \frac{1}{\pi(k - \varepsilon)}\left[(Q' + k\ell)\ \text{arc sin}\ \zeta - Q\ \text{arc sin}\ \frac{\zeta}{\alpha}\right] ,$$

For the equation of the right branch of the free surface (symmetrical with the left):

$$x = \frac{Q}{k - \varepsilon}\left[-\frac{1}{2} + \frac{1}{\pi}\left(1 + \frac{1}{\delta}\right)\text{arc sin}\left(\alpha\ \text{ch}\ \frac{\pi y(k - \varepsilon)}{Q}\right)\right] , \quad (4.9)$$

where

$$\delta = \frac{Q}{(k - \varepsilon)\ell} . \quad (4.10)$$

From (4.9), for $x = \frac{\ell}{2}$, $y = H$ ($H$ is the maximum height on the free surface):

$$H = \frac{Q}{\pi(k - \varepsilon)}\ \ell n\ \text{ctg}\ \frac{\pi\delta}{4(1 + \delta)} . \quad (4.11)$$

To determine the width of the slit $\ell_{dr}$, we notice that in points $M_2$ and $M_8$ the abscissa $x$ reaches an extreme as function of $\xi$. Consequently $\frac{dx}{d\xi} = 0$, and this leads to the condition

$$\ell_{dr} = \frac{Q}{k - \varepsilon}\ f(\delta) , \quad (4.12)$$

in which $f(\delta)$ is

$$f(\delta) = \frac{2}{\pi}\left\{\left(1 + \frac{1}{\delta}\right)\text{arc cos}\left(\frac{1 + \delta}{\sqrt{1 + 2\delta}}\ \cos\frac{\pi\delta}{2(1 + \delta)}\right) - \right.$$

$$\left. - \text{arc cos}\left(\frac{\delta}{\sqrt{1 + 2\delta}}\ \text{ctg}\ \frac{\pi\delta}{2(1 + \delta)}\right)\right\} . \quad (4.13)$$

Where $\delta$ is small (in particular when $\ell$ is large), approximately

$$f(\delta) \approx f(0) ,$$

this leads to

$$f(\delta) \approx \frac{2}{\pi}\left(\sqrt{\frac{\pi^2}{4} - 1} - \text{arc cos}\ \frac{2}{\pi}\right) \approx 0.262 ,$$

and

$$\ell_{dr} \approx \frac{0.262Q}{k - \varepsilon} .$$

For $\varepsilon = 0$ this expression is the same as that obtained for the submerged drain (formula 3.14).

We insert now $x = \frac{1}{2}\ell_{dr}$ in equation (4.9). Designating the ordinate of the free surface in this point by $h_0$, we find

$$h_0 = \frac{Q}{k - \varepsilon}\ F(\delta),\quad F(\delta) = \frac{1}{\pi}\ \text{ch}^{-1}\left(\frac{\sin\frac{\pi(1 + f(\delta))\delta}{2(1 + \delta)}}{\sin\frac{\pi\delta}{2(1 + \delta)}}\right) . \quad (4.14)$$

For small $\delta$ we have

$$F(\delta) \approx F(0) = \frac{1}{\pi}\ \text{ch}^{-1}[1 + f(0)] = 0.226$$

and

$$h_0 \approx \frac{0.226Q}{k - \varepsilon} .$$

These results were obtained by C. N. Numerov [7], who comes to the conclusion that if the drain (horizontal slit) is replaced by a vertical trench of rectangular cross section and width $\ell_{dr}$, then the height of percolation of groundwater at the walls of the trench is smaller than $h_0$, determined by formula (4.14). The case $\varepsilon = 0$ was analyzed by V. I. Aravin [6]. We examine the case of a drain with rectangular cross-section.

The flow net in the case under consideration is such that at some distance downward, the streamlines become almost vertical and the equipotential lines almost horizontal. For a depth $|y|>\ell$, the latter may practically be considered as horizontal. For the head $h$ on such a line, above the drain as datum, we have

(4.15) $$h \approx \frac{Q'y}{kl} + \frac{Q}{k} \Phi(\delta)$$

with an absolute error approximately equal to

$$\frac{Q}{\pi k} \left\{(1 + \delta) \cos^2 \frac{\pi\delta}{2(1 + \delta)} \sin^{2\delta} \frac{\pi\delta}{(1 + \delta)2}\right\} e^{-\frac{2\pi y}{\ell}} .$$

An approximate expression for $\Phi(\delta)$ is given as

$$\Phi(\delta) \approx - \frac{1}{\pi}(1 + \delta) \ln \sin \frac{\pi\delta}{2(1 + \delta)} .$$

From (4.15) we may find $Q'$, knowing $Q$.

When the drains only flow because of infiltration at the free surface, then $Q' = 0$, $Q = Q'' = \varepsilon\ell$ and $\delta = \frac{\varepsilon}{\varepsilon-k}$. If they only flow because of groundwater then $\varepsilon = 0$, $Q = Q'$. From this it follows that if we know the solution without infiltration, we may find the solution with infiltration if we multiply all linear dimensions by $(1 - \frac{\varepsilon}{k})$.

We notice that the above problem is also present when percolation towards the drain is not only due to atmospheric precipitation or irrigation, but also to infiltration from a stratum with confined (artesian) waters through the overlying layer. In other words, the supply to the drain designed to evacuate excess waters occurs from below, on account of flow under head. The possibility of steady flow exists when no infiltration occurs. If there were an impervious boundary below, then unsteady flow in the drain would prevail, with steady flow as a limit case.

V. V. Vedernikov (see literature Chapter V,[10]), gave a solution to the problem of flow towards a system of drains in a region bounded below by an equipotential line (the contour of the drain can be considered as a small circle around a point source).

We notice that application of Zjoukovsky's function is convenient in such cases where besides a free surface we have horizontal equipotential lines and vertical streamlines in the flowregion. F. B. Nelson - Skornjakov analyzed a series of such problems. In his book [15] he examined the flow of groundwaters to a foundation pit or lock, bounded by cut-offs (Fig. 116), drained by lateral horizontal drains (slits). V. I. Aravin, previously had examined the case without these lateral drains [14].

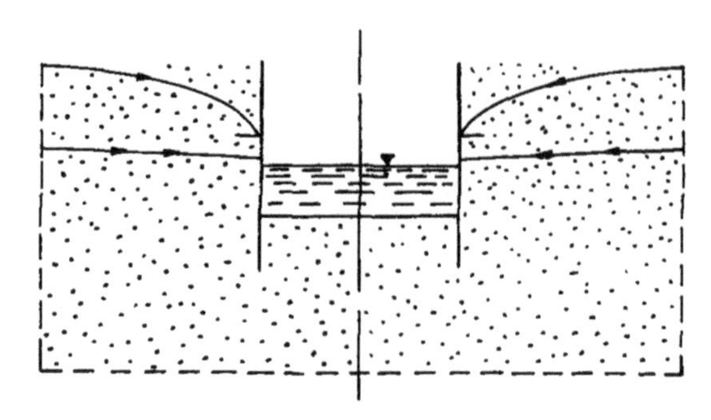

Fig. 116

§5. Seepage From a Canal To Symmetrically Disposed Collectors.

A sketch of the problem is given in Fig. 117. The canal has the single form of an horizontal segment. The drains are horizontal segments AB and GH with slits CB and GF. As in the previous problem, along the complete contour we have:

$$\varphi + \kappa y = 0 \ ,$$

and for the mapping onto the half-plane $\zeta$ (Fig. 117) of the Zjoukovsky function $\theta = \omega - i\kappa z$, we have

$$\theta = a\zeta + b \ . \tag{5.1}$$

We put the velocity potential along the segments ABC and FGH equal to zero. Along BD we have then

$$\varphi = -\kappa T \ ,$$

where $T$ is the elevation of the bottom of the canal above the drains.

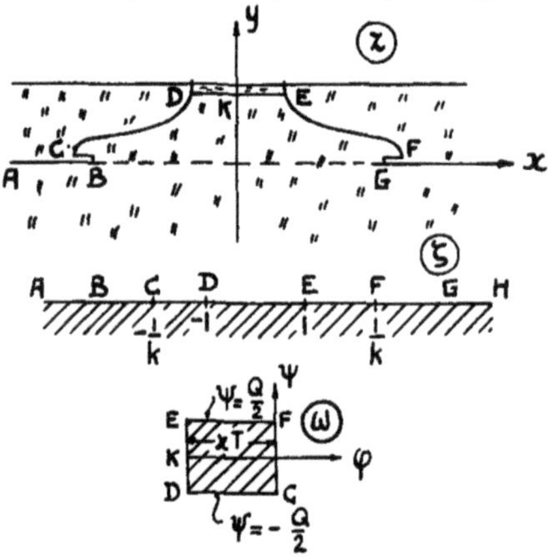

Fig. 117

Designating the full discharge of the canal per unit length by $Q$, we put $\psi = \frac{Q}{2}$ along EF and $\psi = -\frac{Q}{2}$ along CD. The rectangle in the $\omega$-plane (Fig. 117) is mapped onto the lower half-plane $\zeta$ (see a similar problem in Chapter III):

$$\omega = \frac{Qi}{2K}\int_0^\zeta \frac{d\zeta}{\sqrt{(1-\zeta^2)(1-k^2\zeta^2)}} - \kappa T \ . \tag{5.2}$$

For the seepage rate we have

$$Q = \frac{2\kappa TK}{K'} \ . \tag{5.3}$$

We now return to equation (5.1) and rewrite it, considering the value of Zjoukovsky's function:

$$\omega - i\kappa z = a\zeta + b \ . \tag{5.4}$$

For $\zeta = 0$, we have $z = Ti$, $\omega = -\kappa T$ and therefore, after (5.4), $b = 0$.

Furthermore, for $\zeta = 1$

$$z = \frac{B}{2} + Ti, \qquad \omega = -\kappa T + \frac{Qi}{2} ,$$

where $B$, width of canal. After (5.4):

$$a = \frac{i}{2}(Q - \kappa B) ,$$

from which

$$\omega - i\kappa z = \frac{i}{2}(Q - \kappa B)\zeta \quad . \tag{5.5}$$

Putting $\zeta = \frac{1}{k}$, we have:

$$\frac{Qi}{2} - \frac{i\kappa L}{2} = \frac{i(Q - \kappa B)}{2k} ,$$

where $L$, distance between the end points $C$ and $F$ of the drains. From this we find for $Q$ another expression:

$$Q = \frac{\kappa}{1 - k}(B - kL) \quad . \tag{5.6}$$

Equating it with (5.3) we obtain:

$$\frac{B}{T} - k\frac{L}{T} = 2(1 - k)\frac{K}{K'} \quad . \tag{5.7}$$

From this equation, knowing $\frac{B}{T}$ and $\frac{L}{T}$, we may determine $k$, and this allows then to compute the seepage rate by means of formula (5.3). Instead of $L$ we may introduce another quantity, namely the distance $\ell$ from the level in the canal till the level in the collector, or, in other words, the projection of $EF$ on the x-axis (Fig. 117). Between $\ell$, $L$, and $B$:

$$\ell = \frac{1}{2}(L - B), \qquad L = 2\ell + B \quad .$$

Equation (5.7) can be rewritten:

$$\frac{B}{T} = \frac{2k}{1 - k}\frac{\ell}{T} + \frac{2K}{K'} \quad . \tag{5.8}$$

Giving a series of values to $\frac{\ell}{T}$ and the parameter $k$, we find $\frac{B}{T}$ from formula (5.8) and from formula (5.3), written as

$$\frac{Q}{\kappa T} = \frac{2K}{K'} , \tag{5.9}$$

we find the reduced flow rate $\overline{Q} = \frac{Q}{\kappa T}$.

V. V. Vedernikov computed a table for the dependence of $\overline{Q}$ on $\frac{\ell}{T}$ and $\frac{B}{T}$ for some values of these ratios.

Fig. 118 gives a flownet, desired by V. V. Vedernikov for a particular case.

The results thus obtained may with some approximation be applied to elucidate the following problem. Water, percolating from a canal, usually tends to the groundwater, at some depth $T$ below the canal It is of interest to find out, how the seepage from the canal changes for a decrease of $T$, i.e., for a rise of the groundwater level (Fig. 119). The sketch of Fig. 119 is related to the so called "seepage with bedrock concitions at infinity," when water percolates from a canal and spreads along an horizontal surface $AD$, as if the line $AD$ were the asymptote of the free surface. This may not be accomplished in steady flow, since in the treatment of such problems, the lines AB and CD tend to infinity (downward). The steady flow corresponding to the sketch of Fig. 119, occurs when the free surface coincides with an horizontal plane through segment BC.

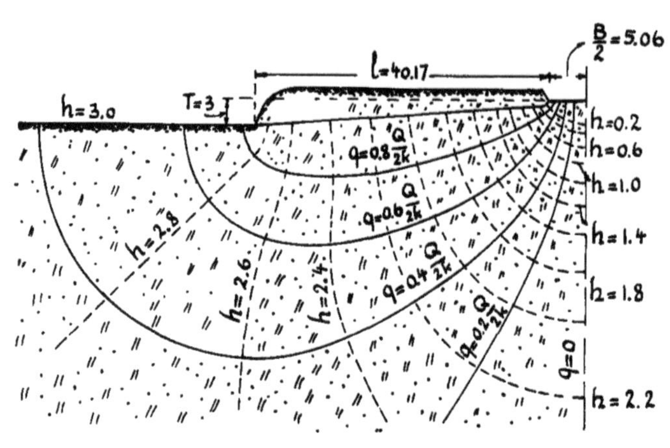


Fig. 118

In order to have the possibility of steady flow when the free surface drops to level AD, one must assume additional characteristics to drain the water; one of these possibilities is considered in the sketch.

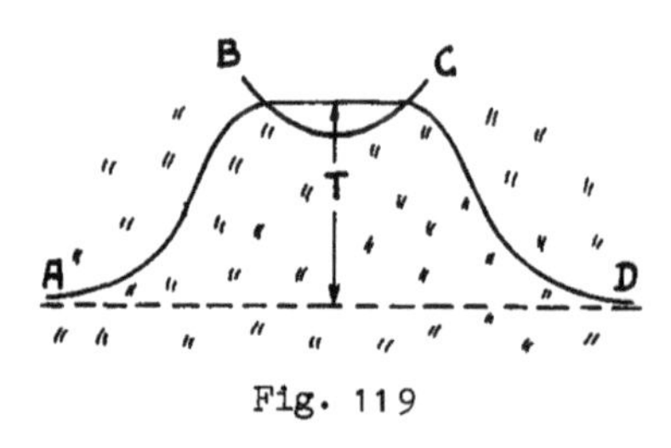


Fig. 119

One may notice that on the segment where $\frac{B}{T}$ changes little, the magnitude $\bar{Q} = \frac{Q}{\kappa T}$ decreases rather slowly with increasing $\frac{\ell}{T}$. For a fixed $\ell$, this means that T decreases, and that

$$\frac{Q}{\kappa} = T\bar{Q}$$

also decreases with decreasing T. Thus the seepage rate from the canal decreases, when the groundwater level approaches the bottom of the canal.

§6. Seepage From Canal With Collector On One Side.

This problem (Fig. 120) is solved in a way analogous to the previous. Here again $\theta = a\zeta$. For $\omega$ we obtain

$$\omega = C\int_0^\zeta \frac{d\zeta}{\sqrt{\zeta(e - \zeta)(1 - \zeta)}} .$$

The substitution $\zeta = et^2$, $e = k^2$ reduces the integral to its normal form; $\omega$ and $\theta$ are related through t:

$$\omega = c'\int_0^t \frac{dt}{\sqrt{(1 - t^2)(1 - k^2t^2)}} , \qquad \theta = a't^2 .$$

Here $a'$, $c'$ are constants, subject to determination.

Further computations, which we skip, give a relation analogous to (5.8):

(6.1) $$\frac{B}{T} = \frac{k^2}{k'^2}\,\frac{\ell}{T} + \frac{K}{K'} \quad ,$$

and for the discharge we obtain

(6.2) $$\frac{Q}{\kappa T} = \frac{B}{T} - \frac{k^2}{k'^2}\,\frac{\ell}{T} = \frac{K}{K'} \quad .$$

Fig. 121 shows a flownet, constructed by V. V. Vedernikov [4] for a particular case. The line $h = 0.6$ is taken for the contour of the canal.

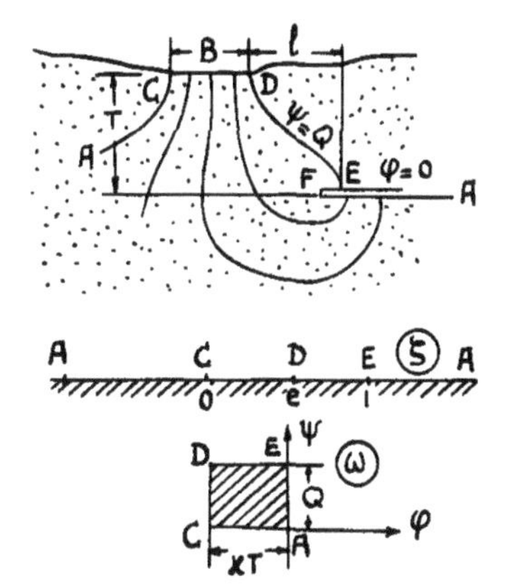


Fig. 120

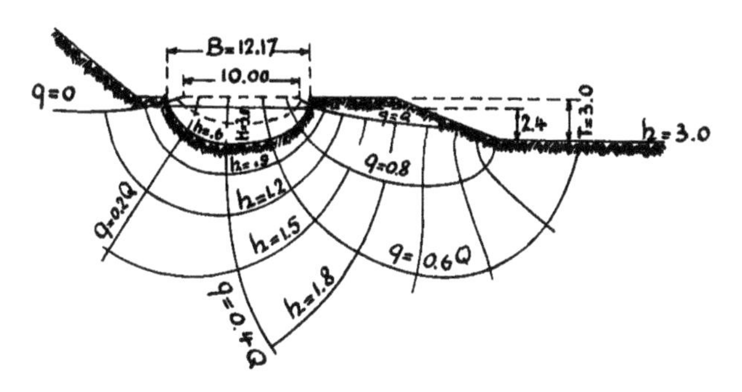


Fig. 121

B. Some Unconfined Flows Derived From Confined Flows.

## §7. Simplified Sketch of Earth Dam On Pervious Layer of Finite Depth.

The calculation of the seepage in the body of an earth dam is very difficult. We have devoted Chapters VI and VII to this problem. Here we consider two simplified schemes, which allow for the computation of an averaging characteristic of the flow in the body of an earth dam on a pervious layer of finite depth, namely the seepage rate under the dam. The first of these problems, treated in the following paragraph, was solved by A. P. Vostchin [8].

An earth dam with low head and tailwater is sketched in Fig. 122. The difference in water levels is H. The pervious layer is bounded at the base by a line the form of which is not known beforehand; it is a streamline that is determined by an artificially imposed condition: along it the expression $\varphi + \kappa y$ must be constant. This condition follows from the circumstances that at the free surface $\varphi + \kappa y$ = constant, and that along the boundaries of the reservoir, at which $y$ = constant, $\varphi$ = constant, also $\varphi + \kappa y$ = constant. Therefore, if we take

$$\varphi + \kappa y = 0 \quad ,$$

at the free surface, then the equation of the impervious boundary AD is

$$\varphi + \kappa y = -\kappa T \quad .$$

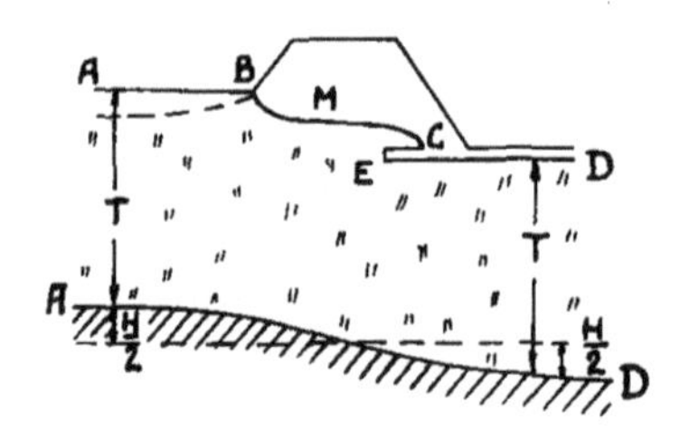


Fig. 122

We obtain a strip of width T (Fig. 123) in the plane of Zjoukovsky's function, which we choose in the form

$$\theta = z + \frac{i}{\kappa}\,\omega = \theta_1 + i\theta_2 \tag{7.1}$$

$$\left(\theta_1 = x - \frac{\psi}{\kappa}\,,\ \theta_2 = y + \frac{\varphi}{\kappa}\right) .$$

Comparing this problem with that of the flow about a flat bottom weir on a layer of depth T (Fig. 124), we see that the flowregion here coincides with the region of Zjoukovsky's function. The region of the complex potential is the same in both cases.

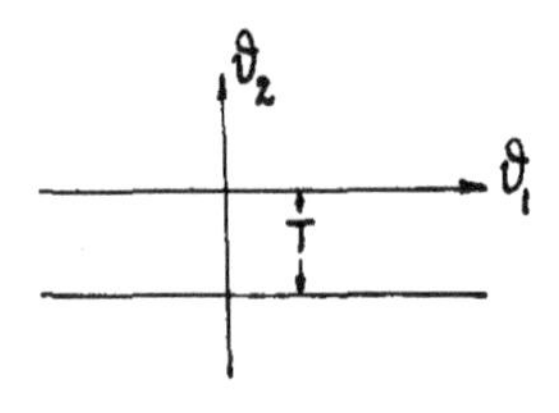


Fig. 123

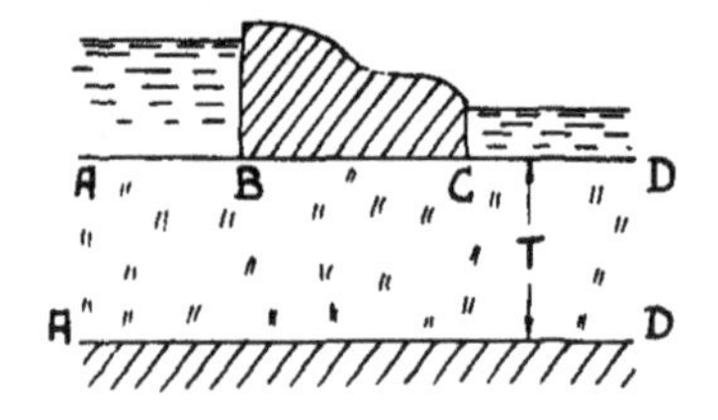


Fig. 124

The comparison leads to the following results. If in the second problem, about the confined flow, we found

$$z = f_1(\zeta), \qquad \omega = f_2(\zeta) \tag{7.2}$$

(where $\zeta$ is the auxiliary complex variable), then for the first problem the solution is immediately written as

$$\theta = z + \frac{i\omega}{\kappa} = f_1(\zeta), \quad \omega = f_2(\zeta) \quad . \tag{7.3}$$

The seepage rate is the same in both cases, for the same H, T and BC = $2\ell$, namely

$$Q = \kappa H\,\frac{K'}{2K} \tag{7.4}$$

where

$$k = \mathrm{th}\,\frac{\pi\ell}{2T} \quad .$$

The equation of the free surface is antisymmetrical about the middle point M (Fig. 122). The impervious bedrock left and right at infinity is at a distance T below head and tailwater reservoirs. The case $T = \infty$ had previously been treated in the book by F. B. Nelson-Skoryakov [10]. Other cases are also solved by means of Zjoukovsky's function.

§8. Simplified Scheme of an Earthdam With Cut-off.

Based on what has been said in §7., we may at once write the solution of the following problem.

Given the simplified scheme of a dam, analogous with the scheme of the previous case, but with a vertical cut-off (diaphragm) of length $S_1$ (Fig. 125) in its body. What is the seepage rate through the dam in this case? We use the formulas for the problem about a flat bottom weir with cut-off. (Fig. 126) (Chapter III, §12.)

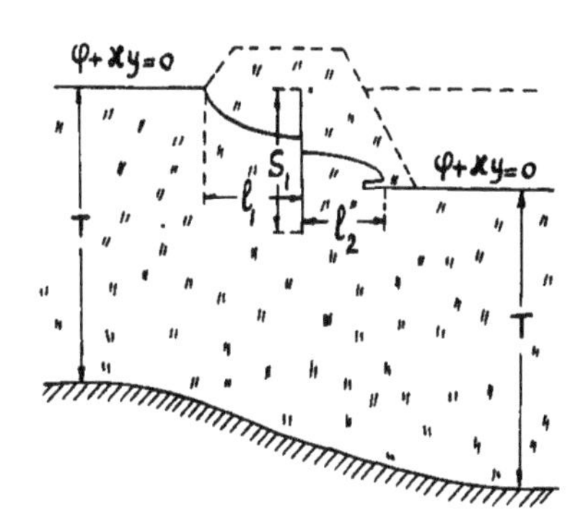


Fig. 125

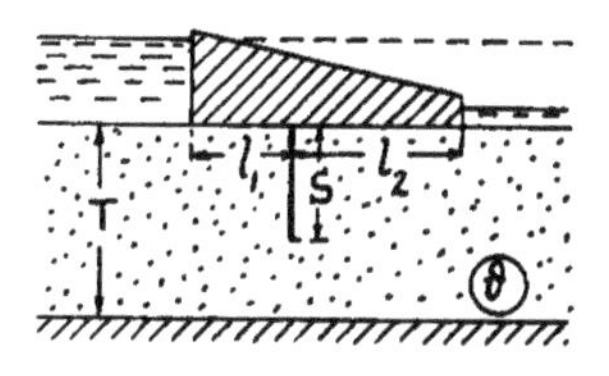


Fig. 126

We designate the unknown value of the velocity potential at the tip of the cut-off by $\varphi_0$. We have the corresponding value of $\theta_2$ [(7.1)] at the tip of the cut-off (the y-axis is oriented upwards along the cut-off)

(8.1) $$\theta_2 = -S_1 + \frac{\varphi_0}{\kappa} = -S \quad .$$

In the formulas of §12 of Chapter III, one must change $z$ into $\theta$ and consider $S$ as the length of the cut-off in the auxiliary $\theta$ plane. We obtain

(8.2) $$\omega = \kappa H - \frac{\kappa H}{K} F\left(\sqrt{\frac{(1 + a_2)(\zeta - 1)}{2(\zeta - a_2)}}, \lambda\right)$$

where

(8.3) $$\zeta = \sqrt{\operatorname{th}^2 \frac{\pi\theta}{2T} + \operatorname{tg}^2 \frac{\pi S}{2T}} \cos \frac{\pi S}{2T}$$

and the quantities $a_1$, $a_2$ and $\lambda$ are determined as

(8.4) $$a_1 = \sqrt{\operatorname{th}^2 \frac{\pi \ell_1}{2T} + \operatorname{tg}^2 \frac{\pi S}{2T}} \cos \frac{\pi S}{2T} \quad ,$$

(8.5) $$a_2 = \sqrt{\operatorname{th}^2 \frac{\pi \ell_2}{2T} + \operatorname{tg}^2 \frac{\pi S}{2T}} \cos \frac{\pi S}{2T} \quad ,$$

(8.6) $$\lambda = \sqrt{\frac{2(a_1 + a_2)}{(1 + a_1)(1 + a_2)}} \quad .$$

The length of the cut-off for the earth dam is determined from (8.1)

(8.7) $$S_1 = S - \frac{\varphi_0}{\kappa} \quad ,$$

and $\varphi_0$ is determined from (8.2), putting

$$\theta = -\,Si, \quad \zeta = 0 \quad .$$

We obtain:

$$\frac{\varphi_0}{\kappa} = H - \frac{H}{K}\,F\left(\sqrt{\frac{1 + a_2}{2a_2}}\,,\ \lambda\right) \quad . \tag{8.8}$$

Thus, in the actual computation of a dam, one has exact values for $\ell_1$ and $\ell_2$ and an approximate value of the length of the cut-off, i.e., the quantity S. The real length of the cut-off $S_1$ is found by formulas (8.7) and (8.8).

To find the equation of the free surface, we notice that on the latter we must satisfy the conditions

$$\psi = 0, \qquad \varphi + \kappa y = 0 \quad .$$

We may write

$$y = -\frac{\varphi}{\kappa}\,, \qquad \theta = x \quad ,$$

and (8.2) gives the equation of the free surface

$$y = \frac{H}{K}\,F\left(\sqrt{\frac{(1 + a_2)(\mp\,\zeta - 1)}{2(\mp\,\zeta - a_2)}}\ ,\ \lambda\right) \ , \tag{8.9}$$

where

$$\zeta = \sqrt{\operatorname{th}^2 \frac{\pi x}{2T} + \operatorname{tg}^2 \frac{\pi S}{2T}}\,\cos\frac{\pi S}{2T} \quad ,$$

in which the upper sign is related to the part of the free surface near the tailwater reservoir, the lower sign to the region near the headwater reservoir.

The intersection of the free surface with the cut-off is found by making $x = 0$ in equations (8.9). We obtain $\zeta = \sin\frac{\pi S}{T}$ and

$$y_1 = -\frac{H}{K}\,F\left(\sqrt{\frac{(1 + a_2)\left(1 + \sin\frac{\pi S}{T}\right)}{2\left(a_2 + \sin\frac{\pi S}{T}\right)}}\ ,\ \lambda\right) \ ,$$

$$y_2 = -\frac{H}{K}\,F\left(\sqrt{\frac{(1 + a_2)\left(1 - \sin\frac{\pi S}{T}\right)}{2\left(a_2 - \sin\frac{\pi S}{T}\right)}}\ ,\ \lambda\right) \ .$$

Here $y_1$ is the ordinate of the intersection of the free surface with the left side of the cut-off, $y_2$ that of the intersection with the right side. The head in the points of intersection is equal to the ordinate $y_1$ and $y_2$ respectively.

This problem was investigated by F. B. Nelson-Skornjakov [18]. A. M. Mkhitarian [9] carried out some computations.

### C. Semi-inverse Method in the Theory of Canal Seepage.

## §9. Seepage From Canals With Curvilinear Perimeter On Infinitely Deep Stratum.

We examine such problems as another application of Zjoukovsky's function. Can a relation between $\theta$ and $\omega = \varphi + i\psi$ of the form

$$\theta = + iz + \frac{\omega}{k} = Ae^{\frac{\omega}{\alpha}} \tag{9.1}$$

give a flow with free surface? Considering $A$ real, we separate $\theta$ into its real and imaginary parts:

$$\begin{cases} - y + \frac{\varphi}{k} = Ae^{\frac{\varphi}{\alpha}} \cos \frac{\psi}{\alpha} , \\ x + \frac{\psi}{k} = Ae^{\frac{\varphi}{\alpha}} \sin \frac{\psi}{\alpha} . \end{cases} \tag{9.2}$$

We notice that equation (9.2) does not change when $\psi$ is replaced by $-\psi$ and $x$ by $-x$; this means that the flow is symmetrical about the y-axis, which is a streamline. Since at the free surface

$$- y + \frac{\varphi}{k} = 0 ,$$

for the value $\psi = \psi_0$, at the free surface one must have

$$\cos \frac{\psi_0}{\alpha} = 0 .$$

For $\psi_0$ we obtain an infinite number of values

$$- q = \psi_0 = (2n + 1)\frac{\pi\alpha}{2} , \tag{9.3}$$

where $n$ an arbitrary integer. From this

$$\frac{1}{\alpha} = - \frac{(2n + 1)\pi}{2q}$$

and

$$\theta = Ae^{-\frac{(2n+1)\pi\omega}{2q}} . \tag{9.4}$$

We take $\psi_0 = \frac{\pi\alpha}{2}$ or, in other words, $n = 0$. This streamline may be the free surface.

Inserting $\varphi = ky$, $\psi = \frac{\pi\alpha}{2}$, in (9.2), we obtain the equation of the free surface

$$x + \frac{\pi\alpha}{2k} = Ae^{\frac{ky}{\alpha}} .$$

This curve has the asymptote

$$x = - \frac{\pi\alpha}{2k} = \frac{q}{k} = x_\infty .$$

For $y = 0$, we obtain the half width of the canal (Fig. 127)

$$x = \frac{B}{2} = \frac{q}{k} + A .$$

We find the line along which $y = 0$; it may be taken for the boundary of the canal. Along its perimeter the head must be constant. From (9.2) we obtain for $\varphi = 0$:

$$- y = A \cos \frac{\psi}{\alpha} , \quad x + \frac{\psi}{k} = A \sin \frac{\psi}{\alpha} . \tag{9.5}$$

When $\psi$ varies from 0 to $\psi_0 = \frac{\pi\alpha}{2}$, the equations give the canal perimeter in parametric form (Fig. 127). For $\psi = 0$, $y = -A = H$, where $H$ is the maximum canal depth. For $\psi = \frac{\pi\alpha}{2} = \psi_0$ we obtain $y = 0$, $x = -\frac{\psi_0}{k} - H = \frac{B}{2}$. From this we find the flowrate through the half cross-section:

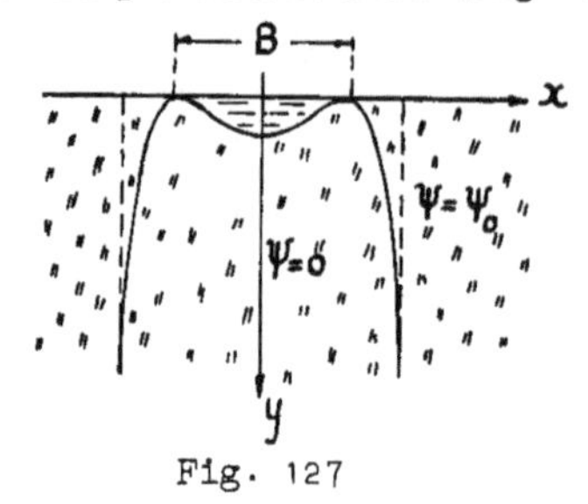


Fig. 127

$$q = -\psi_0 = k\left(H + \frac{B}{2}\right) .$$

The discharge $Q$ through the entire cross-section is:

$$Q = 2q = k(B + 2H) .$$

The width of the flow at infinity is

$$B_\infty = 2x_\infty = B + 2H .$$

From this it follows that the seepage velocity at infinity is equal to the seepage coefficient $k$. The equation of the canal perimeter may be obtained in explicit form, by elimination of $\psi$ from (9.5) and substituting $-H$ for A:

$$\pm x = -\sqrt{H^2 - y^2} + \frac{B + 2H}{\pi}\cos^{-1}\left(\frac{y}{H}\right) .$$

This is a shortened cycloid or trochoid.

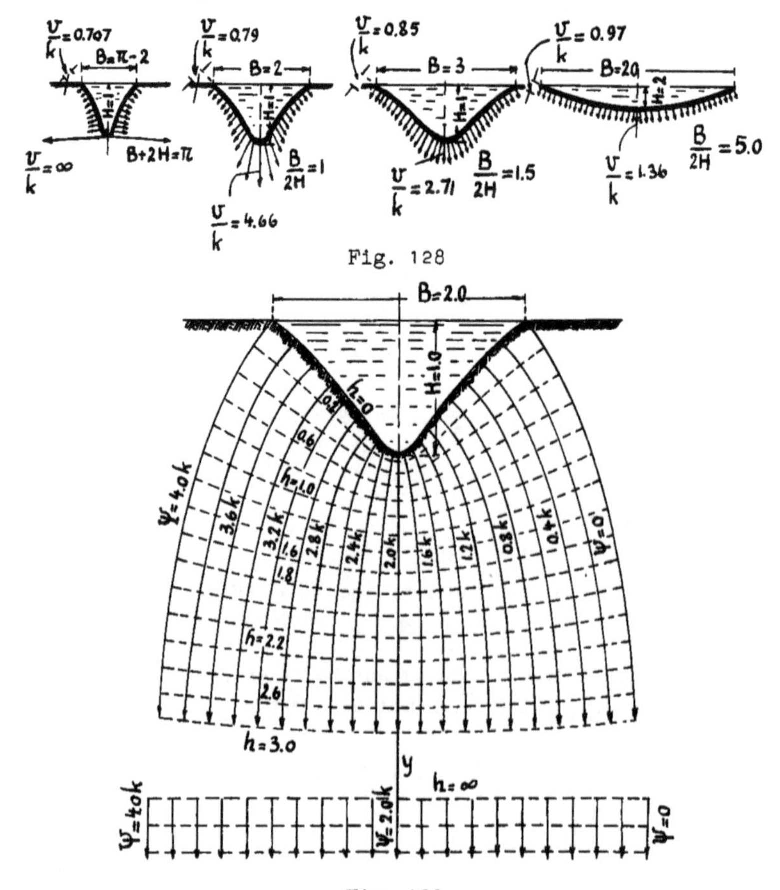


Fig. 128

Fig. 129

Some cross-sections of canals of this kind are sketched in Fig. 128. The flownet and family of isobars are given for a canal with B : H = 2 [4] in Figures 129 and 130.

The solution of this problem was first given by Kozeny, but later proved in another way and developed by N. N. Pavlovsky [2, 3], V. V. Vedernikov [4] and other authors.

We notice that the branches of the free surface tend very rapidly to their asymptote : at a depth $y = \frac{3}{2}(B + 2H)$, the abcissa differs from $x_\infty$ by less than 0.01H. We may consider that at this depth there is already a draining stratum to absorb the water. It is said that seepage takes place without "bedrock conditions at infinity."

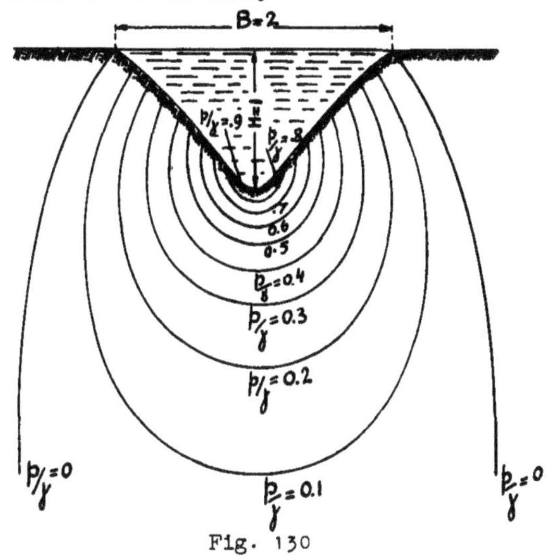


Fig. 130

§10. Seepage With Bedrock Conditions At Infinity.

Now we examine the particular case of equations (9.1) for n = -1. We have, substituting -C for A and putting $\alpha = \frac{Q}{\pi}$ :

$$- iz = \frac{\omega}{k} + Ce^{\frac{\pi\omega}{Q}} . \tag{10.1}$$

Separation of real and imaginary parts gives the equations

$$\left\{ \begin{aligned} - x &= Ce^{\frac{\pi\varphi}{Q}} \sin \frac{\pi\psi}{Q} + \frac{\psi}{k} , \\ y &= Ce^{\frac{\pi\varphi}{Q}} \cos \frac{\pi\psi}{Q} + \frac{\varphi}{k} . \end{aligned} \right. \tag{10.2}$$

It is easy to verify that here C = H, where H is the canal depth, and the equation of the contour of the canal, corresponding to the value $\varphi = 0$, will be:

$$- x = H \sin \frac{\pi\psi}{Q} + \frac{\psi}{k} ,$$

$$y = H \cos \frac{\pi\psi}{Q}$$

or

$$\pm x = \sqrt{H^2 - y^2} + \frac{Q}{k\pi} \cos^{-1}\left(\frac{y}{H}\right) . \tag{10.3}$$

This is also a trochoid, but now the contour of the canal is composed of the wider part of an elongated cycloid.

To obtain the seepage rate, we put $x = \frac{B}{2}$, $y = 0$:

$$\frac{B}{2} = H + \frac{Q}{2k} ,$$

i.e.,

$$Q = k(B - 2H) .$$

Now equation (10.3) for the canal coutour may be written as

$$\pm x = \sqrt{H^2 - y^2} + \frac{B - 2H}{\pi} \cos^{-1}\left(\frac{y}{H}\right) .$$

The equation of the right branch of the free surface is obtained for $\varphi = -\frac{Q}{2}$ in (10.2) and for $C = H$:

$$x = He^{\frac{\pi\varphi}{Q}} + \frac{Q}{2k} , \quad y = \frac{\varphi}{k} ,$$

or

$$x = He^{\frac{\pi ky}{Q}} + \frac{Q}{2k} .$$

Clearly, for $y \to +\infty$, we have $x \to \infty$. The free surface (Fig. 131) has a different character than in the previous case. It is said that seepage from the canal takes place under "bedrock conditions at infinity." In other words, we obtain a flow picture as if there were bedrock at infinity. The velocity at infinity is zero.

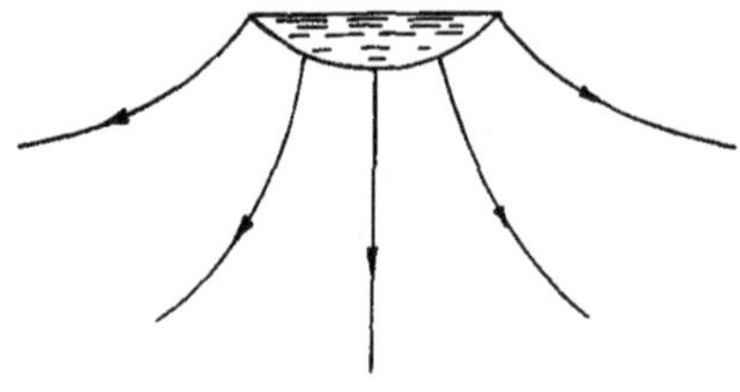

Fig. 131

B. K. Risenkampf [5] emphasized that both cases are obtained from the same flow, and that it is possible to obtain some more cases of seepage with free surface. Indeed, if we examine the flow determined by equations (9.1) and (10.1) in the whole z-plane, then we obtain a family of flowlines as sketched in Fig. 132 and 133. It is possible to obtain other forms of cross-section of the canal, if we assume instead of (10.1) a more general equation

$$z = \frac{i\omega}{k} + F(\omega)$$

and choose the periodic function $F(\omega)$ in an approximate way. N. M. Gersevanov [18] proposed $F(\omega)$ in the form of a series of functions with period $2iQ$:

$$F(\omega) = \sum_{m=-\infty}^{\infty} A_m e^{\frac{m\pi\omega}{Q}} ,$$

and the coefficients $A_m$ are to be determined so that the lower form of the canal is approached.

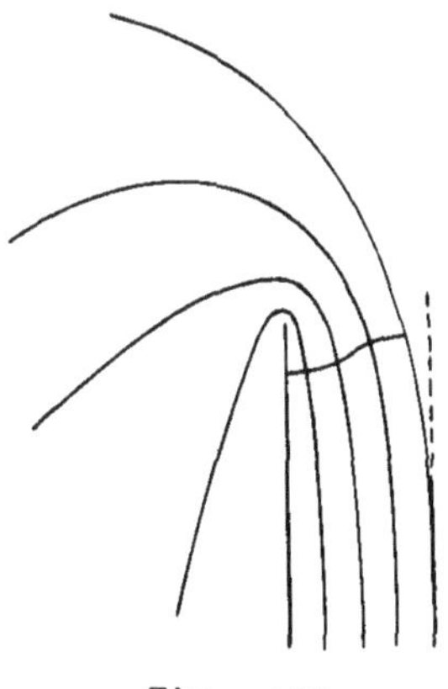

Fig. 132

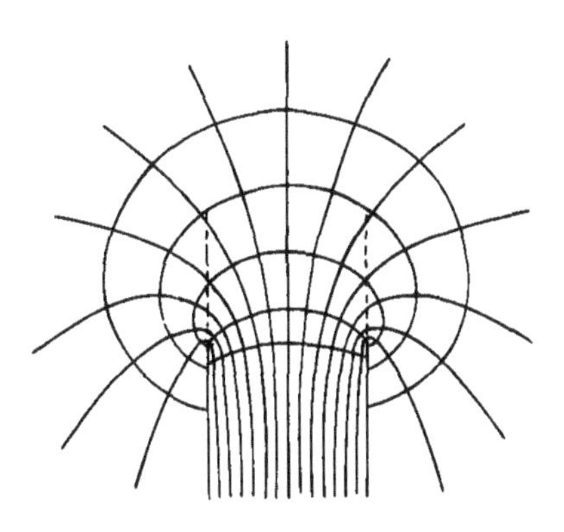

Fig. 133

§11. Flow To a Drainage Ditch With Curvilinear Profile.

M. I. Bazanov [11] solved the problem of the flow from infinity to a ditch with infinitesimally small water depth (strictly spoken, to a empty ditch). Here the entire contour of the ditch is a seepage line, and along the entire contour of the flowregion we have

$$\theta_1 = \text{constant}.$$

The form of the ditch is determined by the condition that in the $\frac{1}{w}$ -plane it is represented by an arc of circle radius S. The author found that the seepage rate Q of the entire ditch is proportional to the depth t of the ditch:

$$Q = k\pi t \quad .$$

A flownet, constructed by M. I. Bazanov for $t = 1$, $b = 2.43$ (b is the half width of the ditch) is represented in Fig. 134.

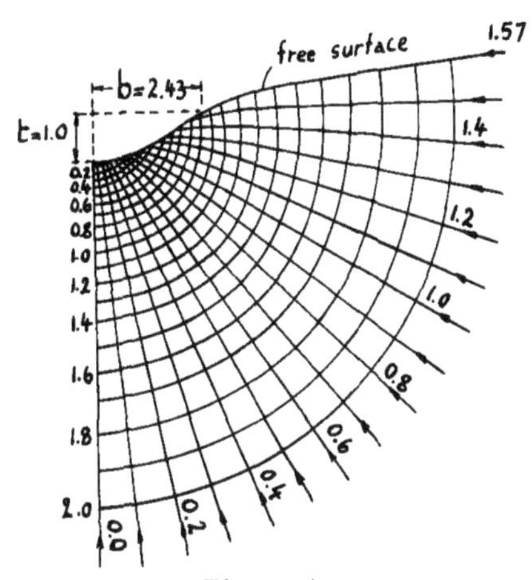


Fig. 134

§12. Simplified Scheme Of Dam With Drain On Soil Of Infinite Depth.

This scheme was examined by V. V. Vedernikov [12]. He proceeded from the condition that in the plane of Zjoukovsky's function, the region was bounded by lines of a form sketched on Fig. 135b. The dam itself is sketched on Fig. 135a. Its inflow surface is curvilinear, but differs little from a straight line.

If we take $\theta = z - \frac{i\omega}{k}$, $\omega = \varphi + i\psi$, then

$$\theta = \frac{H_1}{\pi\alpha} \int_0^\zeta \left( \frac{1 + \zeta}{\zeta} \right)^\alpha d\zeta ,$$

$$\omega = \frac{2kH_2}{\pi} \sin^{-1} \sqrt{\frac{\zeta}{\beta}} .$$

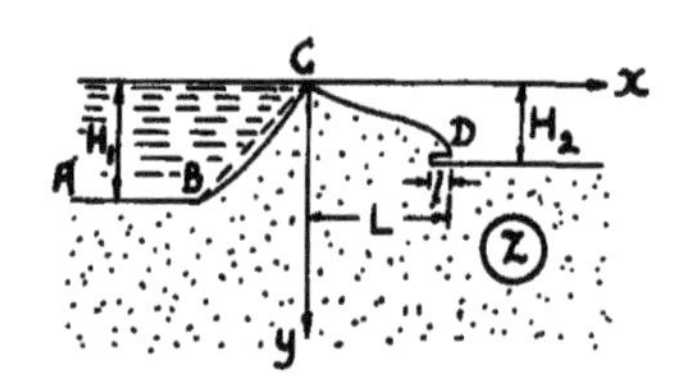


Fig. 135a

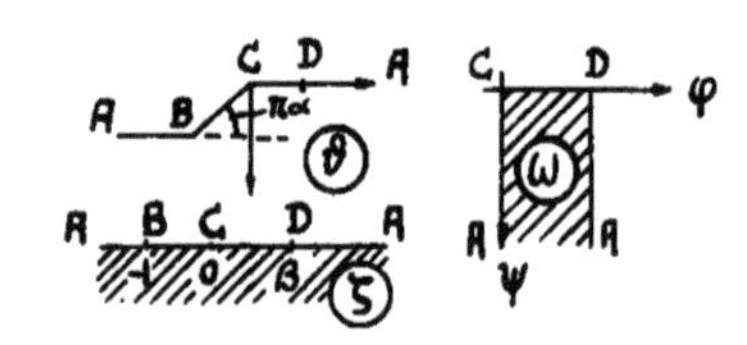


Fig. 135b

We may write the equation of the free surface, considering that there $\varphi - ky = 0$, $\psi = 0$, $\theta = x$, as

$$x = \frac{H_1}{\pi\alpha} \int_0^\zeta \left(\frac{1 + \zeta}{\zeta}\right)^\alpha d\zeta, \quad y = \frac{2H_2}{\pi} \sin^{-1} \sqrt{\frac{\zeta}{\beta}} \quad (0 \leq \zeta \leq \beta) .$$

We find the length $L$ (Fig. 135a) for $\zeta = \beta$

$$L = \frac{H_1}{\pi\alpha} \int_0^\beta \left(\frac{1 + \zeta}{\zeta}\right)^\alpha d\zeta .$$

We find for the inflow surface, where $\varphi = 0$, $-1 < \zeta < 0$, and $\zeta = -t$

$$x = -\frac{H_1 \cos \pi\alpha}{\pi\alpha} \int_0^t \left(\frac{1 - t}{t}\right)^\alpha dt - 2\frac{H_2}{\pi} \ln \frac{\sqrt{t} + \sqrt{\beta + t}}{\sqrt{\beta}} ,$$

$$y = \frac{H_1 \sin \pi\alpha}{\pi\alpha} \int_0^t \left(\frac{1 - t}{t}\right)^\alpha dt \quad (0 \leq t \leq 1) .$$

To find the length $l$ of the drainage segment, we must find the value $\zeta$ where the velocity is infinitely high (this will be for $\frac{d\theta}{d\omega} = -i/k$. For given $H_1$ and $H_2$ one determines $l$ and $L$ by asigning values to the parameter $\beta$.

V. V. Vedernikov also gave an approximate solution of the problem. Other solutions [10] are available for similar schemes.

## * §13. Approximate Solution When The Riverbed Is Clogged.

While water flows in a canal, small particles covered with silt are deposited at the bottom of the canal. When the voids between the sand grains are filled, a layer of compact soil with small seepage coefficient is formed.

If the difference between the values of the seepage coefficients for the clogged layer and the underlying soil is significant, then water may percolate from the edge of the clogged layer in isolated jets and hereby not fill all the soil pores. In this case, the condition of constant pressure holds along the lower boundary $4' - 3 - 4$ of the edge (Fig. 136). At the upper boundary $1' - 2 - 1$, the velocity potential $\varphi$ has a constant value. Along the segments $1 - 4$ and $1' - 4'$, which are streamlines, the pressure is also constant.

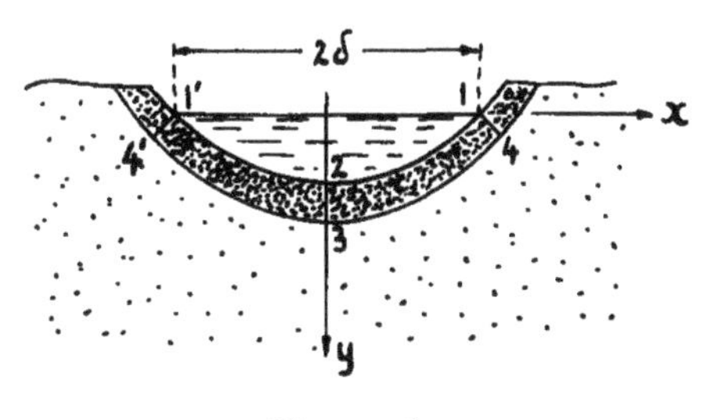


Fig. 136

For reasons of simplicity, we assume further that the seepage coefficient $k = 1$. If points 1 and 4 (and correspondingly $1'$ and $4'$) coincide, then complex $\omega$ and Zjoukovsky's function $\Omega = \omega + iz$, are mapped onto regions as sketched in Fig. 137 and 138, where $1 - 3 - 1'$ and $1 - 2 - 1'$ are curves respectively depending upon the cross-section of the riverbed and the contour of the clogged edge. V. I. Aravin [10] posed the

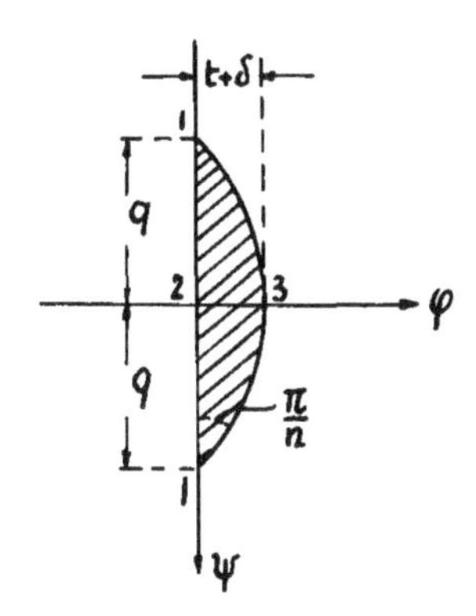


Fig. 137

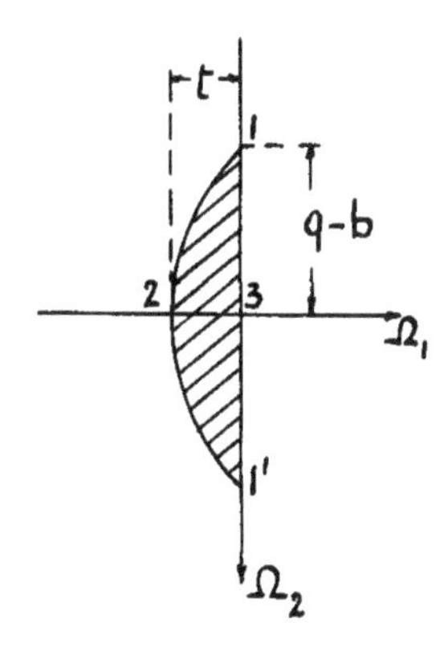


Fig. 138

condition that these contours in the $\omega$ and $\Omega$ planes be arcs of circles making the same angle $\frac{\pi}{n}$ at the vertices of the circular segments (n may be an arbitrary number). From this the form of the contour of the edge may be determined.

Since a circle in one plane transforms into a circle or straight line of another plane by means of a fractional linear mapping, let us consider the mapping

$$\frac{\Omega - A}{\Omega - B} = K\,\frac{\omega - C}{\omega - D} .$$

From the correspondence of the points $1, 3, 1'$ in the planes $\omega$ and $\Omega$, one determines the constants $A, B, C, D, K$ and obtains

$$\frac{\Omega - (q - b)i}{\Omega + (q - b)i} = - \frac{t + \delta + qi}{t + \delta - qi} \frac{\omega - qi}{\omega + qi} . \tag{13.1}$$

Due to the assumption of equal angles for the circular segments, equation (13.1) is also correct for the two other sides, and consequently, represents the conformal mapping of $\omega$ onto $\Omega$.

In (13.1) $q = \frac{Q}{2k}$ is the reduced half seepage rate ($Q$ is the total seepage rate per unit length of canal).

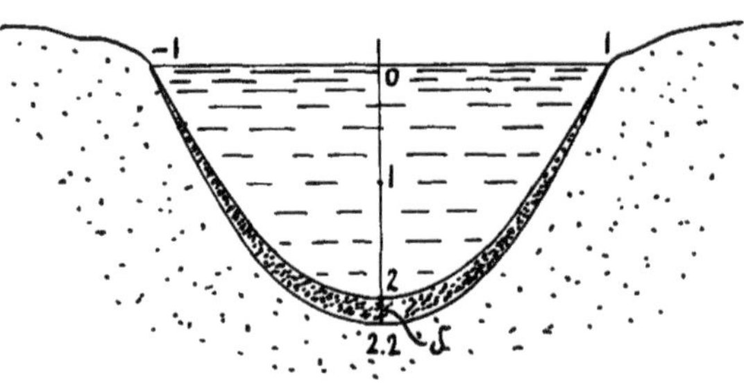


Fig. 139

Along 1 — 2 — 1', $\varphi = 0$, and from (13.1) the equation of the contour of the canal follows:

$$(13.2)\quad \begin{cases} x = - \psi + \dfrac{(q - b)(\psi + q)[\psi(t + \delta)^2 + q^3]}{\psi^2(t + \delta)^2 + q^4} - (q - b) , \\[2ex] y = - \dfrac{(q - b)(\psi + q)(t + \delta)(\psi - q)q}{\psi^2(t + \delta)^2 + q^4} \qquad (- q \le \psi \le q) . \end{cases}$$

Along the exterior surface, $\varphi = y$; therefore equation (13.1) gives (for $t_1 = t + \delta$)

$$x = - \psi + \frac{q - b}{(yt_1 + q^2)^2 + t_1^2\psi^2} \times$$

$$\times \{[(\psi + q)q + yt_1](yt_1 + q^2) + [t_1(\psi + q) - yq]t_1\psi\} -(q - b) ,$$

$$y = - \frac{q^2 - t_1^2}{2t_1} + \frac{1}{2t_1} \sqrt{(q^2 - t_1^2) + 4q^2t_1^2(q^2 - \psi^2)} \qquad (-q \le \psi \le q).$$

From the condition of equal angles in $\omega$ and $\Omega$ planes:

$$\frac{q - b}{t} = \frac{q}{t + \delta}$$

Returning to the arbitrary value of $k$ we find:

$$q = \frac{b(t + \delta)}{\delta}, \qquad Q = 2kq = 2k \frac{b(t + \delta)}{\delta} . \tag{13.3}$$

Curves for $b = 1$, $t = 2$, $\delta = 0.2$ are constructed in Fig. 139.

## D. Application of Functional Analysis.

### §14. Essence of the Method.

The solution of a series of groundwater flow problems by means of functional analysis was obtained by B. K. Risenkampf [5] and N. M. Gersevanov [13].

The essence of the method is this. Let $z = x + iy$ be a function of $\omega = \varphi + i\psi$:

$$z = x + iy = F(\omega) \quad .$$

Then

$$\bar{z} = x - iy = \bar{F}(\bar{\omega}), \quad \bar{\omega} = \varphi - i\psi \quad ,$$

and

$$x = \frac{1}{2}\,[F(\omega) + \bar{F}(\bar{\omega})] \quad , \tag{14.1}$$

$$y = \frac{1}{2}\,[F(\omega) - \bar{F}(\bar{\omega})] \quad . \tag{14.2}$$

Given boundary conditions allow in some cases to find $F(\omega)$. B. K. Risenkampf [5] assumed conditions along two structures and obtained the solution of a free surface flow problem for inclined bedrock, and also of a problem of seepage from a canal, that we already explained in paragraph 10, using other methods.

N. M. Gersevanov [13] analyzed a series of problems where the region of the complex potential is a rectangle.

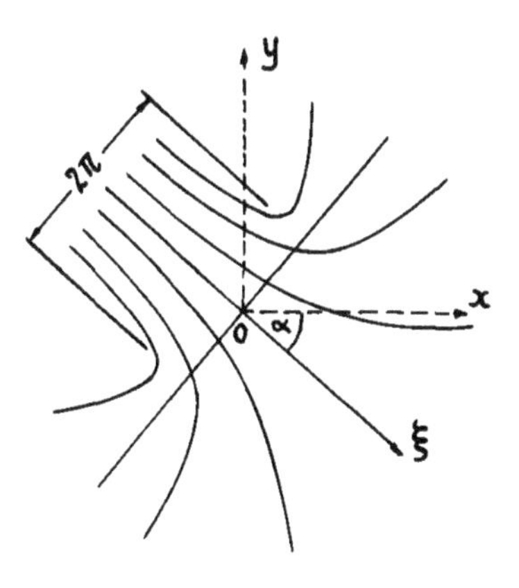


Fig. 140

§15. Groundwater Flow For Inclined Bedrock.

Groundwater flows with a free surface, and has for streamline a straight line, making an angle $\alpha$ with the horizontal x-axis (fig. 140). The equation of the straight line is

$$x \sin \alpha + y \cos \alpha = 0 \; . \tag{15.1}$$

Let $\psi = 0$ along this line. At the free surface we assume $\psi = Q$. In equation (15.1) we replace $x$ and $y$ by their values from (14.1) and (14.2), considering also that $\psi = 0$ along (15.1). Then

$$\sin \alpha[F(\varphi) + \bar{F}(\varphi)] + \frac{\cos \alpha}{i}[F(\varphi) - \bar{F}(\varphi)] = 0 \tag{15.2}$$

and multiplying by $e^{i\alpha}$:

$$e^{2i\alpha}F(\varphi) - \bar{F}(\varphi) = 0 \quad . \tag{15.3}$$

In this equation we replace $\varphi$ by $\omega$ and obtain

$$e^{2i\alpha}F(\omega) = \bar{F}(\omega) \quad .$$

Now we examine the condition at the free surface. Putting $\psi = Q$, we obtain

$$\omega = \varphi + iQ \quad .$$

Moreover, on the free surface

$$\varphi + ky = 0 \quad ,$$

i.e.,

(15.4)
$$y = -\frac{\varphi}{k} .$$

Inserting expression (14.2) and $\psi = Q$ in equation (15.4), we obtain:

(15.5)
$$-\frac{\varphi}{k} = \frac{1}{2i}[F(\varphi + iQ) - \overline{F}(\varphi - iQ)] .$$

We put $\varphi - iQ = \omega$. Then $\varphi + iQ = \omega + 2iQ$, and (15.5) may be written as

$$F(\omega) + 2iQ) - \overline{F}(\omega) = -\frac{2i}{k}(\omega + iQ) .$$

Because of (15.3)

(15.6)
$$F(\omega) + 2iQ) - e^{2i\alpha}F(\omega) = -\frac{2i\omega}{k} + \frac{2Q}{k} .$$

This is a linear equation in finite differences, the solution of which we compose of two parts. We first find a particular solution of non-homogeneous equation (15.6) as a linear function in $\omega$

$$F_*(\omega) = A\omega + B .$$

Substitution in equation (15.6) gives

$$A(\omega + 2iQ) + B - (A\omega + B)e^{2i\alpha} = -\frac{2i\omega}{k} + \frac{2Q}{k} .$$

From which we find

$$A = \frac{e^{-i\alpha}}{k \sin \alpha} , \qquad B = \frac{Qe^{-i\alpha} \operatorname{ctg} \alpha}{k \sin \alpha}$$

and

$$F_*(\omega) = \frac{e^{-i\alpha}}{k \sin \alpha} (\omega + Q \operatorname{ctg} \alpha) .$$

Now we find the general solution of the homogeneous equation

(15.6')
$$F(\omega + 2iQ) - e^{2i\alpha}F(\omega) = 0 .$$

Therefore we replace the unknown function $F(\omega)$ by a new function $f(\omega)$ by the substitution

(15.7)
$$F(\omega) = e^{M\omega}f(\omega) .$$

Substitution of (15.7) in (15.6') leads to

$$e^{M(\omega+2iQ)}f(\omega + 2iQ) - e^{M\omega+2i\alpha}f(\omega) = 0 .$$

We choose the constant $M$ so that for $f(\omega)$ the following equation holds,

$$f(\omega + 2iQ) - f(\omega) = 0 .$$

Hence for $f(\omega)$ one may take any periodic function with period $2iQ$. $M$ has the following value

$$M = \frac{\alpha + n\pi}{Q} ,$$

where $n$ is an arbitrary integer. Thus the general solution of (15.6) is:

(15.8)
$$z = F(\omega) = e^{\frac{\alpha+n\pi}{Q}\omega} f(\omega) + \frac{e^{-i\alpha}}{k \sin \alpha} (\omega + Q \operatorname{ctg} \alpha) ,$$

where $f(\omega)$ is an arbitrary periodic function with period $2iQ$. We put $n = 0$ and

$$f(\omega) = Ce^{-i\alpha} ,$$

where $C$ is a constant, and rewrite (15.8) as

$$ze^{i\alpha} = Ce^{\frac{\alpha\omega}{Q}} + \frac{\omega + Q \operatorname{ctg} \alpha}{k \sin \alpha} . \tag{15.9}$$

Considering $C$ real, we separate (15.9) into its real and imaginary parts. With

$$\zeta = \xi + i\eta = ze^{i\alpha} ,$$

we obtain

$$\begin{cases} \xi = x \cos \alpha - y \sin \alpha = Ce^{\frac{\alpha\varphi}{Q}} \cos \frac{\alpha\psi}{Q} + \frac{\varphi + Q \operatorname{ctg} \alpha}{k \sin \alpha} , \\ \eta = y \cos \alpha + x \sin \alpha = Ce^{\frac{\alpha\varphi}{Q}} \sin \frac{\alpha\psi}{Q} + \frac{\psi}{k \sin \alpha} . \end{cases} \tag{15.10}$$

We examine these equations. We may consider $\xi$, $\eta$ as new coordinates (Fig. 140).

For $C = 0$, we have uniform flow

$$\xi = \frac{\varphi}{k \sin \alpha} + \frac{Q \cos \alpha}{k \sin^2 \alpha} ,$$

$$\eta = \frac{\psi}{k \sin \alpha}$$

and a constant velocity

$$\frac{d\psi}{d\eta} = k \sin \alpha$$

The free surface, for $\psi = Q$, is a straight line

$$\eta = \frac{Q}{k \sin \alpha} .$$

We put

$$\eta_0 = \frac{Q}{k \sin \alpha}$$

and call $\eta_0$ the normal depth.

Returning to the general case, we obtain for $\psi = 0$

$$\eta = y \cos \alpha + x \sin \alpha = 0 ,$$

$$\xi = Ce^{\frac{\alpha\varphi}{Q}} + \frac{\varphi + Q \operatorname{ctg} \alpha}{k \sin \alpha} .$$

We put $\psi = Q$. We have

$$\xi = Ce^{\frac{\alpha\varphi}{Q}} \sin \alpha + \frac{\varphi + Q \operatorname{ctg} \alpha}{k \sin \alpha} ,$$

$$\eta = Ce^{\frac{\alpha\varphi}{Q}} \sin \alpha + \frac{Q}{k \sin \alpha} .$$

We find $\varphi$ from the second equation and substitute it in the first. We obtain the equation of the free surface

$$\xi = \eta \operatorname{ctg} \alpha + \frac{Q}{k\alpha \sin \alpha} \ln \frac{\eta - \frac{Q}{k \sin \alpha}}{C \sin \alpha} . \tag{15.11}$$

The free surface has the asymptote

$$\eta = \frac{Q}{k \sin \alpha} = \eta_0 .$$

Knowing the normal depth $\eta_0$ (i.e., the depth of flow at infinity), we find the discharge

$$Q = k\eta_0 \sin \alpha \quad .$$

If we change the value of the constant $C$, then curve (15.11) is only displaced parallel to itself. Therefore we put

$$C = \pm \frac{\eta_0}{\sin \alpha} \quad . \tag{15.12}$$

By means of (15.12) we rewrite (15.11) as:

$$\xi = \frac{\eta}{\operatorname{tg} \alpha} + \frac{\eta_0}{\alpha} \ln \left| \frac{\eta}{\eta_0} - 1 \right| \ , \tag{15.13}$$

and (15.9) is replaced by

$$ze^{i\alpha} = \pm \frac{\eta_0}{\sin \alpha} e^{\frac{\alpha\omega}{Q}} + \frac{\omega + Q \operatorname{ctg} \alpha}{k \sin \alpha} \quad .$$

The family of streamlines (15.13) is sketched in Fig. 140. If the angle $\alpha$ is small, then

$$\operatorname{tg} \alpha \approx \alpha \approx i \ ,$$

where $i$ is the slope of the streamline $\psi = 0$, and equation (15.13) may be changed to

$$\frac{i\xi}{\eta_0} = \frac{\eta}{\eta_0} + \ln \left| \frac{\eta}{\eta_0} - 1 \right| \quad .$$

This equation of the free surface can be obtained by the hydraulic theory (see Chapter X).

N. M. Gersevanov [13] examined the seepage problem under a flat bottom weir with cut-off on a layer of finite depth, and also the problem of a rectangular cofferdam, when one of its slopes has a particular feature. (See reference [16] for this.) He derives a solution based on the obtained functional relations in the form of a trigonometric series and, in particular, obtains formula (6.7) of Chapter III.

CHAPTER V.

# APPLICATION OF THE METHOD OF INVERSION

## §1. Some Inversion Properties.

Mapping with the inverse radius-vector or inversion in the circle $K$ with radius $R$ and center in point $C$, is the mapping in which point $M$, of the ray $CM$, transforms into point $M'$ of the same ray (Fig. 141) in such a way that

$$CM \cdot CM' = R^2 .$$

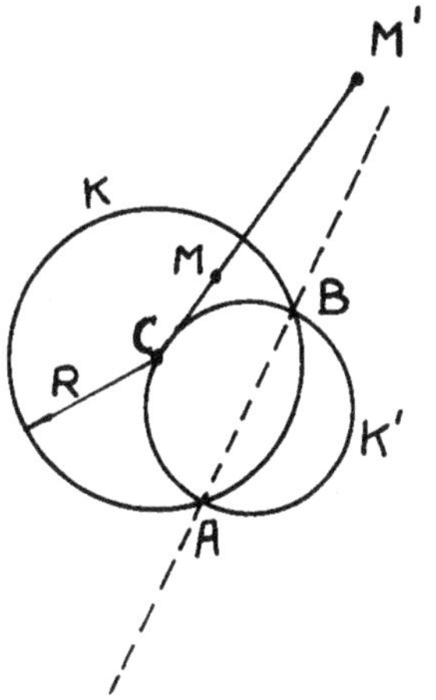

The coordinates of point $M'(x', y')$ are related to the coordinates of point $M(x, y)$ by the equations

$$x' = \frac{R^2 x}{x^2 + y^2}, \quad y' = \frac{R^2 y}{x^2 + y^2} .$$

From this

$$x' + iy' = \frac{R^2(x + iy)}{x^2 + y^2} = \frac{R^2}{x - iy},$$

which may be written by introduction of complex numbers,

$$z = x + iy, \quad z' = x' + iy', \quad \bar{z} = x - iy,$$

as

$$z' = \frac{R^2}{\bar{z}} . \tag{1.1}$$



Fig. 141

The mapping

$$z' = \frac{R^2}{z} \tag{1.2}$$

represents a combination of the inversion with a reflection into the real axis of the z-plane. In this case we have:

$$x' = \frac{R^2 x}{x^2 + y^2}, \quad y' = \frac{R^2 y}{x^2 + y^2} .$$

In an inversion, points on the circle $K$ (Fig. 141) map into themselves. The center of the circle maps into the point at infinity, the point at infinity maps into the center of the circle. A circle or a straight line are mapped into a circle or straight line by inversion.

We observe that we usually construct the region of function $u + iv$

and look for an expression for $1/w = 1/(u - iv)$, i.e., use a mapping of form (1.1), which is a pure inversion of the (u,v) region without reflection in the abscissa axis (see, for example, Fig. 142).

We return to Fig. 141.

Circle K', passing through center C of circle K, transforms into a straight line. Indeed, point C transforms into the point at infinity, points A and B (Fig. 141) remain in their position. Consequently, circle K' transforms into a straight line, passing through points A and B.

From this we have the following consequence.

Assume a region, bounded by arcs of circles (straight lines, which we may consider as circles of infinitely large radius, may also be among the boundaries of the region), so that all these arcs or their continuations intersect in one point. Then, taking this point for center of inversion C and choosing an arbitrary radius R, we apply the inversion transformation to the given region. Because of what was stated above, it is clear that all boundaries of that region transform into rectilinear segments, and, consequently, as a result of the inversion we obtain a rectilinear polygon.

This property of the inversion was used by N. N. Pavlovsky, V. V. Vedernikov, and still other authors in the solution of various seepage problems, when lines passing through one point are found in the velocity hodograph. The essence of the method is explained in examples.

We examine in this chapter problems about seepage from canals of trapezoidal cross-section and their particular cases, rectangular and triangular cross-section. The purpose of the investigation is to elucidate the question of the dependence of the flowrate and form of the free surface on the form of the canal.

Further, the question about the influence of soil capillarity on seepage from canals is important. We examine problems about seepage from a canal with trapezoidal cross-section and from a wide canal.

In this chapter we examine the problem about the flow of groundwater to a drainage ditch of trapezoidal cross-section and with little water in it. This problem is supplemental to that about the flow to a drainage ditch with curvilinear cross-section, communicated at the end of Chapter IV.

Some problems about drainage canals or pipes may be considered simultaneously with problems about irrigation canals and pipes, because the methods for their solution are the same. Such are the problems of §8 and 9. In the latter the influence of evaporation or infiltration on a drainage or irrigation pipe is examined.

The properties of inversion may be used in the solution of some problems by a semi-inverse method: such is the problem about seepage in an earthdam with curvilinear boundaries of upper and lower reservoirs, as explained in §10.

Finally, by means of the method of inversion we examine the problem about smooth contours of foundations of hydraulic structures.

§2. Seepage From a Canal With Trapezoidal Bed.

In Chapter IV., we examined a series of problems about seepage from canals with curvilinear perimeter, in which the form of the canal was not given beforehand, but was obtained as a result of the solution of the problem. This is a semi-inverse method in which part of the conditions are given. Here we examine simple methods, when for a given canal perimeter the seepage net is asked for: from the canal, in case of irrigation, towards the canal, in case of drainage.

Solutions are available for problems about seepage from a canal with trapezoidal cross-section and in particular for rectangular and triangular cross-sections. Then solutions were given by V. V. Vedernikov [3] by means of an application of the inverse transformation to the region of the complex velocity. Recently B. K. Risenkampf [6] represented this solution in series form.

Fig. 142a represents a canal with trapezoidal cross-section, from which water flows downward to infinity, without bedrock conditions (about seepage without these conditions, see Chapter IV, §9). The velocity hodograph is represented in Fig. 142b.

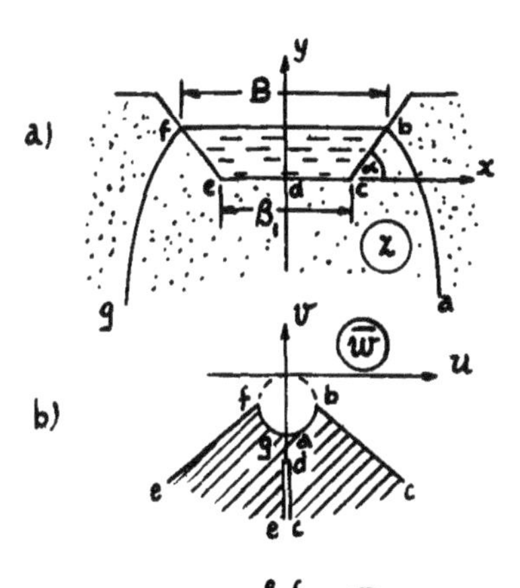


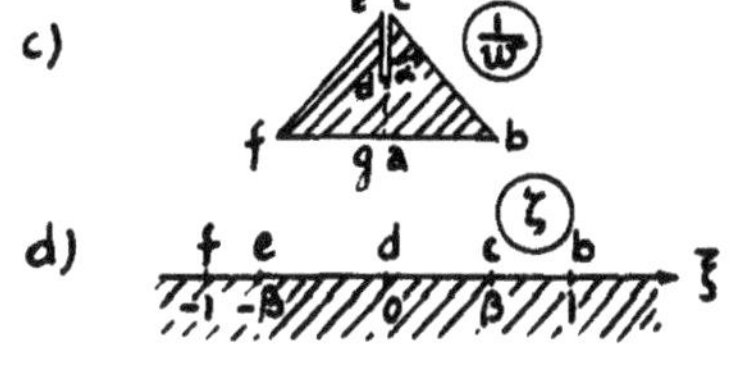


Fig. 142.

Taking the origin of coordinates of the u,v-plane for center of inversion and putting the radius of inversion equal to one, we obtain for $1/w = 1/(u - iv)$ the region represented in Fig. 142c.

The boundaries of the region are rectilinear segments. We call the angle between slope CB and the horizontal axis $\alpha = \pi\sigma$. Then one must have the number

$$\sigma = \frac{\alpha}{\pi} \tag{2.1}$$

in the exponents of the formula which gives a conformal mapping of the polygon of Fig. 142b onto the lower half plane ζ (Fig. 142d).

By means of the Christoffel-Schwartz formula we obtain

$$\frac{1}{w} = M\int_0^\zeta \frac{\zeta\, d\zeta}{(1 - \zeta^2)^{1/2+\sigma}(\beta^2 - \zeta^2)^{1-\sigma}} + N . \tag{2.2}$$

Introducing the notation

$$\int_0^\zeta \frac{\zeta\, d\zeta}{(1 - \zeta^2)^{1/2+\sigma}(\beta^2 - \zeta^2)^{1-\sigma}} = \Phi(\zeta) \quad , \tag{2.3}$$

we may write

$$\frac{1}{w} = M\Phi(\zeta) + N . \tag{2.4}$$

We put

$$\left\{ \begin{aligned} &\int_0^\beta \frac{\zeta \, d\zeta}{(1 - \zeta^2)^{1/2+\sigma}(\beta^2 - \zeta^2)^{1-\sigma}} = J_1 = \Phi(\beta) \\ &\int_\beta^1 \frac{\zeta \, d\zeta}{(1 - \zeta^2)^{1/2+\sigma}(\zeta^2 - \beta^2)^{1-\sigma}} = J_2 . \end{aligned} \right. \tag{2.5}$$

By means of the substitution

$$\frac{\zeta^2 - \beta^2}{1 - \beta^2} = x$$

the second integral is reduced to

$$J_2 = \frac{1}{2\beta'} \int_0^1 x^{\sigma-1}(1 - x)^{-1/2-\sigma} \, dx , \tag{2.6}$$

where

$$\beta' = \sqrt{1 - \beta^2} .$$

Introducing the beta function

$$B(p, q) = \int_0^1 x^{p-1}(1 - x)^{q-1} \, dx \tag{2.7}$$

and recalling the relationship between beta and gamma functions

$$B(p, q) = \frac{\Gamma(p)\Gamma(q)}{\Gamma(p + q)} , \tag{2.8}$$

where

$$\Gamma(p) = \int_0^\infty e^{-x}x^{p-1} \, dx , \tag{2.9}$$

we find

$$J_2 = \frac{1}{2\beta'} B\left(\sigma, \frac{1}{2} - \sigma\right) = \frac{1}{2\beta'} \frac{\Gamma(\sigma)\Gamma\left(\frac{1}{2} - \sigma\right)}{\Gamma\left(\frac{1}{2}\right)} . \tag{2.10}$$

We recall properties of the $\Gamma$-function

$$\Gamma(p + 1) = p\Gamma(p), \quad \Gamma(p)\Gamma(1 - p) = \frac{\pi}{\sin p\pi} , \quad \Gamma\left(\frac{1}{2}\right) = \sqrt{\pi} . \tag{2.11}$$

Using the last of these properties, we write $J_2$ in the form

$$J_2 = \frac{\Gamma(\sigma)\Gamma\left(\frac{1}{2} - \sigma\right)}{2\beta' \sqrt{\pi}} . \tag{2.12}$$

Returning to equation (2.4), we put in it $\zeta = \beta$. We obtain point c, where the velocity is infinite. Consequently (2.4) must give

$$M\Phi(\beta) + N = 0$$

or based on equation (2.5)

$$MJ_1 + N = 0 , \tag{2.13}$$

and (2.4) may be rewritten as

$$\frac{1}{w} = M[\Phi(\zeta) - J_1] . \tag{2.14}$$

We return now to segment cb, determining the slope of the canal banks. Walking around point C where $\zeta = \beta$ in the lower half-plane $\zeta$ over an angle $\pi$ in positive direction; we obtain that

$$(\beta^2 - \zeta^2)^{1-\sigma}$$

transforms into

$$e^{\pi i(1-\sigma)}(\zeta^2 - \beta^2)^{1-\sigma} = - e^{-\pi i\sigma}(\zeta^2 - \beta^2)^{1-\sigma} .$$

We obtain

$$\Phi(\zeta) = J_1 - e^{\pi i\sigma} \int_\beta^\zeta \frac{\zeta \, d\zeta}{(1 - \zeta^2)^{1/2+\sigma}(\zeta^2 - \beta^2)^{1-\sigma}} .$$

Putting $\zeta = 1$ here, we find

$$\Phi(1) = J_1 - e^{\pi i\sigma} J_2 . \tag{2.15}$$

Considering Fig. 142b, we obtain the value of the velocity in point b:

$$u = k \sin \pi\sigma \cos \pi\sigma, \quad v = - k \cos^2 \pi\sigma ,$$
$$w = u - iv = kie^{-\pi i\sigma} \cos \pi\sigma .$$

Therefore formula (2.4) renders for $\zeta = 1$

$$\frac{1}{ki} e^{\pi i\sigma} = M[\Phi(1) - J_1] \cos \pi\sigma ,$$

and this may be rewritten by means of (2.15) as

$$\frac{1}{ki} e^{\pi\sigma i} = - Me^{\pi\sigma i} J_2 \cos \pi\sigma ,$$

from which we find

$$M = \frac{i}{kJ_2 \cos \pi\sigma} . \tag{2.16}$$

Recalling that

$$\omega = \varphi + i\psi ,$$
$$\frac{d\omega}{dz} = w = u - iv$$

and using (2.16), we rewrite (2.14) in the following form

$$\frac{dz}{d\omega} = \frac{1}{w} = \frac{i}{kJ_2 \cos \pi\sigma} [\Phi(\zeta) - J_1] . \tag{2.17}$$

Now we determine the complex potential. We assume that $\varphi = 0$ along the canal perimeter, being an equipotential line. We call $Q$ the total flowrate of the canal per unit length, and obtain the region of the complex potential, sketched in Fig. 143. Its mapping onto the lower half-plane $\zeta$ (Fig. 142d) gives:

$$\omega = \frac{Qi}{\pi} \text{ arc sin } \zeta \quad . \tag{2.18}$$

By means of formulas (2.17) and (2.18), we find the expression

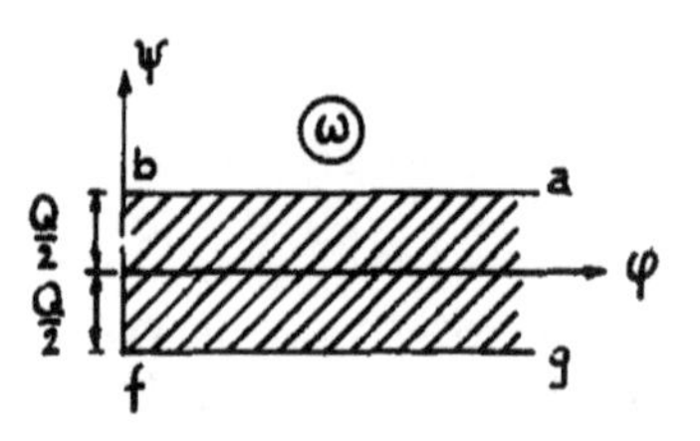


Fig. 143.

$$\frac{dz}{d\zeta} = \frac{dz}{d\omega}\frac{d\omega}{d\zeta} =$$

$$= \frac{Q}{k\pi J_2 \cos \pi\sigma} \frac{1}{\sqrt{1-\zeta^2}}[\Phi(\zeta) - J_1] \ .$$

Integration renders $z$ as function of $\zeta$

$$z = -\frac{Q}{k\pi J_2 \cos \pi\sigma}\left\{\int_0^{\zeta}\frac{\Phi(\zeta)\, d\zeta}{\sqrt{1-\zeta^2}} - J_1 \text{ arc sin } \zeta\right\} \tag{2.19}$$

The coordinate axes $(x, y)$ are chosen in Fig. 142 so that $z = 0$ for $\zeta = 0$ .

We transform the integral occurring in formula (2.19) by means of integration by parts and taking into account (2.3):

$$\int_0^{\zeta}\frac{\Phi(\zeta)\, d\zeta}{\sqrt{1-\zeta^2}} = \Phi(\zeta) \text{ arc sin } \zeta - \int_0^{\zeta}\frac{\zeta \text{ arc sin } \zeta\, d\zeta}{(1-\zeta^2)^{1/2+\sigma}(\beta^2-\zeta^2)^{1-\sigma}} \ . \tag{2.20}$$

Now we rewrite equation (2.19) for different segments of the flow region contour.

Along the canal bottom, for $0 < \zeta < \beta$

$$z = -\frac{Q}{k\pi J_2 \cos \pi\sigma}\left\{\text{arc sin } \zeta \left[\int_0^{\zeta}\frac{\zeta\, d\zeta}{(1-\zeta^2)^{1/2+\sigma}(\beta^2-\zeta^2)^{1-\sigma}} - J_1\right] - \right.$$

$$\left. - \int_0^{\zeta}\frac{\zeta \text{ arc sin } \zeta\, d\zeta}{(1-\zeta^2)^{1/2+\sigma}(\beta^2-\zeta^2)^{1-\sigma}}\right\} \ . \tag{2.21}$$

For $\zeta = \beta$ we obtain $z = \frac{B_1}{2}$ , where $B_1$ the width of the canal bottom, and therefore

$$\frac{B_1}{2} = \frac{Q}{k\pi J_2 \cos \pi\sigma}\int_0^{\beta}\frac{\zeta \text{ arc sin } \zeta\, d\zeta}{(1-\zeta^2)^{1/2+\sigma}(\beta^2-\zeta^2)^{1-\sigma}} \ . \tag{2.22}$$

Along the slope bc for $\beta < \zeta < 1$

$$z = \frac{B_1}{2} + \frac{Q}{k\pi J_2 \cos \pi\sigma} e^{\pi\sigma i}\left\{\text{arc sin } \zeta \int_{\beta}^{\zeta}\frac{\zeta\, d\zeta}{(1-\zeta^2)^{1/2+\sigma}(\zeta^2-\beta^2)^{1-\sigma}} - \right.$$

$$\left. - \int_{\beta}^{\zeta}\frac{\zeta \text{ arc sin } \zeta\, d\zeta}{(1-\zeta^2)^{1/2+\sigma}(\zeta^2-\beta^2)^{1-\sigma}}\right\} \ . \tag{2.23}$$

For $\zeta = 1$, we have $z = \frac{B}{2} + Hi$, where $H$ is the canal depth, $B$ the surface width of the canal, and equality (2.23) gives:

$$(2.24)\quad \begin{cases} \dfrac{B - B_1}{2} = \dfrac{Q}{k\pi J_2}\left\{\dfrac{\pi}{2} J_2 - \displaystyle\int_\beta^1 \dfrac{\zeta \text{ arc sin } \zeta \, d\zeta}{(1 - \zeta^2)^{1/2+\sigma}(\zeta^2 - \beta^2)^{1-\sigma}}\right\} , \\ H = \dfrac{Q}{k\pi J_2} \text{ tg } \pi\sigma \left\{\dfrac{\pi}{2} J_2 - \displaystyle\int_\beta^1 \dfrac{\zeta \text{ arc sin } \zeta \, d\zeta}{(1 - \zeta^2)^{1/2+\sigma}(\zeta^2 - \beta^2)^{1-\sigma}}\right\} . \end{cases}$$

Along the free surface ba, where $1 < \zeta < \infty$, we have

$$(1 - \zeta^2)^{1/2+\sigma} = e^{\pi i(\frac{1}{2}+\sigma)}(\zeta^2 - 1)^{1/2+\sigma} = ie^{\pi\sigma i}(\zeta^2 - 1)^{1/2+\sigma} .$$

For real values of $\zeta$, larger than one, arc sin $\zeta$ transforms as

$$\text{arc sin } \zeta = \int_0^\zeta \frac{d\zeta}{\sqrt{1 - \zeta^2}} = \frac{\pi}{2} + \frac{1}{i}\int_1^\zeta \frac{d\zeta}{\sqrt{\zeta^2 - 1}} = \frac{\pi}{2} - i \, \ln(\zeta + \sqrt{\zeta^2 - 1}) =$$

$$= \frac{\pi}{2} - i \text{ arch } \zeta .$$

We notice that the root $\sqrt{1 - \zeta^2}$ transforms into $+ i\sqrt{\zeta^2 - 1}$, since the contour of the singular point $\zeta = 1$ is followed in the positive direction in the passage from segment cb to segment ba in the lower half-plane $\zeta$ (Fig. 142d).

Considering those circumstances, we obtain from formula (2.23) in the passage to segment $\zeta = \xi > 1$:

$$(2.25)\quad z = \frac{B}{2} + Hi + \frac{Q}{k\pi J_2 \cos \pi\sigma}\left\{- \text{arch } \zeta \int_1^\zeta \frac{\zeta \, d\zeta}{(\zeta^2 - 1)^{1/2+\sigma}(\zeta^2 - \beta^2)^{1-\sigma}} - iJ_2 e^{\pi\sigma i} \text{ arch } \zeta + \int_1^\zeta \frac{\zeta \text{ arch } \zeta \, d\zeta}{(\zeta^2 - 1)^{1/2+\sigma}(\zeta^2 - \beta^2)^{1-\sigma}}\right\} .$$

Separating real and imaginary parts in this equality, we obtain the equations of the free surface ba:

$$(2.26)\quad \begin{aligned} x - \frac{B}{2} &= \frac{Q}{k\pi J_2 \cos \pi\sigma}\left\{\int_1^\zeta \frac{\zeta \text{ arch } \zeta \, d\zeta}{(\zeta^2 - 1)^{1/2+\sigma}(\zeta^2 - \beta^2)^{1-\sigma}} + J_2 \sin \pi\sigma \text{ arch } \zeta - \text{arch } \zeta \int_1^\zeta \frac{\zeta \, d\zeta}{(\zeta^2 - 1)^{1/2+\sigma}(\zeta^2 - \beta^2)^{1-\sigma}}\right\} , \\ y - H &= - \frac{Q}{k\pi} \text{ arch } \zeta \qquad [\text{arch } \zeta = \ln(\zeta + \sqrt{\zeta^2 - 1})] . \end{aligned}$$

We now determine the dependence between the flow rates and dimensions of the canal and the parameters $\sigma$ and $\beta$. We have already obtained the formulas necessary for this: (2.22) and (2.24).

We introduce the notations

$$(2.27)\qquad \left\{\begin{aligned} &\int_0^\beta \frac{\zeta \text{ arc sin } \zeta \, d\zeta}{(1-\zeta^2)^{1/2+\sigma}(\zeta^2-\beta^2)^{1-\sigma}} = f_1(\sigma, \beta) , \\ &\int_\beta^1 \frac{\zeta \text{ arc sin } \zeta \, d\zeta}{(1-\zeta^2)^{1/2+\sigma}(\zeta^2-\beta)^{1-\sigma}} = f_2(\sigma, \beta) . \end{aligned}\right.$$

Then we may replace (2.22) by

$$(2.28)\qquad B_1 = \frac{2Q}{k\pi J_2 \cos \pi\sigma} f_1(\sigma, \beta) ,$$

and (2.24) by

$$H = \frac{Q}{k\pi J_2} \operatorname{tg} \pi\sigma \left\{ \frac{\pi}{2} J_2 - f_2(\sigma, \beta) \right\} ,$$

$$\frac{B - B_1}{2} = \frac{Q}{k\pi J_2} \left\{ \frac{\pi}{2} J_2 - f_2(\sigma, \beta) \right\} .$$

Finally we find for the surface width $B$ of the river

$$(2.29)\qquad B = B_1 + \frac{Q}{k}\left[1 - \frac{2f_2(\sigma,\beta)}{\pi J_2}\right] .$$

Similarly for the depth $H$ of the canal

$$(2.30)\qquad H = \frac{Q}{2k} \operatorname{tg} \sigma\pi \left[1 - \frac{2f_2(\sigma,\beta)}{\pi J_2}\right] .$$

From these formulas we may express the flowrate $Q$ in function of the parameters $\sigma$, $\beta$ and one of three quantities $B$, $b$, $H$ ( the latter are of course correlated as $(B - B_1)/2 = H \operatorname{cotg} \sigma\pi$).

Actually one computes the flowrate $Q$ for given dimensions of the canal and selects the value of $\beta$.

V. V. Vedernikov [3] writes the formula for the flowrate as

$$(2.31)\qquad Q = k(B + AH) .$$

where

$$(2.32)\qquad A = \frac{2}{\operatorname{tg} \sigma\pi} \; \frac{f_2(\sigma,\beta) - \dfrac{1}{\cos \sigma\pi} f_1(\sigma,\beta)}{\dfrac{\pi}{2} J_2 - f_2(\sigma,\beta)} .$$

We determine the formula for the flowrate $Q'$ through the canal bottom.

If in formula (2.18)

$$\omega = \varphi + i\psi = \frac{Qi}{\pi} \text{ arc sin } \zeta$$

we put $\zeta = \beta$ (i.e., we take point $c$ in Fig. 142d), then we find $\varphi = 0$, $\psi = Q'/2$, i.e.,

$$\frac{Q'}{2} = \frac{Q}{\pi} \text{ arc sin } \beta ,$$

from which

$$Q' = \frac{2Q}{\pi} \text{ arc sin } \beta . \tag{2.33}$$

Assigning a series of values to $\beta$ and $\alpha$, V. V. Vedernikov computed the corresponding values of the quantities

$$\frac{kB_1}{Q}, \quad \frac{kB}{Q}, \quad \frac{kH}{Q}, \quad \frac{Q'}{Q}, \quad A \text{ and } \frac{B}{H} .$$

The results of these computations are compiled in Table 7 (all values given by the author have four significant numbers). We notice that, if we call $L$ the width of the percolation flow at infinity,

$$Q = kL ,$$

because the velocity at infinity is equal to the seepage coefficient $k$ .

Table 7

Values of the quantities $kB_1/Q$, $kB/Q$, $kH/Q$, $A$, $B/H$ depending upon $\alpha = \sigma\pi$ (or $m = \text{cotg } \alpha$) and $\beta$ for a canal with trapezoidal cross-section.

| $\alpha$ | $m$ | $\beta^2$ | $\frac{kB_1}{Q}$ | $\frac{kB}{Q}$ | $\frac{kH}{Q}$ | $\frac{Q'}{Q}$ | $A$ | $\frac{B}{H}$ |
|---|---|---|---|---|---|---|---|---|
| 45° | 1 | 0 | 0 | 0.500 | 0.250 | 0 | 2.00 | 2.00 |
| 45° | 1 | 0.25 | 0.167 | 0.545 | 0.189 | 0.333 | 2.40 | 2.88 |
| 45° | 1 | 0.50 | 0.304 | 0.595 | 0.146 | 0.500 | 2.78 | 4.08 |
| 45° | 1 | 0.75 | 0.479 | 0.676 | 0.0986 | 0.667 | 3.29 | 6.86 |
| 45° | 1 | 0.875 | 0.609 | 0.746 | 0.0682 | 0.770 | 3.73 | 10.9 |
| 45° | 1 | 0.9375 | 0.706 | 0.802 | 0.0480 | 0.839 | 4.12 | 16.7 |
| 45° | 1 | 1 | 1 | 1 | 0 | 1 | $\infty$ | $\infty$ |
| 30° | 1.732 | 0 | 0 | 0.656 | 0.189 | 0 | 1.82 | 3.46 |
| 30° | 1.732 | 0.25 | 0.204 | 0.684 | 0.139 | 0.333 | 2.28 | 4.93 |
| 30° | 1.732 | 0.50 | 0.353 | 0.720 | 0.106 | 0.500 | 2.63 | 6.78 |
| 30° | 1.732 | 0.75 | 0.530 | 0.777 | 0.0713 | 0.667 | 3.13 | 10.9 |
| 30° | 1.732 | 0.875 | 0.653 | 0.824 | 0.0494 | 0.770 | 3.55 | 16.7 |
| 30° | 1.732 | 1 | 1 | 1 | 0 | 1 | $\infty$ | $\infty$ |
| 22°30' | 2.414 | 0 | 0 | 0.736 | 0.152 | 0 | 1.74 | 4.82 |
| 22°30' | 2.414 | 0.25 | 0.229 | 0.757 | 0.109 | 0.333 | 2.22 | 6.92 |
| 22°30' | 2.414 | 0.50 | 0.384 | 0.786 | 0.0832 | 0.500 | 2.58 | 9.45 |
| 22°30' | 2.414 | 0.75 | 0.559 | 0.829 | 0.0559 | 0.667 | 3.05 | 14.8 |
| 22°30' | 2.414 | 0.875 | 0.678 | 0.865 | 0.0387 | 0.770 | 3.48 | 22.3 |
| 22°30' | 2.414 | 1 | 1 | 1 | 0 | 1 | $\infty$ | $\infty$ |

It is interesting to compare formula (2.31) for the flowrate with the corresponding formula for the canal with curvilinear (trochoidal) perimeter, which we examined in Chapter IV §9. There we had the quantity $A = 2$ in the notations of formula (2.31). Table 7 shows that for the canal with trapezoidal cross-section A varies for practically interesting cases between limits from 2 to 3-4.

For the canal that corresponds to $\beta = \sin 52°30'$, V. V. Vedernikov constructed the picture of the velocity distribution along the perimeter of the canal, the form of the free surface and also the width of flow at infinity. These results are represented in Fig. 144.

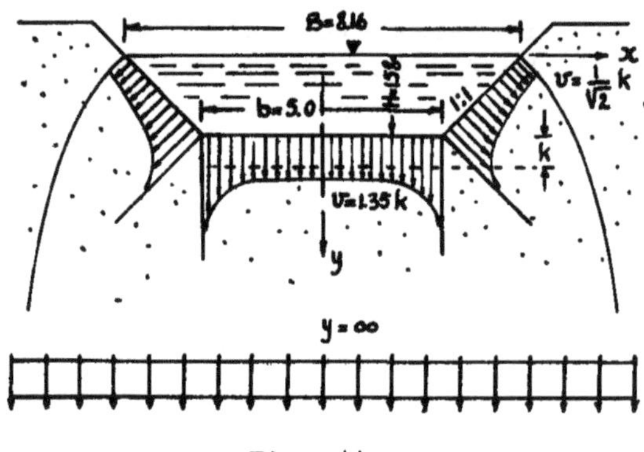


Fig. 144

§3. Canal with Triangular Cross-section.

Separately V. V. Vedernikov investigated a canal with triangular cross-section. This case is derived from the general formulas $\beta = 0$. The quantity A, occurring in the formulae of the flowrate (2.31), varies here between limits given in Table 8.

Table 8

Dependence of A on $\alpha = \sigma\pi$ for canal with triangular cross-section.

| $\alpha$ | 9° | 22°30' | 30° | 45° | 60° | 67°30' | 87° |
|---|---|---|---|---|---|---|---|
| A | 1.58 | 1.74 | 1.82 | 2.00 | 2.20 | 2.31 | 2.53 |

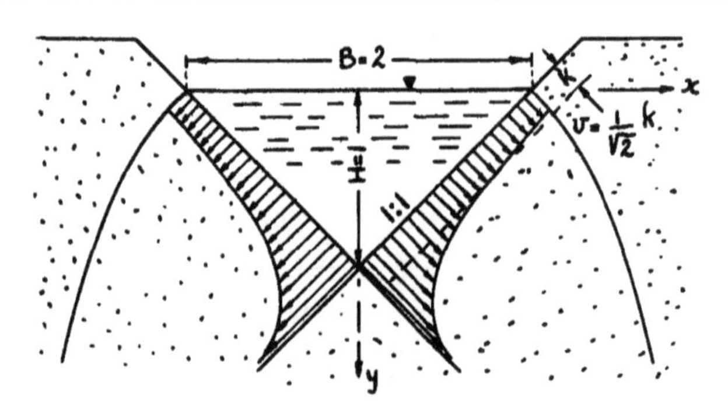


Fig. 145

For $\alpha = 45°$, $A = 2$. Fig. 145 renders a graph, constructed by V. V. Vedernikov, for a canal with triangular cross-section. In the investigations considered here, it was assumed that the seepage velocity at infinity is equal to the seepage coefficient. This means that at a very large depth, theoretically at infinity, perfect drainage conditions are available. J. D. Sokolov [4] solved the problem of the seepage from a trapezoidal canal with draining substratum at finite depth.

## §4. Influence of Soil Capillarity on Seepage from Canals.

We already mentioned that the phenomenon of capillarity plays an important roll in the flow of groundwater with a free surface. Under action of capillary forces, the area of wetting widens and the flowrate increases. We examined the problem of Zhoukovsky's cutoff with influence of capillarity in §2 of Chapter IV. Now we examine the simplest case of seepage from a canal.

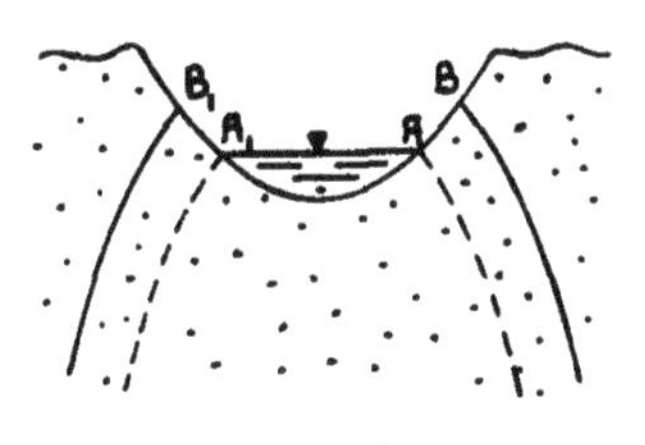


Fig. 146

Consider a canal with arbitrary cross-section (Fig. 146). The dotted line corresponds to the free surface without capillarity, the smooth line to that with capillarity. In the latter case, as we know, the condition [1] prevails that at the free surface the pressure is constant and equal to the atmospheric pressure, decreased by the quantity $\gamma h_\kappa$ (where $h_\kappa$ is the height of capillary rise, $\gamma$ the unit weight of the fluid):

$$p = p_a - \gamma h_\kappa \tag{4.1}$$

at the free surface.

S. F. Averjanov [8] gives an approximate computation of the variable saturation of the soil in the region of the capillary edge. Considering the flow with free surface in the form of Dupuit's parabola, the author assumes that in the region of the capillary fringe the wetting of the soil varies with the depth, according to a parabolic law, decreasing with upward rise. Then the flowrate is determined by a formula, representing the generalized Dupuit formula (Chapter II)

$$Q = k \frac{h_1 - h_2}{x_1 - x_2} \left( \frac{h_1 + h_2}{2} + \alpha h_\kappa \right) ,$$

where $h_\kappa$ is the height of capillary rise, $\alpha$ a number smaller than one, depending on the saturation of the soil. S. F. Averjanov proposes to take $\alpha = 0.3$ for the most occurring cases.

We do not take up the still little developed question about variable soil saturation, and restrict ourselves to the account of capillarity as shown in N. E. Zhoukovsky's formula (4.1).

The question arises as to what condition must be satisfied at the slopes $AB$ and $A_1 B_1$ of the canal (Fig. 146). V. V. Vedernikov, and other authors with him assumed that these segments (up to date all authors considered these segments to be rectilinear) were solid walls, along which the streamfunction assumes a constant value. This condition is observed in the application of our methods to seepage problems. It is clear however, that the solution of these problems is complicated if one accounts for capillarity. We restrict ourselves here to the consideration of the simplest problems, where we may obtain numerical results.

§5. Canal with Trapezoidal Cross-section.

V. V. Vedernikov [7] derived general formulas for the seepage from a canal with trapezoidal cross-section (Fig. 147), considering capillarity. As a particular case he examined that of unit slope, i.e., making an angle

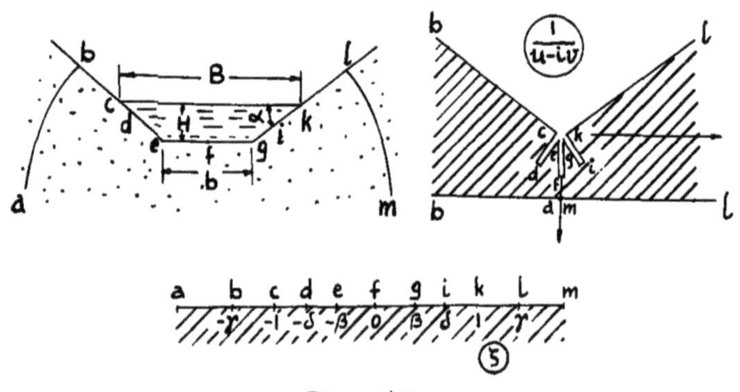


Fig. 147

of 45° with the horizontal.

Formulas for the magnitude of the inverse complex velocity, have the form (Fig. 147)

$$(5.1)\qquad \frac{1}{u - iv} = -\frac{i}{J_1 + J_2}\int_0^{\zeta} \frac{\zeta(\delta^2 - \zeta^2)}{(\beta^2 - \zeta^2)^{1-\sigma}\sqrt{1 - \zeta^2}(\gamma^2 - \zeta^2)^{1+\sigma}} + \frac{iJ_2}{J_1 + J_2}$$

where

$$(5.2)\qquad \delta^2 = \frac{J_3}{J_4}, \quad \sigma = \frac{\alpha}{\pi},$$

and the quantities $J_1, J_2, J_3, J_4$ have the expressions

$$J_1 = \int_0^{\infty} \frac{\eta(\delta^2 + \eta^2)\, d\eta}{(\beta^2 + \gamma^2)^{1-\sigma}\sqrt{1 + \eta^2}(\gamma^2 + \eta^2)^{1+\sigma}},$$

$$J_2 = \int_0^{\beta} \frac{\zeta(\delta^2 - \zeta^2)\, d\zeta}{(\beta^2 - \zeta^2)^{1-\sigma}\sqrt{1 - \zeta^2}(\gamma^2 - \zeta^2)^{1+\sigma}},$$

$$J_3 = \int_{\beta}^{1} \frac{\zeta\, d\zeta}{(\zeta^2 - \beta^2)^{1-\sigma}(\gamma^2 - \zeta^2)^{1+\sigma}\sqrt{1 - \zeta^2}},$$

$$J_4 = \int_{\beta}^{1} \frac{\zeta d\zeta}{(\zeta^2 - \beta^2)^{1-\sigma}(\gamma^2 - \zeta^2)^{1+\sigma}\sqrt{1 - \zeta^2}}.$$

For the complex potential we have

$$\omega = \varphi + i\psi = \frac{Q}{k\pi} \operatorname{arc\,sin} \zeta ,$$

where $Q$ is the flowrate in the canal. The relationship between $z$ and $\zeta$ is given by the formula

$$z = \frac{Q}{\kappa\pi} \int \frac{1}{u - iv} \frac{d\zeta}{\sqrt{1 - \zeta^2}} . \tag{5.3}$$

If we also introduce the integrals

$$J_5 = \int_0^\beta \operatorname{arc\,sin} \zeta \frac{\zeta(\delta^2 - \zeta^2)\, d\zeta}{(\beta^2 - \zeta^2)^{1-\sigma}(\gamma^2 - \zeta^2)^{1+\sigma}\sqrt{1 - \zeta^2}} ,$$

$$J_6 = -\int_\beta^1 \operatorname{arc\,sin} \zeta \frac{\zeta(\delta^2 - \zeta^2)\, d\zeta}{(\zeta^2 - \beta^2)^{1-\sigma}(\gamma^2 - \zeta^2)^{1+\sigma}\sqrt{1 - \zeta^2}} ,$$

$$J_7 = \int_1^\gamma \frac{(\operatorname{arch} \beta - \operatorname{arch} \zeta)(\zeta^2 - \delta^2)\zeta\, d\zeta}{(\zeta^2 - \beta^2)^{1-\sigma}(\gamma^2 - \zeta^2)^{1+\sigma}\sqrt{\zeta^2 - 1}}$$

$$(\operatorname{arch}\ x = \ell n(x + \sqrt{x^2 - 1}))$$

then parameters $\beta$ and $\gamma$, flowrate $Q$ and height of capillary rise $h_k$ along the slope are determined by the following equations:

$$\frac{b}{2} = \frac{Q}{k\pi} \frac{J_5}{J_1 + J_2} \tag{5.4}$$

$$\frac{B}{2} = \frac{Q}{k\pi} \frac{J_5 + J_6 \cos \pi\sigma}{J_1 + J_2} \tag{5.5}$$

$$h = h_k - \frac{Q}{k\pi} \operatorname{arch} \gamma = \frac{J_7 \sin \alpha}{J_1 + J_2} \frac{Q}{k\pi} . \tag{5.6}$$

Here $b$ is the width of the canal bottom, $h$ the height of capillary rise of a given soil.

Table 9

Results of computations for case of unit slope.

| $\frac{B}{H}$ | $\frac{h_k}{H}$ | A | $\frac{h}{h_k}$ | $\gamma^2$ | $\beta^2$ |
|---|---|---|---|---|---|
| 2.0 | 7.3 | 16.0 | 0.246 | 2.25 | 0 |
| 4.1 | 12.6 | 27.1 | 0.24 | | 0.25 |
| 8.3 | 22.0 | 46.8 | 0.233 | | 0.50 |
| 22.8 | 51.2 | 106.8 | 0.225 | | 0.75 |
| 2.0 | 3.0 | 8.8 | 0.285 | 1.44 | 0 |
| 3.8 | 4.8 | 13.9 | 0.273 | | 0.25 |
| 7.2 | 8.1 | 22.9 | 0.254 | | 0.50 |
| 19.7 | 18.6 | 48.2 | 0.248 | | 0.75 |
| 2.0 | 0.97 | 4.7 | 0.32 | 1.1 | 0 |
| 3.4 | 1.46 | 6.6 | 0.315 | | 0.25 |
| 5.8 | 2.2 | 9.5 | 0.306 | | 0.50 |
| 13.2 | 4.1 | 16.5 | 0.287 | | 0.75 |

Calling $H$ the depth of water in the canal and $B$ the width of waterlevel, we have

$$H = \frac{Q}{k\pi} \frac{J_6 \sin \alpha}{J_1 + J_2}, \tag{5.7}$$

$$B = 2H \left( \frac{J_5}{J_6 \sin \alpha} + \text{ctg } \alpha \right). \tag{5.8}$$

The flowrate per unit length of canal is determined by the formula

$$Q = k(B + AH), \tag{5.10}$$

where

$$A = \frac{\pi(J_1 + J_2) - 2(J_5 + J_6 \cos \alpha)}{J_6 \sin \alpha}.$$

(5.10)

We may also determine the flowrate by another formula

$$Q = kNB,$$

where

$$N = \frac{\pi}{2} \frac{J_1 + J_2}{J_5 + J_6 \cos \alpha}.$$

| $\frac{h_k}{H}$ | A | $\frac{h}{h_k}$ |
|---|---|---|
| 0 | 2.8 | 1.0 |
| 0.75 | 3.7 | 0.58 |
| 1.9 | 6.8 | 0.55 |
| 3.2 | 9.6 | 0.48 |
| 27.4 | 43.7 | 0.33 |

Table 10

Results of computations for the case of a vertical slit.

V. V. Vedernikov presented computations for two cases: that of a unit slope (Table 9) and that of a vertical slit ($\alpha = \pi/2$, $\beta = 0$, Table 10).

We notice that the height of capillary rise along the slopes is smaller than the height of static capillary rise.

To supplement this investigation, we consider one more simple case of seepage from a canal, which may be investigated more in detail.

## §6. Canal with Low Waterlevel [6].

When the water level in a canal approaches zero, the flowrate $Q_0$ seeping from the canal is close to

$$Q_0 = \kappa B. \tag{6.1}$$

when there is no capillary rise.

We assume seepage from canal $MM'$ into the soil which has capillary height $h_k$. The velocity hodograph has the form of Fig. 148b, where half of the hodograph is sketched, corresponding to the right half of the flow region. In point $M$ the velocity is finite, in point $L$ zero, at infinity it is equal to the seepage coefficient $k$.

The complex potential region is represented in Fig. 148c. Assuming $\varphi = 0$ along $MM'$, $\psi = 0$ along $MLK$, we find in point $L$ for $\varphi$ the value

$$\varphi = + \kappa h_k.$$

Along the symmetry axis $KN$ we have:

$$\psi = - \frac{Q}{2},$$

where $Q$ is the total flowrate of the canal.

We map the region of the complex potential onto the half-plane $\zeta$. The abscissa of point L is designated in Fig. 148d as $(-a^2)$.

We have for the complex potential

(6.2) $$\omega = -\frac{iQ}{2\pi} \text{ arc cos}(1 - 2\zeta) .$$

Solving this equation for $\zeta$, we find:

(6.3) $$\zeta = \sin^2 \frac{\pi\omega i}{Q} = -\text{sh}^2 \frac{\pi\omega}{Q} .$$

Putting

$$\zeta = -a^2, \quad \omega = +\kappa h_k ,$$

we find:

(6.4) $$a = \text{sh}\,\frac{\pi\kappa h_k}{Q} .$$

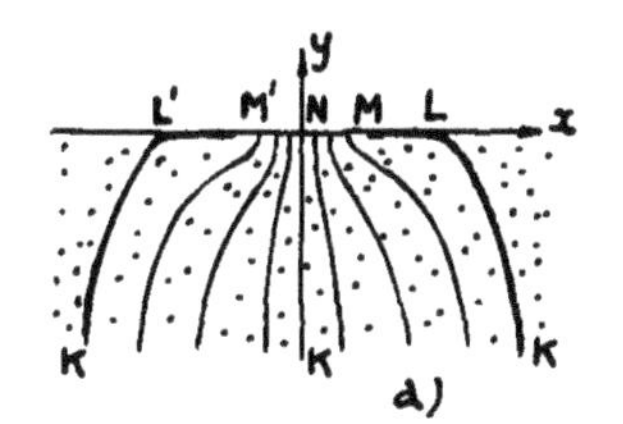


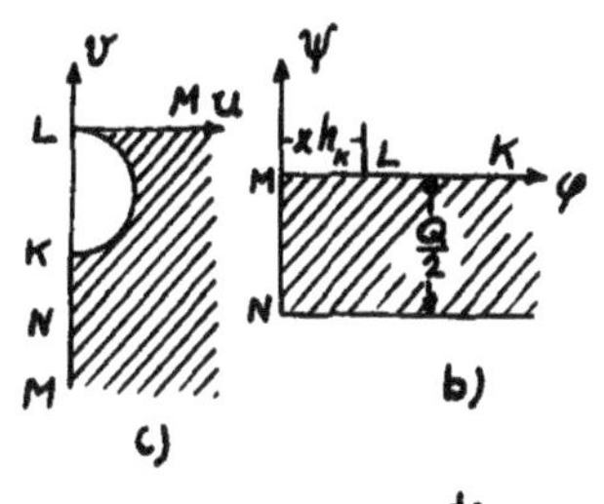


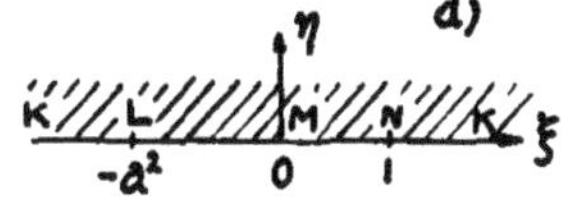


Fig. 148

Returning to the velocity hodograph, we carry out the inversion of the velocity region in a circle with radius one and with center in the origin of coordinates. We obtain a half strip. Its mapping onto the half-plane $\zeta$ is done by the integral

(6.5) $$\frac{1}{u - iv} = \frac{dz}{d\omega} = M \int \frac{d\zeta}{(\zeta + a^2)\sqrt{\zeta}} = \frac{2M}{a} \text{ arc tg } \frac{\sqrt{\zeta}}{a} + N ,$$

where M and N are constants.

The conditions $\zeta = 0$, $u - iv = \infty$, $\zeta = \infty$, $v = -\kappa$ give:

(6.6) $$\frac{dz}{d\omega} = -\frac{2i}{\kappa\pi} \text{ arc tg } \frac{\sqrt{\zeta}}{a} = -\frac{2i}{\kappa\pi} \text{ arc tg } \frac{\sin \frac{\pi\omega i}{Q}}{a} .$$

We abbreviate

(6.7) $$\frac{\pi\omega i}{Q} = W$$

and consider the trigonometric series

(6.8) $$\frac{dz}{d\omega} = -\frac{2i}{\pi\kappa} \text{ arc tg } \frac{\sin W}{a} = -\frac{4i}{\pi\kappa} \sum_{n=1}^{\infty} e^{-(2n-1)\alpha} \frac{\sin(2n - 1)W}{2n - 1} ,$$

where

$$\alpha = \frac{\pi\kappa h_k}{Q} .$$

Series (6.8) is obtained from the known series

$$\text{arc tg } \frac{2p \sin x}{1 - p^2} = 2 \sum_{n=1}^{\infty} p^{2n-1} \frac{\sin(2n - 1)x}{2n - 1} .$$

Therefore, putting $W = U + iV$, we may represent the right part of equation (6.8) in the form

$$- \frac{2}{\pi\kappa} \sum_{n=1}^{\infty} \frac{1}{2n-1} \{e^{-(2n-1)(\alpha - iU + V)} - e^{-(2n-1)(\alpha + iU - V)}\} ,$$

from which it follows that series (6.8) converges in the segment

$$- \alpha < V < \alpha$$

or, in values of $\varphi$, for

$$-\kappa h_k < \varphi < + \kappa h_k .$$

Integrating series (6.8) term by term, we find for $z$

$$z = \frac{4Q}{\kappa\pi^2} \sum_{n=1}^{\infty} e^{-(2n-1)\alpha} \frac{\cos(2n-1) W}{(2n-1)^2} . \tag{6.9}$$

The integration constant is here zero, since $z = 0$ for $\omega = -Qi/2$.

Putting $\zeta = 0$ and $\zeta = -a^2$ in (6.9), we obtain respectively $\omega = 0$, $W = 0$ and $\omega = + \kappa h_k$, $W = i\alpha$. Therefore we find for the canal width $B$:

$$B = \frac{8Q}{\kappa\pi^2} \sum_{n=1}^{\infty} \frac{e^{-(2n-1)\alpha}}{(2n-1)^2} \qquad \left(\alpha = \frac{\pi\kappa h_k}{Q}\right) \tag{6.10}$$

and for the width $B_1$ of the wetted part of the soil

$$B_1 = \frac{8Q}{\kappa\pi^2} \sum_{n=1}^{\infty} e^{-(2n-1)\alpha} \frac{\text{ch}(2n-1)\alpha}{(2n-1)^2} . \tag{6.11}$$

Expanding the expression of the hyperbolic cosine, we rewrite formula (6.11) as:

$$B_1 = \frac{4Q}{\kappa\pi^2} \sum_{n=1}^{\infty} \frac{1 + e^{-2(2n-1)\alpha}}{(2n-1)^2} ,$$

from which follows the equality $\left(\text{if one considers that } \sum_{n=1}^{\infty} \frac{1}{(2n-1)^2} = \frac{\pi^2}{8}\right)$

$$B_1(\alpha) = \frac{Q}{2\kappa} + \frac{1}{2} B(2\alpha) . \tag{6.12}$$

We find $B_1$ and $Q$ for a given canal width from (6.10) and (6.11) or (6.12). Since at infinity the velocity is equal to the seepage coefficient, the width $B_2$ of the seepage flow at infinity is

$$B_2 = \frac{Q}{\kappa} . \tag{6.13}$$

Series (6.10) and (6.11) converge well for large values of $\alpha$, but converge slowly for small $\alpha$. Therefore we transform (6.10) in this form. We call

$$A = e^{-x} + \frac{1}{3^2} e^{-3x} + \frac{1}{5^2} e^{-5x} + \ldots, \tag{6.14}$$

differentiate this series and transform the obtained series

$$\frac{dA}{dx} = - e^{-x} - \frac{1}{3} e^{-3x} - \frac{1}{5} e^{-5x} - \dots =$$

$$= \frac{1}{2} \ln \frac{1 - e^{-x}}{1 + e^{-x}} = \frac{1}{2} \ln \operatorname{sh} \frac{x}{2} - \frac{1}{2} \ln \operatorname{ch} \frac{x}{2} . \tag{6.15}$$

We use the series

$$\ln \frac{\operatorname{sh} u}{u} = \frac{s_2}{\pi^2} u^2 - \frac{1}{2} \frac{s_4}{\pi^4} u^4 + \frac{1}{3} \frac{s_6}{\pi^6} u^6 - \dots,$$

$$\ln \operatorname{ch} u = (2^2 - 1) \frac{s_2}{\pi^2} u^2 - \frac{2^4 - 1}{2} \frac{s_4}{\pi^4} u^4 + \dots,$$

where

$$s_n = 1 + \frac{1}{2^n} + \frac{1}{3^n} + \dots$$

We find for the derivative $dA/dx$ a series that converges for $0 < x < \pi$:

$$\frac{dA}{dx} = \frac{1}{2} \ln \frac{x}{2} -(2 - 1) \frac{s_2}{2^2\pi^2} x^2 + \frac{2^3 - 1}{2} \frac{s_4}{2^2\pi^4} x^4 - \frac{2^5 - 1}{3} \frac{s_6}{2^6\pi^6} x^6 + \dots,$$

Integration of this series leads to this expression

$$A = \frac{1}{2}\left(\ln \frac{x}{2} - 1\right) x - \frac{s_2}{2^2\pi^2} \frac{x^2}{3} + \frac{2^3 - 1}{2} \frac{s_4}{2^4\pi^4} \frac{x^5}{5} - \frac{2^5 - 1}{3} \frac{s_6}{2^6\pi^6} \frac{x^7}{7} + \dots + C.$$

Since for $x = 0$, (6.14) renders $A = 1 + \frac{1}{3^2} + \frac{1}{5^2} + \dots = \frac{\pi^2}{8}$, one must take $C = \frac{\pi^2}{8}$. Returning to expression (6.10) for $B$, we find

$$B = \frac{4Q}{\kappa\pi^2} \left[\frac{\pi^2}{4} + \left(\ln \frac{\alpha}{2} - 1\right) \alpha - \frac{\alpha^3}{36} + \frac{7\alpha^5}{7200} - \frac{31\alpha^7}{635{,}040} + \dots\right] . \tag{6.16}$$

Graphs for the dependence $Q/Q_0$, $B_1/B$ and $B_2/B$ on $h_k/B$ are constructed in Fig. 149. Here $Q_0$ is the flowrate determined by $Q_0 = \kappa B$, i.e., the flowrate from the canal for $h_k = 0$ .

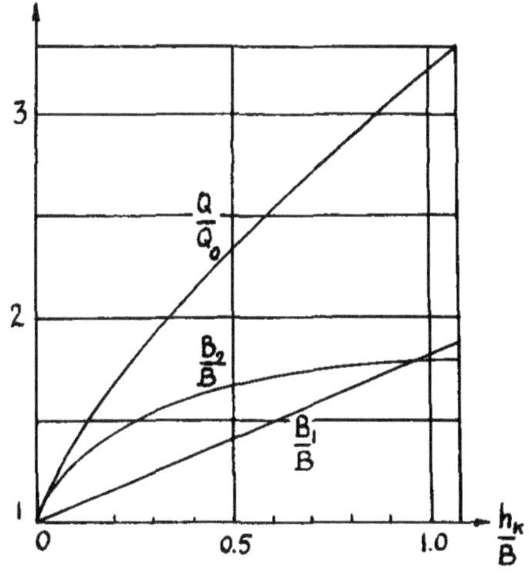


Fig. 149

A study of the graphs reveals that the influence of capillarity on seepage is significant. Therefore the roll of capillarity increases with increasing $h_k/B$ ratio. In real conditions the static capillary rise is often of the order of several tens of centimeters, and sometimes reaches two-three meters. A series of publications in the literature show that one should not take the static capillary rise, but part of it, in the study of seepage. If we take

$h_k = \frac{1}{3} h_{k0}$, where $h_{k0}$ is the height of static rise, then we obtain a value which may reach 1 meter and more. For a canal width of 10 m, we have $h_k : B = 0.1$ and, as apparent from Fig. 149, the flowrate increases by 10% and the difference $B_1 - B$ is equal to 0.9 m. For a smaller canal width the changes of flowrate and the magnitude of spreading at the surface are still significant.

We observe that the dependence of $B_1/B$ on $h_k/B$ is almost linear. For values of $0 < h_k < 0.4B$, we obtain:

$$B_1 = B + 0.883h_k \ ;$$

in a wider interval, for $0 < h_k < 1.1B$ we may assume

$$B_1 = B + 0.8h_k \ .$$

N. N. Verigin [9] examined the influence of capillarity on seepage from an almost semicircular canal; in this canal the spreading in the banks is still stronger than in the case of a wide canal. N. N. Verigin considers the application of this problem to an irrigation system (the width of an irrigation ditch is usually of the order of 20 to 30 cm). We notice that subsoil and cultivation soil have very small capillarity.

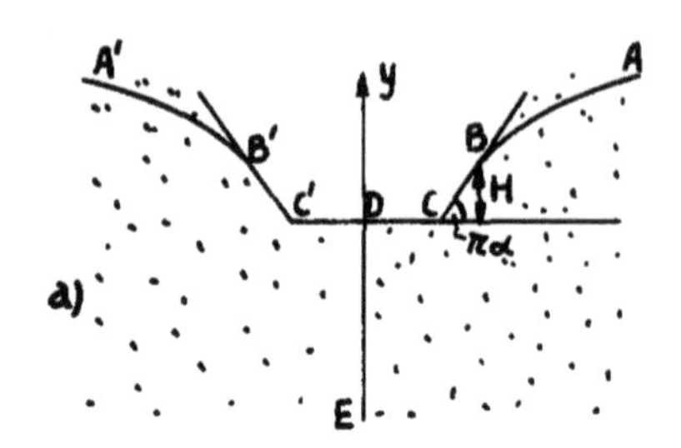


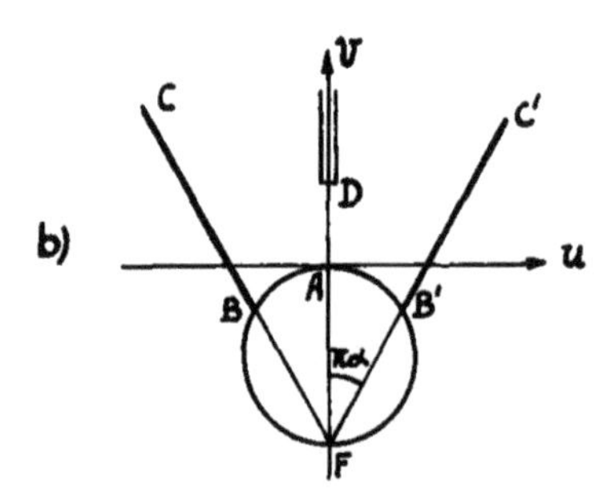


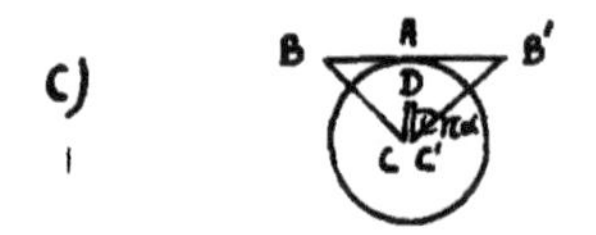


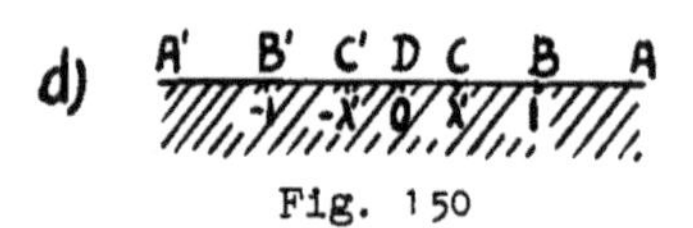


Fig. 150

## §7. Groundwater Flow to a Drainage Ditch of Trapezoidal Cross-section.

The solution of this problem was given by V. V. Vedernikov [3]. Its investigation was also done by J. D. Sokolov [5]. The ditch is sketched in Fig. 150,a. We neglect the water depth in the ditch, since the presence of water in the ditch complicates the solution of the problem strongly.

Along the entire contour of the flowregion we have here the condition that $\varphi + ky = 0$; therefore it is convenient to use Zhoukovsky's function

$$\theta = \omega - ikz = \varphi + ky + i(\psi - kx) \ .$$

The velocity hodograph (Fig. 150b) has arcs of circles and straight lines as boundaries, which all intersect in a point $F$, where $u = 0$, $v = -k$. Inversion in a circle with center in point $F$ gives the polygon of Fig. 150c, which may be mapped easily onto the half-plane $\zeta$ (Fig. 150d).

We observe that

$$\frac{d\theta}{dz} = w - ik ,$$

and see that the inverse quantity

$$U = \frac{dz}{d\theta} = \frac{1}{w - ik} \tag{7.1}$$

exactly corresponds to the function which region is obtained after the above shown inversion in Fig. 150. Mapping the vertices of this polygonal region onto points of plane $\zeta$ as shown in Fig. 150d, we obtain:

$$U = M \int_0^{\zeta} \frac{\zeta \, d\zeta}{(1 - \zeta^2)^{\alpha+1/2}(\lambda'^2 - \zeta^2)^{1-\alpha}} + N .$$

For $\zeta = 0$, we have $w = -iv_0$, where $v_0$ indicates the value of the velocity in point D, smallest velocity along the bottom of the ditch. This gives for $N$ the value

$$N = \frac{1}{-iv_0 - ik} = \frac{i}{v_0 + k} .$$

Further, for $\zeta = \lambda'$ we have $w = \infty$, i.e., $U = 0$. Therefore we may write:

$$U = M \int_{\lambda'}^{\zeta} \frac{\zeta \, d\zeta}{(1 - \zeta^2)^{\alpha+1/2}(\lambda'^2 - \zeta^2)^{1-\alpha}} . \tag{7.2}$$

To find real values of the integrand for $\lambda' < \zeta < 1$, we notice that $(\lambda' - \zeta)^{1-\alpha}$ in the passage from CD to CB, contouring C in the lower half plane anti-clockwise over an angle $\pi$, transforms into

$$(\zeta - \lambda')^{1-\alpha} e^{\pi i(1-\alpha)} = -(\zeta - \lambda')^{1-\alpha} e^{-\pi i\alpha} .$$

Consequently $U$ may be written as

$$U = -Me^{\pi i\alpha} \int_{\lambda'}^{\zeta} \frac{\zeta \, d\zeta}{(1 - \zeta^2)^{\alpha+1/2}(\zeta^2 - \lambda'^2)^{1-\alpha}} . \tag{7.2'}$$

For $\zeta = 1$, as apparent from Fig. 150b,

$$u + iv = - k \sin \pi\alpha e^{\pi i\alpha} ,$$

From this one obtaines in point $\zeta = 1$ the value for $U$

$$U_1 = \frac{1}{u - iv - ki} = \frac{ie^{\pi i\alpha}}{k \cos \pi\alpha} .$$

Inserting $\zeta = 1$, $U = U_1$ in (7.2 ), we obtain after eliminating $e^{\pi i\alpha}$

$$\frac{i}{k \cos \pi\alpha} = - M \int_{\lambda'}^{1} \frac{\zeta \, d\zeta}{(1 - \zeta^2)^{\alpha+1/2}(\zeta^2 - \lambda'^2)^{1-\alpha}} = - MJ ,$$

where

$$J = \int_{\lambda'}^{1} \frac{\zeta \, d\zeta}{(1 - \zeta^2)^{\alpha+1/2}(\zeta^2 - \lambda'^2)^{1-\alpha}} .$$

The substitution

$$1 - \zeta^2 = \lambda^2 x \qquad (\lambda^2 = 1 - \lambda'^2)$$

reduces $\zeta$ to the form

$$J = \frac{1}{2\lambda} \int_0^1 x^{-\alpha-1/2}(1 - x)^{\alpha-1}\, dx \quad .$$

Recalling the value of the beta function

$$B(p, q) = \int_0^1 x^{p-1}(1 - x)^{q-1}\, dx$$

and its connection with the gamma function

$$B(p,q) = \frac{\Gamma(p)\Gamma(q)}{\Gamma(p + q)} \quad ,$$

we write

$$J = \frac{1}{2\lambda} B\left(\frac{1}{2} - \alpha, \alpha\right) = \frac{1}{2\lambda} \frac{\Gamma\left(\frac{1}{2} - \alpha\right)\Gamma(\alpha)}{\Gamma\left(\frac{1}{2}\right)} \quad .$$

Further, by the formula

$$\Gamma(p)\Gamma(1 - p) = \frac{\pi}{\sin \pi p}$$

we find

$$\Gamma\left(\frac{1}{2} - \alpha\right) = \frac{\pi}{\cos \pi\alpha\Gamma\left(\frac{1}{2} + \alpha\right)}$$

and recalling that

$$\Gamma\left(\frac{1}{2}\right) = \sqrt{\pi}$$

we finally find for $J$

$$J = \frac{1}{2\lambda} \frac{\sqrt{\pi}\,\Gamma(\alpha)}{\cos \pi\alpha\Gamma\left(\frac{1}{2} + \alpha\right)} \quad .$$

Inserting the value of $J$ in (7.3) we find for $M$

$$M = -\frac{2\lambda i\Gamma\left(\frac{1}{2} + \alpha\right)}{k\sqrt{\pi}\,\Gamma(\alpha)} \quad .$$

We abbreviate

$$A = \frac{\Gamma\left(\frac{1}{2} + \alpha\right)}{\sqrt{\pi}\,\Gamma(\alpha)} \quad .$$

Then we may write equation (7.1) as

$$\text{(7.4)} \qquad \frac{dz}{d\theta} = U = -\frac{2\lambda iA}{k} \int_0^\zeta \frac{\zeta\, d\zeta}{(1 - \zeta^2)^{\alpha+1/2}(\lambda'^2 - \zeta^2)^{1-\alpha}} + \frac{i}{v_0 + k}$$

Now we find the dependence of $\theta$ on $\zeta$. Since the flowregion corresponds to the half-plane of the $\theta$-plane, this means that $\theta$ and $\zeta$ are connected by means of a linear transformation of the form

$$\theta = M_1\zeta + N_1 \ .$$

The constants $M_1$ and $N_1$ are determined from two conditions. Namely, for $\zeta = 0$, we have $\varphi = 0$, $\psi = Q/2$, $x = 0$; therefore

$$\theta = \frac{iQ}{2} \ .$$

Further, for $\zeta = 1$ we have $\varphi + ky = 0$, $\psi = 0$, $x = B/2$; therefore

$$\theta = -\frac{ikB}{2} \ .$$

As is easy to see, we find for $\theta$:

$$\theta = -\frac{i}{2}(kB + Q)\zeta + \frac{iQ}{2} \tag{7.5}$$

and for $d\theta$:

$$d\theta = -\frac{i}{2}(kB + Q)d\zeta \ .$$

Now integrating equation (7.4) along $\zeta$, we have:

$$z = -\lambda A\left(B + \frac{Q}{k}\right)\int_0^{\zeta} d\zeta_2 \int_0^{\zeta_2} \frac{\zeta_1 \, d\zeta_1}{(1 - \zeta_1^2)^{\alpha+1/2}(\lambda'^2 - \zeta_1^2)^{1-\alpha}} + \frac{kB + Q}{2(v_0 + k)}\zeta \ . \tag{7.6}$$

This equation was given by V. V. Vedernikov, who examined the case $\alpha = \frac{1}{2}$, (vertical banks) more in detail.

J. D. Sokolov subjected (7.6) to further transformations, putting

$$t = \frac{1 - \zeta^2}{\lambda^2} \ ,$$

where

$$\lambda^2 = 1 - \lambda'^2 \ .$$

Then we may write the following expression for $U$:

Along DC $\left(1 \le t \le \frac{1}{\lambda^2}\right)$

$$U = \frac{iA}{k}\int_1^t \frac{dt}{t^{\alpha+1/2}(t - 1)^{1-\alpha}} \ ; \tag{7.7}$$

Along CB $(0 \le t \le 0)$

$$U = \frac{A(i \cos \pi\alpha - \sin \pi\alpha)}{k}\int_t^1 \frac{dt}{t^{\alpha+1/2}(1 - t)^{1-\alpha}} \ ; \tag{7.7'}$$

Along BA $(-\infty < t \le 0)$

$$U = -\frac{A}{k}\int_{-\infty}^{t} \frac{dt}{(-t)^{\alpha+1/2}(1 - t)^{1-\alpha}} + \frac{i}{k} \ . \tag{7.7''}$$

We put $t = 1/\lambda^2$ in formula (7.7) and introduce the notation

$$J_1 = \int_1^{1/\lambda^2} \frac{dt}{t^{\alpha+1/2}(t - 1)^{1-\alpha}} \ . \tag{7.8}$$

Since for this $\zeta = 0$ and $w = u - iv = -iv_0$, we obtain

$$U = \frac{i}{v_0 + k}$$

and

$$\frac{k}{v_0 + k} = AJ_1 \quad . \tag{7.9}$$

The latter equation gives the relationship between parameter $\lambda$ and magnitude of the velocity $v_0$. We introduce the dimensionless quantity $v_0/k = \bar{v}$. Then this equation is written as

$$\frac{1}{\bar{v} + 1} = AJ_1 \quad . \tag{7.9'}$$

We mention two extreme cases of equality (7.9'). First, when $\lambda$ is close to zero, a small value of the velocity $v_0$ goes with it. Indeed, for $\lambda'$ close to one, points C and B in Fig. 150 are close; consequently in the z-plane we have a relatively large width of the bottom b of the ditch, and hence a small value of the velocity $v_0$. But for small values of $\lambda$ we may write:

$$J_1 = \int_1^\infty \frac{dt}{t^{\alpha+1/2}(t-1)^{1-\alpha}} + \int_\infty^{1/\lambda^2} \frac{dt}{t^{3/2}\left(1 - \frac{1}{t}\right)^{1-\alpha}} \quad . \tag{7.10}$$

The first integral of the right side is reduced to the beta function by means of the substitution $t = 1/\tau$. In the brackets of the second one, we neglect $1/\tau$ in comparison with unity. Thus we have

$$J_1 \approx \int_0^1 \tau^{-1/2}(1-\tau)^{\alpha-1}\, d\tau - \int_{1/\lambda^2}^\infty \frac{dt}{t^{3/2}}$$

or

$$J_1 \approx B\left(\tfrac{1}{2}, \alpha\right) - 2\lambda = \frac{\Gamma\left(\frac{1}{2}\right)\Gamma(\alpha)}{\Gamma\left(\frac{1}{2} + \alpha\right)} - 2\lambda \; .$$

Recalling the equality for A, we may abbreviate for small $\lambda$:

$$J_1 \approx \frac{1}{A} - 2\lambda \; .$$

Now we examine the second case, when $\lambda$ is close to one. In this case $\lambda'$ is close to zero, and the bottom of the ditch relatively small, the velocity $v_0$ large. In the integral $J_1$, $t$ now changes between narrow limits 1 to $1/\lambda^2$, close to one. Therefore we put under integral sign (7.8) $t^{\alpha+1/2} \approx 1$. Then we obtain:

$$J_1 \approx \int_1^{1/\lambda^2} \frac{dt}{(t-1)^{1-\alpha}} = \frac{1}{\alpha}\frac{\lambda'^{2\alpha}}{\lambda^{2\alpha}} \approx \frac{\lambda'^{2\alpha}}{\alpha} \quad .$$

Formula (7.9) gives: for $\lambda \approx 0$

$$\lambda \approx \frac{1}{2A\left(1 + \frac{k}{v_0}\right)} \quad ,$$

for $\lambda \approx 1$

$$\lambda' \approx \left[ \frac{\alpha}{A\left(\frac{v_0}{k} + 1\right)} \right]^{1/2\alpha} .$$

We now carry out the transformation of formula (7.6). We apply integration by parts to the double integral. We obtain:

$$J(\zeta) = \int_0^\zeta d\zeta_2 \int_0^{\zeta_2} \frac{\zeta_1 \, d\zeta_1}{(1 - \zeta_1^2)^{1/2+\alpha}(\lambda'^2 - \zeta_1^2)^{1-\alpha}} =$$

$$= \zeta \int_0^\zeta \frac{\zeta_1 \, d\zeta_1}{(1 - \zeta_1^2)^{\alpha+1/2}(\lambda'^2 - \zeta_1^2)^{1-\alpha}} - \int_0^\zeta \frac{\zeta_1^2 \, d\zeta_1}{(1 - \zeta_1^2)^{\alpha+1/2}(\lambda'^2 - \zeta_1^2)^{1-\alpha}}$$

The substitution $t_1 = (1 - \zeta_1^2)/\lambda^2$ (from which follows that $\zeta_1 = \sqrt{1 - \lambda^2 t_1}$) gives, if the two obtained integrals are joined

$$J(\zeta(t)) = -\frac{1}{2\lambda} \int_{1/\lambda^2}^{t} \frac{\sqrt{1 - \lambda^2 t} - \sqrt{1 - \lambda^2 t_1}}{t_1^{\alpha+1/2}(t_1 - 1)^{1-\alpha}} \, dt_1 .$$

Considering the various parts of the contour and recalling (7.9), we have (7.11):

along DC $(1 < t < 1/\lambda^2)$

$$z = -\frac{A(kB + Q)}{2k} \int_1^t \frac{\sqrt{1 - \lambda^2 t_1} - \sqrt{1 - \lambda^2 t}}{t_1^{\alpha+1/2}(t_1 - 1)^{1-\alpha}} \, dt_1 + \frac{b}{2} ,$$

along CB $(0 < t < 1)$

$$z = e^{\pi\alpha i} \frac{A(kB + Q)}{2k} \int_t^1 \frac{\sqrt{1 - \lambda^2 t} - \sqrt{1 - \lambda^2 t_1}}{t_1^{\alpha+1/2}(1 - t_1)^{1-\alpha}} \, dt_1 + \frac{b}{2} ,$$

along BA $(-\infty < t < 0)$

$$z = \frac{iA(kB + Q)}{2k} \int_0^t \frac{\sqrt{1 - \lambda^2 t} - \sqrt{1 - \lambda^2 t_1}}{(-t_1)^{\alpha+1/2}(1 - t_1)^{1-\alpha}} \, dt_1 +$$

$$+ \frac{kB + Q}{2k} (1 + i \operatorname{tg} \pi\alpha)\left(\sqrt{1 - \lambda^2 t} - 1\right) + \frac{B}{2} + iH .$$

From the last equation, separating real and imaginary parts, we obtain the equation of the free surface in parametric form

$$(7.12) \quad \begin{cases} x = \dfrac{B + \overline{Q}}{2} \left(\sqrt{1 - \lambda^2 t} - 1\right) + \dfrac{B}{2} \\ y = \dfrac{A(B + \overline{Q})}{2} \displaystyle\int_0^t \frac{\sqrt{1 - \lambda^2 t} - \sqrt{1 - \lambda^2 t_1}}{(-t_1)^{\alpha+1/2}(1 - t_1)^{1-\alpha}} \, dt_1 + \\ \qquad + \left(x - \dfrac{B}{2}\right) \operatorname{tg} \pi\alpha + H . \end{cases}$$

Here we introduced the notations

$$\overline{Q} = \frac{Q}{k} , \quad \bar{v} = \frac{v_0}{k} .$$

Finally, changing the variable of integration $t_1$ in the last integral by means of

$$t_1 = \frac{\eta_1^\beta}{\eta_1^\beta - 1}, \quad \beta = \frac{1}{\frac{1}{2} - \alpha}$$

and elininating $t$ from equation (7.12), we may write the equation of the free surface in explicit form:

$$y - H = \left(x - \frac{B}{2}\right) \operatorname{tg} \pi\alpha - \frac{A(B + \overline{Q})}{1 - 2\alpha} \int_0^\eta \frac{\frac{2x + \overline{Q}}{B + \overline{Q}} - \sqrt{1 + \frac{\lambda^2 \eta_1^\beta}{1 - \eta_1^\beta}}}{\sqrt{1 - \eta_1^\beta}} \, d\eta_1 ,$$

where

$$\eta = \left[ \frac{\left(\frac{2x + \overline{Q}}{B + \overline{Q}}\right)^2 - 1}{\left(\frac{2x + \overline{Q}}{B + \overline{Q}}\right)^2 - \lambda'^2} \right]^{1/2 - \alpha}$$

Developing the function in the integrand in a series of powers of $\eta_1^\beta$ and limiting oneself only to the free term of the series, we obtain an approximate equality for the free surface near point $B$:

$$y - H \approx \left(x - \frac{B}{2}\right) \left\{ \operatorname{tg} \alpha\pi - \frac{A}{\left(\frac{1}{2} - \alpha\right)\left(\frac{3}{2} - \alpha\right)} \left[ \frac{4}{\lambda^2} \, \frac{x - \frac{B}{2}}{B + \overline{Q}} \right]^{1/2 - \alpha} \right\} .$$

J. D. Sokolov obtains also an approximate expression for the free surface for large $x$ in the following way .

We apply the substitution

$$-t = \frac{1 - \xi^2}{\lambda^2 \xi^2}, \quad t_1 = \frac{1 - \xi_1^2}{\lambda^2 \xi_1^2} \qquad \left(\xi = \frac{1}{\sqrt{1 - \lambda^2 t}}\right) .$$

Then by means of (7.12) we obtain:

$$\left(x - \frac{B}{2}\right) \operatorname{tg} \alpha\pi - (y - H) =$$

$$= A\lambda(B + \overline{Q}) \int_\xi^1 \frac{\xi^{-1} - \xi_1^{-1}}{(1 - \xi_1^2)^{\alpha + 1/2}(1 - \lambda_1^2 \xi_1^2)^{1 - \alpha}} \, d\xi_1 , \tag{7.13}$$

where

$$\xi = \frac{1}{\sqrt{1 - \lambda^2 t}} = \frac{B + \overline{Q}}{2x + Q} . \tag{7.13'}$$

The function in the integrand of the integral of (7.13) behaves as $-1/\xi$ for $\xi = 0$; we add and subtract $-1/\xi$ under the integral sign and divide the integration interval into parts. We obtain

$$I = \int_{\xi}^{1} \frac{\xi^{-1} - \xi_1^{-1}}{(1 - \xi_1^2)^{\alpha+1/2}(1-\xi_1^2\lambda_1^2)^{1-\alpha}} \, d\xi_1 =$$

$$= -\int_{\xi}^{1} \frac{d\xi_1}{\xi_1} + \int_0^1 \left[\frac{\xi^{-1} - \xi_1^{-1}}{(1 - \xi_1^2)^{\alpha+1/2}(1 - \lambda_1^2\xi_1^2)^{1-\alpha}} + \frac{1}{\xi_1}\right] d\xi_1 -$$

$$- \int_0^{\xi} \frac{\xi^{-1} - \xi_1^{-1}}{(1 - \xi_1^2)^{\alpha+1/2}(1 - \lambda_1^2\xi_1^2)^{1-\alpha}} \, d\xi_1 \quad .$$

Developing the function under the integral into a power series of $\lambda_1$, we find:

$$\int_0^1 \frac{d\xi_1}{(1 - \xi_1^2)^{\alpha+1/2}(1 - \lambda_1^2\xi_1^2)^{1-\alpha}} = \frac{\operatorname{tg} \alpha\pi}{2A\lambda} \qquad \left(A = \frac{\Gamma\left(\frac{1}{2} + \alpha\right)}{\sqrt{\pi}\, \Gamma(\alpha)}\right) ,$$

after which we may write:

$$I = \ln\xi + \frac{\operatorname{tg} \alpha\pi}{2A\lambda\xi} - \int_0^{\xi} \frac{\xi^{-1} - \xi_1^{-1}}{(1 - \xi_1^2)^{\alpha+1/2}(1 - \lambda_1^2\xi_1^2)^{1-\alpha}} \, d\xi_1 \quad ,$$

and (7.13) gives:

$$y - H = A\lambda(B + \overline{Q}) \times$$

$$\times \left\{- \ln \xi + \int_0^{\xi}\left[\frac{\xi^{-1} - \xi_1^{-1}}{(1 - \xi_1^2)^{\alpha+1/2}(1 - \lambda_1^2\xi_1^2)^{1-\alpha}} \xi + \frac{1}{\xi_1}\right]d\xi_1 + J_2 - \frac{\operatorname{tg} \alpha\pi}{2A\lambda}\right\} ,$$

where

$$J_2 = \int_0^1 \left[\frac{1}{(1 - \xi_1^2)^{\alpha+1/2}(1 - \lambda_1^2\xi_1^2)^{1-\alpha}} - 1\right] \frac{d\xi_1}{\xi_1} \quad .$$

From this, for large values of $x$ (i.e., small $\xi$), we find because of (7.13'):

$$y - H \approx A\lambda(B + \overline{Q}) \left(\ln \frac{2x + \overline{Q}}{B + \overline{Q}} + E\right) \quad ,$$

where

$$E = \int_0^1 \left\{\frac{1}{(1 - t^2)^{\alpha+1/2}(1 - \lambda_1^2 t^2)^{1-\alpha}} - 1\right\} \frac{1 - t}{t} \, dt \quad .$$

For $\alpha = 1/2$, i.e., for a ditch of rectangular cross section, we have:

$$E = 1 + \ln 2 - \frac{1 + \lambda}{\lambda} \ln(1 + \lambda) \quad .$$

We still must find one basic quantity of interest in the study of the flow, namely the flowrate of the drainage ditch. It must be expressed in function of parameter $\lambda$, also unknown. Thus we must find two equations to determine $\lambda$ and $Q$.

We return to formula (7.11) (for DC) and put in it $t = 1/\lambda^2$. Then $\zeta = 0$, $z = 0$, and we find:

$$b = A(B + \overline{Q})\, I_0 \quad ,$$

where

$$(7.14)\qquad I_0 = \int_1^{1/\lambda^2} \frac{\sqrt{1 - \lambda^2 t}\, dt}{t^{\alpha+1/2}(t - 1)^{1-\alpha}} .$$

Further, putting $t = 0$, i.e., $\zeta = 1$ in formula (7.11) (for CB), and recalling (see fig. 150a, where $C'C = b$) that

$$z = \frac{b}{2} + He^{i\pi\alpha} ,$$

we obtain this equation:

$$\frac{H}{\sin \alpha\pi} = \frac{A(B + \overline{Q})}{2} I_1 ,$$

where

$$(7.15)\quad \left\{ \begin{aligned} I_1 &= \frac{1}{A \cos \pi\alpha} - I_2 = \int_0^1 \frac{1 - \sqrt{1 - \lambda^2 t}}{t^{\alpha+1/2}(1 - t)^{1-\alpha}}\, dt , \\ I_2 &= \int_0^1 \frac{\sqrt{1 - \lambda^2 t}}{t^{\alpha+1/2}(1 - t)^{1-\alpha}}\, dt . \end{aligned} \right.$$

We divide (7.15) by (7.14). Recalling the relationship between $I_0$ and $I_2$,

$$I_0 = I_2 \cos \alpha\pi - \pi\lambda = \frac{1}{A} - I_1 \cos \alpha\pi - \pi\lambda ,$$

we find the connection between $H/b$ and $\lambda$:

$$\frac{2H}{b \sin \alpha\pi} = \frac{I_1}{I_0} = \frac{\frac{1}{A} \cos \alpha\pi - I_2}{I_2 \cos \alpha\pi - \pi\lambda} .$$

Then (7.14) and (7.15) give us the possibility of determining the flowrate $\overline{Q}$ by one of the formulas

$$(7.16)\qquad B + \overline{Q} = \frac{b}{AI_0} \qquad \left(\overline{Q} = \frac{Q}{k}\right)$$

where

$$(7.17)\qquad B + \overline{Q} = \frac{2H}{AI_1 \sin \alpha\pi} = \frac{B}{1 - A\pi\lambda}$$

It is easy to compute the flowrate through the banks, part of the total flowrate $Q$. We call $Q_1'$ the flowrate through both banks, $\overline{Q}_1$ the reduced flowrate, i.e.,

$$\overline{Q}_1 = \frac{Q_1}{k} .$$

To find $Q_1$, we return to equation (7.5), binding $\theta$ and $\zeta$:

$$\theta = -\frac{i}{2}(kB + Q)\zeta + \frac{iQ}{2} .$$

For $\zeta = \lambda'$, we have $z = \frac{b}{2}$, $\psi = Q_1/2$, $\varphi + ky = 0$; therefore we obtain

$$Q_1 = Q + kb - \lambda'(Q + kB) = kb + B\,\frac{\pi A\lambda - \lambda'}{1 - A\pi\lambda} .$$

Using the expressions for $Q$ and $H$, we may obtain these formulas

$$\frac{Q_1}{Q} = \frac{I_2 \cos \alpha\pi - \frac{\lambda'}{A}}{\pi\lambda} \quad , \quad \frac{Q_1}{2kH} = \frac{AI_2 \cos \alpha\pi - \lambda'}{\operatorname{tg} \alpha\pi(1 - AI_2 \cos \alpha\pi)}$$

We now consider the particular use of vertical banks, when $\alpha = 1/2$. All formulas simplify significantly, since all integrals are expressed in finite form.

So, for $U$ we have, instead of (7.7):

along DC $$U = \frac{2i}{k\pi} \operatorname{tg}^{-1} \sqrt{t - 1} \quad ,$$

along CB $$U = \frac{2}{k\pi} \tanh^{-1} \sqrt{1 - t} \quad ,$$

along BA $$U = -\frac{2}{k\pi} \coth^{-1} \sqrt{1 - t} + \frac{i}{k} \quad .$$

Calculating integral (7.8) we find:

$$J_1 = 2\operatorname{tg}^{-1} \frac{\lambda'}{\lambda} \quad .$$

Putting

$$\lambda' = \cos \theta \quad ,$$

we may consider the parameter $\theta$ instead of $\lambda$. It is easy to see that

$$J_1 = 2\operatorname{tg}^{-1}(\operatorname{cotg} \theta) = \pi - 2\theta \quad .$$

Formula (7.9) gives:

$$\frac{k}{v_0 + k} = \frac{\pi - 2\theta}{\pi} \quad ,$$

which may be written as

$$\sin \frac{k\pi}{2(v_0 + k)} = \cos \theta = \lambda' \quad .$$

Further, for $\alpha = 1/2$, integrals $I$ and $I_1$ give:

$$I_0 = \pi(1 - \lambda), \quad I_1 = 2\left(\ln \lambda' + \lambda \ln \frac{1 + \lambda}{\lambda'}\right) \quad .$$

By means of formulas (7.16) and (7.17) we obtain:

$$\bar{Q} = \frac{B\lambda}{1 - \lambda} \, , \quad H = \frac{B}{\pi(1 - \lambda)} \left(\ln \lambda' + \lambda \ln \frac{1 + \lambda}{\lambda'}\right) \quad .$$

These formulas were obtained by V. V. Vedernikov [3]. Assinging values of $\lambda$ in the interval $0 < \lambda < 1$, we compute $H/B$ and $\bar{Q}$.

We notice that the case $\alpha = 1$, when the slopes of the banks become horizontal, gives us Zhukovsky's problem.

Fig. 151 renders another particular case — the entire ditch is reduced to a vertical slit.

## §8. Seepage in Soil Overlying Draining or Waterbearing Stratum.

We communicate the solution of the problem [11] of seepage from a canal of width $B$ with water depth $H$, accounting for soil capillarity, and also for the presence of a draining or waterbearing stratum at some depth $T$ below the water level in the canal. Therefore we assume that the piezometric level of the water in the pervious stratum lies at a distance

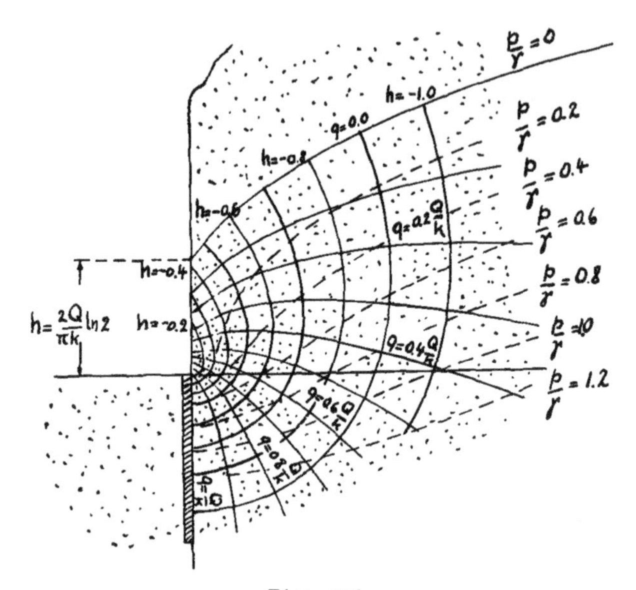


Fig. 151

$H_0$ above the boundary between the soil, through which the canal runs, and the pervious layer (Fig. 152a).

The same solution for an appropriate choice of the meaning of the parameters, is also the solution of the problem of flow of groundwater to a drainage canal from a confined waterbearing stratum.

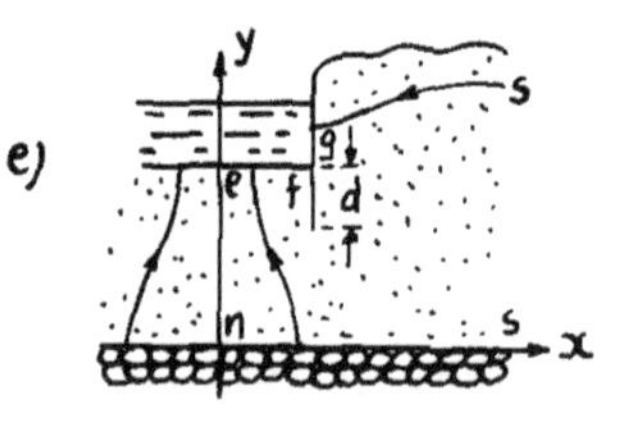


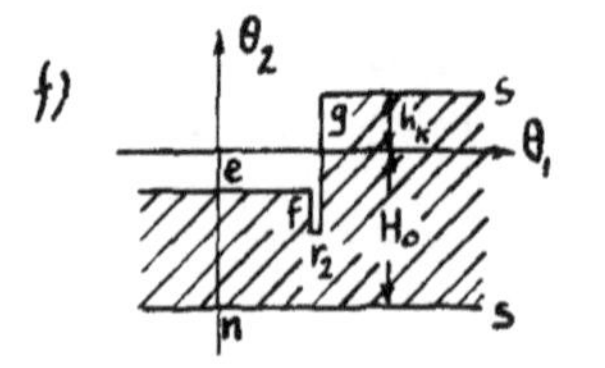


Fig. 152

It is assumed that the vertical walls of the canal are impervious, while the horizontal bottom is pervious.

At the free surface, the velocity potential has the value

$$\varphi = - \kappa(y - h_k) \quad ,$$

where $h_k$ is the height of capillary rise in the soil. Far from the canal for $x = \infty$, the free surface is horizontal and at a distance $H_0 + h_k$ above the waterbearing stratum.

The region of Zhoukovsky's function $\theta = (z + i\omega)/\kappa$ has the form of a polygon with cuts, horizontal or vertical (Fig. 152, e.f.). In the first case the region may consist of two sheets. Mapping onto the $\zeta$ half-plane gives

$$\omega = \frac{iQ}{2K_\rho} F(\text{arc sin } \zeta, \rho) - \kappa T \quad . \tag{8.1}$$

Putting $\zeta = 1/\rho$ and considering that for this $\omega = - \kappa H_0 + \frac{Qi}{2}$, we find:

$$Q = \frac{2\kappa(T - H_0)}{K'_\rho} K_\rho \quad . \tag{8.2}$$

Here $K_\rho$, $K'_\rho$ are complete elliptic integrals with modulus $\rho$. For Zhoukovsky's function, the mapping function is

$$\theta = P \int_0^\zeta \frac{(\alpha^2 - \zeta^2)\, d\zeta}{(1 - \rho^2\zeta^2)\sqrt{(1 - \zeta^2)(1 - k^2\zeta^2}} - iH \quad . \tag{8.3}$$

Recalling that

$$\frac{\alpha^2 - \zeta^2}{1 - \rho^2\zeta^2} = \frac{1}{\rho^2}\left[1 - \frac{1 - \rho^2\alpha^2}{1 - \rho^2\zeta^2}\right] \quad ,$$

we obtain:

$$\theta = \frac{P}{\rho^2}\left[u - (1 - \rho^2\alpha^2) \prod (-\rho^2, \text{arc sin } \zeta, k)\right] \tag{8.4}$$

where $u$ and $\prod$ are elliptic integrals of first and third kind

$$\left\{\begin{aligned} u &= \int_0^\zeta \frac{d\zeta}{\sqrt{(1 - \zeta^2)(1 - k^2\zeta^2)}} \\ \prod (-\rho^2, \text{arc sin } \zeta, k) &= \int_0^\zeta \frac{d\zeta}{(1 - \rho^2\zeta^2)\sqrt{(1 - \zeta^2)(1 - k^2\zeta^2)}} \quad . \end{aligned}\right. \tag{8.5}$$

Computing the residue in point $\zeta = 1/\rho$, we find the value of $P$:

$$P = \frac{2(H_0 + h_k)\sqrt{(1 - \rho^2)(k^2 - \rho^2)\rho}}{\prod (1 - \rho^2\alpha^2)} \quad . \tag{8.6}$$

By means of substitutions

$$\rho = k \text{ sn } a, \quad \zeta = \text{sn } u \tag{8.7}$$

integral $\prod(-\rho^2, \arcsin\zeta, k)$ reduces to an integral expressed in terms of elliptic functions:

$$(8.7') \qquad \prod = \int_0^u \frac{du}{1 - k^2 \operatorname{sn}^2 a \operatorname{sn}^2 u} = \frac{\operatorname{sn} a}{\operatorname{cn} a \operatorname{dn} a} \left\{ \frac{u}{2K} \frac{\theta_1'\left(\frac{a}{2K}\right)}{\theta_1\left(\frac{a}{2K}\right)} - \frac{1}{2} \ln \frac{\theta_0\left(\frac{a+u}{2K}\right)}{\theta_0\left(\frac{a-u}{2K}\right)} \right\}$$

Therefore we obtain for $\theta$:

$$(8.8) \qquad \theta = Su + \frac{H_0 + h_k}{\pi} \ln \frac{\theta_0\left(\frac{a+u}{2K}\right)}{\theta_0\left(\frac{a-u}{2K}\right)} - iH ,$$

where

$$(8.9) \qquad S = \frac{2(H_0 + h_k)}{\pi} \left[ \frac{\alpha^2 \sqrt{(1-\rho^2)(k^2-\rho^2)}}{1 - \rho^2\alpha^2} - \frac{1}{K} \frac{\theta_1'\left(\frac{\pi a}{2K}\right)}{\theta_1\left(\frac{\pi a}{2K}\right)} \right] .$$

Jacoby's function $\theta_0$ is determined by means of the series

$$(8.10) \qquad \theta_0\left(\frac{u}{2K}, k\right) = 1 - 2e^{-\frac{\pi K'}{K}} \cos\frac{\pi u}{K} + 2e^{-\frac{4\pi K'}{K}} \cos\frac{2\pi u}{K} - \dots$$

Further we have:

$$(8.11) \qquad \frac{1}{2} \ln \frac{\theta_0(x+y)}{\theta_0(x-y)} = \sum_{n=1}^{\infty} \frac{\sin 2n\pi x \sin 2n\pi y}{n \operatorname{sh} n\pi\rho} \qquad \left(\rho = \frac{K'}{K}\right) ,$$

$$(8.12) \qquad \frac{\theta_0'(x)}{\theta_0(x)} = 2\pi \sum_{n=1}^{\infty} \frac{\sin 2n\pi x}{\operatorname{sh} n\pi\rho} \qquad \left(\rho = \frac{K'}{K}\right) .$$

We put $\zeta = 1/\rho$. We have:

$$\frac{B - \overline{Q}}{2} - ih_k = \dot{S}(K - iK') - \frac{i(H_0 + h_k)a}{K} + iH ,$$

and this gives after separation of real and imaginary parts:

$$B - \overline{Q} = 2SK, \quad H + h_k = (H_0 + h_k)\frac{a}{K} - SK' .$$

Eliminating $S$ from these formulas, we have:

$$(8.13) \qquad \overline{Q} = \frac{Q}{\kappa} = B + \frac{2K}{K'}(H + h_k) - 2(H_0 + h_k)\frac{a}{K'} .$$

If we eliminate $\overline{Q}$ from formulas (8.2) and (8.11), we find:

$$(T - H_0)\frac{2K_\rho}{K_\rho'} = B + \frac{2K}{K'}(H + h_k) - \frac{2a}{K'}(H_0 + h_k) .$$

To determine the coordinate $\zeta_c$ of the tip of the cut-off, we write

$$\frac{1}{w} = \frac{dz}{d\omega} = \frac{d\theta}{d\omega} - \frac{i}{\kappa} = \frac{2KP}{Qi} \frac{\alpha^2 - \zeta^2}{\sqrt{(1 - k^2\zeta^2)(1 - \rho^2\zeta^2)}} - \frac{i}{\kappa} .$$

Making this expression zero, we find for $\zeta = \zeta_c$ the equation:

$$(8.14) \qquad \frac{\zeta_c^2 - \alpha^2}{\sqrt{(1 - k^2\zeta_c^2)(1 - \rho^2\zeta_c^2)}} = \frac{T - H_0}{K'_\rho} \frac{\pi(1 - \alpha^2\rho^2)}{(H_0 + h_k)\sqrt{(1 - \rho^2)(k^2 - \rho^2)}} .$$

For a non-penetrating cut-off, $d = 0$, and $\zeta_c = 1$. Then equation (8.14) simplifies and becomes:

$$(8.15) \qquad \frac{H_0 + h_k}{T - H_0} = \frac{\pi}{2K'_\rho} \frac{(1 - \rho^2\alpha^2)\sqrt{1 - k^2}}{(1 - \alpha^2)\sqrt{k^2 - \rho^2}} .$$

§9. Drain or Irrigation Canal in the Case of Evaporation or Infiltration. We investigate the case of evaporation to fix the ideas [6].

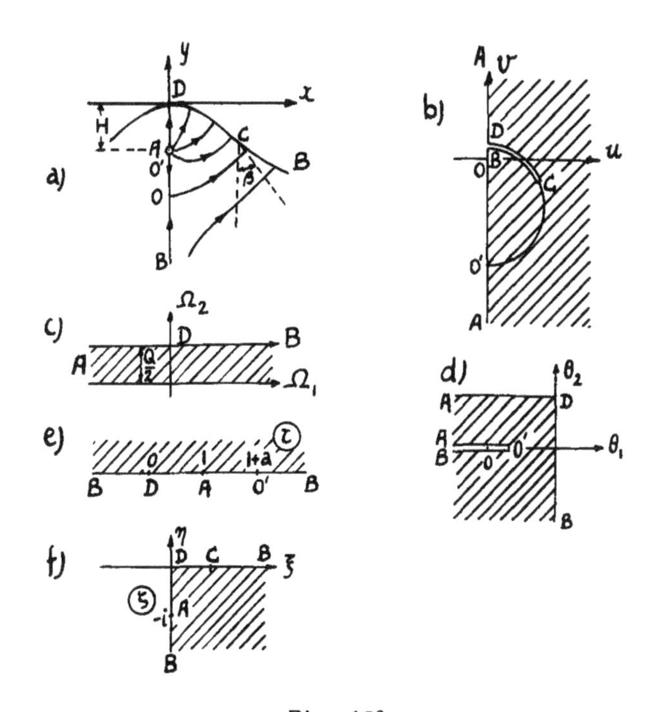

Fig. 153

We put $-\varepsilon = c > 0$ and examine the functions (the y-axis is oriented upward)

$$(9.1) \qquad \theta = \theta_1 + i\theta_2 = \omega - ikz, \qquad \Omega = \Omega_1 + i\Omega_2 = \omega + icz .$$

The flowregion is sketched in Figure 153a, the velocity hodograph in Fig. 153b, to which we might apply inversion in point D. But it is convenient to consider the region $\Omega$, for which we obtain a strip (Fig. 153c). (The region of Zhoukovsky's function is sketched in Fig. 153d. The tip of the cut in this region is in point O'. Indeed, since $\theta_1 = \varphi + ky$, in this point $\partial\theta_1/\partial y = \partial\varphi/\partial y + k = 0$.)

Conformal mapping onto the upper half-plane $\tau$ gives:

$$d\Omega = \frac{Q\, d\tau}{2\pi(\tau - 1)}, \qquad d\theta = B\,\frac{\tau - (1 + a)}{(\tau - 1)\sqrt{\tau}}\, d\tau \quad . \tag{9.2}$$

The substitution $\sqrt{\tau} = i\zeta$ maps the half-plane $\tau$ onto an angle of the $\zeta$ region.

We obtain

$$d\Omega = \frac{Q\zeta\, d\zeta}{\pi(\zeta^2 + 1)}, \qquad d\theta = B\left(2i\, d\zeta + \frac{2ai\, d\zeta}{\zeta^2 + 1}\right) \quad .$$

From this we find

$$dz = -\frac{i}{k + c}\,(d\Omega - d\theta) = -\frac{i}{k + c}\left(-2iB + \frac{\frac{Q}{\pi}\zeta - 2aiB}{\zeta^2 + 1}\right) d\zeta =$$

$$= -\frac{i}{k + c}\left(-2iB + \frac{\frac{Q}{2\pi} + aB}{\zeta + i} + \frac{\frac{Q}{2\pi} - aB}{\zeta - i}\right) d\zeta \tag{9.3}$$

Since point $\zeta = -i$ corresponds to point A of the flowregion, where $z$ has a finite value, we must have $Q/2\pi + aB = 0$, from which

$$B = -\frac{Q}{2a\pi}, \qquad dz = -\frac{i}{k + c}\left(\frac{iQ}{a\pi} + \frac{Q}{\pi}\;\frac{1}{\zeta - i}\right) d\zeta \quad . \tag{9.4}$$

Integrating and considering $z = 0$ for $\zeta = 0$, we obtain:

$$z = \frac{Q}{(k + c)\pi}\left[\frac{\zeta}{a} - i\,\ell n(1 + i\zeta)\right] \quad . \tag{9.5}$$

Since $\zeta = -i$, $z = -iH$, we find:

$$(k + c)\frac{\pi}{Q}\, H = \frac{1}{a} + \ell n\, 2 \quad . \tag{9.6}$$

The equation of the free surface is obtained by putting $\zeta = \xi + i\eta$ and considering that for the free surface $\eta = 0$,

$$x = \frac{Q}{(k + c)\pi}\left(\frac{\xi}{a} + \operatorname{arc\,tg} \xi\right), \quad y = -\frac{Q}{(k + c)\pi}\;\ell n\sqrt{1 + \xi^2} \quad . \tag{9.7}$$

In the inflection point of the free surface, as is easy to see, $\xi = \sqrt{1 + a}$. Calling $\beta$ the angle of the tangent in the inflection point with the vertical, we find:

$$\frac{a}{2\sqrt{1 + a}} = \operatorname{cotg} \beta, \qquad a = \frac{2}{\operatorname{tg} \beta\, \operatorname{tg} \frac{\beta}{2}} \quad . \tag{9.8}$$

The branches of curve (9.7) tend downward to infinity.

Let $h_0$ be the head on the contour of the pipe. By means of formula (9.1) we may write for $\omega$

$$\omega = \varphi + i\psi = -icz + \frac{Q}{2\pi}\,\ell n(1 + \zeta^2) \quad . \tag{9.9}$$

Putting $\zeta = -i + \zeta'$, $z = -iH + z'$, we have on an infinitesimally small circle about point A

$$\omega = -cH + \frac{Q}{2\pi}\,\ell n(-2i\zeta') \quad . \tag{9.10}$$

Because of (9.5) we have with accuracy up to small terms of higher order

$$z' = \frac{Q\zeta'}{2\pi(k + c)\cos\beta} \quad . \tag{9.11}$$

We now put $z' = \delta e^{i\theta}$, where $\delta$ is the radius of the pipe. Substituting this expression in (9.11) and (9.10) and separating real and imaginary part, we find:

$$h_0 = -\frac{Q}{2\pi k}\ \ln\frac{4\pi\delta(k + c)\cos\beta}{Q} + \frac{c}{k}H \ .$$

From the equation

$$h = \frac{p}{\gamma} + y + h_k$$

where $h_k$ is the height of capillary rise, we may write for the pressure $p_0$ on the pipe:

$$\frac{p_0}{\gamma} = \frac{Q}{2\pi k}\ln\frac{Q}{4\pi\delta(k + c)\cos\beta} + \frac{k + c}{k}H - h_k \ . \tag{9.12}$$

Here one assumes that the seepage coefficient of the pipe material is equal to the seepage coefficient of the soil.

When $\beta = 0$, the inflection point moves to infinity, parameter $a$ becomes infinite and we obtain from (9.6)

$$H = \frac{\ln 2}{\pi}\ \frac{Q}{k + c} \ .$$

The free surface has the asympote

$$x = \frac{Q}{2(k + c)}$$

(Fig. 154). The velocity at infinity is $v = -k$, as in general this velocity is $v = c$.

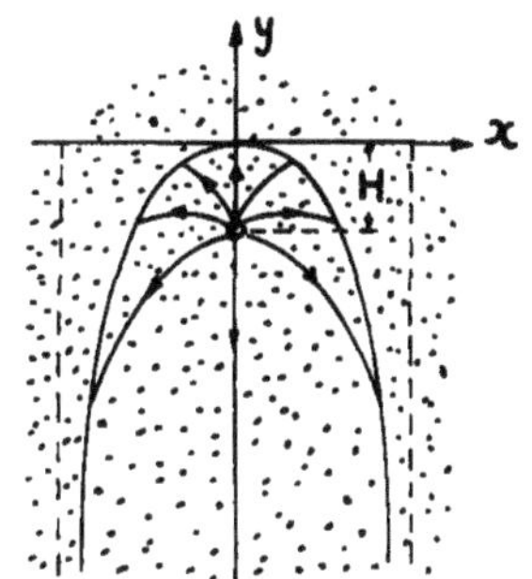


Fig. 154

The problem for the drainage pipe is solved in exactly the same way. We may obtain this solution from the previous one by substituting $-Q$ for $Q$ and putting $-a = b > 0$ .

The equation of the free surface has the form

$$\frac{\pi(k + c)}{Q}x = \frac{\xi}{b} - \text{arc tg}\,\xi, \quad \frac{\pi(k + c)}{Q}y = \frac{1}{2}\ \ln(1 + \xi^2) \quad , \tag{9.13}$$

where

$$b = \frac{2}{1 + \sec\beta} = \frac{2\ \text{tg}\frac{\beta}{2}}{\text{tg}\,\beta} \quad ,$$

where $\beta$ is the angle between vertical and tangent in inflection point $C$ (Fig. 155a). For the depth $H$ of the bedding of the pipe we have

$$\frac{\pi(k + c)}{Q}H = \frac{\sec\beta + 1}{2} - \ln 2.$$

Formula (9.12) for the pressure preserves its form.

When $\beta = 0$, the inflection point disappears and becomes a branch point (Fig. 155b). In formulas (9.13) one must put $b = 1$. The problems

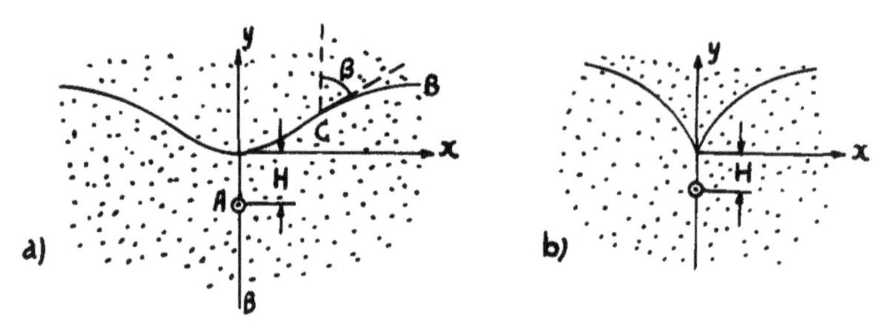


Fig. 155

here examined may be of interest in the theory of the so-called "moleskin drain." Such a drain or irrigation canal is perforated in the following way: a shell with sharp edges is placed at the lower end of a vertically established knife. When the knife cuts through the soil, the shell cuts a cylindrical hole in the soil.

## §10. Semi-inverse Method Applied to Earthdams on Impervious Foundations.

In the previous chapter we already encountered examples of semi-inverse methods, when we solved the problem where part of the boundaries were accurately given and part were determined by some artificial conditions, for the purpose of obtaining much simpler solutions than those obtained when all boundaries were given. Often such conditions are given in the complex potential plane. A more general example of this kind is given by S. N. Numerov [12].

A earthdam with horizontal impervious foundation is sketched in Fig. 156a. The boundaries of the dam with the reservoirs are curvilinear.

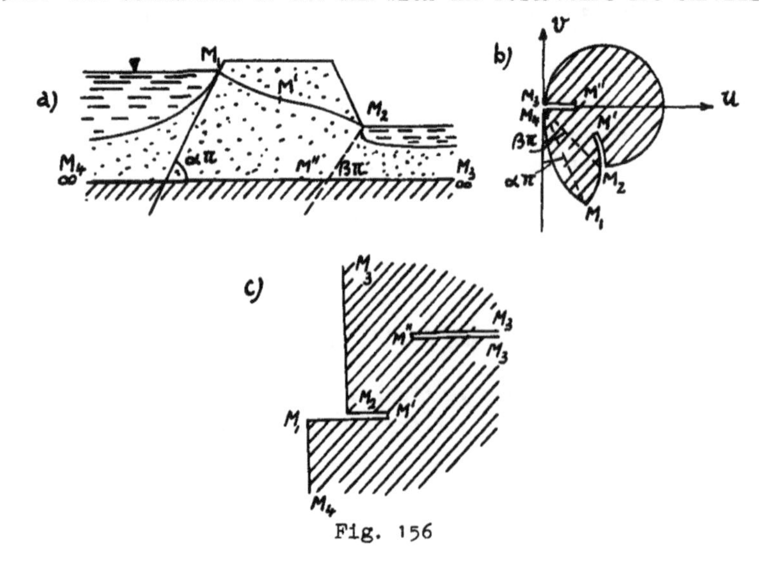


Fig. 156

Their form is not known beforehand, but is determined from the condition that in the hodograph plane the boundaries are represented as arcs of circles, tangent to the ordinate axis in the origin. The tangents to the banks in points $M_1$ and $M_2$, coinciding with the waterlevel in head — and tailwater reservoirs, make angles $\alpha\pi$ and $\beta\pi$ with the horizontal. The water depths in the reservoirs are $H_1$ and $H_2$.

Considering that the vertices $M_1$, $M_2$, $M_3$, $M_4$, $M'$ and $M''$ transform respectively into points $\zeta = -1, +1, 1/k, -1/k, \mu, \nu$ and that there is the rectangle $M_1M_2M_3M_4$ in the $\omega$-plane ($\omega_1 = 0$, $\omega_2 = \kappa H$, $\omega_3 = \kappa H - iQ$, $\omega_4 = -iQ$), we obtain:

$$(10.1) \qquad \frac{\kappa}{u - iv} = \kappa \frac{dz}{d\omega} = M \int \frac{(\zeta - \mu)(\zeta - \nu)d\zeta}{(1 - k^2\zeta^2)\sqrt{(1 - \zeta^2)(1 - k^2\zeta^2)}} + N \quad ,$$

$$(10.2) \qquad \omega = -\frac{\kappa H}{2K}\left[K - \int_0^{\zeta} \frac{d\zeta}{\sqrt{(1 - \zeta^2)(1 - k^2\zeta^2)}}\right] \quad .$$

The substitution $\zeta = \operatorname{sn} u$ $\left(u = \frac{2K\omega}{\kappa H} + K\right)$ reduces (10.1) to the integrals

$$(10.3) \qquad \int \frac{\operatorname{sn} u\, du}{\operatorname{dn}^2 u} = -\frac{1}{k'^2}\frac{\operatorname{cn} u}{\operatorname{dn} u} = \frac{1}{k'^2}\operatorname{sn}(u - K) = \frac{1}{k'^2}\operatorname{sn}\frac{2K\omega}{\kappa H} \quad ,$$

$$(10.4) \qquad \int \frac{du}{1 - k^2\operatorname{sn}^2 u} = pu + q\,\frac{\theta_0'\left(\frac{u - K}{2K}\right)}{\theta_0\left(\frac{u - K}{2K}\right)} \quad .$$

The second of them is an elliptic integral of the third kind of form (8.7') where one must take $\operatorname{sn} a = 1$. Then $a = k$, $\operatorname{dn} a = k'$, $\operatorname{cn} a = 0$ and in the right part of (8.7') one obtains an improper division. Clearing this operation, we find (10.4) where $p$ and $q$ are some constants. The Jacoby theta function $\theta_0$ is determined by means of series (8.10).

Now, introducing new values of the constants, we obtain instead of (10.1):

$$(10.5) \qquad \kappa\frac{dz}{d\omega} = -i + A + 2B\frac{\omega}{\kappa H} + C\,\frac{\theta_0'\left(\frac{\omega}{\kappa H}\right)}{\theta_0\left(\frac{\omega}{\kappa H}\right)} + 2kKD\,\operatorname{sn}\left(\frac{2K\omega}{\kappa H}\right) \quad .$$

We determine the constants A, B, C, D. In point $M_1$, where $\omega = 0$, $u = \kappa \sin \alpha\pi \cos \alpha\pi$, $v = -\kappa\cos^2\alpha\pi$, one must have:

$$(10.6) \qquad A = \operatorname{tg}\alpha\pi \quad .$$

In point $M_2$, where $\omega = \kappa H$, $w = \kappa i \cos\beta\pi e^{-\beta\pi i}$, we find:

$$(10.7) \qquad B = \frac{1}{2}(\operatorname{tg}\beta\pi - \operatorname{tg}\alpha\pi) \quad .$$

Modulus $k$ is the root of the equation

$$(10.8) \qquad \frac{Q}{\kappa H} = \frac{K'}{2K} \quad .$$

Integrating (10.5) along $\omega$, we obtain (see literature to Chapter III [24, 25]):

$$\frac{z}{H} = (A - i)\frac{\omega}{\kappa H} + B\left(\frac{\omega}{\kappa H}\right)^2 + C\,\ln\frac{\theta_0\left(\frac{\omega}{\kappa H}\right)}{\theta_0(0)} -$$

$$- D\,\ln\frac{\mathrm{dn}\left(\frac{2K\omega}{\kappa H}\right) + k\,\mathrm{cn}\left(\frac{2K\omega}{\kappa H}\right)}{1 + k} \quad . \tag{10.9}$$

Applying (10.9) to point $M_2$, where $z = L - iH$ (the x-axis is horizontal, through $M_1$, the y-axis is upward through $M_1$) we find:

$$\frac{L}{H} = A + B + D\,\ln\frac{1 + k}{1 - k} \quad . \tag{10.10}$$

In point $M_3$, $y = -H_1$, $\omega = \kappa H - Qi$; in point $M_4$, $y = -H_1$, $\omega = -Qi$. Therefore we find:

$$-\frac{H_1}{H} = -A\frac{Q}{\kappa H} + \frac{\pi}{2}(C + D) - i \, , \tag{10.11}$$

$$1 + 2B\frac{Q}{\kappa H} - \pi C = 0 \quad . \tag{10.12}$$

Eliminating $C$ and $D$ from the last equations, we find an equation for $k$:

$$\frac{L}{H} = -\frac{2}{\pi}\left(\frac{H_1}{H} + \frac{1}{2}\right)\ln\frac{1 + k}{1 - k} + (B + A)\left(1 - \frac{K'}{\pi K}\,\ln\frac{1 + k}{1 - k}\right) \quad .$$

For given $A$ and $B$ [see (10.6), (10.7)] $C$ and $D$ are determined from (10.12).

In the case treated, it has been assumed that there is no seepage surface. S. N. Numerov shows that acceptable positions for head and tail-water levels are obtained when the inequalities

$$H_1 > \frac{Q}{\kappa}\,\mathrm{tg}\,\alpha\pi, \quad H_2 > \frac{Q}{\kappa}\,\mathrm{tg}\,\beta\pi \tag{10.13}$$

are observed.

Putting $\omega = -\kappa y$, in (10.9), we find the equation of the free surface

$$\frac{\kappa}{H} = -\frac{Ay}{H} + B\left(\frac{y}{H}\right)^2 + C\,\ln\frac{\theta_0\left(\frac{y}{H}, k\right)}{\theta_0(0, k)} -$$

$$- D\,\ln\frac{\mathrm{dn}\left(\frac{2Ky}{H}, k\right) + k\,\mathrm{cn}\left(\frac{2Ky}{H}, k\right)}{1 + k} \quad . \tag{10.14}$$

We consider particular cases.

1) In case of pervious foundation of infinite depth ($T = \infty$), we have: $Q = \infty$, $k = 0$; the relationships (10.9) and (10.14) give ($m = \mathrm{cotg}\,\alpha\pi$; $n = \mathrm{cotg}\,\beta\pi$):

$$\frac{z}{H} = \left(\frac{1}{m} - i\right)\frac{\omega}{\kappa H} + \frac{m-n}{2mn}\left(\frac{\omega}{\kappa H}\right)^2 + \left(\frac{L}{H} - \frac{m+n}{2mn}\right)\sin^2\frac{\pi\omega}{2\kappa H} \tag{10.15}$$

$$(0 \le \varphi \le \kappa H, \quad 0 \le \psi \le \infty)$$

$$\frac{x}{H} = -\frac{1}{m}\frac{y}{H} + \frac{m-n}{2mn}\left(\frac{y}{H}\right)^2 + \left(\frac{L}{H} - \frac{m+n}{2mn}\right)\sin^2\frac{\pi y}{2H} . \tag{10.16}$$

Inequalities (10.13) are replaced by one

$$\frac{L}{H} > \frac{m+n}{2mn} .$$

2) If moreover $n = \infty$, then we have the case considered by F. B. Nelson-Skornjakov (see literature to Chapter IV [10]) when bottom of tail-water reservoir is horizontal. He also obtained the solution for the case $n = \infty$, $H_2 = 0$.

3) If we assume, for finite depth $T$, that the angle $\pi\beta = 0$, i.e., $n = \infty$, then the case of Fig. 157 holds. Changing first the coordinates of the $z$ and $\omega$ planes, we write the equation expressing the relationship between $z$ and $\omega$:

$$z = -\frac{i\omega}{\kappa} - \frac{\kappa}{2Q}\left(\frac{\omega}{\kappa}\right)^2 + \frac{2T}{\pi}\ln \operatorname{ch}\frac{\pi\omega}{2Q} , \tag{10.17}$$

and the equation of the free surface (for it $\psi = 0$, $\omega/\kappa = -y$):

$$x = -\frac{\kappa}{2Q}y^2 - \frac{2T}{\pi}\ln \operatorname{ch}\frac{\pi\kappa y}{2Q} . \tag{10.18}$$

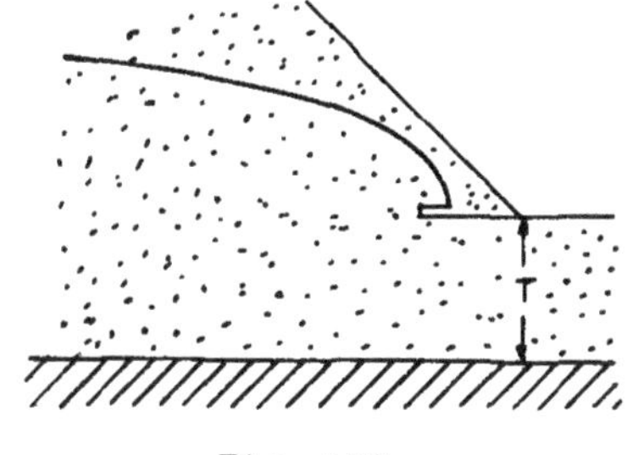


Fig. 157

The width $\ell$ of the filter is determined by the equation that is obtained from (10.17) for $\varphi = 0$, $\psi = \psi_0 = Q$, $y = 0$:

$$\frac{\ell}{T} = \frac{Q}{\kappa T}\left(a + \frac{1}{2}a^2\right) + \frac{2}{\pi}\ln\cos\frac{\pi a}{2} ,$$

in which $a$ is the root of the equation, obtained from the condition $d\omega/dz = \infty$:

$$\frac{Q}{\kappa T} = \frac{\operatorname{tg}\frac{\pi a}{2}}{1+a} = \frac{\pi a}{2} + \ldots .$$

For small $a$ we have $\ln\cos\frac{\pi a}{2} \approx \frac{\pi^2 a^2}{8}$, and this gives the approximate equality

$$\ell \approx \frac{Q^2}{\pi\kappa^2 T} .$$

## §11. Contour of Constant Velocity in Soil of Infinite Depth.

In Fig. 158 we have a typical outline of the subterranean contour of a reinforced concrete weir. Line BC is the footing of the dam. Additional structural elements — apron AB, cut-offs MN, BK and apron with filterblanket CD — have for purpose to secure the stability of the structure and to distribute the headloss $H$ in such a way as to have no large velocities along the contour of the base of the structure.

Using the formula for seepage velocity

$$v = kJ,$$

where $J$ is the gradient of the head, we may make a crude estimate of the average seepage velocity, if we take the average value of $J$ equal to the simple ratio of the effective head $H$ to the total length of the seepage path, i.e., the length $L$ of the subterranean contour:

$$v_{ave} = kJ_{ave} = k\frac{H}{L}.$$

One has to consider the fact [14] that soil erosion and soil deformation do not depend upon the velocities, but on the critical gradient of the head, near one, for which dangerous soil deformation starts, leading to erosion. Thus for sands one may take $J_{crit} = 0.8$. From this consideration one might take a contour length, say one and a half times $H$, so that $v_{ave}$ does not exceed the critical value for the velocity. However local flow velocities may be significantly larger than the average. So at the ends of the flat bottom weir foundation, at the tips of the cut-offs, at the extreme points of the embedded weir foundation, and so on, these velocities are theoretically infinitely large, no matter how long the length of the seepage path.

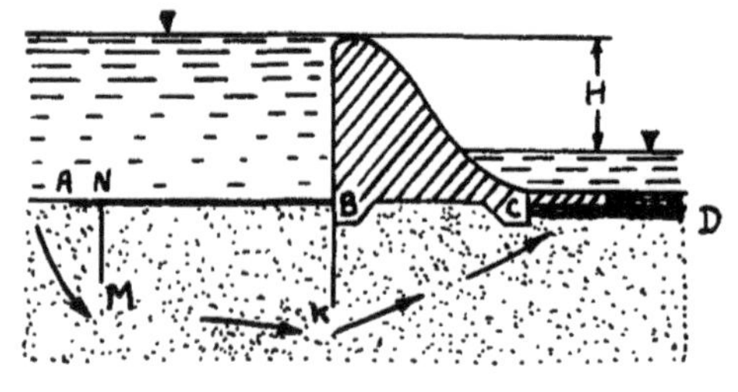

Fig. 158

Why of course these velocities are not infinite and how large they really are, is a question for which the conventionally applied theory has no answer.

A. P. Votschinin introduced the problem of the "ideal" contour, i.e., a contour in all points of which the velocity has the same value.

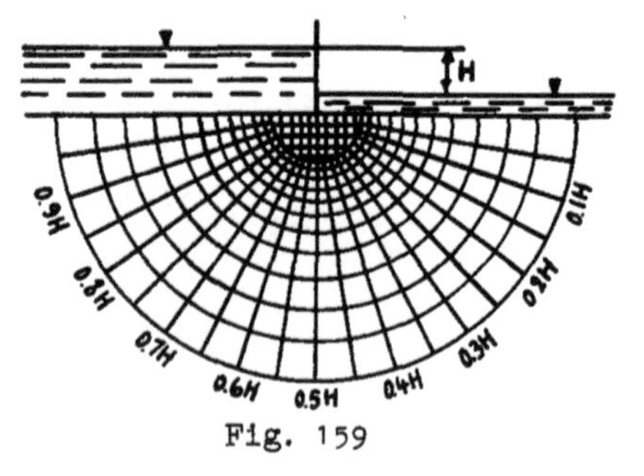

Fig. 159

It is evident that for soil of infinite depth and having horizontal boundaries with head — and tailwater reservoir, the ideal contour is a half circle. Let the radius of this half circle be $R$ (Fig. 159). The complex potential may be written as

$$\omega = \varphi + i\psi = \frac{kH}{\pi i}\ln\left(\frac{z}{R}\right) + C.$$

The complex velocity is

$$w = u - iv = \frac{kH}{\pi i z}$$

and the velocity along the contour

$$v_0 = \frac{kH}{\pi R} = \frac{kH}{L} .$$

Here $L$ is the length of the half circle: $L = \pi R$. The magnitude of gradient $J$ is here exactly $H/L$.

Assuming, for example, $J = 0.8$, we obtain $L = 1.25H$ and this guarantees values of the gradient which nowhere exceed the critical value. However, the half circle is not acceptable from the viewpoint of its stability and method of construction. Therefore we generalize A. P. Votschinin's problem considering a contour consisting of rectilinear segments [15]. First though we fix our attention on the form of the contour of constant velocity in soil of finite depth.

We notice that the form of the contour has a very important meaning in the region of exit of groundwater flow into the lower reservoir, since erosion is often linked with a poor design of the contour in the region on the tailwater reservoir.

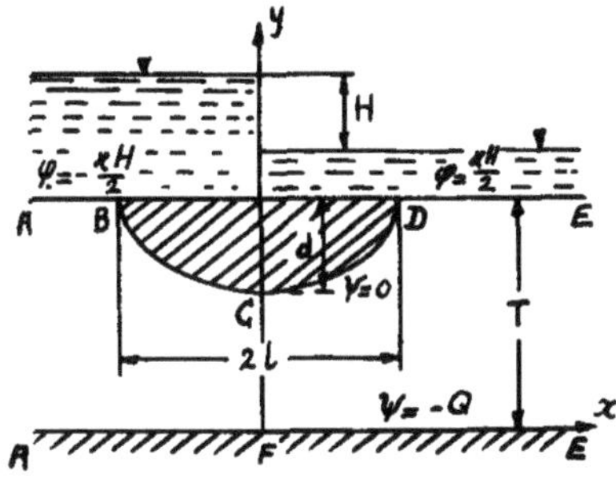


Fig. 160

§12. Contour of Constant Velocity in Soil of Finite Depth.

Suppose we have a horizontal stratum of depth $T$ (Fig. 160). We choose the contour of the base of the impervious structure BCD so that in every point its piezometric slope (or hydraulic gradient) has a given value $J_0 \leq J_{crit}$. $J_{crit}$ is the critical value of the gradient, i.e., the value for which dangerous deformations of the soil are possible [14]. Then the velocity along the contour has a constant value $v_0$, equal to

$$v_0 = \kappa J_0 ,$$

where $\kappa$ is the seepage coefficient.

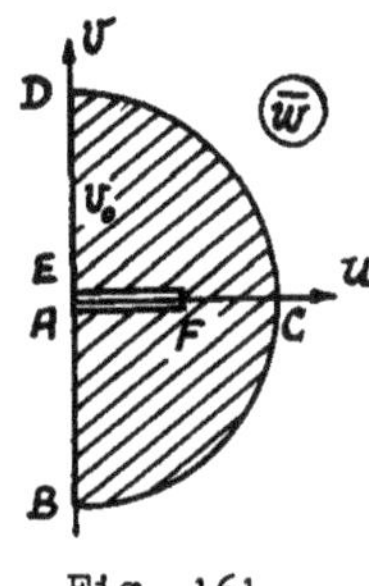


Fig. 161

We take the hodograph as represented in Fig. 161, i.e., along BCD we take the magnitude of the velocity vector equal to $v_0$, so that BCD is a half circle. The cut AFE corresponds to the boundary of impervious bedrock.

In the $\omega = \varphi + i\psi$ plane we have a rectangle, which we choose as sketched in Fig. 162. Mapping the rectangle BDEA onto the half plane $\zeta$, we have:

$$(12.2) \qquad F(\text{arc sin } \zeta, k) = \int_0^{\zeta} \frac{d\zeta}{\sqrt{(1-\zeta^2)(1-k^2\zeta^2)}} = \int_0^{\text{arc sin } \zeta} \frac{d\varphi}{\sqrt{1-k^2 \sin^2 \varphi}} ,$$

$$(12.3) \qquad K = \int_0^1 \frac{d\zeta}{\sqrt{(1-\zeta^2)(1-k^2\zeta^2)}} .$$

For the flowrate $Q$ we have, as usually, the formula

$$Q = \frac{\kappa HK'}{2K} ,$$

where

$$K' = \int_0^{\pi/2} \frac{d\varphi}{\sqrt{1 - k'^2 \sin^2 \varphi}} \qquad (k'^2 = 1 - k^2) \quad .$$

We return to the velocity hodograph. Calling $u - iv = w$, $u + iv = \bar{w}$, we compose function $\bar{W} = \bar{w}^2$. The region of the values of this

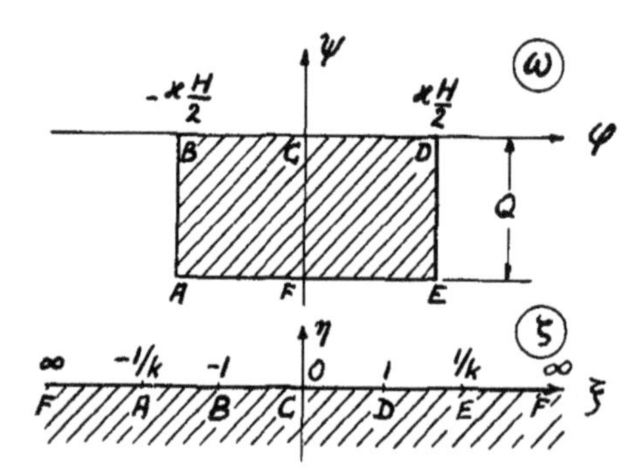


Fig. 162

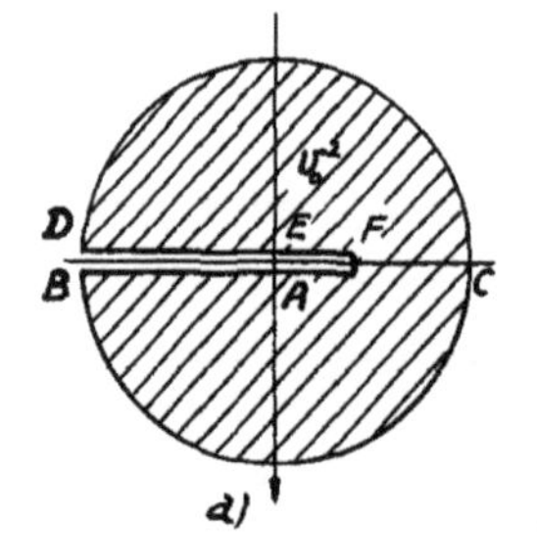


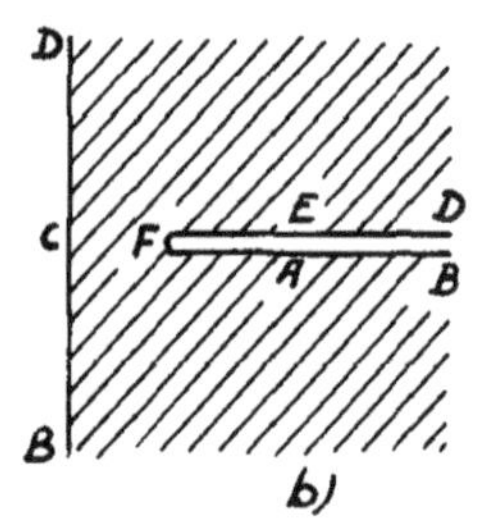


Fig. 163

function is represented in Fig. 163a. The boundary of this region is a circle with radius $v_0^2$ and cut DEFAB. We take point (D,B) for which $W = -v_0^2$, for center of inversion and apply the inverse mapping. Then for function

$$U = \frac{1}{w^2 + v_0^2}$$

we obtain the region represented in Fig. 163b, bounded by rectilinear segments. We map it onto the half plane $\zeta$. We obtain

$$U = \frac{1}{w^2 + v_0^2} = M \int \frac{d\zeta}{(1 - \zeta^2)^{3/2}} = \frac{M\zeta}{\sqrt{1 - \zeta^2}} + N \quad .$$

Constants $M$ and $N$ are determined from the condition that $w = v_0$ for

$\zeta = 0$ and $w = 0$ for $\zeta = 1/k$. We obtain (Fig. 163b)

$$w = - v_0 \frac{ik'\zeta - \sqrt{1 - \zeta^2}}{\sqrt{1 - k^2\zeta^2}} \quad . \tag{12.4}$$

Now, recalling that $w = d\omega/dz$ and that on basis of (12.1) and (12.2)

$$\frac{d\omega}{d\zeta} = \frac{\kappa H}{2K} \frac{1}{\sqrt{(1 - \zeta^2)(1 - k^2\zeta^2)}} \quad , \tag{12.5}$$

We find, dividing (12.5) by (12.4):

$$\frac{dz}{d\zeta} = \frac{\kappa H}{2Kv_0} \frac{1}{(\sqrt{1 - \zeta^2} - ik'\zeta)\sqrt{1 - \zeta^2}} =$$

$$= \frac{\kappa H}{2Kv_0} \left( \frac{1}{1 - k^2\zeta^2} + \frac{ik'\zeta}{(1 - k^2\zeta^2)\sqrt{1 - \zeta^2}} \right) \quad .$$

Integrating, we obtain:

$$z = \frac{\kappa H}{2Kkv_0} \left[ \frac{1}{2} \ln \frac{1 + k\zeta}{1 - k\zeta} - i \operatorname{arc\,tg} \frac{k}{k'} \sqrt{1 - \zeta^2} \right] \quad . \tag{12.6}$$

The integration constant is here taken to be zero, since we must satisfy the condition: for $\zeta = 0$, the quantity $z$ has a purely imaginary value, for $\zeta = 1$, a real value.

Equation (12.6) is convenient for values of $\zeta = \xi$, for which

$$0 < \xi < 1$$

i.e., to determine the form of curve CD, which is obtained when real and imaginary parts are separated in (12.6). First, however, we introduce the depth $T$ of the pervious stratum by considerations as follow. Passing from line DE to line EF (Fig. 160) corresponds to contouring singular point $\xi = 1/k$ over a half circle in positive direction, which gives an increase $\pi i$ to $\ln(1 - k\zeta)$. By formula (12.6) we have for the increase of $z$

$$\frac{\kappa H}{2Kkv_0} \frac{1}{2} (- \pi i) \quad .$$

On the other side, the imaginary part of $z$ in this passage changes from $Ti$ to zero. Therefore we obtain:

$$- \frac{\kappa H \pi i}{4Kv_0k} = - Ti \quad ,$$

from which

$$T = \frac{\kappa H \pi}{4Kv_0k} = \frac{H\pi}{4KkJ_0} \quad . \tag{12.7}$$

Equality (12.6), introducing $T$, may be written as

$$z = \frac{T}{\pi} \left[ \ln \frac{1+k\zeta}{1-k\zeta} - 2i \operatorname{arc\,tg} \frac{k\sqrt{1 - \zeta^2}}{k'} \right] \tag{12.8}$$

Separating real and imaginary parts and considering $\zeta = \xi$, for which $0 < \xi < 1$, we obtain parametric equations for part CD of the contour of the dam foundation

$$x = \frac{T}{\pi} \ln \frac{1 + k\xi}{1 - k\xi}, \quad y = -\frac{2T}{\pi} \operatorname{arc\,tg} \frac{k}{k'} \sqrt{1 - \xi^2} . \tag{12.9}$$

Now we may write the equation of the contour in explicit form:

$$x = \frac{T}{\pi} \ln \frac{1 + \cos \frac{\pi d}{2T} \sqrt{\operatorname{tg}^2 \frac{\pi d}{2T} - \operatorname{tg}^2 \frac{\pi y}{2T}}}{1 - \cos \frac{\pi d}{2T} \sqrt{\operatorname{tg}^2 \frac{\pi d}{2T} - \operatorname{tg}^2 \frac{\pi y}{2T}}} .$$

From this, in particular, we find $\ell$ the half width of the foundation and its depth $d$, putting respectively $\xi = 1$ and $\xi = 0$:

$$\ell = \frac{T}{\pi} \ln \frac{1 + k}{1 - k} ,$$
$$d = \frac{2T}{\pi} \operatorname{arc\,tg} \frac{k}{K'} = \frac{2T}{\pi} \arcsin k . \tag{12.10}$$

From formulas (12.10), we find two expressions for the modulus of the elliptic integral

$$k = \sin \frac{\pi d}{2T} = \operatorname{th} \frac{\pi \ell}{2T} . \tag{12.11}$$

These equalities allow us to determine $k$. They give the relationship between width and depth of the dam foundation. This relationship is expressed in Table 11.

Table 11

Dependence between width and depth $d$ of dam foundation.

| $\frac{d}{T}$ | 0.1 | 0.5 | 0.9 | 0.95 |
|---|---|---|---|---|
| $\frac{\ell}{T}$ | 0.1 | 0.562 | 1.62 | 2.06 |

Fig. 164 gives a schematic graph of a family of dam foundation contours with constant flow around them.

We mention that the flowrate $Q$, computed by means of formula

$$Q = \frac{\kappa H K'}{2K} ,$$

for values of $k$ given by (12.11) is such as if we had a cut-off penetrating a depth $d$ (without dam foundation), or a dam foundation of length $2\ell$.

The depth $T$ of the stratum is hereby not arbitrary; it is related with the head $H$, with the given value $J_0$ of the gradient and with the number $k$.

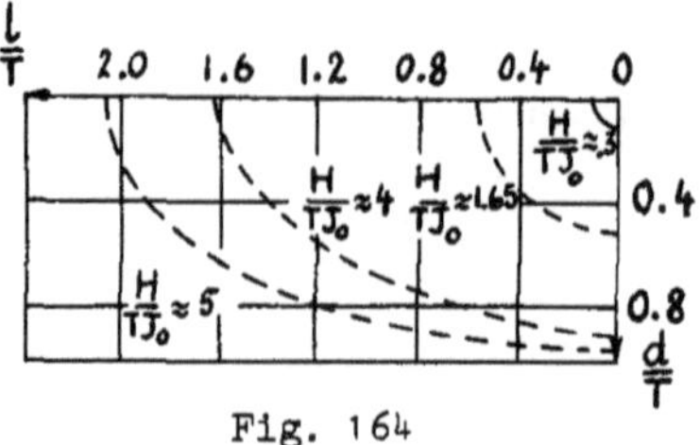


Fig. 164

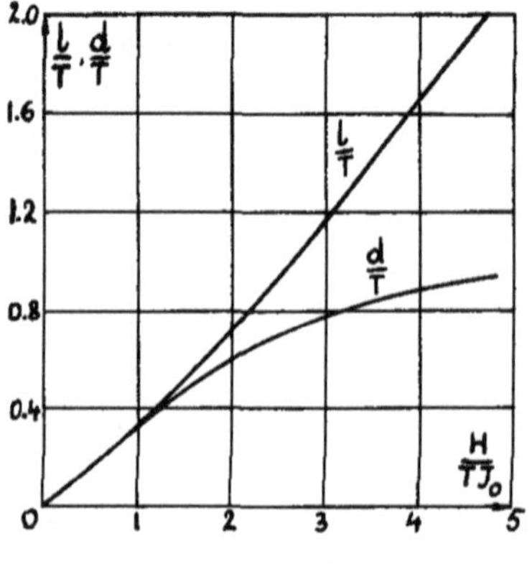


Fig. 165

The direct method to solve the problem requires us to find $k$ from equation (12.7) for given $H$, $T$, $J_0$ and then to find the values of $\ell$ and $d$ corresponding to $k$.

The dependence of $\ell/T$ and $d/T$ on $H/TJ_0$ is given in Fig. 165. We see that for $H/TJ_0 \leq 1$ the contour is close to a half circle and only for relatively large heads it approaches a semi ellips, strongly embedded in the soil.

§13. Contour with Rectilinear Segments and Segments of Constant Velocity.

We assume that the dam foundation has the form of Fig. 166. BC and GF are vertical rectilinear segments, DE is an horizontal rectilinear segment. We impose the condition; at CD and EF, the hydraulic gradient has a given value $J_0$. Then the velocity hodograph is as represented in Fig. 167, and the complex potential plane $\omega$ as in Fig. 168a.

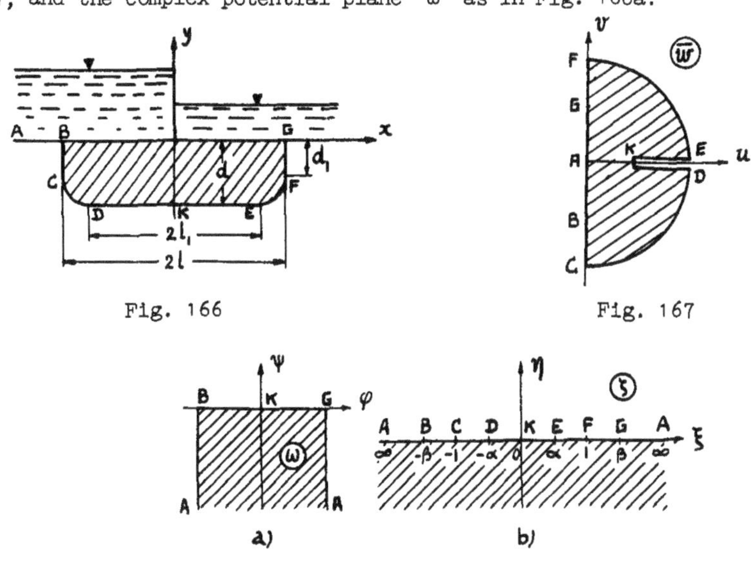


Fig. 166

Fig. 167

Fig. 168

The mapping onto the lower half plane $\zeta$ (Fig. 168b) gives:

$$\omega = \frac{\kappa H}{\pi} \arcsin \frac{\zeta}{\beta}, \quad w = \frac{d\omega}{dz} = \frac{v_0}{\alpha'} \left( \sqrt{1 - \zeta^2} - i \sqrt{\zeta^2 - \alpha^2} \right).$$

From this, as in §12, we find

$$\frac{dz}{d\zeta} = \frac{\kappa H}{\pi v_0 \alpha'} \frac{\sqrt{1 - \zeta^2} + i \sqrt{\zeta^2 - \alpha^2}}{\sqrt{\beta^2 - \zeta^2}}$$

Integrating we have

$$z = \frac{\kappa H}{\pi v_0 \alpha'} \int_1^{\zeta} \frac{\sqrt{1 - \zeta^2} + i \sqrt{\zeta^2 - \alpha^2}}{\sqrt{\beta^2 - \zeta^2}} d\zeta + \ell - d_1 i \quad . \tag{13.1}$$

Here $\ell$ is the half length of the dam foundation, $d_1$ the depth of the vertical segment of the dam foundation (Fig. 166).

Putting $\zeta = \alpha$, and, separating real and imaginary parts, we find the length and depth of the curvilinear part of the foundation (Fig. 166):

$$a = \ell - \ell_1 = \frac{\kappa H}{\pi v_0 \alpha'} \int_\alpha^1 \frac{\sqrt{1 - \zeta^2}}{\sqrt{\beta^2 - \zeta^2}} d\zeta \tag{13.2}$$

$$b = d - d_1 = \frac{\kappa H}{\pi v_0 \alpha'} \int_\alpha^1 \frac{\sqrt{\zeta^2 - \alpha^2}}{\sqrt{\beta^2 - \zeta^2}} d\zeta \quad . \tag{13.3}$$

We take the limits $(0, \alpha)$ in (13.1) and walk around point $\zeta = \alpha$ along a semi circle in negative direction. We obtain the equation for $\ell_1$

$$\ell_1 = \frac{\kappa H}{\pi v_0 \alpha'} \left\{ \int_0^\alpha \sqrt{\frac{1 - \zeta^2}{\beta^2 - \zeta^2}} d\zeta + \int_0^\alpha \sqrt{\frac{\alpha^2 - \zeta^2}{\beta^2 - \zeta^2}} d\zeta \right\} \quad .$$

Comparing this equation with (13.2), we find:

$$\ell = \frac{\kappa H}{\pi v_0 \alpha'} \left\{ \int_0^1 \sqrt{\frac{1 - \zeta^2}{\beta^2 - \zeta^2}} d\zeta + \int_0^\alpha \sqrt{\frac{\alpha^2 - \zeta^2}{\beta^2 - \zeta^2}} d\zeta \right\} \quad . \tag{13.4}$$

We find the equation of the curvilinear segment EF, assigning to $\zeta$ the real values between $\alpha$ and $1$:

$$\left\{ \begin{aligned} x &= \frac{\kappa H}{\pi v_0 \alpha'} \int_\alpha^\zeta \sqrt{\frac{1 - \zeta^2}{\beta^2 - \zeta^2}} d\zeta + \ell_1 \quad , \\ y &= \frac{\kappa H}{\pi v_0 \alpha'} \int_\alpha^\zeta \sqrt{\frac{\zeta^2 - \alpha^2}{\beta^2 - \zeta^2}} d\zeta - d \quad . \end{aligned} \right. \tag{13.5}$$

We examine particular cases.

## §14. "Streamlined" Dam Foundation.

For $\beta = 1$, we obtain a dam foundation with only horizontal framing.

Therefore equation (13.1) gives:

$$z = \frac{\kappa H}{\pi v_0 \alpha'} \left\{ \zeta + i \int_1^\zeta \sqrt{\frac{\zeta^2 - \alpha^2}{1 - \zeta^2}} d\zeta \right\} \quad . \tag{14.1}$$

Since for $\zeta = 1$ we must have $z = \ell$, we obtain from (14.1):

$$\ell = \frac{\kappa H}{\pi v_0 \alpha'} \quad . \tag{14.2}$$

Further, equation (13.2) gives for $\beta = 1$:

$$\ell - \ell_1 = \frac{\kappa H(1 - \alpha)}{\pi v_0 \alpha'} = \frac{\kappa H \alpha'}{\pi v_0 (1 + \alpha)} \quad . \tag{14.3}$$

Finally, from (13.3) we obtain:

$$d = \frac{\kappa H}{\pi v_0 \alpha'} \int_0^1 \sqrt{\frac{\zeta^2 - \alpha^2}{1 - \zeta^2}}\, d\zeta \quad .$$

The substitution

$$t = \frac{\sqrt{1 - \zeta^2}}{\alpha'}$$

reduces the expression for $d$ to

$$d = \frac{\kappa H}{\pi v_0 \alpha'} \int_0^1 \sqrt{\frac{1 - t^2}{1 - \alpha'^2 t^2}}\, dt \quad . \tag{14.4}$$

For $\alpha' \approx 0$, i.e., for sufficiently long dam foundations, we obtain from (14.4) the approximate formula

$$d \approx \frac{\kappa H}{\pi v_0 \alpha'} \cdot \frac{\pi}{4} = \frac{\kappa H}{\pi v_0 \alpha'} \quad . \tag{14.5}$$

We find the equation of the curvilinear part of the dam foundation, assigning real values $\zeta = \xi$ to $\zeta$, between limits from $\alpha$ to $1$. We have:

$$x = \ell \xi \ ,$$

$$y = -\ell \int_{\xi}^{1} \sqrt{\frac{\xi^2 - \alpha^2}{1 - \xi^2}}\, d\xi$$

$$(\alpha < \xi < 1) \quad .$$

Putting

$$\xi = \sqrt{1 - \alpha'^2 t^2} \ ,$$

we obtain the equation of the streamlined part in this form:

$$x = \ell \sqrt{1 - \alpha'^2 t^2}$$

$$y = -\alpha'^2 \ell \int_0^t \sqrt{\frac{1 - t^2}{1 - \alpha'^2 t^2}}\, dt \quad .$$

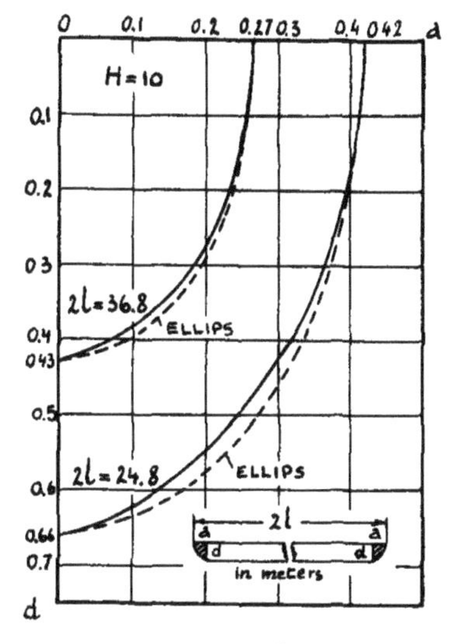


Fig. 169

By means of the substitution $t = \sin \varphi$ the integral in the expression for $y$ is expressed in function of elliptic integrals in the canonic form

$$\int_0^t \sqrt{\frac{1 - t^2}{1 - \alpha'^2 t^2}}\, dt = \frac{1}{\alpha'^2} E(\varphi, \alpha') - \frac{\alpha^2}{\alpha'^2} F(\varphi, \alpha') \quad ,$$

where

$$E(\varphi, k) = \int_0^{\varphi} \sqrt{1 - k^2 \sin^2 \varphi}\, d\varphi \quad ,$$

$$F(\varphi, k) = \int_0^{\varphi} \frac{d\varphi}{\sqrt{1 - k^2 \sin^2 \varphi}} \quad .$$

We finally have the equations of the curvilinear end as:

$$x = \ell\sqrt{1 - \alpha'^2 \sin^2 \varphi} \ ,$$

(14.6)

$$y = -\ell[E(\varphi, \alpha') - \alpha^2 F(\varphi, \alpha')] \ .$$

To represent the form of the curvilinear end, computations were carried out for $J_0 = 1$, $\alpha' = \sin 10°$ and $\alpha' = \sin 15°$. Their results are given in Fig. 169. The curves are close to ellipses with semi axes $a = \ell - \ell_1$ and d.

Assigning a series of other values to $H/\ell J_0$, we may compose a table and graphs (Fig. 170), showing the dependence of the thickness d of the foundation and the length of the curvilinear end on a given half length of the foundation. We see that the shorter the dam foundation is, the thicker it must be to preserve the same gradient $J_0$. Increase of the head H or decrease of the gradient $J_0$ gives rise to a proportional increase of all the dimensions of the foundation.

We introduce the notation

$$\tilde{H} = \frac{H}{\pi J_0 \ell}$$

Then it is easy to obtain for $a/\ell$

$$\frac{a}{\ell} = \frac{\tilde{H}^2}{1 + \sqrt{1 - \tilde{H}^2}} \approx \frac{1}{2}\tilde{H}^2\left(1 + \frac{1}{4}\tilde{H}^2\right) \ .$$

For $d/\ell$ we have the approximate equality

$$\frac{d}{\ell} \approx \frac{\pi}{4}\tilde{H}^2\left(1 + \frac{1}{8}\tilde{H}^2 + \frac{3}{64}\tilde{H}^4\right) \ .$$

Multiplying the obtained equalities term by term with $\ell J_0/H$, we find new relationships

$$\frac{aJ_0}{H} \approx \frac{H}{\pi^2 J_0 \ell}\left(1 + \frac{1}{4}\frac{H^2}{\pi^2 J_0^2 \ell^2}\right) \ ,$$

$$\frac{dJ_0}{H} \approx \frac{H}{2\pi J_0 \ell}\left(1 + \frac{1}{8}\frac{H^2}{\pi^2 J_0^2 \ell^2}\right) \ .$$

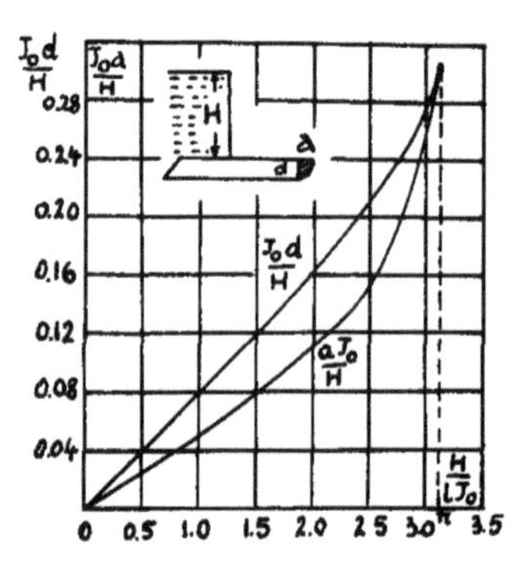


Fig. 170

For sufficiently small ratios $H/\ell$, and namely for

$$\frac{H}{\ell} \leq J_0$$

we have approximate linear relationships of $a/H$, and $d/H$ on $H/\ell$, which we may write as

$$\frac{a}{H} \approx \frac{H}{\pi^2 J_0 \ell} \ ,$$

$$\frac{d}{H} \approx \frac{H}{2\pi J_0 \ell} \ .$$

For given H, $\ell$ and $J_0$ we may thus find a and d.

Fig. 170 gives graphs of $aJ_0/H$, $dJ_0/H$ in function of $H/\ell J_0$ .

For $J_0 = 1$, these graphs express the relationships

$$\frac{a}{H} = f_1\left(\frac{H}{l}\right), \quad \frac{d}{H} = f_2\left(\frac{H}{l}\right) .$$

We emphasize once more that the streamlined ends of the dam foundation were determined from the condition that the hydraulic gradient $J_0$ along them had a given constant value. The velocity along the streamlined parts, as we know, has then the constant value

$$v_0 = kJ_0 .$$

This will be the case, in particular, for the exit velocity, i.e., the velocity in point E of the exit groundwater flow into tailwater.

Along the rectilinear part of the dam foundation, the velocity has values smaller than or equal to $v_0$, with minimum at the center-line of the dam foundation. It is easy to find the value of this velocity

$$v_{min} = v_0 \sqrt{\frac{1 - \alpha}{1 + \alpha}} .$$

§15. "Streamlined" Cut-off.

Another particular case of interest is obtained from the general formula (13.1) for $\alpha = 0$. This is the case of the "streamlined" cut-off or sheetpile (Fig. 171).

Equation (13.1) for $\alpha = 0$ becomes

$$z = \frac{H}{\pi J_0} \int_1^{\zeta} \frac{\sqrt{1 - \zeta^2} + i\zeta}{\sqrt{\beta^2 - \zeta^2}} d\zeta + l - d_1 i ,$$

where $l$ is the half length of the cut-off, $d_1$ the length of its rectilinear part (Fig. 171).

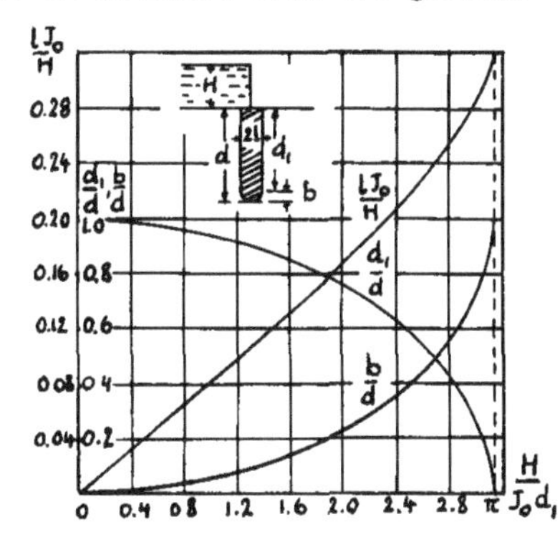


Fig. 171

In the given case, $l_1 = 0$. For $l$, d and $d_1$ we obtain:

$$l = \frac{H}{\pi J_0} \int_0^1 \sqrt{\frac{1 - \zeta^2}{\beta^2 - \zeta^2}} d\zeta ,$$

$$d - d_1 = \frac{H}{\pi J_0} \int_0^1 \frac{\zeta \, d\zeta}{\sqrt{\beta^2 - \zeta^2}} = \frac{H}{\pi J_0} (\beta - \sqrt{\beta^2 - 1}) ,$$

$$d = \frac{H\beta}{\pi J_0}, \quad d_1 = \frac{H}{\pi J_0} \sqrt{\beta^2 - 1} .$$

Given $J_0$, H and d, we find $\beta$, and then we may determine $l$ and all flow elements. It is of interest to find the magnitude of the exit velocity about the cut-off. It is found for $\zeta = \beta$ in the expression of the complex velocity

$$w = u - iv = v_0(\sqrt{1 - \zeta^2} - i\zeta) = - iv_0(\zeta - \sqrt{\zeta^2 - 1}) \quad .$$

We obtain:

$$v_{min} = v_0(\beta - \sqrt{\beta^2 - 1}) = \frac{v_0}{\beta + \sqrt{\beta^2 - 1}} \quad .$$

We transform the formula for $\ell$, expressing the right part in terms of elliptic integrals in canonic form

$$\ell = \frac{H}{\pi J_0 \beta} \int_0^1 \frac{1 - \zeta^2}{1 - \frac{1}{\beta^2}\zeta^2} d\zeta = \frac{H\beta}{\pi J_0}\left[E - \left(1 - \frac{1}{\beta^2}\right)F\right]$$

($E$ and $F$ are complete elliptic integrals with modulus $\frac{1}{\beta}$).

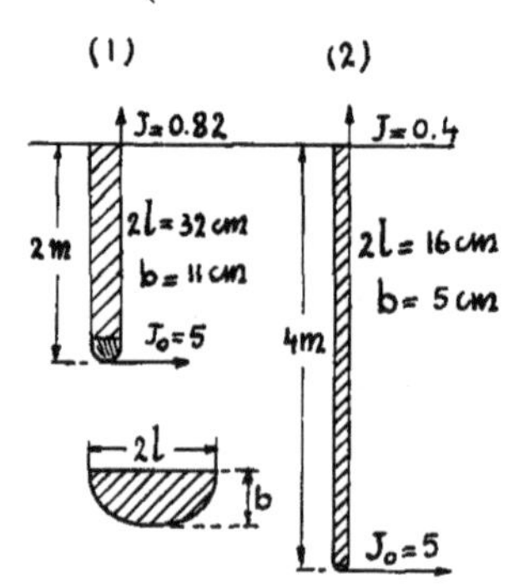


Fig. 172

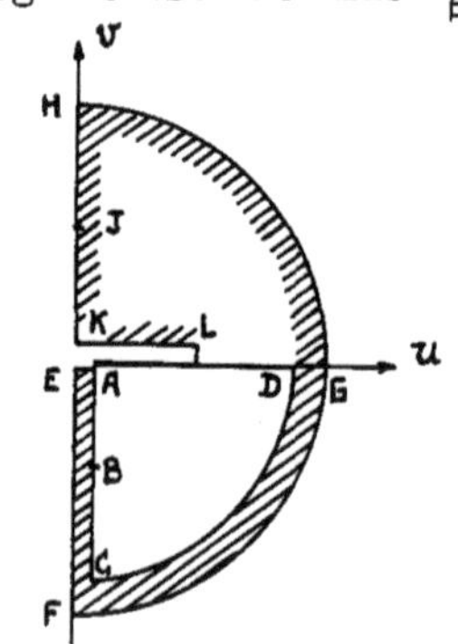


Fig. 173

Assuming $\beta = \pi$, we obtain for $d = H$, $J_0 = 1$:

$$\ell = 0.080d \quad .$$

If, for example, $H = 10$ meters, then in our assumption $d = 10$ meters and we obtain:

$$b = d - d_1 = 0.52d = 52 \text{ cm}, \quad \ell = 80 \text{ cm}.$$

Two other examples are given by Fig. 172, also for $H = 10$ m. and for $J_0 = 5$. The exit gradients prevailing in these conditions are not larger: 0.82 and 0.4. The problem may be generalized by considering a hodograph of the type sketched in Fig. 173. The region covers two sheets and is bounded by arcs of circles of given radius and by rectilinear cuts.

The flowregion is sketched in Fig. 174. For the function $W = \ell n\, w$ one obtains the condition of a constant value $\text{Re}\, W = \ell n\, V$ along the circles, and of a constant $\text{Im}\, W = \arg w$ along

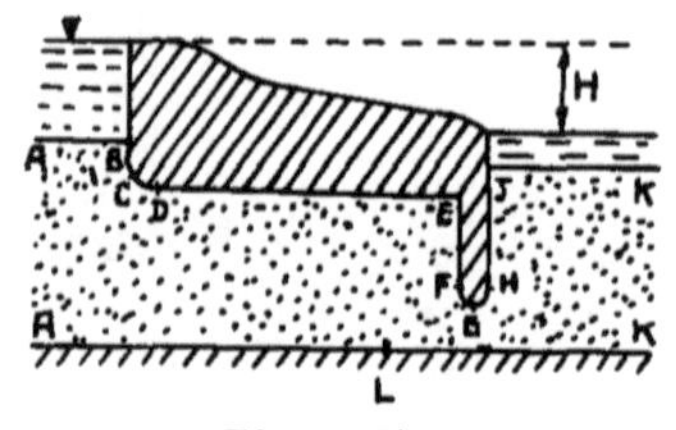


Fig. 174

the edges of the cuts. The relationship between $w$ and $\zeta$ may be found by means of an integral of the Cauchy type.

We notice that the mathematical tool used in the paragraph about streamlined contours is not more complicated than that which we used in Chapter III in the study of similar problems of flow about contours with corner points.

# CHAPTER VI.

## THE MIXED PROBLEM OF THE THEORY OF FUNCTIONS AND ITS APPLICATIONS TO THE SEEPAGE THEORY.

### A. The Mixed Problem Of The Theory Of Functions.

§1 Definition of an Analytic Function By Its Real Part On the Real Axis.

Let a given function $f(\zeta)$ be analytic in the upper half-plane, continuous up to the real axis and having a well-defined value $f_0$ at infinity:

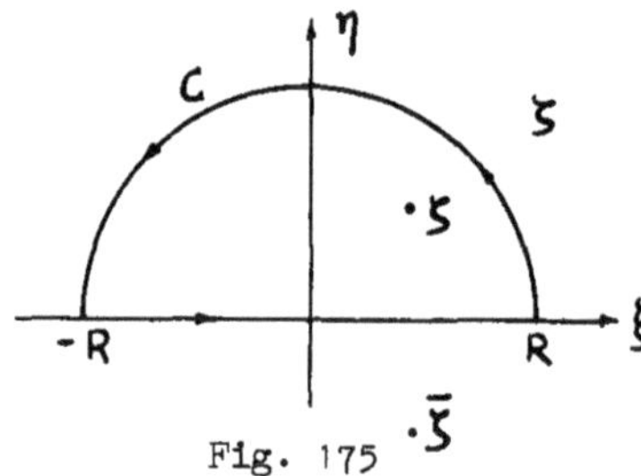

Fig. 175

$$\lim_{\zeta \to \infty} f(\zeta) = f_0 .$$

For its definition we assume that for large values of the modulus of $\zeta$

$$f(\zeta) = f_0 + O\left(\frac{1}{\zeta^\beta}\right) \quad (\beta > 0) .$$

The last equality means that the expression $\zeta^\beta(f - f_0)$ has a limit $A$ that differs from zero:

$$\lim_{\zeta \to \infty} [\zeta^\beta(f - f_0)] = A .$$

We apply Cauchy's theorem to the function $f(\zeta)$, taking a closed contour, composed of the segment $(-R, R)$ of the real axis and a half circle $C$ with radius $R$, walking in a positive direction (Fig. 175). We obtain for point $\zeta$, located inside the contour:

$$f(\zeta) = \frac{1}{2\pi i} \int_{-R}^{R} \frac{f(t)\, dt}{t - \zeta} + \frac{1}{2\pi i} \int_C \frac{f(\tau)\, d\tau}{\tau - \zeta} . \tag{1.1}$$

Let $R$ go to infinity, to obtain:

$$\lim_{R \to \infty} \frac{1}{2\pi i} \int_C \frac{f(\tau) d\tau}{\tau - \zeta} = \lim_{R \to \infty} \frac{1}{2\pi i} \int_C \frac{\left(f_0 + \frac{A}{\tau^\beta}\right) d\tau}{\tau - \zeta} = \frac{f_0}{2} ,$$

because

$$\frac{1}{2\pi i} \int_C \frac{f_0\, d\tau}{\tau - \zeta} = \frac{\pi i f_0}{2\pi i} = \frac{1}{2} f_0 ,$$

and because the second term gives zero. Thus

$$f(\zeta) = \frac{1}{2\pi i} \int_{-\infty}^{\infty} \frac{f(t)\, dt}{t - \zeta} + \frac{1}{2} f_0 . \tag{1.2}$$

We take point $\overline{\zeta} = \xi - i\eta$ in the lower half-plane. It is outside the contour we have chosen; as is well known from the theory of functions of a complex variable, the integral of the right side of equality (1.1) will be equal to zero, and consequently, its limit will be equal to zero. This limit is given by the right side of equation (1.2), and therefore

$$(1.3) \qquad 0 = \frac{1}{2\pi i} \int_{-\infty}^{\infty} \frac{f(t)\,dt}{t - \overline{\zeta}} + \frac{f_0}{2} .$$

In equality (1.3) we replace all complex numbers by their conjugate. We obtain:

$$(1.4) \qquad 0 = -\frac{1}{2\pi i} \int_{-\infty}^{\infty} \frac{\overline{f}(t)\,dt}{t - \zeta} + \frac{\overline{f}_0}{2} .$$

We subtract (1.4) from (1.2) and obtain:

$$f(\zeta) = \frac{1}{2\pi i} \int_{-\infty}^{\infty} \frac{f + \overline{f}}{t - \zeta}\,dt + \frac{1}{2}(f_0 - \overline{f}_0) .$$

Using the notations

$$(1.5) \qquad f = \varphi + i\psi, \quad f_0 = \varphi_0 + i\psi_0 ,$$

we rewrite the latter equality in the following form:

$$(1.6) \qquad f(\zeta) = \frac{1}{\pi i} \int_{-\infty}^{\infty} \frac{\varphi(t)\,dt}{t - \zeta} + i\psi_0 .$$

Exactly in the same way, adding (1.2) and (1.4), we obtain an expression of the analytic function $f(\zeta)$ in the upper half-plane in terms of the value of its imaginary part on the abscissa-axis

$$(1.7) \qquad f(\zeta) = \frac{1}{\pi} \int_{-\infty}^{\infty} \frac{\psi(t)\,dt}{t - \zeta} + \varphi_0 .$$

In front of integrals (1.6) and (1.7) one must write the minus sign if $\zeta$ is a point of the "lower" half-plane.

We consider the integral

$$U(\zeta) = \frac{1}{2\pi i} \int_{-\infty}^{\infty} \frac{\omega(t)\,dt}{t - \zeta}$$

and look for the limit value of $U(\zeta)$, when point $\zeta$ tends to some point $\xi$ of the real axis. Therefore we take two points $N'(-\xi_1, 0)$ and $N''(\xi_2, 0)$ (Fig. 176) at the real axis and look for the limit of expression

$$U(\zeta) = \frac{\omega(\zeta)}{2\pi i} \int_{-\xi_1}^{\xi_2} \frac{dt}{t - \zeta} + \frac{1}{2\pi i} \int_{-\xi_1}^{\xi_2} \frac{\omega(t) - \omega(\zeta)}{t - \zeta}\,dt .$$

This limit will exist for $\xi_1 = \xi_2$, in the sense of principal value [15] (if $\omega$ satisfies, for example, the conditions of Hoelder-Lipschitz, and if conditions of smoothness for $U$ are satisfied).

Therefore, when $\zeta$ tends to point $\xi$ from the upper half-plane,

$$\lim_{\zeta \to \xi} U(\zeta) = \frac{1}{2}\,\omega(\xi) + \frac{1}{2\pi i}\int_{-\infty}^{\infty} \frac{\omega(t) - \omega(\xi)}{t - \xi}\,dt \quad . \tag{1.8}$$

By means of equation (1.8) we obtain from (1.6), with $\omega = 2\varphi$:

$$\lim_{\zeta \to \xi} f(\zeta) = \varphi(\xi) + \frac{1}{\pi i}\int_{-\infty}^{\infty} \frac{\varphi(t) - \varphi(\xi)}{t - \xi}\,dt + i\psi_0 \quad . \tag{1.9}$$

In the right side of (1.9), the real part is equal to $\varphi(\xi)$. This also shows that the function $f(\zeta)$, built according to formula (1.6), really satisfies the required condition.

The same check is applied to equation (1.7).

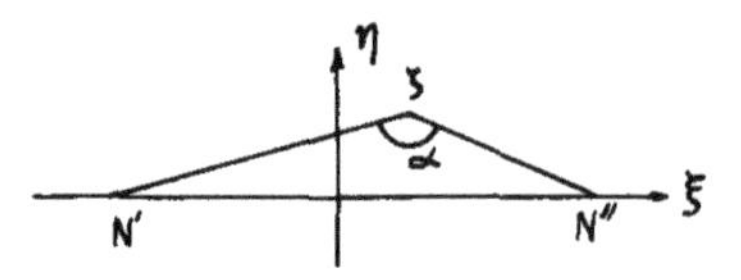


Fig. 176

If at some part of the $\xi$-axis, for example outside of the interval (a, b), the function $\varphi(\xi) = 0$, then one may take the limits a,b in formula (1.6) and write the expression for $f(\zeta)$ in this way:

$$f(\zeta) = \frac{1}{\pi i}\int_{a}^{b} \frac{\varphi(t)\,dt}{t - \zeta} + i\psi_0 \quad .$$

Then in the passage to the $\xi$-axis, we distinguish two cases:

1) if $a < \xi < b$, then

$$\lim_{\zeta \to \xi} f(\zeta) = \varphi(\xi) + \frac{1}{\pi i}\int_{a}^{b} \frac{\varphi(t) - \varphi(\xi)}{t - \xi}\,dt + i\psi_0 \ ;$$

2) if $\xi$ is outside interval, (a, b), then

$$\lim_{\zeta \to \xi} f(\zeta) = \frac{1}{\pi i}\int_{a}^{b} \frac{\varphi(t)}{t - \xi}\,dt + i\psi_0 \quad .$$

This theory is applied to the flow picture of Fig. 177. Rain falls and seeps into a strip of soil of depth d. Groundwater flow is such that line AB may be thought of as being a solid wall. Along segment B'C B we have $v = -\varepsilon$, where $\varepsilon$ is the rainfall intensity, i.e., the amount that falls on a unit area per unit of time. At the remaining boundaries of the flowregion $v = 0$. Mapping the flowregion onto the lower half-plane $\zeta$ (Fig. 177), by means of the Schwarz-Christoffel formula, we find the relation between $\zeta$ and z

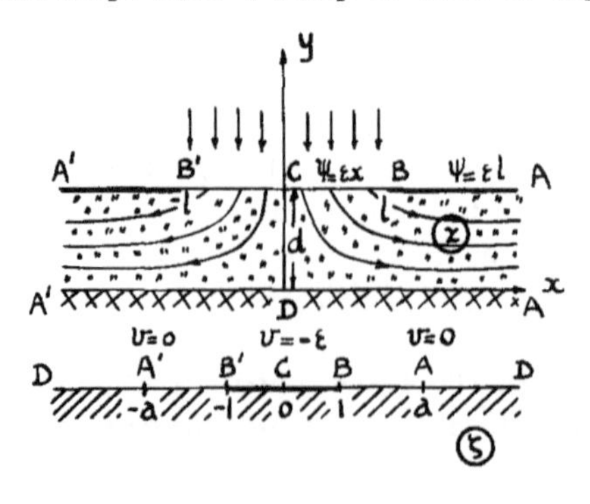


Fig. 177

$$\zeta = a \coth \frac{\pi z}{2d} \quad . \tag{1.10}$$

In order to find the dependence of $w = u - iv$ on $\zeta$, we use formula (1.7) of this chapter, giving the value of the function in the lower half-plane, when its imaginary part is known along the realaxis and when $w = 0$ at infinity:

$$w = \frac{1}{\pi} \int_{-\infty}^{\infty} \frac{v\, d\xi}{\xi - \zeta} = - \frac{\varepsilon}{\pi} \int_{-1}^{1} \frac{d\xi}{\xi - \zeta} = - \frac{\varepsilon}{\pi} \ell n \frac{\zeta - 1}{\zeta + 1} \quad . \tag{1.11}$$

Eliminating $\zeta$ from (1.10) and (1.11) we find $w$ in function of $z$:

$$w = - \frac{\varepsilon}{\pi} \ell n \frac{a \coth \frac{\pi z}{2d} - 1}{a \coth \frac{\pi z}{2d} + 1} \quad .$$

Since for $\zeta = 1$ we have $z = \ell + di$, we find from (1.10):

$$a \coth \frac{\pi \ell}{2d} \quad .$$

We assume (Fig. 178) that the rain falls on a semi-infinite strip. This particular case can be derived from the previous one for $\ell = \infty$, $a = 1$. We have:

$$w = - \frac{\varepsilon}{\pi} \ell n \frac{\coth \frac{\pi z}{2d} - 1}{\coth \frac{\pi z}{2d} + 1} = - \frac{\varepsilon}{\pi} \ell n\, e^{-\frac{\pi z}{d}} = \frac{\varepsilon z}{d} \quad .$$

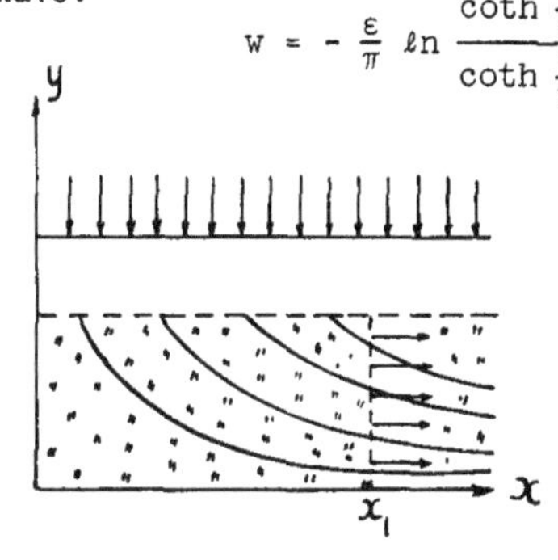


Fig. 178

Integrating this equation over $z$, we obtain:

$$\omega = \varphi + i\psi = \frac{\varepsilon z^2}{2d} + C \quad .$$

Separation of its real and imaginary parts gives

$$\varphi = \frac{\varepsilon}{2d} (x^2 - y^2) + C_1 \ ,$$

$$\psi = \frac{\varepsilon}{d} xy + C_2 \quad .$$

These are families of confocal hyperbolas. In this fow, the seepage velocity does not depend upon $y$ but increases with increasing distance from the $y$-axis:

$$u = \frac{\partial \varphi}{\partial x} = \frac{\varepsilon x}{d} \quad .$$

The seepage rate through a cross-section $x = x_1$, is equal to $q = \varepsilon x_1$, and increases without limit with $x_1$, as this should be because of the conditions of the problem. If we cut the flowregion with a straight line $x = x_1$, we may assume that we have an approximate model of the groundwater flow to a river at a distance $x_1$ from the place where the rain started.

## §2. Mixed Problem of the Theory of Functions.

We consider the problem: to find a function $w = f(\zeta)$, holomorphic in the upper half-plane and continuous along the real axis $L$, except perhaps at some isolated points. In the limit to these points, the order of $f(\zeta)$ increases by less than one, if $w$ satisfies the condition

$$\text{Im}\,[w(\xi)\,e^{-i\omega(\xi)}] = c(\xi)\ \text{Ha L} \quad . \tag{2.1}$$

Here $c(\xi)$ is a piece-wise continuous function, satisfying Gelder's [15, 16] condition along each open segment between points $a_1, a_2, \ldots, a_n$; $\omega(\xi)$ is piece-wise constant.

First let us examine the homogeneous problem, for which $c(\xi) = 0$, and $\omega(\xi) = \pi\alpha_k$ at the segment $(a_k, a_{k+1})$ (Fig. 179).

In §9 of Chapter III, we had an example of such a problem in the conformal mapping of a star-shaped polygon. Indeed, equation (2.1) gives the straight line $v \cos \pi\alpha_k - u \sin \pi\alpha_k = 0$, for $c(\xi) = 0$ and $w = u + iv$. The solution of the problem was given by formula (9.1) of Chapter III [$C$ is a real constant]:

$$w_* = Ce^{\pi\alpha_n i}(a_1 - \zeta)^{\alpha_n - \alpha_1}(a_2 - \zeta)^{\alpha_1 - \alpha_2}\ldots(a_n - \zeta)^{\alpha_{n-1} - \alpha_n} \quad . \tag{2.2}$$

However, the homogeneous condition of (2.1) will also be satisfied if in expression (2.2) each of the exponents is increased or decreased by an integer number.

Regarding the choice of the exponents, we observe this. We designate the jump $\omega(\xi)/\pi$ in the contouring of point $a_k$ by $\gamma_k$ (Fig. 179):

$$\gamma_k = \alpha_{k-1} - \alpha_k \quad (k = 1, 2, \ldots, n), \quad a_{n+1} = a_1 \quad . \tag{2.4}$$

We may choose the integer numbers $\lambda_k$ so as to satisfy the inequality $-1 < \gamma_k + \lambda_k < 1$. If $\gamma_k$ is an integer, then, putting $\lambda_k = -\gamma_k$, we obtain the unique number $\gamma_k + \lambda_k = 0$, satisfying this condition; if $\gamma_k$ is not an integer, there exist two values $\lambda_k$ for which respectively the inequalities

$$\begin{aligned} -1 < \gamma_k + \lambda_k < 0 \quad , \\ 0 < \gamma_k + \lambda_k < 1 \quad , \end{aligned} \tag{2.4}$$

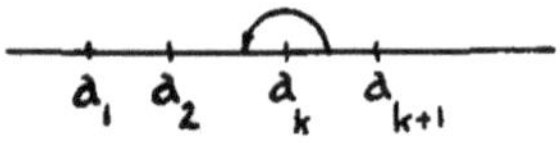

are satisfied.

Fig. 179

If we substitute $\gamma_k + \lambda_k$ from the first inequality (2.4) for $\gamma_k$ in (2.2), the order of infinity of $w_*$ becomes smaller than one in point $a_k$; such a type of solution occurs often in the theory of seepage. But, sometimes one wants a finite solution for $\zeta = a_k$ [17]. Then one must choose $\lambda_k$ according to the second of conditions (2.4).

To determine the choice of $\lambda_k$, we introduce the notations

$$\delta_k = \gamma_k + \lambda_k = \alpha_{k-1} - \alpha_k + \lambda_k \quad . \tag{2.5}$$

The expression

$$T(\zeta) = Ce^{\pi\alpha_n i} \prod_{k=1}^{n} (a_k - \zeta)^{\delta_k} \tag{2.6}$$

is called [16] the "canonic function of class $h(a_1, \ldots, a_q)$, if we have $0 \leq \delta_k < 1$ in points $a_1, a_2 \ldots, a_q$. $C$ is a real constant, for example, one.

The "index" $\kappa$ of the solution of class $h(a_1, \ldots, a_q)$ is the number

$$\kappa = -(\delta_1 + \delta_2 + \ldots + \delta_n) = -(\lambda_1 + \lambda_2 + \ldots + \lambda_n) \quad . \tag{2.7}$$

Returning to the non-homogeneous case (2.1), we rewrite the equation as

$$(2.8) \qquad -\frac{i}{2}\left[w(\xi) - \overline{w}(\xi)e^{2i\omega(\xi)}\right] = c(\xi)e^{i\omega(\xi)} \quad .$$

Function (2.6) satisfies the equation $T(\xi) - \overline{T}(\xi)e^{2i\omega} = 0$. Inserting $e^{2i\omega} = T : \overline{T}$ in (2.8), we obtain:

$$(2.9) \qquad \operatorname{Im}\left(\frac{w}{T}\right) = -\frac{i}{2}\left(\frac{w}{T} - \frac{\overline{w}}{\overline{T}}\right) = \frac{c(\xi)e^{i\omega}}{T(\xi)} \quad .$$

We arrive at the problem of determining a function, holomorphic in the the upper half-plane, by its imaginary part along the real axis (§1). To solve it, we must choose a condition for $c(\xi)$ at infinity. We distinguish two cases.

1) Let the index $\kappa \geq 0$. This means that for $\zeta = \infty$, $T(\zeta)$ tends to zero of order $\kappa$. We require that for $\xi = \pm\infty$, $c(\xi)$ satisfiex the inequality

$$(2.10) \qquad |c(\xi) - A| < \frac{M}{|\xi|^{\kappa+\alpha}}$$

(A, M and $\alpha$ are constants, $\alpha > 0$; $A = 0$ for $\kappa > 0$.

The general solution of class $h(a_1, a_2, \ldots, a_q)$, finite at infinity, has the following form [see formula (1.7), where the arbitrary constant may be written as a polynomial]

$$(2.11) \quad w(\zeta) = \frac{T(\zeta)}{\pi}\int_{-\infty}^{\infty}\frac{c(\xi)e^{i\omega}\,d\xi}{T(\xi)(\xi-\zeta)} + (C_0\zeta^{\kappa} + C_1\zeta^{\kappa-1} + \ldots + C_{\kappa})T(\zeta)$$

($C_0$, $C_1$, ..., $C_\kappa$ are real constants).

2) When $\kappa < 0$, then the solution of class $H(a_1, \ldots, a_q)$, generally speaking, does not exist. Indeed, in this case $T(\zeta)$ tends to infinity of order $(-\kappa)$ for $\zeta \to \infty$. Developing the function under integral (2.11) in powers of $1/\zeta$, we obtain:

$$-\frac{T(\zeta)}{\pi\zeta}\int_{-\infty}^{\infty}\frac{c(\xi)e^{i\omega}}{T(\xi)}\left[1 + \frac{\xi}{\zeta} + \frac{\xi^2}{\zeta^2} + \ldots + \frac{\xi^{-\kappa-1}}{\zeta^{-\kappa-1}} + \ldots\right]d\xi \quad .$$

It is apparent from this that for the existence of a solution with a finite limit for $\zeta = \infty$, all the terms of the series, including that with $\xi^{-\kappa-1}$ must be zero; we require the conditions

$$\int_{-\infty}^{\infty}\frac{C(\xi)e^{i\omega}\xi^m d\xi}{T(\xi)} = 0 \qquad (m = 0, 1. 2. \ldots, -\kappa-1) \quad .$$

When these are satisfied and when we choose $C_0 = \ldots = C_\kappa = 0$, there exists a solution, finite at infinity.

When we look for a solution that may tend to infinity of order $\rho$ for $\zeta = \infty$ ($\rho$ is a positive integer number), it is sufficient to take a polynomial of degree $\kappa + \rho$ instead of $\kappa$ in (2.11). If the solution must go to zero at infinity, we must put $C_0 = 0$ in (2.11).

§3. Particular Case of the Problem.

Let $L'$ be the sum of segments $(a_1, b_1), (a_2, b_2), \ldots, (a_n, b_n)$ of the real axis, $L''$ the remaining part of the real axis (Fig. 180). We must find a function $f(\zeta)$, holomorphic in the upper half-plane, finite at infinity and in points $a_k$, and having an increase of order smaller than one in points $b_k$, with the boundary condition

(3.1) $\operatorname{Re} w = c(\xi)$ at $L'$,

$\operatorname{Im}(w) = c(\xi)$ at $L''$,

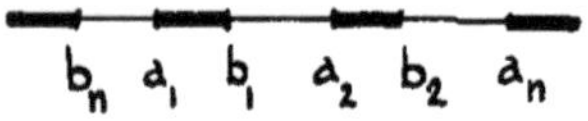

Fig. 180

where $c(\xi)$ satisfies the Hoelder condition in the intervals between points $a_k$ and $b_k$, and condition (2.10) at infinity. We take $\alpha_n = 0$, $\omega = 0$ at $L''$ and $\omega = -\frac{\pi}{2}$ in (2.1), and the jumps in points $a_k$ and $b_k$ respectively equal to 1/2 and - 1/2. The index $\kappa$ of class $h(a_1, \ldots, a_n)$ is zero and function (2.6) is:

$$T(\zeta) = \prod_{k=1}^{n} (\sqrt{a_k - \zeta} : \sqrt{b_k - \zeta})$$

The solution of class $h(a_1, \ldots, a_n)$ is:

$$f(\zeta) = \frac{1}{\pi} \prod_{k=1}^{n} \sqrt{\frac{\zeta - a_k}{\zeta - b_k}} \int_{-\infty}^{\infty} \prod_{1}^{n} \sqrt{\frac{\xi - b_k}{\xi - a_k}} \frac{ce^{i\omega}}{\xi - \zeta} d\xi + C_0 \prod_{1}^{n} \sqrt{\frac{\zeta - a_k}{\zeta - b_k}} . \tag{3.2}$$

$C_0$ is a real constant.

If in points $a_k$ we only require that the condition be satisfied that in these points the order of increase be smaller than one, then the solution, finite at infinity, is [16 - 18]:

$$f(\zeta) = \frac{1}{\pi} \prod_{k=1}^{n} \sqrt{\frac{\zeta - a_k}{\zeta - b_k}} \int_{-\infty}^{\infty} \prod_{1}^{n} \sqrt{\frac{\xi - b_k}{\xi - a_k}} \frac{ce^{i\omega}}{\xi - \zeta} d\xi + \frac{C_0 + C_1\zeta + \ldots + C_n\zeta^n}{\sqrt{\prod_{1}^{n} (\zeta - a_k)(\zeta - b_k)}} \tag{3.3}$$

where $C_0, C_1, \ldots, C_n$ are real constants, $\omega = 0$ at $L''$ and $\omega = -\frac{\pi}{2}$ at $L'$.

As an example, we examine the problem of the confined flow under n weirs on a stratum of arbitrary form (Fig. 181), for which it is easy to construct the complex potential as a function of $\zeta$ [18].

Let the points $a, b_1, c_1, \ldots, b_n, c_n, d$ correspond to the points $A, B_1, C_1, \ldots, B_n, C_n, D$ on the real axis of the $\zeta$-plane.

The real part of function $\omega = \varphi + i\psi$ assumes constant values $(-\kappa H_m)$ along segments $ab_1, cb_2, \ldots, c_nd$, its imaginary part assumes constant values $Q_m$ along segments $b_1c_1, \ldots, b_nc_n$. The derivative, taken along the real axis of the $\zeta$-plane, has its real part equal to zero along the segments $c_mb_{m+1}$, its imaginary part equal to zero along $b_mc_m$.

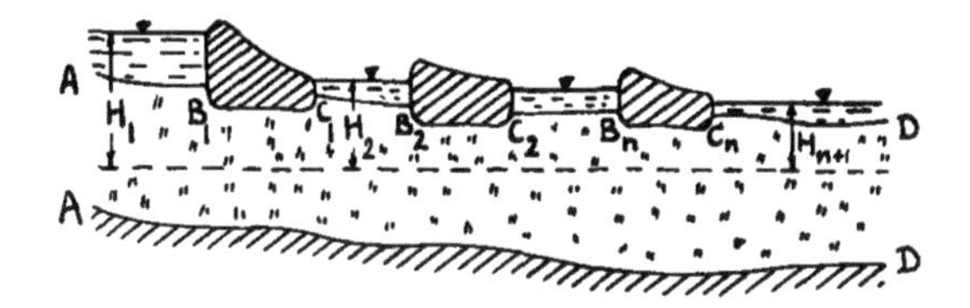


Fig. 181

Therefore $d\omega/d\zeta$ may be written as

$$\frac{d\omega}{d\zeta} = \frac{P(\zeta)}{\sqrt{R(\zeta)}} , \tag{3.4}$$

where $R(\zeta) = (\zeta - a)(\zeta - b_1)(\zeta - c_1)(\zeta - b_2) \ldots (\zeta - b_n)(\zeta - c_n)(\zeta - d)$, if $A \neq D$, and $R(\zeta) = -(\zeta - b_1)(\zeta - c_1) \ldots (\zeta - b_n)(\zeta - c_n)$, if $A = D$.

The polynomial $P(\zeta)$ must not be of a degree higher than $n + 1$ when the stratum rests on bedrock and not higher than $n - 1$ when there is no bedrock. Indeed, in the first case we have for $\zeta = \infty$, assuming a smooth contour AD, a finite velocity in the point of the z-plane corresponding to the point at infinity of the $\zeta$-plane; in the second case $d\omega/d\zeta = O(1/\zeta)$.

Integrating $d\omega/d\zeta$ between limits $b_m$, $c_m$ and $c_m, b_{m+1}$, we find:

$$\int_{b_m}^{c_m} \frac{P(\zeta)\,d\zeta}{\sqrt{R(\zeta)}} = k(H_m - H_{m+1}), \quad \frac{1}{i} \int_{c_m}^{b_{m+1}} \frac{P(\zeta)}{\sqrt{R(\zeta)}}\,d\zeta = Q_{m+1} - Q_m . \tag{3.5}$$

Given the values of some of the quantities $H_1, H_2, \ldots, H_{n+1}$, $Q_1, Q_2, \ldots, Q_{n+1}$, and namely $n + 1$ of them for $A \neq D$ and $n$ of them for $A = D$, equations (3.5) enable us to determine the coefficients of the polynomial $P(\zeta)$; the remaining unknowns $H_m$ and $Q_m$ are determined from the rest of the equations (3.5).

We notice that the determinant of the system of equations to determine the coefficients of the polynomial $P(\zeta)$ is different from zero [16,18];

## B. Problems About Drains and Canals.

### §4. Drain in Stratum on Bedrock with Non-symmetrical Flow.

In paragraphs 2 and 3 we had a particular case of the Riemann-Hilbert [16] problem. Its solution by means of the multiplication of the unknown function with function $T(\zeta)$ (2.6) is reduced to the solution of Dirichlet's problem by means of formulas (1.6), (1.7). This method was applied by S. N. Numerov to a series of problems in the theory of seepage. For each problem a function is chosen that satisfies well determined conditions at infinity, function $T(\zeta)$ is constructed to suit the first function, and one of formulas (1.6) and (1.7) is applied.

In §1 we dealt with a very simple example of such a problem. In this and in the following paragraphs, we analyze a series of more complicated problems, in which significant computation difficulties arise. In some examples (§§4, 5) we give detailed computations. In the next examples, we omit part of the numerical work.

Fig. 182 represents the case of a drain, receiving a non-symmetrical flow of groundwater in a stratum resting on bedrock.

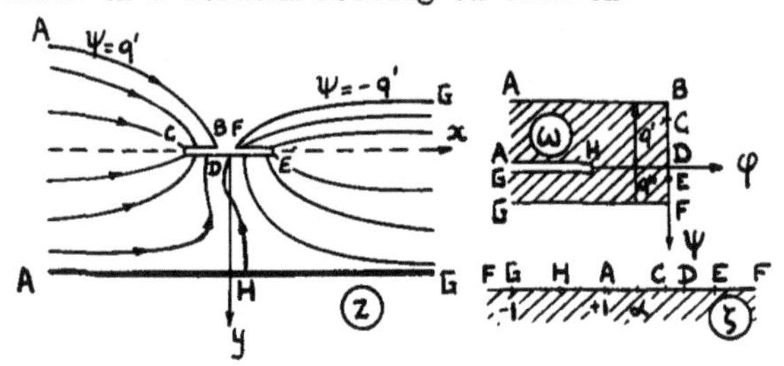

Fig. 182

Water flows to the drainage slit BCDEF from two sides at infinity [9]. Line DH is a streamline, separating the flowregion in two parts, left and right. The y-axis is oriented downward. The depth of the pervious stratum is designated by T.

The flowrate from the left side to the drain is $q'$, that from the right side is $q''$. The velocity potential along the drain is assumed to be zero. In the $\omega$-plane we have the polygon of Fig. 182. We examine the values of Zhukovsky's function $\theta = z - i\omega/k$ along the boundaries of the flow-region. Along the free surface and along the contour of the drain

$$\varphi - ky = 0, \quad \operatorname{Im} \theta = \theta_2 = 0 \quad . \tag{4.1}$$

Along the impervious bedrock AHG, the value of $\varphi$ is unknown. But if we map the regions $z$ and $\omega$ onto the half-plane $\zeta$ and determine $\omega$ in function of $\zeta$, we have the value $\varphi(\zeta)$ and also $\operatorname{Im} \theta(\zeta)$ along the entire real axis of the $\zeta$-plane. Since we are especially interested in the value of $\operatorname{Im} \theta$ along the segment AHG, it is convenient to map points A and G in points $\pm 1$ of the $\zeta$-plane. Moreover let point B transform into point $\zeta = \alpha$, and F into the point at infinity of the $\xi$-axis. For $\omega$ we obtain this expression [see further, §6, for derivation of analogous formula (6.7)]:

$$\omega = -\frac{q'}{\pi} \ln \frac{1 + \sqrt{\dfrac{\alpha - 1}{\alpha - \xi}}}{1 - \sqrt{\dfrac{\alpha - 1}{\alpha - \xi}}} - \frac{q''}{\pi} \ln \frac{1 + \sqrt{\dfrac{\alpha - \xi}{\alpha + 1}}}{1 - \sqrt{\dfrac{\alpha - \xi}{\alpha + 1}}} \quad . \tag{4.2}$$

Along segment $|\xi| < 1$, where $\psi = 0$, function $\omega = \varphi(\xi)$. The imaginary part of Zjoukovsky's function is equal to

$$\operatorname{Im} \theta = \theta_2 = y - \frac{\varphi}{k} = T - \frac{\varphi(\xi)}{k} \quad ,$$

i.e.,

$$\theta_2 = T + \frac{q'}{k\pi} \ln \frac{1 + \sqrt{\dfrac{\alpha - 1}{\alpha - \xi}}}{1 - \sqrt{\dfrac{\alpha - 1}{\alpha - \xi}}} + \frac{q''}{k\pi} \ln \frac{1 + \sqrt{\dfrac{\alpha - \xi}{\alpha + 1}}}{1 - \sqrt{\dfrac{\alpha - \xi}{\alpha + 1}}} \quad (|\xi| < 1). \tag{4.3}$$

As we saw

$$\theta_2(\xi) = 0 \quad \text{for} \quad |\xi| > 1 \quad . \tag{4.4}$$

Thus along the entire $\xi$-axis, the values of function $\theta_2$, imaginary part of function $\theta$, are known. By means of (1.6) we find:

$$\theta(\zeta) = \frac{T}{\pi}\, \ell n\, \frac{1 - \zeta}{-1 - \zeta} + \frac{q'}{\pi^2 k} \int_{-1}^{1} \frac{\ell n \dfrac{\sqrt{\alpha - t} + \sqrt{\alpha - 1}}{\sqrt{\alpha - t} - \sqrt{\alpha - 1}}}{t - \zeta}\, dt + \tag{4.5}$$

$$+ \frac{q''}{\pi^2 k} \int_{-1}^{1} \frac{\ell n \dfrac{\sqrt{\alpha + 1} + \sqrt{\alpha - t}}{\sqrt{\alpha + 1} - \sqrt{\alpha - t}}}{t - \zeta}\, dt + \frac{q''}{k} \quad .$$

The constant term must be $q''/k$ in order to satisfy the condition

$$\theta(\infty) = \frac{q''}{k} \quad .$$

For $\zeta = \alpha$ we obtain $\theta(\alpha) = - q'/k$ and therefore

$$- \frac{q'}{k} = \frac{T}{\pi}\, \ell n\, \frac{\alpha - 1}{\alpha + 1} + \frac{q'}{\pi^2 k}\, J_1 + \frac{q''}{\pi^2 k}\, J_2 + \frac{q''}{k} \quad . \tag{4.6}$$

Here

$$J_1 = \int_{-1}^{1} \frac{\ell n\left[\left(1 + \sqrt{\dfrac{\alpha - 1}{\alpha - t}}\right) : \left(1 - \sqrt{\dfrac{\alpha - 1}{\alpha - t}}\right)\right]}{t - \alpha}\, dt \quad ,$$

$$J_2 = \int_{-1}^{1} \frac{\ell n\left[\left(1 + \sqrt{\dfrac{\alpha - \tau}{\alpha + 1}}\right) : \left(1 - \sqrt{\dfrac{\alpha - \tau}{\alpha + 1}}\right)\right]}{\tau - \alpha}\, d\tau \quad .$$

Through the substitution of variables

$$\begin{cases} 2 \tanh^{-1} \sqrt{\dfrac{\alpha - 1}{\alpha - t}} = \ell n \left[\left(1 + \sqrt{\dfrac{\alpha - 1}{\alpha - t}}\right) : \left(1 - \sqrt{\dfrac{\alpha - 1}{\alpha - t}}\right)\right] = u_1 \; , \\ 2 \tanh^{-1} \sqrt{\dfrac{\alpha - \tau}{\alpha + 1}} = \ell n \left[\left(1 + \sqrt{\dfrac{\alpha - \tau}{\alpha + 1}}\right) : \left(1 - \sqrt{\dfrac{\alpha - \tau}{\alpha + 1}}\right)\right] = v_1 \; , \end{cases} \tag{4.7}$$

we reduce both integrals to the same expression:

$$J_1 = 2 \int_c^\infty \frac{u_1\, du_1}{\operatorname{sh} u_1} \; , \qquad J_2 = 2 \int_c^\infty \frac{v_1\, dv_1}{\operatorname{sh} v_1} \quad ,$$

where

$$c = \ell n\, \frac{1 + \sqrt{\dfrac{\alpha - 1}{\alpha + 1}}}{1 - \sqrt{\dfrac{\alpha - 1}{\alpha + 1}}} \quad . \tag{4.8}$$

Since $J_1 = J_2$, we determine the sum $q' + q''$ from (4.6)

$$\frac{q' + q''}{k} = \frac{- \dfrac{T}{\pi}\, \ell n\, \dfrac{\alpha - 1}{\alpha + 1}}{1 + \dfrac{1}{\pi^2}\, J_1} \quad .$$

We transform $J_1$ by integration by parts

$$J_1 = 2\int_c^\infty u_1 d \,\ell n \text{ th} \frac{u_1}{2} = 2\left(-c \,\ell n \text{ th} \frac{c}{2} - \int_c^\infty \ell n \text{ th} \frac{u_1}{2} du_1\right) .$$

Further, the integral between brackets may be written as

$$J_3 = \int_c^\infty \ell n \frac{1 - e^{-u}}{1 + e^{-u}} du ;$$

and substitution $e^{-u} = t$ gives

$$J_3 = \int_0^{e^{-c}} \frac{\ell n(1 - t)dt}{t} - \int_0^{e^{-c}} \frac{\ell n(1 + t)dt}{t} .$$

For $J_1$ we obtain:

$$J_1 = 2\left(\ell n \coth \frac{c}{2} + \int_0^{e^{-c}} \frac{\ell n(1 + t)}{t} dt - \int_0^{e^{-c}} \frac{\ell n(1 - t)}{t} dt\right) . \tag{4.9}$$

Knowing that

$$\tanh \frac{c}{2} = \sqrt{\frac{\alpha - 1}{\alpha + 1}} , \tag{4.10}$$

we rewrite the expression for the total flowrate $q = q' + q''$

$$\frac{q}{kT} = \frac{\frac{2}{\pi} \ell n \coth \frac{c}{2}}{1 - \frac{2}{\pi^2}\left(c \,\ell n \coth \frac{c}{2} + \int_0^{e^{-c}} \frac{\ell n(1 + t)}{t} dt - \int_0^{e^{-c}} \frac{\ell n(1 - t)}{t} dt\right)} . \tag{4.11}$$

Designating by $f_1(q/kT)$ the inverse of (4.11), we obtain:

$$c = f_1\left(\frac{q}{kT}\right) .$$

We look for the equation of the left branch of the free surface. Values of $1 < \xi < \alpha$ in the $\zeta$-plane correspond to this branch. Therefore we have for the free surface:

$$\frac{\omega}{k} = -\frac{q'i}{k} + y .$$

On the other side, inserting $\zeta = \xi$ in (4.2), with $1 < \xi < \alpha$, we separate the imaginary part in the logarithm and obtain:

$$\frac{\omega}{k} = -\frac{q'i}{k} - \frac{q'}{\pi k} \ell n \frac{\sqrt{\frac{\alpha - 1}{\alpha - \xi}} + 1}{\sqrt{\frac{\alpha - 1}{\alpha - \xi}} - 1} - \frac{q''}{\pi k} \ell n \frac{1 + \sqrt{\frac{\alpha - \xi}{\alpha + 1}}}{1 - \sqrt{\frac{\alpha - \xi}{\alpha + 1}}} . \tag{4.12}$$

We introduce the notations

$$\ell n \frac{\sqrt{\frac{\alpha - 1}{\alpha - \xi}} + 1}{\sqrt{\frac{\alpha - 1}{\alpha - \xi}} - 1} = u, \quad \ell n \frac{1 + \sqrt{\frac{\alpha - \xi}{\alpha + 1}}}{1 - \sqrt{\frac{\alpha - \xi}{\alpha + 1}}} = v . \tag{4.13}$$

Between $u$ and $v$, there exists the relation

$$\tanh \frac{v}{2} = \tanh \frac{c}{2} \tanh \frac{u}{2} \quad . \tag{4.14}$$

The intervals of the variables $u$ and $v$ are:

$$0 < u < \infty, \quad 0 < v < c \quad .$$

Now we may write $y$ along the free surface in function of the parameters $u$ and $v$:

$$y = - \frac{q'}{\pi k} u - \frac{q''}{\pi k} v \quad . \tag{4.15}$$

In order to find $x$ in function of $u, v$ along the free surface, we return to expression (4.5) for Zjoukovsky's function. At the free surface $\theta = \theta_1 = x + \frac{\psi}{k} = x - \frac{q'}{k}$, and therefore we obtain from (4.5), with $\zeta = \xi$:

$$x = \frac{q}{k} + \frac{T}{\pi} \ln \frac{\xi - 1}{\xi + 1} + \frac{q'}{\pi^2 k} J_1(\xi) + \frac{q''}{\pi^2 k} J_2(\xi) \quad . \tag{4.16}$$

Here

$$J_1(\xi) = \int_{-1}^{1} \frac{u_1 dt}{t - \xi} , \quad J_2(\xi) = \int_{-1}^{1} \frac{v_1 d\tau}{\tau - \xi} ,$$

where $u_1$ and $v_1$ have values (4.7)

To compute the integral

$$J_1(\xi) = 2 \int_{-1}^{1} \frac{\tanh^{-1} \sqrt{\frac{\alpha - 1}{\alpha - t}}}{\xi - t} dt ,$$

we change variables according to (4.7) and (4.13)

$$t = \alpha - (\alpha - 1) \coth^2 \frac{u_1}{2} , \quad \xi = \alpha - (\alpha - 1) \tanh^2 \frac{u}{2}$$

and put

$$e^{-u_1} = \tau \quad .$$

We obtain:

$$J_1(\xi) = - e^{-u}(e^u + 1)^2 \int_0^{e^{-c}} \frac{(1 + \tau) \ln \tau \, d\tau}{(1 - \tau)(\tau + e^u)(\tau + e^{-u})} \quad .$$

Separation of the multiplier of $\ln \tau$ in partial fractions gives:

$$J_1(\xi) = - 2 \int_0^{e^{-c}} \frac{\ln \tau}{1 - \tau} d\tau - \int_0^{e^{-c}} \frac{\ln \tau \, d\tau}{e^u + \tau} - \int_0^{e^{-c}} \frac{\ln \tau \, d\tau}{e^{-u} + \tau} \quad .$$

In the second integral we substitute $\tau = e^u t$, in the third $\tau = e^{-u} t$. We obtain:

$$J_1(\xi) = - 2 \int_0^{e^{-c}} \frac{\ln \tau}{1 - \tau} d\tau - \int_0^{e^{-c-u}} \frac{\ln t}{1 + t} dt - \int_0^{e^{-c+u}} \frac{\ln t}{1 + t} dt -$$

$$- u \ln \frac{1 + e^{-c-u}}{1 + e^{-c+u}} \quad .$$

Now we integrate all three integrals of the right side of the last equation by parts. Thus we generate the term

$$2c \ln (1 - e^{-c}) + (c + u) \ln (1 + e^{-c-u}) + (c + u) \ln (1 + e^{-c+u}) .$$

Therefore $J_1(\xi)$ becomes

$$J_1(\xi) = 2 \int_0^{e^{-c}} \frac{\ln(1 - t)}{t} dt - \int_0^{e^{-c-u}} \frac{\ln(1 + t)}{t} dt - \int_0^{e^{-c+u}} \frac{\ln(1 + t)}{t} dt +$$

$$+ c \ln \frac{(1 + e^{-c-u})(1 + e^{-c+u})}{(1 - e^{-c})^2} .$$

We return to the computation of the integral

$$J_2(\xi) = 2 \int_{-1}^{1} \frac{\tanh^{-1} \sqrt{\frac{\alpha - t}{\alpha + 1}}}{\xi - t} dt .$$

Considering that

$$\xi = \alpha - (\alpha + 1) \tanh^2 \frac{v}{2} , \qquad \tau = \alpha - (\alpha + 1) \tanh^2 \frac{v_1}{2} ,$$

taking into account (4.7) and (4.13) and substituting

$$e^{-v_1} = t_1 = e^{v} t ,$$

we reduce this integral in the form

$$J_2(\xi) = 2 \int_0^{e^{-c}} \frac{\ln(1 + t)}{t} dt - \int_0^{e^{-c-v}} \frac{\ln(1 - t)}{t} dt - \int_0^{e^{-c+v}} \frac{\ln(1 - t)}{t} dt - \tag{4.17}$$

$$- c \ln \frac{(1 - e^{-c-v})(1 - e^{-c+v})}{(1 + e^{-c})^2} .$$

From

$$\coth \frac{v}{2} = \coth \frac{c}{2} \coth \frac{u}{2}$$

we obtain

$$\frac{(1 + e^{-c+u})(1 + e^{-c-u})}{(1 - e^{-c})^2} = \coth^2 \frac{c}{2} \frac{\cosh^2 \frac{u}{2}}{\cosh^2 \frac{v}{2}} .$$

Finally we obtain:

$$\frac{J_1(\xi)}{\pi^2} = \frac{2}{\pi^2} \int_{-1}^{1} \frac{\tanh^{-1} \sqrt{\frac{\alpha - 1}{\alpha - t}}}{\xi - t} dt = \frac{2}{\pi^2} c \ln\left(\coth \frac{c}{2} \frac{\cosh \frac{u}{2}}{\cosh \frac{v}{2}} \right) + \tag{4.18}$$

$$+ f_5(e^{-c}) + \frac{1}{2} f_4(e^{-c-u}) + \frac{1}{2} f_4(e^{-c+u}) ,$$

$$\frac{J_2(\xi)}{\pi^2} = \frac{2}{\pi^2} \int_{-1}^{1} \frac{\tanh^{-1} \sqrt{\frac{\alpha - t}{\alpha + 1}}}{\xi - t} dt = \frac{2}{\pi^2} c \ln \frac{\coth \frac{c}{2} \cosh \frac{u}{2}}{\cosh \frac{v}{2}} + \tag{4.19}$$

$$+ f_4(e^{-c}) + \frac{1}{2} f_5(e^{-c-v}) + \frac{1}{2} f_5(e^{-c+v}) .$$

Where

$$(4.20)\qquad f_4(t) = \frac{2}{\pi^2}\int_0^t \frac{\ln(1+t)}{t}\,dt\ , \qquad f_5(t) = -\frac{2}{\pi^2}\int_0^t \frac{\ln(1-t)}{t}\,dt\ .$$

We transform $\ln\frac{\xi - 1}{\xi + 1}$ by means of formulas

$$\xi - 1 = (\alpha - 1)\left(1 - \text{th}^2\,\frac{u}{2}\right), \qquad \xi + 1 = (\alpha + 1)\left(1 - \text{th}^2\,\frac{v}{2}\right),$$

to find

$$\ln\frac{\xi - 1}{\xi + 1} = 2\ln\frac{\cosh\frac{v}{2}}{\cosh\frac{u}{2}}\ .$$

We may now write (4.16) in its final form

$$(4.21)\qquad \begin{aligned} x = \frac{2}{\pi}\left(T + \frac{qc}{\pi k}\right)\ln\frac{\cosh\frac{v}{2}}{\cosh\frac{u}{2}} + \frac{q'}{k}\Big\{f_4(e^{-c}) - \frac{1}{2}f_4(e^{-c+u}) - \\ - \frac{1}{2}f_4(e^{-c-u})\Big\} + \frac{q''}{k}\Big\{f_5(e^{-c}) - \frac{1}{2}f_5(e^{-c+v}) - \frac{1}{2}f_5(e^{-c-v})\Big\} \end{aligned}$$

We add to this the equation for $y$:

$$(4.22)\qquad y = -\frac{q'}{\pi k}u - \frac{q''}{\pi k}v\ .$$

We obtained a system of equations to determine the coordinates $x$ and $y$ of points at the free surface in function of two parameters linked by equation (4.14).

S. N. Numerov made a table for functions $f_4(t)$ and $f_5(t)$. By means of these functions, equation (4.11) for the flowrate is written as

$$q = \frac{\frac{2kT}{\pi}\ln\coth\frac{c}{2}}{1 - \frac{2}{\pi^2}c\ln\coth\frac{c}{2} - f_4(e^{-c}) - f_5(e^{-c})}\ .$$

The equation of the right branch of the free surface is obtained from the equation of the left branch, by replacing $x$ by $-x$, $q'$ by $q''$ and $q''$ by $q'$.

For large values of $x/T(|x|/T > 3)$, parameter $u$ becomes large and $v$ approaches zero.

S. N. Numerov gives for this case an approximate equation of the free surface.

When the flowrates $q'$ and $q''$ are known, $c$ is determined by means of a special table [9], and then the branches of the free surface can be constructed.

S. N. Numerov analyzed separately the case of unilateral flow to a drain, when $q'' = 0$, and to a system of two symmetrical drains.

## §5. Flow to a Rectangular Trench with Sloping Free Surface in a Stratum of Infinite Depth.

We reproduce the solution of a problem given by S. N. Numerov [11] (see Figure 183). The rectangular drainage ditch does not contain water, and consequently the pressure along its contour is considered to be constant. Here it is convenient to apply Zjoukovsky's function, which may be chosen so that its imaginary part is equal to zero along the entire contour of the flowregion. In this case one may assume that the half-plane of the auxiliary variable $\zeta$ coincides with the region of Zjoukovsky's function, and take $\zeta$ for Zjoukovsky's function:

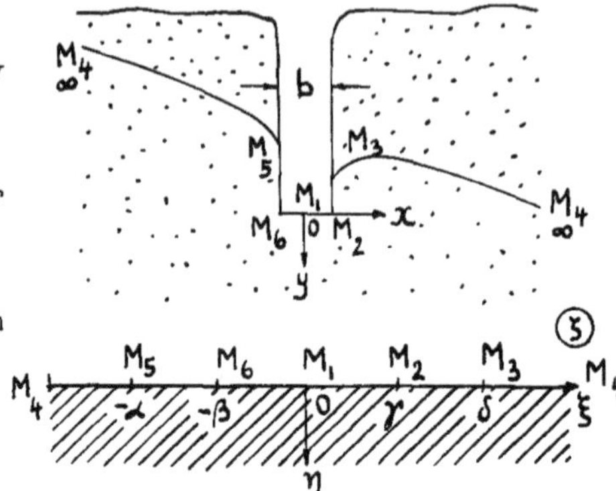


(5.1)
$$\zeta = \xi + i\eta = z - i\omega \quad ,$$

in which

Fig. 183

(5.2)
$$\begin{cases} \xi = x + \psi, \quad \eta = y - \varphi \quad , \\ u - iv = k \dfrac{d\omega}{dz} \quad , \end{cases}$$

Here $\omega$, $\varphi$, $\psi$ are reduced functions, i.e., related to a seepage coefficient of unit magnitude.

At the center of the bottom of the trench $M_1(0, 0)$ we assume $\psi = 0$, we designate the flowrate through line $M_5M_6M_1$ by $q'$, and that through $M_1M_2M_3$ through $q''$. The total reduced flowrate of the drain is $q$:

(5.3)
$$q = q' + q'' \quad .$$

Since the drain receives a flow from infinity, we may single out a logarithmic term in the expression of the complex potential as a function of $\zeta$

(5.4)
$$\omega = -\frac{q}{\pi}\,\ell n\,\zeta + iq'' + W \quad .$$

Function $W$ is analytic in the upper half-plane and continuous up to its boundaries, except for points $\zeta = 0$ and $\zeta = \infty$. To determine its values at the boundaries where $\zeta = \xi$, we write it as

(5.5)
$$\begin{cases} W = U + iV = -iq'' + \dfrac{q}{\pi}\,\ell n\,\zeta + \omega \quad , \\ U = \dfrac{q}{\pi}\,\ell n|\zeta| + \varphi, \quad V = -q'' + \dfrac{q}{\pi}\arg\zeta + \psi \quad , \end{cases}$$

in which $\arg\zeta = 0$ for $\zeta = \xi > 0$ and $\arg\zeta = \pi$ for $\zeta = \xi < 0$ .

Along segment $M_4M_5$, where $\psi = -q'$, we have $V = 0$. Along $M_5M_6$, where $x = -\frac{b}{2}$, $\psi = \xi - x$, we have $V = \xi + q' + \frac{b}{2}$. Continuing in this way, we find the e conditions at the boundaries:

(5.6)
$$\begin{cases} V(\xi) = 0, & -\infty < \xi < -\alpha, \quad \delta < \xi < \infty , \\ V(\xi) = \alpha + \xi, & -\alpha \leq \xi \leq -\beta, \\ V(\xi) = \xi - \delta, & \gamma \leq \xi \leq \delta, \\ U(\xi) = \dfrac{q}{\pi}\,\ell n\,|\xi|, & -\beta \leq \xi \leq \gamma. \end{cases}$$

From condition $V = 0$ for $\zeta = -\alpha$ and $\zeta = \delta$, we find:

$$\alpha = q' + \frac{b}{2}, \quad \delta = q'' + \frac{b}{2} \quad . \tag{5.7}$$

Function

$$W : \sqrt{(\gamma - \zeta)(-\beta - \zeta)}$$

satisfies the conditions of applicability of formula (1.7). Therefore we may write:

$$\frac{W}{\sqrt{(\gamma - \zeta)(-\beta - \zeta)}}$$

$$= \frac{1}{\pi}\int_{-\alpha}^{-\beta} \frac{(\alpha + t)\,dt}{(t - \zeta)\sqrt{\gamma - t)(-\beta - t)}} + \frac{q}{\pi^2}\int_{-\beta}^{\gamma} \frac{\ln|t|\,dt}{(t - \zeta)\sqrt{(\gamma - t)(\beta + t)}} +$$

$$+ \frac{1}{\pi}\int_{\gamma}^{\delta} \frac{(\delta - t)\,dt}{(t - \zeta)\sqrt{(t - \gamma)(\beta + t)}} + C \quad , \tag{5.8}$$

where $C$ is the limit of the real part of this function at infinity.

Carrying out the computations and substituting the result in formula (5.4), we find the following expression for $\omega$:

$$\omega = -iq' + C\sqrt{(-\beta - \zeta)(\gamma - \zeta)} + \frac{2}{\pi}(\alpha - \zeta)\tanh^{-1}\frac{A}{f} +$$

$$+ \frac{2}{\pi}(\delta - \zeta)\tanh^{-1}(Bf) - \frac{2}{\pi}q\tanh^{-1} f \quad , \tag{5.9}$$

where the following notations are used:

$$\begin{cases} f = \sqrt{\dfrac{-\beta - \zeta}{\gamma - \zeta}}, \quad A = \sqrt{\dfrac{\alpha - \beta}{\alpha + \gamma}}, \quad B = \sqrt{\dfrac{\delta - \gamma}{\delta + \beta}}, \\ C = -J + \dfrac{2}{\pi}\tanh^{-1} A - \dfrac{2}{\pi}\tanh^{-1} B, \quad \tanh^{-1} x = \dfrac{1}{2}\ln\dfrac{1 + x}{1 - x} \end{cases} \tag{5.10}$$

The correctness of formula (5.9) may be checked immediately in an example. The constant $C$, as shown further, is the slope of the surface at infinity.

There are six constants in equation (5.9): $J$, $q$, $\alpha$, $\beta$, $\gamma$, $\delta$. Quantities $q'$ and $q''$ are expressed in terms of $\alpha$ and $\delta$. We have four conditions, allowing us to determine four of the constants:

1) From (5.7) we have

$$\alpha + \delta = b + q \; ; \tag{5.11}$$

2) In points $M_6$, $M_2$ where the velocity is infinite, $d\zeta/d\omega = dz/d\omega - i$ gives $d\omega/d\zeta = i$, i.e.,

$$\omega'(-\beta) = i, \quad \omega'(\gamma) = i \; : \tag{5.12}$$

3) For $\zeta = 0$, we take

$$\omega(0) = 0 \; . \tag{5.13}$$

From (5.12) we have:

$$C = -J + \frac{2}{\pi}\tanh^{-1} A - \frac{2}{\pi}\tanh^{-1} B = 0 \quad ,$$

$$q = \sqrt{(\alpha - \beta)(\alpha + \gamma)} + \sqrt{(\delta - \gamma)(\delta + \beta)}$$

We have from (5.11) and (5.14):

$$\alpha - \beta = \frac{b}{4} e^{-\frac{\pi}{2} J} \operatorname{ch}^{-1} \frac{\pi J}{2} \left(e^{\frac{\pi J}{2}} \sqrt{1 + \frac{2q}{b}} - 1\right)^2 ,$$

$$\alpha + \gamma = \frac{b}{4} e^{-\frac{\pi J}{2}} \operatorname{ch}^{-1} \frac{\pi J}{2} \left(e^{\frac{\pi J}{2}} \sqrt{1 + \frac{2q}{b}} + 1\right)^2 ,$$

$$\delta - \gamma = \frac{b}{4} e^{\frac{\pi J}{2}} \operatorname{ch}^{-1} \frac{\pi J}{2} \left(e^{\frac{\pi J}{2}} \sqrt{1 + \frac{2q}{b}} - 1\right)^2 ,$$

$$\delta + \beta = \frac{b}{4} e^{\frac{\pi J}{2}} \operatorname{ch}^{-1} \frac{\pi J}{2} \left(e^{-\frac{\pi J}{2}} \sqrt{1 + \frac{2q}{b}} + 1\right)^2 .$$

From this we find the following expressions for $A$ and $B$:

$$A = \sqrt{\frac{\alpha - \beta}{\alpha + \gamma}} = \left(e^{\frac{\pi J}{2}} \sqrt{1 + \frac{2q}{b}} - 1\right) : \left(e^{\frac{\pi J}{2}} \sqrt{1 + \frac{2q}{b}} + 1\right) , \tag{5.16}$$

$$B = \sqrt{\frac{\delta - \gamma}{\delta + \beta}} = \left(e^{-\frac{\pi J}{2}} \sqrt{1 + \frac{2q}{b}} - 1\right) : \left(e^{-\frac{\pi J}{2}} \sqrt{1 + \frac{2q}{b}} + 1\right) .$$

<u>Equations of free surface</u>. At the left branch we have

$$\varphi = y, \quad \psi = -q', \quad -\alpha - \xi = -\frac{b}{2} - x .$$

From equation (5.9), considering that $C = 0$, we have:

$$y = \frac{2}{\pi}\left(\frac{b}{2} + x\right) \tanh^{-1} \frac{A}{f} + \frac{2}{\pi}\left(\frac{b}{2} + q - x\right) \tanh^{-1} (Bf) - \frac{2}{\pi} q \tanh^{-1} f , \tag{5.17}$$

where

$$f^2 = \frac{b\left(e^{\frac{\pi J}{2}} \sqrt{1 + \frac{2q}{b}} - 1\right)^2 - 4e^{\frac{\pi J}{2}} \left(\frac{b}{2} + x\right) \cosh \frac{\pi J}{2}}{b\left(e^{\frac{\pi J}{2}} \sqrt{1 + \frac{2q}{b}} + 1\right)^2 - 4e^{\frac{\pi J}{2}} \left(\frac{b}{2} + x\right) \cosh \frac{\pi J}{2}} , \tag{5.18}$$

and for which $-\infty < x \leq -\frac{b}{2}$ .

Putting $x = -b/2$, we find the height of the seepage surface at the left side of the ditch:

$$\frac{h'}{q} = \frac{J}{2} + \frac{1}{\pi}\left(1 + \frac{b}{2q}\right) \ell n \left(1 + \frac{2q}{b}\right) + \frac{1}{\pi}\left(1 + \frac{b}{q}\right) \ell n \frac{\cosh \frac{\pi J}{2}}{1 + \frac{q}{b}} . \tag{5.19}$$

For sufficiently far points on the free surface, namely for $-x \gg q$, after separating the terms with $\ell n(-x)$ and developing the remaining expression in powers of $1/x$, we find an asymptotic expression, of order $O(q/x)$:

$$\frac{y}{q} \sim \frac{Jx}{q} - \frac{1}{\pi} \ell n \left(-\frac{x}{q}\right) + D , \tag{5.20}$$

where

$$D = -\frac{1}{\pi}\left[1 + \ell n(1 + e^{\pi J}) + \ell n \frac{2q}{b} - \left(1 + \frac{b}{2q}\right) \ell n \left(1 + \frac{b}{2q}\right)\right] .$$

It follows from (5.20) that $\zeta$ is equal to the slope of the free surface at infinity.

For the right branch of the free surface we have:

$$(5.21)\quad y = \frac{2}{\pi}\left(\frac{b}{2} + q + x\right)\tanh^{-1}(Af) - \frac{2}{\pi}\left(x - \frac{b}{2}\right)\tanh^{-1}\frac{B}{f} - \frac{2q}{\pi}\tanh^{-1} f \quad ,$$

where

$$f^2 = \frac{b\left(e^{-\frac{\pi J}{2}}\quad 1 + \frac{2q}{b} - 1\right)^2 + 4e^{-\frac{\pi J}{2}}\cosh\frac{\pi J}{2}\left(x - \frac{b}{2}\right)}{b\left(e^{-\frac{\pi J}{2}}\quad 1 + \frac{2q}{b} + 1\right)^2 + 4e^{-\frac{\pi J}{2}}\cosh\frac{\pi J}{2}\left(x - \frac{b}{2}\right)}$$

Putting $x = b/2$, we find the height of the seepage surface at the right side

$$(5.22)\quad \frac{h''}{q} = -\frac{1}{2}J + \frac{1}{\pi}\left(1 + \frac{b}{2q}\right)\ell n\left(1 + \frac{b}{2q}\right) + \frac{1}{\pi}\left(1 + \frac{b}{q}\right)\ell n\,\frac{\cosh\frac{\pi J}{2}}{1 + \frac{q}{b}} \quad .$$

Comparing (5.19) and (5.22), we find the difference in seepage heights

$$h' - h'' = Jq = J\,\frac{Q}{k} \quad ,$$

where $Q$ — total flowrate per unit length of drain.

For $x >> q$, we have the asymptotic formula

$$(5.23)\quad \frac{y}{q} \approx J\,\frac{x}{q} - \frac{1}{\pi}\,\ell n\,\frac{x}{q} + D + J \quad .$$

The coordinates of the maximum of the right branch of the free surface are determined by the condition $dy/dx = 0$. Because of (5.21) we have:

$$(5.24)\quad \frac{x_{max}}{q} = \frac{1}{e^{\pi J} - 1}, \quad \frac{y_{max}}{q} = \frac{2}{\pi}\left(\frac{b}{q} + 1\right)\tanh^{-1}\sqrt{AB} - \frac{2}{\pi}\tanh^{-1}\sqrt{\frac{B}{A}} \quad .$$

For small $J$ we obtain from this the approximate equalities

$$(5.25)\quad \frac{x_{max}}{q} \approx \frac{1}{\pi J}\,, \quad \frac{y_{max}}{q} \approx \frac{1}{\pi}\,\ell n\,\frac{\pi J b}{4q} + \frac{1}{\pi}\left(1 + \frac{b}{2q}\right)\ell n\left(1 + \frac{2q}{b}\right) \quad .$$

To solve practical problems, the slope of the natural free surface and the coordinates of point $M_0(x_0, y_0)$ of the free surface, sufficiently far from the ditch, must be given. Inserting the coordinates of this point in the equation of one branch, either (5.20) or (5.23), we find $q$ and thereafter one finds the remaining characteristics of the flow. The particular case $b = 0$ of the problem has been treated by N. N. Verigin [14]. If, moreover, $J = 0$, we have the problem that was treated by V. V. Vedernikov (see literature to Chapter IV [4]). We have considered this problem in Chapter V, §7.

## §6. About Seepage Towards a Drain or Canal In the Case of Inclined Bedrock.

A sketch of the flow is given in Fig. 184. The drain has the form of an horizontal slit. The boundary of the bedrock makes the angle $\alpha\pi$ with the horizontal [10].

In this paragraph we designate by $\omega = \varphi + i\psi$ the reduced complex potential, i.e., the usual potential divided by the seepage coefficient, so that the complex seepage velocity is expressed as:

$$w = u - iv = k\frac{d\omega}{dz} \quad . \tag{6.1}$$

We designate by $Q$ and $Q'$ the seepage rates respectively at infinity to the left and to the right, by $q$ and $q'$ the reduced flowrates:

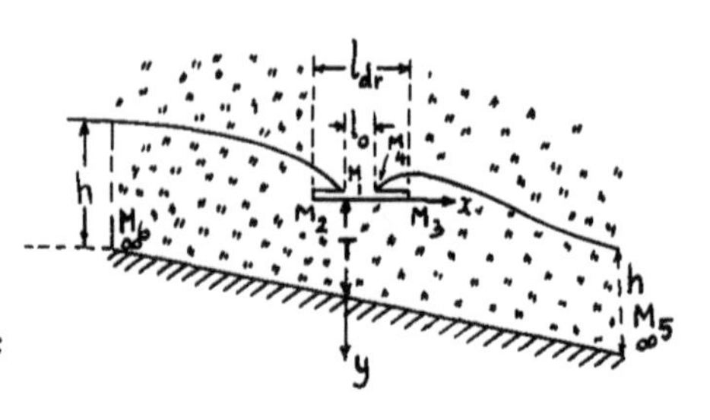


Fig. 184

$$Q = kq, \quad Q' = kq' \quad . \tag{6.2}$$

We have three cases:

1) Flow from all directions to the drain, when the flowrate $q$ is larger than $q'$. In this case the right branch of the free surface has a maximum.

2) Seepage from the drain into the soil, i.e., seepage from a canal. In this case the left branch of the free surface has a minimum.

3) In the upper part of the drain influx of groundwater takes place, in the lower part seepage from the drain into the soil occurs.

All three cases are sketched in Figs. 185-187 together with the corresponding regions of the $\omega = \varphi + i\psi$ plane.

Construction of the solution. Conformal mapping of these regions onto the half-plane $\zeta$ (Fig. 188) gives:

$$\omega = iq - \frac{2}{\pi} q \cosh^{-1}\frac{1}{\sqrt{\beta\zeta}} + \frac{2}{\pi} q' \sinh^{-1}\sqrt{\frac{1-\beta}{\beta(1-\zeta)}}$$

$$\left(\sinh^{-1} x = \ell n\,(x + \sqrt{x^2+1}), \quad \cosh^{-1} x = \ell n\,(x + \sqrt{x^2-1})\right) \quad . \tag{6.3}$$

Indeed, let the end of the cut in the $\omega$-plane, i.e., point $M_0$ transform into point $a$ of the $\zeta$-plane. The Christoffel-Schwarz formula, applied to any of the three figures of the $\omega$-plane gives:

$$\omega = M\int \frac{(\zeta - a)\,d\zeta}{\zeta(\zeta - 1)\sqrt{\beta - \zeta}} =$$

$$= Ma\int\frac{d\zeta}{\zeta\sqrt{\beta-\zeta}} + M(1-a)\int\frac{d\zeta}{(\zeta-1)\sqrt{\beta-\zeta}} \quad .$$

Considering that the residue of the first integral of the right side at point $\zeta = 0$ is $-iq$, and that the residue of the second at $\zeta = 1$ is $iq'$, and performing the integrations, we find formula (6.3).

Now we consider Zjoukovsky's function (see §5)

$$W = U + iV = z - i\omega \quad . \tag{6.4}$$

For the intervals $(-\infty, 0)$, $(1, \infty)$ i.e., along the free surface and along the contour of the drain, the pressure is atmospheric, and therefore

$$V(\xi) = y - \varphi = 0 \quad . \tag{6.5}$$

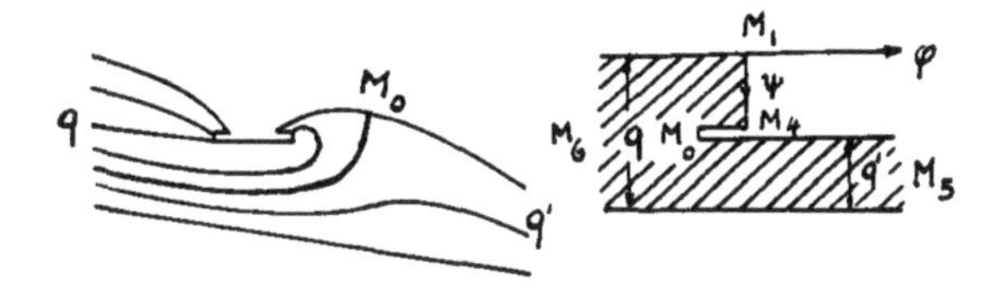


Fig. 185

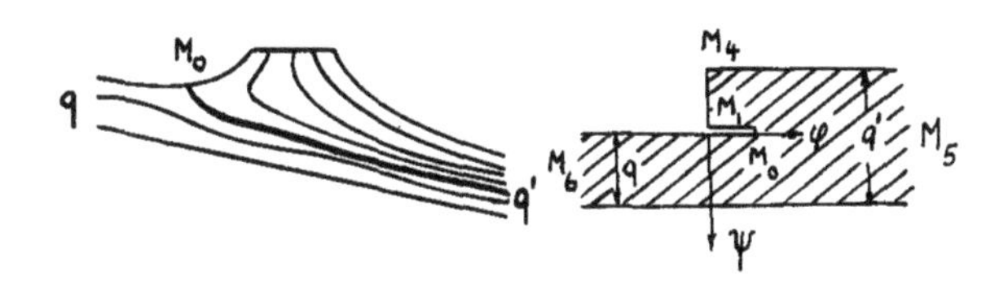


Fig. 186

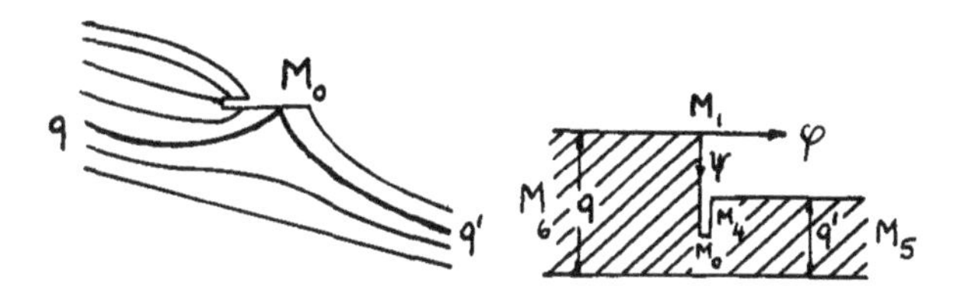


Fig. 187

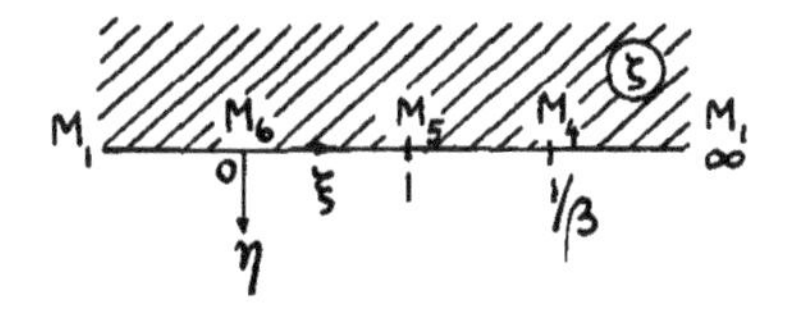


Fig. 188

In the interval $0 < \xi < 1$, we have the conditions

$$x \sin \alpha\pi - y \cos \alpha\pi = - T \cos \alpha\pi, \quad \psi = q \quad . \tag{6.6}$$

Because of (6.4), $U = x + \varphi$, $V = y - \varphi$, and therefore (6.6) is rewritten as:

$$U \sin \alpha\pi - V \cos \alpha\pi = - T \cos \alpha\pi + q \sin \alpha\pi + \varphi(\xi) \cos \alpha\pi \quad .$$

Considering that because of (6.3)

$$\varphi(\xi) = - \frac{2q}{\pi} \cosh^{-1} \frac{1}{\sqrt{\beta\xi}} + \frac{2q'}{\pi} \sinh^{-1} \sqrt{\frac{1 - \beta}{\beta(1 - \xi)}} \quad , \tag{6.7}$$

We find that $V = \operatorname{Im} W = 0$ outside the segment $(0, 1)$ of the real axis; inside the interval $(0, 1)$ we have:

$$- \operatorname{Im}[e^{-\alpha\pi i} W] = - T \cos \alpha\pi + q \sin \alpha\pi + \varphi(\xi) \cos \alpha\pi = f(\xi) \quad .$$

The point at infinity of the $\zeta$-plane corresponds to zero for $W$. Applying formula (1.7) to the function

$$(W + T \operatorname{cotg} \alpha\pi - q)\zeta^{-\alpha}(1 - \zeta)^{\alpha-1} \quad ,$$

we have:

$$z(\zeta) = - T \operatorname{cotg} \alpha\pi + q + i\omega - \frac{\cos \alpha\pi}{\pi}\zeta^{\alpha}(\zeta - 1)^{1-\alpha}\int_0^1 \frac{t^{-\alpha}(1 - t)^{\alpha-1}\varphi(t)dt}{t - \zeta} \tag{6.8}$$

<u>Determination of the constants</u>. Equations (6.3) and (6.8) contain two unknown constants: $q'$ and $\beta$. The quantity $q$ is given by means of the depth of flow $h$ at infinity to the left:

$$q = h \sin \alpha\pi \cos \alpha\pi \quad . \tag{6.9}$$

In exactly the same way, designating by $h'$ the depth of flow at infinity to the right side

$$q' = h' \sin \alpha\pi \cos \alpha\pi \quad . \tag{6.9'}$$

The conditions $z(\infty) = 0$, $\varphi = \psi = 0$ and $z\left(\frac{1}{\beta}\right) = \ell_0$, $\varphi = 0$, $\psi = q - q'$ give:

$$0 = - T \operatorname{cotg} \alpha\pi + q + \frac{\cos \alpha\pi}{\pi} \int_0^1 t^{-\alpha}(1 - t)^{\alpha-1}\varphi(t)\, dt \quad ,$$

$$\ell_0 = - T \operatorname{cotg} \alpha\pi + q' + \frac{\cos \alpha\pi}{\pi} (1 - \beta)^{1-\alpha} \int_0^1 t^{-\alpha}(1 - t)^{\alpha-1} \frac{\varphi(t)\, dt}{1 - \beta t} \quad .$$

*We may consider* $\beta$ *as given, and* $q$ *and* $q'$ *unknown, and write the latter* equalities as a system of equations

$$a_{11}q + a_{12}q' = T, \quad a_{21}q + a_{22}q' = J\ell_0 \quad (J = \operatorname{tg} \alpha\pi) \quad , \tag{6.10}$$

where

$$a_{11}(\beta, \alpha) = \operatorname{tg} \alpha\pi - \frac{2}{\pi^2} \sin \alpha\pi \int_0^1 t^{-\alpha}(1 - t)^{\alpha-1} \cosh^{-1} \frac{1}{\sqrt{\beta t}}\, dt \quad , \tag{6.11}$$

$$a_{12}(\beta, \alpha) = \frac{2}{\pi^2} \sin \alpha\pi \int_0^1 t^{-\alpha}(1 - t)^{\alpha-1} \sinh^{-1} \sqrt{\frac{1 - \beta}{\beta(1 - t)}}\, dt \quad , \tag{6.12}$$

$$a_{21}(\beta, \alpha) = - a_{11}(\beta, \alpha) - a_{12}(\beta, 1 - \alpha) \quad , \tag{6.13}$$

$$a_{22}(\beta, \alpha) = - a_{11}(\beta, 1 - \alpha) - a_{12}(\beta, \alpha) = a_{21}(\beta, 1 - \alpha) \quad . \tag{6.14}$$

The latter equalities may be obtained by the substitution $\tau = \frac{1 - t}{1 - \beta t}$, in the integral of $\ell_0$. This gives:

$$(1 - \beta)^{1-\alpha} \int_0^1 t^{-\alpha}(1 - t)^{\alpha-1}\left(q \cosh^{-1} \frac{1}{\sqrt{\beta t}} - q' \sinh^{-1} \sqrt{\frac{1 - \beta}{\beta(1 - t)}}\right)\frac{dt}{1 - \beta t} =$$

$$= \int_0^1 \tau^{\alpha-1}(1 - \tau)^{-\alpha} \left(q \sinh^{-1} \sqrt{\frac{1 - \beta}{\beta(1 - \tau)}} - q' \cosh^{-1} \frac{1}{\sqrt{\beta\tau}} \right) d\tau.$$

To simplify the expressions for the quantities $a_{11}$, $a_{12}$, ..., S. N. Numerov applies different methods for two intervals, including the number $\beta$.

1) Interval $0 \leq \beta \leq \frac{1}{2}$. Differentiating $a_{11}$ with respect to $\beta$, we find

$$\beta\frac{\partial a_{11}}{\partial \beta} = \frac{1}{\pi^2} \sin \alpha\pi \int_0^1 t^{-\alpha}(1 - t)^{\alpha-1}(1 - \beta t)^{-\frac{1}{2}} dt = \frac{1}{\pi} F\left(\frac{1}{2}, 1 - \alpha, 1 ; \beta\right) ,$$

where $F(\frac{1}{2}, 1 - \alpha, 1 ; \beta)$ is the hypergeometric function (see Chapter VII, §5). Integrating the obtained equation, we have

$$a_{11} = \frac{1}{\pi}\int_0^\beta \frac{1}{\beta}\left[F\left(\frac{1}{2}, 1 - \alpha, 1 ; \beta\right) - 1\right] d\beta + \frac{1}{\pi} \ln \beta + c ,$$

where the constant $c$ does not depend upon $\beta$. It is determined by passing to the limit in (6.11) for $\beta \to 0$ considering that $\lim\left(2 \cosh^{-1} \frac{1}{\sqrt{\beta t}} + \ln \beta\right) = \ln \frac{2}{\sqrt{t}}$:

$$c = \lim_{\beta \to 0} \left(a_{11} - \frac{1}{\pi} \ln \beta\right) = \operatorname{tg} \alpha\pi + \frac{\sin \alpha\pi}{\pi^2} \int_0^1 t^{-\alpha}(1 - t)^{\alpha-1}\ln \frac{t}{4} dt =$$

$$= \operatorname{tg} \alpha\pi + \frac{1}{\pi} [- \ln 4 - \psi(1) + \psi(1 - \alpha)] .$$

Here $\psi(a)$. is the logarithmic derivative of the $\Gamma$-function. To obtain the latter formula we may differentiate the beta-function with respect to one of its parameters. We find:

$$a_{11} = \operatorname{tg} \alpha\pi \left\{ \psi(1 - \alpha) - \psi(1) - \ln \frac{4}{\beta} + \right.$$

$$\left. + \int_0^\beta \frac{1}{\beta} \left[F\left(\frac{1}{2}, 1 - \alpha, 1 ; \beta\right) - 1\right] d\beta\right\}$$

and similarly

$$a_{12} = \operatorname{cotg} \alpha\pi + \frac{1}{\pi} \left\{\psi(1) - \psi(1 - \alpha) + 2 \cosh^{-1} \frac{1}{\sqrt{\beta}} - \int_0^\beta \frac{F\left(\frac{1}{2}, 1 - \alpha, 1 ; \beta\right)}{\beta \sqrt{1 - \beta}} d\beta \right\}$$

or, developing the hypergeometric function in power series in $\beta$:

$$a_{11}(\beta, \alpha) = \operatorname{tg} \alpha\pi + \frac{1}{\pi} \left[\psi(1 - \alpha) - \psi(1) - \ln \frac{4}{\beta} + \right.$$

$$(6.15) \qquad \left. + \frac{\frac{1}{2}(1 - \alpha)}{1 \cdot (1!)^2} \beta + \frac{\frac{1}{2} \cdot \frac{3}{2} (1 - \alpha)(2 - \alpha)}{2\cdot(2!)^2} \beta^2 + \dots \right] ,$$

$$a_{12}(\beta, \alpha) = \text{cotg}\ \alpha\pi + \frac{1}{\pi}\left[\psi(1) - \psi(1-\alpha) + 2\cosh^{-1}\frac{1}{\sqrt{\beta}} - \frac{\frac{1}{2}(1-\alpha)}{(1!)^2}\int_0^\beta \frac{d\beta}{\sqrt{1-\beta}} - \frac{\frac{1}{2}\cdot\frac{3}{2}(1-\alpha)(2-\alpha)}{(2!)^2}\int_0^\beta \frac{\beta\, d\beta}{\sqrt{1-\beta}} + \dots\right].$$

We now examine the second interval for $\beta$.

2) Interval $\frac{1}{2} \leq \beta < 1$. We use the auxiliary substitution for the hypergeometric function (see Chapter VII, §5):

$$F(a, b, c; x) = \frac{\Gamma(c)\,\Gamma(c-a-b)}{\Gamma(c-a)\,\Gamma(c-b)} F(a, b, a+b+1-c, 1-x) + \frac{\Gamma(c)\,\Gamma(a+b-c)}{\Gamma(a)\,\Gamma(b)}(1-x)^{c-a-b} \times F(c-a, c-b, c+1-a-b, 1-x).$$

Applied to the function $F(\frac{1}{2}, 1-\alpha, 1; \beta)$ it gives the equality:

$$\beta\frac{\partial a_{11}}{\partial\beta} = \frac{1}{\pi}F\left(\frac{1}{2}, 1-\alpha, 1; \beta\right) = \frac{\text{tg}\ \alpha\pi\Gamma(1-\alpha)}{\pi^{3/2}\Gamma\left(\frac{3}{2}-\alpha\right)}F\left(\frac{1}{2}, 1-\alpha, \frac{3}{2}-\alpha; 1-\beta\right) + \frac{\Gamma\left(\frac{1}{2}-\alpha\right)}{\pi^{3/2}\Gamma(1-\alpha)}(1-\beta)^{-\frac{1}{2}+\alpha}F\left(\frac{1}{2}, \alpha, \frac{1}{2}+\alpha; 1-\beta\right).$$

Integrating this over $\beta$ and determining the value of the constant of integration, we find:

$$a_{11}(\beta, \alpha) = \text{tg}\ \alpha\pi - \frac{2}{\pi^2}\sin\alpha\pi\int_0^1 t^{-\alpha}(1-t)^{\alpha-1}\cosh^{-1}\frac{1}{\sqrt{\beta t}}\,dt + \frac{\text{tg}\ \alpha\pi\Gamma(1-\alpha)}{\pi^{3/2}\Gamma\left(\frac{3}{2}-\alpha\right)}\int_\beta^1 \frac{1}{\beta}F\left(\frac{1}{2}, 1-\alpha, \frac{3}{2}-\alpha; 1-\beta\right)d\beta - \frac{\Gamma\left(\frac{1}{2}-\alpha\right)}{\pi^{3/2}\Gamma(1-\alpha)}\int_\beta^1 \frac{F\left(\frac{1}{2}, \alpha, \frac{1}{2}+\alpha; 1-\beta\right)}{\beta(1-\beta)^{1/2-\alpha}}d\beta$$

and similarly

$$a_{12}(\beta, \alpha) = -\frac{\text{tg}\ \alpha\pi\Gamma(1-\alpha)}{\pi^{3/2}\Gamma\left(\frac{3}{2}-\alpha\right)}\int_\beta^1 \frac{F\left(\frac{1}{2}, 1-\alpha, \frac{3}{2}-\alpha; 1-\beta\right)}{\beta\sqrt{1-\beta}}d\beta + \frac{\Gamma\left(\frac{1}{2}-\alpha\right)}{\pi^{3/2}\Gamma\left(\frac{3}{1}-\alpha\right)}\int_\beta^1 \frac{F\left(\frac{1}{2}, \alpha, \frac{1}{2}+\alpha; 1-\beta\right)}{\beta(1-\beta)^{1-\alpha}}d\beta.$$

These equalities may be written in this form:

$$a_{11} = \operatorname{tg} \alpha\pi - \frac{4}{\pi^2} \sin \alpha\pi \int_0^{\pi/2} \operatorname{tg}^{1-2\alpha} t \, \ln \operatorname{ctg} \frac{t}{2} \, dt +$$

$$+ \frac{\operatorname{tg} \alpha\pi \Gamma(1-\alpha)}{\pi^{3/2}\Gamma\left(\frac{3}{2}-\alpha\right)} \left[ \ln \frac{1}{\beta} + \frac{\frac{1}{2}(1-\alpha)}{1!\left(\frac{3}{2}-\alpha\right)} \int_0^{1-\beta} \frac{t\,dt}{1-t} + \right.$$

$$\left. + \frac{\frac{1}{2}\cdot\frac{3}{2}(1-\alpha)(2-\alpha)}{2!\left(\frac{3}{2}-\alpha\right)\left(\frac{5}{2}-\alpha\right)} \int_0^{1-\beta} \frac{t^2\,dt}{1-t} + \ldots \right] -$$

$$- \frac{\Gamma\left(\frac{1}{2}-\alpha\right)}{\pi^{3/2}\Gamma(1-\alpha)} \left[ \int_0^{1-\beta} \frac{t^{\alpha-\frac{1}{2}}}{1-t}\,dt + \frac{\frac{1}{2}\alpha}{1!\left(\frac{1}{2}+\alpha\right)} \int_0^{1-\beta} \frac{t^{\alpha+\frac{1}{2}}}{1-t}\,dt + \right.$$

$$(6.16) \qquad \left. + \frac{\frac{1}{2}\cdot\frac{3}{2}\cdot\alpha(1+\alpha)}{2!\left(\frac{1}{2}+\alpha\right)\left(\frac{3}{2}+\alpha\right)} \int_0^{1-\beta} \frac{t^{\alpha+\frac{3}{2}}}{1-t}\,dt + \ldots \right],$$

$$a_{12} = - \frac{\operatorname{tg} \alpha\pi \Gamma(1-\alpha)}{\pi^{3/2}\Gamma\left(\frac{3}{2}-\alpha\right)} \left[ \int_0^{1-\beta} \frac{dt}{(1-t)\sqrt{t}} + \frac{\frac{1}{2}(1-\alpha)}{1!\left(\frac{3}{2}-\alpha\right)} \int_0^{1-\beta} \frac{\sqrt{t}}{1-t}\,dt + \right.$$

$$\left. + \frac{\frac{1}{2}\cdot\frac{3}{2}(1-\alpha)(2-\alpha)}{2!\left(\frac{3}{2}-\alpha\right)\left(\frac{5}{2}-\alpha\right)} \int_0^{1-\beta} \frac{t^{\frac{3}{2}}\,dt}{1-t} + \ldots \right] +$$

$$+ \frac{\Gamma\left(\frac{1}{2}-\alpha\right)}{\pi^{3/2}\Gamma(1-\alpha)} \left[ \int_0^{1-\beta} \frac{dt}{t^{1-\alpha}(1-t)} + \frac{\frac{1}{2}\alpha}{1!\left(\frac{1}{2}+\alpha\right)} \int_0^{1-\beta} \frac{t^{\alpha}\,dt}{1-t} + \right.$$

$$\left. + \frac{\frac{1}{2}\cdot\frac{3}{2}\alpha(1+\alpha)}{2!\left(\frac{1}{2}+\alpha\right)\left(\frac{3}{2}+\alpha\right)} \int_0^{1-\beta} \frac{t^{\alpha+1}}{1-t}\,dt + \ldots \right].$$

Extreme of the free surface. In the case of flow from all sides towards a drain, as we said before, the right branch of the free surface has a maximum. To determine this, we rewrite for the right branch, by means of (6.3):

$$y = \varphi = - \frac{2q}{\pi} \cosh^{-1} \frac{1}{\sqrt{\beta\xi}} + \frac{2q'}{\pi} \cosh^{-1} \sqrt{\frac{1-\beta}{\beta(\xi-1)}} \left(1 \le \xi \le \frac{1}{\beta}\right).$$

Equating the derivative with respect to $\xi$

$$\frac{dy}{d\xi} = - \frac{1}{2\pi} (1-\beta\xi)^{-\frac{1}{2}} \left( \frac{q}{\xi} - \frac{q'\sqrt{1-\beta}}{\xi-1} \right)$$

to zero, and considering that $q' : q = h' : h$, we find

$$\xi_{max} = \frac{1}{1 - \frac{h'}{h}\sqrt{1 - \beta}} .$$

The excess of the maximum above the horizontal water level in the drain is:

$$-y_{max} = \frac{2q}{\pi} \cosh^{-1} \sqrt{\frac{h - h'\sqrt{1 - \beta}}{h\beta}} - \frac{2q'}{\pi} \cosh^{-1} \sqrt{\frac{\sqrt{1 - \beta}(h - h'\sqrt{1 - \beta})}{h'\beta}} . \tag{6.17}$$

The condition that $\xi_{max}$ be in the interval $(1, 1/\beta)$ requires:

$$\beta < 1 - \frac{h'}{h} \sqrt{1 - \beta} < 1 ,$$

from which we obtain the inequality

$$\frac{h}{h'} > \frac{1}{\sqrt{1 - \beta}} .$$

In the same way, for the left branch of the free surface, in the case of seepage from a canal, we obtain the value of the minimum ordinate:

$$y_{min} = - \frac{2q}{\pi} \cosh^{-1} \sqrt{\frac{h'\sqrt{1 - \beta} - h}{h\beta}} + \frac{2q'}{\pi} \cosh^{-1} \sqrt{\frac{\sqrt{1 - \beta}(h'\sqrt{1 - \beta} - h)}{h'\beta}} \tag{6.18}$$

with the condition

$$\frac{h}{h'} < \sqrt{1 - \beta} .$$

<u>Case of small slope</u>. When the angle $\alpha\pi$ is small, the expressions for $a_{ik}$ may be simplified. Putting $\mathrm{tg}\, \alpha\pi = J$, we again consider two cases

1) $0 \leq \beta \leq \frac{1}{2}$. From equation (6.15) we find:

$$a_{11} \approx - \frac{1}{\pi}\left(\ln \frac{4}{\beta} - \frac{\frac{1}{2}}{1.1!}\beta - \frac{\frac{1}{2}\cdot\frac{3}{2}}{2.2!}\beta^2 - \dots\right) \approx - \frac{2}{\pi} \cosh^{-1} \frac{1}{\sqrt{\beta}} ,$$

$$a_{12} \approx \frac{1}{J} + \frac{1}{\pi}\left(2 \cosh^{-1} \frac{1}{\sqrt{\beta}} - \frac{\frac{1}{2}}{1!} \int_0^\beta \frac{d\beta}{\sqrt{1 - \beta}} - \frac{\frac{1}{2}\cdot\frac{3}{2}}{2!} \int_0^\beta \frac{\beta\, d\beta}{\sqrt{1 - \beta}} - \dots\right) \approx \frac{1}{J} + \frac{1}{\pi} \ln \frac{4(1 - \beta)}{\beta} ,$$

$$a_{21} \approx - J + \frac{1}{\pi}\left(\ln \frac{4}{\beta} - 2 \cosh^{-1} \frac{1}{\sqrt{\beta}} - \frac{\frac{1}{2}}{1.1!}\beta - \frac{\frac{1}{2}\cdot\frac{3}{2}}{2.2!}\beta^2 - \dots\right) \approx - J$$

$$a_{22} \approx J + \frac{1}{\pi}\left(\ln\frac{4}{\beta} - 2\cosh^{-1}\frac{1}{\sqrt{\beta}} + \right.$$

$$\left. + \frac{\frac{1}{2}}{1!}\int_0^\beta \frac{dt}{\sqrt{1-t}} - \frac{\frac{1}{2}\cdot\frac{3}{2}}{2!}\int_0^\beta \frac{t\,dt}{\sqrt{1-t}} + \dots\right) \approx J - \frac{1}{\pi}\ln(1-\beta) \quad .$$

Instead of (6.10) we may write

$$-\frac{2}{\pi}\cosh^{-1}\frac{1}{\sqrt{\beta}}\cdot q + \left(\frac{1}{J} + \frac{1}{\pi}\ln\frac{4(1-\beta)}{\beta}\right)q' \approx J \quad ,$$

$$\text{(6.19)} \qquad -Jq + \left[J - \frac{\ln(1-\beta)}{\pi}\right]q' \approx J\ell_0 \quad .$$

We designate by $q_{dr}$ the reduced seepage rate per unit length of drain

$$q_{dr} = q - q'$$

and put

$$\text{(6.20)} \qquad \gamma = \frac{\pi}{2}\,\frac{\ell_0 + q_{dr}}{T} \quad .$$

Equation (6.19) may be written in the following form:

$$\frac{1}{J}q' \approx T, \qquad -\frac{\ln(1-\beta)}{\pi}q' \approx J(\ell_0 + q_{dr}) \quad .$$

From this we find the approximate equalities

$$\text{(6.21)} \qquad q' \approx JT, \qquad \beta \approx 1 - e^{-2\gamma}$$

and also

$$\text{(6.22)} \qquad h' = \frac{q'}{\sin\alpha\pi\,\cos\alpha\pi} \approx T, \qquad q_{dr} \approx J(h-t) \quad .$$

2) $\frac{1}{2} \leq \beta \leq 1$. From equation (6.16) we obtain the approximate equalities

$$a_{11} \approx \frac{1}{\pi}\int_0^{1-\beta}\frac{dt}{(1-t)\sqrt{t}} \approx -\frac{2}{\pi}\cosh^{-1}\frac{1}{\sqrt{\beta}} \quad ,$$

$$a_{12} \approx \frac{1}{\pi}\int_0^{1-\beta}\frac{dt}{t^{1-\alpha}} \approx \frac{1}{J}(1-\beta)^\alpha, \qquad \alpha_{21} \approx 1 \quad ,$$

$$a_{22} \approx \frac{4}{\pi}\sin\alpha\pi\int_0^{\pi/2}\operatorname{tg}^{2\alpha-1}t\,\ln\operatorname{cotg}\frac{t}{2}\,dt - \frac{1}{\pi}\int_0^{1-\beta}\frac{dt}{t^{1-\alpha}} \approx \frac{1}{J}[1-(1-\beta)^\alpha].$$

Instead of (6.10) we have:

$$-\frac{2}{\pi}q\cosh^{-1}\frac{1}{\sqrt{\beta}} + \frac{1}{J}(1-\beta)^\alpha q' \approx T, \qquad q + \frac{1}{J}[1-(1-\beta)^\alpha]q' \approx J\ell_0 \quad .$$

From this we find for $q'$ and $\beta$ approximate equalities (6.21). Consequently, the approximate equalities (6.21) and (6.22) hold for all values of $0 \leq \beta \leq 1$.

Further, we may find approximate values of the extreme ordinates of the free surface for small $\alpha\pi$

$$- y_{max} \approx \frac{2q}{\pi}\left\{\cosh^{-1}\sqrt{\frac{\bar{h} - e^{-\gamma}}{\bar{h}(1 - e^{-2\gamma})}} - \frac{1}{\bar{h}}\cosh^{-1}\sqrt{\frac{e^{-\gamma}(\bar{h} - e^{-\gamma})}{1 - e^{-2\gamma}}}\right\} ,$$

in which

$$\bar{h} = \frac{h}{T} > e^{\gamma} ,$$

$$y_{min} \approx \frac{2q}{\pi}\left\{\frac{1}{\bar{h}}\cosh^{-1}\sqrt{\frac{e^{-\gamma}(e^{-\gamma} - \bar{h})}{1 - e^{-2\gamma}}} - \cosh^{-1}\sqrt{\frac{e^{-\gamma} - \bar{h}}{\bar{h}(1 - e^{-2\gamma})}}\right\} ,$$

in which

$$\bar{h} = \frac{h}{T} < e^{-\gamma} .$$

From this it follows that for $\bar{h} > e^{\gamma}$ we have the case of flow from all sides to a drain, for $\bar{h} < e^{-\gamma}$ we have the case of seepage from a canal into the soil. For $e^{-\gamma} \leq \bar{h} < e^{\gamma}$ we have the case of Fig. 187.

About the magnitude of the drainage slit. By means of equations (6.3) and (6.8) $x$ varies along the slit according to

$$x = - T \operatorname{cotg} \alpha\pi + q + \frac{2q}{\pi}\cos^{-1}\frac{1}{\sqrt{\beta\xi}} - \frac{2q'}{\pi}\cos^{-1}\sqrt{\frac{1 - \beta}{\beta(\xi - 1)}} -$$

$$(6.23) \qquad - \frac{2\cos\alpha\pi}{\pi^2}\xi^{\alpha}(\xi - 1)^{1-\alpha} \times$$

$$\times \int_0^1 t^{-\alpha}(1 - t)^{\alpha-1}\left(q\cosh^{-1}\frac{1}{\sqrt{\beta t}} - q'\sinh^{-1}\sqrt{\frac{1 - \beta}{\beta(1 - t)}}\right)\frac{dt}{\xi - t}$$

$$\left(\frac{1}{\beta} \leq \xi \leq \infty\right) .$$

Introducing

$$M(t) = q\cosh^{-1}\frac{1}{\sqrt{\beta t}} - q'\sqrt{\frac{1 - \beta}{\beta(1 - t)}}$$

and considering that for $\xi = \infty$, $x = 0$, we have

$$(6.24) \quad 0 = - T\operatorname{cotg}\alpha\pi + q + (q - q') - \frac{2\cos\alpha\pi}{\pi^2}\int_0^1 t^{-\alpha}(1 - t)^{\alpha-1}M(t)\, dt.$$

Subtracting (6.24) from (6.23), we write the equation for $x$ as:

$$x = \frac{2q}{\pi}\cos^{-1}\frac{1}{\sqrt{\beta\xi}} - \frac{2q'}{\pi}\cos^{-1}\sqrt{\frac{1 - \beta}{\beta\xi\left(1 - \frac{1}{\xi}\right)}} + q' - q +$$

$$+ \frac{2\cos\alpha\pi}{\pi^2}\int_0^1 t^{-\alpha}(1 - t)^{\alpha-1}M(t)\left[1 - \frac{\left(1 - \frac{1}{\xi}\right)^{-\alpha}}{1 + \frac{1 - t}{\xi - 1}}\right] dt .$$

We consider the case of small values of $\beta$, corresponding to a small value of the width of the slit (compared with the depth $T$). In this case $\xi$ assumes large values, and the approximate equalities hold:

$$\frac{1-\beta}{\beta\xi\left(1-\frac{1}{\xi}\right)} \approx \frac{1}{\beta\xi}, \quad 1 - \frac{\left(1-\frac{1}{\xi}\right)^{-\alpha}}{1+\frac{1-t}{\xi-1}} \approx \frac{1-\alpha-t}{\xi} .$$

We obtain, considering that $q - q' = q_{dr}$ :

$$x \approx \frac{2}{\pi} q_{dr} \cos^{-1} \frac{1}{\sqrt{\beta\xi}} + \frac{2 \cos \alpha\pi}{\pi^2 \xi} \int_0^1 t^{-\alpha}(1-t)^{\alpha-1} M(t)(1-\alpha-t)\, dt ,$$

or finally,

$$x \approx \frac{2}{\pi} q_{dr} \cos^{-1} \frac{1}{\sqrt{\beta\xi}} + [\ell_0 + q_{dr}] \frac{1}{\beta\xi} . \tag{6.25}$$

In points $M_2$, $M_3$ the velocity becomes infinite; consequently

$$\frac{dx}{d\xi} = \frac{q_{dr}}{\pi\xi\sqrt{1-\beta\xi}} - \frac{\ell_0 + q_{dr}}{\beta\xi^2} \approx 0 .$$

We introduce the notation

$$\delta = \frac{2q_{dr}}{\pi(\ell_0 + q_{dr})} . \tag{6.26}$$

Let the roots of the equation

$$\frac{1}{(\beta\xi)^2} - \frac{1}{\beta\xi} - \frac{\delta^2}{4} = 0$$

be $\xi_2$, $\xi_3 (\xi_2 > \xi_3)$. Inserting these in the approximate formula for $x$ (6.25) and carrying out the computations, we find:

$$\ell_{dr} \approx \frac{2}{\pi} q_{dr}\left(\cos^{-1} \frac{1}{\sqrt{\beta\xi_3}} - \cos^{-1} \frac{1}{\sqrt{\beta\xi_2}}\right) + (\ell_0 + q_{dr}) \left(\frac{1}{\beta\xi_3} - \frac{1}{\beta\xi_2}\right)$$

or

$$\ell_{dr} \approx -\frac{2}{\pi} q_{dr} \cos^{-1}\left(\sqrt{\frac{1}{\beta\xi_2} \cdot \frac{1}{\beta\xi_3}} + \sqrt{1 - \frac{1}{\beta\xi_2} - \frac{1}{\beta\xi_3} - \frac{1}{\beta\xi_2} \cdot \frac{1}{\beta\xi_3}}\right) + (\ell_0 + q_{dr}) \left(\frac{1}{\beta\xi_3} - \frac{1}{\beta\xi_2}\right) .$$

Using the properties of the roots of a quadratic equation, we find:

$$\ell_{dr} \approx \frac{2q_{dr}}{\pi} \left(\sqrt{\frac{1}{\delta^2} - 1} - \cos^{-1} \delta\right) . \tag{6.27}$$

This expression is compatible with the formula for a single drain (Chapter IV, §1).

As is easy to see, $\delta \leq \frac{2}{\pi}$. For $\ell_0 = 0$, we have $\delta = \frac{2}{\pi}$ and

$$\ell_{dr} \approx \frac{2}{\pi} \left(\sqrt{\frac{\pi^2}{4} - 1} - \cos^{-1} \frac{2}{\pi}\right) q_{dr} \approx 0.262 q_{dr} .$$

By formula (6.27) we find the difference

$$\ell_{dr} - \ell_0 \approx q_{dr} \left[1 - \frac{2\delta}{\pi(\sqrt{1-\delta^2} + 1)} - \frac{2}{\pi} \cos^{-1}\delta\right] .$$

The derivative of this expression is positive:

$$\frac{\partial(\ell_{dr} - \ell_0)}{\partial\delta} = q_{dr}\,\frac{1 - \sqrt{1 - \delta^2}}{\Delta} > 0 \quad .$$

Consequently for a given reduced flowrate in the drain, the difference between its inner and outer width decreases with increasing outer width. The largest value of the difference, say $0.262q_{dr}$, occurs for $\ell_0 = 0$.

In the case of seepage from the drain into the soil, we have $\ell_{dr} = \ell_0$

We examine a particular case: seepage from a canal when no natural flow takes place (Fig. 189). In this case $q = 0$. We determine the dis-

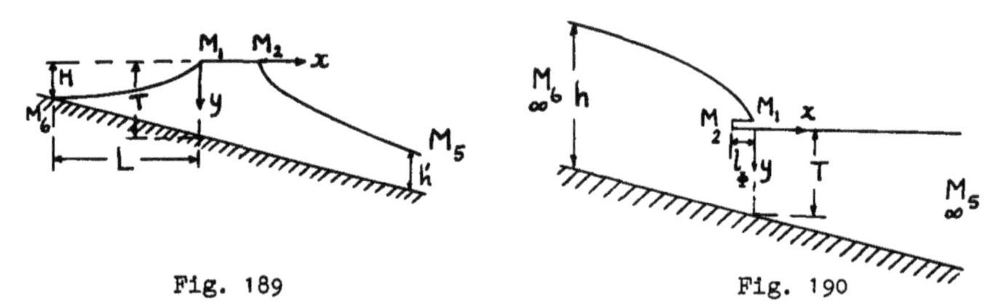


Fig. 189 Fig. 190

tance $L$ of the spreading from the canal and the depth $H$ of the free surface at its intersection with bedrock. We find the boundary. Considering the interval $-\infty < \xi \leq 0$, we write by means of (6.8) and because $\varphi = y$:

$$x = -T\cot\!g\,\alpha\pi + $$
$$+ \frac{2q'}{\pi}\cos\alpha\pi(-\xi)^{\alpha}(1-\xi)^{1-\alpha}\int_0^1 t^{-\alpha}(1-t)^{\alpha-1}\sinh^{-1}\sqrt{\frac{1-\beta}{\beta(1-t)}}\,\frac{dt}{t-\xi}$$

$$y = \frac{2q'}{\pi}\cosh^{-1}\sqrt{\frac{1-\beta}{\beta(1-\xi)}} \quad .$$

Passing to the limit for $\xi = 0$, we find:

$$-L = \frac{T}{J} - \frac{2q'}{\pi J}\cosh^{-1}\frac{1}{\sqrt{\beta}}\,, \qquad H = \frac{2q'}{\pi}\cosh^{-1}\sqrt{\frac{1-\beta}{\beta}}\;,$$

where $J = \text{tg}\,\alpha\pi$.

We treat the flow to an horizontal filter (Fig. 190). Here we have $\ell_0 = \infty$, $q' = 0$, $\beta = 1$. Therefore formulas (6.3) and (6.8) give:

$$\omega = \frac{2\xi q}{\pi}\sinh^{-1}\frac{1}{\sqrt{\zeta}}\,, \qquad \zeta = \frac{1}{\sinh^2\frac{\pi\omega}{2\xi q}}\;,$$

$$z = -\frac{T}{J} + q + i\omega + \left(\frac{T}{J} - q + Aq\right)\cosh^{-2\alpha}\frac{\pi\omega}{2q} -$$

(6.28)

$$-\frac{4q}{\pi^2}\cos\alpha\pi\int_0^{\frac{\pi}{2}}\text{tg}^{1-2\alpha}t\,\sinh^{-1}\!\left(\cosh\frac{\pi\omega}{q}\cot\!g\,t\right)dt \quad ,$$

$$A = \frac{4\cos\alpha\pi}{\pi^2}\int_0^{\frac{\pi}{2}}\text{tg}^{1-2\alpha}t\cdot\ln\cot\!g\,\frac{t}{2}\,dt \quad .$$

We designate by $\ell_\Phi$ the width of the upper part of the slit; we call this the filter. On the boundary of the filter $\omega = i\psi$ and because of (6.28) we have:

$$x = -\frac{T}{J} + q - \psi + \left(\frac{T}{J} - q + Aq\right)\cos^{-2\alpha}\frac{\pi\psi}{2q} -$$
$$- \frac{4q}{\pi^2}\cos\alpha\pi\int_0^{\pi/2} tg^{1-2\alpha}t \sinh^{-1}\left(\cos\frac{\pi\psi}{2q}\operatorname{cotg} t\right) dt \quad .$$

In point $M_2$ of the filter, x reaches an extreme and therefore:

$$\frac{dx}{d\psi} = -1 + \frac{\alpha\pi}{q}\left(\frac{T}{J} - q + Aq\right)\cos^{-1-2\alpha}\frac{\pi\psi}{2q}$$
$$+ \frac{2\cos\alpha\pi}{\pi}\int_0^{\pi/2} B(t)\, dt = 0 \quad .$$

$$B(t) = \frac{\sin\frac{\pi\psi}{2q}}{tg^{2\alpha} t\sqrt{1 + \cos^2\frac{\pi\psi}{2q}\cdot\operatorname{cotg}^2 t}} \quad .$$

The value of $\psi$ in $M_2$ is determined by the equation

$$\operatorname{cotg}\frac{\pi\psi}{2q} = \frac{\alpha\pi}{\sin\alpha\pi}\left(\frac{T}{q}\cos\alpha\pi - \sin\alpha\pi + \sin\alpha\pi A\right)\cos^{-2\alpha}\frac{\pi\psi}{2q} +$$
$$+ \frac{2\cos\alpha\pi}{\pi}\int_0^{\pi/2} C(t)\, dt,$$

where the notation for C is introduced

$$C(t) = \frac{\cos\frac{\pi\psi}{2q}\sin^{1-2\alpha} t\cos^{2\alpha}t}{\sqrt{1 - \sin^2\frac{\pi\psi}{2q}\cos^2 t}} \quad .$$

S. N. Numerov found the approximate formula

$$\ell_\Phi \approx \frac{q^2}{\pi T} \quad .$$

The above explained method has been applied by Numerov to the problem of flow to a drain when part of the free surface was subject to infiltration or evaporation [2].

## C. Earth Dams

### §7. Drained Earth Dam With Trapezoidal Profile on Impervious Bedrock.

Now we consider the seepage through an earth dam, built on bedrock, with drain or tailwater filter in prismatic form [7]. The face of the drain CGB is represented by a sloping straight line (Fig. 191). The velocity hodograph for such a dam has the shape of Fig. 192, with a great number of singular points. The existence of a seepage surface introduces a complication in the hodograph, and is represented by the line CG, not passing through the origin of coordinates.

The seepage surface may not exist in particular cases. So, when

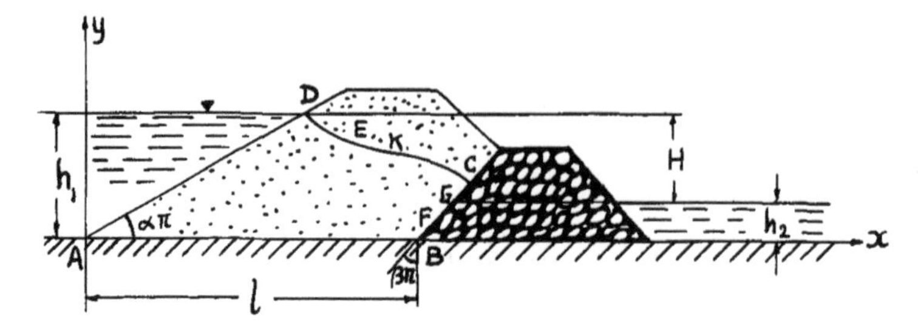


Fig. 191

the cut GFB is extended and point F assumes the position $F_1$ the lower part of the hodograph plane disappears and the shaded area of Fig. 192 remains. In this case the free surface is orthogonal to the lower segment.

To construct the hodograph of Fig. 192, we assumed that the free surface has one inflexion point E. But, we may assume that there are two inflexion points, K and E, at the free surface. In this case, proceeding along the circular arc DE (Fig. 193), we return along the cut to point K, then we rise again along a circular arc, passing on to the other Riemann sheet, up to point C (lying under point A in the second sheet). From C we proceed along the straight line CB. In the case of hodograph degeneration, where there is no seepage surface, the exit of the free surface in the exit surface has zero velocity in point C and the tangent to the free surface is horizontal.

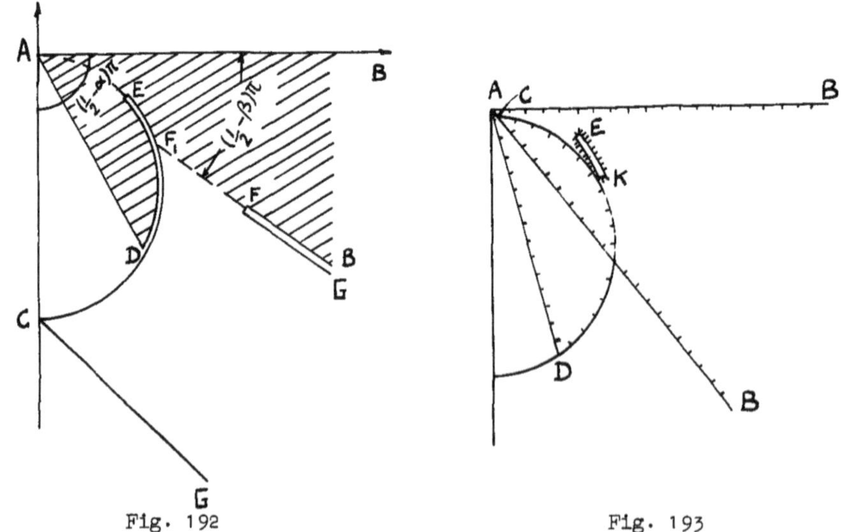


Fig. 192 Fig. 193

From this example, it is obvious how complicated the hodograph can be and how many parameters the Christoffel-Schwarz formula may contain.

S. N. Numerov applied his method, shown by us in the previous paragraphs, assuming that there was no seepage surface. He reduces the solution in table-form, allowing computation.

In this assumption, the flowregion is ABCD (Fig. 191). The boundaries of this region are streamlines and equipotential lines, We put, along AB

$$\psi = 0 \quad ,$$

and along CD

$$\psi = Q \quad .$$

Velocity potential and pressure are related

$$\varphi = -\kappa\left(\frac{p}{\rho g} + y\right) + C \quad .$$

Choosing the constant $C$ as

$$C = \kappa\frac{h_1 + h_2}{2} + \frac{\kappa p_0}{\rho g} \quad ,$$

i.e., considering

$$\varphi = -\kappa\left(\frac{p - p_0}{\rho g} + y\right) + \kappa\frac{h_1 + h_2}{2} \quad , \tag{7.1}$$

we have also assumed that the atmospheric pressure is $p_0$, and that the velocity potential has the value

$$\varphi = -\frac{\kappa H}{2}$$

at the boundary of the headwater reservoir and

$$\varphi = \frac{\kappa H}{2}$$

along the boundary of the tailwater reservoir, where $H = h_1 - h_2$ (see Chapter II, §12).

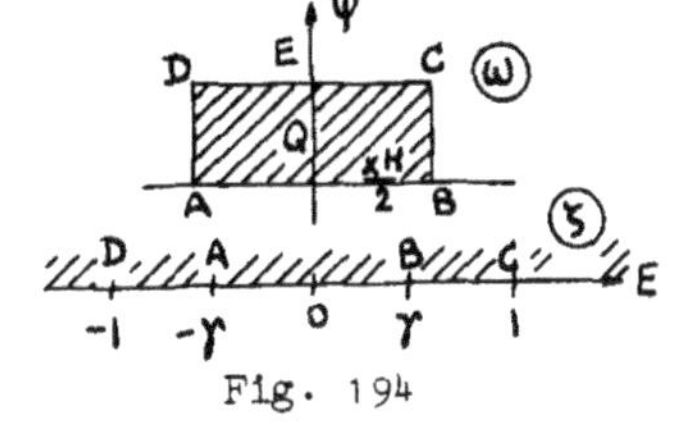

Fig. 194

The region of the complex potential is represented in Fig. 194. Mapping it onto the upper half-plane $\zeta$, as shown in Fig. 194, we obtain (as before, in a series of cases),

$$\omega = \frac{\kappa H}{2K}\int_0^{\zeta}\frac{d\zeta}{\sqrt{(1 - \zeta^2)(\gamma^2 - \zeta^2)}} \quad , \tag{7.2}$$

where

$$K = \int_0^{\frac{\pi}{2}}\frac{d\varphi}{\sqrt{1 - \gamma^2\sin^2\varphi}} \qquad (k = \gamma) \quad .$$

To bring integral (7.2) into its canonical form,, we introduce the substitution

$$\zeta = \gamma t \quad .$$

We obtain:

$$\omega = \frac{\kappa H}{2K}\int_0^{\frac{\zeta}{\gamma}}\frac{dt}{\sqrt{(1 - t^2)(1 - \gamma^2 t^2)}} \quad . \tag{7.3}$$

Putting $\zeta = \xi$, we proceed along the real axis of the $\zeta$-plane.

Expression (7.3) is suitable for $\zeta = \xi$, when $|\xi| < \gamma$. For the segment $\gamma < \xi < 1$ we write it in the form

$$\omega = \frac{\kappa H}{2} + \frac{\kappa Hi}{2K} \int_1^{\zeta/\gamma} \frac{dt}{\sqrt{(t^2 - 1)(1 - \gamma^2 t^2)}} \quad . \tag{7.4}$$

The substitution

$$\gamma^2 t^2 + \gamma'^2 t'^2 = 1 \quad ,$$

where

$$\gamma' = \sqrt{1 - \gamma^2} \quad ,$$

gives:

$$J(\zeta) = \int_1^{\frac{\zeta}{\gamma}} \frac{dt}{\sqrt{(t^2 - 1)(1 - \gamma^2 t^2)}} = \int_{\frac{\sqrt{1-\zeta^2}}{\gamma'}}^{1} \frac{dt'}{\sqrt{(1 - t'^2)(1 - \gamma'^2 t'^2)}} =$$

$$= K' - \int_0^{\frac{\sqrt{1-\zeta^2}}{\gamma'}} \frac{dt}{\sqrt{(1 - t^2)(1 - \gamma'^2 t^2)}} \quad , \tag{7.5}$$

where

$$K' = \int_0^{\frac{\pi}{2}} \frac{d\varphi}{\sqrt{1 - \gamma'^2 \sin^2 \varphi}}$$

Using the notation

$$\int_0^x \frac{dt}{\sqrt{(1 - t^2)(1 - k^2 t^2)}} = F\,(\sin^{-1} x,\ k) \quad , \tag{7.6}$$

we may write

$$J(\xi) = K' - F\left(\sin^{-1} \frac{\sqrt{1 - \xi^2}}{\gamma'}\ ,\ \gamma'\right) \quad . \tag{7.7}$$

(7.4) is reduced for $\gamma < \xi < 1$ to

$$\omega(\xi) = \frac{\kappa H}{2} + \frac{\kappa Hi}{2K} \left[K' - F\left(\sin^{-1} \frac{\sqrt{1 - \xi^2}}{\gamma'},\ \ \gamma'\right)\right] \quad . \tag{7.8}$$

In particular, for $\xi = 1$, we find the flowrate

$$Q = \frac{\kappa HK'}{2K} \quad . \tag{7.9}$$

Using this formula, we rewrite (7.8) for $\gamma < \xi < 1$ as

$$\omega(\xi) = \frac{\kappa H}{2} + iQ \left[1 - \frac{F(\lambda,\ \gamma')}{K'}\ \right] \quad . \tag{7.10}$$

Here

$$\lambda = \sin^{-1} \frac{\sqrt{1 - \xi^2}}{\gamma'} \quad . \tag{7.11}$$

In a similar way, for the interval $-1 < \xi < -\gamma$, we obtain this value

$$\omega(\xi) = -\frac{\kappa H}{2} + iQ\left[1 - \frac{F(\lambda, \gamma')}{K'}\right] , \tag{7.12}$$

where $\lambda$ has the value (7.11).

Now we consider $z$, following S. N. Numerov, as a function of $\omega$. It is convenient to deal with another function instead of $z(\omega)$, based on these considerations. If instead of the real free surface we adopted the Kozeny parabola for the flow with horizontal drainage slit beginning in point B, we would determine the corresponding flow by means of the equation

$$z = -\frac{1}{2\kappa Q}\left[\frac{\kappa(h_1 + h_2)}{2} - \omega\right]^2 .$$

For the case under consideration we introduce the function $Z(\omega)$ by means of

$$Z(\omega) = X + iY = z(\omega) + \frac{1}{2\kappa Q}\left[\frac{\kappa(h_1 + h_2)}{2} - \omega\right]^2 . \tag{7.13}$$

For this function we obtain the following conditions at the contour of the flowregion.

Along AB, where $\psi = 0$ and $y = 0$

$$Y = 0 . \tag{7.14}$$

Along BC, with equation

$$x \sin \beta\pi - y \cos \beta\pi = \ell \sin \beta\pi$$

and along which $\varphi = \frac{1}{2}\kappa H$, we obtain:

$$X \sin \beta\pi - Y \cos \beta\pi = \left(\ell + \frac{\kappa h_2^2}{2Q}\right) \sin \beta\pi + \frac{h_2 \cos \beta\pi}{Q}\psi - \frac{\sin \beta\pi}{2\kappa Q}\psi^2 . \tag{7.15}$$

Along the free surface CD, where

$$\varphi = \frac{1}{2}\kappa(h_1 + h_2) - \kappa y$$

and $\psi = Q$, we obtain again

$$Y = 0 . \tag{7.16}$$

Finally, along the entrance surface DA, where $\varphi = -\frac{\kappa H}{2}$, and with equation

$$x \sin \alpha\pi - y \cos \alpha\pi = 0 ,$$

we have the equation

$$X \sin \alpha\pi - Y \cos \alpha\pi = \frac{\kappa h_1^2 \sin \alpha\pi}{2Q} + \frac{h_1 \cos \alpha\pi}{Q}\psi - \frac{\sin \alpha\pi}{2\kappa Q}\psi^2 . \tag{7.17}$$

We see that our problem is reduced to the following: along the sides of a rectangle in the $\omega$-plane, a linear combination of the functions $X$, $Y$ is given; find the function $X + iY$ inside the rectangle $\omega$. This is a so-called Hilbert problem. However, S. N. Numerov substitutes the half-plane $\zeta$ for the region $\omega$. He maps the flowregion onto the half plane $\zeta$, and then by methods, shown in §2 of this chapter, reduces the problem to that of Dirichlet.

We already have the functional relation between $\omega$ and $\xi$, expressed by formula's (7.10) and (7.12), for two segments of the $\xi$-axis, along which we obtained non-zero conditions (7.15) and (7.17). Considering that along these segments

$$\varphi = \pm \frac{\kappa H}{2} \quad ,$$

we find by means of formulas (7.10) and (7.12), in both cases for $\gamma < \xi <$ and for $-1 \leq \xi \leq -\gamma$,

$$\psi = Q\left[1 - \frac{F(\lambda, \gamma')}{K'}\right] \quad ,$$

where

$$\lambda = \sin^{-1} \frac{\sqrt{1 - \xi^2}}{\gamma'} \quad .$$

We insert this value of $\psi$ in equations (7.15) and (7.17). After simplifications we obtain:

$$\begin{cases} \text{for } \gamma < \xi < 1 \\ X(\xi) \sin \beta\pi - Y(\xi) \cos \beta\pi = b_0 - \frac{b_1}{K'} F(\lambda, \gamma') - \frac{b_2}{K'^2} F^2(\lambda, \gamma') = x_1(\xi) \; ; \\ \text{for } -1 < \xi < -\gamma \\ X(\xi) \sin \alpha\pi - Y(\xi) \cos \alpha\pi = a_0 - \frac{a_1}{K'} F(\lambda, \gamma') - \frac{a_2}{K'^2} F^2(\lambda, \gamma') = x_2(\xi) \; ; \end{cases}$$

$$\text{for } |\xi| < \gamma \quad \text{and} \quad |\xi| > 1$$

$$Y = 0 \quad , \tag{7.18}$$

in which

$$\begin{cases} b_0 = \ell \sin \beta\pi + h_2 \cos \beta\pi + \frac{\kappa h_2^2 \sin \beta\pi}{2Q} - \frac{Q \sin \beta\pi}{2\kappa} \quad , \\ b_1 = h_2 \cos \beta\pi - \frac{Q \sin \beta\pi}{\kappa} \; , \quad b_2 = \frac{Q \sin \beta\pi}{2\kappa} \quad , \\ a_0 = h_1 \cos \alpha\pi + \frac{\kappa h_1^2 \sin \alpha\pi}{2Q} - \frac{Q \sin \alpha\pi}{2\kappa} \quad , \\ a_1 = h_1 \cos \alpha\pi - \frac{Q \sin \alpha\pi}{\kappa} \; ; \quad a_2 = \frac{Q \sin \alpha\pi}{2\kappa} \quad . \end{cases} \tag{7.19}$$

From this it is obvious that we obtain conditions at segments of the real axis as we have seen in §2.

Indeed let us, for example, examine the expression

$$e^{\pi i(\frac{1}{2} - \alpha)} Z \quad .$$

Obviously we have

$$\mathrm{Re}[e^{\pi i(\frac{1}{2} - \alpha)} Z] = X \sin \alpha\pi - Y \cos \alpha\pi \quad .$$

We compose the function

$$T(\zeta) = i(\gamma - \zeta)^{\beta}(\gamma + \zeta)^{1-\alpha}(1 - \zeta)^{1-\beta}(1 + \zeta)^{\alpha} \quad . \tag{7.20}$$

This function assumes purely imaginary values for real values of $\zeta = \xi$, for which $-\gamma < \xi < \gamma$. We take its argument equal to $\pi/2$ for $0 < \xi < \gamma$. At different segments of the real axis, we have the following expressions for $T(\zeta)$:

$$(7.21)\quad \begin{cases} \text{for } |\xi| < \gamma \\ T(\xi) = i(\gamma - \xi)^{\beta}(\gamma + \xi)^{1-\alpha}(1 - \xi)^{1-\beta}(1 + \xi)^{\alpha} \quad ; \\ \text{for } \gamma < \xi < 1 \\ T(\xi) = (\xi - \gamma)^{\beta}(\gamma + \xi)^{1-\alpha}(1 - \xi)^{1-\beta}(1 + \xi)^{\alpha}e^{\pi i(\frac{1}{2} - \beta)} \quad ; \\ \text{for } -1 < \xi < -\gamma \\ T(\xi) = (\gamma - \xi)^{\beta}(-\gamma - \xi)^{1-\alpha}(1 - \xi)^{1-\beta}(1 + \xi)^{\alpha}e^{\pi i(\frac{3}{2} - \alpha)} \quad ; \\ \text{for } -\infty < \xi < -1 \\ T(\xi) = -i(\gamma - \xi)^{\beta}(-\gamma - \xi)^{1-\alpha}(1 - \xi)^{1-\beta}(-1 - \xi)^{\alpha} \quad ; \\ \text{for } 1 < \xi < \infty \\ T(\xi) = -i(\xi - \gamma)^{\beta}(\gamma + \xi)^{1-\alpha}(\xi - 1)^{1-\beta}(1 + \xi)^{\alpha} \quad . \end{cases}$$

The product

$$T(\zeta)\, Z(\zeta)$$

is purely imaginary for $|\xi| < \gamma$ and $|\xi| > 1$, i.e., its real part is zero in these intervals.

In the intervals $\gamma < \xi < 1$ and $-1 < \xi < -\gamma$, the real part of the function

$$T(\xi)\, Z(\xi)$$

is known.

To apply Cauchy's formula to find $T(\zeta)\, Z(\zeta)$, we must investigate this product at $\zeta = \infty$.

Since $Z(\zeta)$ is an analytic function, finite in the whole half-plane $\zeta$, it can be developed in a Laurent series at infinity of the form

$$(7.22)\qquad Z(\zeta) = \sum_{n=0}^{\infty} \frac{c_n}{\zeta^n} \quad .$$

Therefore, noticing that the imaginary part of $Z$ vanishes for $|\xi| > 1$, we obtain the following property for function $Z(\zeta)$ by Schwarz's principle of symmetry: it takes on conjugate values in conjugate points $\zeta$ and $\overline{\zeta}$. From this it follows that the coefficients $c_n$ must be real.

As far as $Z(\zeta)$ is concerned, equations (7.21) show that about $\zeta = \infty$ we have:

$$T(\zeta) = -i\zeta^2\left(1 - \frac{\gamma}{\zeta}\right)^{\beta}\left(1 + \frac{\gamma}{\zeta}\right)^{1-\alpha}\left(1 - \frac{1}{\zeta}\right)^{1-\beta}\left(1 + \frac{1}{\zeta}\right)^{\alpha} =$$

$$= -i\zeta^2\left(1 + \sum_{n=1}^{\infty} \frac{d_n}{\zeta^n}\right) \quad ,$$

in which all coefficients $d_n$ are real.

Product $T(\zeta)\,Z(\zeta)$ at infinity has the form

$$T(\zeta)\,Z(\zeta) = -\,i\zeta^2\left(1 + \frac{d_1}{\zeta} + \frac{d_2}{\zeta^2} + \dots\right)\left(c_0 + \frac{c_1}{\zeta} + \frac{c_2}{\zeta^2} + \dots\right) =$$

$$= -\,ic_0\zeta^2 - i(c_1 + c_0d_1)\zeta - i(c_0 + c_1d_1 + c_0d_2) + \dots\,.$$

Function

$$W(\zeta) = T(\zeta)\,Z(\zeta) + ic_0\zeta^2 + i(c_1 + c_0d_1)\zeta \tag{7.23}$$

is finite at infinity and it is convenient to use it as an auxiliary function. In the intervals where $Y = 0$, the new function has the property that its real part vanishes. As far as the other intervals are concerned, it is not difficult to find the value of the real part of this function, there. We obtain:

$$f(\xi) = \operatorname{Re} W(\xi) = \begin{cases} X_1(\xi)\lambda_1(\xi) & \text{for } \gamma < \xi < 1\,, \\ -X_2(\xi)\lambda_2(\xi) & \text{for } -1 < \xi < -\gamma\,, \\ 0 & \text{for } |\xi| < \gamma \text{ and } |\xi| > 1\,, \end{cases} \tag{7.24}$$

where $X_1(\xi)$, and $X_2(\xi)$ have values (7.18), for which

$$\lambda_1(\xi) = (\xi - \gamma)^{\beta}(\gamma + \xi)^{1-\alpha}(1 - \xi)^{1-\beta}(1 + \xi)^{\alpha}\,,$$

$$\lambda_2(\xi) = (\gamma - \xi)^{\beta}(-\gamma - \xi)^{1-\alpha}(1 - \xi)^{1-\beta}(1 + \xi)^{\alpha}\,.$$

The problem is reduced to that of Dirichlet and for its solution formula (1.6) of this chapter gives:

$$(c_0d_1 + c_1)\zeta + c_0\zeta^2 + (\gamma - \zeta)^{\beta}(\gamma + \zeta)^{1-\alpha}(1 - \zeta)^{1-\beta}(1 + \zeta)^{\alpha}\,Z(\zeta) =$$

$$= -\frac{1}{\pi}\int_{-1}^{1}\frac{f(t)}{t - \zeta}\,dt + C \tag{7.26}$$

The equation so obtained contains unknown constants $c_0$, $c_1$, $d_1$ and $C$. They may be determined, by making respectively $\zeta = \pm\gamma, \pm 1$ in (7.26). For $\zeta = \gamma$, we have from (7.26):

$$(c_0d_1 + c_1)\gamma + c_0\gamma^2 = -\frac{1}{\pi}\int_{-1}^{1}\frac{f(t)\,dt}{t - \gamma} + C\,.$$

Subtracting this expression from (7.26) and dividing by $\gamma - \zeta$, we obtain:

$$c_0d_1 + c_1 + c_0(\gamma + \zeta) - (\gamma - \zeta)^{\beta-1}(\gamma + \zeta)^{1-\alpha}(1 - \zeta)^{1-\beta}(1 + \zeta)^{\alpha}\,Z(\zeta) =$$

$$= -\frac{1}{\pi}\int_{-1}^{1}\frac{f(t)\,dt}{(t - \gamma)(t - \zeta)}\,.$$

Subtracting from this equality an analogous correlation, corresponding to $\zeta = -\gamma$, we obtain:

$$c_0 - (\gamma - \zeta)^{\beta-1}(\gamma + \zeta)^{-\alpha}(1 - \zeta)^{1-\beta}(1 + \zeta)^{\alpha} Z(\zeta) = - \frac{1}{\pi}\int_{-1}^{1} \frac{f(t)\,dt}{(t - \gamma)(t + \gamma)(t - \zeta)} .$$

We make $\zeta = 1$, subtract the corresponding equations, eliminate $c_0$ and fine:

$$(\gamma - \zeta)^{\beta-1}(\gamma + \zeta)^{-\alpha}(1 - \zeta)^{-\beta}(1 + \zeta)^{\alpha} Z(\zeta) = - \frac{1}{\pi}\int_{-1}^{1} \frac{f(t)\,dt}{(t^2 - \gamma^2)(t - 1)(t - \zeta)} . \tag{7.27}$$

Finally we make $\zeta = -1$ in this equation. Since the left part becomes zero for this value, the right part gives the following equation:

$$\int_{-1}^{1} \frac{f(t)\,dt}{(\gamma^2 - t^2)(1 - t^2)} = 0 . \tag{7.28}$$

Subtracting from the right part of (7.27) the left part of (7.28), we reduce (7.27) to a more symmetrical form. Solving (7.27) for $Z(\zeta)$ we write:

$$Z(\zeta) = - \frac{1}{\pi} (\gamma - \zeta)^{1-\beta}(\gamma + \zeta)^{\alpha}(1 - \zeta)^{\beta}(1 + \zeta)^{1-\alpha} \times \int_{-1}^{1} \frac{f(t)\,dt}{(\gamma^2 - t^2)(1 - t^2)(t - \zeta)} . \tag{7.29}$$

Finally, bearing in mind the dependence (7.13) between $Z$ and $z$, we write the equation for $z(\zeta)$

$$z(\zeta) = - \frac{1}{2\kappa Q}\left[\kappa\frac{h_1 + h_2}{2} - \omega\right]^2 - \frac{T_1(\zeta)}{\pi}\int_{-1}^{1} \frac{f(t)\,dt}{(\gamma^2 - t^2)(1 - t^2)(t - \zeta)} , \tag{7.30}$$

where

$$T_1(\zeta) = (\gamma - \zeta)^{1-\beta}(\gamma + \zeta)^{\alpha}(1 - \zeta)^{\beta}(1 + \zeta)^{1-\alpha} .$$

Equation (7.28) contains an unknown parameter $\gamma$ and may serve to determine the value of this parameter. When $\gamma$ is found, the flowrate $Q$ is determined by formula (7.9). We rewrite equation (7.28) in a more convenient form, using (7.24) and (7.18):

$$\int_{-1}^{-1}\left[a_0 - \frac{a_1}{K'} F - \frac{a_2}{K'^2} F^2\right] \frac{\lambda_2(t)\,dt}{(t^2 - \gamma^2)(1 - t^2)} - \int_{\gamma}^{1}\left[b_0 - \frac{b_1}{K'} F - \frac{b_2}{K'^2} F^2\right] \frac{\lambda_1(t)\,dt}{(t^2 - \gamma^2)(1 - t^2)} = 0 . \tag{7.31}$$

Here

$$F = F\left(\sin^{-1} \frac{\sqrt{1 - t^2}}{\gamma'} , \gamma'\right) .$$

Equation (7.31) may be written in another way, designating the integrals:

$$J_0 = \int_{-1}^{-\gamma} \frac{\lambda_2(t)\,dt}{(t^2 - \gamma^2)(1 - t^2)} =$$

$$= \int_{-1}^{-\gamma} (\gamma - t)^{\beta-1}(-\gamma - t)^{-\alpha}(1 - t)^{-\beta}(1 + t)^{\alpha-1}\,dt \quad .$$

Replacing $t$ by $-t$, we may bring $J_0$ in the form

$$J_0 = \int_{\gamma}^{1} \Phi(t)\,dt \quad ,$$

$$\Phi(t) = (\gamma + t)^{\beta-1}(t - \gamma)^{-\alpha}(1 + t)^{-\beta}(1 - t)^{\alpha-1} \quad .$$

Furthermore, we put:

$$J_1 = \int_{-1}^{-\gamma} \frac{\lambda_2(t)\,F\,dt}{(t^2 - \gamma^2)(1 - t^2)} = \int_{\gamma}^{1} \Phi(t)\,F\left(\sin^{-1}\frac{\sqrt{1 - t^2}}{\gamma'},\ \gamma'\right)dt \ ,$$

$$J_2 = \int_{-1}^{-\gamma} \frac{\lambda_2(t)\,F^2\,dt}{(t^2 - \gamma^2)(1 - t^2)} = \int_{\gamma}^{1} \Phi(t)\,F^2\left(\sin^{-1}\frac{\sqrt{1 - t^2}}{\gamma'},\ \gamma'\right)dt \ ,$$

and we introduce three analogous integrals:

$$J'_0 = \int_{\gamma}^{1} \Phi_1(t)\,dt \quad ,$$

$$J'_1 = \int_{\gamma}^{1} \Phi_1(t)\,F\left(\sin^{-1}\frac{\sqrt{1 - t^2}}{\gamma'},\ \gamma'\right)dt \quad ,$$

$$J'_2 = \int_{\gamma}^{1} \Phi_1(t)\,F^2\left(\sin^{-1}\frac{\sqrt{1 - t^2}}{\gamma'},\ \gamma'\right)dt \quad ,$$

where

$$\Phi_1(t) = (t - \gamma)^{\beta-1}(t + \gamma)^{-\alpha}(1 - t)^{-\beta}(1 + t)^{\alpha-1} \quad .$$

We rewrite (7.31) in the form

$$a_0J_0 - \frac{a_1}{K'}J_1 - \frac{a_2}{K'^2}J_2 - b_0J'_0 + \frac{b_1}{K'}J'_1 + \frac{b_2}{K'^2}J'_2 = 0 \quad . \tag{7.32}$$

In practical cases, $\ell/H > 10$ usually. Then the parameter $\gamma$ is very close to unity, $\gamma'$ very close to zero. The integrals $J_n$ and $J'_n$ may then be simplified, assuming

$$F(\theta,\ \gamma') \approx \theta, \quad \gamma \approx 1 \quad .$$

We consider integral $J_0$. We put:

$$(\gamma + t)^{\beta-1} = (1 + t)^{\beta-1}\left[1 + \frac{\gamma - 1}{1 + t}\right]^{\beta-1} =$$

$$= (1 + t)^{\beta-1} + (\beta - 1)(\gamma - 1)(1 + t)^{\beta-2} + \ldots \approx (1 + t)^{\beta-1} \quad . \tag{7.33}$$

The function under the integral sign in $J_0$ is simplified

$$(1 + t)^{-1}(t - \gamma)^{-\alpha}(1 - t)^{\alpha-1} \approx \frac{1}{2}(t - \gamma)^{-\alpha}(1 - t)^{\alpha-1} \quad ,$$

since

$$(7.34) \qquad (1 + t)^{-1} = \frac{1}{2}\left(1 + \frac{t - 1}{2}\right)^{-1} = \frac{1}{2}\left(1 - \frac{t - 1}{2} + \ldots\right) \approx \frac{1}{2} \quad .$$

Then

$$J_0 \approx \frac{1}{2}\int_\gamma^1 (t - \gamma)^{-\alpha}(1 - t)^{\alpha-1}\, dt \quad .$$

The substitution

$$1 - t = (1 - \gamma)x$$

reduces the integral of the right part to the Beta function

$$J_0 \approx \frac{1}{2}\int_0^1 x^{\alpha-1}(1 - x)^{-\alpha}\, dx = \frac{1}{2}\, B(\alpha, 1 - \alpha) = \frac{1}{2}\, \frac{\Gamma(\alpha)\Gamma(1 - \alpha)}{\Gamma(1)} = \frac{\pi}{2 \sin \alpha\pi} \quad .$$

Exactly in the same way we find the approximate equality

$$J_0' \approx \frac{\pi}{2 \sin \beta\pi} \quad .$$

By means of series (7.33), (7.34) we may estimate the approximations that have been made.

Returning to the integral $J_1$, we introduce in it:

$$F\left(\sin^{-1} \frac{\sqrt{1 - t^2}}{\gamma'}, \gamma'\right) \approx \sin^{-1} \frac{\sqrt{1 - t^2}}{\gamma'} \quad .$$

Applying the same simplification as for $J_0$, we write:

$$J_1 \approx \frac{1}{2}\int_\gamma^1 (t - \gamma)^{-\alpha}(1 - t)^{\alpha-1} \sin^{-1} \frac{\sqrt{1 - t^2}}{\gamma'} \quad .$$

We make the substitution

$$\sin^{-1} \frac{\sqrt{1 - t^2}}{\gamma'} = \frac{\pi}{2}\tau \quad .$$

From this

$$t = \sqrt{1 - \gamma'^2 \sin^2 \frac{\pi\tau}{2}} \approx 1 - \frac{1}{2}\gamma'^2 \sin^2 \frac{\pi\tau}{2} \quad ,$$

$$dt \approx -\frac{\pi}{2}\gamma'^2 \sin \frac{\pi\tau}{2} \cos \frac{\pi\tau}{2}\, d\tau \quad ,$$

$$1 - t \approx \frac{1}{2}\gamma'^2 \sin^2 \frac{\pi\tau}{2} \quad ,$$

$$t - \gamma \approx (1 - \gamma)\left[1 - \frac{1}{2}(1 + \gamma)\sin^2 \frac{\pi\tau}{2}\right] \approx (1 - \gamma)\cos^2 \frac{\pi\tau}{2} \quad .$$

Now we obtain for $J_1$:

$$J_1 \approx \frac{\pi^2}{4}\int_0^1 \tau\, \mathrm{cotg}^{1-2\alpha} \frac{\pi\tau}{2}\, d\tau = \frac{\pi^2}{4} f_1(\alpha) \quad .$$

In the same way we obtain:

$$J_2 \approx \frac{\pi^3}{8}\int_0^1 \tau^2 \operatorname{cotg}^{1-2\alpha}\frac{\pi\tau}{2}\,d\tau = \frac{\pi^3}{4} f_3(\alpha) \quad .$$

Passing on to integral $J_1'$, we write:

$$J_1' \approx \frac{1}{2}\int_\gamma^1 (t-\gamma)^{\beta-1}(1-t)^{-\beta}\sin^{-1}\frac{\sqrt{1-t^2}}{\gamma'}\,dt =$$

$$= \frac{\pi^3}{4}\int_0^1 \tau \operatorname{cotg}^{2\beta-1}\frac{\pi\tau}{2}\,d\tau = \frac{\pi^3}{4} f_1(1-\beta) \quad .$$

Similarly

$$J_2' \approx \frac{\pi^3}{8}\int_0^1 \tau^2 \operatorname{cotg}^{2\beta-1}\frac{\pi\tau}{2}\,d\tau = \frac{\pi^3}{4} f_3(1-\beta) \quad .$$

Equality (7.32) contains the quantity $K'$. For $\gamma'$, close to zero, we may take

$$K' \approx \frac{\pi}{2}$$

It is interesting to observe that parameter $\gamma$ completely disappears in all approximate formulas.

We insert now the approximate expressions for the integrals $J_0$, $J_1$, ... in (7.32). Simplifying for the common factor $\pi/2$, we obtain:

$$\frac{a_0}{\sin\alpha\pi} - \frac{b_0}{\sin\beta\pi} - a_1 f_1(\alpha) + b_1 f_1(1-\beta) - 2a_2 f_3(\alpha) + 2b_2 f_3(1-\beta) = 0 \quad .$$

Finally, assuming equality (7.19), we write:

$$h_1 \operatorname{cotg}\alpha\pi - h_2 \operatorname{cotg}\beta\pi + \frac{\kappa h_1^2}{2Q} - \frac{\kappa h_2^2}{2Q} - \ell -$$

$$- \left(h_1 \cos\alpha\pi - \frac{Q}{\kappa}\sin\alpha\pi\right) f_1(\alpha) + \left(h_2 \cos\beta\pi - \frac{Q}{\kappa}\sin\beta\pi\right) f_1(1-\beta) -$$

$$- \frac{Q}{\kappa}\sin\alpha\pi f_3(\alpha) + \frac{Q}{\kappa}\sin\beta\pi f_3(1-\beta) = 0$$

From this we may find the approximate equality for $\ell$

$$\ell \approx \frac{\kappa\left(h_1^2 - h_2^2\right)}{2Q} + h_1[\operatorname{cotg}\alpha\pi - \cos\alpha\pi f_1(\alpha)] +$$

$$+ h_2[-\operatorname{cotg}\beta\pi + \cos\beta\pi f_1(1-\beta)] - \frac{Q}{\kappa}[-\sin\alpha\pi f_1(\alpha) +$$

$$+ \sin\beta\pi f_1(1-\beta) + \sin\alpha\pi f_3(\alpha) - \sin\beta\pi f_3(1-\beta)] \; .$$

We introduce the notation:

$$f_2(\alpha) = f_1(\alpha) - f_3(\alpha) = \int_0^1 t\left(1 - \frac{1}{2}t\right)\operatorname{cotg}^{1-2\alpha}\frac{\pi t}{2}\,dt \quad .$$

Then we may rewrite the equation for $\ell$ as:

$$\ell \approx \frac{\kappa\left(h_1^2 - h_2^2\right)}{2Q} + h_1[\operatorname{cotg} \alpha\pi - \cos \alpha\pi f_1(\alpha)] - h_2[\operatorname{cotg} \beta\pi - \cos \beta\pi f_4(1 - \beta)] -$$

$$- \frac{Q}{\kappa}[- \sin \alpha\pi f_2(\alpha) + \sin \beta\pi f_2(1 - \beta)] \quad .$$

We use one more transformation, using the property of the functions $f_1(1 - \beta)$ and $f_2(1 - \beta)$, which we obtained by putting $t = 1 - \tau$ in the equality

$$f_1(1 - \beta) = \int_0^1 t \operatorname{cotg}^{2\beta - 1} \frac{\pi t}{2} \, dt \quad .$$

We have:

$$f_1(1 - \beta) = \int_0^1 \operatorname{cotg}^{1-2\beta} \frac{\pi\tau}{2} \, d\tau - \int_0^1 \tau \operatorname{cotg}^{1-2\beta} \frac{\pi\tau}{2} \, d\tau = \frac{1}{\sin \beta\pi} - f_1(\beta) \quad .$$

Similarly we find:

$$f_3(1 - \beta) = \frac{1}{2}\int_0^1 (1 - \tau)^2 \operatorname{cotg}^{1-2\beta} \frac{\pi\tau}{2} \, d\tau = \frac{1}{2 \sin \beta\pi} - f_1(\beta) + f_3(\beta) \quad ,$$

$$f_2(1 - \beta) = \int_0^1 (1 - \tau)\left(\frac{1}{2} + \frac{\tau}{2}\right) \operatorname{cotg}^{1-2\beta} \frac{\pi\tau}{2} \, d\tau = \frac{1}{2 \sin \beta\pi} - \frac{1}{2} f_3(\beta) \quad .$$

Inserting the obtained expressions in the formula for $\ell$, we finally obtain after some simplifications:

$$\ell \approx \frac{\kappa\left(h_1^2 - h_2^2\right)}{2Q} + h_1[\operatorname{cotg} \alpha\pi - \cos \alpha\pi f_1(\alpha)] - h_2 \cos \beta\pi f_1(\beta) -$$

$$- \frac{Q}{\kappa}\left[\frac{1}{2} - \sin \alpha\pi f_2(\alpha) - \sin \beta\pi f_3(\beta)\right] \quad . \tag{7.35}$$

This approximate formula has a high degree of accuracy for wide dams. Tables [8, 12] have been computed for $f_1$, $f_2$, $f_3$.

In the special case of a rectangular dam with horizontal drain, i.e., for $\alpha = \frac{1}{2}$, $\beta = 0$, and

$$h_1 = H, \qquad h_2 = 0 \quad ,$$

we obtain from (7.35):

$$\ell \approx \frac{\kappa H^2}{2Q} - \frac{Q}{6\kappa} \quad ,$$

from which we find the approximate equation for $Q$

$$Q \approx \frac{\kappa H^2}{\ell + \sqrt{\ell^2 + \frac{1}{3} H^2}} \quad .$$

P. A. Shankin [13] established graphs for the dependence of $f_1$, $f_2$, $f_3$ on $\alpha$, given in Fig. 195.

Equation of the free surface. To obtain the equation of the free surface in equation (7.30) we must put:

$$\omega = \varphi + iQ = \frac{1}{2}\kappa(h_1 + h_2) - \kappa y + iQ \quad .$$

We put:

$$x = -\frac{\kappa y^2}{2Q} + \frac{Q}{2\kappa} + \frac{1}{\pi}(\gamma - \xi)^{1-\beta}(-\gamma - \xi)^{\alpha}(1 - \xi)^{\beta}(-1 - \xi)^{1-\alpha} \times$$

$$\times \left\{ \int_{-1}^{-\gamma} \frac{\left(a_0 - \frac{a_1}{K'}F - \frac{a_2}{K'^2}F^2\right)dt}{(\gamma - t)^{1-\beta}(-\gamma - t)^{\alpha}(1 - t)^{\beta}(1 + t)^{1-\alpha}(t - \xi)} - \right.$$

$$(7.36) \qquad \left. - \int_{1}^{\gamma} \frac{\left(b_0 - \frac{b_1}{K'}F - \frac{b_2}{K'^2}F^2\right)dt}{(t - \gamma)^{1-\beta}(\gamma + t)^{\alpha}(1 - t)^{\beta}(1 + t)^{1-\alpha}(t - \xi)} \right\} \quad .$$

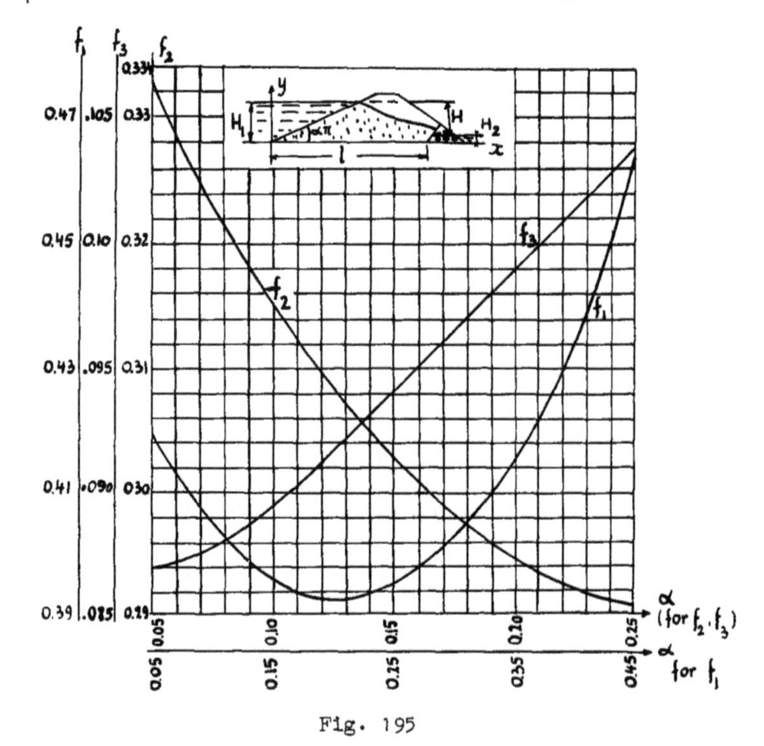


Fig. 195

We examine the left part of the free surface for which

$$-\infty < \xi < -1 \quad .$$

For $\gamma \approx 1$, equation (7.36) may be simplified, as we did before for the computations of integrals $J_1$, $J_2$. Assuming

$$K' \approx \frac{\pi}{2} \quad ,$$

we obtain:

$$x \approx \frac{\kappa y^2}{2Q} + \frac{Q}{2\kappa} + \frac{1}{\pi}(1 - \xi)(-\gamma - \xi)^{\alpha}(-1 - \xi)^{1-\alpha} \times$$

$$\times \left\{ \frac{1}{2} \int_{-1}^{-\gamma} \frac{\left(a_0 - \frac{2a_1}{\pi} F - \frac{4a_2}{\pi^2} F^2\right) dt}{(-\gamma - t)^{\alpha}(1 + t)^{1-\alpha}(t - \xi)} - \right.$$

$$\left. - \frac{1}{2} \int_{\gamma}^{1} \frac{\left(b_0 - \frac{2b_1}{\pi} F - \frac{4b_2}{\pi^2} F^2\right) dt}{(t - \gamma)^{1-\beta}(1 - t)^{\beta}(1 - \xi)} \right\} . \tag{7.37}$$

Here we may write the approximate equality

$$F\left(\sin^{-1} \frac{\sqrt{1 - t^2}}{\gamma'} , \gamma'\right) \approx \sin^{-1} \frac{\sqrt{1 - t^2}}{\gamma'} .$$

We notice that the difference $t - \xi$ in the second integral is replaced by $1 - \xi$. This is possible because in this case in the series

$$\frac{1}{t - \xi} = \frac{1}{1 - \xi}\left(1 - \frac{t - 1}{1 - \xi} + \dots\right)$$

for

$$\gamma \approx 1, \quad 1 < t < \gamma, \quad 1 - \xi > 2$$

we may take

$$\frac{1}{t - \xi} \approx \frac{1}{1 - \xi} .$$

After this simplification, equation (7.31) becomes:

$$\int_{-1}^{-\gamma} \frac{a_0 - \frac{2a_1}{\pi} F - \frac{4a_2}{\pi^2} F^2}{(-\gamma - t)^{\alpha}(1 + t)^{1-\alpha}} dt \approx \int_{\gamma}^{1} \frac{b_0 - \frac{2b_1}{\pi} F - \frac{4b_2}{\pi^2} F^2}{(t - \gamma)^{1-\beta}(1 - t)^{\beta}} dt .$$

Multiplying both parts of this equality with

$$\frac{1}{\pi}(-t - \xi)^{\alpha}(-1 - \xi)^{1-\alpha} ,$$

we subtract them from (7.37). We recall that

$$\frac{1 - \xi}{t - \xi} - 1 = \frac{1 - t}{t - \xi} .$$

Under the first integral sign, where $t \approx -1$, we may take $1 - t \approx 2$. Therefore, from (7.37) we obtain the following equation:

$$x \approx -\frac{\kappa y^2}{2Q} + \frac{Q}{2\kappa} +$$

$$+ \frac{1}{\pi}(-\gamma - \xi)^{\alpha}(-1 - \xi)^{1-\alpha} \int_{-1}^{-\gamma} \frac{a_0 - \frac{2a_1}{\pi} F - \frac{4a_2}{\pi^2} F^2}{(-\gamma - t)^{\alpha}(1 + t)^{1-\alpha}(t - \xi)} dt .$$

Now we introduce the new variable of integration by means of the substitution

$$t = -\frac{-\xi - \gamma + (1-\gamma)\xi \sin^2 \frac{\pi\tau}{2}}{-\xi - \gamma - (1-\gamma) \sin^2 \frac{\pi\tau}{2}} \quad . \tag{7.38}$$

We obtain:

$$x \approx -\frac{\kappa y^2}{2Q} + \frac{Q}{2\kappa} + \int_0^1 \operatorname{cotg}^{1-2\alpha} \frac{\pi\tau}{2}\left(a_0 - \frac{2a_1}{\pi} F - \frac{4a_2}{\pi^2} F^2\right) d\tau \quad . \tag{7.89}$$

In this we may take:

$$F \approx \sin^{-1}\sqrt{\frac{1-t^2}{1-\gamma^2}} \approx \sin^{-1}\sqrt{\frac{1+t}{1-\gamma}} \quad , \tag{7.40}$$

since

$$\frac{1-t}{1+\gamma} \approx \frac{2}{2} = 1 \quad .$$

Replacing $\sin^{-1} x$ by $\operatorname{tg}^{-1}\frac{x}{\sqrt{1-x^2}}$, we obtain:

$$F \approx \operatorname{tg}^{-1}\sqrt{\frac{1+t}{-t-\gamma}} \quad .$$

Furthermore, by means of (7.38) we find:

$$F \approx \operatorname{tg}^{-1}\left(\sqrt{\frac{-1-\xi}{-\gamma-\xi}} \operatorname{tg}\frac{\pi\tau}{2}\right) \quad .$$

Because of equation (7.3) we have:

$$\zeta = \gamma \operatorname{sn} \frac{2K\omega}{\kappa H}$$

and for $-\infty < \xi < -1$

$$\xi = \gamma \operatorname{sn} \frac{2K}{\kappa H}(\varphi + iQ) \quad .$$

Based on (7.9) we obtain:

$$\xi = \gamma \operatorname{sn}\left[\frac{K'}{Q}(\varphi + iQ),\ \gamma\right] = \gamma \operatorname{sn}\left(\frac{K'\varphi}{Q} + iK',\ \gamma\right) \quad .$$

For the left branch of the free surface, for which $y$ is close to $h_1$, we may write:

$$\frac{\varphi}{Q} = \frac{\kappa(h_1 + h_2)}{2Q} - \frac{\kappa y}{Q} = \frac{\kappa(h_1 - y)}{Q} - \frac{\kappa H}{2Q} = \frac{\kappa(h_1 - y)}{Q} - \frac{K}{K'} \quad .$$

Using formulas (see literature to Chapter III [24])

$$\gamma \operatorname{sn}(x + iK',\ \gamma) = \frac{1}{\operatorname{sn}(x,\ \gamma)} \quad ,$$

$$\operatorname{sn}(x - K,\ \gamma) = -\frac{\operatorname{cn}(x,\ \gamma)}{\operatorname{dn}(x,\ \gamma)} \quad ,$$

we obtain:

$$\xi = \frac{1}{\operatorname{sn}\left(\frac{K'\varphi}{Q}, \gamma\right)} = -\frac{\operatorname{dn}\left[K' \frac{\kappa(h_1 - y)}{Q}, \gamma\right]}{\operatorname{cn}\left[K' \frac{\kappa(h_1 - y)}{Q}, \gamma\right]} . \tag{7.41}$$

Returning to equation (7.38) we obtain:

$$u = \sqrt{\frac{-1-\xi}{-\gamma-\xi}} \approx \sqrt{\frac{\xi^2 - 1}{\xi^2 - \gamma^2}} .$$

The latter approximation may be made because for $\gamma \approx 1$

$$\frac{\xi + 1}{\xi + \gamma} \approx 1 .$$

Because of (7.41) we obtain:

$$u = \sqrt{\frac{\operatorname{dn}^2 x - \operatorname{cn}^2 x}{\operatorname{dn}^2 x - \gamma^2 \operatorname{cn}^2 x}} = \operatorname{sn}\left[K' \frac{\kappa(h_1 - y)}{Q}, \gamma\right] \approx \tanh \frac{\pi\kappa(h_1 - y)}{2Q} . \tag{7.42}$$

Therefore (7.40) becomes:

$$F \approx \operatorname{tg}^{-1}\left(\sqrt{\frac{1-\xi}{-\gamma-\xi}} \operatorname{tg}\frac{\pi\tau}{2}\right) \approx \operatorname{tg}^{-1}\left(\sqrt{\frac{\xi^2 - 1}{\xi^2 - \gamma^2}} \operatorname{tg}\frac{\pi\tau}{2}\right) = \operatorname{tg}^{-1}\left(u \operatorname{tg}\frac{\pi\tau}{2}\right) ,$$

where $u$ has value (7.42).

Now, recalling the value of (7.19) for $a_0$, and the equality $(0 < \alpha < \frac{1}{2})$

$$\int_0^1 \operatorname{cotg}^{1-2\alpha} \frac{\pi\tau}{2}\, d\tau = \frac{1}{\sin \alpha\pi} ,$$

we write the equation of the last part of the free surface in the form

$$x \approx \frac{\kappa(h_1^2 - y^2)}{2Q} + h_1 \operatorname{cotg} \alpha\pi - \\ - \frac{2a_1}{\pi}\int_0^1 F \operatorname{cotg}^{1-2\alpha}\frac{\pi\tau}{2}\, d\tau - \frac{4a_2}{\pi^2}\int_0^1 F^2 \operatorname{cotg}^{1-2\alpha}\frac{\pi\tau}{2}\, d\tau .$$

Because of (7.19)

$$a_1 = h_1 \cos \alpha\pi - \frac{Q \sin \alpha\pi}{\kappa}, \quad a_2 = \frac{Q \sin \alpha\pi}{2\kappa} .$$

Therefore it is convenient to introduce the notations

$$F_1(u, \alpha) = \frac{2 \cos \alpha\pi}{\pi} \int_0^1 \operatorname{cotg}^{1-2\alpha}\frac{\pi\tau}{2} \operatorname{tg}^{-1}\left(u \operatorname{tg}\frac{\pi\tau}{2}\right) d\tau ,$$

$$F_2(u, \alpha) = \\ = \frac{2 \cos \alpha\pi}{\pi}\int_0^1 \operatorname{cotg}^{1-2\alpha}\frac{\pi\tau}{2} \operatorname{tg}^{-1}\left(u \operatorname{tg}\frac{\pi\tau}{2}\right)\left[1 - \frac{1}{\pi}\operatorname{tg}^{-1}\left(u \operatorname{tg}\frac{\pi\tau}{2}\right)\right] d\tau .$$

Finally we write the equation of the left branch of the free surface in the

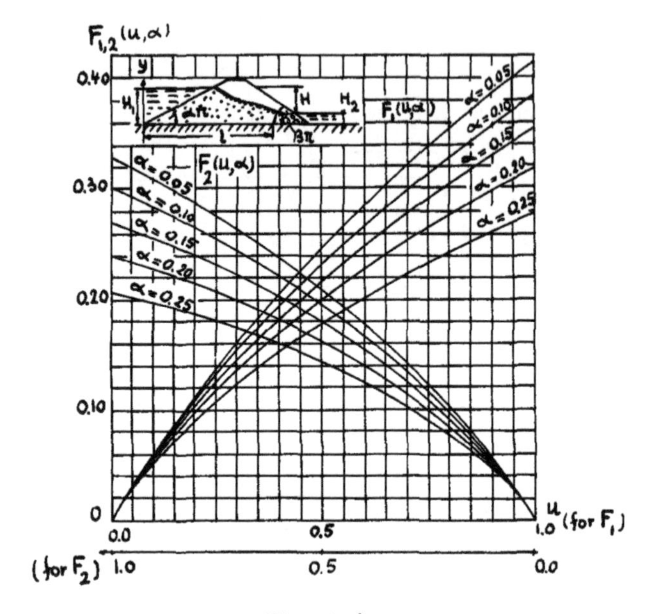


Fig. 196

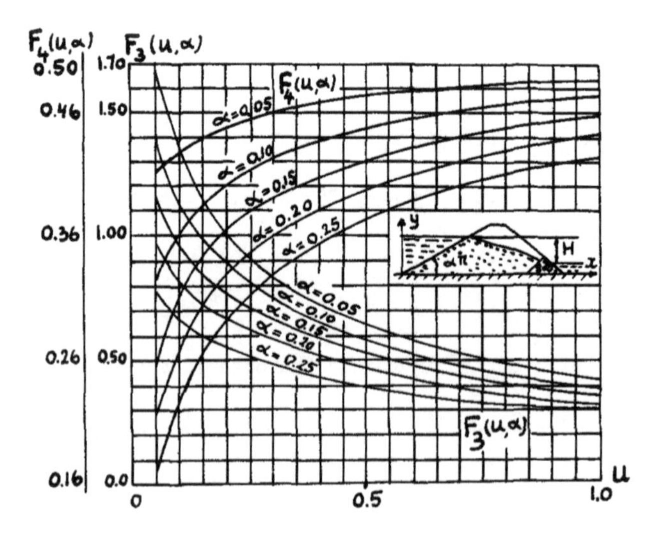


Fig. 197

form

$$x \approx \frac{\kappa(h_1^2 - y^2)}{2Q} + h_1\left[\operatorname{cotg} \alpha\pi - F_1\left(\tanh \frac{\kappa\pi(h_1 - y)}{2Q}, \alpha\right)\right] +$$

$$(7.43) \qquad + \frac{Q}{\kappa} \operatorname{tg} \alpha\pi\, F_2\left(\tanh \frac{\kappa\pi(h_1 - y)}{2Q}, \alpha\right).$$

The equation of the right branch of the free surface is obtained in a similar way

(7.44)
$$x \approx \ell + \frac{\kappa(h_2^2 - y^2)}{2Q} + h_2F_3\left[\tanh \frac{\kappa\pi(y - h_2)}{2Q}, \beta\right] + \frac{Q}{\kappa} F_4\left[\tanh \frac{\kappa\pi(y - h_2)}{2Q}, \beta\right].$$

where

$$F_3(u, \beta) = \frac{2 \cos \beta\pi}{\pi} \int_0^1 \operatorname{tg}^{1-2\beta} \frac{\pi\tau}{2} \operatorname{cotg}^{-1}\left(u \operatorname{tg} \frac{\pi\tau}{2}\right) d\tau ,$$

$$F_4(u, \beta) = \frac{1}{2} - \frac{2}{\pi^2} \sin \beta\pi \int_0^1 \operatorname{tg}^{1-2\beta} \frac{\pi\tau}{2} \operatorname{arc\,cotg}^2\left(u \operatorname{tg} \frac{\pi\tau}{2}\right) d\tau$$

S. N. Numerov composed tables of values of the function $F_1(u, \alpha)$, $F_2(u, \alpha)$, $F_3(u, \alpha)$, $F_4(u, \alpha)$ [8, 12]. Graphs of these functions [13] are given in Fig. 196 and 197.

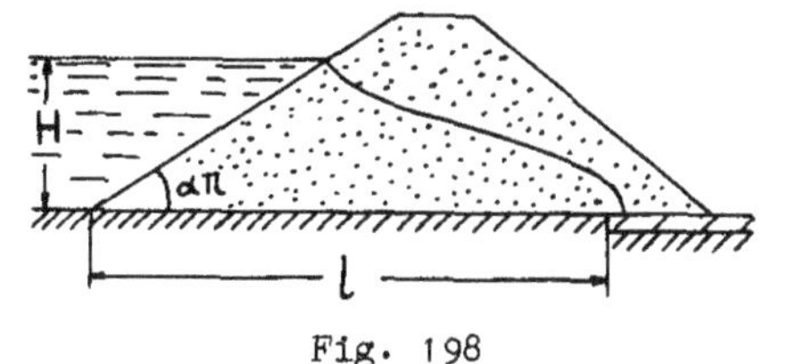


Fig. 198

For a dam with horizontal drain [1, 6], (Fig. 198), the position of the entire free surface may be determined by one equation, compatible with (7.43):

$$x \approx \kappa \frac{H^3 - y^2}{2Q} + H\left[\operatorname{cotg} \alpha\pi - F_1\left(\tanh \kappa\pi \frac{H - y}{2Q}, \alpha\right)\right] +$$

$$+ \frac{Q}{\kappa} F_2\left(\tanh \kappa\pi\frac{H - y}{2Q}, \alpha\right) \operatorname{tg} \alpha\pi .$$

Two examples of free surfaces are given in Fig. 199, for $Q = 0.2\kappa H$, $\ell \approx 3.16H$ [1].

For rectangular dams with horizontal drain $(\alpha = \frac{1}{2}, \beta = 0)$ the equation of the free surface has the form

$$x = \kappa \frac{H^2 - y^2}{2Q} + \frac{Q}{\pi} F\left(\tanh \kappa\pi \frac{H - y}{2Q}\right) ,$$

where

$$F(u) = \frac{2}{\pi}\int_0^1 \mathrm{tg}^{-1}\left(u\ \mathrm{tg}\ \frac{\pi\tau}{2}\right)\left[1 - \frac{1}{\pi}\ \mathrm{tg}^{-1}\left(u\ \mathrm{tg}\ \frac{\pi\tau}{2}\right)\right] d\tau \quad .$$

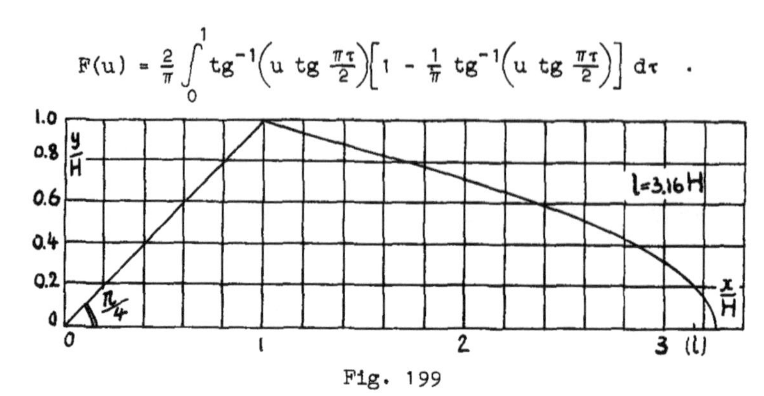


Fig. 199

The graph giving F in function of u [12] is sketched in Fig. 200.

By means of (7.44) we may find for the neighborhood of point C — exit of free surface in tailwater — the expression of the derivative

$$\frac{dx}{dy} = -\frac{\kappa y}{Q} + \left[\left(\sin\beta\pi - \frac{\kappa h_2}{Q}\cos\beta\pi\right)\int_0^1 \frac{\mathrm{tg}^{2-2\beta}\ \frac{\pi\tau}{2}\ d\tau}{1 - u^2\ \mathrm{tg}^2\ \frac{\pi\tau}{2}} - \right.$$

$$\left. - \frac{2\sin\beta\pi}{\pi}\int_0^1 \frac{\mathrm{tg}^{2-2\beta}\ \frac{\pi\tau}{2}\ \mathrm{cotg}^{-1}\left(u\ \mathrm{tg}\ \frac{\pi\tau}{2}\right) d\tau}{1 + u^2\ \mathrm{tg}^2\ \frac{\pi\tau}{2}}\right](1 - u^2)$$

For $u \to 0$ both integrals tend to infinity and, generally speaking

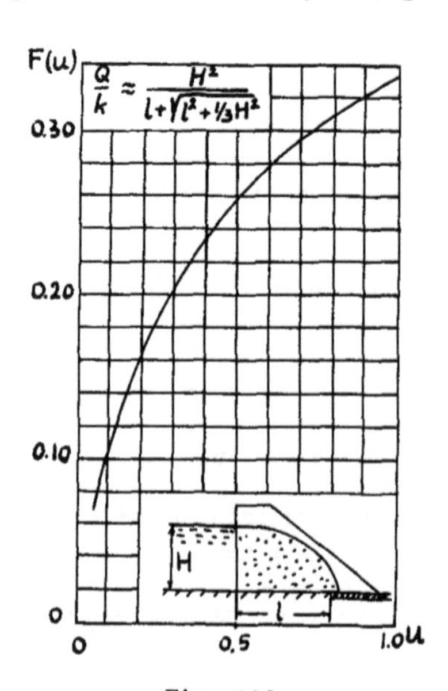


Fig. 200

$\frac{dx}{dy} \to \pm \infty$ . To elucidate the sign question, we make the substitution

$$u \operatorname{tg} \frac{\pi\tau}{2} = \operatorname{tg} \frac{\pi s}{2} ,$$

which gives:

$$\frac{dx}{dy} = -\frac{\kappa y}{Q} + u^{2\beta-1}(1 - u^2)\left[\left(\sin \beta\pi - \frac{\kappa h_2}{Q} \cos \beta\pi\right)\int_0^1 \frac{\operatorname{cotg}^{2\beta} \frac{\pi s}{2} \, ds}{1 + u^2 \operatorname{cotg}^2 \frac{\pi s}{2}} - \right.$$

$$\text{(7.45)} \qquad \left. - \sin \beta\pi \int_0^1 \frac{s \operatorname{cotg}^{2\beta} \frac{\pi s}{2} \, ds}{1 + u^2 \operatorname{cotg}^2 \frac{\pi s}{2}} \right] .$$

*From this we find the limit of the expression*

$$\lim_{u \to 0} \left(u^{1-2\beta} \frac{dx}{dy}\right) = \left(\sin \beta\pi - \frac{\kappa h_2}{Q} \cos \beta\pi\right) \cdot \frac{1}{\cos \beta\pi} - \sin \beta\pi f_1\left(\frac{1}{2} - \beta\right) .$$

*We notice that the sign of the derivative* $dx/dy$ depends on the sign of the quantity

$$A = \frac{\kappa h^2}{Q} - \operatorname{tg} \beta\pi + \sin \beta\pi f_1\left(\frac{1}{2} - \beta\right) .$$

For $A = 0$, $dx/dy$ is finite and non-zero.

If $A > 0$, then

$$\frac{dx}{dy} \to -\infty ,$$

i.e., the free surface approaches the exit surface horizontally.

If $A < 0$, i.e., $dx/dy \to +\infty$, the free surface approaches the exit surface with a tangent oriented along the surface towards its foot. It is evident that the latter form of flow has no physical meaning and that this case corresponds to flow with a seepage surface.

Indeed, for the absence of a seepage surface it is necessary and sufficient that $A \geq 0$, i.e.,

$$\text{(7.46)} \qquad h_2 \geq \sin \pi\beta\left[\frac{1}{\cos \pi\beta} - f_1\left(\frac{1}{2} - \beta\right)\right] \frac{Q}{k} .$$

Methods of approximate computations of the magnitude of the seepage surface are shown in the book quoted before [12].

## *§8. Earth Dam With Horizontal Drain on Pervious Stratum of Infinite Depth [3].

In many cases earth dams are erected on pervious soil, and sometimes the seepage coefficient of the earth dam differs only slightly from the seepage coefficient of the stratum. A filter (often called "tailwater filter") is established to receive the outflow in the tailwater basin. The purpose of the filter is to prevent the outwash of soil particles. Usually the filter is made of several soil layers with different coarseness, increasingly coarse in the seepage direction.

The simplest case of this type is that of an earth dam with

horizontal drain on a pervious stratum of infinite depth (Fig. 201).

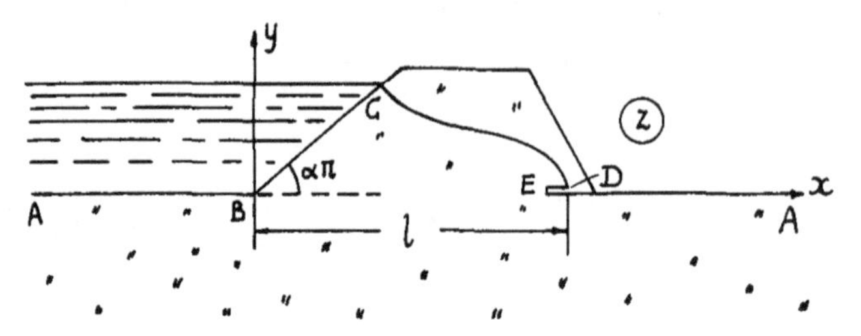


Fig. 201

We introduce the complex potential

(8.1) $$\omega = \varphi + i\psi, \quad \varphi = -k\left(\frac{p}{\rho g} + y\right)$$

(see Chapter II, §1.)

We introduce the reduced complex potential $\Omega$, by means of the equality

(8.2) $$\Omega = \frac{\omega}{kh} = \Phi + i\Psi$$

where

$$\Phi = \frac{\varphi}{kh}, \quad \Psi = \frac{\psi}{kh} \quad .$$

Here $k$ is the seepage coefficient, $h$ the water level in the head reservoir. There is not tailwater in this scheme. $\Omega$ is a dimensionless quantity. The boundary conditions for $\Omega$ are the following.

Along the free surface CD, considering the atmospheric pressure to be zero, we obtain $\varphi = - ky$; consequently

$$\Phi = - \frac{y}{h}$$

or

(8.3) $$y = - h\Phi \quad .$$

Along the contour of the headwater reservoir ABC, where

$$p = \rho g(h - y) \quad ,$$

we have $\varphi = - kh$; consequently

$$\Phi = - 1 \quad .$$

Along the segment DE and the boundary of tailwater EA, we have $p = 0$, $y = 0$, and therefore

$$\Phi = 0 \quad .$$

The region of the reduced complex potential, if we take

$$\Psi = 0$$

along the free surface, has the shape of a half strip (Fig. 202). Mapping this region onto the upper half-plane $\zeta$ (Fig. 202), we have

(8.4) $$\begin{cases} \zeta = - \cos \pi\Omega \quad , \\ \Omega = \frac{1}{\pi} \cos^{-1}(- \zeta) \quad . \end{cases}$$

We consider $z$ as a function of $\Omega$. Along the various boundaries of the region $\Omega$ we have the following conditions for $z(\Omega)$. Along AB and DEA

$$y = 0 \quad . \tag{8.5}$$

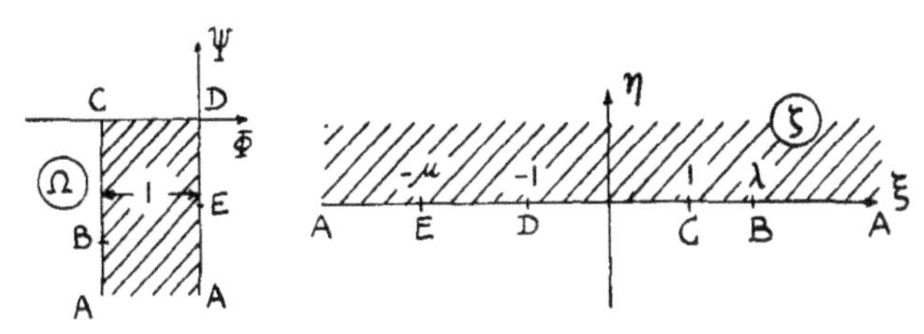


Fig. 202

Along segment BC we may write the equation

$$x \sin \alpha\pi - y \cos \alpha\pi = 0 \tag{8.6}$$

or in complex form

$$\operatorname{Re}(ize^{-\alpha\pi i}) = 0 \quad . \tag{8.7}$$

Along the free surface

$$y = - h\Phi \tag{8.8}$$

Considering now $z$ as a function of $\zeta$, we obtain at the various segments of the $\xi$-axis, instead of (8.5), (8.7) and (8.8) the following conditions

$$\begin{cases} y = \operatorname{Im} z = 0 & \text{for } \xi > \lambda \text{ and } \xi < -1 \quad , \\ \operatorname{Im}(ze^{-\alpha\pi i}) = 0 & \text{for } 1 < \xi < \lambda \quad , \\ y = \operatorname{Im} z = - \frac{h}{\pi} \cos^{-1}(- \xi) & \text{for } -1 < \xi < 1 \quad . \end{cases} \tag{8.9}$$

The latter condition is obtained from (8.8), if we assume that for $-1 < \xi < 1$ we have $\Omega = \Phi$ .

Conditions (8.9) are of the same form as those we considered in §2. This permits us to reduce the problem to Dirichlet's. But first we have to investigate the behaviour of $z(\zeta)$ at infinity.

Formulas (8.9) show that $z$ assumes real values for $\xi > \lambda$ and $\xi < -1$ on the real axis. Because of the Riemann-Schwarz principle of symmetry, when a function is analytic in a region containing a segment of the abscissa axis, and assumes continuous real values at this segment, then this function may be analytically continued in the lower half-plane, assuming conjugate values in conjugate points.

Thus $z(\zeta)$ is analytic in the whole plane with a cut along the real axis from point $\xi = -1$ to point $\xi = \lambda$. Function $1/z$ may be developed in a Laurent series in the neighborhood of the point at infinity; since $z(\infty) = \infty$, $1/z$ may only contain negative powers of $\zeta$:

$$\operatorname{Im} = \sum_{n=1}^{\infty} \frac{a_n}{\zeta^n} \quad . \tag{8.10}$$

In this series $a_1 \neq 0$, since at infinity

$$z \approx a_1 \zeta \quad .$$

All coefficients $a_n$ are real. This follows from the fact that in infinity along the $\xi$-axis, the variable $z$ has real values (the $\xi$-axis transforms into the x-axis) and we may write:

$$\operatorname{Im} = \sum_{n=1}^{\infty} \frac{a_n}{\zeta^n} = 0$$

for sufficiently large $\xi$. But multiplying this expression with $\xi$ and putting $\xi = -\infty$, we obtain:

$$\operatorname{Im} a_1 = 0 \quad .$$

Continuing this process further, we convince ourselves that all $a_n$ are real

From (8.10) it follows that

$$z(\zeta) = b_{-1}\,\zeta + \sum_{n=0}^{\infty} \frac{b_n}{\zeta^n} \quad , \tag{8.11}$$

where all coefficients of the series are real.

The problem of finding function $z(\zeta)$ is reduced to the problem analyzed in §2. As function $T(\zeta)$ we take

$$T(\zeta) = \ell(1 - \zeta)^{\alpha}(\lambda - \zeta)^{1-\alpha} \quad . \tag{8.12}$$

We choose the branch of this multivalued function, for which

$$\arg w = \frac{\pi}{2} \quad \text{for} \quad \zeta = \xi < 1 \quad .$$

Then, for $-\infty < \xi < 1$

$$T(\xi) = e^{\frac{\pi i}{2}}(1 - \xi)^{\alpha}(\lambda - \xi)^{1-\alpha} \quad . \tag{8.13}$$

Going around point $\xi = 1$ over an angle $\pi$ in the negative sense, argument $(1 - \zeta)^{\alpha}$ receives an increment $-\alpha\pi$, and for $1 < \xi < \lambda$ we may write:

$$T(\xi) = (\xi - 1)^{\alpha}(\lambda - \xi)^{1-\alpha} e^{\pi i(1/2 - \alpha)} \quad . \tag{8.14}$$

For the interval $\lambda < \xi < \infty$ we have

$$T(\xi) = (\xi - 1)^{\alpha}(\xi - \lambda)^{1-\alpha} e^{-\frac{\pi i}{2}} \quad . \tag{8.15}$$

To investigate product $T(\zeta)$ at infinity, we bring $\xi$ out of the brackets of equality (8.15), rewritten as

$$T(\xi) = -i\xi\left(1 - \frac{1}{\xi}\right)^{\alpha}\left(1 - \frac{\lambda}{\xi}\right)^{1-\alpha} = -i\xi - i\sum_{n=0}^{\infty} \frac{c_n}{\xi^n} \quad ,$$

where $c_n$ are real coefficients. But also

$$T(\zeta) = -i\zeta - i\sum_{n=0}^{\infty} \frac{c_n}{\zeta^n} \quad . \tag{8.16}$$

Product $z(\zeta)\,T(\zeta)$ because of (8.11) and (8.16) has the following form near the point at infinity

$$z(\zeta)\,T(\zeta) = -\,b_{-1}i\zeta^2 - i(b_0 + b_{-1}c_0)\zeta + i\sum_{n=0}^{\infty} \delta_n\zeta^{-n} \quad ,$$

where coefficients $\delta_n$ are real.

We compose the function

$$W(\zeta) = z(\zeta)\,T(\zeta) + b_{-1}i\zeta^2 + i(b_0 + b_{-1}c_0)\zeta \quad . \tag{8.17}$$

It is analytic near the point at infinity. Conditions for $W(\zeta)$ have the following form because of (8.9), (8.13), (8.14) and (8.15):

$$\operatorname{Re} W(\xi) = f^*(\xi) = \begin{cases} -\frac{h}{\pi}\cos^{-1}(-\xi)(1-\xi)^{\alpha}(\lambda-\xi)^{1-\alpha} & \text{for } |\xi| < 1 \; , \\ 0 & \text{for } |\xi| > 1 \; . \end{cases} \tag{8.18}$$

To determine function $W(\zeta)$, we may now apply formula (1.6). We obtain:

$$W(\zeta) = -\frac{i}{\pi}\int_{-1}^{1} \frac{f^*(t)\,dt}{t-\zeta} + ic \quad , \tag{8.19}$$

where $c$ is a real constant, subject to determination.

Taking into account (8.17) and (8.12), we may rewrite after simplifying by $i$:

$$(1-\zeta)^{\alpha}(\lambda-\zeta)^{1-\alpha}z(\zeta) + b_{-1}\zeta^2 + (b_0 + b_{-1}c_0)\zeta = -\frac{1}{\pi}\int_{-1}^{1} \frac{f^*(t)\,dt}{t-\zeta} + c \quad . \tag{8.20}$$

A series of unknown constants is introduced in equation (8.20). The following method is used to eliminate these. We put $\zeta = 1$. We obtain:

$$b_{-1} + (b_0 + b_{-1}c_0) = -\frac{1}{\pi}\int_{-1}^{1} \frac{f^*(t)\,dt}{t-1} + c \quad .$$

This equality is subtracted from (8.20) and the result is divided by $1-\zeta$. We have:

$$(1-\zeta)^{\alpha-1}(\lambda-\zeta)^{1-\alpha}z(\zeta) - b_{-1}(\zeta+1) - (b_0 + b_{-1}c_0) = \frac{1}{\pi}\int_{-1}^{1} \frac{f^*(t)\,dt}{(t-1)(t-\zeta)} \; .$$

Now we make $\zeta = \lambda$. Carrying out the subtraction and dividing by $\lambda - \zeta$, we obtain:

$$(1-\zeta)^{\alpha-1}(\lambda-\zeta)^{-\alpha}z(\zeta) + b_{-1} = -\frac{1}{\pi}\int_{-1}^{1} \frac{f^*(t)\,dt}{(1-t)(\lambda-t)(t-\zeta)} \quad . \tag{8.21}$$

We may choose the coordinate axes so that for $\zeta = \lambda$, $z = 0$ (Fig. 201, 202). Therefore since contouring of the point $\zeta = \lambda$ by $\pi$ in the $\zeta$-plane corresponds to contouring over an angle $\pi(1+\alpha)$ in the z-plane, we have this expression for $z$:

$$z = (\lambda-\zeta)^{1+\alpha}[A + B(\zeta-\lambda) + \ldots] \quad .$$

From this it is apparent that the product

$$(\lambda - \zeta)^{-\alpha} z(\zeta) = (\lambda - \zeta)[A + B(\zeta - \lambda) + \dots]$$

vanishes for $\zeta = \lambda$. Therefore, making $\zeta = \lambda$ in (8.21) we obtain:

$$b_{-1} = \frac{1}{\pi} \int_{-1}^{1} \frac{f^*(t)\, dt}{(1 - t)(\lambda - t)^2} .$$

We insert this value of $b_{-1}$ in (8.21) and rewrite this equation in the following form:

$$(8.22) \qquad z(\zeta) = -\frac{1}{\pi} (1 - \zeta)^{1-\alpha} (\lambda - \zeta)^{\alpha+1} \int_{-1}^{1} \frac{f^*(t)\, dt}{(1 - t)(\lambda - t)^2 (t - \zeta)} .$$

Recalling value (8.18) for $f^*(t)$, we obtain:

(8.23)

$$z(\zeta) = \frac{h}{\pi^2}(1 - \zeta)^{1-\alpha}(\lambda - \zeta)^{1+\alpha} \int_{-1}^{1} \frac{\cos^{-1}(-t)\, dt}{(1 - t)^{1-\alpha}(\lambda - t)^{1+\alpha}(t - \zeta)} .$$

Adding to this equation (8.4)

$$\Omega = \frac{1}{\pi} \cos^{-1}(-\zeta) ,$$

we obtain the solution of the problem in parameter form.

Equation (8.23) contains an unknown parameter $\lambda$. To determine it, we use the condition that $z = \ell$ for $\zeta = -1$ (segment $\ell$ in Fig. 201 is the distance between points B and D). From (8.23) we obtain:

$$(8.24) \qquad \frac{\ell}{h} = \frac{1}{\pi^2} 2^{1-\alpha} (\lambda + 1)^{1+\alpha} \int_{-1}^{1} \frac{\cos^{-1}(-t)\, dt}{(1 - t)^{1-\alpha}(\lambda - t)^{1+\alpha}(1 + t)} .$$

This transcendental equation serves to determine $\ell$ for given $\ell/h$. Since the function

$$\frac{\lambda + 1}{\lambda - t}$$

decreases monotonously with $\lambda$ for $-1 < t < 1$, the function

$$\left(\frac{\lambda + 1}{\lambda - t}\right)^{1+\alpha}$$

has the same behaviour and so has the right part of equation (8.24). For each given value $\ell/h$ there is one value of $\lambda$.

<u>Transformation of basic formulas</u>. We insert in formula (8.23)

$$\zeta = -\cos \pi\Omega$$

and

$$t = -\cos \pi\tau .$$

We obtain:

$$z = h(1 + \cos \pi\Omega)^{1-\alpha}(\lambda - \cos \pi\Omega)^{1+\alpha} \times$$

$$\times \int_{0}^{1} \frac{\tau \sin \pi\tau\, d\tau}{(1 + \cos \pi\tau)^{1-\alpha}(\lambda + \cos \pi\tau)^{1+\alpha}(\cos \pi\Omega - \cos \pi\tau)} .$$

We use a transformation in terms of trigonometric functions of the half angle, introducing the notation

$$(8.25) \qquad f(\tau) = \frac{\left(1 + \frac{2}{\lambda - 1} \cos^2 \frac{\pi\tau}{2}\right)^{1+\alpha}}{\cos^{2\alpha} \frac{\pi\tau}{2}} .$$

We obtain:

$$z = h \cos^2 \frac{\pi\Omega}{2} f(\Omega) \int_0^1 \frac{\tau \sin \pi\tau \, d\tau}{\cos^2 \frac{\pi\tau}{2} f(\tau)(\cos \pi\Omega - \cos \pi\tau)} .$$

For $\tau = 1$ the function under the integral becomes infinite. Applying the transformation

$$\frac{\cos^2 \frac{\pi\Omega}{2}}{\cos^2 \frac{\pi\tau}{2} (\cos \pi\Omega - \cos \pi\tau)} = \frac{1}{2 \cos^2 \frac{\pi\tau}{2}} + \frac{1}{\cos \pi\Omega - \cos \pi\tau} ,$$

we split the integral in two terms

$$(8.26) \qquad z = hf(\Omega) \left\{c + \int_0^1 \frac{\tau \sin \pi\tau \, d\tau}{f(\tau)(\cos \pi\Omega - \cos \pi\tau)}\right\} .$$

Here

$$c = \frac{1}{2} \int_0^1 \frac{\tau \sin \pi\tau \, d\tau}{\cos^2 \frac{\pi\tau}{2} f(\tau)} = \int_0^1 \frac{\tau \operatorname{tg} \frac{\pi\tau}{2} \, d\tau}{f(\tau)} .$$

Putting

$$\tau = 1 - t$$

we obtain:

$$f(\tau) = f(1 - t) = \left(\frac{\lambda + 1}{\lambda - 1}\right)^{1+\alpha} \frac{\left(1 - \frac{2}{\lambda + 1} \sin^2 \frac{\pi t}{2}\right)^{1+\alpha}}{\sin^{2\alpha} \frac{\pi t}{2}} ,$$

and the integral for $c$ is split into two:

$$c = \left(\frac{\lambda - 1}{\lambda + 1}\right)^{1+\alpha} \int_0^1 \frac{\sin^{2\alpha-1} \frac{\pi t}{2} \cos \frac{\pi t}{2} \, dt}{\left(1 - \frac{2}{\lambda + 1} \sin^2 \frac{\pi t}{2}\right)^{1+\alpha}} - \int_0^1 \frac{t \operatorname{cotg} \frac{\pi t}{2}}{f(1 - t)} .$$

The integral in the first term is reduced to the integral

$$\int_0^1 x^{\alpha-1}(1 - \mu x)^{-\alpha-1} \, dx = \frac{1}{\alpha} (1 - \mu)^{-\alpha}$$

by means of the substitution $\sin^2 \frac{\pi t}{2} = x$ and is equal to

$$\frac{1}{\pi\alpha}\left(\frac{\lambda - 1}{\lambda + 1}\right)^{1\alpha} .$$

Therefore, introducing the notation

$$(8.27) \qquad J_1(\alpha, \lambda) = \int_0^1 \frac{t \operatorname{cotg} \frac{\pi t}{2} \, dt}{f(1 - t)} ,$$

we find that

$$c(\alpha,\ \lambda) = \frac{1}{\pi\alpha}\left(\frac{\lambda - 1}{\lambda + 1}\right) - J_1 \quad . \tag{8.28}$$

Equation (8.24) serves to determine parameter $\lambda$ and may be written in the form

$$\frac{\ell}{h} = \frac{1}{\pi\alpha}\ {\frac{\lambda + 1}{\lambda - 1}}^{\alpha} + {\frac{\lambda + 1}{\lambda - 1}}^{1+\alpha}\ (J_2 - J_1) \quad . \tag{8.29}$$

where

$$J_2(\alpha,\ \lambda) = \int_0^1 \frac{t \operatorname{cotg} \frac{\pi t}{2}\, dt}{f(t)} \quad . \tag{8.30}$$

To obtain an expression of the complex velocity, we recall that

$$\frac{d\omega}{dz} = kh\,\frac{d\Omega}{dz} \quad .$$

It is simpler to take the inverse of the complex velocity, namely

$$\frac{dz}{d\omega} = \frac{1}{u - iv} = \frac{1}{kh}\ \frac{dz}{d\Omega} \quad .$$

We have

$$\frac{dz}{d\omega} = \frac{1}{kh}\frac{dz}{d\zeta} : \frac{d\Omega}{d\zeta}$$

Because of (8.4)

$$\frac{d\Omega}{d\zeta} = \frac{1}{\pi\sqrt{1 - \zeta^2}} \quad .$$

Therefore

$$\frac{dz}{d\omega} = \frac{\pi}{kh}\ \sqrt{1 - \zeta^2}\ \frac{dz}{d\zeta} \tag{8.31}$$

To compute the derivative $z'(\zeta)$, we transform expression (8.23) for $z(\zeta)$. We notice that

$$\frac{1 - \zeta}{t - \zeta} = \frac{1 - t}{t - \zeta} + 1$$

and

$$\lambda - \zeta = (\lambda - t) + (t - \zeta) \quad .$$

From this

$$\frac{(\lambda - \zeta)(1 - \zeta)}{t - \zeta} = \frac{(\lambda - t)(1 - t)}{t - \zeta} + (1 - t) + (\lambda - \zeta) \quad .$$

and integral (8.23), after insertion in it of the multiplier $(1 - \zeta)(\lambda - \zeta)$ may be developed into three terms:

$$z(\zeta) = \frac{h}{\pi^2}\left(\frac{\lambda - \zeta}{1 - \zeta}\right)^{\alpha}\left\{ \int_{-1}^{1} \frac{\cos^{-1}(-t)\,dt}{(1 - t)^{-\alpha}(\lambda - t)^{\alpha}(t - \zeta)} + \right.$$

$$\left. + \int_{-1}^{1} \frac{\cos^{-1}(-t)\,dt}{(1 - t)^{-\alpha}(\lambda - t)^{1+\alpha}} + (\lambda - \zeta)\int_{-1}^{1} \frac{\cos^{-1}(-t)\,dt}{(1 - t)^{1-\alpha}(\lambda - t)^{1+\alpha}} \right\} . \tag{8.32}$$

Putting

$$T(\zeta) = \left(\frac{\lambda - \zeta}{1 - \zeta}\right)^{\alpha} ,$$

we find:

$$T'(\zeta) = \alpha(\lambda - 1)\left(\frac{\lambda - \zeta}{1 - \zeta}\right)^{\alpha-1} .$$

Differentiating $z(\zeta)$, we write:

$$z'(\zeta) = \frac{h}{\pi^2} T'(\zeta) \int_{-1}^{1} \frac{\cos^{-1}(-t)\, dt}{(1 - t)^{1-\alpha}(\lambda - t)^{1+\alpha}(t - \zeta)} +$$

$$(8.33) \qquad + \frac{h}{\pi^2} T(\zeta) \left\{ \int_{-1}^{1} \frac{\varphi(t)\, dt}{(t - \zeta)^2} - \int_{-1}^{1} \frac{\cos^{-1}(-t)dt}{(1 - t)^{1-\alpha}(\lambda - t)^{1+\alpha}} \right\}$$

where

$$\varphi(t) = \frac{\cos^{-1}(-t)}{(1 - t)^{-\alpha}(\lambda - t)^{\alpha}} .$$

Here we substitute three terms in the multiplier of $T'(\zeta)$ for one expression.

We integrate by parts

$$\int_{-1}^{1} \frac{\varphi(t)\, dt}{(t - \zeta)^2} = - \frac{\varphi(t)}{t - \zeta}\Bigg|_{-1}^{1} + \int_{-1}^{1} \frac{\varphi'(t)\, dt}{t - \zeta} .$$

It is easy to see that the term resulting from the integration by parts vanishes. (Development of the integral into three terms was done to split off this term.) Therefore

$$\varphi'(t) = \frac{1}{(1 - t)^{1/2-\alpha}(1 + t)^{1/2}(\lambda - t)^{\alpha}} - \alpha(\lambda - 1)\frac{\cos^{-1}(-t)}{(1 - t)^{1-\alpha}(\lambda - t)^{1+\alpha}} ,$$

and therefore

$$\int_{-1}^{1} \frac{\varphi(t)\, dt}{(t - \zeta)^2} = \int_{-1}^{1} \frac{dt}{(1 - t)^{1/2-\alpha}(1 + t)^{1/2}(\lambda - t)^{\alpha}(t - \zeta)} - \alpha(\lambda - 1)\int_{-1}^{1} \frac{\varphi(t)\, dt}{t - \zeta} .$$

Substitution of these terms in equation (8.33) causes all terms to vanish, except two, giving:

$$z'(\zeta) = \frac{h}{\pi^2}\left(\frac{\lambda - \zeta}{1 - \zeta}\right)^{\alpha} \left\{ \int_{-1}^{1} \frac{dt}{(1 - t)^{1/2-\alpha}\sqrt{1 + t}(\lambda - t)^{\alpha}(t - \zeta)} - \int_{-1}^{1} \frac{\cos^{-1}(-t)\, dt}{(1 - t)^{1-\alpha}(\lambda - t)^{1+\alpha}} \right\} .$$

Now taking into account (8.31) and recalling that

$$\zeta = -\cos \pi\Omega ,$$

we find the inverse of the complex velocity:

$$(8.34) \qquad \frac{1}{u - iv} = - \frac{1}{k\pi} \sin \pi\Omega(\lambda + \cos \pi\Omega)^{\alpha}(1 + \cos \pi\Omega)^{-\alpha} \left\{ - A + \int_{-1}^{1} \frac{\Phi(t)\, dt}{t + \cos \pi\Omega} \right\} ,$$

where

$$(8.35)\qquad \begin{cases} A = \displaystyle\int_{-1}^{1} \frac{\cos^{-1}(-t)\,dt}{(1-t)^{1-\alpha}(\lambda - t)^{1+\alpha}} , \\ \Phi(t) = \dfrac{1}{(1-t)^{1/2-\alpha}(1+t)^{1/2}(\lambda - t)^{\alpha}} . \end{cases}$$

Introducing the variable of integration t,

$$t = -\cos \pi\tau ,$$

we rewrite (8.34) as:

$$(8.3\ \ )\qquad \frac{1}{u - iv} = \frac{\sin \pi\Omega}{k\pi} F(\Omega) \left\{ \frac{\pi c}{\lambda - 1} - \int_0^1 \frac{d\tau}{F(\tau)(\cos \pi\Omega - \cos \pi\tau)} \right\} ,$$

where

$$F(\tau) = \frac{\left(1 + \frac{2}{\lambda - 1} \cos^2 \frac{\pi\tau}{2}\right)^{\alpha}}{\cos^{2\alpha} \frac{\pi\tau}{2}} .$$

To find the equation of the free surface, we must pass to the limit for $\Omega \to \Phi$ in (8.26). Therefore integral (8.26) becomes improper. As we did before, it is convenient for limit purposes to write the integral as

$$J = \int_0^1 \frac{\tau \sin \pi\tau \, d\tau}{f(\tau)(\cos \pi\Omega - \cos \pi\tau)} =$$

$$= \int_0^1 \frac{\left(\frac{\tau}{f(t)} - \frac{\Omega}{f(\Omega)}\right) \sin \pi\tau}{\cos \pi\Omega - \cos \pi\tau} d\tau + \frac{\Omega}{f(\Omega)} \int_0^1 \frac{\sin \pi\tau \, d\tau}{(\cos \pi\Omega - \cos \pi\tau)} .$$

The second integral gives:

$$\lim_{\Omega \to \Phi} \int_0^1 \frac{\sin \pi\tau \, d\tau}{\cos \pi\Omega - \cos \pi\tau} = i + \frac{1}{\pi} \ln \frac{1 + \cos \pi\Phi}{1 - \cos \pi\Phi} .$$

Thus, from (8.26) we obtain for the free surface

$$z = hf(\Phi)c + hf(\Phi) \int_0^1 \frac{\left[\frac{\tau}{f(\tau)} - \frac{\Phi}{f(\Phi)}\right] \sin \pi\tau}{\cos \pi\Phi - \cos \pi\tau} d\tau - h\left(\Phi i + \frac{\Phi}{\pi} \ln \frac{1 + \cos \pi\Phi}{1 - \cos \pi\Phi}\right)$$

Equating the imaginary parts of this equation gives:

$$y = -h\Phi ,$$

and this is compatible with equation (8.3). Separation of the real parts by means of the substitution

$$\Phi = -\frac{y}{h}$$

gives the equation of the free surface:

$$(8.37)\qquad x = \frac{2y}{\pi} \ln \operatorname{cotg} \frac{\pi y}{2h} + \cosh f\left(\frac{y}{h}\right) + J\left(\frac{y}{h}\right)$$

$$(8.38)\qquad J(u) = u \int_0^1 \frac{\frac{\tau f(u)}{u f(\tau)} - 1}{\cos \pi u - \cos \pi\tau} \sin \pi\tau \, d\tau ,$$

in which $f(u)$ has value (8.25).

To determine the length of the drainage segment ED, we must use the condition that the velocity becomes infinite in point E. It would be possible to make the right part of expression (8.36) zero and to solve it for $\Omega = i\psi$, when $-\infty < \psi < 0$. Below an approximate formula for the computation of $b = ED$ is given.

Approximate formulas. Equation (8.37) is not convenient for the computation of $x$ for $y$ values near $h$. Therefore S. N. Numerov made the following transformation. He computes the derivative

$$[\sqrt{1-\zeta^2}z'(\zeta)]' = -\frac{h}{\pi^2}\left\{\frac{2^\alpha\pi^2 c}{(\lambda-1)^{1+\alpha}}\frac{d}{d\zeta}[(1-\zeta)^{1/2-\alpha}(1+\zeta)^{1/2}(\lambda-\zeta)^\alpha] - \right.$$

$$\left. - 2^\alpha(\lambda-1)^{1-\alpha}B(1-\zeta)^{-1/2-\alpha}(1+\zeta)^{-1/2}(\lambda-\zeta)^{\alpha-1}\right\}, \tag{8.39}$$

where

$$B = \left(\frac{\alpha}{\lambda-1} - \alpha^2\right)c - \frac{2}{\pi(\lambda-1)}\int_0^1 \frac{\sin^{2+2\alpha}\frac{\pi\tau}{2}\,d\tau}{\left(1+\frac{2}{\lambda-1}\sin^2\frac{\pi\tau}{2}\right)^\alpha}.$$

To obtain formula (8.39) we may act in an analogous way as for the computation of $z'(\zeta)$. We write:

$$z'\sqrt{1-\zeta^2} = -\frac{h}{\pi^2}P(\zeta)\left\{A - \int_{-1}^{1}\frac{\Phi(t)\,dt}{t-\zeta}\right\}, \tag{8.40}$$

where

$$P(\zeta) = (\lambda-\zeta)^\alpha(1-\zeta)^{1/2-\alpha}(1+\zeta)^{1/2}.$$

Before differentiating the integral of the right part of equality (8.40), we introduce in it the product $(1-\zeta)(1+\zeta)$ and develop it in partial fractions

$$\frac{(1-\zeta)(1+\zeta)}{t-\zeta} = \frac{(1+t)(1-t)}{t-\zeta} + (1+t) - (1-\zeta).$$

Designating

$$P_1(\zeta) = (\lambda-\zeta)^\alpha(1-\zeta)^{-1/2-\alpha}(1+\zeta)^{-1/2},$$

we may write (8.40) as

$$z'\sqrt{1-\zeta^2} = -\frac{h}{\pi^2}\left\{AP(\zeta) - P_1(\zeta)\left[\int_{-1}^{1}\frac{(1-t^2)\varphi(t)\,dt}{t-\zeta} + \right.\right.$$

$$\left.\left. + \int_{-1}^{1}(1+t)\varphi(t)\,dt - (1-\zeta)\int_{-1}^{1}\varphi(t)\,dt\right]\right\}$$

and proceed further as for the computation of $z'(\zeta)$.

We now integrate (8.39), recalling that (Fig. 203)

$$\sqrt{1-\zeta^2}z'(\zeta)\Bigg|_{\zeta=1} = -\frac{kh}{\pi}\frac{1}{u-iv}\Bigg|_{\zeta=1} = -\frac{he^{\pi i(\alpha-1/2)}}{\pi\cos\alpha\pi} = \frac{he^{\pi i(\alpha+1/2)}}{\pi\cos\alpha\pi}.$$

We find for $z'(\zeta)$ a new expression

$$z'(\zeta) = \frac{he^{\pi i(\alpha+1/2)}}{\pi \cos \alpha\pi} \frac{1}{\sqrt{1 - \zeta^2}} - \frac{ch}{(\lambda - 1)^{\alpha+1} 2^{-\alpha}} \left(\frac{\lambda - \zeta}{1 - \zeta}\right)^\alpha +$$

(8.41)

$$+ 2^\alpha (\lambda - 1)^{1-\alpha} \frac{Bh}{\sqrt{1 - \zeta^2}} \times \int_1^\zeta (1 - t)^{-1/2-\alpha} (1 + t)^{-1/2} (\lambda - t)^{\alpha-1} dt .$$

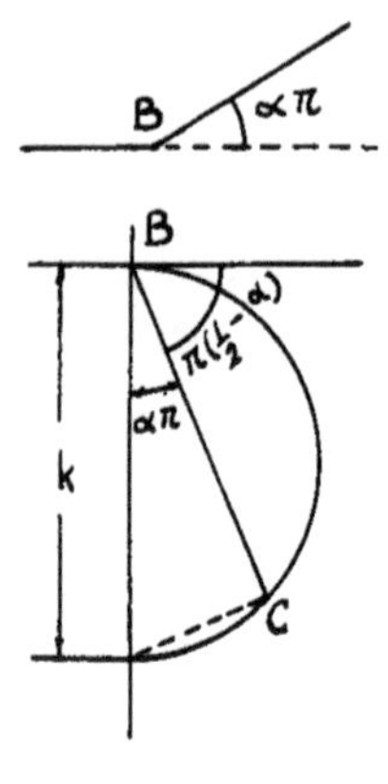

Fig. 203

We introduce a new variable of integration

(8.42)
$$\eta = \frac{1 - t}{2}$$

and compute the integral by means of a power series in $\eta$. We obtain:

$$J = \int_1^\zeta (1 - t)^{-1/2-\alpha} (1 + t)^{-1/2} (\lambda - t)^{\alpha-1} dt =$$

$$= -2^{-\alpha} (\lambda - 1)^{\alpha-1} \int_0^{\frac{1-\zeta}{2}} \eta^{-1/2-\alpha} (1 - \eta)^{-1/2} \left(1 + \frac{2}{\lambda - 1} \eta\right)^{\alpha-1} d\eta =$$

$$= -2^{-\alpha} (\lambda - 1)^{\alpha-1} \left\{ \frac{1}{2^{1/2-\alpha} (1 - \zeta)^{\alpha-1/2}} \left[ \frac{1}{\frac{1}{2} - \alpha} - \frac{D}{\frac{3}{2} - \alpha} \frac{1 - \zeta}{2} + \ldots \right] \right\} ,$$

where

$$D = \frac{2(1 - \alpha)}{1 - \lambda} - \frac{1}{2} .$$

Now we have for $z'(\zeta)$:

$$z'(\zeta) = \frac{he^{\pi i(\alpha+1/2)}}{\pi \cos \alpha\pi} \frac{1}{\sqrt{1 - \zeta^2}} - \frac{hc}{2^{1-\alpha} (\lambda - 1)^{\alpha+1}} \left(\frac{\lambda - \zeta}{1 - \zeta}\right)^\alpha -$$

$$- \frac{Bh}{2^{1/2-\alpha} (1 - \zeta)^\alpha (1 + \zeta)^{1/2}} \left\{ \frac{1}{\frac{1}{2} - \alpha} - \frac{D}{\frac{3}{2} - \alpha} \frac{1 - \zeta}{2} + \ldots \right\}$$

We integrate this expression between limits from 1 to $\zeta$, and apply again substitution (8.42) under the integral. We obtain, recalling that for $\zeta = 1$

$$z = \frac{he^{\pi i\alpha}}{\sin \alpha\pi} ,$$

the following equation

$$\frac{z}{h} = \frac{e^{\pi i\alpha}}{\sin \alpha\pi} - \frac{e^{i\pi(1/2+\alpha)}}{\pi \cos \alpha\pi} \cos^{-1} \zeta + \frac{2c}{1 - \lambda} \int_0^{\frac{1-\zeta}{2}} \eta^{-\alpha} \left(1 + \frac{2}{\lambda - 1} \eta\right)^\alpha d\eta +$$

$$+ B \int_0^{\frac{1-\zeta}{2}} \eta^{-\alpha} (1 - \eta)^{-1/2} \left\{ \frac{1}{\frac{1}{2} - \alpha} - \frac{D}{\frac{3}{2} - \alpha} \eta + \ldots \right\} d\eta$$

Developing the function under the integral in a power series of $\eta$, we find in the neighborhood of $\zeta = 1$:

$$\frac{z}{h} = \frac{e^{i\alpha\pi}}{\sin\alpha\pi} - \frac{e^{i\pi(1/2+\alpha)}}{\pi\cos\alpha\pi}\cos^{-1}\zeta +$$

(8.43)
$$+\frac{1}{1-\alpha}\left(\frac{B}{\frac{1}{2}-\alpha}+\frac{2c}{1-\lambda}\right)\left(\frac{1-\zeta}{2}\right)^{1-\alpha} - \frac{E}{2-\alpha}\left(\frac{1-\zeta}{2}\right)^{2-\alpha} + \dots ,$$

where

$$E = B\left\{\frac{1}{\frac{3}{2}-\alpha}\left[\frac{2(1-\alpha)}{1-\lambda}-\frac{1}{2}\right] - \frac{1}{2}\,\frac{1}{\frac{1}{2}-\alpha}\right\} - \frac{4c\alpha}{(1-\lambda)^2} .$$

The free surface corresponds to values of $\zeta$ between $-1$ and $1$, for which $\zeta = -\cos\pi\Phi$, since here $\psi = 0$.

Besides, at the free surface

$$\Phi = -\frac{y}{h} ,$$

so that

(8.44)
$$\zeta = -\cos\frac{\pi y}{h} .$$

We separate the real part in equation (8.43). Taking into account (8.44) and considering that $1 - \zeta = 2\cos^2\frac{\pi y}{2h}$, we have this equation of the free surface in the neighborhood of the tailwater basin:

$$\frac{x}{h} = \operatorname{cotg}\alpha\pi + \operatorname{tg}\alpha\pi\left(1-\frac{y}{h}\right) + a_1\cos^{2-2\alpha}\frac{\pi y}{2h} + a_2\cos^{4-2\alpha}\frac{\pi y}{2h} + \dots ,$$

where

$$a_1(\alpha,\lambda) = \frac{1}{(1-\alpha)\left(\frac{1}{2}-\alpha\right)(\lambda-1)}\left\{c(1-\alpha) - \alpha^2 c(\lambda-1) - \frac{2}{\pi}\int_0^1 \frac{\cos^2\frac{\pi t}{2}}{F(t)}\,dt\right\} ,$$

$$a_2(\alpha,\lambda) = \frac{1}{(2-\alpha)\left(\frac{3}{2}-\alpha\right)\left(\frac{1}{2}-\alpha\right)(\lambda-1)^2} \times$$

$$\times\left\{c\alpha[(1-2\alpha)(2-\alpha) + (1-\alpha)^2(1+2\alpha)(\lambda-1) - \alpha(1-\alpha)(\lambda-1)^2] + \right.$$

$$\left. + \frac{2}{\pi}(1-\alpha)[1-2\alpha-(\lambda-1)]\int_0^1 \frac{\cos^2\frac{\pi t}{2}}{F(t)}\,dt\right\} .$$

To determine the width of the drainage slit, S. N. Numerov gives $z(\zeta)$ in the neighborhood of $\zeta = -1$:

$$z(\zeta) = \ell + \frac{hi}{\pi}\left(\frac{\pi}{2}+\sin^{-1}\zeta\right) - h\left\{\frac{2c}{1-\lambda}\left(\frac{\lambda+1}{\lambda-1}\right)^{\alpha} - 2B\left(\frac{\lambda-1}{\lambda+1}\right)^{1-\alpha}\right\}\frac{1+\zeta}{2} + \dots$$

Since at the boundary of the drain $\Phi = 0$, $\Omega = i\Psi$, the following equation for $x$ is obtained:

(8.45)
$$x(\Psi) = \ell - h\Psi + \left\{\frac{2c}{1-\lambda}\left(\frac{\lambda+1}{\lambda-1}\right)^{\alpha} - 2B\left(\frac{\lambda-1}{\lambda+1}\right)^{1-\alpha}\right\}\sinh^2\frac{\pi\Psi}{2} + \dots .$$

In point E of the boundary of the drain $x'(\Psi) = 0$. We find $\Psi$ from this condition and substitute it in equation (8.45), to find for the drainage segment b the formula

$$\frac{b}{h} = \frac{(\lambda + 1)^{1-\alpha}(\lambda - 1)^{1+\alpha}}{2\pi^2 c[\lambda + 1 - \alpha(\lambda - 1) + \alpha^2(\lambda - 1)^2] + 4\pi(\lambda - 1)\int_0^1 \frac{\cos^2 \frac{\pi t}{2}\, dt}{F(t)}}$$

S. N. Numerov took the values $\lambda = 2$, $\alpha + \frac{1}{4}$ and carried out the computation of the drainage segment and the length $\ell$. He found

$$\frac{\ell}{h} = 3.76, \quad \frac{b}{h} = 0.046 \quad .$$

Computation of coordinates of points of the free surface gives a graph, represented in Fig. 204.

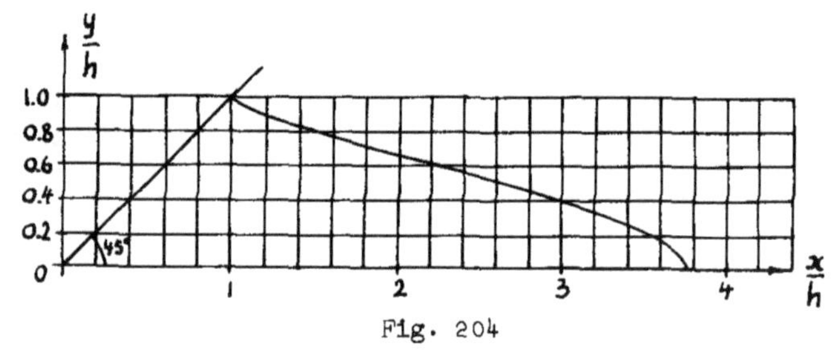


Fig. 204

We notice that S. N. Numerov gave the solution of the problem about the seepage through an earthdam on a pervious layer of infinite depth with inclined drain [4] .

CHAPTER VII

# APPLICATION OF THE ANALYTIC THEORY OF LINEAR DIFFERENTIAL EQUATIONS.

## A. General Theory.

### §1. Introductory Remarks.

As indicated by the author of the present book [1], N. E. Kochin initiated the idea of applying the analytic theory of linear differential equations to problems about groundwater flow. This idea was the basis of a method, developed in the works of the author [1-3], and also of B. K. Risenkampf [8]. The method was applied to the solution of two problems about seepage in earthdams: in a rectangular batardeau [5,8], and dam with rectangular trapezoidal cross-section [4,7,10,11]. In the case of more complicated dams [3], we arrive at differential equations with a large number of parameters, subject to determination, and this complicates the effectiveness of the application of the method. This difficulty is quite similar to that encountered in the application of Christoffel's formula to a polygon with a large number of sides. By means of the analytic theory of differentail equations, some problems about seepage in two soils of different permeability have been solved. (See Chapter VIII.)

### §2. Conditions at the Real Axis for Two Basic Functions.

In Chapter II we dealt with boundary conditions for flows with boundaries consisting only of straight lines and free surfaces. We notice (Chapter III, §2) that all these conditions either consist of linear functions of the coordinates (equations of straight lines) equal to zero, or of linear functions of the coordinates and of the velocity potential and the stream function. In other words, along each boundary the conditions may be represented as two equations:

$$(2.1)\qquad \begin{cases} k_1x + \ell_1y + m_1\varphi + n_1\psi = p\ , \\ k_2x + \ell_2y + m_2\varphi + n_2\psi = q\ , \end{cases}$$

where the coefficients are constant.

Equations (2.1) may be written as follows, using the symbol Im for the imaginary part:

$$(2.2)\qquad \operatorname{Im}(kz + \ell\omega) = p, \qquad \operatorname{Im}(mz + n\omega) = q\ .$$

Here $k,\ell,m,n$ are complex, $p$ and $q$ real numbers.

We may verify that for all boundaries which we treated the determinant

$$\begin{vmatrix} k & \ell \\ m & n \end{vmatrix}$$

is different from zero, except for one particular case of infiltration when $\varepsilon = \kappa$ (Chapter II, §2).

Further we map the region of flow and the region of the complex velocity onto the half plane of the auxiliary complex variable $\zeta$. In this mapping the boundaries of the regions are mapped onto segments of the real axis of the $\zeta$-plane; therefore the values of $\zeta$, corresponding to the contour of the flow region, are real.

We differentiate (2.1) or (2.2) with respect to the real variable $\zeta$ and introduce the notations

$$F = \frac{d\omega}{d\zeta}, \quad Z = \frac{dz}{d\zeta}. \tag{2.3}$$

Instead of conditions (2.2) we obtain homogeneous conditions of the following form:

$$\operatorname{Im}(kZ + \ell F) = 0, \quad \operatorname{Im}(mZ + nF) = 0. \tag{2.4}$$

Using these, one succeeds in building the whole theory of this chapter.

Equations (2.4) show that the quantities

$$kZ + \ell F \quad \text{and} \quad mZ + nF$$

take on real values at the corresponding segment of the real axis. The quotient of these quantities assumes only real values, so that we may write:

$$\operatorname{Im}\left(\frac{kZ + \ell F}{mZ + nF}\right) = 0,$$

or, dividing numerator and denominator by $Z$

$$\operatorname{Im}\left(\frac{k + \ell \frac{F}{Z}}{m + n\frac{F}{Z}}\right) = 0.$$

Observing that

$$\frac{F}{Z} = \frac{d\omega}{d\zeta} : \frac{dz}{d\zeta} = \frac{d\omega}{dz} = w,$$

we obtain the equation

$$\operatorname{Im}\left(\frac{k + \ell w}{m + nw}\right) = 0. \tag{2.5}$$

This equation determines a circle in the w-plane. Indeed the linear fractional mapping

$$w_1 = \frac{k + \ell w}{m + nw}$$

maps a circle in the w-plane onto a circle in the $w_1$-plane and vice-versa.

But this mapping transforms equation (2.5) into equation

$$\operatorname{Im} w_1 = 0,$$

i.e., into the equation of the real axis of the $w_1$-plane, which is a circle in the broad sense of the word and, consequently, maps onto some circle in the w-plane.

It is immediately evident that the circle (2.5) passes through points $w = -k/\ell$ and $w = -m/n$.

## §3. Problem of Determining Two Functions by Conditions at the Real Axis.

We examine the problem of finding two functions F and Z of the complex variable $\zeta$, when these functions satisfy conditions of this form at segments of the real axis:

$$\operatorname{Im}(k_s Z + \ell_s F) = 0, \quad \operatorname{Im}(m_s Z + n_s F) = 0 \qquad (s = 1, 2, \ldots, n) \tag{3.1}$$

We take two neighboring segments $M_1M_2$ and $M_2M_3$ of the $\zeta$ axis and write above them the conditions of Fig. 205.

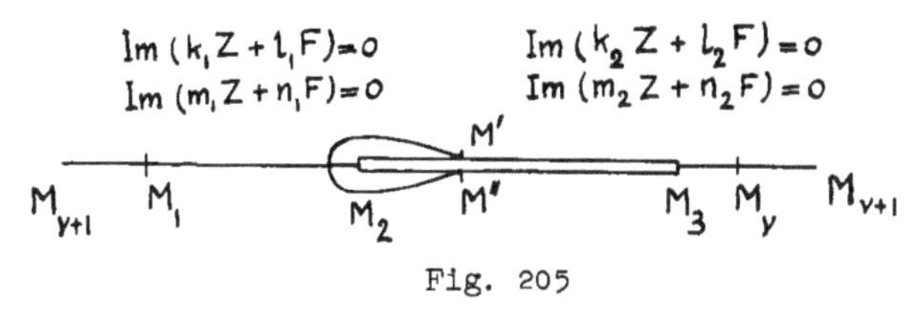


Fig. 205

We show that F and Z, while contouring a singular point, say $M_2$, *undergo a linear transformation.*

We continue the functions in the lower half plane and apply Schwarz' principle of symmetry in the simplest form: when a function $f(z)$ is analytic in a region G, adjacent to the real axis, and continuous up to segment AB, — along segment AB $\operatorname{Im} f(z) = 0$ — then $f(z)$ may be continued in the lower half-plane so that in conjugate points the function assumes conjugate values (Fig. 206). In other words, when we take point $z^*$ in the lower half-plane, for which

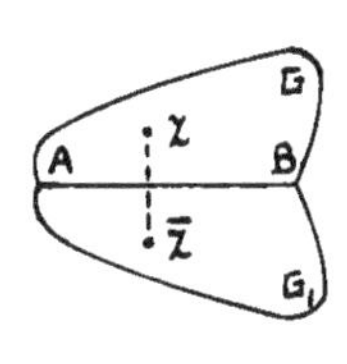


Fig. 206

$$z^* = \bar{z} = x + iy,$$

we obtain

$$f^*(z^*) = \overline{f(z)} \; . \tag{3.2}$$

where $f^*$ is the value of the analytic continuation.

We apply this rule to the linear combinations written above segment $M_1M_2$. We make a cut along segment $M_2M_3$ and call the points on upper and lower sides of the cut M′ and M″ (Fig. 205). The values of functions F and Z, after contouring the singular point $M_2$, when we are in point M″ are called $F^*$ and $Z^*$. Based on equations (3.2) we have:

$$(3.3)\qquad \begin{cases} k_1 Z^* + \ell_1 F^* = \bar{k}_1 \bar{Z} + \bar{\ell}_1 \bar{F} , \\ m_1 Z^* + n_1 F^* = \bar{m}_1 \bar{Z} + \bar{n}_1 \bar{F} . \end{cases}$$

We now return to conditions at segment $M_2M_3$. We notice that the equality $\operatorname{Im} w = 0$ is equivalent with

$$w - \bar{w} = 0$$

and write the conditions on $M_2M_3$ in the form

$$(3.4)\qquad \begin{cases} k_2 Z + \ell_2 F - \bar{k}_2 \bar{Z} - \bar{\ell}_2 \bar{F} = 0 , \\ m_2 Z + n_2 F - \bar{m}_2 \bar{Z} - \bar{n}_2 \bar{F} = 0 . \end{cases}$$

$\bar{F}$ and $\bar{Z}$ may be eliminated from (3.4) and (3.3). We obtain $F^*$ and $Z^*$ depending on $F$, $Z$,

$$(3.5)\qquad \begin{cases} Z^* = \alpha Z + \beta F, \\ F^* = \gamma Z + \delta F. \end{cases}$$

So we see that indeed functions $F$ and $Z$ undergo a linear transformation when a singular point is contoured.

By means of the characteristic equation

$$(3.6)\qquad \begin{vmatrix} \alpha - \lambda & \beta \\ \gamma & \delta - \lambda \end{vmatrix} = 0$$

we may transform (3.5) into the canonical form [21]. This means that there are functions $U, V$, which undergo the transformation

$$(3.7)\qquad U^* = \lambda' U, \quad V^* = \lambda'' V,$$

when contouring particular points, where $\lambda', \lambda''$ are the roots of the characteristic equation, for which $F$ and $Z$ are linear combinations of these functions.

Equation (3.7) holds when $\lambda'$ is not equal to $\lambda''$. In the case of a double root of equation (3.6), the canonical transformation has the form

$$(3.8)\qquad U^* = \lambda' U, \quad V^* = \alpha U + \lambda' V.$$

We introduce the numbers $\alpha'$, $\alpha''$,

$$(3.9)\qquad \alpha' = \frac{\ln \lambda'}{2\pi i}, \quad \alpha'' = \frac{\ln \lambda''}{2\pi i} .$$

Let the value $\zeta = \zeta_0$ correspond to point $M_2$. Functions

$$(\zeta - \zeta_0)^{\alpha'}, \quad (\zeta - \zeta_0)^{\alpha''}$$

in contouring the singular point $\zeta_0$ in the positive sense over an ahgle $2\pi$ become respectively

$$e^{2\pi i \alpha'} (\zeta - \zeta_0)^{\alpha'} = \lambda' (\zeta - \zeta_0)^{\alpha'}$$

$$e^{2\pi i \alpha''} (\zeta - \zeta_0)^{\alpha''} = \lambda'' (\zeta - \zeta_0)^{\alpha''}$$

Therefore the eqotients

$$\frac{U}{(\zeta - \zeta_0)^{\alpha'}} , \quad \frac{V}{(\zeta - \zeta_0)^{\alpha''}}$$

do not undergo any change, i.e., remain single valued functions. Consequently, they may be developed into Laurent series about $\zeta = \zeta_0$. For U and B we therefore obtain:

$$\left\{\begin{aligned} U &= (\zeta - \zeta_0)^{\alpha'} \sum_{k=-\infty}^{\infty} a_k(\zeta - \zeta_0)^k , \\ V &= (\zeta - \zeta_0)^{\alpha''} \sum_{k=-\infty}^{\infty} b_k(\zeta - \zeta_0)^k . \end{aligned}\right. \tag{3.10}$$

When $\lambda' = \lambda''$, the first expression remains, the second generally contains $\ln(\zeta - \zeta_0)$:

$$V = \frac{\alpha U}{2\pi i \alpha'} \ln(\zeta - \zeta_0) + U \sum_{k=-\infty}^{\infty} c_k(\zeta - \zeta_0)^k . \tag{3.11}$$

When the Laurent series in formulas (3.10), (3.11) contains only a finite number of terms with negative powers of $\zeta - \zeta_0$, point $\zeta_0$ is called a "regular singular point." Since the exponents $\alpha'$ and $\alpha''$ are only determined within an integer by means of formulas (3.9), one may change their meaning so that in this case the power series do not contain terms with negative powers of $\zeta - \zeta_0$.

We may represent the canonic system of functions U,V in the neighborhood of a regular singular point $\zeta = \zeta_0$ as

$$\left\{\begin{aligned} U &= (\zeta - \zeta_0)^{\rho_1} \sum_{k=0}^{\infty} a_k(\zeta - \zeta_0)^k , \\ V &= (\zeta - \zeta_0)^{\rho_2} \sum_{k=0}^{\infty} b_k(\zeta - \zeta_0)^k + \alpha U \ln(\zeta - \zeta_0) , \end{aligned}\right. \tag{3.12}$$

in which $a_0$, $b_0$ differ from zero. When $\alpha' - \alpha''$ differs from zero or an integer, then one must consider $\alpha = 0$. To simplify characteristic equation (3.6) one may proceed in this way. We replace functions Z and F by their linear combination, putting

$$\left\{\begin{aligned} Z_1 &= k_2 Z + \ell_2 F , \\ F_1 &= m_2 Z + n_2 F . \end{aligned}\right. \tag{3.13}$$

We have along segment $M_2M_3$:

$$\operatorname{Im} Z_1 = 0, \quad \operatorname{Im} F_1 = 0 . \tag{3.14}$$

Along segment $M_1M_2$ we obtain conditions of the same form as before, but with other coefficients. We put along $M_1M_2$

$$\operatorname{Im} (kZ_1 + \ell F_1) = 0, \quad \operatorname{Im} (mZ_1 + nF_1) = 0 . \tag{3.15}$$

Equations (3.3) and (3.4) are simplified now; we obtain

$$k_1 Z_1^* + \ell_1 F_1^* = \bar{k}Z_1 + \bar{\ell}F_1 \quad ,$$
$$mZ_1^* + nF_1^* = \bar{m}Z_1 + \bar{n}F_1 \quad ,$$
$$Z_1 = \bar{Z}_1, \quad F_1 = \bar{F}_1$$

from which

$$(3.16) \qquad \begin{cases} Z_1^* = \dfrac{\bar{k}n - \bar{m}\ell}{kn - m\ell} Z_1 + \dfrac{\bar{\ell}n - \ell\bar{n}}{kn - m\ell} F_1 , \\[2ex] F_1^* = \dfrac{k\bar{m} - \bar{k}m}{kn - m\ell} Z_1 + \dfrac{k\bar{n} - m\bar{\ell}}{kn - m\ell} F_1 . \end{cases}$$

The characteristic equation has the form

$$\begin{vmatrix} \dfrac{\bar{k}n - \bar{m}\ell}{kn - m\ell} - \lambda & \dfrac{\bar{\ell}n - \ell\bar{n}}{kn - m\ell} \\[2ex] \dfrac{k\bar{m} - \bar{k}m}{kn - m\ell} & \dfrac{k\bar{n} - m\bar{\ell}}{kn - m\ell} - \lambda \end{vmatrix} = 0$$

or

$$(3.17) \qquad (kn - \ell m)\lambda^2 - (k\bar{n} - \bar{m}\ell + \bar{k}n - m\bar{\ell})\lambda + \bar{k}\bar{n} - \bar{\ell}\bar{m} = 0 .$$

The roots of this equation $\lambda'$ and $\lambda''$ must coincide with the roots of equation (3.6), since $Z_1$ and $F_1$ as linear combinations of $Z$ and $F$ must belong to the same exponents. For $\lambda'$ and $\lambda''$ determined, one finds exponents $\alpha'$ and $\alpha''$ by means of (3.9).

Considering the behaviour of the functions $\omega(\zeta)$, $z(\zeta)$ about singular points of the flowregion (see §8 below), we may convince ourselves of the fact that in all encountered cases of groundwater flow the singular points are regular singular points of the function $F = d\omega/d\zeta$, $Z = dz/d\zeta$.

From the analytic theory of differential equations [21], the following is known.

Two linearly independent functions $U$ and $V$, having a regular singular point $\zeta_0$, i.e., represented by equations of form (3.12) near this point, constitute a system of solutions of the linear differential equation

$$(3.18) \qquad u'' + p(\zeta)u' + q(\zeta)u = 0,$$

where $p(\zeta)$ and $q(\zeta)$ are

$$(3.19) \qquad p = \frac{p_1(\zeta)}{\zeta - \zeta_0} , \quad q = \frac{q_1(\zeta)}{(\zeta - \zeta_0)^2} .$$

Here $p_1(\zeta)$ and $q_1(\zeta)$ are regular functions in point $\zeta_0$, i.e., may be developed into a series of positive integer powers of $\zeta - \zeta_0$.

As will be shown later, functions $Z$ and $F$ have only regular singular points. Consequently, they may be represented as linear combinations of two linearly independent solutions of equation (3.18) with coefficients of form (3.19). We notice that exponents $\alpha'$ and $\alpha''$ are only single-valuedly determined when the geometrical shape of hodograph and flow regions are completely known [8,9].

## §4. Equation with Three Regular Singular Points.

Linear equation (3.18), when all its singular points are regular, is called an equation of the "Fuchs" type. We first examine the case where we have three regular singular points. Then the linear differential equation of the problem may be completely constructed.

We may assume, according to further statements, that the singular points in the plane are $0, 1$ and $\infty$.

We assume that one of the exponents for each of the finite singular points vanishes. In other words, let the exponents about the singular points be:

$$0, \alpha \text{ about } \zeta = 0,$$
$$0, \beta \text{ about } \zeta = 1,$$
$$\gamma, \gamma' \text{ about } \zeta = \infty.$$

The branch of the analytic function $Y$, corresponding to these exponents, is indicated by Riemann as

$$Y = P\left\{\begin{matrix} 0 & 1 & \infty & \\ 0 & 0 & \gamma & \zeta \\ \alpha & \beta & \gamma' & \end{matrix}\right\}. \tag{4.1}$$

The differential equation for $Y$ is written as [21-24]:

$$Y'' + \left(\frac{1-\alpha}{\zeta} \quad \frac{1-\beta}{\zeta-1}\right)Y' + \frac{\gamma\gamma'}{\zeta(\zeta-1)}Y = 0. \tag{4.2}$$

In this case the sum of all the exponents must be one (Fuch's correlation)

$$\alpha + \beta + \gamma + \gamma' = 1. \tag{4.3}$$

We call $U, V$ a fundamental canonical system of solutions of equation (4.2) about point $\zeta = 0$. It may be written as

$$\left\{\begin{aligned} U &= 1 + a_1\zeta + a_2\zeta^2 + \ldots, \\ V &= \zeta^\alpha(1 + a_1'\zeta + a_2'\zeta^2 + \ldots). \end{aligned}\right. \tag{4.4}$$

We take the system of linearly independent integrals about $\zeta = 1$ as

$$\left\{\begin{aligned} U_1 &= 1 + b_1(\zeta - 1) + \ldots, \\ V_1 &= (1-\zeta)^\beta[1 + b_1'(\zeta - 1) + \ldots]. \end{aligned}\right. \tag{4.5}$$

We take the fundamental system of solutions about $\zeta = \infty$ as

$$\left\{\begin{aligned} U &= \left(\frac{1}{\zeta}\right)^\gamma\left[1 + \frac{c_1}{\zeta} + \frac{c_2}{\zeta} + \ldots\right], \\ V &= \left(\frac{1}{\zeta}\right)^{\gamma'}\left[1 + \frac{c_1'}{\zeta} + \frac{c_2'}{\zeta^2} + \ldots\right]. \end{aligned}\right. \tag{4.6}$$

Coefficients $a_1, a_2, \ldots, b_1, b_2, \ldots, c_1, c_2, \ldots$ of series (4.4)-(4.6) are real numbers. Therefore series (4.4) converge inside circle $C_0$ (Fig. 207) with center in point $\zeta = 0$ and going through the nearest singular point, i.e., $\zeta = 1$; series (4.5) converge inside circle $C_1$, for which $|\zeta - 1| < 1$;

series (4.6) converges for $|\zeta| > 1$, i.e., outside circle $C_0$.

Equation (4.2) is linked to the problem of the conformal mapping of a curvilinear triangle onto a half-plane. To show this, we compose the function

$$w = \mu \frac{V}{U} = \mu \frac{\zeta^\alpha(1 + a_1\zeta + \dots)}{1 + b_1\zeta + \dots}, \tag{4.7}$$

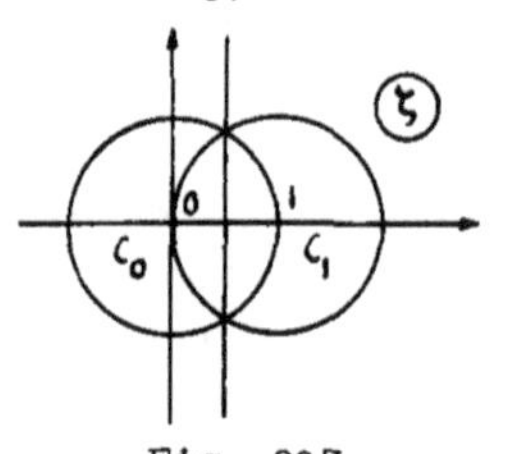


Fig. 207

where $\mu$ is a given constant. To determine it, we consider $\alpha$, $\beta$ positive.

For $\zeta = \xi$, for which $0 < \xi < 1$, function $w$ assumes real values; for these values $w$ covers some segment of the abscissa axis. For $\zeta = \xi < 0$, we have (assuming $\arg \zeta = 0$ for $\zeta = \xi > 0$):

$$\zeta^\alpha = e^{\pi i \alpha}(-\xi)^\alpha .$$

Consequently, $w$ covers the points of the straight segment AC (Fig. 208).

The linear fractional mapping

$$w_1 = \frac{A + Bw}{C + Dw} = \frac{AU + BV}{CU + DV} \tag{4.8}$$

maps straight lines into circles. To study circles about the singular point $\zeta = 1$, we pass from the system of function U, V to the system $U_1$, $V_1$ by means of the auxiliary transformation

$$U = pU_1 + qV_1, \quad V = rU_1 + sV_1. \tag{4.9}$$

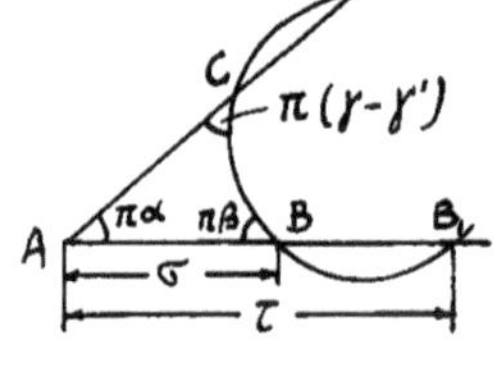


Fig. 208

(There must be a linear dependence between the systems of functions U, V and $U_1$, $V_1$, since they are fundamental systems of solutions of the linear equation.)

By means of (4.9) we may rewrite (4.7) as:

$$w = \frac{\mu r U_1 + \mu s V_1}{pU_1 + qV_1} .$$

Reasoning similar to that above shows that also in points $\zeta = 1$ and $\zeta = \infty$ we obtain intersecting arcs of circles. We have the theorem [21]: functions of kind (4.7) or (4.8) map conformally the upper half-plane onto a curvilinear triangle, without branch points inside.

## §5. The Hypergeometric Functions and their Properties.

Hypergeometric is the name of the function

$$F(a,b,c,x) = 1 + \frac{ab}{c} x + \frac{a(a+1)b(b+1)}{1 \cdot 2 \cdot c(c+1)} x^2 + \dots \tag{5.1}$$

It is a particular solution of this equation

$$x(1 - x)Y'' + [c - (a + b + 1)x]Y' - abY = 0 . \tag{5.2}$$

Equation (5.2) is derived from (4.2) when one puts

$$\gamma = a, \quad \gamma' = b, \quad \alpha = 1 - c, \quad \beta = c - a - b$$

[this satisfies correlation (4.3)].

Every solution of this equation is expressed by means of the Riemann symbol as

$$Y = P \left\{ \begin{matrix} 0 & \infty & 1 & \\ 0 & a & 0 & x \\ 1 - c & b & c - a - b & \end{matrix} \right\}. \tag{5.3}$$

We notice the property of the Riemann symbol for an arbitrary number of singular points. Let

$$Y = P \left\{ \begin{matrix} a_1 & a_2 & \dots & a_n & \infty & \\ \alpha_1 & \alpha_2 & \dots & \alpha_n & \alpha & x \\ \beta_1 & \beta_2 & \dots & \beta_n & \beta & \end{matrix} \right\}. \tag{5.4}$$

If, in the expression for $Y$ we single out the factor

$$(x - a_1)^p ,$$

then the exponents about $x = a_1$ in the second factor are decreased by $p$, and consequently become

$$\alpha_1 - p, \quad \beta_1 - p,$$

but the exponents about the point at infinity are increased by $p$, so that instead of $\alpha$ and $\beta$ we have

$$\alpha + p, \quad \beta + p .$$

In other words:

$$P \left\{ \begin{matrix} a_1 & a_2 & \dots & a_n & \infty & \\ \alpha_1 & \alpha_2 & \dots & \alpha_n & \alpha & x \\ \beta_1 & \beta_2 & \dots & \beta_n & \beta & \end{matrix} \right\} =$$

$$= (x - a_1)^n P \left\{ \begin{matrix} a_1 & a_2 & \dots & a_n & \infty & \\ \alpha_1 - p & \alpha_2 & \dots & \alpha_n & \alpha + p & x \\ \beta_1 - p & \beta_2 & \dots & \beta_n & \beta + p & \end{matrix} \right\}. \tag{5.5}$$

In particular, by means of this transformation we may obtain a Riemann function, having one vanishing exponent about each finite singular point. Namely

$$P \left\{ \begin{matrix} a_1 & a_2 & \dots & a_n & \infty & \\ \alpha_1 & \alpha_2 & \dots & \alpha_n & \alpha & x \\ \beta_1 & \beta_2 & \dots & \beta_n & \beta & \end{matrix} \right\} = (x - a_1)^{\alpha_1} (x - a_2)^{\alpha_2} \dots (x - a_n)^{\alpha_n'} \times$$

$$\times P \left\{ \begin{matrix} a_1 & a_2 & \dots & a_n & \infty & \\ 0 & 0 & \dots & 0 & \alpha + \alpha_1 + \dots + \alpha_n & x \\ \beta_1 - \alpha_1 & \beta_2 - \alpha_2 & \dots & \beta_n - \alpha_n & \beta + \alpha_1 + \dots + \alpha_n & \end{matrix} \right\}. \tag{5.6}$$

Using Riemann's notation, we may find other solutions of the hypergeometric equation. Thus to obtain besides $F(a,b,c,x)$, another solution of equation (5.2) about $x = 0$, we single out the factor $x^{1-c}$ in (5.3). We have:

$$Y = x^{1-c}P\begin{Bmatrix} 0 & \infty & 1 & \\ c-1 & a+1-c & 0 & x \\ 0 & b+1-c & a+b+1-c & \end{Bmatrix} =$$

$$= x^{1-c}P\begin{Bmatrix} 0 & \infty & 1 & \\ 0 & a+1-c & 0 & x \\ c-1 & b+1-c & a+b+1-c & \end{Bmatrix} . \tag{5.7}$$

The expression with factor $x^{1-c}$ is an hypergeometric function in which the first arguments have the exponents at infinity

$$a + 1 - c, \quad b + 1 - c,$$

while the third argument corresponds to the addition (up to one) of a non vanishing exponent for $x = 0$. Thus we obtain

$$Y = x^{1-c}F(a + 1-c,\ b + 1-c,\ 2 - c,\ x) .$$

This function is linearly independent of $F(a,b,c,x)$ and therefore we have as fundamental system of solutions about $x = 0$ for equation (5.2):

$$\left\{\begin{aligned} U &= F(a,b,c,x) , \\ V &= x^{1-c}F(a + 1 - c,\ b + 1 - c,\ 2 - c,\ x) . \end{aligned}\right. \tag{5.8}$$

Similarly we may find systems of solutions about singular points $x = 1$ and $x = \infty$:

$$\left\{\begin{aligned} U_1 &= F(a,b,\ a + b - c + 1,\ 1 - x) , \\ V_1 &= (1 - x)^{c-a-b}F(c - a,\ c - b,\ c - a - b + 1,\ 1 - x) , \end{aligned}\right. \tag{5.9}$$

$$\left\{\begin{aligned} U_\infty &= x^{-a}F(a,\ a-c + 1,\ a - b + 1,\ \tfrac{1}{x}) , \\ V_\infty &= x^{-b}F(b,\ b-c + 1,\ b - a + 1,\ \tfrac{1}{x}) . \end{aligned}\right. \tag{5.10}$$

Series (5.8) converge for $|x| < 1$, series (5.9) for $|x-1| < 1$, series (5.10) for $|x| > 1$.

Replacing $x$ respectively by

$$x' = x,\quad 1 - x,\quad \frac{1}{x},\quad \frac{x}{x - 1},\quad \frac{x - 1}{x}$$

changes the coefficients of the equation, but retains singular points $0, 1$ and $\infty$.

From this we may find new forms of the integrals of equation (5.2). Together with those already obtained, twenty four forms are given by Kummer [22].

Among the various branches of the hypergeometric function, there exists a linear relation, expressed by means of an auxiliary transformation of form (4.9). Thus we have:

$$
(5.11)\quad
\begin{aligned}
F(a,b,c,z) &= pF(a,b,\ a+b+1-c,\ 1-z) + \\
&\quad + q(1-z)^{c-a-b}F(c-a,\ c-b,\ c+1-a-b,\ 1-z), \\
z^{1-c}F(a+1-c,\ b+1-c,\ 2-c,\ z) &= \\
&= rF(a,b,\ a+b+1-c,\ 1-z) + \\
&\quad + s(1-z)^{c-a-b}F(c-a,\ c-b,\ c+1-a-b,\ 1-z)
\end{aligned}
$$

where

$$
(5.12)\quad
\left\{
\begin{aligned}
p &= \frac{\Gamma(c)\Gamma(c-a-b)}{\Gamma(c-a)\Gamma(c-b)}, \qquad q = \frac{\Gamma(c)\Gamma(a+b-c)}{\Gamma(a)\Gamma(b)}, \\
r &= \frac{\Gamma(2-c)\Gamma(c-a-b)}{\Gamma(1-a)\Gamma(1-b)}, \qquad s = \frac{\Gamma(2-c)\Gamma(a+b-c)}{\Gamma(a+1-c)\Gamma(b+1-c)} .
\end{aligned}
\right.
$$

when $c - a - b > 0$, the hypergeometric series converge for $z = 1$. Therefore, from (5.11) we obtain for this case

$$
(5.13)\qquad F(a,b,c,1) = \frac{\Gamma(c)\Gamma(c-a-b)}{\Gamma(c-a)\Gamma(c-b)} .
$$

We still notice the representation of the hypergeometric function as an Eulerian integral [24].

$F(a,b,c;z)$ may be written as a definite integral

$$
(5.14)\qquad F(a,b,c;z) = \frac{\Gamma(c)}{\Gamma(b)\Gamma(c-b)} \int_0^1 t^{b-1}(1-t)^{c-b-1}(1-zt)^{-a}\,dt .
$$

This equality may be checked by decomposing

$$(1-zt)^{-a}$$

into a power series of $t$ and using the properties of the gamma- and beta-functions in the integration of (5.14).

Since $F(a,b,c;z)$ is symmetrical with respect to $a$ and $b$ (i.e., does not change by permutation of these arguments), we have another form of the function:

$$
(5.15)\qquad F(a,b,c;z) = \frac{\Gamma(c)}{\Gamma(a)\Gamma(c-a)} \int_0^1 t^{a-1}(1-t)^{c-a-1}(1-zt)^{-b}\,dt .
$$

These representations are convenient when $b$ and $c - b$ in (5.14) and $a$, $c - a$ in formula (5.15) have positive real parts and when $z$ does not lie within the interval $(1, \infty)$ of the real axis.

Each of the integrals (5.14), (5.15) corresponds to a system of twenty four integrals, obtained from the basic integrals by means of linear transformations of the integration variable and the parameter $z$, for which the singular points $t = 0, \infty, 1, 1/z$, are rearranged among themselves. So, equations

$$
\begin{aligned}
zt\tau - t - \tau + 1 &= 0 , \\
zt\tau - zt - z\tau + 1 &= 0 , \\
zt\tau - 1 &= 0
\end{aligned}
$$

transform integral (5.14), if we take $\tau$ for new variable of integration, respectively to these forms:

$$(1 - z)^{c-a-b} \int_0^1 \tau^{c-b-1}(1 - \tau)^{b-1}(1 - z\tau)^{a-c}\, d\tau \quad ,$$

$$z^{1-c}(1 - z)^{c-a-b} \int_{1/z}^{\infty} \tau^{-a}(\tau - 1)^{a-c}(z\tau - 1)^{b-1}\, d\tau \quad ,$$

$$z^{1-c} \int_{1/z}^{\infty} \tau^{a-c}(\tau - 1)^{-a}(z\tau - 1)^{c-b-1}\, d\tau \quad .$$

Replacing $z$ by one of the expressions

$$\frac{1}{z},\quad 1 - z,\quad \frac{1}{1 - z},\quad \frac{z - 1}{z},\quad \frac{z}{z - 1}\ ,$$

we obtain all 24 forms of Eulerian integrals of the equation. Thanks to the development of a mathematical tool to solve equations of the Fuchs type with three singular points one may analyze seepage problems where the number of singular points is three. The problem becomes more complicated when the number of singular points exceeds three. In seepage problems the number of singular points is usually 5,7 etc. In order to consider the difficulties that arise in this case, we analyze a curvilinear polygon with number of vertices exceeding three.

## *§6. Curvilinear Polygons, with Number of Vertices Exceeding Three.

We suppose that we have a curvilinear polygon, i.e., a polygon bounded by arcs of a circle

$$\operatorname{Im}\left(\frac{k_s + \ell_s w}{m_s + n_s w}\right) = 0 \quad (s = 1,2,\ldots,\nu,\ \nu + 1)\ . \tag{6.1}$$

The angles in the vertices of the polygon are indicated as

$$\pi\alpha_1, \pi\alpha_2, \ldots, \pi\alpha_{\nu+1}\ .$$

We map the polygon conformally onto the upper half-plane of an auxiliary complex variable $\zeta$. Let the vertices of the polygon map into points $a_s(S = 1,2,\ldots,\ \nu + 1)$ of the real axis and let $a_{\nu+1} = \infty$.

If moreover, we choose two points $a_s$, of arbitrary subscript, then the location of the remaining points is completely determined (but not known before hand) for a given polygon.

Function $w$ may be represented as the ratio of two linearly independent particular solutions of the differential equation

$$Y'' + \left(\frac{1 - \alpha_1}{\zeta - a_1} + \frac{1 - \alpha_2}{\zeta - a_2} + \ldots + \frac{1 - \alpha_\nu}{\zeta - a_\nu}\right) Y' +$$

$$+ \frac{\alpha'_{\nu+1}\ \alpha''_{\nu+1}\ (\zeta - \lambda_1)\ \ldots(\zeta - \lambda_{\nu-2})}{(\zeta - a_1)(\zeta - a_2)\ \ldots\ (\zeta - a_\nu)}\ Y = 0\ . \tag{6.2}$$

Here $\alpha_1, \alpha_2, \ldots, \alpha_\nu$ be the angles in the vertices of the polygon divided by $\pi$), corresponding by conformal mapping to finite points on the real axis; $\alpha'_{\nu+1}$ and $\alpha''_{\nu+1}$ are such that

$$\alpha'_{\nu+1} - \alpha''_{\nu+1} = \alpha_{\nu+1} \tag{6.3}$$

($\pi\alpha_{\nu+1}$ is the angle for the vertex that is mapped into $\zeta = \infty$).

Therefore, between the numbers $\alpha_s (S = 1, 2, \ldots, \nu), \alpha'_{\nu+1}$ and $\alpha''_{\nu+1}$, the Fuchs correlation must be satisfied

$$\alpha_1 + \alpha_2 + \ldots + \alpha_\nu + \alpha'_{\nu+1} + \alpha''_{\nu+1} = \nu - 1. \tag{6.4}$$

Here $\nu$ is the number of singular points located at a finite distance from the origin.

Finally, the quantities $\lambda_1 \lambda_2, \ldots, \lambda_{\nu-2}$ in equation (6.2) are the so-called "additional parameters," not known before hand. In the case of a curvilinear triangle, these parameters, as we say, do not exist and this facilitates the solution of the problem.

We call the linearly independent integrals of equation (6.2) about points $a_s, U_s$ and $V_s$. Then $w$ is a linear fractional function of these

$$w = \frac{AU_s + BV_s}{CU_s + DV_s}, \tag{6.5}$$

where $A, B, C, D$ are constants.

Some properties of the conformal mapping of quadrangles and pentagons are given in works [3,6].

## §7. Case of Real Exponents. Second Derivation of the Characteristic Equation.

We return to the question of determining the exponents about singular points (see §3). The case where $\alpha'$ and $\alpha''$ are real is of special interest. It is easy to see that the necessary and sufficient condition for $\alpha'$ and $\alpha''$ to be real is the condition of intersection of circles:

$$\operatorname{Im}\left(\frac{k + \ell w_1}{m + n w_1}\right) = 0, \qquad \operatorname{Im} w_1 = 0, \tag{7.1}$$

obtained from equations (3.14) and (3.15) where

$$w_1 = \frac{F_1}{Z_1} .$$

Indeed, when circles (7.1) intersect, $\alpha' - \alpha''$ is a real number since $\pi(\alpha' - \alpha'')$ is the angle between these circles. The ratio $\lambda'/\lambda''$ must then have a unit modulus. But equation (3.17) has the form

$$a\lambda^2 - b\lambda + \bar{a} = 0,$$

where $b$ is a real number, $a$ and $\bar{a}$ complex conjugate numbers. Therefore

$$\frac{\lambda'}{\lambda''} = \frac{b + \sqrt{b^2 - 4a\bar{a}}}{b - \sqrt{b^2 - 4a\bar{a}}} .$$

In order to have $|\lambda'/\lambda''| = 1$, it is sufficient and necessary that

$$b^2 - 4a\bar{a} < 0$$

or

$$(\bar{k}n + k\bar{n} - \bar{m}\ell - m\bar{\ell})^2 - 4(kn - \ell m)(\bar{k}\bar{n} - \bar{\ell}\bar{m}) \leq 0 , \tag{7.2}$$

and this is the condition for which circles (7.1) intersect (or be tangent). Indeed, in the latter case the coordinates of the points of intersection are the roots of equation

$$w^2 + \frac{\ell\bar{m} - \bar{k}n + k\bar{n} - \bar{\ell}m}{\ell\bar{n} - \bar{\ell}n} w + \frac{k\bar{m} - \bar{k}m}{\ell\bar{n} - \bar{\ell}n} = 0,$$

and these roots are real for

$$(\bar{k}n - k\bar{n} + \bar{\ell}m - \ell\bar{m})^2 - 4(k\bar{m} - \bar{k}m)(\bar{\ell}n - \ell\bar{n}) \leq 0 ,$$

what corresponds with equation (7.2).

Since the product $\lambda'\lambda'' = \bar{a}/a$ always has a unit modulus, it is easy to see that $\alpha' - \alpha''$ real implies $\alpha'$ and $\alpha''$ real.

Assuming now that the condition for $\alpha'$ and $\alpha''$ to be real about $\xi = 0$ is satisfied, we return to the condition on the segments of form (3.1). We assume that

$$Z = AU + BV, \quad F = CU + DV , \tag{7.3}$$

where

$$U = \zeta^{\alpha'} \sum_{n=0}^{\infty} a_n \zeta^n, \quad V = \zeta^{\alpha''} \sum_{n=0}^{\infty} b_n \zeta^n , \tag{7.4}$$

in which we assume for simplicity that the terms with $\ln \zeta$ are absent.

The first condition along the segment $M_2M_3$ may then be represented according to (7.3), as

$$\operatorname{Im}(k_2Z + \ell_2F) = \operatorname{Im}[(k_2A + \ell_2C)U + (k_2B + \ell_2D)V] = 0 . \tag{7.5}$$

If the exponents about all singular points are real, what we assume, then the coefficients in series (7.4) are real. Therefore (7.5) may be represented as:

$$U \operatorname{Im}(k_2A + \ell_2C) + V \operatorname{Im}(k_2B + \ell_2D) = 0 .$$

Since functions $U$ and $V$ are linearly independent, their respective multipliers must be zero:

$$\operatorname{Im}(k_2A + \ell_2C) = 0, \quad \operatorname{Im}(k_2B + \ell_2D) = 0 . \tag{7.6}$$

In the same way we obtain two more equations:

$$\operatorname{Im}(m_2A + n_2C) = 0, \quad \operatorname{Im}(m_2B + n_2D) = 0. \tag{7.7}$$

We now go from segment $M_2M_3$ to segment $M_1M_2$ (Fig. 205), turning around point $\zeta = 0$ along an half circle in the upper half-plane. If we assume $\arg \zeta = 0$ on segment $M_2M_3$, then $\arg \zeta = \pi$ on segment $M_1M_2$ and consequently, along segment $M_1M_2$ for $\zeta = \xi < 0$,

$$\zeta^{\alpha'} = |\zeta|^{\alpha'} e^{\pi i \alpha'}, \quad \zeta^{\alpha''} = |\zeta|^{\alpha''} e^{\pi i \alpha''}.$$

For $\zeta = \xi < 0$

$$U = e^{\pi i\alpha'}U', \quad V = e^{\pi i\alpha''}V',$$

where $U'$ and $V'$ are real for $\xi < 0$. Along $M_1M_2$ we obtain:

$$\operatorname{Im}(k_1Z + \ell_1F) = \operatorname{Im}[(k_1A + \ell_1C)e^{\pi i\alpha'}U' + (k_1B + \ell_1D)e^{\pi i\alpha''}V'] = 0 .$$

From this:

$$(7.8) \qquad \operatorname{Im}[(k_1A + \ell_1C)e^{\pi i\alpha'}] = 0, \quad \operatorname{Im}[(k_1B + \ell_1D)e^{\pi i\alpha''}] = 0 .$$

In a similar way we find:

$$(7.9) \qquad \operatorname{Im}[(m_1A + n_1C)e^{\pi i\alpha'}] = 0, \quad \operatorname{Im}[(m_1B + n_1D)e^{\pi i\alpha''}] = 0 .$$

The eight equations (7.6)-(7.9) are divided into two groups, one to which $A$ and $C$ belong, the other related to $B$ and $C$. We rewrite these equations, introducing complex conjugate numbers ($\bar{A}$ — conjugate of $A$ and so on):

$$(7.10) \qquad \begin{cases} k_2A + \ell_2C - \bar{k}_2\bar{A} - \bar{\ell}_2\bar{C} = 0 , \\ m_2A + n_2C - \bar{m}_2\bar{A} - \bar{n}_2\bar{C} = 0 , \\ (k_1A + \ell_1C)e^{2\pi i\alpha'} - \bar{k}_1\bar{A} - \bar{\ell}_1\bar{C} = 0 , \\ (m_1A + n_1C)e^{2\pi i\alpha'} - \bar{m}_1\bar{A} - \bar{n}_1\bar{C} = 0 , \end{cases}$$

$$(7.11) \qquad \begin{cases} k_2B + \ell_2D - \bar{k}_2\bar{B} - \bar{\ell}_2\bar{D} = 0 , \\ m_2B + n_2D - \bar{m}_2\bar{B} - \bar{n}_2\bar{D} = 0 , \\ (k_1B + \ell_1D)e^{2\pi i\alpha''} - \bar{k}_1\bar{B} - \bar{\ell}_1\bar{D} = 0 , \\ (m_1B + n_1D)e^{21i\alpha''} - \bar{m}_1\bar{B} - \bar{n}_1\bar{D} = 0 . \end{cases}$$

In order to have a solution different from zero for system (7.10) its determinant must be zero.

$$\begin{vmatrix} k_1e^{2\pi i\alpha'} & \ell_1e^{2\pi i\alpha'} & \bar{k}_1 & \bar{\ell}_1 \\ m_1e^{2\pi i\alpha'} & n_1e^{2\pi i\alpha'} & \bar{m}_1 & \bar{n}_1 \\ k_2 & \ell_2 & \bar{k}_2 & \bar{\ell}_2 \\ m_2 & n_2 & \bar{m}_2 & \bar{n}_2 \end{vmatrix} = 0 .$$

If we substitute $\alpha''$ for $\alpha'$ in this determinant, we obtain the condition that system (7.11) has a non zero solution for $B$ and $D$. This shows that $e^{2\pi i\alpha'}$ and $e^{2\pi i\alpha''}$ are roots of the quadratic equation in $\lambda$:

$$(7.12) \qquad \begin{vmatrix} k_1\lambda & \ell_1\lambda & \bar{k}_1 & \bar{\ell}_1 \\ m_1\lambda & n_1\lambda & \bar{m}_1 & \bar{n}_1 \\ k_2 & \ell_2 & \bar{k}_2 & \bar{\ell}_2 \\ m_2 & n_2 & \bar{m}_2 & \bar{n}_2 \end{vmatrix} = 0 .$$

Thus we obtained the characteristic equation for $\lambda = e^{2\pi i\alpha}$. Although in its derivation we used the condition that $\alpha'$ and $\alpha''$ be real, equation (7.12) holds also for complex $\alpha', \alpha''$. We may convince ourselves of this by transforming equation (3.6) into form (7.12).

## §8. Exponents for Basic Cases of the Seepage Theory.

We communicate results concerning the method of determining the exponents about singular points for different combinations of boundary conditions, which are encountered in the study of groundwater flow.

We examine [8,6] some combinations of boundaries, which show up in problems about earthdams. Boundaries of seepage regions may be:

a) rectilinear impervious walls,

b) rectilinear boundaries of reservoirs,

c) rectilinear seepage surfaces,

d) free surfaces, along which evaporation may occur (or to which water may flow from outside)

All together we have six possible combinations of boundaries (ab), (ac), (ad), (bc), (bd), (cd) shown below.

1. Case ab. Let the flow surface of the dam make an angle $\pi\alpha$ with a linear impervious basis (Fig. 209).

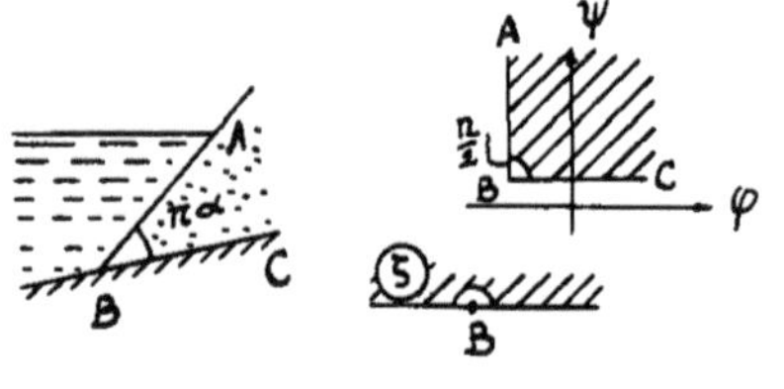


Fig. 209

The conformal mapping of region $z$ onto the half-plane $\zeta$, in the neighborhood of B, if we assume that point B maps onto the origin of coordinates of the $\zeta$ plane, has the form

$$z = \zeta^{\alpha}(a_0 + a_1\zeta + \dots) .$$

For function Z we obtain

$$Z = \frac{dz}{d\zeta} = \zeta^{\alpha-1}(b_0 + b_1\zeta + \dots) .$$

Consequently Z belongs to the exponent $\alpha - 1$.

Lines $\varphi$ = constant and $\psi$ = constant concur in point B of the $\omega = \varphi + i\psi$ plane; therefore we have:

$$\omega = \zeta^{1/2}(c_0 + c_1\zeta + \dots) ,$$

$$F = \frac{d\omega}{d\zeta} = \zeta^{-1/2}(\ell_0 + \ell_1\zeta + \dots) ,$$

i.e., function F corresponds to exponent $-1/2$.

So we have the exponents $(\alpha - 1, -1/2)$ for functions Z and F.

When angle $\pi\alpha$ is 90°, the exponents are $(-1/2, -1/2)$. In this case point B on the velocity hodograph is an ordinary point. The substitution

$$Z = \frac{Z_1}{\sqrt{\zeta}} , \quad F = \frac{F_1}{\sqrt{\zeta}}$$

leads to functions $Z_1$ and $F_1$, for which point B is also an ordinary point, i.e., the development of these functions have the form

$$Z_1 = a_0 + a_1\zeta + \dots, \quad F_1 = b_0 + b_1\zeta + \dots$$

Points having a singularity that can be removed by a substitution of the form

$$U = \zeta^m U_1 ,$$

may be called "removable singular points [8,9]. In our case we may construct a linear combination of functions $Z_1$" and $F_1$, corresponding to exponent +1, and consider that functions $Z_1$ and $F_1$ correspond to exponents 0 and 1, but functions Z and F to exponents -1/2 and 1/2.

2. Case ac. The impervious wall joins a seepage surface (Fig. 210). As in the previous case, we have for Z

$$Z = \zeta^{\alpha-1}(a_0 + a_1\zeta + \dots) .$$

We consider the plane of the complex velocity. In this plane we have straight lines intersecting in point B under the angle $\pi(\frac{1}{2} - \alpha)$.

Consequently we have for u - iv:

$$u - iv = \frac{F}{Z} = \zeta^{1/2 - \alpha}(c_0 + c_1\zeta + \dots) .$$

From this

$$F = Z\zeta^{1/2 - \alpha}(c_0 + \dots) = \zeta^{-1/2}(a_0c_0 + \dots) .$$

Thus functions Z and F must correspond to exponents $\alpha - 1$, $-1/2$. Here contrary to the previous case for $\alpha = 1/2$, the series for Z and F contains a logarithmic term. Indeed, in this case the straight lines in the u - iv plane become parallel, and therefore

$$u - iv = \frac{F}{Z} = m \ln \zeta + \dots$$

Consequently

$$F = Z \ln \zeta + \dots$$

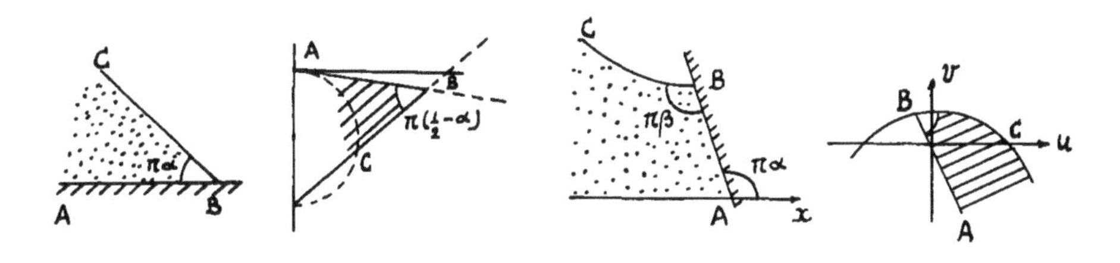


Fig. 210 Fig. 211

3. Case ad. The impervious wall joins with the free surface. (Fig. 211). Evaporation with intensity C takes place from CB, or there is infiltration at the surface CB, so that $\psi + cx$ = constant. Moreover, along CB the pressure is constant, i.e., $\varphi + ky$ = constant.

Let the equation of wall AB be

$$y \cos \alpha\pi - x \sin \alpha\pi = 0 .$$

Along the wall $\psi$ = constant.

The conditions along segments AB and BC of the $\zeta$-plane are represented in Fig. 212.

| A | B | C |
|---|---|---|
| $\mathrm{Im}(F) = 0$ | | $\mathrm{Im}(\ell F + kZ) = 0$ |
| $\mathrm{Im}(Ze^{-\pi i\alpha}) = 0$ | | $\mathrm{Im}(F + ciZ) = 0$ |

Fig. 212

The corresponding characteristic equation is

$$\begin{vmatrix} \lambda e^{-\pi i\alpha} & 0 & e^{\pi i\alpha} & 0 \\ 0 & \lambda & 0 & 1 \\ k & i & k & -i \\ ci & 1 & -ci & 1 \end{vmatrix} = 0$$

This equation is brought in the form

$$\lambda^2 - 2\lambda \frac{k - c}{k + c} e^{\pi i\alpha} \cos \pi\alpha + e^{2\pi i\alpha} = 0 .$$

Each time $((k - c)/(k + c))^2 \cos^2\pi\alpha < 1$ (in particular, this inequality exists for $c > 0$, i.e., in the case of evaporation), we may put

$$\frac{k - c}{k + c} \cos \pi\alpha = \cos \pi\delta$$

in the expression for $\lambda$

$$\lambda = e^{\pi i\alpha} \left[ \frac{k - c}{k + c} \cos \pi\alpha \pm i \sqrt{1 - \left(\frac{k - c}{k + c}\right)^2 \cos^2 \pi\alpha} \right]$$

we have

$$\lambda + e^{\pi i(\alpha \pm \delta)} .$$

The unknown exponents about singular point B, determined by formula $r_1 = \frac{\ell n \lambda'}{2\pi i}$, $r_2 = \frac{\ell n \lambda''}{2\pi i}$, are respectively $\frac{1}{2}(\alpha + \delta) + m$, $\frac{1}{2}(\alpha - \delta) + m'$, where $m$ and $m'$ are integer numbers and $0 < \delta < 1$ .

A study of the velocity hodograph reveals that $m = -1$, $m' = 0$ and that $\pi\delta$ is the complement of the angle ABC in the velocity hodograph (i.e., the sum of these angles is $\pi$). The angle $\pi\beta$ between the wall and the free surface line is determined as

$$\pi\beta = \pi \frac{\alpha + \delta}{2} .$$

In this case the exponents of the system of functions Z and F have the form

$$\frac{\alpha - \delta}{2} , \frac{\alpha + \delta}{2} - 1 .$$

4. Case bc (Fig. 213). The rectilinear boundary of the reservoir is also the rectilinear boundary of the seepage surface.

Here in the z-plane and consequently also for the function Z we have an ordinary point:

$$Z = a_0 + a_1 \zeta + \ldots$$

In the u - iv plane, point B is at infinity as the intersection of two parallel lines, perpendicular to the straight line ABC of the z-plane. Therefore about point B

$$u - iv = c_0 \ \ell n \zeta + \ldots$$

The exponent of the system of functions Z and F is twofold zero (0,0).

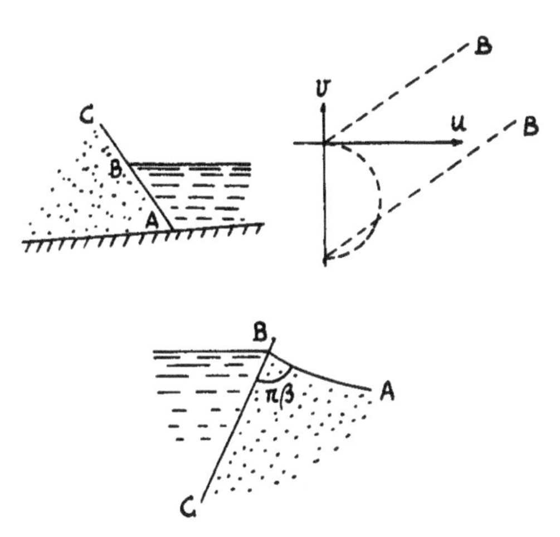


Fig. 214

5. Case bd (Fig. 214). The free surface joins the boundary of the reservoir, infiltration or evaporation take place. The conditions on the line AB are the same as in the third case, but the conditions on segment BC, making the angle $\pi\alpha$ with the horizontal, are

$$\operatorname{Im}(Ze^{-\pi i\alpha}) = 0, \quad \operatorname{Im}(iF) = 0 .$$

The corresponding characteristic equation

$$\begin{vmatrix} k\lambda & i\lambda & k & -i \\ ci\lambda & \lambda & -ci & 1 \\ e^{-\pi i\alpha} & 0 & e^{\pi i\alpha} & 0 \\ 0 & i & 0 & -i \end{vmatrix} = 0$$

can be written as

$$\lambda^2 + 2\lambda \frac{k - c}{k + c} e^{\pi i(1/2-\alpha)} \cos \pi\left(\frac{1}{2} - \alpha\right) + e^{2\pi i(1/2-\alpha)} = 0 .$$

From this

$$\lambda = e^{\pi i(1/2-\alpha)}\left[- \frac{k - c}{k + c} \cos \pi\left(\frac{1}{2} - \alpha\right) \pm \sqrt{\left(\frac{k - c}{k + c}\right)^2 \cos^2\pi\left(\frac{1}{2} - \alpha\right) - 1}\right]$$

Putting

$$- \frac{k - c}{k + c} \cos \pi \left(\frac{1}{2} - \alpha\right) = \cos \pi\delta ,$$

we obtain:

$$\lambda = e^{\pi i(1/2-\alpha \pm \delta)} .$$

from which we obtain the values for the exponents

$$\frac{1}{2}\left(\frac{1}{2} - \alpha + \delta\right) - 1, \quad \frac{1}{2}\left(\frac{1}{2} - \alpha - \delta\right)$$

or

$$-\frac{3}{4} - \frac{1}{2}(\alpha - \delta), \quad \frac{1}{4} - \frac{1}{2}(\alpha + \delta) .$$

If there is no evaporation, i.e., $c = 0$, then $\delta = \alpha + 1/2$, and the exponents are

$$-\frac{1}{2}, \quad -\alpha .$$

6. Case cd. (Fig. 215). The free surface joins with a rectilinear seepage surface. Conditions on segments AB and BC are represented in Fig. 216.

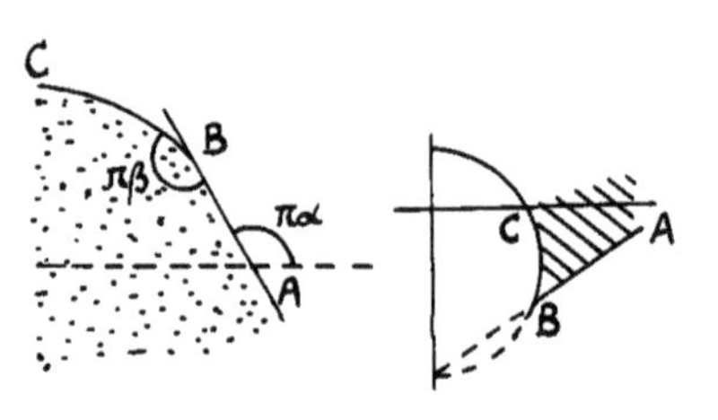


Fig. 215

| $\operatorname{Im}(Ze^{-\pi i\alpha}) = 0$ | $\operatorname{Im}(F + ciZ) = 0$ |
|---|---|
| $\operatorname{Im}(iF + kZ) = 0$ | $\operatorname{Im}(iF + kZ) = 0$ |

A B C

Fig. 216

The characteristic equation

$$\begin{vmatrix} k\lambda & i\lambda & k & -i \\ \lambda e^{-\pi i\alpha} & 0 & e^{\pi i\alpha} & 0 \\ k & i & k & -i \\ ci & 1 & -ci & 1 \end{vmatrix} = 0$$

has the roots

$$\lambda_1 = 1, \quad \lambda_2 = -e^{2\pi i\alpha} = e^{2\pi i(1/2+\alpha)}$$

From this it is not difficult to arrive at the conclusion that the exponents are $0$ and $\alpha - 1/2$ about vertex B. When the slope makes an obtuse angle with the horizontal, then $\pi\beta = \pi$ and when the angle is sharp, then $\pi\beta = \pi(\alpha + 1/2)$.

Finally we make a remark concerning the exponents about the point at infinity [8].

If we assume that the point at infinity in the $\zeta$-plane corresponds to a given vertex B, then the values of the exponents previously obtained must be increased by two. Indeed, we put

$$\zeta = \frac{1}{\tau} .$$

Point $\zeta = \infty$ transforms into point $\tau = 0$; after a change of variables for the functions

$$Z = \frac{dz}{d\zeta}, \quad F = \frac{d\omega}{d\zeta}$$

We obtain

$$Z = -\frac{dz}{d\tau}\tau^2, \quad F = -\frac{d\omega}{d\tau}\tau^2 ,$$

from which the expressed statement follows.

## B. Problem of Seepage Through a Vertical Dam

### §9. Seepage Rate Through a Dam and Discharge of a Well.

Fig. 217 illustrates a dam with slopes BC and AD on impervious horizontal foundation CD of length L. The depth of water in the lower reservoir is h, in the upper reservoir H; one has to determine the flow elements and also the length $h_0$ of the seepage surface AE (see §2, Chapter II).

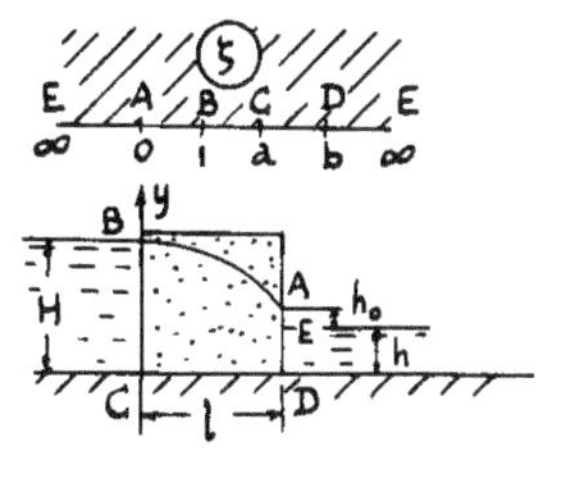


Fig. 217

First we examine how to determine the total flowrate.

We show that the flowrate through the vertical dam of Fig. 217, i.e., the flowrate through segment BC or the equivalent flowrate through AED, is determined by Dupuit's formula, if we trace Dupuit's parabola through the endpoints of the ordinates H and h. The proof of this theorem is given by Charnii [12].

It was already established long ago, by considering a series of investigations, that Dupuit's formula gave good results for flowrates through vertical dams. The available hydrodynamic solutions of B. B. Davison [13] and Hamel [14] were in close agreement with Dupuit's formula. Computations were carried out for six cases of vertical dams [15], for which the flowrate appeared to be close to that given by Dupuit's formula, but not completely the same. In was unknown why there was a discrepancy: because of incorrect computations (due to their difficulty) or because Dupuit's formula was not exact for vertical dams.

We assume steady flow with a velocity potential $\varphi(x,y)$ for which at the free surface AB the condition of constant pressure is satisfied

$$\varphi(x,y) + ky = 0 \ . \tag{9.1}$$

If we designate the ordinate of the free surface by $Y = Y(x)$, we have this identity at the free surface

$$\varphi(x,Y(x)) + kY(x) = 0 \ . \tag{9.2}$$

Along segment BC we have for $\varphi$

$$\varphi(0,y) = - kH \ , \tag{9.3}$$

and along ED

$$\varphi(\ell,y) = - kh. \tag{9.4}$$

Condition (9.1) must be satisfied along seepage segment AE.

We compute the flowrate through an arbitrary vertical cross-section of the dam. Evidently

$$q = \int_0^Y \frac{\partial\varphi}{\partial x}\, dy \ . \tag{9.5}$$

We examine the integral

$$J(x) = \int_0^{Y(x)} \varphi(x,y)\, dy \tag{9.6}$$

and take its derivative with respect to $x$

$$\frac{dJ}{dx} = \int_0^{Y} \frac{\partial\varphi}{\partial x}\, dy + \varphi(x,Y)\, \frac{dY}{dx} \ . \tag{9.7}$$

Accounting for (9.2), we may write:

$$q = \frac{dJ}{dx} - \varphi(x,Y)\, \frac{dY}{dx} = \frac{dJ}{dx} + kY\, \frac{dY}{dx} \ . \tag{9.8}$$

Integrating this equation, we obtain:

$$qx = \int_0^{Y} \varphi(x,y)\, dy + \frac{kY^2}{2} + C.$$

Considering that for $x = 0$, $Y = H$, we have:

$$0 = \int_0^{H} \varphi(0,y)\, dy + \frac{kH^2}{2} + C.$$

By means of (9.3) we find from the last equation

$$C = \frac{kH^2}{2} \ ,$$

and consequently,

$$qx = \int_0^{Y} \varphi(x,y)\, dy + \frac{k(Y^2 + H^2)}{2} \ . \tag{9.9}$$

Now we make $x = \ell$. For $x = \ell$, $Y = h_0$, and we split the integral of the right side into two parts using relation (9.4):

$$\int_0^{h_0} \varphi(x,y)\, dy = - \int_0^{h} kh\, dy - k \int_h^{h_0} y\, dy = - kh^2 - \frac{k}{2}(h_0^2 - h^2) \ .$$

Now equation (9.9) renders:

$$q\ell = - kh^2 - \frac{k}{2}(h_0^2 - h^2) + k\, \frac{h_0^2 + H^2}{2} \ .$$

We see that $h_0^2$ disappears from the right side and that

$$q = \frac{k(H^2 - h^2)}{2\ell} \ , \tag{9.10}$$

i.e., Dupuit's formula is indeed obtained. We encountered it in §10 of Chapter II and derive it in §2 of Chapter X, where we put $\ell = x_2 - x_1$ .

Exactly in the same way the formula is derived for the axi-symmetrical problem in the case of a fully penetrating well.

We consider a well of radius $\delta$, in which water stands at a level h. Let the cylindrical surface $r = R$ be a surface of equipotential (Fig. 218).

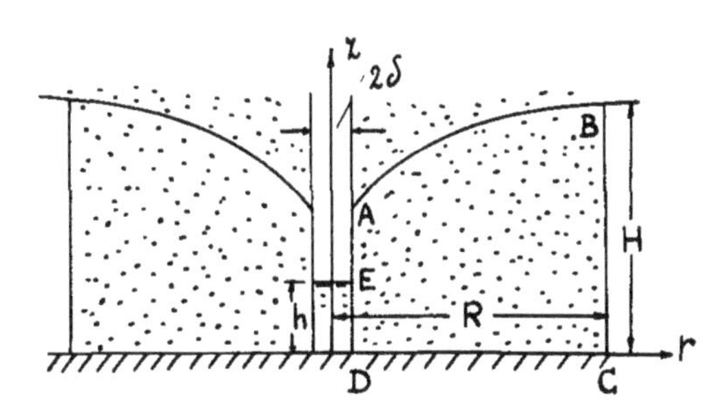


Fig 218

Let the depth of the free surface be

$$z = h + h_0 \quad \text{for} \quad r = \delta, \quad z = H \quad \text{for} \quad r = R.$$

In cylindrical coordinates $(r,z)$, we have the velocity potential $\varphi(r,z)$ so that the projections of the velocity are

$$v_r = \frac{\partial\varphi}{\partial r}, \quad v_z = \frac{\partial\varphi}{\partial z}.$$

At the free surface we have the relation stating that the pressure is constant:

$$\varphi(r,z) + kz = 0 . \tag{9.11}$$

The flowrate through an arbitrary cylinder of radius $r$ and axis coinciding with the z-axis, is expressed as

$$q = 2\pi r \int_0^z \frac{\partial\varphi}{\partial r}\, dz , \tag{9.12}$$

where $z$ is the ordinate of the free surface. If we put

$$\frac{\ln r}{2\pi} = x ,$$

then the problem is reduced to the previous case of vertical dam. To obtain the final result it is sufficient to substitute the difference

$$\frac{1}{2\pi}(\ln R - \ln \delta) ,$$

for $\ell = x_2 - x_1$ in the expression for the flowrate (9.10). This renders Dupuit's formula for a well (see Chapter X, §8):

$$q = \frac{\pi k(H^2 - h^2)}{\ln \frac{R}{\delta}} . \tag{9.13}$$

Thus at present we do not have an exact and complete solution of the problem of unconfined flow to a well, but we have an exact formula for the flowrate of the well.

We notice that in 1937, S. M. Proskurnikov [15] carried out a series of experiments by means of the electrodynamic analog method (see further

Chapter XI). The results obtained for the discharge through a dam were very close to those given by Dupuit's formula and S. M. Proskurnikov derived this formula for a dam, starting from an artificial hypothesis.

§10. Construction of the Solution for a Vertical Dam.

Using the results of §6, we find the exponents of the system of functions Z,F for our problem (Fig. 217, 219).

About point $A(\zeta = 0)$, the sixth case takes place, and therefore the exponents are 0,0. Here $\alpha = 1/2$.

About point $B(\zeta = 1)$, the fifth case takes place. Here we find that $-1/2$ is a double root of the characteristic equation, i.e., that the exponents are $-1/2$, $-1/2$.

About point $C(\zeta = a)$, corresponding to the first case, we may consider the exponents to be $-1/2$, $1/2$, so that here the singularity is removed.

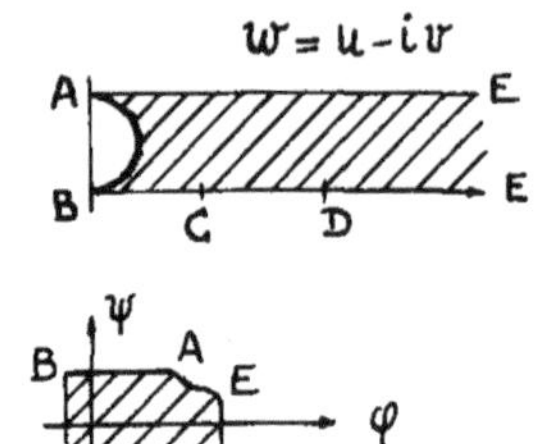


Fig. 219

About point $D(\zeta = b)$, we obtain the same result as about C, i e., the exponents are $-1/2$, $1/2$.

About point $E(\zeta = \infty)$, we have the fourth case, to which corresponds a double zero root of the characteristic equation. Since this is a singular point, at infinity, the exponents about it are 2,2 (see end of §8).

Functions F and Z are different branches of the function that may be represented by means of Riemann's symbol:

$$\tilde{Y} = P\begin{Bmatrix} 0 & 1 & a & b & \infty & \\ 0 & -\frac{1}{2} & -\frac{1}{2} & -\frac{1}{2} & 2 & \zeta \\ 0 & -\frac{1}{2} & \frac{1}{2} & \frac{1}{2} & 2 & \end{Bmatrix} . \tag{10.1}$$

We notice that the sum of all exponents is three and consequently Fuchs' condition (6.4) is here satisfied (the number of singular points $\nu + 1$ is five here).

The identity (5.6) holds for Riemann's function. Applying it to function $\tilde{Y}$, we may write

$$\tilde{Y} = (1-\zeta)^{-1/2}(\zeta - a)^{-1/2}(\zeta - b)^{-1/2} P\begin{Bmatrix} 0 & 1 & a & b & \infty & \\ 0 & 0 & 0 & 0 & \frac{1}{2} & \zeta \\ 0 & 0 & 1 & 1 & \frac{1}{2} & \end{Bmatrix} = \tag{10.2}$$

$$= \frac{Y}{\sqrt{(1-\zeta)(\zeta - a)(\zeta - b)}} .$$

From this it follows that the new function Y that we introduced, has the exponents 0 and 1 in points $\zeta = a$ and $\zeta = b$ and as shown above, does not contain logarithmic terms in its series developments.

Therefore $\zeta = a$ and $\zeta = b$ are ordinary points and may be excluded from the consideration, i.e., function $Y$ has only three singular points and may be represented as

$$Y = P \left\{ \begin{matrix} 0 & 1 & \infty & \\ 0 & 0 & \frac{1}{2} & \zeta \\ 0 & 0 & \frac{1}{2} & \end{matrix} \right\} \tag{10.3}$$

We have seen that function

$$P \left\{ \begin{matrix} 0 & 0 & \infty & \\ 0 & 0 & \gamma & \zeta \\ \alpha & \beta & \gamma' & \end{matrix} \right\} \tag{10.4}$$

is a solution of equation (5.2)

$$Y'' + \left( \frac{1 - \alpha}{\zeta} \quad \frac{1 - \beta}{\zeta - 1} \right\} Y' + \frac{\gamma\gamma'}{\zeta(\zeta - 1)} = 0 . \tag{10.5}$$

In the case under consideration

$$\alpha = \beta = 0, \quad \gamma = \gamma' = \frac{1}{2} ,$$

and consequently, differential equation (10.5) assumes the form

$$\zeta(1 - \zeta) Y'' + (1 - 2\zeta) Y' - \frac{1}{4} Y = 0 , \tag{10.6}$$

for which, if $U$ and $V$ are its linearly independent integral,

$$F = \frac{AU + BV}{\sqrt{(1 - \zeta)(\zeta - a)(\zeta - b)}} , \qquad Z = \frac{CU + DV}{\sqrt{(1 - \zeta)(\zeta - a)(\zeta - b)}} . \tag{10.7}$$

About $\zeta = 0$ we have the following integral of this equation written as a hypergeometric function (here $a = 1/2$, $b = 1/2$, $c = 1$):

$$F\left(\frac{1}{2}, \frac{1}{2}, 1, \zeta\right) = 1 + \left(\frac{1}{2}\right)^2 \zeta + \left(\frac{1\cdot 3}{2\cdot 4}\right)^2 \zeta^2 + \left(\frac{1\cdot 3\cdot 5}{2\cdot 4\cdot 6}\right)^2 \zeta^3 + \ldots \tag{10.8}$$

We consider the definite integral

$$\int_0^1 \frac{dt}{\sqrt{t(1 - t)(1 - \zeta t)}} . \tag{10.9}$$

The substitution $t = \sin^2\varphi$ gives:

$$\int_0^{\pi/2} \frac{d\varphi}{\sqrt{1 - \zeta \sin^2\varphi}} = \frac{\pi}{2}\left[1 + \left(\frac{1}{2}\right)^2 \zeta^2 + \left(\frac{1.3}{2.4}\right)^2 \zeta^2 + \ldots\right] .$$

It is apparent from this that hypergeometric function (10.8) differs from the complete elliptic integral of the first kind (considered as a function of the square of the modulus $k^2 = \zeta$) only by a constant multiplying factor. Consequently, we may take for one of the particular solutions of equation (10.6):

$$U = K(\zeta) = \int_0^{\pi/2} \frac{d\varphi}{\sqrt{1 - \zeta \sin^2 \varphi}} \qquad (k^2 = \zeta) . \tag{10.10}$$

Substitution of $1 - \zeta$ for $\zeta$ does not change equation (10.7) and therefore

$$V = K(1 - \zeta) = K' \qquad (k'^2 = 1 - \zeta) \tag{10.11}$$

is also a solution of our equation, regular about point $\zeta = 1$.

The series for $K(\zeta)$ converges for $|\zeta| < 1$, i.e., in a circle $C_0$ (Fig. 207), the series for $K(1 - \zeta)$ in a circle $C_1$, for which $|\zeta - 1| = 1$. We show how $K(1 - \zeta)$ behaves for $\zeta = 0$.

Between $K$ and $K'$ exists a relation (obtained by means of the theorem about the Wronskian determinant) [21]

$$K \frac{dK'}{d\zeta} - K' \frac{dK}{d\zeta} = - \frac{\pi}{4\zeta(1 - \zeta)} .$$

It may be rewritten as

$$\frac{d}{d\zeta}\left(\frac{K'}{K}\right) = - \frac{\pi}{4K^2\zeta(1 - \zeta)} = - \frac{1}{\pi}\left\{\frac{1}{\zeta} + \frac{1}{2} + P_1(\zeta)\right\} ,$$

where $P_1(\zeta)$ is a series in positive integer powers of $\zeta$.

From this, by integration, we find:

$$K' = K(1 - \zeta) = - \frac{K}{\pi} \ln \zeta + P(\zeta) ,$$

where $P(\zeta)$ has the form

$$P(\zeta) = - 2 \sum_{n=1}^{\infty} \left[\frac{1\cdot 3\cdot 5\ldots(2n - 1)}{2\cdot 4\cdot 6\ldots 2n}\right]^2 \sum_{m=1}^{2n} \frac{(-1)^{m-1}}{m} \zeta^n .$$

Thus $K(1 - \zeta)$ represents a solution with logarithmic singularity about $\zeta = 0$; consequently $K(\zeta)$ and $K(1 - \zeta)$ are linearly independent and form a fundamental system of solutions about $\zeta = 0$:

$$U = K(\zeta), \quad V = K(1 - \zeta) . \tag{10.12}$$

By means of formulas (5.10) and analogous formulas, we find that

$$U_1 = \frac{1}{\sqrt{\zeta}} K\left(\frac{1}{\zeta}\right), \quad V_1 = \frac{1}{\sqrt{\zeta}} K\left(\frac{\zeta - 1}{\zeta}\right) \tag{10.13}$$

is a fundamental system of integrals of equation (10.7). The series for functions (10.13) converge for $|\zeta| > 1$.

Substituting $1 - \zeta$ for $\zeta$, we find a new system of integrals from (10.13)

$$U_2 = \frac{1}{\sqrt{1 - \zeta}} K\left(\frac{1}{1 - \zeta}\right), \quad V_2 = \frac{1}{\sqrt{1 - \zeta}} K\left(\frac{\zeta}{1 - \zeta}\right) . \tag{10.14}$$

Their series converge outside circle $C_1$ (Fig. 207).

Conformal mapping of the "modular" velocity triangle $w = u - iv$

(Fig. 219, where $AB = k$) onto the upper half-plane $\zeta$ (Fig. 217) is carried out by the formula

$$w = \frac{AK(\zeta) + BK(1-\zeta)}{CK(\zeta) + DK(1-\zeta)} .$$

Considering the segment $(0,1)$ for $\zeta$, we have this correspondence for points $\zeta$ and $w$: for $\zeta = 0$, $w = ki$; for $\zeta = 1$, $w = 0$ and for $\zeta = 1/2$, $w = k/2 + (k/2)i$. From this we find, $A, B, C, D$ and obtain

$$w = k \frac{K(1-\zeta)}{K(\zeta) - iK(1-\zeta)} . \tag{10.15}$$

To pass on to other sides of the triangle, one must know auxiliary substitutions, linking $U, V$ with other integrals. If point $\zeta$ is in the upper half-plane, then the following relations hold

$$\left\{\begin{aligned} K(\zeta) - iK(1-\zeta) &= \frac{1}{\sqrt{\zeta}} K\left(\frac{1}{\zeta}\right) = -\frac{i}{\sqrt{1-\zeta}} K\left(\frac{1}{1-\zeta}\right) , \\ K(1-\zeta) &= \frac{1}{\sqrt{\zeta}} K\left(\frac{\zeta-1}{\zeta}\right) = \frac{1}{\sqrt{1-\zeta}} \left[K\left(\frac{1}{1-\zeta}\right) - iK\left(\frac{\zeta}{\zeta-1}\right)\right] . \end{aligned}\right. \tag{10.16}$$

By means of these we find for the segment $1 < \zeta < \infty$

$$w = k \frac{K\left(\frac{\zeta-1}{\zeta}\right)}{K\left(\frac{1}{\zeta}\right)} , \tag{10.17}$$

and for the segment $-\infty < \zeta < 0$

$$w = ki \frac{K\left(\frac{\zeta}{1-\zeta}\right) - iK\left(\frac{\zeta}{\zeta-1}\right)}{K\left(\frac{1}{1-\zeta}\right)} . \tag{10.18}$$

The exactness of (10.17) and (10.18) is immediately checked from consideration of Fig. 217 and 219.

Now it is easy to compose expressions for functions $F$ and $Z$, starting from equations (10.7) and (10.16) and considering that the numerators of these expressions must be proportional to the numerators of formulas (10.15), (10.17) and (10.18). We obtain:

$$\left\{\begin{aligned} Z &= \frac{dx}{d\zeta} + i\frac{dy}{d\zeta} = -A\frac{K(\zeta) - iK(1-\zeta)}{\sqrt{(1-\zeta)(a-\zeta)(b-\zeta)}} , \\ F &= \frac{d\varphi}{d\zeta} = -kA\frac{K(1-\zeta)}{\sqrt{(1-\zeta)(a-\zeta)(b-\zeta)}} ; \end{aligned}\right. \tag{10.19}$$

for $-\infty < \zeta < 0$

$$Z = i\frac{dy}{d\zeta} = \frac{AiK\left(\frac{1}{1-\zeta}\right)}{(1-\zeta)\sqrt{(a-\zeta)(b-\zeta)}} ,$$

$$F = \frac{d\varphi}{d\zeta} + \frac{id\psi}{d\zeta} = -\frac{kdy}{d\zeta} + \frac{id\psi}{d\zeta} = -\frac{kA\left[K\left(\frac{1}{1-\zeta}\right) - iK\left(\frac{\zeta}{\zeta-1}\right)\right]}{(1-\zeta)\sqrt{(a-\zeta)(b-\zeta)}} ; \tag{10.20}$$

for $1 < \zeta < a$

$$(10.21)\quad \begin{cases} Z = i\dfrac{dy}{d\zeta} = -A\dfrac{iK\left(\frac{1}{\zeta}\right)}{\sqrt{\zeta(\zeta-1)(\zeta-a)(\zeta-b)}}, \\ F = i\dfrac{d\psi}{d\zeta} = -kA\dfrac{iK\left(\frac{\zeta-1}{\zeta}\right)}{\sqrt{\zeta(\zeta-1)(a-\zeta)(b-\zeta)}}; \end{cases}$$

for $a < \zeta < b$

$$(10.22)\quad \begin{cases} Z = \dfrac{dx}{d\zeta} = A\dfrac{K\left(\frac{1}{\zeta}\right)}{\sqrt{\zeta(\zeta-1)(\zeta-a)(b-\zeta)}}, \\ F = \dfrac{d\varphi}{d\zeta} = kA\dfrac{K\left(\frac{\zeta-1}{\zeta}\right)}{\sqrt{\zeta(\zeta-1)(\zeta-a)(b-\zeta)}}; \end{cases}$$

for $b < \zeta < \infty$

$$(10.23)\quad \begin{cases} Z = i\dfrac{dy}{d\zeta} = A\dfrac{iK\left(\frac{1}{\zeta}\right)}{\sqrt{\zeta(\zeta-1)(\zeta-a)(\zeta-b)}}, \\ F = i\dfrac{d\psi}{d\zeta} = kA\dfrac{iK\left(\frac{\zeta-1}{\zeta}\right)}{\sqrt{\zeta(\zeta-1)(\zeta-a)(\zeta-b)}}. \end{cases}$$

Here multiplier $A$ has a constant magnitude. Carrying out the integration over $\zeta$ between corresponding limits, we find formulas for the length of segments $H$, $\ell$, $h$, $h_0$ and for flowrates $Q$, $Q_h$, $Q_{h_0}$ through segments $H$, $h$, $h_0$. We obtain:

$$(10.24)\quad \ell = A\int_a^b \frac{K\left(\frac{1}{\zeta}\right)d\zeta}{\zeta\sqrt{(\zeta-1)(\zeta-a)(b-\zeta)}} = AJ_1,$$

$$(10.25)\quad H = A\int_1^a \frac{K\left(\frac{1}{\zeta}\right)d\zeta}{\zeta\sqrt{(\zeta-1)(\zeta-a)(\zeta-b)}} = AJ_2,$$

$$(10.26)\quad h = A\int_b^\infty \frac{K\left(\frac{1}{\zeta}\right)}{\zeta\sqrt{(\zeta-1)(\zeta-a)(\zeta-b)}}\,d\zeta = AJ_3,$$

$$(10.27)\quad h_0 = A\int_{-\infty}^0 \frac{K\left(\frac{1}{1-\zeta}\right)d\zeta}{(1-\zeta)\sqrt{(a-\zeta)(b-\zeta)}} = AJ_4,$$

$$(10.28)\quad Q = kA\int_1^a \frac{K\left(\frac{\zeta-1}{\zeta}\right)}{\sqrt{\zeta(\zeta-1)(a-\zeta)(b-\zeta)}}\,d\zeta = kAJ_5,$$

$$(10.29)\quad Q_h = kA\int_b^\infty \frac{K\left(\frac{\zeta-1}{\zeta}\right)}{\sqrt{\zeta(\zeta-1)(\zeta-a)(\zeta-b)}}\,d\zeta = kAJ_6,$$

$$(10.30)\quad Q_{h_0} = kA\int_{-\infty}^0 \frac{K\left(\frac{\zeta}{\zeta-1}\right)}{(1-\zeta)\sqrt{(a-\zeta)(b-\zeta)}}\,d\zeta = kAJ_7.$$

The full flowrate, as we know, is determined by formula (9.10). Therefore from (10.24) — (10.26) and (10.28) this relation between the integrals must follow:

$$2J_1J_5 = J_2^2 - J_3^2 .$$

For convenience of computation, we introduce the notations

$$\alpha = \frac{1}{b}, \quad \beta = \frac{1}{a}, \qquad 0 \leq \alpha \leq \beta \leq 1 \tag{10.31}$$

and substitute in the integrals

$$\frac{1}{\zeta} = \tau . \tag{10.32}$$

Further, we substitute $\tau$ by new expressions, different for the various integrals, which make the function in the integrand finite for the integration limits. We put:

$$\begin{cases} \tau = \alpha + (\beta - \alpha) \sin^2 \psi & (a < \zeta < b) , \\ \tau = \beta + (1 - \beta) \sin^2 \psi & (1 < \zeta < a) , \\ \tau = \alpha \sin^2 \psi & (b < \zeta < \infty) . \end{cases} \tag{10.33}$$

Introducing still the notation

$$\frac{2A}{\sqrt{ab}} = C ,$$

we find the following formulas:

$$\ell = C \int_0^{\pi/2} \frac{K[\alpha + (\beta - \alpha) \sin^2\psi]}{\sqrt{1 - \alpha - (\beta - \alpha) \sin^2 \psi}} \, d\psi , \tag{10.34}$$

$$H = C \int_0^{\pi/2} \frac{K[\beta + (1 - \beta) \sin^2\psi] \, d\psi}{\sqrt{\beta - \alpha - (1 - \beta) \sin^2 \psi}} , \tag{10.35}$$

$$h = C \sqrt{\alpha} \int_0^{\pi/2} \frac{K(\alpha \sin^2 \psi) \sin \psi \, d\psi}{\sqrt{(1 - \alpha \sin^2\psi)(\beta - \alpha \sin^2 \psi)}} . \tag{10.36}$$

In the integrals for $h_0$ and $Q_{h_0}$ we introduce the substitution

$$\frac{1}{1 - \zeta} = \tau = \cos^2\psi ,$$

and this gives:

$$h_0 = C \int_0^1 \frac{K(\cos^2\psi) \sin \psi \cos \psi \, d\psi}{\sqrt{(1 - \alpha_1 \sin^2\psi)(1 - \beta_1 \sin^2 \psi)}} , \tag{10.37}$$

$$Q_{h_0} = kC \int_0^1 \frac{K(\sin^2\psi) \sin \psi \cos \psi \, d\psi}{\sqrt{(1 - \alpha_1 \sin^2\psi)(1 - \beta_1 \sin^2\psi)}} . \tag{10.38}$$

Here we have put

$$\alpha_1 = 1 - \alpha, \quad \beta_1 = 1 - \beta . \tag{10.39}$$

Integral $Q_h$ remains to be transformed. Putting

$$\frac{\zeta - 1}{\zeta} = \tau = 1 - \alpha \sin^2\psi \quad ,$$

we have

$$Q_h = kC\sqrt{\alpha} \int_0^{\pi/2} \frac{K(1 - \alpha \sin^2\psi) \sin \psi \, d\psi}{\sqrt{(1 - \alpha \sin^2\psi)(\beta - \alpha \sin^2\psi)}} \quad . \tag{10.40}$$

Finally, we find the equation of the free surface in parametric form. Therefore we separate the real part from the imaginary in the first equation (10.19) and introduce the substitution

$$\zeta = \sin^2\psi \; .$$

We obtain:

$$\left\{ \begin{aligned} x &= \ell - C\int_0^{\psi} \frac{K(\sin^2\psi) \sin \psi \, d\psi}{\sqrt{(1 - \alpha \sin^2\psi)(1 - \beta \sin^2\psi)}} \quad , \\ y &= h + h_0 + C\int_0^{\psi} \frac{K(\cos^2\psi) \sin \psi \, d\psi}{\sqrt{(1 - \alpha \sin^2\psi)(1 - \beta \sin^2\psi)}} \quad . \end{aligned} \right. \tag{10.41}$$

Constant C remains arbitrary; it may, for example, be taken equal to one.

Making $\psi = \pi/2$ in the equations thus obtained, we find another expression for $\ell$, and also an equality for $H - h - h_0$:

$$\ell = C \int_0^{\pi/2} \frac{K(\sin^2\psi) \sin \psi \, d\psi}{\sqrt{(1 - \alpha \sin^2\psi)(1 - \beta \sin^2\psi)}} \quad , \tag{10.42}$$

$$H - h - h_0 = C \int_0^{\pi/2} \frac{K(\cos^2\psi) \sin \psi \, d\psi}{\sqrt{(1 - \alpha \sin^2\psi)(1 - \beta \sin^2\psi)}} \quad . \tag{10.43}$$

In computations of the derived formulas, we may use developments of the functions in the integrand in power series of parameters $\alpha, \beta$ and so on.

We designate by $I_n$, $J_n$ the integrals which converge in the coefficients of these developments:

$$I_n = \int_0^{\pi/2} K(\sin^2\psi) \sin^n\psi \, d\psi \quad ,$$

$$J_n = \int_0^{\pi/2} K(\cos^2\psi) \sin^n\psi \, d\psi \quad .$$

We communicate the values of some of these:

$$I_1 = \int_0^1 K(1 - t^2)dt = \frac{\pi^2}{4}\,, \quad I_3 = \frac{3\pi^2}{16}, \quad I_5 = \frac{41\pi^2}{256}, \quad I_7 = \frac{147\pi^2}{1024} \quad ,$$

$$J_1 = \int_0^1 K(t^2)dt = 2G = 2\left(1 - \frac{1}{3^2} + \frac{1}{5^2} - \ldots\right) = 1.8319 \; ,$$

$$J_3 = 1.124, \quad J_5 = 0.8764, \quad J_7 = 0.7478 .$$

Here $G$ is Catalan's constant.

We now compute $\ell$. For small values of $\alpha$ and $\beta$, coresponding to slender dams, we have a series development for integral (10.42):

$$\ell(\alpha,\beta) = \frac{\pi^2}{4} \{1 + 0.375(\alpha + \beta) + 0.24023(\alpha^2 + \beta^2) + $$
$$+ 0.16016\alpha\beta + 0.1794(\alpha^3 + \beta^3) + 0.1075\alpha\beta(\alpha + \beta) +$$
$$+ 0.1443(\alpha^4 - \beta^4) + \ldots\}. \tag{10.44}$$

From formula (10.34) putting

$$\int_0^\psi \frac{d\psi}{\sqrt{1 - \frac{\beta - \alpha}{1 - \alpha}\sin^2\psi}} = u ,$$

we find

$$\ell = \frac{1}{\sqrt{\alpha_1}} \int_0^K K(1 - \alpha_1 \operatorname{dn}^2 u)\, du \quad , \tag{10.45}$$

where $\alpha_1 = 1 - \alpha$ and where the elliptic function $\operatorname{dn} u$ has a quadratic modulus

$$k^2 = \frac{\beta - \alpha}{1 - \alpha} .$$

The function in the integrand is close to linear for small $\alpha_1$.

We compute $H$ and $Q$. For these values we have power series in $\beta_1$ and $\gamma_1$ :

$$H = \frac{\pi}{2\sqrt{\alpha_1}} \{2.0794 + 0.07243\beta_1 + 0.39486\gamma_1 +$$
$$+ 0.01737\beta_1^2 + 0.02326\beta_1\gamma_1 - 0.21039\gamma_1^2 -$$
$$- \frac{\ln \beta_1}{2} (1 + \frac{1}{8}\beta_1 + \frac{1}{4}\gamma_1 + \frac{27}{512}\beta_1^2 + \frac{3}{64}\beta_1\gamma_1 + \frac{9}{64}\gamma_1^2 + \ldots)\} , \tag{10.46}$$

$$Q = \frac{k\pi^2}{4\sqrt{\alpha_1}} \{1 + \frac{1}{8}\beta_1 + \frac{1}{4}\gamma_1 + \frac{27}{512}\beta_1^2 + \frac{3}{64}\beta_1\gamma_1 + \frac{9}{64}\gamma_1^2 + \ldots\} . \tag{10.47}$$

These series converge well for $\beta \geq 0.9$ and $\alpha < \beta$ .

For $h$ and $Q_h$, in case of small $\alpha$ together with values of $\beta$, close to one, we have the series

$$h = \frac{\pi}{2}\sqrt{\gamma}(1 + \frac{1}{2}\alpha + \frac{1}{3}\gamma + 0.3417\alpha^2 + 0.2\alpha\gamma + 0.2\gamma^2 +$$
$$+ 0.2625\alpha^3 + 0.1464\alpha^2\gamma + 0.1286\alpha\gamma^2 + 0.1429\gamma^3 + \ldots) , \tag{10.48}$$

$$Q_h = k\sqrt{\gamma}(E_0 + \frac{1}{2}\gamma E_1 + \frac{3}{8}\gamma^2 E_2 + \ldots) , \tag{10.49}$$

where

$$E_0 = (1 + \frac{1}{2}\alpha + \frac{41}{120}\alpha^2 + \ldots)\ln\frac{2}{\sqrt{\alpha}} + 1 + \frac{1}{4}\alpha + \frac{817}{7200}\alpha^2 + \ldots,$$

$$E_1 = (\frac{2}{3} + \frac{2}{5}\alpha + \frac{41}{140}\alpha^2 + \ldots)\ln\frac{2}{\sqrt{\alpha}} + \frac{5}{9} + \frac{9\alpha}{50} + \frac{5309}{58800}\alpha^2 + \ldots,$$

$$E_2 = \left(\frac{8}{15} + \frac{12}{35}\alpha + \frac{82}{315}\alpha^2 + \ldots\right)\ln\frac{2}{\sqrt{\alpha}} + \frac{94}{225} + \frac{179}{1225}\alpha + \frac{30\,419}{396900}\alpha^2 + \ldots$$

Computations for $\beta = 0.9;\ 0.99\ldots;\ 1 - 10^{-8}$ and $\alpha = 0;\ 0,1;\ldots;\ 0.9;\ldots;\ 1 - 10^{-8}$, and subsequent graphical interpolation lead to graphs represented in Fig. 220-226.

Concerning the graphs, those for $h = 0$ (no tailwater) are accurate where the computations are simpler and where there is no necessity for interpolation. Those for non-zero $h$ have a smaller degree of accuracy when interpolations must be made. In especially doubtful cases, dotted lines replace the continuous lines.

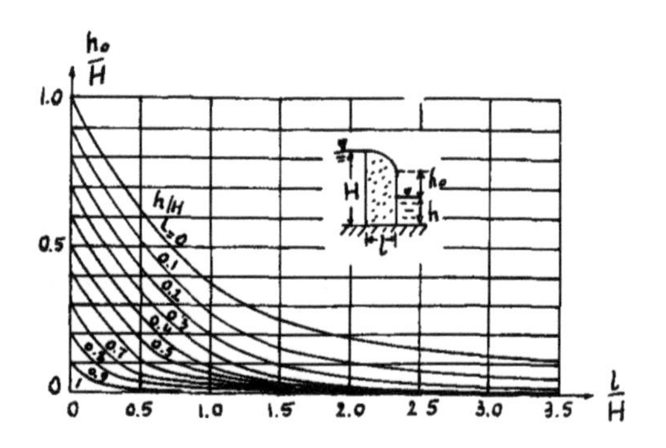


Fig. 220

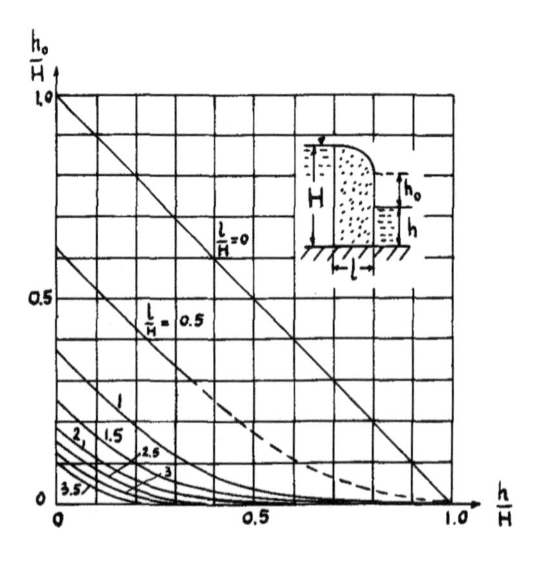


Fig. 221

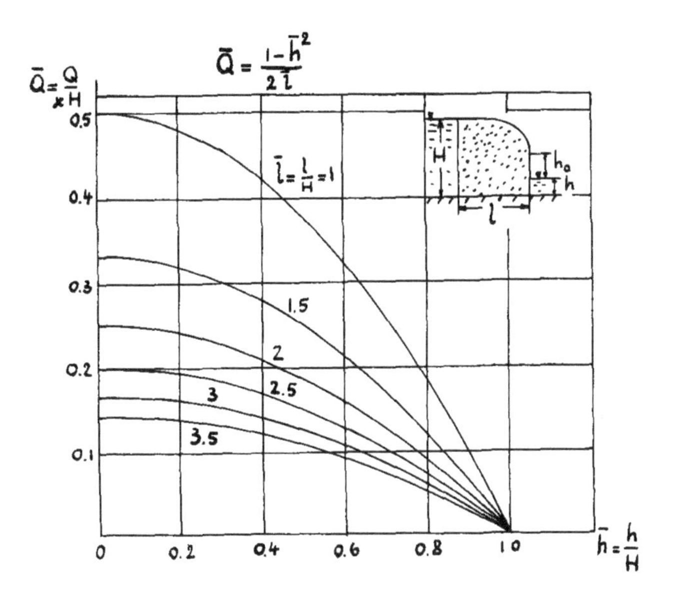


Fig. 222

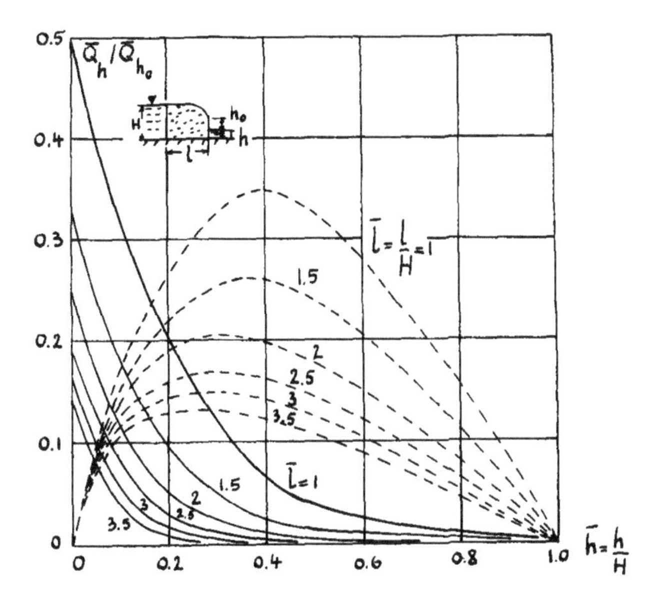


Fig. 223

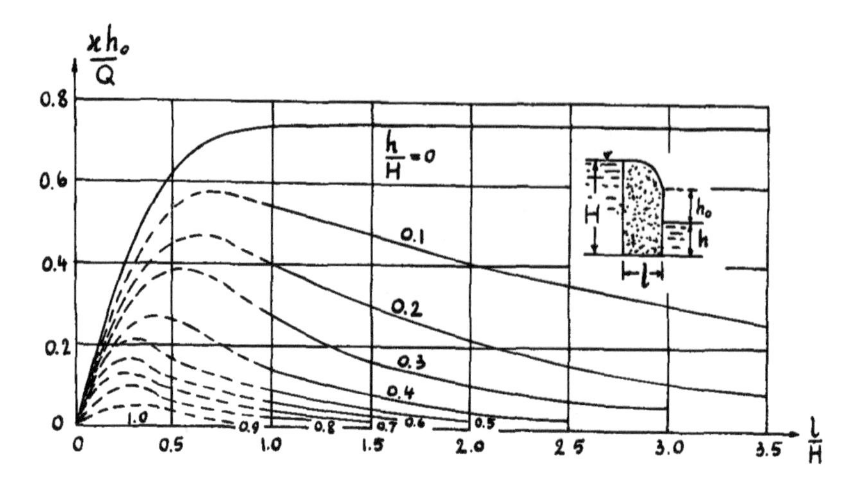


Fig. 224

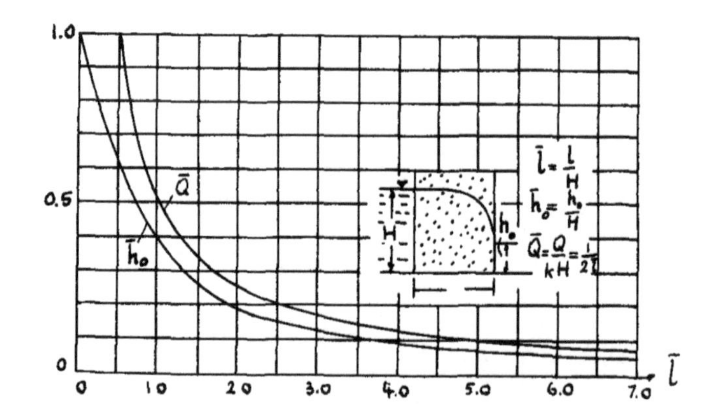


Fig. 225

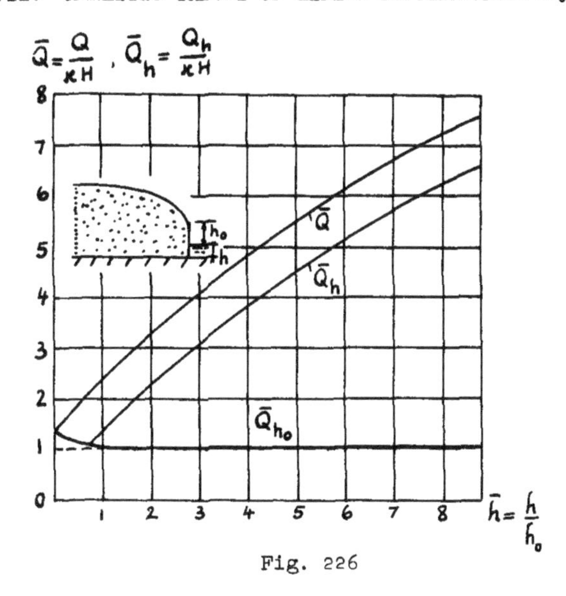


Fig. 226

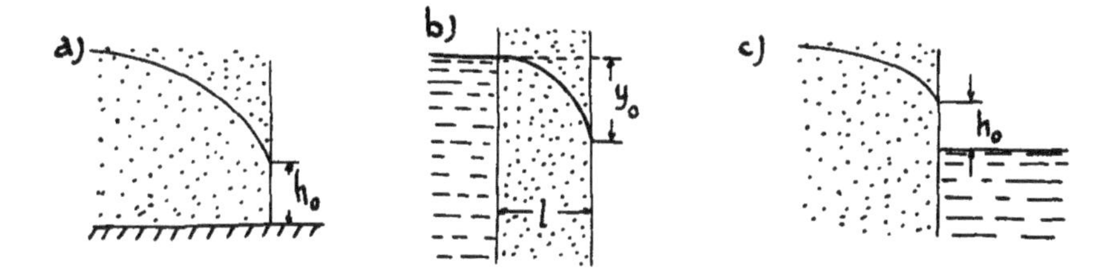


Fig. 227

We notice some particular cases.

1. Case $\alpha = 0$, $\beta = 1$. This is the case when there is no tail-water and when the base of the dam is infinitely long (Fig. 227 a). In this case from (10.37) we obtain for the length of the seepage segment:

$$h_0 = C \int_0^1 K(\cos^2\psi) \sin\psi \, d\psi = CJ_1 = 1.8319C \ . \tag{10.50}$$

The flowrate $Q_{h_0}$ is determined from (10.38):

$$Q_{h_0} = kC \int_0^1 K(\sin^2\psi) \sin\psi \, d\psi = CJ_1 = \frac{\pi^2}{4} C \ .$$

From this we find the dependence between $Q_{h_0}$ and $\ell$:

$$Q_{h_0} = \frac{k\pi^2}{8G} h_0 = 1.3469 \, kh_0 \ . \tag{10.51}$$

The equations of the free surface are simplified and may be written as

$$(10.52)\qquad x = \frac{C}{2}\int_0^m \frac{K(m)\,dm}{1-m}\,, \qquad y - h_0 = y_1 = \frac{C}{2}\int_0^m \frac{K(1-m)\,dm}{1-m}\,.$$

We notice that for $\ell = \infty$, but $h$ non-zero, i.e., for existing tailwater, equality (10.51) is approximately satisfied. In fact $Q_{h_0}/kh_0$ varies between limits 1.346 to 1, when $h/h_0$ varies from 0 to $\infty$. In the case of a finite dam but with sufficiently long base, equality (10.51) is also approximately satisfied. Namely for $\ell/H \geq 1$

$$Q \approx 1.35\ kh_0\ .$$

For $\ell/H < 1$ this equality becomes less accurate the more $\ell/H$ tends to zero (Fig. 226).

2. Case $\alpha = \beta = 0$. This is the case of an infinitely high dam with no tailwater. We designate by $y_0$ the segment that is shown in Fig. 227 b. Formula (10.44) gives:

$$\ell = C\,\frac{\pi^2}{4}\ .$$

In formula (10.41) we obtain for $y_0$

$$y_0 = C\int_0^{\pi/2} K(\cos^2\psi)\ \sin\psi\ d\psi = 2CG.$$

From this we find the relation

$$\frac{y_0}{\ell} = \frac{8G}{\pi^2}$$

or

$$y_0 = 0.7425\ell\ .$$

3. Case $\alpha = \beta = 1$. Fig. 227 c corresponds to this case. Formulas (10.38) and (10.37) for $Q_{h_0}$ and $h_0$ render:

$$h_0 = C\int_0^1 K(t^2)\ t\ dt = C, \qquad Q_{h_0} = kC\int_0^1 K(1-t^2)\,t\,dt = kC\ .$$

Consequently

$$Q_{h_0} = kh_0\ .$$

Table 12 renders the dependence of the coordinates of points of the free surface on parameter $m$, for the case of no tailwater for an infinitely long dam or, what is the same, for the case of water flowing from infinity towards a ditch with vertical walls and no water in it [by equation (10.52) for $C = 1$].

To obtain ordinates of the free surface $y = y_1 + h_0$, one must add $h_0 = 1.8319$ to the values of $y_1$ of Table 12. In the graph of Fig. 228, $h_0$ was assumed to be one, therefore all the values $x_1$, $y_1$ of Table 12 are divided by 1.8319. Therefore according to (10.51) the flowrate $Q_{h_0}$

TABLE 12

Dependence of the coordinates of the free surface on parameter m in the case of an infinitely long dam with no tailwater.

| m | 0 | 0.05 | 0.10 | 0.25 | 0.5 | 0.6 | 0.7 |
|---|---|---|---|---|---|---|---|
| x | 0 | 0.036 | 0.073 | 0.250 | 0.593 | 0.805 | 1.094 |
| $y_1$ | 0 | 0.084 | 0.160 | 0.445 | 0.773 | 0.976 | 1.226 |
| m | 0.8 | 0.9 | 0.99 | 0.999 | 0.9999 | $1-10^{-5}$ | $1-10^{-6}$ |
| x | 1.5833 | 2.370 | 5.977 | 10.890 | 17.126 | 24.686 | 33.572 |
| $y_1$ | 1.568 | 2.133 | 3.961 | 5:771 | 7.579 | 9.388 | 11.196 |
| m | $1-10^{-7}$ | $1-10^{-8}$ | $1-10^{-9}$ | $1-10^{-12}$ | $1-10^{-15}$ | $1-10^{-18}$ | $1-10^{-24}$ |
| x | 43.784 | 55.321 | 68.184 | 114.72 | 173.19 | 243.59 | 420.18 |
| $y_1$ | 13.005 | 14.813 | 16.622 | 22.047 | 22.472 | 32.898 | 43.748 |

will be 1.3469k. The flowrate through horizontal segment d of the drain is given by a parabola with parameter $p \approx 2d = 1.3469$ [see formula (10.5), Chpater II]. In Fig. 228 it is shown in dotted form; it is the asymptotic parabola for the constructed free surface.

We notice that if we move this parabola along its axis, we obtain again parabolas which tend asymptotically towards the free surface of Fig. 228. The parabola with vertex in the origin of coordinates is the most interesting. The free surface lies between two parabolas.

We already mentioned the electro-hydrodynamical analog experiments of S. M. Proskurnikov[5] in §9. Besides the flowrates for a series of dams, he also determined the length of the seepage surface. The results were much more accurate for slender dams than for wide dams, as may be ascertained by comparison of experiment and theory.

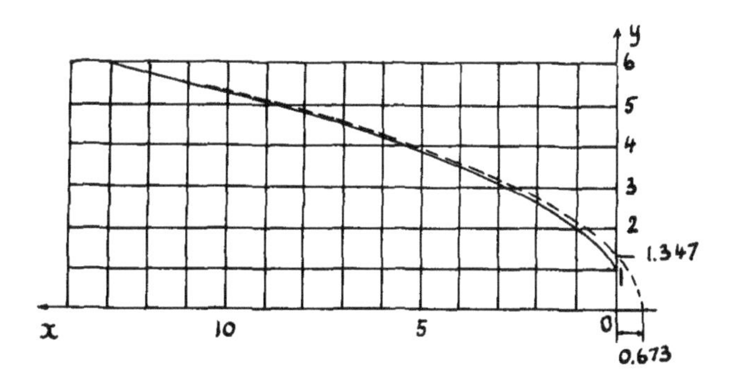


Fig. 228

## C. Dams With Rectangular Trapezoidal Cross-Section.

### §11. Seepage in a Trapezoidal Dam when Evaporation Occurs.

We call cofferdam (batardeau) not only a dam with vertical faces but also a dam with vertical inflow face and sloping down stream face, making an angle $\alpha\pi$, smaller than $\pi$, with the horizontal.

Let the dam be on impervious foundation (Fig. 229 a). Headwater depth is $H$; there is no tailwater so that water percolates through the outflow surface, leaving a seepage surface. Evaporation from the free surface takes place in such a way that on CB we have:

$$\psi + cx = \text{constant} \tag{11.1}$$

where $c > 0$. The velocity hodograph is represented in Fig. 228 b. We notice that CB is not a streamline and that angle $\theta_1 = \pi\varepsilon$ made by the free surface where the horizontal is not zero. As results from the hodograph, segment OC is equal to $\sqrt{kc}$ and also to $k \text{ tg } \theta_1$, so that, with $\theta_1 = \pi\varepsilon$,

$$\text{tg } \theta_1 = \text{tg } \pi\varepsilon = \sqrt{\frac{c}{k}} \; . \tag{11.2}$$

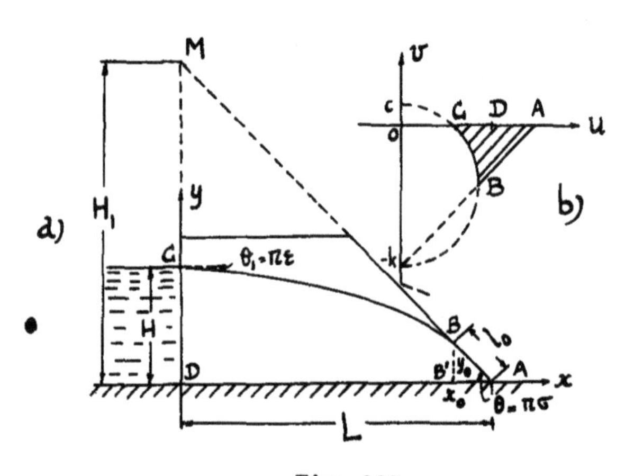


Fig. 229

Let the points A, B, C, D map respectively into $\zeta = 0, 1, \infty, -a$ of the real axis of the auxiliary plane $\zeta$. For $\zeta = -a$ we have a removed singularity with exponent $- 1/2$ for functions F and Z. Putting

$$F = \frac{F_1}{\sqrt{\zeta + a}} , \qquad Z = \frac{Z_1}{\sqrt{\zeta + a}} , \tag{11.3}$$

we find these exponents for functions $F_1$ and $Z_1$ about the singular points :

$$-\frac{1}{2}, \quad 1 - \sigma \quad \text{for} \quad \zeta = 0 ,$$

$$1 + \varepsilon, \quad 1 - \varepsilon \quad \text{for} \quad \zeta = \infty ,$$

$$0, \quad \frac{1}{2} - \sigma \quad \text{for} \quad \zeta = 1 .$$

Functions $Z_1$ and $F_1$ represent linear combinations of two branches of Riemann's function

$$Y = P\begin{Bmatrix} 0 & \infty & 1 & \\ -\frac{1}{2} & 1+\varepsilon & 0 & \zeta \\ \sigma - 1 & 1 - \varepsilon & \frac{1}{2} - \sigma & \end{Bmatrix} = \zeta^{-1/2} P \begin{Bmatrix} 0 & \infty & 1 & \\ 0 & \frac{1}{2}+\varepsilon & 0 & \zeta \\ \sigma - \frac{1}{2} & \frac{1}{2} - \varepsilon & \frac{1}{2} - \sigma & \end{Bmatrix} = \bar{\zeta}^{1/2} Y_1$$

About point $\zeta = 0$, the linearly independent branches of function $Y_1$ are expressed by means of hypergeometric functions $U$ and $V$:

$$U = F(a,b,c,\zeta) = F\left(\frac{1}{2} + \varepsilon,\ \frac{1}{2} - \varepsilon,\ \frac{3}{2} - \sigma,\ \zeta\right),$$

$$(11.4) \qquad V = \zeta^{1-c}F(a - c + 1,\ b - c + 1,\ 2 - c,\ \zeta) =$$

$$= \zeta^{\sigma - 1/2}F\left(\sigma + \varepsilon,\ \sigma - \varepsilon,\ \frac{1}{2} + \sigma,\ \zeta\right) .$$

Functions $F$ and $Z$ which satisfy all the conditions on the segments of the real axis, assume the form

$$F = \frac{B}{\sqrt{\zeta(\zeta + a)}} \left\{ \frac{ir}{p}(k\ \mathrm{tg}^2\theta - c)\ U + k\ \mathrm{tg}\ \theta(1 - i\ \mathrm{tg}\ \theta)\ V \right\} ,$$

$$(11.5) \qquad Z = B(1 - i\ \mathrm{tg}\ \theta)\ \frac{V}{\sqrt{\zeta(\zeta + a)}} .$$

Here $B$ is a constant subject to determination, $r$ and $p$ are coefficients of the auxiliary substitution [see formula (5.12)]. The ratio $r/p$ is expressed by means of $\Gamma$ functions:

$$\frac{r}{p} = \frac{\Gamma(2 - c)\Gamma(c - a)\Gamma(c - b)}{\Gamma(1 - a)\Gamma(1 - b)\Gamma(c)} = \frac{\cos \pi\varepsilon}{\pi\Gamma\left(\frac{1}{2} + \delta\right)}\ \Gamma\left(\frac{3}{2} - \delta\right)\Gamma(\delta - \varepsilon)\Gamma(\delta + \varepsilon).$$

Formulas (11.5) are easy to check in the passage from one segment to the other.

The length $L$ of the dam base is found by means of formula

$$(11.6) \qquad L = B(1 - i\ \mathrm{tg}\ \theta) \int_{-a}^{0} \frac{V\ d\zeta}{\sqrt{\zeta(\zeta + a)}} .$$

The expression for the flowrate is

$$(11.7) \qquad Q = Bk \int_0^1 \mathrm{tg}^2\theta \left(\frac{rq}{p} - s\right)(1 - \zeta)^{\sigma - 1/2}\ d\zeta + \frac{rBc}{p} \int_0^1 \frac{U\ d\zeta}{\sqrt{\zeta(\zeta + a)}} .$$

We notice the special case where $\theta = \pi$. This means that we have an horizontal drainage strip. Formulas (11.4) become:

$$U = F\left(\frac{1}{2} + \varepsilon,\ \frac{1}{2} - \varepsilon,\ \frac{1}{2},\ \zeta\right) ,$$

$$(11.8) \qquad V = \zeta^{1/2}F\left(1 + \varepsilon,\ 1 - \varepsilon,\ \frac{3}{2},\ \zeta\right) .$$

Gauss made these expressions available

$$F\left(\frac{1}{2} + \varepsilon, \frac{1}{2} - \varepsilon, \frac{1}{2}, \sin^2 t\right) = \frac{\cos 2\varepsilon t}{\cos t} ,$$

$$F\left(1 + \varepsilon, 1 - \varepsilon, \frac{3}{2}, \sin^2 t\right) = \frac{\sin 2\varepsilon t}{2\varepsilon \sin t \cos t} .$$

Putting $\sin^2 t = \zeta$, we find for our case:

$$(11.9)\quad \begin{cases} F\left(\frac{1}{2} + \varepsilon, \frac{1}{2} - \varepsilon, \frac{1}{2}, \zeta\right) = \dfrac{(\sqrt{1 - \zeta} + \sqrt{-\zeta})^{2\varepsilon} + (\sqrt{1 - \zeta} - \sqrt{-\zeta})^{2\varepsilon}}{2\sqrt{1 - \zeta}} , \\ F\left(1 + \varepsilon, 1 - \varepsilon, \frac{3}{2}, \zeta\right) = \dfrac{(\sqrt{1 - \zeta} + \sqrt{-\zeta})^{2\varepsilon} - (\sqrt{1 - \zeta} - \sqrt{-\zeta})^{2\varepsilon}}{2\varepsilon\sqrt{\zeta(1 - \zeta)}} . \end{cases}$$

Restricting ourselves to the case of an infinitely long dam, or, otherwise, considering the flow to an horizontal drain from infinity, we obtain for $\omega$ and $z$

$$\omega = \frac{i\sqrt{kc}}{2\varepsilon} A \int \frac{(\sqrt{1 - \zeta} + \sqrt{-\zeta})^{2\varepsilon} + (\sqrt{1 - \zeta} - \sqrt{-\zeta})^{2\varepsilon}}{\sqrt{\zeta(1 - \zeta)}} d\zeta ,$$

$$z = -\frac{Ai}{4\varepsilon} \int \frac{(\sqrt{1 - \zeta} + \sqrt{-\zeta})^{2\varepsilon} - (\sqrt{1 - \zeta} - \sqrt{-\zeta})^{2\varepsilon}}{\sqrt{\zeta(1 - \zeta)}} d\zeta .$$

Substitution $\zeta = \sin^2 t$ leads to integrals, expressed in finite form

$$(11.10)\quad \begin{cases} z = -\dfrac{Ai}{2\varepsilon} \displaystyle\int_0^t \sin 2\varepsilon t \, dt = -\dfrac{A}{4\varepsilon^2}(1 - \cos 2\varepsilon t) , \\ \omega = \dfrac{i\sqrt{kc}}{\varepsilon} A \displaystyle\int_0^t \cos 2\varepsilon t \, dt = \dfrac{i\sqrt{kc}}{2\varepsilon^2} A \sin 2\varepsilon t . \end{cases}$$

We shift the origin of coordinates to the point of exit of the free surface in the drain (on the tailwater segment). The length $\ell_0$ of the drainage strip is then found from (11.10) for $t = \pi/2$, when $x = -\ell_0$, $y = 0$:

$$\ell_0 = \frac{A}{4\varepsilon^2}(1 - \cos \pi\varepsilon) .$$

If we insert in the second of equations (11.10) $t = \pi/2$, $\varphi = 0$, $\psi = Q$, we find the relation between flowrate $Q$ and drainage strip

$$Q = \frac{\sqrt{kc}}{2\varepsilon^2} A \sin \pi\varepsilon = + \sqrt{k}(\sqrt{c + k} + \sqrt{k})\ell_0 .$$

To find the equation of the free surface, we put $\varphi + ky = 0$, $\psi + cx = 0$; for this $t = \pi/2 + i\lambda$, where $\lambda > 0$. We obtain the hyperbola

$$y^2 = \frac{c}{k} x^2 - 2\ell_0 \left(1 + \sqrt{\frac{k + c}{k}}\right) x .$$

§12. Seepage Through a Trapezoidal Dam without Evaporation.

Here we examine the particular case of the previous problem when $\varepsilon = 0$, i.e., no evaporation takes place. We find the following formulas for the elements of the dam:

$$
(12.1)\quad \begin{cases}
H = B \displaystyle\int_0^{1/(1+a)} \frac{F\left(\sigma, \frac{1}{2}, \sigma + \frac{1}{2}, 1 - x\right)}{(1 - x)^{1-\sigma} \sqrt{x}\sqrt{1 - (1 + a)x}}\, dx \,, \\[2ex]
L = B \displaystyle\int_0^1 \frac{F\left(\sigma, \sigma, \sigma + \frac{1}{2}, -ax\right)}{a^{1/2-\sigma} x^{1-\sigma} \sqrt{1 - x}}\, dx \,, \\[2ex]
Q = kB \dfrac{\Gamma^2(1 - \sigma)\left(\frac{1}{2} - \sigma\right) \sin^2\theta}{\Gamma^2\left(\frac{3}{2} - \sigma\right) \cos\theta} \displaystyle\int_0^1 \frac{F\left(1 - \sigma, 1 - \sigma, \frac{3}{2} - \sigma, 1 - x\right)}{(1 - x)^{\sigma - 1/2} \sqrt{x(x + a)}}\, dx, \\[2ex]
\ell_0 = B \displaystyle\int_0^1 \frac{F\left(\sigma, \sigma, \sigma + \frac{1}{2}, x\right)}{x^{1-\sigma} \sqrt{x(x + a)}}\, dx \,.
\end{cases}
$$

In the case of an infinitely long dam ($L = \infty$), S. V. Falkovich [7] derived an expression for $k\ell_0/Q$ in function of the angle of the tailwater slope with the horizontal

$$
(12.2)\qquad \frac{k\ell_0}{Q} = \frac{4}{\pi^2} \int_0^{\pi/2} (\operatorname{cotg} x)^{1-2\sigma} \ln \operatorname{cotg} \frac{x}{2}\, dx \,.
$$

G. K. Mikhailov [11] reduced formulas (12.1) to a form suitable for computations. Therefore he developed the hypergeometric functions in series and then carried out the integration. The results were these, suitable for $\alpha \gg 1$:

$$
H = \frac{\sqrt{\pi t}}{\sqrt{1+a}} \sum_{n=0} \frac{\mu_n + 2\nu_n\left[\ln 2 + \sum_{m=1}^{2n} \frac{(-1)^m}{m} + \ln\sqrt{1 + a}\right]}{\Gamma(n + 1)(1 + a)^n}\, \Gamma(n + \tfrac{1}{2}) \,,
$$

$$
L = \sum_{n=0} t(\nu_n I^n + \mu_n I^n) +
$$

$$
+ \frac{(-1)^n}{\sqrt{a - 1}} \frac{\Gamma^2(\sigma + n)\Gamma\left(\sigma + \frac{1}{2}\right)}{\Gamma^2(\sigma)\Gamma\left(\sigma + n + \frac{1}{2}\right)\Gamma(n + 1)(n + \sigma)}\, F\left(\tfrac{1}{2}, 1, n+\sigma+1, \tfrac{-1}{a-1}\right),
$$

(12.3)

$$Q = k\ell_0 \frac{\sin^2\theta}{\cos\theta} - k\frac{\Gamma^2(1-\sigma)\left(\frac{1}{2}-\sigma\right)}{\Gamma^2\left(\frac{3}{2}-\sigma\right)\operatorname{ctg}^2\theta} \frac{1}{\pi\sqrt{1+a}} \times$$

$$\times \sum_{n=0}^{\infty} \frac{\Gamma^2\left(n+\frac{1}{2}\right)\Gamma\left(\frac{3}{2}-\sigma\right)}{\Gamma\left(n+\frac{3}{2}-\sigma\right)\Gamma(n+1)\left(n+\frac{1}{2}\right)} F\left(\frac{1}{2}, 1, n+\frac{3}{2}, \frac{1}{1+a}\right),$$

$$\ell_0 = \frac{1}{\sqrt{1+a}} \sum_{n=0}^{\infty} \frac{\Gamma^2(\sigma+n)\Gamma\left(\sigma+\frac{1}{2}\right)}{\Gamma^2(\sigma)\Gamma\left(\sigma+n+\frac{1}{2}\right)\Gamma(n+1)(n+\sigma)} \times$$

$$\times F\left(\frac{1}{2}, 1, \sigma+n+1, \frac{1}{1+a}\right).$$

For simplicity one has assumed $B = 1$ and the following notations were introduced:

$$\bar{I}^n = -\int_{\frac{1}{1+a}}^{1/2} \frac{x^n \ln x\, dx}{\sqrt{x}\sqrt{(a+1)x-1}}, \qquad \bar{\bar{I}}^n = \int_{\frac{1}{1+a}}^{1/2} \frac{x^n\, dx}{\sqrt{x}\sqrt{(a+1)x-1}},$$

$$\frac{F\left(\sigma, \frac{1}{2}, \sigma+\frac{1}{2}, 1-x\right)}{(1-x)^{1-\sigma}} \equiv \frac{-t\left[(\ln x - R)F\left(\sigma, \frac{1}{2}, 1, x\right) + d_n x^n\right]}{(1-x)^{1-\sigma}} \equiv$$

$$\equiv t\sum_{n=0} (\mu_n x^n - \nu_n x^n \ln x), \quad \text{where} \quad t = \frac{\sigma}{\sqrt{\pi}} \frac{\Gamma\left(\sigma+\frac{1}{2}\right)}{\Gamma(\sigma+1)}.$$

The first part of the last identity is based on Gauss' formula to transform function $F(\alpha, \beta, \alpha+\beta, 1-x)$ into the function of argument $x$, the second is based on the usual development in power series.

Some computations were made by means of these formulas. By means of a passage to the limit for $\sigma \to 0$, G. K. Mikhailov obtained a simpler formula

$$\tilde{Q} = \frac{\tilde{h}^2}{1+\sqrt{1-\tilde{h}^2}}. \tag{12.4}$$

For $\sigma \to 1/2$, in the same way he derived

$$\tilde{Q} = \tilde{h} - (1-\tilde{h}) \ln\frac{1}{1-\tilde{h}}. \tag{12.5}$$

Here $\tilde{h} = H/H_1$ and $\tilde{Q} = Q/Q_0$, where $Q_0 = kH_1^2/L$, flowrate for maximum head.

We communicate briefly the derivation of formula (12.4). It is evident that for $\sigma \to 0$ also $a \to \infty$. However, in order to obtain finite values for $\tilde{h}$ and $\tilde{Q}$, it is necessary to put $\sigma \ln a = \alpha$ = constant. We examine the behaviour of $H\sqrt{a}$ for $a \to \infty$ and $\sigma \ln a = \alpha$. Therefore we reduce the corresponding formula of (12.3) to

$$H\sqrt{a+1} = t\pi(\mu_0 + 2\nu_0 \ln 2\sqrt{1+a}) + $$
$$+ t\sqrt{\pi} \sum_1^\infty \frac{\mu_n + 2\nu_n\left[\ln 2 + \sum\limits_{m=1}^{2n} \frac{(-1)^m}{m} + \ln\sqrt{1+a}\right]}{\Gamma(n+1)(1+a)^n} \Gamma\left(n+\frac{1}{2}\right)$$

We notice that

$$\mu_0 = R = \frac{1}{\sigma} - \psi(\sigma) + \psi(0) + \ln 4, \quad \nu_0 = 1 .$$

Here $\psi(x)$ is the logarithmic derivative of the $\Gamma$-function. Passing to the limit we have

$$\lim H\sqrt{a} = \pi(1 + 2\sigma \ln\sqrt{a}) = \pi(1+\alpha) , \tag{12.6}$$

since the sum on the right side tends to zero with increasing $a$. We examine now the behaviour of $\ell_0\sigma\sqrt{a}$:

$$\ell_0\sigma\sqrt{a} = \sqrt{\frac{a}{a+1}}\left\{F\left(\frac{1}{2}, 1, 1+\sigma, \frac{1}{1+a}\right) + \right.$$
$$\left. + \sigma^3 \sum_1^\infty \frac{\Gamma^2(\sigma+n)\Gamma\left(\sigma+\frac{1}{2}\right)}{\Gamma\left(\sigma+n+\frac{1}{2}\right)\Gamma(n+1)\Gamma^2(\sigma+1)(n+\sigma)} F\left(\frac{1}{2}, 1, \sigma+n+1, \frac{1}{1+a}\right)\right\}$$

The sum on the right side tends to zero for $\sigma \to 0$, and we have:

$$\lim \ell_0\sigma\sqrt{a} = 1 .$$

We now consider the quantity $Q\sqrt{a}/\sigma k$. Evidently

$$\lim \frac{Q\sqrt{a}}{\sigma k} = \lim \pi^2\sigma\ell_0\sqrt{a} -$$
$$- \lim 2\sigma \sum_{n=0}^\infty \frac{\Gamma^2\left(n+\frac{1}{2}\right)\Gamma\left(\frac{3}{2}-\sigma\right)}{\Gamma\left(n+\frac{3}{2}-\sigma\right)\Gamma(n+1)\left(n+\frac{1}{2}\right)} F\left(\frac{1}{2}, 1, n+\frac{3}{2}, \frac{1}{1+a}\right) ;$$

the second limit of the right side goes to zero for $\sigma \to \infty$ and $\lim Q\sqrt{a}/\sigma k = \pi^2$.

We finally study the quantity $2L\sigma\sqrt{a}$. We separate $L$ into two parts $(L = L_1 + L_2)$ corresponding to the two terms of the expression under the summation sign in the corresponding formula (12.3):

$$\lim 2L_1\sigma\sqrt{a} = \lim 2\sigma\sqrt{a}\, t \sum_0^\infty (\nu_n I^n + \mu_n \bar{\bar{I}}^n) =$$
$$= \lim 2\sigma\sqrt{a}\, t(I^0 + R\bar{\bar{I}}^0) + \lim 2\sigma t\sqrt{a} \sum_1^\infty (\nu_n I^n + \mu_n \bar{\bar{I}}^n) .$$

The second limit on the right side is zero, since $\bar{I}^n$ and $\bar{\bar{I}}^n$ go to zero as $a^{-1/2}$. Consequently

$$\lim 2L_1\sigma\sqrt{a} = \lim 2\sigma\sqrt{a}\, t(\bar{I}^0 + R\bar{\bar{I}}^0) = \lim 2t\sqrt{a}(\sigma\bar{I}^0 + \bar{\bar{I}}^0) \ .$$

But

$$\bar{\bar{I}}^0 = \frac{\ln a + \sqrt{a^2 - 1}}{\sqrt{1 + a}} \quad \text{and} \quad \lim 2t\sqrt{a}\,\bar{\bar{I}}^0 = 2\alpha \ .$$

Developing $\bar{I}^0$ into a converging series, we may show that

$$\lim 2t\sqrt{a}\,\bar{I}^0\sigma = \alpha^2$$

Consequently, $\lim 2L_1\sigma\sqrt{a} = \alpha^2 + 2\alpha$.

Passing on to the part with $L_2$, we notice that

$$L_2\sigma\sqrt{a} = \sqrt{\frac{a}{a-1}}\left\{F\left(\frac{1}{2}, 1, 1 + \sigma, \frac{-1}{a-1}\right) + \right.$$

$$\left. + \sigma^3\sum_{n=1}^{\infty} \frac{(-1)^n\Gamma^2(\sigma + n)\Gamma\left(\sigma + \frac{1}{2}\right)F\left(\frac{1}{2}, 1, n + \sigma + 1, \frac{-1}{a-1}\right)}{\Gamma^2(\sigma + 1)\Gamma\left(\sigma + n + \frac{1}{2}\right)\Gamma(n + 1)(n + \sigma)}\right\}$$

and $\lim 2L_2\sigma\sqrt{a} = 2$. Finally $\lim 2L\sigma\sqrt{a} = \alpha^2 + 2\alpha + 2$.

Consequently:

$$(12.7)\quad \lim \tilde{h} = \lim \frac{H \operatorname{cotg}\theta}{L} = \lim \frac{H}{\pi\sigma L} = \lim \frac{H\sqrt{a}}{\pi\sigma L\sqrt{a}} = \frac{2(1 + \alpha)}{\alpha^2 + 2\alpha + 2} \ ,$$

$$(12.8)\quad \lim \tilde{Q} = \lim \frac{Q \operatorname{cotg}^2\theta}{Lk} = \lim \frac{Q}{\pi^2\sigma^2 Lk} = \frac{1}{\pi^2}\lim \frac{Q\sqrt{a}}{k\sigma} \cdot \frac{1}{\sigma L\sqrt{a}} =$$

$$= \frac{2}{\alpha^2 + 2\alpha + 2} \ ..$$

Eliminating the parameter $\alpha$ from these equations and dropping the limit signs for $\tilde{h}$ and $\tilde{Q}$, we find (12.4).

In a similar but more tedious way, formula (12.5) is derived.

One must notice that for $\sigma = 0$ and $\sigma = 1/2$, the computation scheme for the dam looses its meaning, but formulas (12.4) and (12.5) preserve their significance. They may be considered as approximations for $\sigma \approx 0$ and $\sigma \approx 1/2$.

Graphs of formulas (12.4) and (12.5) and also results of computations by means of formulas (12.3) for $\sigma = 0.1$ and $\sigma = 0.25$ are given in Fig. 230. Letter m in Fig 230

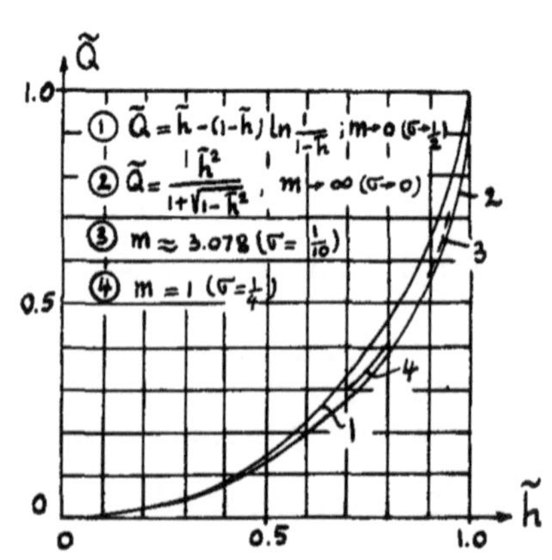


Fig. 230

designates the coefficient corresponding to the tailwater slope i.e., $m = \cot g\ \theta$. As is apparent from Fig. 230, formulas (12.4) and (12.5) give approximate curves between which lie the curves for all the slopes of the tailwater surface $(\sigma < 1/2)$.

We may arrive at the same limiting curves in another way. Therefore we consider first a dam built from anisotropic horizontally layered soil, for which $k_y = 0$ (see Chapter VIII, §9). The flow in such soil is horizontal, and

$$v_x = -k\frac{dh}{dx}.$$

We take an elemental horizontal strip at height $y$ above the dam foundation. The length of this strip is $\ell = L - y \cot g\ \theta = L(1 - y/H_1)$. The head difference at its ends is $H - y$. Therefore

$$v_x = k\frac{H-y}{\ell} = \frac{kH_1}{L} \cdot \frac{H-y}{H_1-y}.$$

We find the flowrate from formula

$$(12.9) \qquad Q = \frac{kH_1}{L}\int_0^H \frac{H-y}{H_1-y}\,dy = \frac{kH_1}{L}\left[H - (H_1 - H)\,\ell n\,\frac{H_1}{H_1-H}\right].$$

Formula (12.9) is compatible for the notations $\tilde{Q}$ and $\tilde{h}$ with formula (12.5). Formula (12.9) was given and recommended by N. T. Meletchenko [16] as approximate for isotropic dams.

We now consider seepage in a vertically layered soil for $k_y = \infty$. In this case vertical lines $x$ = constant are lines of equipotential (see Chapter VIII). As shown before, this scheme is the exact interpretation of Dupuit's equation, for which the free surface has the form

$$y^2 = ax + b.$$

We find the constants $a$ and $b$ from the condition that the parabola passes through point $C(0,H)$ and intersects the tailwater slope in some point $B(x_0,y_0)$ (Fig. 229). Apparently

$$y^2 = H^2 - \frac{H^2 - y_0^2}{x_0}x.$$

The flowrate $Q$ through section $CD$ (Fig. 229) is

$$(12.10) \qquad Q = k\frac{H^2 - y_0^2}{2x_0}.$$

The horizontal seepage velocity is constant in triangle $ABB'$. The difference in head between points $B'$ and $A$ is $y_0$, the length of the seepage path between them is $L - x_0$. Therefore the seepage rate in section $BB'$, equal to the flowrate in section $CD$, is:

$$(12.11) \qquad Q = ky_0\frac{y_0}{L - x_0} = ky_0 \operatorname{tg}\theta.$$

If we add to (12.10) and (12.11) the condition that point $(x_0, y_0)$ lies on the tailwater slope, then we finally obtain this formula for the seepage rate:

$$Q = k \frac{H^2}{L + \sqrt{L^2 - H^2 \cot g^2 \theta}} \quad . \tag{12.12}$$

In notations $\tilde{Q}$ and $\tilde{h}$ formula (12.12) assumes the form of (12.4). Formula (12.12) was recommended by N. N. Pavlovsky as approximate for isotropic dams [17].

Since the curves $\tilde{Q} = \tilde{Q}(\tilde{h})$ for isotropic soil must lie between the curves of formulas (12.9) and (12.12), this consideration about seepage in maximum — anisotropic soils allow us to estimate much better the quantity of the seepage flowrate in a dam. Application of the model of maximum — anisotropic soils proves also to be useful in some other problems of the seepage theory (see Chapter XIII).

Curves of the flowrate $\tilde{Q} = \tilde{Q}(\tilde{h})$ in real dams, as shown in Fig. 230, are very close to the curve that is drawn by means of formula (12.4) [or, what is the same, by means of formulas (12.12)]. Therefore, for practical computations we may recommend N. N. Pavlovsky's formula (12.12).

Concerning the dependence of the ratio of the length of the seepage surface to the seepage rate $(k\ell_0/Q)$ on the angle of the down stream slope, this dependence is, as shown by computations, practically the same for all real dams. For values H/L not exceeding those encountered in practice, the ratio $k\ell_0/H$ depends only on the angle of the down stream slope with

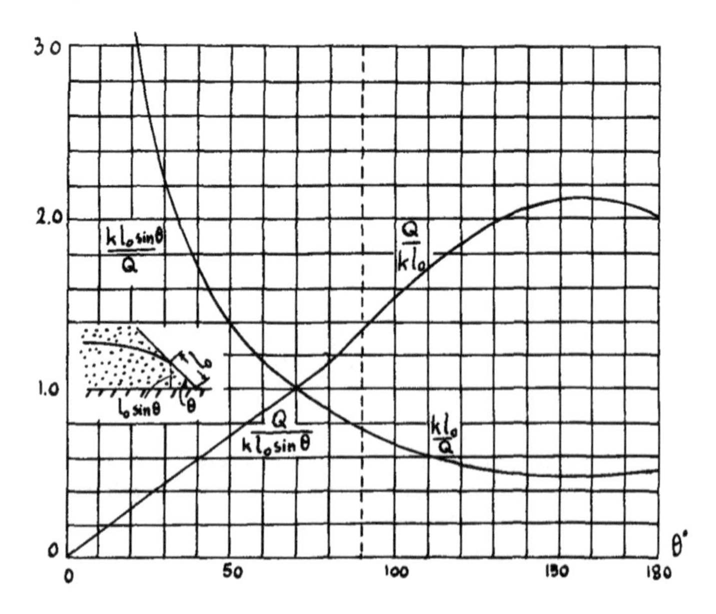


Fig. 231

the horizontal. Therefore, for this ratio one may use formula (12.12) by S. V. Falkovich, reduced to this form by G. K. Mikhailov

$$\frac{k\ell_0}{Q} = \frac{4^\sigma}{\pi^2} \sum_{n=0}^{\infty} \sum_{m=0}^{n} \frac{(-1)^{n+m}\,\Gamma(m - 1 + 2\sigma)}{(n + \sigma)^2\,\Gamma(-1 + 2\sigma)m!} . \tag{12.13}$$

Graphs of these dependence are given in Fig. 231, in which for convenience for $\theta < 90°$ values of $k\ell_0 \sin\theta/Q$ instead of $k\ell_0/Q$ are used, as they are more often encountered in practical computations.

For $\theta < \pi/2$, formula (12.13) may be represented, with error less than 3%, by the approximate formula [11]:

$$\frac{k\ell_0}{Q} = \frac{\alpha \operatorname{cotg}\theta + 6 - \alpha}{4 \sin\theta} , \tag{12.14}$$

where $\alpha = 4$ for $\operatorname{cotg}\theta > 1$ and $\alpha = 3$ for $\operatorname{cotg}\theta < 1$ .

In order to pass from the schemes that we studied to real dam profiles, it is necessary to consider various slopes of the inflow surface, different from the vertical. It is known that the slope of the inflow of a dam shows little effect on the quantity of seepage. In the last decade many authors have used the method for a vertical slope and applied it to a non-vertical slope. The real dam was replaced by an equivalent one and the flowrate was computed for a vertical upstream slope.

Several semiempirical formulas were proposed for the length of the base $\Delta L$ of the equivalent rectangular that replaces the triangle of the upstream face. From an analysis of the existing hydrodynamic investigations, G. K. Mikhailov found that for $\operatorname{tg}\psi < 4/3$ this quantity is expressed with sufficient accuracy by

$$\Delta L = \frac{H}{2 + \operatorname{tg}\psi} , \tag{12.15}$$

where $\psi$ is the angle of the upstream face with the horizontal. For $\operatorname{tg}\psi > 4/3$, formula (12.15) may be replaced more accurately for this interval by

$$\Delta L = \frac{H}{2} \cos\psi . \tag{12.16}$$

Influence of tailwater. To obtain a guiding estimate of the influence of tailwater on seepage through the dam, G. K. Mikhailov considered a triangular dam for maximum pool level, reaching to the top of the triangle. For such a dam with gentle slopes, when the angle at the top is larger than 90°, for maximum pool level, the free surface in the dam exists; an exact solution of the problem is difficult in this case. If the angle at the top is smaller than 90°, then groundwater flow fills the entire region of the dam. In this case an exact solution may be obtained by means of conformal mapping of the triangular dam onto a polygon of the hodograph plane. A simpler problem, when one face of the dam is vertical, has been treated by A. M. Mkhitarian in the seepage computation of a trapezoidal dam on impervious foundation. In this case the hodograph region degenerates, into a triangle.

Fig. 232 renders some graphs, characterizing the influence of tailwater on the seepage rate. $Q_{max}$ means here the seepage flow when there is no tailwater.

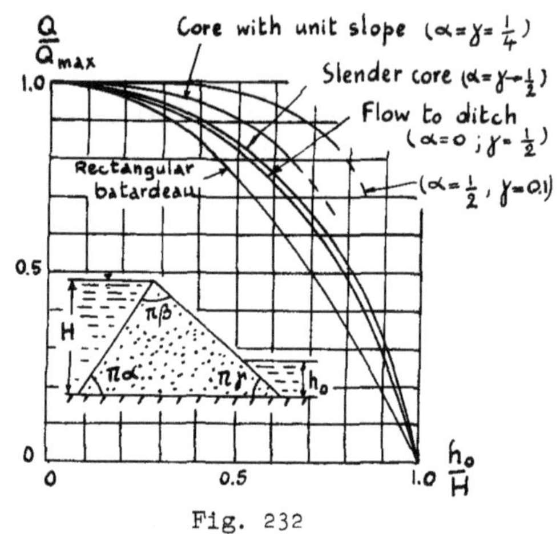


Fig. 232

The problem of seepage in similar triangular dams without tailwater was treated by M. M. Morgulis [19].

Curve $\alpha = 1/2$, $\gamma = 0.1$ was constructed by means of computations by A. M. Mkhitarian [18], the remaining ones by means of computations by G. K. Mikhailov.

As is apparent from Fig. 232, the presence of tailwater does not affect very much the drop in seepage. The effect decreases with increasing slope of the downstream dam face.

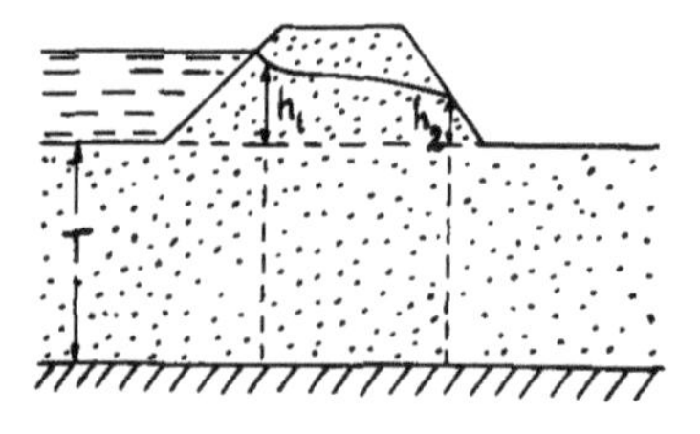


Fig. 233

We notice that for the case of a dam with trapezoidal profile, on a pervious stratum of finite depth (Fig. 233) and with tailwater level exactly coinciding with the surface of the stratum, A. M. Mkhitarian carried out computations of flowrates. He used the method of fragments, dividing the dam in three parts by means of two verticals. Solutions for each fragment were carried out by means of hydromechanical analysis and for some cases put in graphical form.

CHAPTER VIII.

# SEEPAGE IN HETEROGENEOUS AND ANISOTROPIC SOILS.

# SEEPAGE OF TWO FLUIDS.

## A. Heterogeneous Soils.

### §1. Weir on Two Strata of Same Thickness.

At present the following results are available in the seepage theory of heterogeneous soils. The exact solution for seepage in two horizontal strata of the same thickness is given for: the flow about a flat bottom weir and the flow about a cut-off [1]. There is an exact solution for the problem of flow about two weirs, symmetrical with respect to a vertical separation line, and also for the flow about one flat bottom weir on soil consisting of two strata. Besides these, there exists a series of exact solutions about seepage in multilayered soils with a singular feature, such as a point vortex or source, both in the plane [2] and in three dimensions. Choosing the intensity of such singularities, distributed along the contours or in the flow regions of hydraulic structures, we may calculate the solution for various contours of hydraulic structures.

Since in nature the ratio of the seepage coefficients of adjacent strata is very high — usually more than 10, but sometimes of the order of $10^2 - 10^4$ and higher — the development of approximate theories is of interest.

We examine the following problem. A flat bottom weir (or the foundation of an hydraulic structure, having the shape of an horizontal strip) of length $2\ell$ rests on a stratum of depth $h$ with seepage coefficient $\kappa_1$. The lower stratum, also of depth $h$, has a seepage coefficient $\kappa_2$ and rests on impervious bedrock (Fig. 234). All the flow elements must be determined.

Quantities related to the first stratum are indicated by the index 1, those related to the second region by index 2. We consider only the right half of the figure. We write down the conditions that must be satisfied along the contour of regions $A_1DCBA$ and $ABEA_2$.

The boundaries of the reservoirs are equipotential lines. Consequently, along these we have the conditions:

along the boundary of the head reservoir

$$\varphi_1 = -\kappa_1 \frac{H}{2}, \tag{1.1}$$

along $DA_1$

$$\varphi_1 = \kappa_1 \frac{H}{2}, \tag{1.2}$$

where $H = H_1 - H_2$, the acting head.

2l
y
H1
H2
A1
C
D
h
A
B
x
h
A2
E


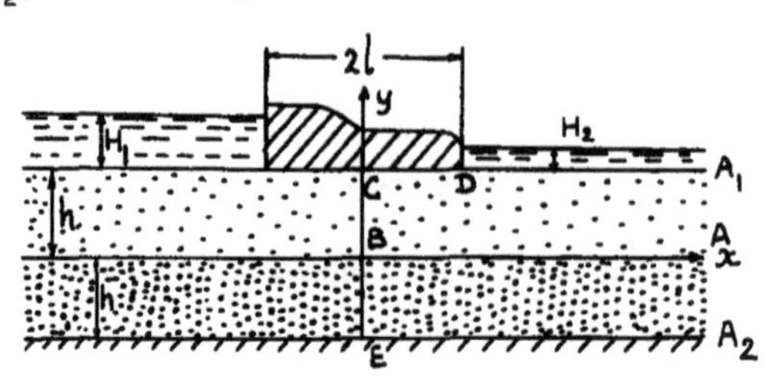

Fig. 234.

Due to the symmetry of the flow, along CB, which is a line of constant potential, we have

$$\varphi_1 = 0 \quad . \tag{1.3}$$

Line BE must also be an equipotential line, so that there $\varphi_2 =$ constant. We put for BE

$$\varphi_2 = 0 \quad . \tag{1.4}$$

Furthermore, CD and $EA_2$ are streamlines. We put:

along $EA_2$

$$\psi_2 = 0 \quad , \tag{1.5}$$

along CD

$$\psi_1 = Q \quad . \tag{1.6}$$

At the boundary AB of the two media, these conditions (see Chapter II, §3) must be satisfied

$$\frac{\varphi_1}{\kappa_1} = \frac{\varphi_2}{\kappa_2} , \quad \psi_1 = \psi_2 \quad . \tag{1.7}$$

We map the region $A_1DCBEA_2$ of the $z = x + iy$ plane onto the region of the complex variable $\zeta$, which has a cut, so that region $A_1DCBA$ maps

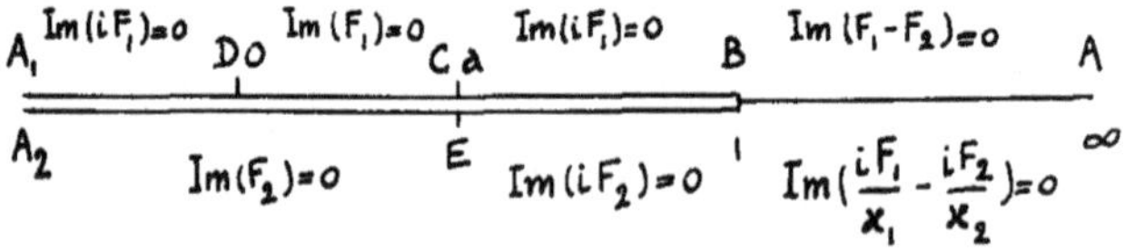

A1
Im(iF1)=0
D 0
Im(F1)=0
C a
Im(iF1)=0
B
Im(F1-F2)=0
A
A2
Im(F2)=0
E
Im(iF2)=0
1
Im(iF1/x1 - iF2/x2)=0
∞


Fig. 235.

onto the upper half-plane, region $ABEA_2$ onto the lower.

The relationship between z and $\zeta$ has the form

$$\zeta = \frac{1 + a}{2} + \frac{1 - a}{2} \cosh \frac{\pi z}{h} \quad , \tag{1.8}$$

where a corresponds to points C and E in the $\zeta$-plane. Making $\zeta = 0$, $z = \ell + hi$ in equation (1.8) we find

$$a = \tanh^2 \frac{\pi \ell}{2h} \quad .$$

Thus it is easy to compute $a$ for given $\ell$ and $h$.

We consider the complex potentials $\omega_1 = \varphi_1 + i\psi_1$ and $\omega_2 = \varphi_2 + i\psi_2$ of the first and second flow region. In the $\zeta$-plane, $\omega_1(\zeta)$ is considered for the upper half-plane, $\omega_2(\zeta)$ for the lower. We may express the boundary conditions by means of the functions $\omega_1$ and $\omega_2$:

along $DA_1$ $\qquad \mathrm{Im}(i\omega_1) = \kappa_1 \frac{H}{2}$ ,

along CD $\qquad \mathrm{Im}\,(\omega_1) = Q$ ,

along CB $\qquad \mathrm{Im}(i\omega_1) = 0$ ,

along BE $\qquad \mathrm{Im}(i\omega_2) = 0$ ,

along $EA_2$ $\qquad \mathrm{Im}\,(\omega_2) = 0$ ,

along AB $\qquad \mathrm{Im}\left(\frac{i\omega_1}{\kappa_1} - \frac{i\omega_2}{\kappa_2}\right) = 0 \quad , \qquad \mathrm{Im}(\omega_1 - \omega_2) = 0.$

In order to obtain homogeneous conditions at the contour, we introduce the functions

$$F_1 = \frac{d\omega_1}{d\zeta} , \quad F_2 = \frac{d\omega_2}{d\zeta} .$$

For $F_1$ and $F_2$ we obtain conditions written near the corresponding segments of Fig. 235.

In point C (Fig. 235), the lines $\varphi_1$ = constant and $\psi_1$ = constant converge and consequently in the complex potential plane we have here a 90° angle. Therefore the series for $\omega_1(\zeta)$ about point $\zeta = a$ must have the form (see Chapter II, §5)

$$\omega_1(\zeta) = \sqrt{\zeta - a}\,[a_0 + a_1(\zeta - a) + \dots] .$$

In the same way, in point E $\omega_2(\zeta)$ has the form

$$\omega_2(\zeta) = \sqrt{\zeta - a}\,[b_0 + b_1(\zeta - a) + \dots] .$$

From this we find for the functions $F_1(\zeta)$ and $F_2(\zeta)$

$$F_1(\zeta) = \frac{c_0 + c_1(\zeta - a) + \dots}{\sqrt{\zeta - a}} = \frac{\Phi_1(\zeta)}{\sqrt{\zeta - a}} ,$$

$$F_2(\zeta) = \frac{d_0 + d_1(\zeta - a) + \dots}{\sqrt{\zeta - a}} = \frac{\Phi_2(\zeta)}{\sqrt{\zeta - a}} , \tag{1.10}$$

where $\Phi_1(\zeta)$ and $\Phi_2(\zeta)$ are functions for which $\zeta = a$ is an ordinary point. We consider the functions $\Phi_1$ and $\Phi_2$ instead of $F_1$ and $F_2$. Along the real axis of the half plane $\zeta$ for $\zeta = \xi > a$, the root $\sqrt{\zeta - a}$ has a real value. Consequently the conditions for functions $\Phi_1$ and $\Phi_2$ to the right of $a$ will be the same as for $F_1$ and $F_2$. For $\zeta = \xi < a$, the root $\sqrt{\zeta - a}$ has an imaginary value, and therefore the conditions on $A_1D$, DC and $A_2E$ are changed as shown in Fig. 236.

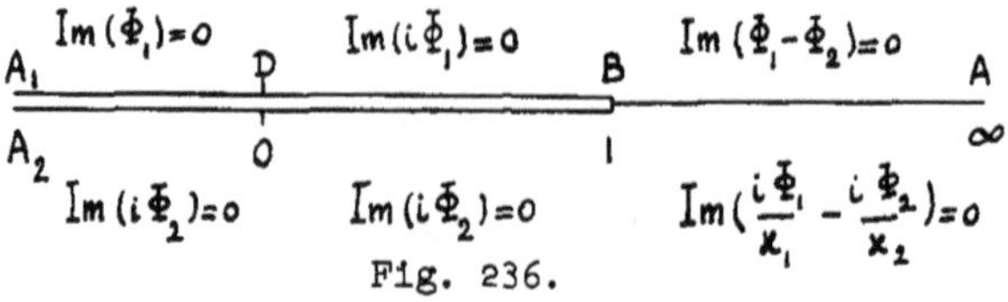

$A_1$ $\operatorname{Im}(\Phi_1)=0$ $D$ $\operatorname{Im}(i\Phi_1)=0$ $B$ $\operatorname{Im}(\Phi_1-\Phi_2)=0$ $A$

$A_2$ $0$ $1$ $\infty$

$\operatorname{Im}(i\Phi_2)=0$ $\operatorname{Im}(i\Phi_2)=0$ $\operatorname{Im}\left(\frac{i\Phi_1}{\varkappa_1}-\frac{i\Phi_2}{\varkappa_2}\right)=0$

Fig. 236.

Functions $F_1$ and $F_2$ must be analytic functions, respectively in the upper and lower half-planes, having regular singular points in $\zeta = 0$, $1$, $a$, $\infty$. We assume that each of these functions may be continued (analytically) in the other half-plane and we try to find $F_1$ and $F_2$ as solutions of a linear differential equation of the $2^{nd}$ order.

We examine the behaviour of the functions about the singular points.

About $\zeta = 0$. Function $\omega_1(\zeta)$ must have the form

$$\omega_1(\zeta) = \zeta^{1/2}(a_1 + a_1\zeta + \dots) \quad ,$$

since here streamline and equipotential line converge. Then

$$F_1(\zeta) = \zeta^{-1/2}(a_0' + \dots) \quad ,$$

i.e., $F_1(\zeta)$, and also $\Phi_1(\zeta)$, belong to exponent $-1/2$. Point $\zeta = 0$ is ordinary for $F_2(\zeta)$, i.e., the exponent for it is zero as well for $F_2(\zeta)$ as for $\Phi_2(\zeta)$.

About $\zeta = 1$. We assume that functions $\Phi_1$ and $\Phi_2$ may be continued through segment BA. We make a cut along segment BA. We change some notations, i.e., we limit the notations $\Phi_1$, $\Phi_2$ to the values of the functions under consideration at the lower side of the cut and call $\Phi_1^*$, $\Phi_2^*$ the values of the same functions at the upper side of the cut. Then the difference $\Phi_1 - \Phi_2$ goes over into $\Phi_1^* - \Phi_2^*$. But since along

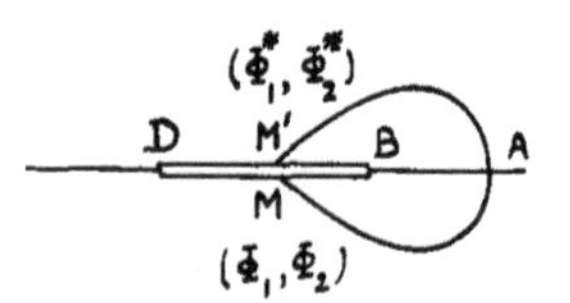


Fig. 237

BA the difference $\Phi_1 - \Phi_2$ assumes real values, in point M', which we consider as the complex conjugate of M, this difference changes into the complex conjugate,

$$\Phi_1^* - \Phi_2^* = \bar{\Phi}_1 - \bar{\Phi}_2 \quad . \tag{1.11}$$

Exactly in the same way, the second condition on AB gives:

$$\frac{i\Phi_1^*}{\kappa_1} - \frac{i\Phi_2^*}{\kappa_2} = -\frac{i\bar{\Phi}_1}{\kappa_1} + \frac{i\bar{\Phi}_2}{\kappa_2} \quad . \tag{1.12}$$

Solving these equations for $\Phi_1^*$ and $\Phi_2^*$, we obtain:

$$\begin{aligned} \Phi_1^* &= \frac{\kappa_1 + \kappa_2}{\kappa_1 - \kappa_2}\bar{\Phi}_1 - \frac{2\kappa_1}{\kappa_1 - \kappa_2}\bar{\Phi}_2 , \\ \Phi_2^* &= \frac{2\kappa_2}{\kappa_1 - \kappa_2}\bar{\Phi}_1 - \frac{\kappa_1 + \kappa_2}{\kappa_1 - \kappa_2}\bar{\Phi}_2 \end{aligned} \tag{1.13}$$

The conditions on segment CB, in the new notations, must be represented as

$$\mathrm{Im}(i\Phi_1^*) = 0, \quad \mathrm{Im}(i\Phi_2) = 0$$

or

$$\Phi_1^* + \bar{\Phi}_1^* = 0, \quad \Phi_2 + \bar{\Phi}_2 = 0 \quad , \tag{1.14}$$

from which

$$\bar{\Phi}_1^* = -\Phi_1^* , \quad \bar{\Phi}_2 = -\Phi_2 \quad .$$

Substituting all quantities by their conjugate in equality (1.13) for $\Phi_1^*$, we obtain:

$$\bar{\Phi}_1^* = \frac{\kappa_1 + \kappa_2}{\kappa_1 - \kappa_2}\Phi_1 - \frac{2\kappa_1}{\kappa_1 - \kappa_2}\Phi_2 \quad .$$

Adding the last equality with equality (1.13) for $\Phi_1^*$ and accounting for (1.14), we find:

$$\Phi_1 + \bar{\Phi}_1 = 0 \quad .$$

Therefore functions $\bar{\Phi}_1^*$ and $\Phi_2^*$ may be expressed in $\Phi_1$ and $\Phi_2$:

$$\begin{aligned} \Phi_1^* &= \frac{\kappa_1 + \kappa_2}{\kappa_2 - \kappa_1}\Phi_1 - \frac{2\kappa_1}{\kappa_2 - \kappa_1}\Phi_2 \quad , \\ \Phi_2^* &= \frac{2\kappa_2}{\kappa_2 - \kappa_1}\Phi_1 - \frac{\kappa_2 + \kappa_1}{\kappa_2 - \kappa_1}\Phi_2 \quad , \end{aligned} \tag{1.15}$$

The characteristic equation

$$\begin{vmatrix} \dfrac{\kappa_2 + \kappa_1}{\kappa_2 - \kappa_1} - \lambda & -\dfrac{2\kappa_1}{\kappa_2 - \kappa_1} \\ \dfrac{2\kappa_2}{\kappa_2 - \kappa_1} & -\dfrac{\kappa_2 + \kappa_1}{\kappa_2 - \kappa_1} - \lambda \end{vmatrix} = 0$$

has roots $\lambda = 1$, $\lambda' = -1$. The exponents for which functions $\omega_1$ and $\omega_2$ remain finite in point B, are respectively:

$$\gamma = \frac{\ln \lambda}{2\pi i} = 0, \qquad \gamma' = \frac{\ln \lambda'}{2\pi i} = -\frac{1}{2} \; .$$

About $\zeta = \infty$. A positive contour of the point at infinity may be replaced by a negative contour of points D and B. Therefore, the functions in the analysis must be continued through segment BA. Consequently, formulas (1.13) hold also in this case. If we call $\Phi_1$, $\Phi_2$ the values of the function up to the contour, i.e., at the upper side of the cut (Fig. 232), and $\Phi_1^*$, $\Phi_2^*$ their values after the contour, i.e., at the lower side of the cut, then the conditions at AD are

$$\mathrm{Im}(\Phi_1) = 0, \quad \mathrm{Im}(i\Phi_2^*) = 0 \quad ,$$

or

$$\bar{\Phi}_1 = \Phi_1 \; , \quad \bar{\Phi}_2^* = -\Phi_2^* \quad .$$

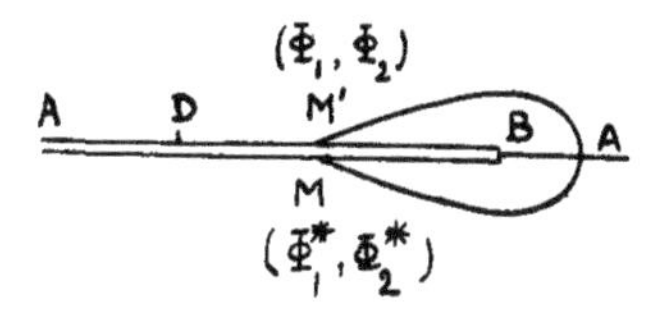


Fig. 238

In the second equation (1.13) we replace all quantities by their complex conjugate. We obtain:

$$\bar{\Phi}_2^* = -\Phi_2^* = \frac{2\kappa_2}{\kappa_1 - \kappa_2}\Phi_1 - \frac{\kappa_1 + \kappa_2}{\kappa_1 - \kappa_2}\Phi_2 \quad .$$

Adding this equality to the second of (1.13), we find:

$$0 = \frac{4\kappa_2}{\kappa_1 - \kappa_2}\Phi_1 - \frac{\kappa_1 + \kappa_2}{\kappa_1 - \kappa_2}(\Phi_2 + \bar{\Phi}_2) \quad ,$$

from which

$$\bar{\Phi}_2 = -\Phi_2 + \frac{4\kappa_2}{\kappa_1 + \kappa_2}\Phi_1 \quad . \tag{1.16}$$

Substituting $\bar{\Phi}_1$ and $\bar{\Phi}_2$ in equation (1.13) we obtain:

$$\Phi_1^* = -\frac{\kappa_1^2 - 6\kappa_1\kappa_2 + \kappa_2^2}{(\kappa_1 - \kappa_2)(\kappa_1 + \kappa_2)}\Phi_1 + \frac{2\kappa_1}{\kappa_1 - \kappa_2}\Phi_2 \quad ,$$

$$\Phi_2^* = \frac{2\kappa_2}{\kappa_1 - \kappa_2}\Phi_1 + \frac{\kappa_1 + \kappa_2}{\kappa_1 - \kappa_2}\Phi_2 \quad .$$

The characteristic equation

$$\begin{vmatrix} \dfrac{\kappa_1^2 - 6\kappa_1\kappa_2 + \kappa_2^2}{(\kappa_1 - \kappa_2)(\kappa_1 + \kappa_2)} - \lambda & \dfrac{2\kappa_1}{\kappa_1 - \kappa_2} \\ -\dfrac{2\kappa_2}{\kappa_1 - \kappa_2} & \dfrac{\kappa_1 + \kappa_2}{\kappa_1 - \kappa_2} - \lambda \end{vmatrix} = \lambda^2 - 2\frac{\kappa_1 - \kappa_2}{\kappa_1 + \kappa_2}\lambda + 1 = 0$$

has the roots

$$\lambda = \frac{\sqrt{\kappa_1} + i\sqrt{\kappa_2}}{\sqrt{\kappa_1} - i\sqrt{\kappa_2}}, \qquad \lambda' = \frac{\sqrt{\kappa_1} - i\sqrt{\kappa_2}}{\sqrt{\kappa_1} + i\sqrt{\kappa_2}} \quad .$$

We put:

$$\operatorname{tg} \varepsilon\pi = \sqrt{\frac{\kappa_2}{\kappa_1}}, \qquad \sin \varepsilon\pi = \sqrt{\frac{\kappa_2}{\kappa_1 + \kappa_2}}, \qquad \cos \varepsilon\pi = \sqrt{\frac{\kappa_1}{\kappa_1 + \kappa_2}} \quad .$$

Then

$$\lambda = e^{2\pi\varepsilon i}, \qquad \lambda' = e^{-2\pi\varepsilon i},$$

from which

$$\beta = \frac{\ln \lambda}{2\pi i} = \varepsilon + n, \quad \beta' = \frac{\ln \lambda'}{2\pi i} = -\varepsilon + n' \quad .$$

The condition of finiteness of the functions $\omega_1$ and $\omega_2$ for $\zeta = 0$ gives for $n$ and $n'$ the values

$$n = n' = 1 \quad ,$$

so that

$$\beta = 1 + \varepsilon, \quad \beta' = 1 - \varepsilon \quad .$$

Assuming that $\Phi_1$ and $\Phi_2$ do not have any other singular points, we find that $\Phi_1$ and $\Phi_2$ represent branches of the Riemann function

$$Y = P \begin{Bmatrix} 0 & \infty & 1 & \\ 0 & a & 0 & \zeta \\ 1 - c & b & c - a - b & \end{Bmatrix} = P \begin{Bmatrix} 0 & \infty & 1 & \\ 0 & 1 + \varepsilon & 0 & \zeta \\ -\frac{1}{2} & 1 - \varepsilon & -\frac{1}{2} & \end{Bmatrix} . \tag{1.17}$$

We have:

$$a = 1 + \varepsilon, \quad b = 1 - \varepsilon, \quad c = \frac{3}{2} .$$

$\Phi_1$ and $\Phi_2$ are linear combinations of the integrals of equation

$$\zeta(1 - \zeta)Y'' + [c - (a + b + 1)\zeta]Y' - abY = 0$$

or

$$\zeta(1 - \zeta)Y'' + \left(\frac{3}{2} - 3\zeta\right)Y' - (1 - \varepsilon^2)Y = 0 . \tag{1.18}$$

Indeed, about $\zeta = 0$ we may take a fundamental system of solutions

$$U = F(a,b,c,\zeta) = F\left(1 + \varepsilon, 1 - \varepsilon, \frac{3}{2}, \zeta\right) ,$$

$$V = \zeta^{1-\sigma}F(a - c + 1, b - c + 1, 2 - c, \zeta) = \tag{1.19}$$

$$= \zeta^{-1/2}F\left(\frac{1}{2} + \varepsilon, \frac{1}{2} - \varepsilon, \frac{1}{2}, \zeta\right) . \tag{1.19}$$

where the hypergeometric function is determined by the series

$$F(a,b,c,\zeta) = 1 + \frac{ab}{c}\zeta + \frac{a(a + 1)b(b + 1)}{1\cdot 2\cdot c(c + 1)}\zeta^2 + \ldots$$

The obtained hypeometric functions may be represented in the form (see literature to Chapter VII, [25])

$$F\left(\frac{1}{2} + \varepsilon, \frac{1}{2} - \varepsilon, \frac{1}{2}, \zeta\right) = \frac{(\sqrt{1 - \zeta} + \sqrt{-\zeta})^{2\varepsilon} + (\sqrt{1 - \zeta} - \sqrt{-\zeta})^{2\varepsilon}}{2\sqrt{1 - \zeta}}$$

(1.20)

$$F\left(1 + \varepsilon, 1 - \varepsilon, \frac{3}{2}, \zeta\right) = \frac{(\sqrt{1 - \zeta} + \sqrt{-\zeta})^{2\varepsilon} - (\sqrt{1 - \zeta} - \sqrt{-\zeta})^{2\varepsilon}}{4\varepsilon i\sqrt{\zeta(1 - \zeta)}}$$

Functions $\Phi_1$ and $\Phi_2$ are expressed linearly in U and V, i.e.,

$$\Phi_1 = AU + BV ,$$

$$\Phi_2 = CU + DV ,$$

where A, B, C, D are complex constants.

Since on segment DB, where $0 < \zeta < 1$, functions U and V as-- sume real values, it is necessary for the satisfaction of the conditions

$$\mathrm{Im}(i\Phi_1) = \mathrm{Im}(i\Phi_2) = 0$$

that the constants A, B, C, D be imaginary.

Therefore, changing the notations, we have

$$\Phi_1 = AiU + BiV ,$$

$$\Phi_2 = CiU + DiV ,$$

where A, B, C, D are real constants. In order that the function $\Phi_1$ be real on the segment AB, where $\zeta < 0$, and the function $\Phi_2$ be imaginary, it is necessary that

$$A = D = 0 .$$

Therefore

$$\Phi_1 = BiV = \frac{Bi}{\sqrt{\zeta}} F\left(\frac{1}{2} + \varepsilon, \frac{1}{2} - \varepsilon, \frac{1}{2}, \zeta\right) ,$$

$$\Phi_2 = CiU = CiF\left(1 + \varepsilon, 1 - \varepsilon, \frac{3}{2}, \zeta\right)$$

or

$$\Phi_1 = \frac{Bi}{2\sqrt{\zeta(1-\zeta)}} \{(\sqrt{1-\zeta} + \sqrt{-\zeta})^{2\varepsilon} + (\sqrt{1-\zeta} - \sqrt{-\zeta})^{2\varepsilon}\} ,$$

$$\Phi_2 = \frac{C}{4\varepsilon\sqrt{\zeta(1-\zeta)}} \{(\sqrt{1-\zeta} + \sqrt{-\zeta})^{2\varepsilon} - (\sqrt{1-\zeta} - \sqrt{-\zeta})^{2\varepsilon}\} .$$

We assume now that $\zeta > 1$. Then, progressing around point $\zeta = 1$ along a half circle, we obtain:

$$(1 - \zeta)^{1/2} = \mp i(\zeta - 1)^{1/2} ,$$

depending upon which half-plane the function is considered, upper or lower. For function $\Phi_1$, one has to take the minus sign, for function $\Phi_2$ the plus sign.

Exactly in the same way, when going from $\zeta < 0$ to values $\zeta > 0$, we obtain:

$$\sqrt{-\zeta} = \mp i\sqrt{\zeta} ,$$

in which the minus sign is related to function $\Phi_1$, the plus sign to $\Phi_2$.

For $\zeta > 1$, functions $\Phi_1$ and $\Phi_2$ may be represented as

$$\Phi_1 = \frac{Bi}{2} \frac{e^{-\pi\varepsilon i}(\sqrt{\zeta - 1} + \sqrt{\zeta})^{2\varepsilon} + e^{\pi\varepsilon i}(\sqrt{\zeta} - \sqrt{\zeta - 1})^{2\varepsilon}}{-i\sqrt{\zeta(\zeta - 1)}} , \tag{1.21}$$

$$\Phi_2 = \frac{C}{4\varepsilon i} \frac{e^{\pi\varepsilon i}(\sqrt{\zeta - 1} + \sqrt{\zeta})^{2\varepsilon} - e^{-\pi\varepsilon i}(\sqrt{\zeta} - \sqrt{\zeta - 1})^{2\varepsilon}}{\sqrt{\zeta(\zeta - 1)}} . \tag{1.22}$$

Condition $Im(\Phi_1 - \Phi_2) = 0$ for $\zeta > 1$ leads to equality

$$Im\left(- Be^{-\pi\varepsilon i} + \frac{iC}{2\varepsilon} e^{\pi\varepsilon i}\right) = B \sin \varepsilon\pi + \frac{C}{2\varepsilon} \cos \varepsilon\pi = 0 ,$$

from which

$$C = - 2\varepsilon \operatorname{tg} \varepsilon\pi B .$$

The second condition on segment AB is identically satisfied.

For functions $F_1$ and $F_2$, we obtain, based on (1.10), (1.21) and (1.22):

$$F_1 = \frac{Bi \{(\sqrt{1-\zeta} + \sqrt{-\zeta})^{2\varepsilon} + (\sqrt{1-\zeta} - \sqrt{-\zeta})^{2\varepsilon}\}}{2\sqrt{\zeta(1-\zeta)(\zeta - a)}} ,$$

(1.23)

$$F_2 = - \frac{B \operatorname{tg} \varepsilon\pi \{(\sqrt{1-\zeta} + \sqrt{-\zeta})^{2\varepsilon} - (\sqrt{1-\zeta} - \sqrt{-\zeta})^{2\varepsilon}\}}{2\sqrt{\zeta(1-\zeta)(\zeta - a)}} .$$

Functions $\omega_1$ and $\omega_2$ are determined by means of the integrals

$$(1.24)\quad \begin{cases} \omega_1 = \dfrac{Bi}{2}\displaystyle\int \dfrac{(\sqrt{1-\zeta}+\sqrt{-\zeta})^{2\varepsilon}+(\sqrt{1-\zeta}-\sqrt{-\zeta})^{2\varepsilon}}{\sqrt{\zeta(1-\zeta)(\zeta-a)}}\,d\zeta\ , \\ \omega_2 = -\dfrac{B\,\mathrm{tg}\,\varepsilon\pi}{2}\displaystyle\int \dfrac{(\sqrt{1-\zeta}+\sqrt{-\zeta})^{2\varepsilon}-(\sqrt{1-\zeta}-\sqrt{-\zeta})^{2\varepsilon}}{\sqrt{\zeta(1-\zeta)(\zeta-a)}}\,d\zeta\ . \end{cases}$$

To determine constant B, we take the integral for $\omega_1$ between limits $\zeta = a$, $\zeta = 0$. Since

$$\omega_1(0) = \frac{\kappa_1 H}{2} + Qi, \quad \omega_1(a) = Qi\ ,$$

we have

$$(1.25)\qquad Bi\int_a^0 \frac{(\sqrt{1-\zeta}+\sqrt{-\zeta})^{2\varepsilon}+(\sqrt{1-\zeta}-\sqrt{-\zeta})^{2\varepsilon}}{\sqrt{\zeta(1-\zeta)(\zeta-a)}}\,d\zeta = \kappa_1 H\ .$$

Parameter a, as shown, is determined by equality (1.9). We introduce the notations:

$$(1.26)\qquad \tanh\frac{\pi\ell}{2h} = k, \quad \cosh\frac{\pi\ell}{2h} = \frac{1}{k'}\ .$$

Then

$$(1.27)\qquad a = k^2\ .$$

Under integral sign (1.26) we substitute

$$\zeta = a\sin^2\alpha = k^2\sin^2\alpha\ .$$

We obtain:

$$(1.28)\qquad B = \frac{\kappa_1 H}{2J}\ ,$$

where

$$(1.29)\qquad J = \int_0^{\pi/2}\frac{\cos(2\ \mathrm{arc\ sin}\ k\sin\alpha)\,d\alpha}{\sqrt{1-k^2\sin^2\alpha}} = \int_0^{\mathrm{arc\ sin}\ k}\frac{\cos 2\varepsilon\alpha\,d\alpha}{\sqrt{k^2-\sin^2\alpha}}\ .$$

Taking the integral for $\omega_1(\zeta)$ between limits $\zeta = -\infty$, $\zeta = 0$, we find the flowrate

$$(1.30)\qquad Q = \frac{\kappa_1 HJ'}{2J\cos\varepsilon\pi}$$

where

$$(1.31)\qquad J' = \int_0^{\pi/2}\frac{\cos(2\varepsilon\ \mathrm{arc\ sin}\ k'\sin\alpha)\,d\alpha}{\sqrt{1-k'^2\sin^2\alpha}} = \int_0^{\mathrm{arc\ sin}\ k'}\frac{\cos 2\varepsilon\alpha\,d\alpha}{\sqrt{k'^2-\sin^2\alpha}}\ .$$

Since

$$\frac{d\omega_s}{dz} = \frac{d\omega_s}{d\zeta}\cdot\frac{dz}{d\zeta} = F_s\frac{d\zeta}{dz} \qquad (s = 1,\ 2)\ ,$$

and using formulas (1.8) and (1.23), we find expressions for the complex velocities in the last and second medium:

(1.32)‡
$$u_1 - iv_1 = \frac{\kappa_1 H\pi}{4HJ} \times$$

$$\times \frac{(-\sin\lambda iz + i\sqrt{\mathrm{ch}^2\lambda\ell - \sin^2\lambda iz})^{2\varepsilon} + (-\sin\lambda iz - i\sqrt{\mathrm{ch}^2\lambda\ell - \sin^2\lambda iz})^{2\varepsilon}}{\sqrt{\mathrm{ch}^2\lambda\ell - \sin^2\lambda iz}} ,$$

$$u_2 - iv_2 = \frac{\kappa_1 H\pi i\ \mathrm{tg}\ \pi\varepsilon}{4HJ} \times$$

$$\times \frac{(+\sin\lambda iz + i\sqrt{\mathrm{ch}^2\lambda\ell - \sin^2\lambda iz})^{2\varepsilon} - (+\sin\lambda iz - i\sqrt{\mathrm{ch}^2\lambda\ell - \sin^2\lambda iz})^{2\varepsilon}}{\sqrt{\mathrm{ch}^2\lambda\ell - \sin^2\lambda iz}} ,$$

where

$$\lambda = \frac{\pi}{2h} .$$

We show that all boundary conditions are satisfied for $u_1 - iv$ and $u_2 - iv_2$, i.e., that the constructed solution is correct.

We introduce the reduced velocity $\tilde{u} - i\tilde{v}$, where

$$u - iv = \frac{\kappa_1 H}{2h} (\tilde{u} - i\tilde{v}) .$$

Along the boundary line BA, where $z = x$, we have after separation of real and imaginary parts,

$$\tilde{u}_1 = \frac{\pi}{J} \frac{M + N}{P} \cos \varepsilon\pi, \quad \tilde{v}_1 = \frac{\pi}{J} \frac{N - M}{P} \sin \varepsilon\pi ,$$

where $M = (P - \mathrm{sh}\ \lambda x)^{2\varepsilon}$, $N = (P + \mathrm{sh}\ \lambda x)^{2\varepsilon}$, $P = \sqrt{\mathrm{ch}^2\lambda\ell + \mathrm{sh}^2\lambda x}$

$$\left(\lambda = \frac{\pi}{2h},\quad 0 \le x \le \infty\right) .$$

Along segment CD of the base

$$\tilde{v}_1 = 0, \quad \tilde{u}_1 = \frac{\pi}{J} \frac{(\mathrm{ch}\ \lambda\ell)^{2\varepsilon} \cos 2\varepsilon\theta}{\sqrt{\mathrm{ch}^2\lambda\ell - \mathrm{ch}^2\lambda x}} ,$$

where

$$\theta = \arccos \frac{\mathrm{ch}\ \lambda x}{\mathrm{ch}\ \lambda\ell} \quad \left(\lambda = \frac{\pi}{2h},\quad 0 \le x \le \ell\right) .$$

Along boundary $DA_1$ of tailwater

$$\tilde{u}_1 = 0, \quad \tilde{v}_1 = \frac{\pi}{J} \frac{\mathrm{ch}\ 2\varepsilon\theta}{(\mathrm{ch}\ \lambda\ell)^{1-2\varepsilon} \mathrm{sh}\ \theta} ,$$

where

$$\mathrm{ch}\ \theta = \frac{\mathrm{ch}\ \lambda x}{\mathrm{ch}\ \lambda\ell} \quad \left(\lambda = \frac{\pi}{2h}\right) .$$

Along segment BC, where $x = 0$, $z = yi$, we have

$$\tilde{v}_1 = 0, \quad \tilde{u}_1 = \frac{\pi}{J} \frac{\mathrm{ch}\ \lambda\ell \cos 2\varepsilon\theta}{\sqrt{\mathrm{ch}^2\lambda\ell - \sin^2\lambda y}} ,$$

‡ Note: In this paragraph sh, ch, th stand for sinh, cosh, tanh.

where

$$\theta = \text{arc tg}\,\frac{\sqrt{\text{ch}^2\lambda\ell - \sin^2\lambda y}}{\sin\lambda y} \qquad \left(\lambda = \frac{\pi}{2h}\right) .$$

Along segment BE we have

$$\tilde{v}_2 = 0, \quad \tilde{u}_2 = \frac{\text{tg}\,\pi\varepsilon}{J}\,\frac{\text{ch}\,\lambda\ell\,\sin 2\varepsilon\theta}{\sqrt{\text{ch}^2\lambda\ell - \sin^2\lambda y}} ,$$

where $\theta$ and $\lambda$ have the same expressions as in the previous case. Along the impervious boundary $EA_2$

$$\tilde{v}_2 = 0, \quad \tilde{u}_2 = \frac{\text{tg}\,\varepsilon\pi}{J}\,\frac{\text{ch}\,\lambda\ell\,\sin 2\varepsilon\theta}{\sqrt{\text{ch}^2\lambda\ell - \text{ch}^2\lambda x}} ,$$

where

$$\theta = \text{arc tg}\,\frac{\sqrt{\text{ch}^2\lambda\ell - \text{ch}^2\lambda x}}{\text{ch}\,\lambda x} \qquad \left(\lambda = \frac{\pi}{2h}\right) .$$

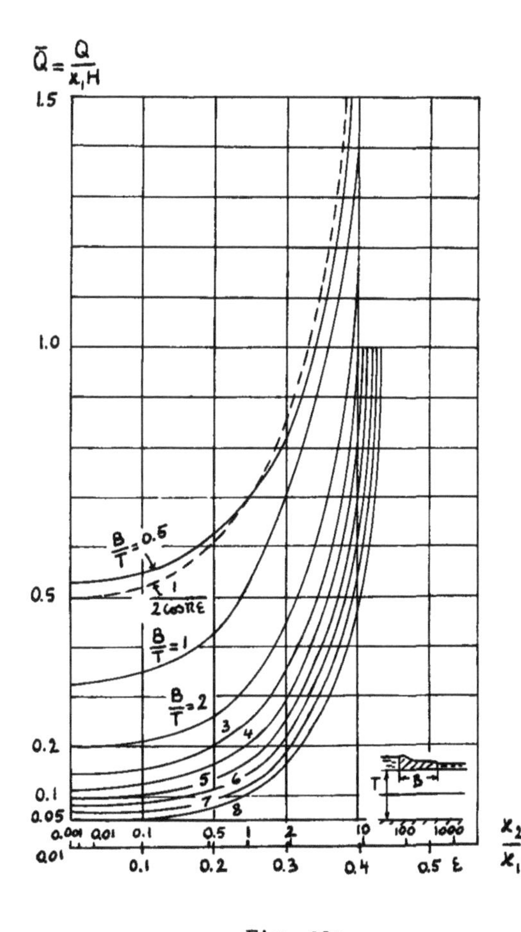


Fig. 239

We designate by $\bar{Q}$ the reduced flowrate, i.e., the flowrate for $\kappa_1 = 1$, $H = 1$. Then

$$Q = \kappa_1 H\bar{Q} \ .$$

Fig. 239 renders graphs for the dependence of $\bar{Q}$ on

$$\varepsilon = \frac{1}{\pi} \text{arc tg} \sqrt{\frac{\kappa_2}{\kappa_1}} \ .$$

We notice that for $\varepsilon = 0$, i.e., $\kappa_2 = 0$, we have flow in an homogeneous upper stratum of depth h with seepage coefficient $\kappa_1$.

For $\varepsilon = 1/4$ we have $\kappa_1 = \kappa_2$, i.e., we have seepage in a homogeneous stratum of depth 2h. For $\varepsilon = 1/2$, we have $\kappa_2/\kappa_1 = \infty$, i.e., there is no resistance in the lower stratum.

The flow takes place as if there were only water (no soil) in this stratum. Line BA (Fig. 234) is a line of equal head for which

$$\varphi = -\kappa_1 \frac{H_1 + H_2}{2} \ .$$

For $\varepsilon = 1/2$, $\bar{Q} = \infty$. For $\varepsilon$ close to $1/2$,

$$\bar{Q} \approx \frac{1}{\cos \varepsilon\pi} \ ,$$

exactly

$$\lim_{\varepsilon \to 1/2} \bar{Q} \cos \varepsilon\pi = 1 \ .$$

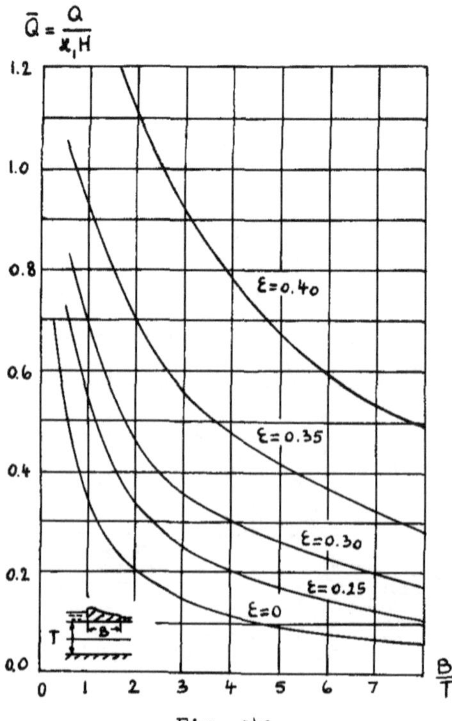


Fig. 240

Graphs of the dependence of $\bar{Q}$ on $B/T$ are given in Fig. 240. Here $B = 2\ell$ stands for the length of the dam, $T = 2h$ is the depth of the stratum (all soil).

Putting $\varepsilon = 0$ in formulas (1.30) and (1.31), we obtain: the flow-rate in homogeneous medium of depth $h$

$$Q = \frac{\kappa_1 HK'}{2K} ,$$

where $K$ is the complete elliptic integral of the first kind with modulus

$$k = \operatorname{th} \frac{\pi B}{2T} ,$$

and $K'$ is the complete elliptic integral of the second kind of the complementary modulus.

## §2. Cut-off in Two Strata of Same Depth.

A cut-off penetrates over a depth $d$ in two layered soil composed of two horizontal strata (Fig. 241) of identical thickness $h$. The seepage coefficients of these strata are respectively $\kappa_1$ and $\kappa_2$.

In this problem, we distinguish three cases.

1°. Let $d < h$.

As in problem of §1, we map region $ABCDA_1$ onto the upper and region $ABEA_2$ onto the lower half-plane of the complex variable $\zeta$ as shown in Fig. 242.

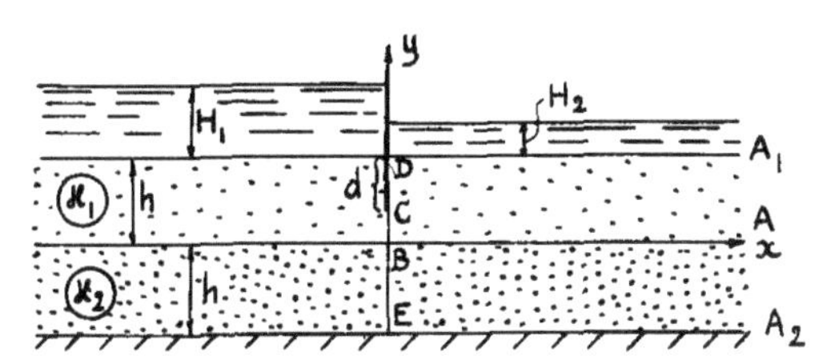


Fig. 241

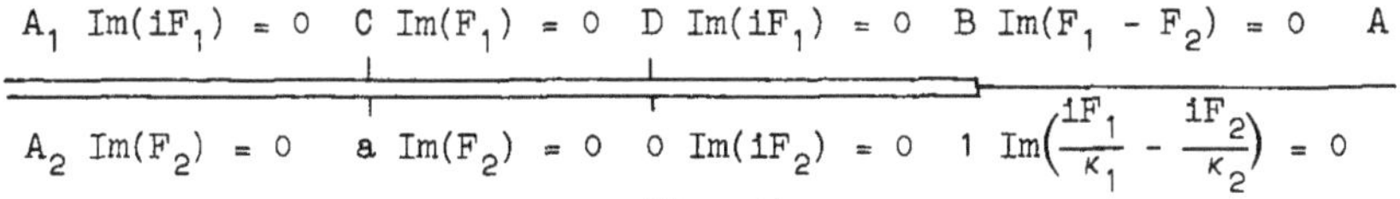

$A_1$ $\mathrm{Im}(iF_1) = 0$ $C$ $\mathrm{Im}(F_1) = 0$ $D$ $\mathrm{Im}(iF_1) = 0$ $B$ $\mathrm{Im}(F_1 - F_2) = 0$ $A$

$A_2$ $\mathrm{Im}(F_2) = 0$ $a$ $\mathrm{Im}(F_2) = 0$ $0$ $\mathrm{Im}(iF_2) = 0$ $1$ $\mathrm{Im}\left(\frac{iF_1}{\kappa_1} - \frac{iF_2}{\kappa_2}\right) = 0$

Fig. 242

The dependence between $z$ and $\zeta$ again is expressed as

$$\zeta = \frac{1 + a}{2} + \frac{1 - a}{2} \operatorname{ch} \frac{\pi z}{h} ,$$

but here $a$ has another value. Indeed, assuming here $\zeta = 0$, $z = (h-d)i$, we find

$$a = - \operatorname{tg}^2 \frac{\pi d}{2h} . \tag{2.1}$$

We introduce the notations

$$\sin\frac{\pi d}{2h} = k, \qquad \cos\frac{\pi d}{2h} = k' \; ; \tag{2.2}$$

where

$$a = -\frac{k^2}{k'^2} \; .$$

We also put here:

$$F_1 = \frac{\Phi_1}{\sqrt{\zeta - a}}, \qquad F_2 = \frac{\Phi_2}{\sqrt{\zeta - a}} \; .$$

Then for functions $\Phi_1$ and $\Phi_2$ the same conditions as in the previous problem are satisfied on the segments $(-\infty, 0)$, $(0, 1)$ and $(1, \infty)$. Consequently, the solution is expressed by the same functions (1.20) and for $F_1$ and $F_2$ we obtain the same equations (1.24) and (1.25).

But the constant $B$ here has another value. Indeed, integrating formula (1.25) from $a$ to $0$, we find:

$$\frac{B}{2}\int_0^a \frac{(\sqrt{1-\zeta} + \sqrt{-\zeta})^{2\varepsilon} + (\sqrt{1-\zeta} - \sqrt{-\zeta})^{2\varepsilon}}{\sqrt{-\zeta(1-\zeta)(\zeta - a)}}\, d\zeta = \frac{\kappa_1 H}{2} \; .$$

The substitution

$$\zeta = a \sin^2\varphi$$

gives

$$B = \frac{\kappa_1 H}{J} \; ,$$

where

$$J = 2\int_0^{\pi/2} \frac{(\sqrt{1 + b\sin^2\varphi} + \sqrt{b}\sin\varphi)^{2\varepsilon} + (\sqrt{1 + b\sin^2\varphi} - \sqrt{b}\sin\varphi)^{2\varepsilon}}{\sqrt{1 + b\sin^2\varphi}}\, d\varphi$$

in which

$$b = -a = \frac{k^2}{k'^2} \; .$$

Putting

$$\varphi = \frac{\pi}{2} - \alpha \; ,$$

we find:

$$J = 2(k')^{1-2\varepsilon}(J_1 + J_2) \; ,$$

where

$$\begin{aligned} J_1 &= \int_0^{\pi/2} \frac{(\sqrt{1 - k^2\sin^2\varphi} + k\cos\varphi)^{2\varepsilon}}{\sqrt{1 - k^2\sin^2\varphi}}\, d\varphi = \int_0^K (\mathrm{dn}\, u + k\, \mathrm{cn}\, u)^{2\varepsilon} du \; , \\ J_2 &= \int_0^{\pi/2} \frac{(\sqrt{1 - k^2\sin^2\varphi} - k\cos\varphi)^{2\varepsilon}}{\sqrt{1 - k^2\sin^2\varphi}}\, d\varphi = \int_0^K (\mathrm{dn}\, u - k\, \mathrm{cn}\, u)^{2\varepsilon} du \end{aligned} \tag{2.3}$$

($\mathrm{dn}\, u$, $\mathrm{cn}\, u$ are elliptic functions).

Further, introducing limits $-\infty$ and $a$ in expression (1.25), we find for the flowrate $Q$:

$$Q = \frac{B}{2} \int_{-\infty}^{a} \frac{(\sqrt{1-t} + \sqrt{-t})^{2\varepsilon} + (\sqrt{1-t} - \sqrt{-t})^{2\varepsilon}}{\sqrt{t(1-t)(t-a)}} dt$$

Putting

$$t = a \operatorname{cosec}^2\varphi = -\frac{k^2}{k'^2 \sin^2\varphi} ,$$

we obtain:

$$Q = Bk'^{1-2\varepsilon} \int_0^{\pi/2} \frac{(\sqrt{1 - k'^2\sin^2\varphi} + k)^{2\varepsilon} + (\sqrt{1 - k'^2\sin^2\varphi} - k)^{2\varepsilon}}{\cos^2\varphi \sqrt{1 - k'^2\sin^2\varphi}} d\varphi . \tag{2.4}$$

We may compute the flowrate $Q$ in another way, as the sum of the flowrates through segments EB and BD. A formula, obtained in this way, and used for numerical computation of $Q$, is given below. We do not enter into the details of the computations but restrict ourselves to the communication of the final results.

1°. For $d < h$, we put

$$Q = \kappa_1 H \bar{Q} ,$$

and have

$$\bar{Q} = \frac{1}{J_1 + J_2} \left[ \frac{k'^{2\varepsilon}}{\cos \varepsilon\pi} J_3 + \frac{1}{2} \operatorname{tg} \varepsilon\pi (J_1 - J_2) \right] , \tag{2.5}$$

where

$$J_3 = \int_0^{\pi/2} \frac{\cos 2\varepsilon\varphi \, d\varphi}{\sqrt{1 - k'^2\sin^2\varphi}} , \tag{2.6}$$

and $J_1$ and $J_2$ are determined by formulas (2.3) and (2.4). Further, the expressions for the complex velocities in the first and second medium have the form

$$u_1 - iv_1 = -\frac{\kappa_1 H\pi i}{4h(J_1 + J_2)} \times$$

$$\times \frac{(-\sin \lambda iz + \sqrt{\sin^2\lambda iz - k'^2})^{2\varepsilon} + (-\sin \lambda iz - \sqrt{\sin^2\lambda iz - k'^2})^{2\varepsilon}}{\sqrt{\sin^2\lambda iz - k'^2}} ,$$

$$u_2 - iv_2 = \frac{\kappa_1 H\pi \operatorname{tg} \varepsilon\pi}{4h(J_1 + J_2)} \times$$

$$\times \frac{(\sin \lambda iz + \sqrt{\sin^2\lambda iz - k'^2})^{2\varepsilon} - (\sin \lambda iz - \sqrt{\sin^2\lambda iz - k'^2})^{2\varepsilon}}{\sqrt{\sin^2\lambda iz - k'^2}}$$

where

$$\lambda = \frac{\pi}{2h} .$$

In particular, for the exit velocity $v_0$, i.e., for the velocity in point C of the cut-off, we have:

$$v_0 = \frac{\kappa H \bar{v}}{2h} ,$$

where $\bar{v}$ is the reduced velocity, determined by the formula

$$\bar{v} = \frac{(1 + k)^{2\varepsilon} + (1 - k)^{2\varepsilon}}{2k(J_1 + J_2)} .$$

For $\varepsilon = 0$, i.e., for $\kappa_2 = 0$, we obtain one stratum of depth h. In this case

$$\bar{Q} = \frac{K'}{2K}, \quad \bar{v} = \frac{\pi}{2kK} ,$$

where K is the complete elliptic integral of the first kind with modulus

$$k = \sin \frac{\pi d}{2h} ,$$

and K' is the same integral with complementary modulus. For $\varepsilon = 1/4$, we have one stratum of depth 2h. For $\varepsilon = 1/2$, we have $\kappa_2 = \infty$, i.e., we obtain the problem of Chapter III, §11 of the cut-off above a draining stratum. There

$$\bar{Q} = \infty, \quad \bar{v} = \frac{1}{k} . \tag{2.8}$$

2°. For $d > h$ we have the formulas

$$u_1 - iv_1 = - \frac{\kappa_1 H\pi i}{4hJ_0} \frac{(-\sin \lambda iz + \sqrt{\sin^2 \lambda iz - k'^2})^{2\varepsilon} - (-\sin \lambda iz - \sqrt{\sin^2 \lambda iz - k'^2})^{2\varepsilon}}{\sqrt{\sin^2 \lambda iz - k'^2}}$$

$$u_2 - iv_2 = \frac{\kappa_1 H\pi \operatorname{tg} \varepsilon\pi}{4hJ_0} \frac{(\sin \lambda iz + \sqrt{\sin^2 \lambda iz - k'^2})^{2\varepsilon} + (\sin \lambda iz - \sqrt{\sin^2 \lambda iz - k'^2})^{2\varepsilon}}{\sqrt{\sin^2 \lambda iz - k'^2}} .$$

Here

$$J_0 = J_1 - J_2 + \frac{2k'^{2\varepsilon}}{\sin \varepsilon\pi} J_3, \quad \lambda = \frac{\pi}{2h} .$$

For the reduced flowrate and the reduced exit velocity we have:

$$\bar{Q} = \frac{\operatorname{tg} \varepsilon\pi (J_1 + J_2)}{2J_0}, \quad \bar{v} = \frac{\pi}{2k} \frac{(1 + k)^{2\varepsilon} - (1 - k)^{2\varepsilon}}{J_0} \tag{2.9}$$

We mention some particular cases. For $\varepsilon = 0$, i.e., $\kappa_2 = 0$, we must evidently have (since there is no flow in this case):

$$\bar{Q} = 0, \quad \bar{v} = 0 .$$

For $\varepsilon = 1/4$, we have one stratum of depth 2h.

3°. Finally, we put $d = h$, $\bar{Q} = \infty$, $\bar{v} = 1$ .

This case was examined by N. K. Girinsky [3]. It may be obtained from the case $d < h$ and also from the case $d > h$ for $k = 1$, $k' = 0$ and this renders:

$$u_1 - iv_1 = -\frac{\kappa_1 Hi\sqrt{\pi}}{2h\,\Gamma(\varepsilon)}\,\frac{\Gamma\left(\frac{1}{2}+\varepsilon\right)}{(-\sin\lambda iz)^{1-2\varepsilon}},$$

$$u_2 - iv_2 = \frac{\sqrt{\kappa_1\kappa_2}\,H\sqrt{\pi}}{2h\,\Gamma(\varepsilon)}\,\frac{\Gamma\left(\frac{1}{2}+\varepsilon\right)}{(-\sin\lambda iz)^{1-2\varepsilon}},$$

where $\Gamma(\varepsilon)$ is the gamma function, $\lambda = \pi/2h$. Further

$$Q = \frac{1}{2}\kappa_1 H \operatorname{tg}\varepsilon\pi = \frac{1}{2}\sqrt{\kappa_1\kappa_2}H, \qquad v_0 = \frac{\kappa_1 H\sqrt{\pi}}{2h}\,\frac{\Gamma\left(\frac{1}{2}+\varepsilon\right)}{\Gamma(\varepsilon)},$$

i.e.,

$$\overline{Q} = \frac{1}{2}\operatorname{tg}\varepsilon\pi = \frac{1}{2}\sqrt{\frac{\kappa_2}{\kappa_1}}, \qquad \bar{v} = \frac{\sqrt{\pi}\,\Gamma\left(\frac{1}{2}+\varepsilon\right)}{\Gamma(\varepsilon)}$$

Figures 243-246 render graphs for the dependence of $\overline{Q}$ and $\bar{v}$ on the quantity $\varepsilon = 1/\pi \operatorname{arc\,tg}\sqrt{\kappa_2/\kappa_1}$ for $d/2h = 1/8, 1/4, \ldots, 7/8$ and graphs for the dependence of $\overline{Q}'$, $\bar{v}'$ on $d/2h$ for different values of $\varepsilon$.

For this problem we derive once more the formulas for the velocities along the contour of the region,

$$\tilde{u} - i\tilde{v} = \frac{2h}{\kappa_1 H}(u - iv), \quad \lambda = \frac{\pi}{2h}.$$

1° Case $d < h$.

Along the lower basin

$$\tilde{u}_1 = 0, \quad \tilde{v}_1 = \frac{\pi}{2(J_1 + J_2)}\,\frac{(\operatorname{ch}\lambda x + M)^{2\varepsilon} + (\operatorname{ch}\lambda x - M)^{2\varepsilon}}{M},$$

$$M = \sqrt{\operatorname{ch}^2\lambda x - k'^2} \qquad (0 \leq x < \infty).$$

Along the cut-off

$$\tilde{u}_1 = 0, \quad \tilde{v}_1 = \frac{\pi}{2(J_1 + J_2)}\,\frac{(\sin\lambda y + N)^{2\varepsilon} + (\sin\lambda y - N)^{2\varepsilon}}{N},$$

$$N = \sqrt{\sin^2\lambda y - k'^2} \qquad (h - d < y \leq h).$$

Along the vertical below the cut-off

$$\tilde{v}_1 = 0, \quad \tilde{u}_1 = \frac{\pi k'^{2\varepsilon}}{J_1 + J_2}\,\frac{\cos 2\varepsilon\theta}{\sqrt{k'^2 - \sin^2\lambda y}},$$

where

$$\theta = \arcsin\frac{\sqrt{k'^2 - \sin^2\lambda y}}{k'} \qquad (0 \leq y < h - d).$$

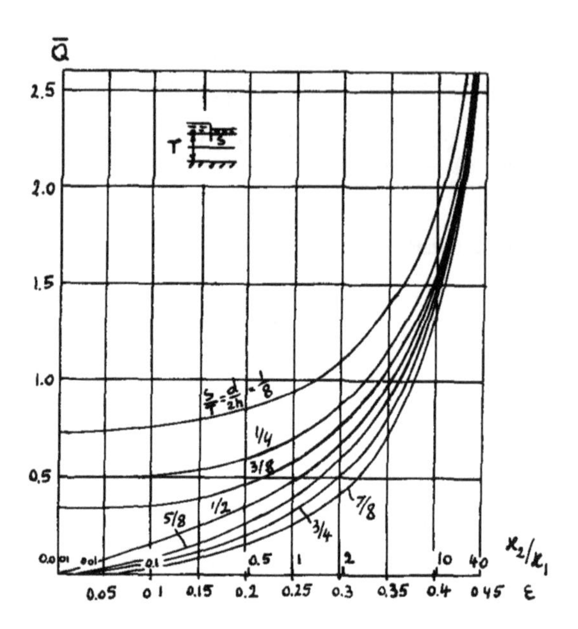

Fig. 243

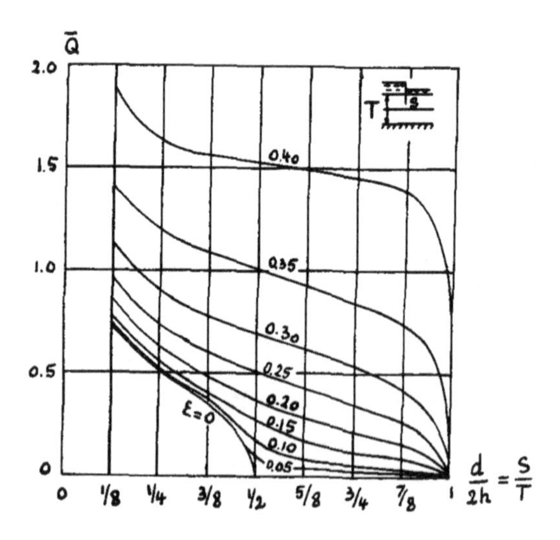

Fig. 244.

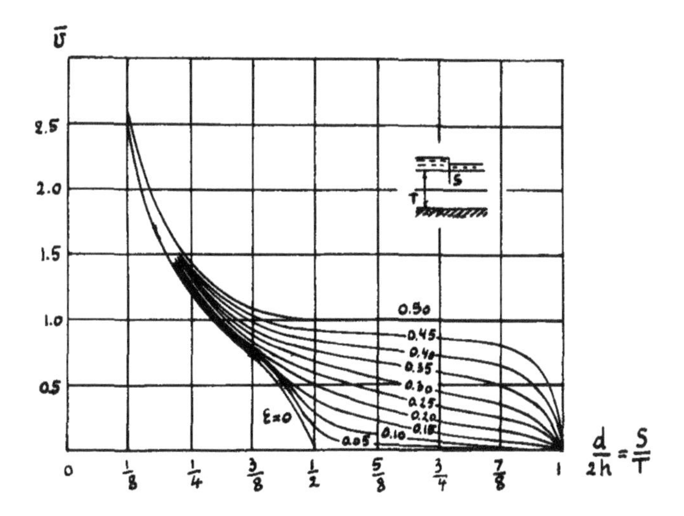


Fig. 245.

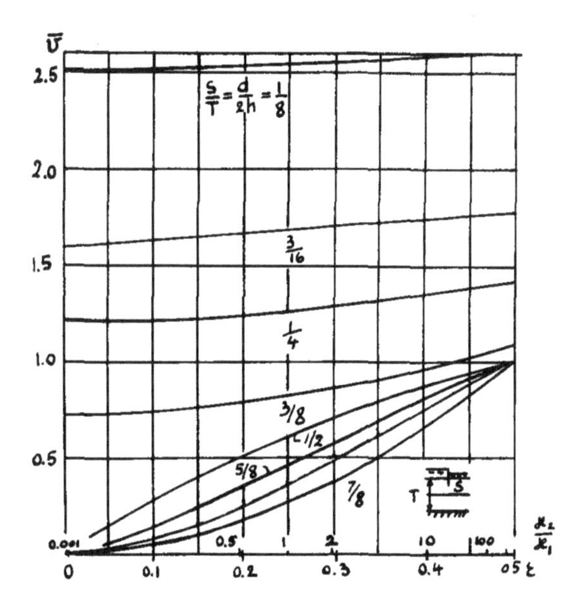


Fig. 246.

Along the boundary line

$$\tilde{u}_1 = \frac{\pi}{2(J_1 + J_2)} \frac{(P + \operatorname{sh} \lambda x)^{2\varepsilon} + (P - \operatorname{sh} \lambda x)^{2\varepsilon}}{P} \cos \varepsilon\pi ,$$

$$\tilde{v}_1 = \frac{\pi}{2(J_1 + J_2)} \frac{(P + \operatorname{sh} \lambda x)^{2\varepsilon} - (P - \operatorname{sh} \lambda x)^{2\varepsilon}}{P} \sin \varepsilon\pi ,$$

where

$$P = \sqrt{\operatorname{sh}^2 \lambda x + k'^2} \qquad (0 \leq x < \infty) .$$

Along the impervious foundation of the lower stratum

$$\tilde{v}_2 = 0, \qquad \tilde{u}_2 = \frac{\pi \operatorname{tg} \varepsilon\pi}{2(J_1 + J_2)} \frac{(R + \operatorname{ch} \lambda x)^{2\varepsilon} - (\operatorname{ch} \lambda x - R)^{2\varepsilon}}{R} ,$$

where

$$R = \sqrt{\operatorname{ch}^2 \lambda x - k'^2} \qquad (0 \leq x < \infty) .$$

Along the vertical in the lower stratum

$$\tilde{v}_2 = 0, \qquad \tilde{u}_2 = \frac{k'\pi \operatorname{tg} \varepsilon\pi}{2(J_1 + J_2)} \frac{\sin 2\varepsilon\theta}{\sqrt{k'^2 - \sin^2 \lambda y}} ,$$

where

$$\theta = \operatorname{arc} \operatorname{tg} \frac{\sqrt{k'^2 - \sin^2 \lambda y}}{\sin \lambda y} .$$

2° Case $d > h$.

Along the boundary line

$$\tilde{u}_1 = \frac{\pi}{2J_0} \frac{(M + \operatorname{sh} \lambda x)^{2\varepsilon} - (M - \operatorname{sh} \lambda x)^{2\varepsilon}}{M} \cos \varepsilon\pi ,$$

$$\tilde{v}_1 = \frac{\pi}{2J_0} \frac{(M + \operatorname{sh} \lambda x)^{2\varepsilon} + (M - \operatorname{sh} \lambda x)^{2\varepsilon}}{M} \sin \varepsilon\pi ,$$

where

$$M = \sqrt{\operatorname{sh}^2 \lambda x + k'^2} .$$

Along the boundaries of the reservoir

$$\tilde{u}_1 = 0, \qquad \tilde{v}_1 = \frac{\pi}{2J_0} \frac{(\operatorname{ch} \lambda x + N)^{2\varepsilon} - (\operatorname{ch} \lambda x - N)^{2\varepsilon}}{N} ,$$

$$N = \sqrt{\operatorname{ch}^2 \lambda x - k'^2} \qquad (0 \leq x < \infty) .$$

Along the cut-off in the upper layer

$$\tilde{u}_1 = 0, \qquad \tilde{v}_1 = \frac{\pi}{2J_0} \frac{(\sin \lambda y + P)^{2\varepsilon} - (\sin \lambda y - P)^{2\varepsilon}}{P} \qquad (0 \leq y < h)$$

Along the cut-off in the lower layer

$$\tilde{u}_2 = 0, \qquad \tilde{v}_2 = \frac{\pi k' \operatorname{tg} \varepsilon\pi}{2J_0} \frac{\cos 2\varepsilon\theta}{\sqrt{k'^2 - \sin^2 \lambda y}} ,$$

where

$$\theta = \text{arc tg} \frac{\sqrt{k'^2 - \sin^2 \lambda y}}{\sin \lambda y} \qquad (d - h < y < 0) .$$

Below the cut-off in the lower stratum

$$\tilde{v}_2 = 0, \quad \tilde{u}_2 = \frac{\pi \text{ tg } \varepsilon\pi}{2J_0} \frac{(-\sin \lambda y + R)^{2\varepsilon} + (-\sin \lambda y - R)^{2\varepsilon}}{R} ,$$

$$R = \sqrt{\sin^2 \lambda y - k'^2} \qquad (h - d < y < -h) .$$

3° Case $d = h$.

Along the boundaries of the reservoir

$$\tilde{u}_1 = 0, \quad \tilde{v}_1 = \frac{\pi\Gamma\left(\frac{1}{2} + \varepsilon\right)}{\Gamma(\varepsilon)} \frac{1}{(\text{ch } \lambda x)^{1-2\varepsilon}} \qquad (0 \leq x < \infty) .$$

Along the cut-off

$$\tilde{u}_1 = 0, \quad \tilde{v}_1 = \frac{\sqrt{\pi}\,\Gamma\left(\frac{1}{2} + \varepsilon\right)}{\Gamma(\varepsilon)} \frac{1}{(\sin \lambda y)^{1-2\varepsilon}} \qquad (0 < y \leq h) .$$

Along the boundary line

$$\tilde{u}_1 = \frac{\sqrt{\pi}\,\Gamma\left(\frac{1}{2} + \varepsilon\right)}{\Gamma(\varepsilon)} \frac{\sin \varepsilon\pi}{(\text{sh } \lambda x)^{1-2\varepsilon}} , \quad \tilde{v}_1 = \frac{\sqrt{\pi}\,\Gamma\left(\frac{1}{2} + \varepsilon\right)}{\Gamma(\varepsilon)} \frac{\cos \varepsilon\pi}{(\text{sh } \lambda x)^{1-2\varepsilon}}$$

$$(0 \leq x < \infty) .$$

Along the vertical under the cut-off

$$\tilde{v}_2 = 0, \quad \tilde{u}_2 = \frac{\sqrt{\pi}\,\Gamma\left(\frac{1}{2} + \varepsilon\right)}{\Gamma(\varepsilon)} \frac{\text{tg } \varepsilon\pi}{(-\sin \lambda y)^{1-2\varepsilon}} \qquad (-h \leq y < 0).$$

Along the impervious foundation

$$\tilde{v}_2 = 0, \quad \tilde{u}_2 = \frac{\sqrt{\pi}\,\Gamma\left(\frac{1}{2} + \varepsilon\right)}{\Gamma(\varepsilon)} \frac{1}{(\text{ch } \lambda x)^{1-2\varepsilon}} \qquad (0 \leq x < \infty) .$$

We communicate here the formulas for the velocities in the case of a cut-off "with zero penetration," on two-layered medium with identical thickness of both strata. We may use formulas (2.7) in which we make $k = 0$, $k' = 1$. It follows that

$$J_1 = \frac{\pi}{2}, \qquad J_2 = 0.$$

Therefore we obtain:

$$u_1 - iv_1 = -\frac{\kappa_1 H}{2h} [e^{\pi\varepsilon i} e^{\pi\varepsilon z/h} + e^{-\pi\varepsilon i} e^{-\pi\varepsilon z/h}] \sec \frac{\pi z}{2hi} =$$

$$= -\frac{\kappa_1 H}{h} \sec \frac{\pi z}{2hi} \cos \frac{\pi\varepsilon(hi - z)}{hi} ,$$

$$u_2 - iv_2 = \frac{\kappa_1 H \text{ tg } \varepsilon\pi}{h} \sec \frac{\pi z}{2hi} \sin \frac{\pi\varepsilon(hi + z)}{hi} .$$

By means of these formulas we obtain the following expressions for the velocities. Along the boundary line

$$\tilde{u}_1 = \frac{2 \text{ ch } 2\lambda\varepsilon x}{\text{ch } \lambda x} \cos \varepsilon\pi, \quad \tilde{v}_1 = \frac{2 \text{ sh } 2\lambda\varepsilon x}{\text{ch } \lambda x} \sin \varepsilon\pi \qquad (0 \le c < \infty) .$$

Along the vertical in the upper stratum

$$\tilde{v}_1 = 0, \quad \tilde{u}_1 = \frac{2 \cos \dfrac{\pi\varepsilon(h - y)}{h}}{\cos \lambda y} \qquad (0 < y < h).$$

Along the vertical in the lower stratum

$$\tilde{v}_2 = 0, \quad \tilde{u}_2 = 2 \text{ tg } \varepsilon\pi \frac{\sin \pi\varepsilon\left( \dfrac{y}{h} + \dfrac{\pi}{2} \right)}{\cos \lambda y} \qquad (0 \le y < h) .$$

Along the boundary of the reservoir

$$\tilde{u}_1 = 0, \quad \tilde{v}_1 = \frac{2 \text{ ch } 2\lambda\varepsilon x}{\text{sh } \lambda x} \qquad (0 \le x < \infty) .$$

Along the impervious boundary

$$\tilde{v}_2 = 0, \quad \tilde{u}_2 = \frac{2 \text{ sh } 2\lambda\varepsilon x}{\text{sh } \lambda x} \qquad (0 \le x < \infty) .$$

By means of the method applied in this chapter we may also solve a series of symmetrical problems of seepage, in two-layered medium: for example, we might examine a dam with cut-off in the middle of its base, or two symmetrically located weirs. But in these problems the number of singular points increases and one has to study the presence of additional singularities.

We may consider the ratio:

$$\frac{\Phi_1}{\Phi_2} = W(\zeta)$$

and construct the region of function $W$. We examine what we obtain in the problems that are of interest to us.

We separate equation

$$\text{Im} \left( \frac{i\Phi_1}{\kappa_1} - \frac{i\Phi_2}{\kappa_2} \right) = 0$$

into equation

$$\text{Im} \ (\Phi_1 - \Phi_2) = 0 .$$

On line $AB$ we have

$$\text{Im} \left( \frac{iW - \dfrac{i\kappa_1}{\kappa_2}}{W - 1} \right) = 0 . \tag{2.10}$$

This equation of a circle passing through points

$$W = 1 \quad \text{and} \quad W = \frac{\kappa_1}{\kappa_2} .$$

It is impossible to consider that in the formula for $W$, functions $\Phi_1$ and $\Phi_2$ satisfy those conditions on the remaining segments as written in Fig. 236. One must, as shown above, replace function $\Phi_1$ by $\Phi_1^*$ for example at the upper side of the cut. Then, based upon formula (1.15) the condition for function $\Phi_2$ may be represented as

$$(2.11)\quad \begin{cases} \operatorname{Im}[(\kappa_2 + \kappa_1)\Phi_1 - 2\kappa\,\Phi_2] = 0, & \operatorname{Im}\ \Phi_2 = 0 \quad \text{for } -\infty < \xi < 0\ , \\ \operatorname{Im}[(\kappa_2 + \kappa_1)\Phi_1 i - 2\kappa\,\Phi_2 i = 0, & \operatorname{Im}(i\Phi_2) = 0 \quad \text{for } 0 < \xi < 1\ . \end{cases}$$

From these two equations, we obtain for $-\infty < \xi < 0$:

$$\operatorname{Im}[(\kappa_2 + \kappa_1)W - 2\kappa_1] = 0$$

or

$$\operatorname{Im} W = 0\ .$$

Similarly for $0 < \xi < 1$

$$\operatorname{Im}[(\kappa_1 + \kappa_2)Wi - 2\kappa_1 i] = 0$$

or

$$\operatorname{Im}\left(iW - \frac{2\kappa_1 i}{\kappa_2 + \kappa_1}\right) = 0\ .$$

If we put $W = U + iV$ Then the equations obtained render two straight lines: the real $W$-axis and

$$U = \frac{2\kappa_1}{\kappa_1 + \kappa_2}\ ,$$

and in the $W$ plane a triangle is formed by these straight lines and by circle (2.10). The theory points to the analogy between the problems in two-layered soil which we examined and problems of seepage with free surface, in which function $W$ corresponds to the complex velocity [5].

## §3. Cut-offs without Penetration on Soil with Two Strata.

We analyze here two schemes [2].

1. Two cut-offs without penetration and without symmetry in their disposition. This case is represented in Fig. 247.

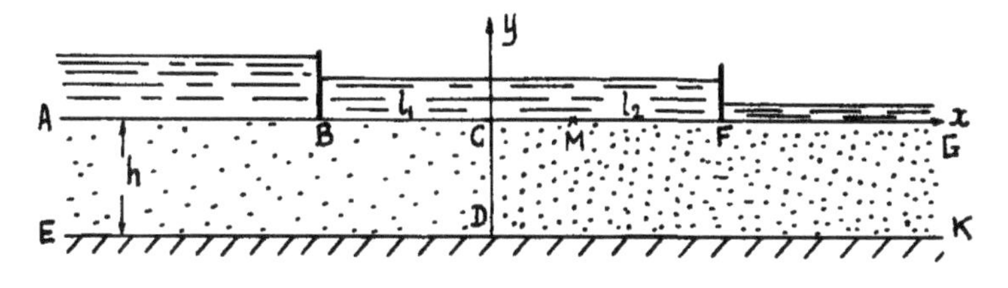


Fig. 247

If we map the flow region conformally onto the plane with the branch cut, as was done in the case of the two cut-offs, then we find that the singularity on the upper side of the cut is in point $\zeta = -a$, and on the lower side of the cut in point $\zeta = -b$. In the contour of point C we obtain different sheets of the Riemann surface and if we take for example $\zeta = -a$ in the upper sheet of this surface, then $\zeta = -b$ lies in the lower sheet.

Analyzing the solution of the problem for two cut-offs without penetration, sketched in Fig. 247, but in homogeneous medium, i.e., for $\kappa_1 = \kappa_2$, we may arrive at the conclusion that functions $F_1$ and $F_2$ for the problem with two different seepage coefficients must be represented as

$$F_1 = \frac{1}{\zeta + a}\left(\frac{Ai}{\sqrt{1 - \zeta}} + \frac{B}{\sqrt{\zeta}}\right) + \frac{1}{\zeta + b}\left(\frac{Ci}{\sqrt{1 - \zeta}} + \frac{D}{\sqrt{\zeta}}\right) ,$$

$$F_2 = \frac{1}{\zeta + a}\left(\frac{Ki}{\sqrt{1 - \zeta}} + \frac{L}{\sqrt{\zeta}}\right) + \frac{1}{\zeta + b}\left(\frac{Mi}{\sqrt{1 - \zeta}} + \frac{N}{\sqrt{\zeta}}\right) .$$

Conditions, which must be satisfied on the segments of the $\zeta$ plane with cut, lead to these ratios between the constants:

$$K = A, \quad M = C, \quad L = \frac{\kappa_2}{\kappa_1} B, \quad N = \frac{\kappa_2}{\kappa_1} D ,$$

so that the expressions for $F_1$ and $F_2$ become

$$F_1 = \frac{Ai\sqrt{\zeta} + B\sqrt{1 - \zeta}}{(\zeta + a)\sqrt{\zeta(1 - \zeta)}} + \frac{Ci\sqrt{\zeta} + D\sqrt{1 - \zeta}}{(\zeta + b)\sqrt{\zeta(1 - \zeta)}} ,$$

$$F_2 = \frac{Ai\sqrt{\zeta} + B\frac{\kappa_2}{\kappa_1}\sqrt{1 - \zeta}}{(\zeta + a)\sqrt{\zeta(1 - \zeta)}} + \frac{Ci\sqrt{\zeta} + \frac{\kappa_2}{\kappa_1}D\sqrt{1 - \zeta}}{(\zeta + b)\sqrt{\zeta(1 - \zeta)}} .$$

Computing the values of the functions $F_1$ and $F_2$ about points $\zeta = -a$ and $\zeta = -b$, we find ($H_1, H_2$ are the differences in waterlevel of the reservoirs):

$$\pi\frac{-A\sqrt{a} + B\sqrt{1 + a}}{\sqrt{a(1 + a)}} = \kappa_1 H_1 , \qquad \pi\frac{C\sqrt{b} + D\frac{\kappa_2}{\kappa_1}\sqrt{1 + b}}{\sqrt{b(1 + b)}} = -\kappa_2 H_2 ,$$

$$\pi\frac{A\sqrt{a} + B\frac{\kappa_2}{\kappa_1}\sqrt{1 + a}}{\sqrt{a(1 + a)}} = 0 , \qquad \pi\frac{-C\sqrt{b} + D\sqrt{1 + b}}{\sqrt{b(1 + b)}} = 0 .$$

The mapping of the regions onto the $\zeta$-plane is given, as before, by the formula

$$\zeta = \sin^2\frac{\pi z}{2hi} .$$

Therefore we find:

$$a = \mathrm{sh}^2\frac{\pi l_1}{2h} , \qquad b = \mathrm{sh}^2\frac{\pi l_2}{2h} ,$$

where $\ell_1$ and $\ell_2$ are the lengths of segments CB and CF ($\ell_1, \ell_2 > 0$).

We obtain for the constants the values

$$A = -\frac{\kappa_1\kappa_2 H_1}{\pi(\kappa_1 + \kappa_2)} \operatorname{ch} \lambda\ell_2, \qquad C = -\frac{\kappa_1\kappa_2 H_2}{\pi(\kappa_1 + \kappa_2)} \operatorname{ch} \lambda\ell_1 ,$$

$$B = -\frac{\kappa_1\kappa_2 H_1}{\pi(\kappa_1 + \kappa_2)} \operatorname{sh} \lambda\ell_2, \qquad D = \frac{\kappa_1\kappa_2 H_2}{(\kappa_1 + \kappa_2)} \operatorname{sh} \lambda\ell_1 ,$$

$$\left(\lambda = \frac{\pi}{2h}\right).$$

The final formulas for the velocities are

$$(3.1)\quad \begin{cases} u_1 - iv_1 = \\ = -\dfrac{\kappa_1\kappa_2 i}{h(\kappa_1 + \kappa_2)}\left[\dfrac{H_1(-\operatorname{ch}\lambda\ell_1 \operatorname{sh}\lambda z + \operatorname{sh}\lambda\ell_1 \operatorname{ch}\lambda z)\frac{\kappa_1}{\kappa_2}}{\operatorname{sh}\lambda(\ell_1 + z)\operatorname{sh}\lambda(\ell_1 - z)} - \dfrac{H_2}{\operatorname{sh}\lambda(\ell_2 - z)}\right] \\ u_2 - iv_2 = \\ = -\dfrac{\kappa_1\kappa_2 i}{h(\kappa_1 + \kappa_2)}\left[\dfrac{H_1}{\operatorname{sh}\lambda(\ell_1 + z)} - \dfrac{H_2(\operatorname{ch}\lambda\ell_2 \operatorname{sh}\lambda z + \operatorname{sh}\lambda\ell_2 \operatorname{ch}\lambda z)\frac{\kappa_2}{\kappa_1}}{\operatorname{sh}\lambda(\ell_2 + z)\operatorname{sh}\lambda(\ell_2 - z)}\right]. \end{cases}$$

When $M$, the point at the boundary of the *middle reservoir where* the velocity becomes zero, lies in the second region (where $x > 0$), then its abscissa $x_0$ is determined by the equation

$$H_1 \operatorname{sh}\lambda(\ell_2 + x_0)\operatorname{sh}\lambda(\ell_2 - x_0) =$$

$$(3.2)\qquad = H_2\left(\operatorname{ch}\lambda\ell_2 \operatorname{sh}\lambda x_0 + \frac{\kappa_2}{\kappa_1}\operatorname{sh}\lambda\ell_2 \operatorname{ch}\lambda x_0\right)\operatorname{sh}\lambda(\ell_1 + x_0).$$

When this point lies in the first region, then in equation (3.2) one must substitute $H_2, \ell_2, \kappa_2$ by $H_1, \ell_1, \kappa_1$ and $x_0$ by $-x_0$.

Abbreviating

$$\lambda\ell_1 = \frac{\pi\ell_1}{2h} = \alpha_1, \quad \lambda\ell_2 = \frac{\pi\ell_2}{2h} = \alpha_2, \quad \frac{H_2}{H_1} = H, \quad \frac{\kappa_2}{\kappa_1} = \kappa,$$

we find from equation (3.2):

$$\operatorname{th}\frac{\pi x_0}{2h} = -\frac{H(\operatorname{ch}\alpha_2 \operatorname{sh}\alpha_1 + \kappa \operatorname{sh}\alpha_2 \operatorname{ch}\alpha_1)}{2\operatorname{ch}\alpha_2(\operatorname{ch}\alpha_2 + H\operatorname{ch}\alpha_1)} +$$

$$+\frac{\sqrt{[H(\operatorname{ch}\alpha_2 \operatorname{sh}\alpha_1 - \kappa \operatorname{sh}\alpha_2 \operatorname{ch}\alpha_1) + \operatorname{sh} 2\alpha_2]^2 + 2H(\kappa + 1)\operatorname{sh} 2\alpha_2 \operatorname{sh}(\alpha_2 - \alpha_1)}}{2\operatorname{ch}\alpha_2(\operatorname{ch}\alpha_2 + H\operatorname{ch}\alpha_1)}$$

For $\ell_1 = \ell_2 = \ell$, i.e., for $\alpha_1 = \alpha_2$, this equation becomes

$$\operatorname{th}\frac{\pi x_0}{2h} = \frac{\kappa_1 H_1 + \kappa_2 H_2}{\kappa_1(H_1 + H_2)}\operatorname{th}\frac{\pi\ell}{2h}.$$

2. Cut-off without penetration in infinite half-plane.

We put $H_2 = 0$ in the formulas of the previous case. Then we have a cut-off only in point $z = -\ell_1$. We call

$$-\ell_1 = \ell, \quad H_1 = H.$$

Formulas (3.1) become

$$(3.3) \quad \begin{cases} u_1 - iv_1 = \dfrac{\kappa_1\kappa_2 H}{ih(\kappa_1 + \kappa_2)} \dfrac{\text{ch}\,\lambda\ell\ \text{sh}\,\lambda z + \dfrac{\kappa_1}{\kappa_2}\text{sh}\,\lambda\ell\ \text{ch}\,\lambda z}{\text{sh}\,\lambda(z-\ell)\ \text{sh}\,\lambda(z+\ell)}, \\ u_2 - iv_2 = \dfrac{\kappa_1\kappa_2 H}{ih(\kappa_1 + \kappa_2)} \dfrac{1}{\text{sh}\,\lambda(z-\ell)} \qquad \left(\lambda = \dfrac{\pi}{2h}\right). \end{cases}$$

We now consider the case of soil of infinite depth, i.e., we make $h = \infty$. Equations (3.3), after simplifications, become

$$(3.4) \quad \begin{cases} u_1 - iv_1 = \dfrac{\kappa_1 H}{\pi i}\left(\dfrac{1}{z-\ell} + \dfrac{\lambda}{z+\ell}\right), \\ u_2 - iv_2 = \dfrac{\kappa_2 H}{\pi i}\dfrac{1-\lambda}{z-\ell}, \qquad \left(\lambda = \dfrac{\kappa_1 - \kappa_2}{\kappa_1 + \kappa_2}\right). \end{cases}$$

When the boundary line between the two soils is the abscissa axis of the xy-plane and when a point vortex in the region 1 is moved to point $z = z_0$, then the complex velocities in regions 1 and 2 are respectively

$$(3.5) \quad \begin{cases} u_1 - iv_1 = \dfrac{\kappa_1\Gamma}{2\pi i}\left(\dfrac{1}{z-z_0} + \dfrac{\lambda}{z-\bar{z}_0}\right), \\ u_2 - iv_2 = \dfrac{\kappa_2\Gamma}{2\pi i}\dfrac{1-\lambda}{z-z_0}, \end{cases}$$

Here $\Gamma$ is the intensity of the point vortex. In this case it is impossible to speak about a cut-off without penetration, since amongst the equipotential lines there is no horizontal line, which could be taken for boundary of the water reservoir. However, the analysis of the point vortex is a useful concept in the theory of groundwater flow, since by means of a system of vortices of suitable intensity, one may solve problems of flow about hydraulic structures.

If the first parts of formulas (3.4) are multiplied by $i$, then, instead of a point vortex, we obtain a point sink or source in the half-plane with seepage coefficient $\kappa_1$. (In the other half-plane there is soil with seepage coefficient $\kappa_2$.)

Streamlines and equipotential lines for a cut-off without penetration in media 1 and 2, for $\kappa_1/\kappa_2 = 1/3$, are given in Fig. 248.

Integrating equations (3.4), we obtain for the complex potential of regions 1 and 2 respectively:

$$\omega_1(z) = \frac{\kappa_1 H}{\pi i}\,[\ln(z-\ell) + \lambda\ln(z+\ell) + C_1,$$

$$\omega_2(z) = \frac{\kappa_2 H}{\pi i}\,(1-\lambda)\ln(z-\ell) + C_2 .$$

It is easy to see that the streamlines in the first region are

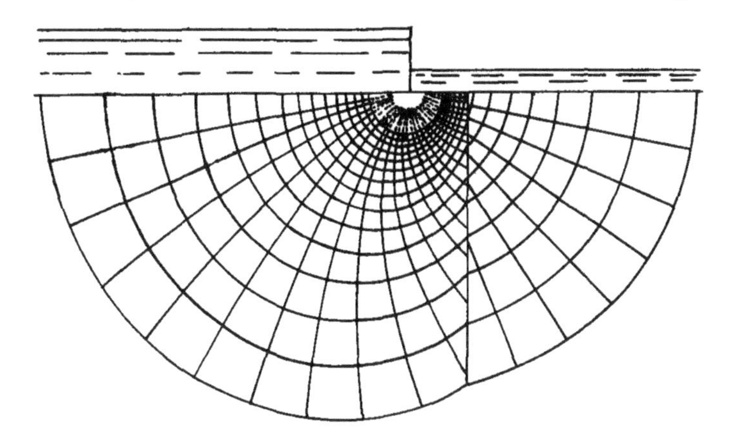

Fig. 248

"generalized lemniscates," i.e., lines with equation

$$|z - \ell| \cdot |z + \ell|^{\lambda} = \text{constant.}$$

The streamlines in the second region are arcs of circles.

§4. Point Vortex in Multilayered Region.

To solve problems of seepage under hydraulic structures in a medium that consists of horizontal layers of soil of different porosity, we may use the hydrodynamical approach of flow about a body. The contour of the base of the hydrotechnical structure is replaced by a system of point vortices and the intensity distribution of these vortices is so chosen that the boundary conditions are satisfied, i.e., the normal components of the velocities are zero.

To construct the integral equation, which must be satisfied by the intensity of the singularities, one must find the expression of the complex potential of a point vortex in a multilayered region [2].

Fig. 249 represents a horizontal stratum of thickness $h_1$ with seepage coefficient $\kappa_1$, then a stratum of thickness $h_2$ with seepage coefficient $\kappa_2$, and so on, and finally, the bottom stratum of thickness $h_n$ with seepage coefficient $\kappa_n$.

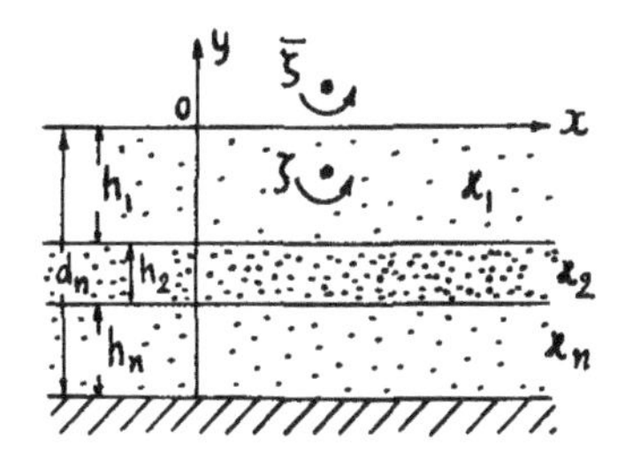


Fig. 249

Parallel with the quantities $h_r$, we examine their sums $d_r$, so that

$$d_1 = h_1 \ ,$$

$$d_2 = h_1 + h_2 \ , \ldots ,$$

$$d_n = h_1 + h_2 + \ldots + h_n \ .$$

The line $y = - d_n$ represents an impervious boundary.

We place a point vortex of intensity one in point $\zeta = \xi + i\eta$ of the upper stratum.

We designate by $w_r = u_r - iv_r$ the complex velocity in point $z$, of the $r^{th}$ layer, due to the existence of a point vortex in $\zeta$. We write function $w_r$ as a Fourier integral

$$(4.1) \qquad w_r(z) = \int_0^\infty [A_r(\alpha)e^{i\alpha z} + B_r(\alpha)e^{-i\alpha z}]d\alpha \qquad (r = 2,3,\ldots,n),$$

where $A_r(\alpha)$ and $B_r(\alpha)$ — complex functions of the real variable $\alpha$.

Concerning the velocity in the point of the upper stratum, which contains the point vortex, we isolate in its expression the term

$$\frac{1}{2\pi i}\frac{1}{z-\zeta} .$$

Along the x-axis, corresponding to the boundary of the water reservoir, the horizontal component of the velocity is zero, so that we may write

$$\operatorname{Im}(iw_1) = 0 .$$

From this it follows that the function of the complex variable $iw_1$ may be continued in the upper half-plane (in the region of the strip between the straight lines $y = 0$ and $y = +h_1$), so that in point $\bar{z}$ the function $iw_1$ transforms into $-i\bar{w}_1$. This allows us to write $w_1$ as:

$$(4.2) \qquad w_1 = \frac{1}{2\pi i}\frac{1}{z-\zeta} + \frac{1}{2\pi i}\frac{1}{z-\bar{\zeta}} + \int_0^\infty [A_1(\alpha)e^{i\alpha z} + B_1(\alpha)e^{-i\alpha z}]d\alpha ,$$

where

$$(4.3) \qquad B_1(\alpha) = -\bar{A}_1(\alpha) .$$

Along the impervious boundary of the bottom $n^{th}$ stratum, the vertical velocity must be zero. This determines the relationship between $A_n(\alpha)$ and $B_n(\alpha)$ as

$$(4.4) \qquad B_n(\alpha) = e^{2d_n\alpha} A_1(\alpha) .$$

At the boundary between the $r^{th}$ and $(r+1)^{th}$ strata, the normal velocity component must be continuous, the tangential velocity components must be proportional to the corresponding seepage coefficients. Therefore, for $y = -d_r$, we have

$$v_{r+1} = v_r, \qquad \frac{u_{r+1}}{\kappa_{r+1}} = \frac{u_r}{\kappa_r}$$

or, in other words,

$$\operatorname{Im}(w_{r+1}) = \operatorname{Im} w_r, \qquad \kappa_r \operatorname{Re}(w_{r+1}) = \kappa_{r+1} \operatorname{Re}(w_r) .$$

Finally, the last condition may be written as:

$$(4.5) \qquad \begin{aligned} w_{r+1} - \bar{w}_{r+1} &= w_r - \bar{w}_r , \\ \kappa_r(w_{r+1} + \bar{w}_{r+1}) &= \kappa_{r+1}(\bar{w}_r + w_r) . \end{aligned}$$

For $z = x - d_r i$ we have:

$$w_{r+1}(z) = \int_0^\infty [A_{r+1}(\alpha)e^{d_r\alpha+i\alpha x} + B_{r+1}(\alpha)e^{-d_r\alpha-i\alpha x}]d\alpha ,$$

$$w_r(z) = \int_0^\infty [A_r(\alpha)e^{d_r\alpha+i\alpha x} + B_r(\alpha)e^{-d_r\alpha-i\alpha x}]d\alpha .$$

Conditions (4.5) must not only be satisfied for the functions themselves but also for their expressions as integrands. Therefore we obtain:

$$(A_{r+1} - A_r)e^{d_r\alpha+i\alpha x} + (B_{r+1} - B_r)e^{-d_r\alpha-i\alpha x} =$$
$$= (\bar{A}_{r+1} - \bar{A}_r)e^{d_r\alpha-i\alpha x} + (\bar{B}_{r+1} - \bar{B}_r)e^{-d_r\alpha+i\alpha x} ,$$

$$(\kappa_r A_{r+1} - \kappa_{r+1}A_r)e^{d_r\alpha+i\alpha x} + (\kappa_r B_{r+1} - \kappa_{r+1}B_r)e^{-d_r\alpha-i\alpha x} =$$
$$= - (\kappa_r\bar{A}_{r+1} - \kappa_{r+1}\bar{A}_r)e^{d_r\alpha-i\alpha x} - (\kappa_r\bar{B}_{r+1} - \kappa_{r+1}\bar{B}_r)e^{-d_r\alpha+i\alpha x} .$$

These equations must be satisfied identically in $\kappa$, and therefore we may equate amongst each other the coefficients for $e^{i\alpha x}$ and $e^{-i\alpha x}$. This leads to four equations, from which we write two:

$$(4.6)\quad \begin{aligned} &(A_{r+1} - A_r)e^{d_r\alpha} - (\bar{B}_{r+1} - \bar{B}_r)e^{-d_r\alpha} = 0 , \\ &(\kappa_r A_{r+1} - \kappa_{r+1}A_r)e^{d_r\alpha} + (\kappa_r\bar{B}_{r+1} - \kappa_{r+1}\bar{B}_r)e^{-d_r\alpha} = 0 . \end{aligned}$$

We obtain the two other equations, if we replace all complex quantities by their conjugates in (4.6).

Equations (4.6) hold for $r = 2,3,\dots,n-1$. To satisfy the condition at the lines $y = -h_1 = -d_1$, we represent the principal part of the expression of $w_1$ as a determined integral

$$\frac{1}{z - \zeta} = i\int_0^\infty e^{-i\alpha(z-\zeta)}d\alpha \quad \text{for } \operatorname{Im}(z - \zeta) < 0 .$$

Along line $y = -d_1$, condition $\operatorname{Im}(z - \zeta) < 0$, equivalent to $\operatorname{Im}(z - \bar{\zeta}) < 0$, is satisfied. Therefore function $w_1(z)$ may be represented as

$$w_1(z) = \int_0^\infty \left\{A_1(\alpha)e^{i\alpha z} + B_1(\alpha)e^{-i\alpha z} + \frac{1}{\pi} e^{-i\alpha(z-\xi)} \operatorname{ch} \alpha\eta \right\} d\alpha$$
$$(y - \eta < 0) ,$$

and for $z = x - d_1 i$ as

$$w_1(z) = \int_0^\infty \left\{A_1(\alpha)e^{d_1\alpha+i\alpha x} + B_1(\alpha)e^{-d_1\alpha-i\alpha x} + \frac{1}{\pi} e^{-d_1\alpha-i\alpha(x-\xi)} \operatorname{ch} \alpha\eta\right\} d\alpha .$$

The conditions at the boundary line give:

$$(4.7)\quad \begin{aligned} &(A_1 - A_2)e^{d_1\alpha} - (\bar{B}_1 - \bar{B}_2)e^{-d_1\alpha} = \frac{1}{\pi}\, e^{-d_1\alpha} \cdot e^{-i\alpha\xi} \operatorname{ch} \alpha\eta \;, \\ &(\kappa_2 A_1 - \kappa_1 A_2)e^{d_1\alpha} + (\kappa_2 \bar{B}_1 - \kappa_1 \bar{B}_2)e^{-d_1\alpha} = -\frac{\kappa_2}{\pi} e^{-d_1\alpha - i\alpha\xi} \operatorname{ch} \alpha\eta \;. \end{aligned}$$

We introduce the notations

$$e^{2d_r\alpha} = \alpha_r, \quad \frac{1}{\pi}\, e^{i\alpha\xi} \operatorname{ch} \alpha\eta = b, \quad \frac{\kappa_r}{\kappa_{r+1}} = \lambda_r \;,$$

so that the system of equations (4.3), (4.4), (4.6) and (4.7) may be written as

$$(4.8)\quad \begin{aligned} A_1 + \bar{B}_1 &= 0 \;, \\ \alpha_1 A_1 - B_1 - \alpha_1 A_2 + \bar{B}_2 &= b \;, \\ \alpha_1 A_1 + \bar{B}_1 - \lambda_1 \alpha_1 A_2 - \lambda_1 \bar{B}_2 &= -b \;, \\ \alpha_2 A_2 - \bar{B}_2 - \alpha_2 A_3 + \bar{B}_3 &= 0 \;, \\ \alpha_2 A_2 + \bar{B}_2 - \lambda_2 \alpha_2 A_3 - \lambda_2 \bar{B}_3 &= 0 \;, \\ \cdots\cdots\cdots\cdots\cdots\cdots & \\ \alpha_n A_n - \bar{B}_n &= 0 \;. \end{aligned}$$

Particularly, for $n = 2$ we obtain:

$$\bar{B}_1 = -A_1, \quad \bar{B}_2 = \alpha_2 A_2 \;,$$

$$(\alpha_1 + 1)A_1 + (\alpha_2 - \alpha_1)A_2 = b \;,$$

$$(\alpha_1 - 1)A_1 - \lambda(\alpha_2 + \alpha_1)A_2 = -b \;,$$

from which we find

$$A_1 = \frac{1}{2\pi}\, \frac{\operatorname{ch} \alpha\eta(\kappa_1 \operatorname{ch} h_2\alpha - \kappa_2 \operatorname{sh} h_2\alpha)}{\kappa_2 \operatorname{sh} h_1\alpha \operatorname{sh} h_2\alpha + \kappa_1 \operatorname{ch} h_1\alpha \operatorname{ch} h_2\alpha}\, e^{-i\alpha\xi - h_1\alpha} \;,$$

$$A_2 = \frac{1}{2\pi}\, \frac{\operatorname{ch} \alpha\eta}{\kappa_2 \operatorname{sh} h_1\alpha \operatorname{sh} h_2\alpha + \kappa_1 \operatorname{ch} h_1\alpha \operatorname{ch} h_2\alpha}\, e^{-i\alpha\xi - (h_1 + h_2)\alpha} \;.$$

We have for the complex velocities in the first and second flow regions:

$$w_1(z) = \frac{1}{2\pi i}\, \frac{1}{z - \zeta} + \frac{1}{2\pi i}\, \frac{1}{z - \bar{\zeta}} + \frac{2i}{\pi} \int_0^\infty M(\alpha) \sin \alpha(z - \zeta) e^{-h_1\alpha}\, d\alpha,$$

$$w_2(z) = \frac{2}{\pi} \int_0^\infty N(\alpha) \cos \alpha(z - \bar{\zeta})\, d\alpha \;,$$

where

$$M(\alpha) = \frac{\operatorname{ch} \alpha\eta(\kappa_1 \operatorname{ch} h_2\alpha - \kappa_2 \operatorname{sh} h_2\alpha)}{(\kappa_1 + \kappa_2)\operatorname{ch} h\alpha + (\kappa_1 - \kappa_2)\operatorname{ch}(h_1 - h_2)\alpha} ,$$

$$N(\alpha) = \frac{\operatorname{ch} \alpha\eta}{(\kappa_1 + \kappa_2)\operatorname{ch} h\alpha + (\kappa_1 - \kappa_2)\operatorname{ch}(h_1 - h_2)\alpha}$$

$$(h = h_1 + h_2) .$$

We mention the particular cases:

1) $\kappa_2 = 0$, i.e., the lower stratum is impervious. Then

$$w(z) = \frac{1}{4h_1 i}\left[\operatorname{csch} \frac{\pi(z - \zeta)}{2h_1} + \operatorname{csch} \frac{\pi(z - \bar{\zeta})}{2h_1}\right] .$$

2) $\kappa_2 = \infty$, i.e., the line $y = -h_1$ represents a equipotential line. Then

$$w(z) = \frac{1}{4h_1 i}\left[\operatorname{cth} \frac{\pi(z - \zeta)}{2h_1} + \operatorname{cth} \frac{\pi(z - \bar{\zeta})}{2h_1}\right] .$$

3) The lower stratum is infinitely deep, $h_2 = \infty$. This case was examined by B. K. Risenkampf [3]. We may write (considering that the vortex approaches the x-axis, i.e., $\zeta = \xi$):

$$w_1(z) = \frac{1}{2\pi i}\,\frac{1}{z - \xi} + \frac{1}{2\pi i}\int_0^\infty \frac{(\kappa_2 - \kappa_1)\sin \alpha(z - \xi)}{\kappa_1 \operatorname{ch} h_1\alpha + \kappa_2 \operatorname{sh} h_1\alpha}\, e^{-h_1\alpha}\, d\alpha$$

or in the form of a power series of $\lambda = \dfrac{\kappa_1 - \kappa_2}{\kappa_1 + \kappa_2}$:

$$w_1(z) = \frac{1}{2\pi i}\,\frac{1}{z - \xi} - \frac{\lambda(z - \xi)}{\pi i[4h_1^2 + (z - \xi)^2]} + \frac{\lambda^2(z - \xi)}{\pi i[9h_1^2 + (z - \xi)^2]} + \cdots$$

We now come to the derivation of the integral equation. Let the foundation of the hydraulic structure be bounded by the line AB (Fig. 250). Reflecting this line into the abscissa axis, we obtain a closed contour C. We assume that contour C has a tangent in each point. We distribute along contour C vortices with intensity $\Gamma(s)\,ds$ per element of arc ds. We call $w(z,\zeta)$ the complex velocity due to a vortex with unit intensity in point $\zeta$. Then the arc element ds calls for a velocity

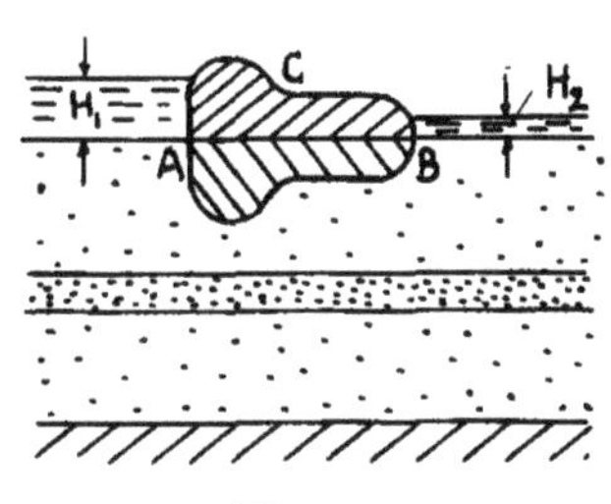

Fig. 250

$$\Gamma(s)\, w(z,\zeta)\, ds$$

in point z.

Therefore $\zeta = \xi(s) + i\eta(s)$ is a function of arc length s. The velocity $W(z)$, due to the influence in point z of the system of all vortices, distributed along contour $C$, is given by

$$W(z) = \int_C \Gamma(s)\, w(z,\zeta)\, ds. \tag{4.9}$$

We put

$$w(z,\zeta) = \frac{i}{2\pi i} \frac{1}{z - \zeta} + \Phi(z,\zeta) ,$$

where $\Phi(z,\zeta)$ is an holomorphic function. We let point $z$ approach the contour point $\zeta_0 = \zeta(\sigma)$, where $\sigma$ is the length of the arc in point $\zeta_0$. In the limit we obtain for the integral of the Cauchy type ($\theta(\sigma)$ is the angle of the tangent with the x-axis):

$$\lim_{z\to\zeta} \frac{1}{2\pi i}\int_C \frac{\Gamma(s)\, ds}{z - \zeta} = \frac{1}{2}\Gamma(\sigma)e^{-i\theta(\sigma)} + \frac{1}{2\pi i}\,\text{P. V.}\int_C \frac{\Gamma(s)\, ds}{\zeta(\sigma) - \zeta(s)} . \tag{4.10}$$

Multiplying (4.9) term by term with $e^{i\theta(\sigma)}$ and passing to the limit for $z \to \zeta_0$, we obtain:

$$W(\zeta_0)e^{i\theta(\sigma)} = \frac{1}{2}\Gamma(\sigma) + \frac{1}{2\pi i}\quad \text{P. V.}\int_C \Gamma(s)\, w(z,\zeta)e^{i\theta(\sigma)}\, ds . \tag{4.11}$$

But $W(\zeta_0)e^{i\theta(\sigma)} = V_t - iV_n$, where $V_t$ is the projection of the velocity on the tangent to the contour, $V_n$ the projection on the normal. Therefore, separating real and imaginary parts in (4.11) and considering that $V_t = \Gamma(\sigma)$, $V_n = 0$ on the contour, we obtain two integral equations: the equation of the second kind with ordinary kernel

$$\frac{1}{2}\Gamma(\sigma) = \int_C \Gamma(s)\, E(\sigma,s)\, ds ,$$

where

$$E(\sigma,s) = u(\sigma,s)\sin\theta(\sigma) - v(\sigma,s)\cos\theta(\sigma) ,$$

and the equation of the first kind with degenerate kernel

$$\int_C \Gamma(s)\, G(\sigma,s)\, ds = 0 ,$$

where

$$G(\sigma,s) = u(\sigma,s)\cos\theta(\sigma) + v(\sigma,s)\sin\theta(\sigma) .$$

N. K. Kalinin [4] and B. K. Risenkampf [3] and also the author of this book [2] have shown some way to find the solution of the last equation in the case of a flat bottom dam.

§5. Simplest Flows in Stratified Soils.

To conclude this subdivision, we consider the simplest, one-dimensional flows in heterogeneous soils [19]. We need these problems further in the interpretation of anisotropic soils.

We examine two cases.

1. Horizontal confined seepage. The scheme of such a flow is sketched in Fig. 251: the soil is composed of horizontal strata (with thickness $\ell_1, \ell_2, \dots \ell_n$) separated from each other by horizontal planes. The seepage coefficients are called respectively $k_1, k_2, \dots, k_n$. The upper and bottom stratum are bounded by impervious walls.

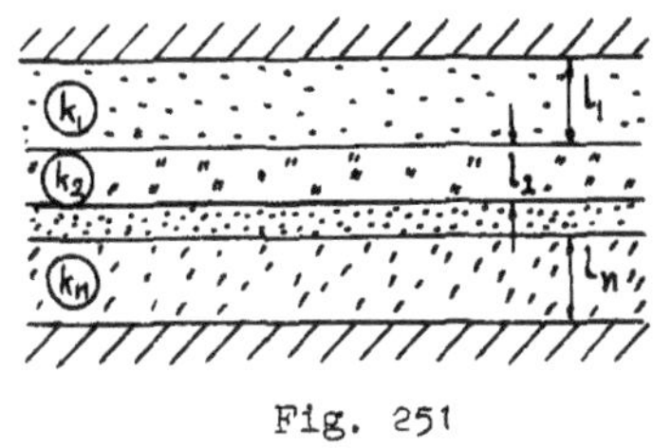

Fig. 251

The flowrate through each stratum is computed by the formula

$$Q_i = -k\frac{H_2 - H_1}{L}\ell_i . \tag{5.1}$$

Here $H_2 - H_1$ is the difference in head between the end points of the stratum, $L$ the length of the seepage path.

For the total flowrate through the entire soil strip, we obtain:

$$Q = \sum_{\nu=1}^{n} k_\nu \ell_\nu \frac{H_1 - H_2}{L} . \tag{5.2}$$

Dividing the flowrate by the cross-section, we find the average seepage velocity $u$. Taking the thickness of the strata perpendicular to the paper to be one, we find that the area $S$ is numerically equal to the total depth of flow:

$$S = 1 \cdot \sum_{\nu=1}^{n} \ell_\nu = \ell .$$

Therefore we obtain for the average velocity of flow:

$$u = \frac{\sum k_\nu \ell_\nu}{\ell}\frac{H_1 - H_2}{L} .$$

Introducing the notation

$$k = \frac{\sum_{\nu=1}^{n} k_\nu \ell_\nu}{\ell} \tag{5.3}$$

we have

$$u = k\frac{H_1 - H_2}{L} . \tag{5.4}$$

We may call $k$ the equivalent or reduced seepage coefficient: this is the average value, weighted over the depths of strata, of the seepage coefficient of the individual strata. The flow proceeds as if we had homogeneous soil with seepage coefficient equal to the average seepage coefficient of all strata.

2. Vertical seepage in stratified soils. We assume that flow occurs in the same soil as above, but in vertical direction. (Fig. 252.)

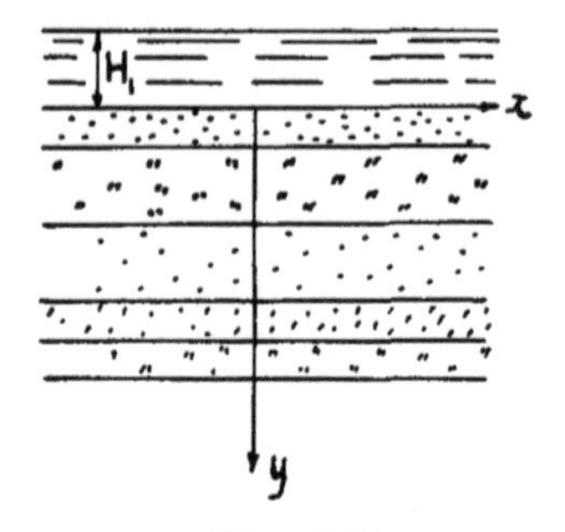

Fig. 252

For the velocity potential and head, with the ordinate axis oriented downward (see Chapter II, §1):

$$\varphi_i = -k_i\left(\frac{p_i}{\rho g} - y\right), \quad h_i = \frac{p_i}{\rho g} - y$$

(5.5) $(i = 1,2,\ldots,n)$.

We assume zero atmospheric pressure, so that the head at $y = 0$ is $H_1$. Along the lines

$$y = \ell_1, \quad y = \ell_1 + \ell_2, \ldots, \qquad y = \ell_1 + \ell_2 + \ldots + \ell_{n-1}$$

the values of the head are unknown. We call them respectively $h^{(1)}, h^{(2)}, \ldots, h^{(n-1)}$. Along the line $y = \ell = \ell_1 + \ldots + \ell_n$, the head is $H_2$.

Applying Darcy's formula (Chapter I, §6) to all of the strata, we call $v_i$ the velocity in the $i^{th}$ stratum:

$$v_1 = -k_1 \frac{h^{(1)} - H_1}{\ell_1}, \quad v_2 = -k_2 \frac{h^{(2)} - h^{(1)}}{\ell_2}, \ldots,$$

(5.6)

$$v_n = -k_n \frac{H_2 - h^{(n-1)}}{\ell_n} .$$

All these velocities are equal, since the fluid is incompressible and the flow one-dimensional. The continuity equation therefore gives $\partial v/\partial y = 0$, i.e., a constant value for the velocity. Equating the expressions (5.6) mutually, we obtain $n-1$ equations for $n-1$ unknowns $h^{(1)}, h^{(2)}, \ldots, h^{(n-1)}$:

$$\frac{h^{(1)} - H_1}{\alpha_1} = \frac{h^{(2)} - h^{(1)}}{\alpha_2} = \ldots = \frac{H_2 - h^{(n-1)}}{\alpha_n}, \tag{5.7}$$

where

$$\alpha_i = \frac{\ell_i}{k_i} \qquad (i = 1,2,\ldots,n) . \tag{5.8}$$

We rewrite them as the following system (the first equation is an identity and is introduced for convenience of computation):

$$\alpha_1(h^{(1)} - H_1) = \alpha_1(h^{(1)} - H_1) ,$$

$$\alpha_2(h^{(1)} - H_1) = \alpha_1(h^{(2)} - h^{(1)}) ,$$

$$\alpha_3(h^{(1)} - H_1) = \alpha_1(h^{(3)} - h^{(2)}) ,$$

$$\cdots\cdots\cdots\cdots\cdots\cdots$$

$$\alpha_n(h^{(1)} - H_1) = \alpha_1(H_2 - h^{(n-1)}) .$$

Adding the left and right sides of these equations, we have

$$(\alpha_1 + \alpha_2 + \dots + \alpha_n)(h^{(1)} - H_1) = \alpha_1(H_2 - H_1) ,$$

from which we have for $v_1$ and consequently for all $v_i$:

$$v_1 = -\frac{h^{(1)} - H_1}{\alpha_1} = -\frac{H_2 - H_1}{\alpha_1 + \alpha_2 + \dots + \alpha_n}$$

Calling $v$ the general expression for the velocity, we have

$$v = \frac{H_1 - H_2}{\frac{\ell_1}{k_1} + \frac{\ell_2}{k_2} + \dots + \frac{\ell_n}{k_n}} . \tag{5.9}$$

We may apply this derivation of this velocity to any stratum, for example the upper, and write:

$$v = k_1 \frac{H_1 - H_2}{L} ,$$

where

$$L = \ell_1 + \frac{k_1}{k_2}\ell_2 + \dots + \frac{k_1}{k_n}\ell_n .$$

$L$ may be called the equivalent or reduced length.

We may give another interpretation to formula (5.9). We put

$$\frac{\ell_1}{k_1} + \frac{\ell_2}{k_2} + \dots + \frac{\ell_n}{k_n} = \frac{\ell}{k} ; \tag{5.10}$$

so that we may write

$$v = k\frac{H_1 - H_2}{\ell} . \tag{5.11}$$

Here $k$ is the equivalent or reduced seepage coefficient, rendering the same velocity as the heterogeneous soil for a depth of stratum equal to that of the heterogeneous soil.

## B. Anisotropic Soils.

### §6. Equations of Motion. Boundary Conditions.

It is well known that in loess-soil, the permeability in the vertical direction is higher than in the horizontal. In a series of other soils we have the opposite picture: the permeability in the horizontal direction is higher than in the vertical.

The case of plane flow in soil with few strata, bounded by parallel planes, was briefly examined by Dachler, then to some extent by V. I Aravin [7] and more in detail by N- N- Pavlovsky in his recent work.

V. I. Aravin investigates the general law of stratified two-dimensional flow. Examples of flow under hydraulic structures in anisotropic soils [9] and of flow with free surface and with boundary line (interphase) between two fluids in such soils [8] were also given.

We do not consider three-dimensional cases, but analyze only plane flows.

Anisotropic is the name for soil in which the magnitude of the seepage coefficient in a given point of the flow region depends upon the direction of the seepage velocity.

To analyze the general case, we assume that we have families of curves $\xi$ = constant and $\eta$ = constant, where

$$\xi = f_1(x,y), \quad \eta = f_2(x,y) \ , \tag{6.1}$$

mutually orthogonal and having the property that the seepage coefficients $k_\xi$, $k_\eta$ along the tangents to these curves are respectively maximum and minimum (Fig. 253).

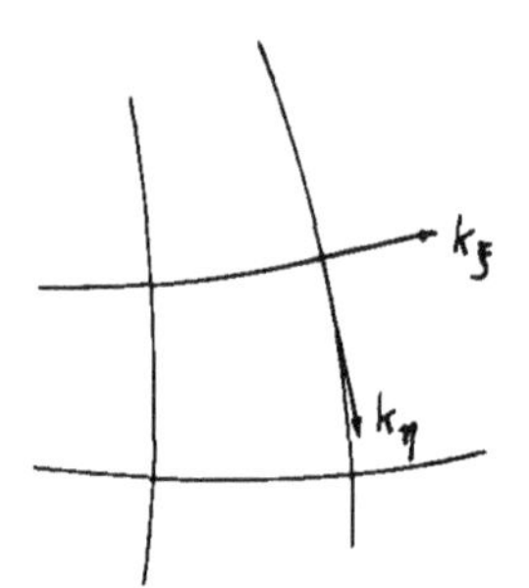


Fig. 253

We call the elements of arc along these axes respectively $ds_1$ and $ds_2$, the Lame' coefficients $H_1$, $H_2$:

$$H_1 = \sqrt{\left(\frac{\partial x}{\partial \xi}\right)^2 + \left(\frac{\partial y}{\partial \xi}\right)^2}, \qquad H_2 = \sqrt{\left(\frac{\partial x}{\partial \eta}\right)^2 + \left(\frac{\partial y}{\partial \eta}\right)^2} \ .$$

The equation of continuity can be written as [21]:

$$\frac{\partial(H_2 v_\xi)}{\partial \xi} + \frac{\partial(H_1 v_\eta)}{\partial \eta} = 0 \ , \tag{6.2}$$

where

$$v_\xi = - k_\xi \frac{\partial h}{\partial s_1} = - k_\xi \frac{1}{H_1} \frac{\partial h}{\partial \xi} \ , \qquad v_\eta = - k_\eta \frac{\partial h}{\partial s_2} = - k_\eta \frac{1}{H_2} \frac{\partial h}{\partial \eta} \ , \tag{6.3}$$

$h$ is the piezometric head (see Chapter II, §1).

On account of (6.3) we may write for (6.2):

$$\frac{\partial}{\partial \xi}\left(k_\xi \frac{H_2}{H_1} \frac{\partial h}{\partial \xi}\right) + \frac{\partial}{\partial \eta}\left(k_\eta \frac{H_1}{H_2} \frac{\partial h}{\partial \eta}\right) = 0 \tag{6.4}$$

If $k_\xi$ and $k_\eta$ are constant, then they may be put before the differentiation sign.

V. I. Aravin [7] considered particular cases of stratification such as in concentric circles or in the form of a bunch of straight lines. We restrict ourselves in the sequel to "homogeneous" anisotropy, when the family of lines (6.1) consists of parallel straight lines and where the magnitudes of $k_\xi$ and $k_\eta$ are constant.

For isotropic soil we had the basic relationship for the linear seepage law

$$V = - k \text{ grad } h \ ,$$

where $k$ is the seepage coefficient, $h$ is linked with the pressure by the formula

$$h = \frac{p}{\rho g} + y + \text{const.}$$

In the case of anisotropic soil, we have a symmetric seepage tensor. Considering only two-dimensional flow, we may write the seepage tensor as

$$\begin{Vmatrix} k_{11} & k_{12} \\ k_{21} & k_{22} \end{Vmatrix} .$$

We take the principal axes of this tensor along the $x_1, y_1$ axes of a rectangular coordinate system. Then the seepage tensor may be written as

$$\begin{Vmatrix} k_{x_1} & 0 \\ 0 & k_{y_1} \end{Vmatrix} .$$

The quantities $k_{x_1}$, $k_{y_1}$ are seepage coefficients in the direction of the principal axes.

The equation of motion in $x_1, y_1$ coordinates are written as (see Chapter I, §11):

$$0 = -\frac{1}{\rho}\frac{\partial p}{\partial x_1} - \frac{g}{k_{x_1}} u_1 - g \sin\alpha, \qquad 0 = -\frac{1}{\rho}\frac{\partial p}{\partial y_1} - \frac{g}{k_{y_1}} v_1 - g\cos\alpha . \tag{6.5}$$

Here $\alpha$ is the angle measured from the horizontal x-axis (Fig. 254).

Putting

$$\Phi(x_1, y_1) = -\frac{p}{\rho g} - x_1 \sin\alpha - y_1 \cos\alpha , \tag{6.6}$$

where

$$u_1 = -k_{x_1}\frac{\partial\Phi}{\partial x_1}, \qquad v_1 = -k_{y_1}\frac{\partial\Phi}{\partial y_1}, \tag{6.7}$$

We may call $\Phi$ the reduced head. We also introduce function $\Psi(x_1, y_1)$ by means of the equalities

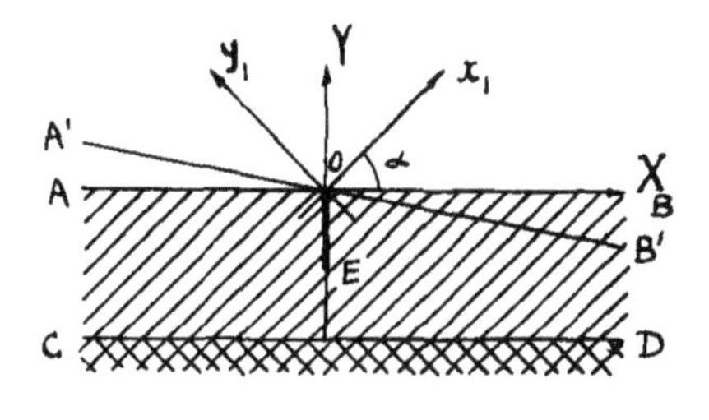


Fig. 254

$$u_1 = \sqrt{k_{x_1} k_{y_1}}\,\frac{\partial\Psi}{\partial y_1}, \qquad v_1 = -\sqrt{k_{x_1} k_{y_1}}\,\frac{\partial\Psi}{\partial x_1} .$$

From the continuity equation

$$\frac{\partial u_1}{\partial x_1} + \frac{\partial v_1}{\partial y_1} = 0$$

it follows that $\Phi$ satisfies the equation

$$k_{x_1}\frac{\partial^2\Phi}{\partial x_1^2} + k_{y_1}\frac{\partial^2\Phi}{\partial y_1^2} = 0 .$$

If we make one more transformation

$$(6.8) \qquad x = x_1 , \quad y_1 = \nu y , \quad \nu^2 = k_{y_1} : k_{x_1}$$

Then $\Phi$ satisfies Laplace's equation in the new $x,y$ coordinates

$$\frac{\partial^2 \Phi}{\partial x^2} + \frac{\partial^2 \Phi}{\partial y^2} = 0 \quad .$$

The combination $\Phi + i\Psi$, where $\Psi$ is the previously introduced reduced streamfunction, is a function of the complex variable $z = x + iy$.

The flowrate through the contour bounded by the arc $MN$, is expressed by means of the streamfunction as:

$$(6.9) \qquad Q = \sqrt{k_{x_1} k_{y_1}} \; (\Psi_N - \Psi_M) \quad .$$

The three coordinate systems that we introduced are linked by the equations:

$$(6.10) \qquad \begin{cases} X = x_1 \cos\alpha - y_1 \sin\alpha = x\cos\alpha - \nu y \sin\alpha , \\ Y = x_1 \sin\alpha + y_1 \cos\alpha = x \sin\alpha + \nu y \cos\alpha , \end{cases}$$

$$(6.11) \qquad \begin{cases} x_1 = x = X\cos\alpha + Y \sin\alpha , \\ y_1 = \nu y = -X \sin\alpha + Y\cos\alpha . \end{cases}$$

Here we have introduced the notation

$$\nu = \sqrt{k_{y_1}} : \sqrt{k_{x_1}} \quad .$$

In particular, the equations of the horizontal lines $AB, CD$ and of the vertical line $DE$ in the $x,y$ coordinate system are:

$$(6.12) \qquad y = -\frac{1}{\nu} x \operatorname{tg} \alpha ,$$

$$(6.13) \qquad x \sin\alpha + \nu y \cos\alpha + h = 0 ,$$

$$(6.14) \qquad y = \frac{1}{\nu} x \operatorname{ctg} \alpha .$$

In Fig. 254 $A'B'$ is the position of the straight line $AB$ in the $x,y$ system.

In the $x,y$ coordinate system, we may consider some fictitious flow in the region that is obtained from that in the $X,Y$ plane through the affine transformation (6.10), in which straight lines transform into straight lines and in which, generally speaking, the angles between straight lines are not preserved.

On the boundaries of the flowregion in the $x,y$ plane, these conditions must be satisfied:

1) On the boundary of the water reservoir

$$\Phi(x,y) = \text{const.},$$

since on the corresponding boundary of the $x_1, y_1$ plane we have

$$\Phi(x_1, y_1) = \text{const.}$$

2) On the impervious boundary:

$$\Psi(x,y) = \text{const.}$$

3) On the free surface, when the pressure is constant, we have, according to (6.6) and (6.11)

$$\Phi + x \sin \alpha + \nu y \cos \alpha = \text{const.} \tag{6.15}$$

Moreover, on the free surface, as streamline

$$\Psi(x,y) = \text{const.}$$

We differentiate (6.15) along $s$ on the free surface (in the $x,y$ plane), multiply term by term with $\partial\Phi/\partial s$ and obtain:

$$\left( \frac{\partial \Phi}{\partial s} \right)^2 + \sin \alpha \frac{\partial \Phi}{\partial s}\frac{dx}{ds} + \nu \cos \alpha \frac{\partial \Phi}{\partial s}\frac{dy}{ds} = 0 \quad .$$

Introducing the components $u,v$ of the fictitious dimensionless velocity,

$$\frac{\partial \Phi}{\partial s}\frac{dx}{ds} = u, \quad \frac{\partial \Phi}{\partial s}\frac{dy}{ds} = v ,$$

we have the equation of the circle

$$u^2 + v^2 + u \sin \alpha + \nu u \cos \alpha = 0 \tag{6.16}$$

in the plane of the velocity $u,v$. This circle passes through the origin of coordinates and has its center in the point $(-1/2 \sin \alpha, -\nu/2 \cos \alpha)$.

4) Along the seepage surface, equation (6.15) holds. Let the seepage segment be a straight line, making the angle $\beta$ with the x-axis. Differentiating (6.15) along the arc $s$ on the seepage surface, we obtain:

$$\frac{\partial \Phi}{\partial s} + \sin \alpha \frac{dx}{ds} + \nu \cos \alpha \frac{dy}{ds} = 0 \ .$$

Considering that

$$\frac{\partial \Phi}{\partial s} = \frac{\partial \Phi}{\partial x}\frac{dx}{ds} + \frac{\partial \Phi}{\partial y}\frac{dy}{ds} = \frac{\partial \Phi}{\partial x} \cos \beta + \frac{\partial \Phi}{\partial y} \sin \beta = u \cos \beta + v \sin \beta,$$

we have found the equation of the straight line

$$u \cos \beta + v \sin \beta + \sin \alpha \cos \beta + \nu \cos \alpha \sin \beta = 0 \ . \tag{6.17}$$

This straight line passes through the point $(- \sin \alpha, - \nu \cos \alpha)$ and is perpendicular to the seepage surface.

We see that for flow in anisotropic soil the methods investigated in the previous chapter hold with some slight alteration only.

We mention the particular case of anisotropy when $\alpha = 0$. If therefor $\nu < 1$, then we have horizontal stratification, if $\nu > 1$, we have vertical.

Formulas (6.10) now become

$$X = x_1 = x ,$$

$$Y = y_1 = \nu y .$$

The directions of the coordinate axes of the systems coincide, but in the region of the fictitious flow $(x,y)$, this change in scale along the y-axis holds: the dam foundation preserves its length, but the vertical cut-off is shortened for horizontal stratification and stretched for vertical.

The relationship between the reduced flowrate and the pressure, according to (6.6) must be

$$\Phi(x_1,y_1) = \frac{p}{\rho g} - y_1 .$$

In Chapter XI we communicate examples of some confined flows in anisotropic soil for $\alpha$ different from zero. In the auxiliary plane $(x,y)$, inclined cut-offs are obtained instead of the vertical cut-offs of the real flow region. If one has constructed the flownet in the $(x,y)$ plane by some method, then by means of formulas (6.10) one may return to the $(X,Y)$ plane.

We do not examine confined flows here, but describe an example of unconfined flow in homogeneous anisotropic soil.

## §7. Flow to a Drain on Impervious Foundation in Anisotropic Soil [8].

We consider the simplest case of flow with a free surface, the flow of groundwater towards an horizontal filter resting on horizontal impervious bedrock (Fig. 255). This problem for isotropic soil has been studied in Chapter II.

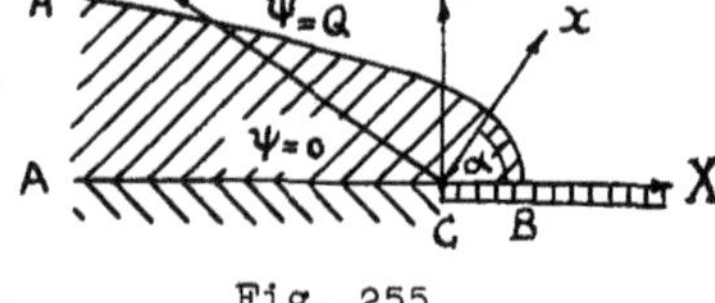

Fig. 255

We repeat the solution found before, introducing other symbols, namely capital letters for the coordinates of the flowregion in isotropic soil:

$$Z = X + iY$$

and $Q_0$ for the flowrate in the drain. We have

$$Z = -\frac{\omega^2}{2kQ_0} = -\frac{(\varphi + i\psi)^2}{2kQ_0} . \tag{7.1}$$

We write the equation of the free surface, along which $\psi = Q_0$, as

$$X_0 = -\frac{\varphi^2 - Q_0^2}{2kQ_0}, \quad Y_0 = -\frac{\varphi}{kQ_0}$$

or, considering that $\varphi = -kY$ along the free surface,

$$Y_0^2 = \frac{Q_0^2}{k^2} - \frac{2Q_0}{k} X_0 . \tag{7.2}$$

The length of the drainage sugment $\ell_0$ is related to the flowrate $Q_0$ as

$$Q_0 = 2k\ell_0 . \tag{7.3}$$

We put $k = 1$ and use this solution in the passage to anisotropic soil.

For anisotropic soil we have in the $\Phi,\Psi$ plane the same figure (half strip) as for isotropic soil.

The circle corresponding to the free surface in the $u,v$ plane, is moved (Fig. 256) so that its center lies in the point $(-1/2 \sin\alpha, -\nu/2 \cos\alpha)$.

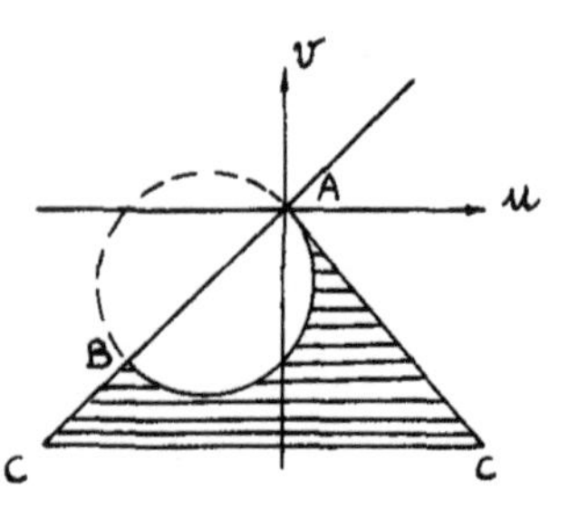

Fig. 256

The equation of line BC in the flow region has the form

$$Y = x \sin \alpha + \nu y \cos \alpha = 0 .$$

The line BC in the hodograph plane must be perpendicular to this straight line, i.e., must have the equation

$$v = \nu u \operatorname{cotg} \alpha = u \operatorname{cotg} \gamma .$$

The angle between the straight line ABC and the v-axis is called $\gamma$, for which

$$\operatorname{cotg} \gamma = \nu \operatorname{cotg} \alpha .$$

The equation of the straight line AC in the coordinates $x,y$ has the same form

$$Y = x \sin \alpha + \nu y \cos \alpha = 0 .$$

The line AC of the hodograph is parallel with line AC, i.e., has the equation

$$u \sin \alpha + \nu v \cos \alpha = 0$$

or

$$v = - u \operatorname{tg} \gamma ;$$

consequently, it is orthogonal to ABC.

If we rotate the velocity hodograph,in the uv plane over an angle $\gamma$ anti-clockwise, then we obtain the velocity hodograph, in the isotropic case, for which we call $v_x, v_y$ the components of the velocity. Then

$$u + iv = e^{-i\gamma}(v_x + iv_y),$$

or

$$u - iv = e^{i\gamma}(v_x - iv_y), \quad \frac{d\omega}{dz} = e^{i\gamma} \frac{d\omega}{dZ} .$$

From this, since function $\omega$ assumes unique values in corresponding points of planes $z$ and $Z$, we have:

$$z = e^{-i\gamma} Z . \tag{7.4}$$

Suppose we found the equation of the free surface for isotropic soil:

$$Z = Z_0(\zeta) = X_0(\zeta) + iY_0(\zeta) .$$

$\zeta$ is a parameter with real values.

Because of formulas (6.11) and (7.4) we have:

$$x + iy = X \cos \alpha + Y \sin \alpha + i \frac{-X \sin \alpha + Y \cos \alpha}{\nu} = e^{-i\gamma}(X_0 + iY_0) .$$

Separating real from imaginary parts, we find

$$\begin{aligned} X &= mX_0 + nY_0 , \\ Y &= pY_0 , \end{aligned} \tag{7.5}$$

where

$$m = \frac{\nu}{\sqrt{\sin^2\alpha + \nu^2\cos^2\alpha}} , \quad n = \frac{(1-\nu^2)\sin \alpha \cos \alpha}{\sqrt{\sin^2\alpha + \nu^2 \cos^2\alpha}} , \quad p = \sqrt{\sin^2\alpha + \nu^2\cos^2\alpha} .$$

Such is the relationship between coordinates of the flowregion for anisotropic ans isotropic soil in the case where the regions $\Phi + i\Psi$ are identical in both cases, but the region $u + iv$ is rotated over an angle $\gamma$

relative to the region $u_x + iv_y$.

In particular, the coordinates of the points of the free surface in anisotropic soil $X, Y$ are related to the coordinates of the free surface in isotropic soil $X_0, Y_0$ by means of (7.5).

Applying these equations to the particular flow case that we consider, we obtain, after substitution of (7.5) into (7.2), the equation of the free surface as a parabola.

$$\frac{Y^2}{p^2} = \bar{Q}_0^2 - \frac{2\bar{Q}_0}{m}\left(X - \frac{n}{p}Y\right) ,$$

where

$$\bar{Q}_0 = 2\ell_0 .$$

The flowrate through CB based on formula (6.9) is

$$Q = \sqrt{k_{x_1} k_{y_1}}\ \bar{Q}_0$$

The relationship between flowrate $Q$ and drainage segment CB, with length $\ell$, is determined as

$$Q = \frac{2\ell}{m} \nu \sqrt{k_{x_1}\ k_{y_1}}$$

$$Q = 2k_{x_1} \ell \sqrt{\sin^2\alpha + \nu^2\cos^2\alpha}.$$

## §8. Interpretation of Anisotropic Soil by Means of Two Alternating Soils.

We may represent anisotropic soil as composed of thin alternating strata of constant thickness, and with seepage coefficients $k_1$ and $k_2$ (Fig. 257). Then the flow parallel to these strata will be similar to that of the first case of §5, and for the velocity along the strata (direction of the x-axis) we may write:

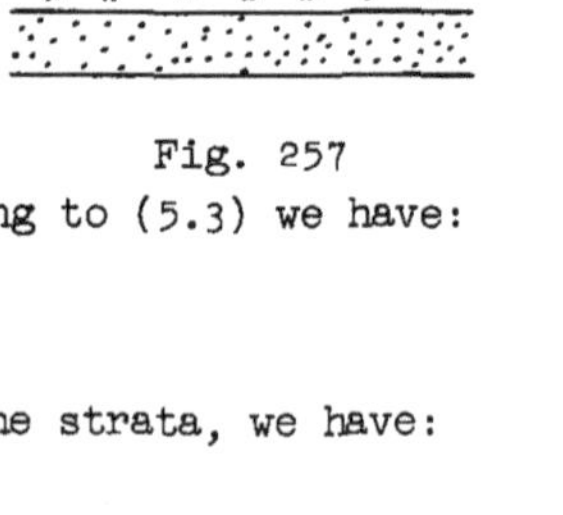

$$v_x = - k_x \frac{\partial h}{\partial x} , \tag{8.1}$$

Fig. 257

for which, considering $\ell_1 = \ell_2$, $n = 2$, according to (5.3) we have:

$$k_x = \frac{k_1 + k_2}{2} . \tag{8.2}$$

For the flow in the direction normal to the strata, we have:

$$v_y = - k_y \frac{\partial h}{\partial y} , \tag{8.3}$$

where, according to (5.10) we have for $\ell_1 = \ell_2$, $n = 2$

$$\frac{\ell}{k_1} + \frac{\ell}{k_2} = \frac{2\ell}{k_y} , \tag{8.4}$$

from which

$$k_y = \frac{1}{2} \frac{1}{\frac{1}{k_1} + \frac{1}{k_2}} . \tag{8.5}$$

We mention special values of $k_2$.

1) Let $k_2 = 0$; this means that the pervious strata alternate with impervious ones. In this case we have

$$k_x = \frac{k_1}{2}, \quad k_y = 0 .$$

Only flow along the strata (theoretically we must assume that the thickness of these strata is unfinitesimally small) is possible, with rectilinear flowlines.

2) $k_2 = \infty$; this means that in the $k_2$ layers the flow proceeds as in a continuous fluid without soil. Therefore (8.2) and (8.5) give:

$$k_x = \infty, \quad k_y = \frac{k_1}{2} .$$

The formula $v_x = - k_x \frac{\partial h}{\partial x}$ shows that one must have

$$\frac{\partial h}{\partial x} = 0 ,$$

because otherwise we would obtain an infinitely large value of $v_x$. But from the last condition follows that the head does not depend on $x$, i.e., that the lines $y$ = constant are equipotential lines.

Now we assume that we have a vertical alternation of strata with seepage coefficients $k_1$ and $k_2$. We obtain:

$$k_x = \frac{1}{2} \frac{1}{\frac{1}{k_1} + \frac{1}{k_2}} , \quad k_y = \frac{k_1 + k_2}{2} .$$

Putting $k_2 = \infty$, we find

$$k_x = \frac{k_1}{2} , \quad k_y = \infty .$$

This is the case of flow in which the equipotential lines are vertical lines $x$ = constant, which follows from the equation

$$v_y = - k_y \frac{\partial h}{\partial y}$$

from the condition that $k_y = \infty$ and that $v_y$ remains finite.

Such a scheme is a model of Dupuit flow, assuming that for a slightly changing free surface the equipotential lines are vertical lines. This assumption is approximately satisfied in homogeneous soil (slightly varying flow), and exactly in anisotropic soil with $k_y = \infty$.

It was already mentioned that there are anisotropic soils in nature with seepage coefficient in the vertical direction larger than in the horizontal. For these Dupuit's hypothesis must be well satisfied.

## C. Two Fluids of Different Density.

### §9. About Flow of Two Liquids of Different Density.

The problem of seepage of two fluids of different density arises in the case where there is a zone of salty water at some depth in the seepage of water under a dam. Such a zone is formed when the pervious stratum lies on a layer of salty rocks; sea waters, percolating in the soil, may be found under fresh ground waters; at different depths, groundwaters may have a different mineral content, while waters percolating from a canal may be fresh.

We examine the case [2,10] where a stratum of salty water A'D'D"A" lies at some depth under a hydraulic structure. The salty water overlies a stratum of impervious salts (Fig. 258,a). The whole stratum A D D" A" is homogeneous, i.e., is composed of one and the same material.

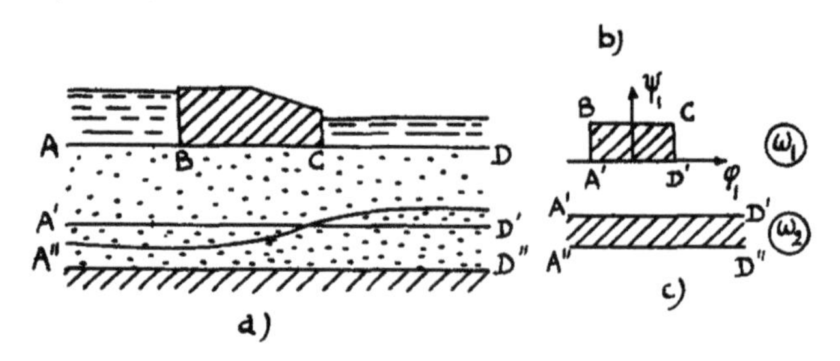


Fig. 258

It is assumed that after the construction of the hydraulic structure, the salty solution is squeezed so that the left part of the line A'D' sinks and the right part rises.

How does this phenomenon proceed, i.e., how does the boundary line between fresh and salt water change in time?

The second part of this book is devoted to unsteady flows of this kind. Here we analyze the case of steady flow.

We consider both fluids incompressible, having respectively densities $\rho_1$ and $\rho_2$, where $\rho_2 > \rho_1$. Considering that the seepage coefficient depends not only on the soil, but also on the fluid, and in particular on its viscosity, we assume that the seepage coefficients of the two flowregions are respectively $k_1$ and $k_2$.

We write the equations for the velocities $u_i, v_i$ $(i = 1,2)$ for each of the two fluids:

$$(9.1) \qquad u_i = \frac{\partial \varphi_i}{\partial x}, \quad v_i = \frac{\partial \varphi_i}{\partial y}, \qquad \Delta\varphi_i = 0, \qquad \varphi_i = -k_i\left(\frac{p_i}{\rho_i g} + y\right) \quad (i = 1,2).$$

The condition of equal pressure on the boundary line (interface) renders the equation

$$(9.2) \qquad \frac{\rho_1}{k_1}\varphi_1 - \frac{\rho_2}{k_2}\varphi_2 = (\rho_2 - \rho_1)\, y$$

on the boundary line (interface).

Moreover, the boundary line is a streamline, i.e., on this line

$$\psi_1 = \text{const.} \quad \psi_2 = \text{const.}$$

If we consider the plane of complex potential $\omega_1 = \varphi_1 + i\psi_1$ for the first region, then we obtain a rectangle A'D'CB (Fig. 258,b). Consequently, $\varphi_1$ is finite in the whole region 1.

We pass on to region 2. If flow would be possible in this region, then we would obtain the strip A'D'A"D" (Fig. 258,b) in the plane of its complex potential $\omega_2 = \varphi_2 + i\psi_2$. From this it follows that $\varphi_2 = \pm \infty$, when $x = \pm \infty$. But it would then follow from equation (9.2), that also $\varphi_1 = \pm \infty$ for $x = \pm \infty$, which is impossible. Consequently, in the case of steady flow, the fluid in region 2 must remain at rest.

Thus the problem is reduced to that of determining the flow in one region, and for which the boundary line is subject to a condition exactly analogous to that prevailing at free surface lines. Indeed, making the value $\varphi_2$ constant in equation (9.2), we rewrite this equation as

$$\varphi_1 - \mu y = \text{const.}, \tag{9.3}$$

where

$$\mu = \left( \frac{\rho_2}{\rho_1} - 1 \right) k_1 . \tag{9.4}$$

We would have obtained the same expression, if we considered $k_1 = k_2$. In other words, the magnitude of the seepage coefficient of the second medium does not enter in the conditions of the problem.

It is evident that a circle corresponds to equation (9.3) (see Chapter II, §4) in the velocity hodograph

$$u^2 + v^2 - \mu v = 0 . \tag{9.5}$$

Consider a hydraulic structure of arbitrary shape. Along the boundaries of the reservoirs AB and CD (Fig. 258), the velocity potential has constant values. We put:

$$\varphi_1 = -\frac{1}{2} k_1 H \quad \text{on AB} ,$$

$$\varphi_2 = \frac{1}{2} k_1 H \quad \text{on CD} ,$$

where

$$H = H_1 - H_2 .$$

The depths of the flowing fluid at $x = -\infty$ and $x = +\infty$ are respectively h' and h". Writing equation (9.3) in the form

$$y = \frac{\varphi_1}{\mu} + C ,$$

we have: for $x = -\infty$

$$-h' = -\frac{k_1 H}{2\mu} + C ,$$

for $x = +\infty$

$$-h'' = \frac{kH}{2\mu} + C .$$

From this

$$C = -\frac{h' + h''}{2} .$$

For steady flow in a finite region, one would have to satisfy the condition of equal area occupied initially and finally by the salty water. In our case these areas are infinite and therefore we replace the above condition by the following: we assume that $1/2(h' + h'')$ be equal to the depth of the upper stratum in the equilibrium state

$$\frac{h' + h''}{2} = h_1 .$$

Then

$$C = -h_1 ,$$

$$(9.6) \qquad h' = h_1 + \frac{k_1 H}{2\mu} = h_1 + \frac{H}{2\left(\frac{\rho_2}{\rho_1} - 1\right)} ,$$

$$h'' = h_1 - \frac{k_1 H}{2\mu} = h_1 - \frac{H}{2\left(\frac{\rho_2}{\rho_1} - 1\right)} .$$

The difference of depth is equal to

$$h' - h'' = \frac{H}{\frac{\rho_2}{\rho_1} - 1} .$$

Thus the displacement depth becomes larger the smaller the difference is in densities $\rho_1$ and $\rho_2$.

If

$$h_1 \leq \frac{H}{2\left(\frac{\rho_2}{\rho_1} - 1\right)} ,$$

then the right end of the boundary line must rest on the boundary of the tailwater reservoir.

If the depth $h_2$ of the bottom stratum satisfies the inequality

$$h_2 \leq \frac{H}{2\left(\frac{\rho_2}{\rho_1} - 1\right)}$$

then the boundary line (interface) rests on the boundary of the impervious bottom. Four forms of interface are possible, as rendered by Fig. 259.

For a series of particular forms of the base of the hydraulic structure, it is easy to obtain a solution by means of conformal mapping onto the half-plane of the potential regions and of the regions of the velocity hodograph (or its inverse). For a cut-off we may obtain a solution, using Zhukovsky's function as was done for the flow with free surface.

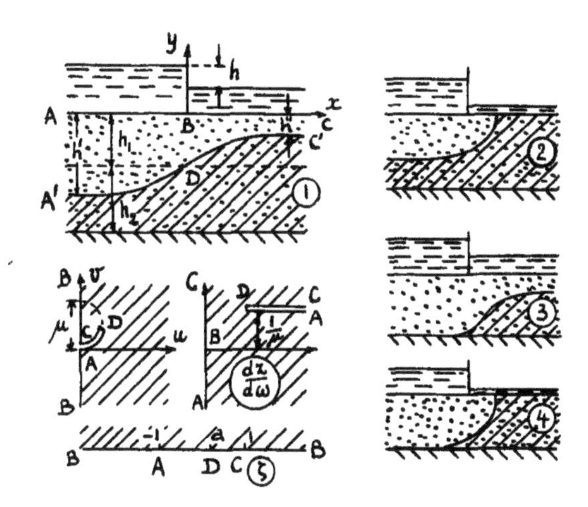

Fig. 259

§10. Flow About a Cut-off without Penetration.

We examine case 1 of Fig. 259. Assuming $\varphi = -\frac{kH}{2}$ along AB, $\varphi = \frac{kH}{2}$ along BC, $\psi = 0$ along A'DC', we obtain for $\omega = \varphi + i\psi$

$$\omega = \frac{kH}{\pi} \text{ arc sin } \zeta \ . \tag{10.1}$$

In the u,v plane, the flow region is a half plane with cut along circle (9.5). Inversion in the circle of radius $\mu$ with center in the origin of coordinates gives the region of the function $\mu^2 \frac{dz}{d\omega}$. After simplification by $\mu^2$, we have

$$\frac{dz}{d\omega} = C \int \frac{(\zeta - a)d\zeta}{(1 - \zeta^2)^{3/2}} + D = C \frac{1 - a\zeta}{(1 - \zeta^2)^{3/2}} + D \ . \tag{10.2}$$

For $|\xi| < 1$ we have $\text{Im} \frac{dz}{d\omega} = \frac{1}{\mu}$, for $|\xi| > 1$, $\text{Re} \frac{dz}{d\omega} = 0$, for $\zeta = \infty$, $\frac{dz}{d\omega} = 0$. This gives:

$$C = \frac{1}{\mu a} , \quad D = \frac{\ell}{\mu} \ . \tag{10.3}$$

We multiply (10.2) by $\frac{d\omega}{d\zeta} = \frac{kH}{\pi\sqrt{1 - \zeta^2}}$ and integrate over $\zeta$.

Considering that contouring of the points $\zeta = \pm 1$ gives increases for $z$, respectively equal to $h''i$, $h'i$, for which the equalities (9.6) hold, we find:

$$z = \frac{k_1 H - 2\mu h_1}{2\mu\pi} \ \ell n(1 - \zeta) + \frac{k_1 H + 2\mu h_1}{2\mu\pi} \ \ell n(1 + \zeta) +$$

$$+ \frac{ik_1 H}{\mu\pi} \text{ arc sin } \zeta + \frac{k_1 H}{\mu\pi} \ell n \ 2 - ih_1 \ . \tag{10.4}$$

The equations of the interface are obtained for $-1 < \zeta < 1$:

$$X_0 = \frac{H}{\pi\left(\frac{\rho_2}{\rho_1} - 1\right)} \ln(2\sqrt{1-\zeta^2}) + \frac{h_1}{\pi} \ln\frac{1+\zeta}{1-\zeta} \,,$$

(10.5)

$$Y_0 = \frac{H}{\pi\left(\frac{\rho_2}{\rho_1} - 1\right)} \arcsin\zeta - h_1 \,.$$

If there exists a relationship between $h_1$ and $H$

$$h_1 \approx \frac{k_1 H}{2\mu} = \frac{H}{2\left(\frac{\rho_2}{\rho_1} - 1\right)} \,,$$

then the equation of the interface becomes

$$(10.6) \qquad X_0 = \frac{2h_1}{\pi} \ln[2(1+\zeta)], \quad Y_0 = \frac{2h_1}{\pi} \arcsin\zeta - h_1 \,.$$

The case of a cut-off with penetration and flat bottom dam is analyzed in reference [12]. The same paroblem for anisotropic soil is treated in reference [11].

## *§11. Diffusion of Dissoluble Matter in Bases of Hydraulic Structures.

In previous paragraphs we neglected diffusion of salts, considering that we had two fluids with densities remaining constant in time. Actually between these fluids a diffusion process of dissoluble matter takes place (rock salts, etc.), traces of which may be found in bases of hydraulic structures. As a result of this process, soil deformations may occur.

We analyze the problem, neglecting the change in density of the fluids in the seepage region and the deformation of this region.

We take the differential equation for diffusion of two fluids [17], in which $v_x, v_y$ are the components of the velocity of the fluid particles

$$(11.1) \qquad D\,\Delta c = \frac{\partial c}{\partial x} v_x + \frac{\partial c}{\partial y} v_y + \frac{\partial c}{\partial t} \,.$$

For steady flow this equation becomes [18]

$$(11.2) \qquad D\,\Delta c = \frac{\partial c}{\partial x}\frac{u}{m} + \frac{\partial c}{\partial y}\frac{v}{m} \,.$$

Here $c$ is the concentration of dissoluble matter in a given point of the seepage region (gram/cm$^2$), $D$ the diffusion coefficient, i.e., a parameter characterizing the diffusive properties of the medium and of the dissoluble matter, expressed in cm$^2$/sec, $u, v$ the components of the seepage velocity, $m$ the porosity.

We assume that $\omega$ is the reduced complex potential, and the usual complex potential is

$$kH\omega = kH(\varphi + i\psi) \,.$$

Consequently, the complex seepage velocity is

$$u - iv = kH \frac{d\omega}{dz} ,$$

where $H = H_1 - H_2$, the effective head; $H_1$, $H_2$ are the levels in headwater and tailwater reservoirs. Therefore

$$\varphi(x,y) = -\frac{1}{H}\left(\frac{p - p_0}{\rho g} - y - H_1\right) ,$$

(11.3)

$$u = kH \frac{\partial\varphi}{\partial x} , \quad v = kH \frac{\partial\varphi}{\partial y} ,$$

$p_0$ is a constant, for example the atmospheric pressure. The y-axis is oriented downward.

We assume that we found the relationship between the complex coordinate $z$ and the reduced complex potential $\omega$

(11.4)

$$z = f(\omega) .$$

In equation (11.2) it is convenient to pass from the variables $x,y$ to the variables $\varphi$, $\psi$. It is easy to see that

$$\frac{\partial c}{\partial x} u + \frac{\partial c}{\partial y} v = \frac{\partial c}{\partial \varphi}\left(\frac{\partial\varphi}{\partial x} u + \frac{\partial\varphi}{\partial y} v\right) + \frac{\partial c}{\partial \psi}\left(\frac{\partial\psi}{\partial x} u + \frac{\partial\psi}{\partial y} v\right) .$$

The expression between the second brackets is zero, that between the first is

$$\frac{V^2}{kH} ,$$

where $V$ is the magnitude of the velocity, say $\sqrt{u^2 + v^2}$. Therefore

(11.5)

$$\frac{\partial c}{\partial x} u + \frac{\partial c}{\partial y} v = \frac{V^2}{kH} \frac{\partial c}{\partial \varphi} .$$

Relating Laplace's operator

$$\Delta c = \frac{\partial^2 c}{\partial x^2} + \frac{\partial^2 c}{\partial y^2}$$

to the variables $\varphi$, $\psi$, we find

(11.6)

$$\Delta c = \left(\frac{V}{kH}\right)^2 \left(\frac{\partial^2 c}{\partial \varphi^2} + \frac{\partial^2 c}{\partial \psi^2}\right) .$$

If we insert equations (11.5) and (11.6) into equation (11.2), then we obtain, after division by $V^2/kH$,

(11.7)

$$\frac{\partial^2 c}{\partial \varphi^2} - 2\lambda \frac{\partial c}{\partial \varphi} + \frac{\partial^2 c}{\partial \psi^2} = 0 .$$

Here $\lambda$ is a dimensionless parameter

$$\lambda = \frac{kH}{mD} .$$

We may find a solution of equation (11.7) in points of the rectangle $(\varphi,\psi)$ for the following boundary conditions. Along the bottom of the headwater and tailwater reservoirs, the concentration is zero. Along the contour of the impervious foundation of the dam, we have condition $\frac{\partial c}{\partial n} = 0$ [20].

This gives

$$\frac{\partial c}{\partial n} = \frac{\partial c}{\partial \psi}\frac{\partial \psi}{\partial \psi} + \frac{\partial c}{\partial \varphi}\frac{\partial \varphi}{\partial n} = \frac{\partial c}{\partial \psi}\frac{\partial \varphi}{\partial n} = V\frac{\partial c}{\partial \psi} = 0 , \tag{11.8}$$

from which $\frac{\partial c}{\partial \psi} = 0$ along segment $\psi = 0$ .

Finally, along the boundary of the impervious bedrock, which, by assumption, has the capacity of holding a layer of dissoluble material, we have $c = c_0$. There $c_0$ is a well determined degree of saturation with dissoluble matter.

The solution is found by means of a Fourier series as:

$$\frac{c}{c_0} = 1 - \frac{4}{\pi} e^{\lambda\varphi} \sum_{n=0}^{\infty} \frac{(-1)^n}{(2n+1)(1 - e^{-2\mu_n})} [(e^{-\lambda} - e^{-\mu_n})e^{-\mu_n(1-\varphi)} +$$

$$+ (1 - e^{-\lambda-\mu_n})e^{-\mu_n\varphi}]\cos (2n+1)\frac{\pi\psi}{2q} , \tag{11.9}$$

where

$$\mu_n = \sqrt{\lambda^2 + \left[(2n+1)\frac{\pi}{2q}\right]^2} , \tag{11.10}$$

$q$ is the reduced flowrate, i.e., $q = Q/kH$, where $Q$ is the total flowrate. The quantity $c/c_0$ depends only on two parameters

$$\lambda = \frac{kH}{mD} , \quad q = \frac{Q}{kH} .$$

The computation of the flowrate of dissoluble matter from the segment $MN$ of the impervious bedrock is carried out by means of the formula

$$Q_{MN} = -\int_M^N D \left.\frac{\partial c}{\partial n}\right|_{\psi=q} ds = -D\int_M^N v \left.\frac{\partial c}{\partial \psi}\right|_{\psi=q} ds = -D\int_m^n \left.\frac{\partial c}{\partial \psi}\right|_{\psi=q} d\varphi .$$

We may estimate [18] that, for a depth of stratum larger then twice the length of the subterranean contour of the base, because of the slowness of the diffusion process, the deformation of the boundaries of the bedrock practically does not have an influence upon the flownet.

Equation (11.9) together with (11.4), or with a flownet constructed in one way or the other (by the electro-hydrodynamic analogy for example, see Chapter XI), allow for the determination of the concentration of dissolved matter in any point of the flowregion.

## §12. Heterogeneous Soil with Seepage Coefficient a Continuous Function of Coordinates.

In the analysis of two-dimensional flow in heterogeneous medium with variable seepage coefficient $k(x,y)$, we introduce the dimensionless quantity

$$K(x,y) = \frac{k(x,y)}{k_0} ,$$

where $k_0$ is some constant value of the seepage coefficient (for example, the average over the entire flowregion). Then we may write for vector $\vec{V}$ of the seepage velocity (see literature [27] to Chapter III):

$$\text{(12.1)} \qquad \operatorname{div} \vec{V} = 0, \quad \vec{V} = K\left(\frac{\partial\varphi}{\partial x} + i\frac{\partial\varphi}{\partial y}\right), \quad \varphi = k_0 h \ .$$

We introduce function $\omega = \varphi + i\psi$, so that

$$\vec{V} = \frac{\partial\psi}{\partial y} - i\,\frac{\partial\psi}{\partial x} = K\left(\frac{\partial\varphi}{\partial x} + i\,\frac{\partial\varphi}{\partial y}\right) \ .$$

Then we obtain the equations

$$\text{(12.2)} \qquad \frac{\partial\varphi}{\partial x} = \frac{1}{K}\,\frac{\partial\psi}{\partial y}\,, \quad \frac{\partial\varphi}{\partial y} = -\,\frac{1}{K}\,\frac{\partial\psi}{\partial x} \ .$$

There is a theorem about the conservation of region for functions $\varphi, \psi$, satisfying the system (12.2) of the elliptic type: the region of the plane $z = x + iy$ maps onto the region of the plane $\varphi + i\psi$; an infinitesimally small circle of the z-plane corresponds to an infinitesimally small ellips of the $\omega$-plane with half-axes $a, b$, parallel to the axes $\varphi, \psi$ for which $b/a = K$.

To the flow here discussed, one may apply a series of variational theorems, examples of which are given at the end of Chapter III. (See literature [27, 28] to Chapter III.)

## CHAPTER IX. NATURAL AND MAN-MADE WELLS
## HORIZONTAL DRAINS

### A. Fully Penetrating Wells

§1. Fully Penetrating Well in the Center of a Stratum

We consider the motion of a fluid - water or oil - in a pervious layer (usually sand), bordered by two horizontal impervious strata. We consider a man-made well of cylindrical form penetrating through the whole layer - such a well is called "fully or completely penetrating" as distinguished from "partially penetrating" to a depth only part of the pervious stratum. We may consider the movement in such a stratum to be plane parallel. We explain here the theory of a circular fully penetrating well. It is possible to use the results also as an approximation for wells of other forms, for example rectangular (see Chapter X, Par. 9). We consider radially symmetrical flow towards a well of radius $r_o$ (Fig. 260 and 261). We compute the flowrate Q through a cylindrical surface with height M and arbitrary radius r. It is equal to

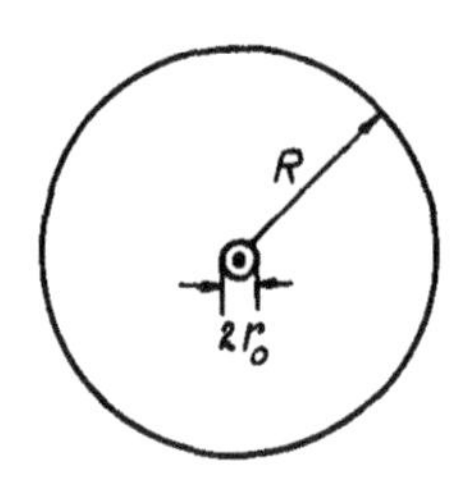

Fig. 260

$$Q = 2\pi rMv \tag{1.1}$$

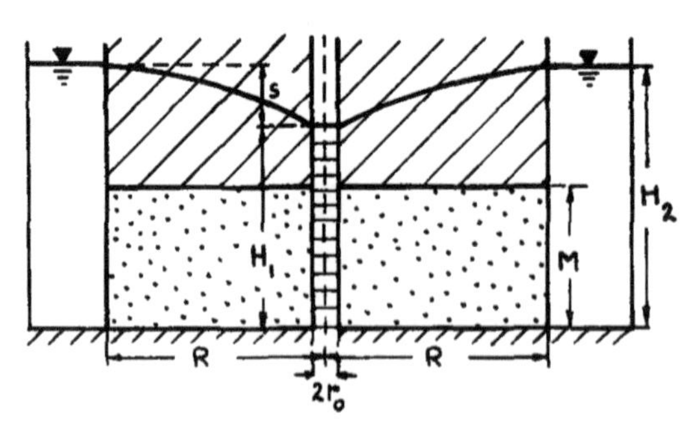

Fig. 261

where v is the radial seepage velocity, depending only upon r because of the proposed radial symmetry. The fluid is considered to be incompressible, the movement steady. We consider the movement confined, i.e., without free surface. In these conditions, the discharge Q through a cylinder of any radius $r$ must be constant. Consequently, the velocity $v$ will be inversely proportional to the distance $r$. Taking the direction of this velocity towards the center of the well, which we move into the origin of coordinates, we introduce the concept of radial velocity

$$v_r = -v = \partial\varphi/\partial r \tag{1.2}$$

and obtain

$$v_r = -Q/2\pi Mr \tag{1.3}$$

When we integrate this equation, we find the velocity potential or the head proportional to the latter, $h = -\varphi/k$ :

$$h = -\frac{\varphi}{k} = \frac{Q}{2\pi kM} \ln r + C \quad .$$

In order to determine the constant $C$, one must have some condition. We will assume that the head $H_2$ is known at a distance $R$ from the center of the well. This gives:

$$H_2 = \frac{Q}{2\pi kM} \ln R + C \quad .$$

We also assume that we know the head $H_1$ at the well itself, i.e., for $r = r_o$. Then we obtain:

$$H_1 = \frac{Q}{2\pi kM} \ln r_o + C \quad .$$

From the last two equations one may eliminate $C$ to find the equation for the flowrate of the well

$$Q = 2\pi kM(H_2 - H_1)/\ln \frac{R}{r_o} \quad . \tag{1.4}$$

This is the basic formula to compute the discharge of the well.

For the head $h$ we obtain the equation

$$h = \frac{Q}{2\pi kM} \ln \frac{r}{r_o} + H_1 \quad . \tag{1.5}$$

We notice that earlier the concept of "radius of influence" of the well has been introduced, at a distance where the influence of a given well dies out (in the case of unconfined flow, this concept is now also used). However, this concept is not well-defined, since the dependence of head $h$ on the radius vector $r$ is logarithmic and $h$ increases till infinity for $r \longrightarrow \infty$ .

V N Tsjelkatchev and G. V. Pichatchev [1] proposed instead of the notion of region of influence another concept - that of region of intake of the well. This concept has a value in the study of an oil-bearing stratum, which usually is in so called "water-confined regimen", surrounded by a region of water. The border between the region of soil, saturated with oil or water, and the region of water (without soil) is called the "contour of intake", which is taken for an isobar, or even, in the case of horizontal movement, for a line of constant head.

We recall that in the study of the flow of oil towards a fully penetrating well, the pressure is studied instead of the head $h$; both are related through

$$p = \rho g h = -\rho g \varphi / k \tag{1.6}$$

and instead of the seepage coefficient $k$ the permeability $k_0$ of the stratum is used; the latter are related as follows

$$k = k_0 g/\nu \tag{1.7}$$

Putting $p = p_c$, $r = r_0$ at the contour of the well, $r = R$ and $p = p_0$ at the contour of intake (we consider here the region of intake to be circular, concentric with the contour of the well), we obtain the formula of the discharge of the well in the form

$$Q = 2\pi k_0 M(p_0 - p_c)/\mu \ \ell n \frac{R}{r_0} \tag{1.8}$$

where $\mu$ - viscosity of the oil, $M$ - thickness of stratum. Returning to formula (1.4), we notice that it is used in hydrogeology to compute the seepage coefficient in a constant pumping test. Therefore it is necessary to know the thickness of the pervious layer $M$ and the values of the head $H_1$ and $H_2$ at the well and at a distance $R$ from it. The quantity $H_2$ is the head at one of the observation wells, which are usually established round the actual well and the comparison of the heads in these serves as a check of the applicability of formula (1.4) to the conditions of a given well.

Sometimes as a result of observation it is established that in the confined layer a constant flow of ground water takes place. Then a linear function of the coordinates, corresponding to the constant flow, must be added to the terms of the right side in equation (1.4). When we choose the direction of the $x$ - axis parallel to the direction of the flow, having a velocity $U$, where $U > 0$, then we obtain (Fig, 262)

$$w = -Uz - \frac{Q}{2\pi} \ell n \ z + C \tag{1.9}$$

from where we have for the velocity potential and stream function:

$$\varphi = -Ux - \frac{Q}{2\pi} \ell n \ r + C', \qquad \psi = -Uy - \frac{Q}{2\pi} \operatorname{arc\,tg} \frac{y}{x} + C \ .$$

The streamline $\psi = 0$ disintegrates in two parts: $y = 0$ and

$$y = -x \operatorname{tg} \frac{2\pi U y}{Q} \ .$$

The latter has the asymptotes $y = \pm \frac{Q}{2U}$ and intersects the ordinate axis in the points $y = \pm \frac{Q}{4\pi U}$, and the abscissa axis in the point $x = -\frac{Q}{2\pi U}$, in which point the flow branches and the velocity is zero. The region that is bordered by the line, $\psi = 0$ is called the region of intake of the well (Fig. 262).

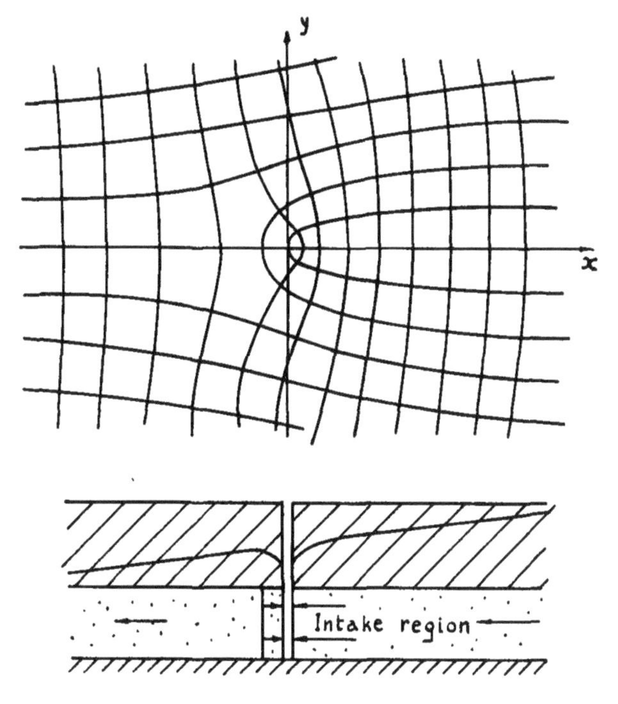


Fig. 262

§2. Well Eccentrically Established in a Circular Contour.

Let the region of intake be a circle with radius $R$ and center in the origin of coordinates (Fig, 263). We take the thickness of the stratum equal to unity. The well moves to the point $M_1(X_1, Y_1)$. We take an arbitrary point $M$ in the region and designate its complex coordinate by $Z$, by $Z_1$ that of the point $M_1$. By $q_1$ we designate the discharge of the well with center in the point $M_1$.

Then the complex potential in the point $Z$

$$W(Z) = \varphi(X, Y) + i\,\psi(X, Y) \tag{2.1}$$

may be obtained by putting a sink of intensity $q_1$ in point $M_1$ and a source of the same intensity in point $M_1'$ (with complex coordinate $Z_2 = R^2/\bar{Z}_1$), representing the inverse of point $M_1$ with respect to the circle of radius $R$ (Chapter V, Par. 1):

$$W(Z) = -\frac{q_1}{2\pi}\,\ell n\,\frac{Z-Z_1}{Z-\frac{R^2}{\bar{Z}_1}} + C = -\frac{q_1}{2\pi}\left\{\ell n(Z - Z_1) - \ell n\,\frac{R^2-Z\bar{Z}_1}{R}\right\} + C_1 \tag{2.2}$$

where $\bar{Z}_1 = X_1 - iY_1$. By $q_1$ we designate the discharge of the sink, placed in point $M_1$, which we consider as the center of the well, by $C_1$ - a constant, for the present arbitrary. The function $\varphi(X, Y)$ is the real part of equation (2.2) and therefore may be written in the form

$$\varphi(X, Y) = -\frac{q_1}{2\pi}\left\{\ell n|Z - Z_1| - \ell n\left|\frac{R^2-Z\bar{Z}_1}{R}\right|\right\} + C_1' \tag{2.3}$$

After formula (1.6), taking into account (1.7), we have for the pressure $p$:

$$p(X, Y) = \frac{q_1\mu}{2\pi k_0}\ \ell n \frac{R|Z-Z_1|}{|R^2-Z\bar{Z}_1|} + C_2 \tag{2.4}$$

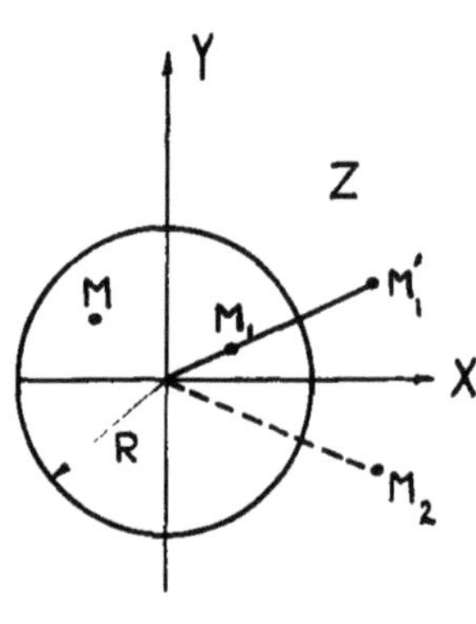

Fig. 263

Near the well, the lines of equal pressure are very close to circles, and therefore for the contour of the well we may take one of these isobars, close to a circle of radius $\delta$ (radius of well with center in the point $Z_1$). Along the contour of the well the pressure in this condition is equal to $p_1$. In particular, $p = p_1$ in the point $Z = Z_1 + \delta$, and this leads to (2.5)

$$p_1 = \frac{q_1 M}{2\pi k_0}\ \ell n \frac{\delta R}{R^2-|Z_1|^2} + C_2$$

using eq. (2.4). Let the pressure at the intake contour be equal to $p_0$. In particular, $p = p_0$ in the point (Fig. 263) for which $Z = R$. This gives the equality

$$p_0 = C_2 \quad . \tag{2.6}$$

We subtract (2.5) from (2.6) and put

$$|Z_1|^2 = X_1^2 + Y_1^2 = R_1^2 \quad .$$

We obtain

$$p_0 - p_1 = \frac{q_1\mu}{2\pi k_0}\ \ell n \frac{R^2-R_1^2}{\delta R} \quad .$$

From this, we obtain the discharge of the well

$$q_1 = \frac{2\pi k_0(p_0-p_1)}{\mu[\ell n(R^2 - R_1^2) - \ell n(R\delta)]} \quad .$$

When the well is located in the center of the circle, then its discharge after (1.8) is equal to

$$q_0 = \frac{2\pi k_0(p_0-p_1)}{\mu(\ell n\ R-\ell n\ \delta)} \quad .$$

We form the ratio

$$\frac{q_1}{q_0} = \frac{\ln \bar{R}}{\ln\left(\bar{R} - \frac{\bar{R}_1^2}{\bar{R}}\right)}$$

where

$$\bar{R} = \frac{R}{\delta} , \ \bar{R}_1 = \frac{R_1}{\delta} .$$

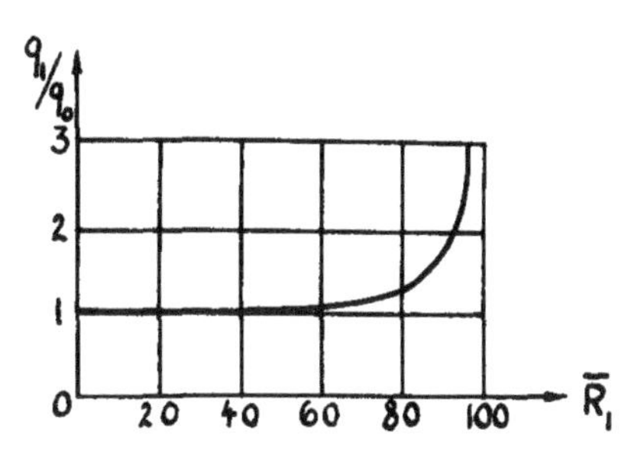

Fig. 264

In Fig. 264, the dependence of the quantity $q_1/q_0$ on $\bar{R}_1$ for a given $\bar{R} = 100$ is given. We see that the discharge of the well abruptly increases when the well approaches the contour of intake.

§3. The Case of the Arbitrary Contour. [2].

We perform the conformal mapping of the inside of a circle $|Z| = R$ onto the interior of some region S of the z plane (Fig. 265) by means of the mapping function

$$Z = F(z) \tag{3.1}$$

Let the contour of the circle, represented in Fig. 263, correspond to the contour L, sketched in Fig. 265, and the point $M_1$ correspond to the point with the same letter in Fig. 265. We designate the complex coordinate of this point by $z_1$. Then the distribution of the pressure in the new region will be expressed by a formula, which we obtain from (2.4), replacing Z by means of (3.1)

$$p(x, y) = \frac{q_1 \mu}{2\pi k_0} \ln \frac{R|F(z)-F(z_1)|}{|R^2-F(z)\overline{F(z_1)}|} + C . \tag{3.2}$$

The condition that in the point $z = z_1 + \delta$ the pressure $p = p_1$, gives (*), if we neglect the small terms of higher order in $\delta$:

$$p_1 = \frac{q_1 \mu}{2\pi k_0} \ln \frac{\delta |F'(z_1)| R}{R^2 - |F(z_1)|^2} + C . \tag{3.3}$$

Since along the circle of Fig. 263 $|Z| = R$, the equation of the contour L according to equation (3.1) will be:

$$|F(z)| = R .$$

(*) Indeed

$$F(z_1+\delta) - F(z_1) = \delta F'(z_1) + \frac{\delta^2}{2} F''(z_1) + \dots \approx \delta F'(z_1)$$

$$F(z_1+\delta)\overline{F(z_1)} = |F(z_1)|^2 + \delta F'(z_1)\overline{F)z_1)} + \dots \approx |F(z_1)|^2$$

We take an arbitrary point $z_0$ at the contour L. In this point, as everywhere along the contour L, let the pressure be equal to $p_0$. Then we have from (3.2):

$$p_0 = \frac{q_1 \mu}{2\pi k_0} \ln \frac{R|F(z_0) - F(z_1)|}{|R^2 - F(z_0)\overline{F(z_1)}|} + C. \tag{3.4}$$

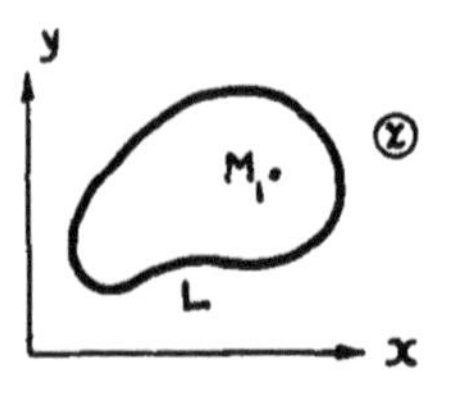


Fig. 265

Taking into account that $|F(z_0)| = R$, we find

$$\frac{R|F(z_0) - F(z_1)|}{|R^2 - F(z_0)\overline{F(z_1)}|} = 1 \quad , \quad C = p_0 \quad . \tag{3.5}$$

Subtracting (3.3) from (3.4) we find

$$p_0 - p_1 = \frac{q_1 \mu}{2\pi k_0} \ln \frac{R^2 - |F(z_1)|^2}{R\,\delta\,|F'(z_1)|} \quad . \tag{3.6}$$

From this we find the general formula for the discharge of a well inside a contour that we have obtained from a circular contour by conformal mapping (3.1):

$$q_1 = \frac{2\pi k_0 (p_0 - p_1)}{\mu(\ln(R^2 - |F(z_1)|^2) - \ln(R\,\delta\,|F'(z_1)|))} \quad . \tag{3.7}$$

If we take $R = 1$, then we obtain:

$$q_1 = \frac{2\pi k_0 (p_0 - p_1)}{\mu(\ln(1 - |F(z_1)|^2) - \ln(\delta\,|F'(z_1)|))} \quad . \tag{3.8}$$

It is natural to take an ellipse as example of intake region. The function that gives the conformal mapping of the interior of an ellipse onto the interior of a unit circle, has the form:

$$Z = F(z) = \sqrt{k}\ \mathrm{sn}\left(\frac{2K}{\pi} \arcsin \frac{z}{c}\right) \quad . \tag{3.9}$$

Where $k$ - the modulus of the elliptic function, determined by means of the equality

$$\frac{K'}{K} = \frac{4}{\pi} \tanh^{-1} \frac{b}{a} = \frac{2}{\pi} \ln \frac{a + b}{a - b} \tag{3.10}$$

$$\left( K = \int_0^{\frac{\pi}{2}} \frac{d\varphi}{\sqrt{1 - k^2 \sin^2 \varphi}} \quad , \quad K' = \int_0^{\frac{\pi}{2}} \frac{d\varphi}{\sqrt{1 - k'^2 \sin^2 \varphi}} \quad , \quad k'^2 = 1 - k^2 \right)$$

Here $c^2 = a^2 - b^2$, $a$ and $b$ - semi axes of the ellipse. We take the derivative:

$$F'(z) = \frac{2K \sqrt{k}\ \mathrm{cn}\, u\ \mathrm{dn} u}{\pi \sqrt{c^2 - z^2}} \quad , \quad u = \frac{2K}{\pi} \arcsin \frac{z}{c} \quad . \tag{3.11}$$

By means of formula (3.9) we find for the discharge of the well, which is located in point $z_1$, the expression

$$q_1 = \frac{2\pi k_0(p_0 - p_1)}{\mu\{\ell n[(1 - k\, sn^2 u_1)\sqrt{c^2 - z_1^2}] - \ell n[\alpha|cn\, u_1\, du\, u_1|]\}} .$$

Here

$$u_1 = \frac{2K}{\pi} \arcsin \frac{z_1}{c} , \quad \alpha = \frac{2K\delta\sqrt{k}}{\pi} ,$$

where $\delta$, as always, is the radius of the well.

If the well is located in the center of the ellipse, then

$$q_1 = \frac{2\pi k_0(p_0 - p_1)}{\mu[\ell n(\pi c) - \ell n(2\delta K\sqrt{k})]}$$

By means of this formula, ratios have been computed that give the discharge of a well in an elliptic region compared to the discharge $q_0$ of a well in a circular region with radius $R$, and with center coinciding with that of the ellipse:

$$\frac{q_1}{q_0} = \ell n \frac{R}{\delta} : \ell n \frac{\pi c}{2\delta K\sqrt{k}} .$$

If we take ellipses of the same magnitude as the circle, then $R^2 = ab$. We put $a = nb$ and assign to $n$ a series of integer values $n = 1,2,3$---

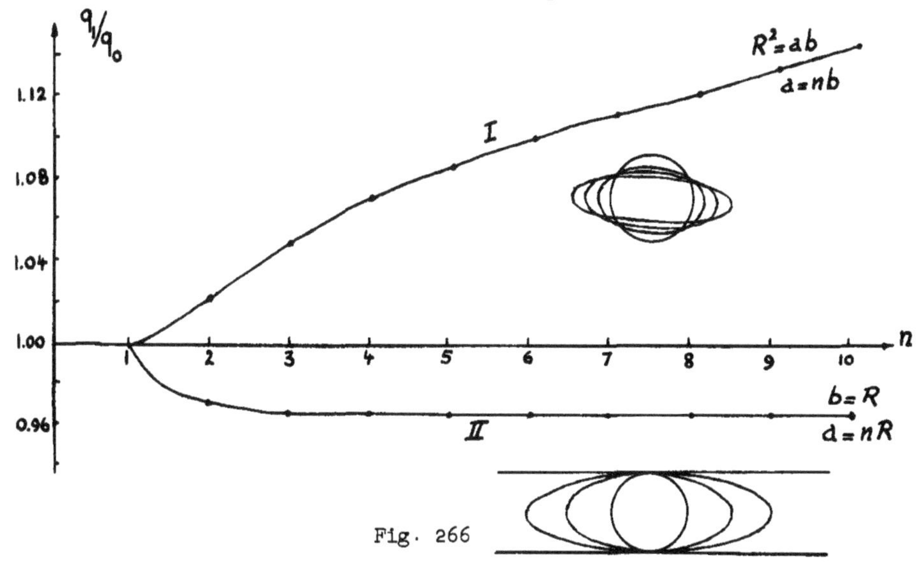


Fig. 266

Along with calculations of this kind, curve I of Fig. 266 has been constructed It shows that the discharge increases with increased stretching of the ellipse under conservation of its area; however, the increase is very slow.

If we take $b = R$, i.e., keep the small semiaxis of the ellipse constant, and put $a = nR$, where $n$ an integer, then we obtain curve II of Fig. 266. From this it appears that, when the region is stretched out in one direction, the discharge of the central well decreases, but very slowly. In the limit for $n \to \infty$, when instead of the elliptical region we obtain an infinite strip, bordered by two parallel straight lines, its discharge is 96.6% of the discharge for a circular region.

Calculations have confirmed the general assumption that a change in the form of the region shows little influence on the magnitude of the discharge of the well.

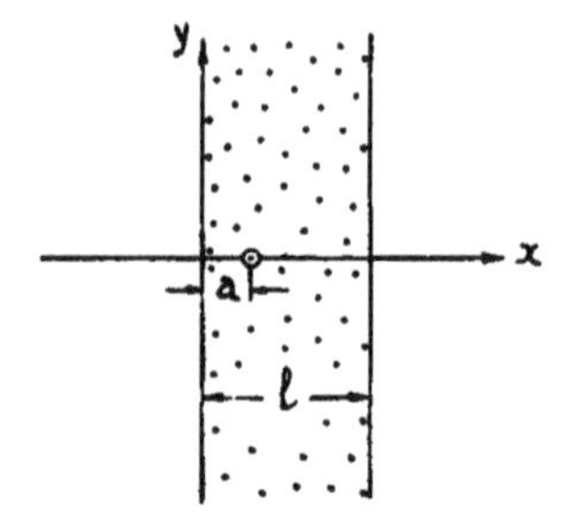

Fig. 267

We consider the well in an infinite strip (between two canals) In this case (fig. 267), the complex potential has the form

$$f(z) = -\frac{q_1}{2\pi}\left\{\ell n \sin\frac{\pi}{2\ell}(z - a) - \ell n \sin\frac{\pi}{2\ell}(z + a)\right\} + C \quad .$$

For the pressure we have the expression

$$p = \frac{q_1 \mu}{2\pi k_0}\left\{\ell n\left|\sin\frac{\pi}{2\ell}(z - a)\right| - \ell n\left|\sin\frac{\pi}{2\ell}(z + a)\right|\right\} + C \ .$$

Near the well, in the point $z = a + \delta$, $p = p_1$, and therefore

$$p_1 \approx \frac{q_1 \mu}{2\pi k_0}\left\{\ell n(\pi\delta) - \ell n\left(2\ell \sin\frac{\pi a}{\ell}\right)\right\} + C \quad .$$

In the point $z = \ell$ we have $p_0 = C$ (we assume $p = p_0$ for $x = 0$ and $x = \ell$). From this the discharge of the well

$$q_1 = \frac{2\pi k_0 (p_0 - p_1)}{\mu \ \ell n\left(\frac{2\ell}{\pi\delta}\sin\frac{\pi a}{\ell}\right)}$$

## §4. About the Interference of Wells.

With the increase of the number of fully penetrating wells drilled in the stratum, the total discharge of the wells increases. However, this increase is not proportional to the number of wells, and namely the total discharge of $n$ single wells is smaller that $n$-times the discharge of one well (it is assumed that all the wells are found in single conditions). This phenomenon is called well-interference. In this case, the larger the number of wells to be found in a given place, the smaller the discharge of each well will be. For some value $n = N$, further increase of the number of wells does not pay anymore, as the expenditures for the well do not justify the small increment in the total discharge, which is contributed by the $(N + 1)^{th}$ well. Problems about the most advantageous number of wells have an important meaning in the theory of oil seepage [3]. Here we limit ourselves to considerations of the simplest cases.

Two wells. Putting sinks in the points $z = \pm a$, each of intensity $Q$, we obtain the scheme of two wells in a stratum. The complex potential of this flow is:

$$w = \varphi + i\psi = -\frac{Q}{2\pi} \ln[(z - a)(z + a)] + C \quad , \tag{4.1}$$

where $Q$ - discharge of each well. The velocity potential in plane, horizontal movement is related to the pressure by the equation

$$\varphi = -\frac{kp}{\rho g} = -\frac{k_o}{\mu} p \tag{4.2}$$

(see Chapter I, Par. 7). Therefore, separating in (4.1) the real part and applying (4.2), we may write

$$p = \frac{\mu Q}{2\pi k_o} \ln|(z - a)(z + a)| + C_1 \quad . \tag{4.3}$$

To determine the constant $C_1$ and the discharge $Q$ we need two conditions.

Considering that the radius of the well $\delta$ is small in comparison with the distance $2a$ between the wells, we take for the contour of the cross-section of the well the isobar that passes through the point

$$z = a + \delta \quad .$$

In the indicated conditions, this isobar is close to a circle of radius $\delta$ with center in the point $a$. Let the pressure at the contour of the well be $p_o$. Then formula (4.3) gives, if one neglects the quantity $\delta^2$,

$$p_o = \frac{\mu Q}{2\pi k_o} \ln(2 a \delta) + C_1 \quad . \tag{4.4}$$

We assume now that we know the isobar

$$r = R \ .$$

Actually, in the flow of two wells, the isobars will not take the form of circles, since the isobars are represented by a family of lemniscates. However, at sufficiently large distance from the origin, these lemniscates become close to circles. Considering $R$ sufficiently large with respect to $a$, we assume that for $r = R$, the pressure is equal to $p_k$ (the contour pressure, as called in the theory of oil seepage). Inserting $r = R$ and $p = p_k$ in (4.3), we obtain:

$$p_k = \frac{\mu Q}{2\pi k_o} \ln(R^2 - a^2) + C_1 \quad . \tag{4.5}$$

Subtracting (4.4) from (4.5), we obtain

$$p_k - p_o = \frac{\mu Q}{2\pi k_o} \ln \frac{R^2 - a^2}{2 a \delta} \quad .$$

From this, we find the discharge of each well:

$$Q = \frac{2\pi k_o}{\mu} \frac{p_k - p_o}{\ln \frac{R^2 - a^2}{2a\delta}} . \tag{4.6}$$

We compare this discharge with that of a single well of radius $\delta$, with center in the origin of coordinates, for which the pressure for $r = \delta$ and $r = R$ is equal respectively to $p_o$ and $p_k$. As we have seen [see (1.8)], this discharge, which we designate by $Q_1$, is equal to

$$Q_1 = \frac{2\pi k_o}{\mu} \frac{p_k - p_o}{\ln \frac{R}{\delta}} . \tag{4.7}$$

The ratio $\frac{Q}{Q_1}$ is equal to

$$\frac{Q}{Q_1} = \frac{\ln \frac{R}{\delta}}{\ln \frac{R^2 - a^2}{2a\delta}} = \frac{\ln \frac{R}{\delta}}{\ln \frac{R}{\delta} + \ln \left(\frac{R}{2a} - \frac{a}{2R}\right)} .$$

Since we consider $R/2a$ sufficiently large (otherwise the circle $r = R$ in the case of the two wells may not be taken for an isobar), the second term of the denominator is signigicantly positive, and

$$\frac{Q}{Q_1} < 1 .$$

Thus, our statement in the case of two wells is justified. To describe the dependence of the total discharge of the wells on their number, in the case of more than two wells, we examine the following example [3]

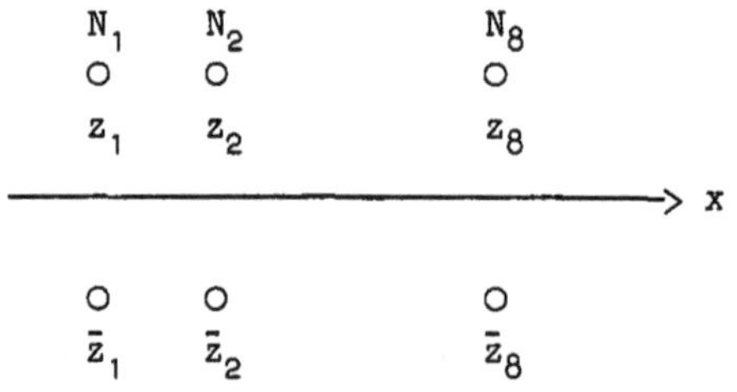

We assume that initially the stratum is tapped by one well No. 1, and that then subsequently the wells No. 2, 3. and so on till well No. 8 included are introduced. (Fig. 268). The distance of all wells from a straight linear contour of intake be L, the distance between wells is designated by 2h. Numerical values are adopted: the radius of the well $\delta = 10^{-1}$m, the distance between wells $2h = 100$m, $L = 10^4$m. The intake contour may be the bank of a river or canal.



Fig. 268

After putting sources in points conjugate to the centers of the wells, we find for the complex potential

$$w = -\frac{Q}{2\pi} \sum_{s=1}^{n} \ln \frac{\frac{\pi}{h}(z - z_s)}{\frac{\pi}{h}(z - \bar{z}_s)} \qquad (n = 1, 2, \ldots, 8) ,$$

where $z_s$ - the coordinate of the $s^{th}$ well, and $Q$ - the discharge of each well. Designating by $Q_o$ the discharge of a single well (No. 1), we find the quantity

$$q_n = \frac{nQ}{Q_o} \cdot 100\%$$

giving the total discharge of $n$ wells expressed in percentage of the discharge of a single well.

| $n$ | $q_n$ | $\frac{q_n - q_1}{n-1}$ | $\Delta q_n$ |
|---|---|---|---|
| 1 | 100 | 0 | |
| 2 | 145 | 45 | 45 |
| 3 | 175 | 37.5 | 30 |
| 4 | 197 | 32 | 22 |
| 5 | 215.5 | 29 | 18.5 |
| 6 | 231 | 26 | 15.5 |
| 7 | 245 | 24 | 14 |
| 8 | 248 | 22.5 | 13 |

Table 13

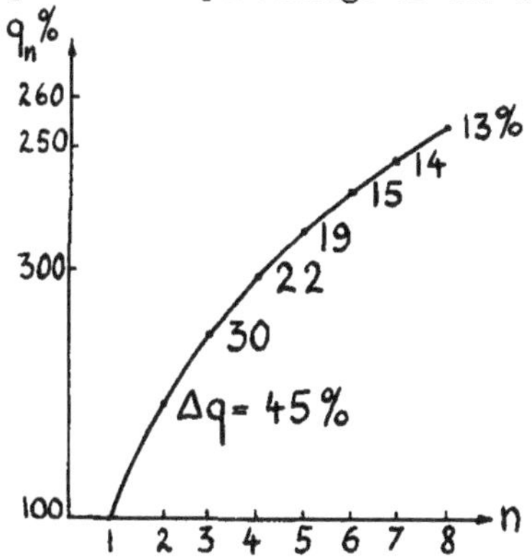

Fig. 269

Table 13

Dependence of the total discharge upon the number of wells and values, in percentages, of the average increment of the total discharge caused by the addition of each well.

In Table 13 are given the computations of V. N. Tsjelkatchev [3]: values of the quantity $q_n$ depending upon the number $n$ of wells, and also values of $(q_n - q_1)/(n - 1)$ - the average increment of the total discharge caused by the addition of each well, in percentages. In Fig. 269 is given the graph of the dependence of the total discharge on the number of wells, in percentages. The numbers in the vicinity of the various points are the increments $\Delta q_n$:

$$\Delta q_n = q_n - q_{n-1} \quad .$$

We see that with increasing number of wells, the increment of the total discharge decreases, and therefore a dense grid of wells may prove not to be remunerative.

## §5. Flow to a Fully Penetrating Well in Nonhomogeneous Medium [4].

One may interpret the second problem of Chapter VIII, Par. 4 otherwise, by replacing the point vortex by a point source. Therefore it suffices to mutate the roles of $\varphi$ and $\psi$. Let us namely assume that the confined stratum, in which the movement of fluid towards a fully-penetrating well takes place, consists of two parts: one half plane in which the soil has a seepage coefficient $k_1$ and another half plane with seepage coefficient $k_2$ (Fig. 270).

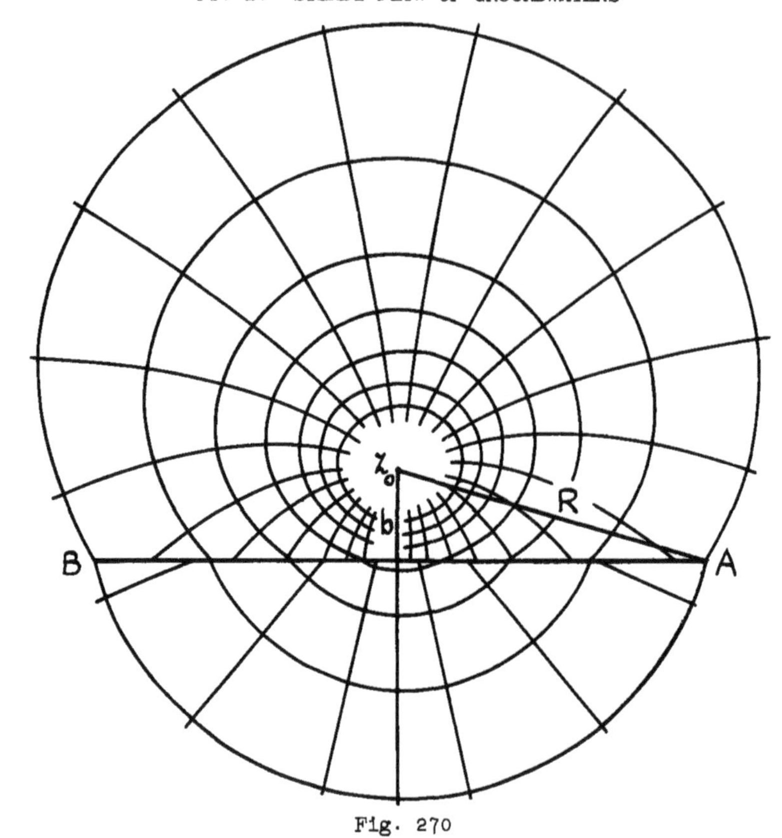


Fig. 270

Then, designating by $Q$ the discharge of the well, one may write for the first and second region, if the well is situated in the first regions, in the point $z_0$:

$$
(5.1) \qquad \begin{aligned} \omega_1(z) &= -\frac{Q}{2\pi}\,[\ln(z - z_0) + \lambda \ln(z - \bar{z}_0)] + C_1 \\ \omega_2(z) &= -\frac{Q}{2\pi}\,(1-\lambda) \ln(z - z_0) + C_2 \qquad \left(\lambda = \frac{k_1 - k_2}{k_1 + k_2}\right) \end{aligned}
$$

For the discharge of the well, one obtains the formula

$$
(5.2) \qquad Q = 2\pi k_1(\varphi_k - \varphi_0)/[\ln(R/\delta) + \lambda \ln R/2b]
$$

where $\delta$ - radius of the well, $b$ - distance of the well from the abscissa axis, $R$ - distance from the well to the point of intersection $A$ of the borderline $AB$ with the isobar at which the pressure is equal to

$$
p_k = -\frac{\rho g}{k_1}\,\varphi_k \;.
$$

Formula (5.1) is made so that along the abscissa axis the conditions (3.2) and (3.3) of Chapter II are satisfied:

$$\frac{\varphi_1}{k_1} = \frac{\varphi_2}{k_2} , \quad \psi_1 = \psi_2 .$$

If the well is found in a region, bordered by a circle of radius $r = 1$, where the soil has a seepage coefficient $k_1$, but if in the annular region between the circles with radii $1$ and $R$ soil with seepage coefficient $k_2$ is available, for which the outside circle is an isobar, then we have these formulas for the velocity (fig. 271).

$$(5.3)\quad \begin{aligned} u_1 - iv_1 &= -\frac{Q}{2\pi}\left\{\frac{1}{z - z_0} + \frac{\lambda}{z - \frac{1}{\bar z_0}} + (1-\lambda^2)\sum_{n=0}^{\infty}\frac{z^n \bar z_0^{\,n+1}}{R^{2n+2}-\lambda}\right\}, \\ u_2 - iv_2 &= -\frac{Q}{2\pi}\left\{\frac{1-\lambda}{z - z_0} + \frac{\lambda}{z} + (1-\lambda)\sum_{n=0}^{\infty}\frac{z^n \bar z_0^{\,n+1}}{R^{2n+2}-\lambda} + \right. \\ &\qquad \left. + \lambda(1-\lambda)\sum_{n=1}^{\infty}\frac{z_0^n}{z^{n+1}(R^{2n}-\lambda)}\right\}. \end{aligned}$$

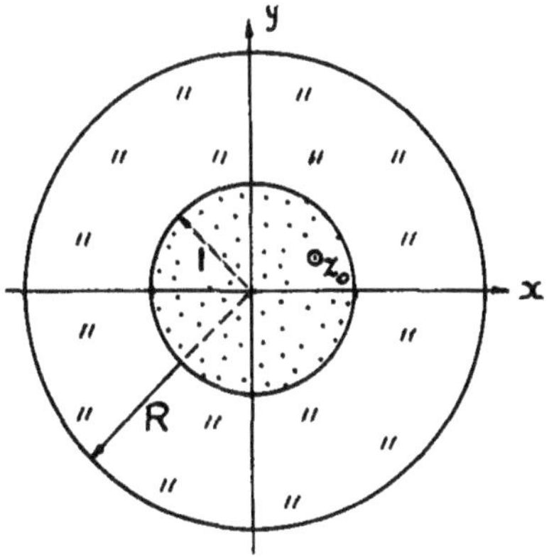

Fig. 271

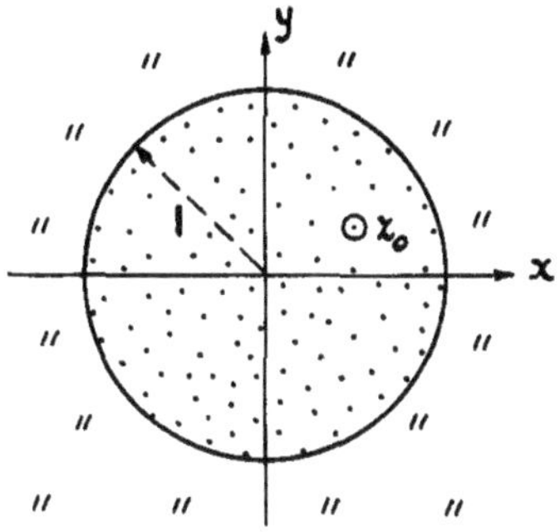

Fig. 272

When the outside region is infinite, i.e., $R = \infty$, then we obtain Fig. 272).

$$(5.4)\quad \begin{aligned} u_1 - iv_1 &= -\frac{Q}{2\pi}\left\{\frac{1}{z - z_0} + \frac{\lambda}{z - \frac{1}{z_0}}\right\} \\ u_2 - iv_2 &= -\frac{Q}{2\pi}\left\{\frac{1-\lambda}{z - z_0} + \frac{\lambda}{z}\right\} \end{aligned}$$

Formulas (5.3) may also be written in this form

$$(5.5)\quad \begin{aligned} u_1 - iv_1 &= -\frac{Q}{2\pi}\left\{\frac{1}{z - z_0} + \frac{\lambda}{z - \frac{1}{\bar z_0}} - \sum_{n=1}^{\infty}\frac{\lambda^{n-1}}{z - \frac{R^{2n}}{\bar z_0}}\right\} \\ u_2 - iv_2 &= -\frac{Q(1-\lambda)}{2\pi}\left\{\frac{1}{z - z_0} + \sum_{n=1}^{\infty}\frac{\lambda^n}{z - \frac{z_0}{R^{2n}}} - \sum_{n=1}^{\infty}\frac{\lambda^{n-1}}{z - \frac{R^{2n}}{\bar z_0}}\right\} \end{aligned}$$

## B. Horizontal Drains

### §6. Application of the Method of Sources to Problems About Horizontal Drains.

The surface of the ground may be marsh-ridden for various reasons: due to the excess flow of ground water to a given plot, under influence of atmospheric precipitation, during flooding to desalt brackish soil, and so on. We examine the case [5] where a water layer of depth $H$ covers the surface of the ground. At a certain depth below the surface, a horizontal pipe is laid that drains the water and carries it to some collector. (Fig. 273, 274)

We will assume that a sink is located in the center of the pipe with discharge $Q$ per unit length of pipe. The coordinates of the center of the sink will be $(0,0,h_1)$. In the point $(0,0,-h_1)$, we put a source of the same intensity as the sink. Then we have for the complex velocity potential:

$$(6.1) \qquad \varphi = \frac{Q}{2\pi} \ln \left| \frac{z + h_1 i}{z - h_1 i} \right| - kH = \frac{Q}{2\pi} \ln \frac{\sqrt{x^2 + (y + h_1)^2}}{\sqrt{x^2 + (y - h_1)^2}} - kH$$

The constant is chosen equal to $-kH$ in order to satisfy the condition $\varphi = -kH$ along the x-axis.

We assume that at the contour of the drainage pipe the head is equal to $h_{dr}$. We take for contour of the drain a line of constant potential, passing through the point $(\delta, h_1)$, where $\delta$ is a small number in comparison with $h_1$, so that the equipotential under consideration is almost a circle. We have $x = \delta$, $y = h_1$:

$$\varphi = -kh_{dr} = \frac{Q}{2\pi} \ln \frac{\sqrt{\delta^2 + 4h_1^2}}{\delta} - kH \approx \frac{Q}{2\pi} \ln \frac{2h_1}{\delta} - kH$$

and we find the expression of the discharge:

$$(6.2) \qquad Q = \frac{2\pi k(H - h_{dr})}{\ln (2h_1/\delta)}$$

In Fig. 274 a flownet, constructed by V. V. Vedernikov [5] for a particular case is represented. He solved this problem and also the following one, by a conformal mapping of the flow region and complex potential region onto a half plane. His notations differ from ours: he designates by $h_1$ the distance from the surface of the ground to the lowest point of the drainage pipe and by $2\delta$ the diameter of the pipe.

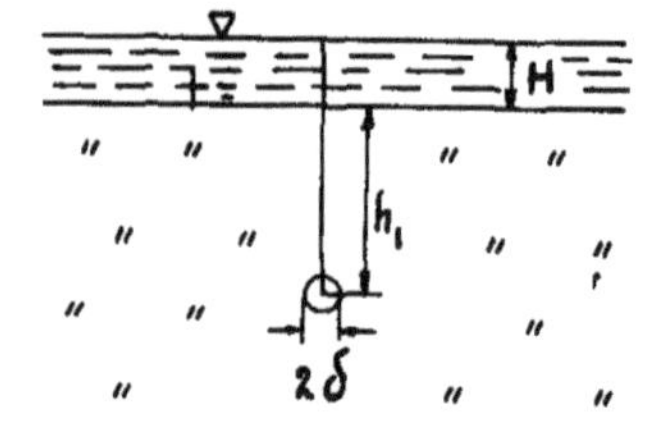


Fig. 273

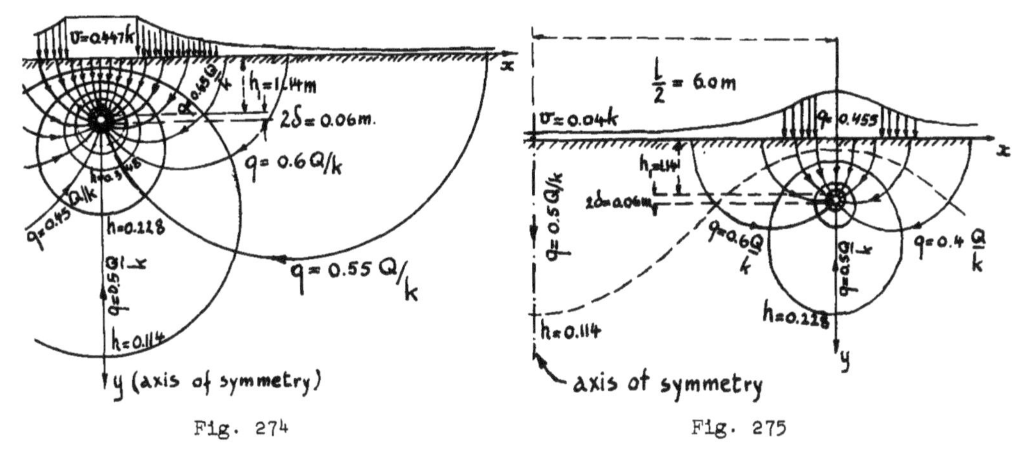


Fig. 274 Fig. 275

In the case of an infinite number of parallel drainage pipes, at a distance $\ell$ from each other (Fig. 275), we have:

$$\varphi = \frac{Q}{2\pi} \ell n \left| \frac{\sin \frac{\pi}{\ell} (z + h_1 i)}{\sin \frac{\pi}{\ell} (z - h_1 i)} \right| - kH$$

Putting $x = o$, $y = h_1 + \delta$, $\varphi = - kh_{dr}$, we obtain

$$-kh_{dr} = \frac{Q}{2\pi} \ell n \frac{\operatorname{sh} \frac{2\pi h_1}{\ell}}{\operatorname{sh} \frac{\ell\delta}{\ell}} - kH$$

from which we find the expression of the discharge:

$$Q = \frac{2\pi k(H - h_{dr})}{\ell n \operatorname{sh}^2 \frac{2\pi h_1}{\ell} - \ell n \operatorname{sh} \frac{\pi\delta}{\ell}} \tag{6.3}$$

We notice that we consider $\frac{\delta}{\ell}$ so small, that instead of $\operatorname{sh} \frac{\pi\delta}{\ell}$ we may take $\frac{\pi\delta}{\ell}$. This is shown in a different-looking form of formula by various authors in works of this kind.

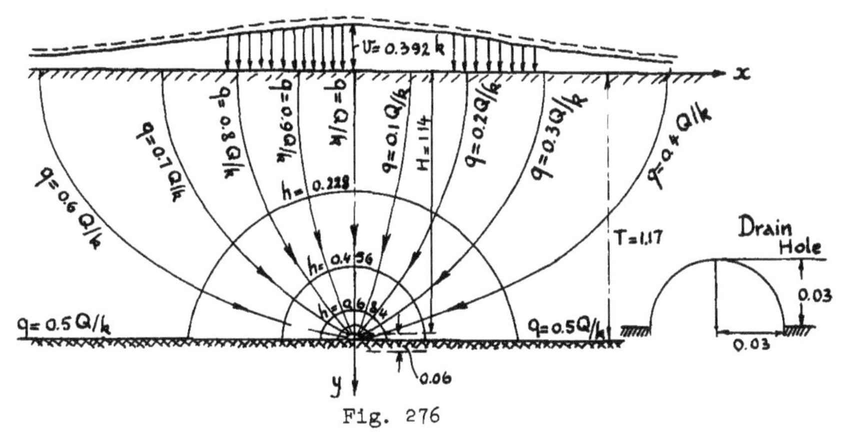


Fig. 276

The case of a drain on an impervious bottom is treated in an analogous way (Fig. 276). Here, if we designate by $T$ the thickness of the pervious layer and if we alternately distribute sources and sinks at distances $\pm 2nT$ from the basic drain, we have:

$$\varphi = -\frac{q}{\pi}\,\ell n\left|\text{th}\,\frac{\pi(z - iT)}{4T}\right| - kH \quad .$$

Therefore $q$, the discharge per unit length of drain, is equal to one-half of the discharge of a source put in the point $(o, T)$:

$$q = \frac{k\pi(H - h_{dr})}{\ell n\ \text{th}\,\frac{\pi\delta}{2T}} \approx \frac{k\pi(H - h_{dr})}{\ell n\,\frac{\pi\delta}{2T}} \quad .$$

## §7. Underground Layer of Weak Transmissivity, Transmitting Waters of the Adjacent Water-bearing Stratum [6].

We assume that the confined water-bearing stratum B (Fig. 277), reaching deep till infinity, has a piezometric head $H_o$ and a seepage coefficient $k_o$. At the top, this stratum is bordered by a weakly pervious layer of rock of thickness $T$, with seepage coefficient $k_1$. Above this layer lies an artesian stratum with piezometric head $H_1$. In the point $\zeta$ of the lower layer a point sink is established. Its equipotential, close to a circle, may be taken for the contour of the drain. Considering $H_1$ to be constant, we assume that along the x-axis the vertical component of the velocity is determined by Darcy's formula which in the treated case may be written as (see below Par. 8):

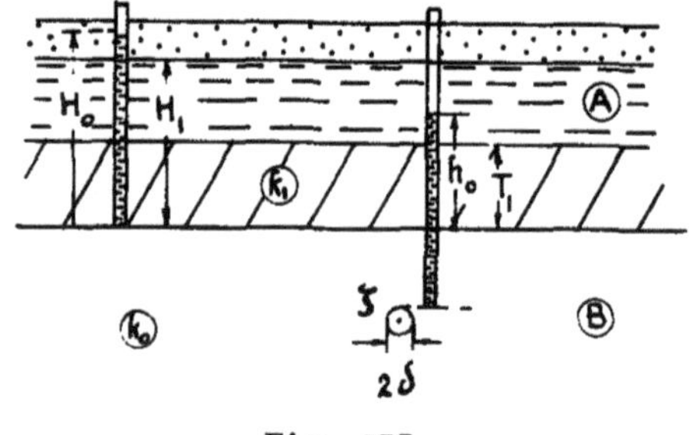


Fig. 277

$$v_y = -k_1\,\frac{H_1 - h}{T_1} = -\frac{k_1}{T_1}\left(H_1 + \frac{\varphi}{k_o}\right) \quad . \tag{7.1}$$

We construct the function

$$W = \frac{d(\varphi + i\psi)}{dz} - \frac{ik_1}{T_1}\left(H_1 + \frac{\omega}{k_o}\right) \quad . \tag{7.2}$$

If we find $W = F(z)$, then we obtain for $\omega$ after integration:

$$\omega = e^{-az}\int e^{az}[F(z) - b]\,dz + Ce^{-az} \tag{7.3}$$

where

$$a = -\frac{ik_1}{k_0T_1} \quad , \quad b = -\frac{ik_1H_1}{T_1} \quad .$$

It is easy to see that in the case of one sink, established in the point $z = \zeta$, we have:

$$F(z) = -\frac{Q}{2\pi}\left(\frac{1}{z-\zeta} + \frac{1}{z-\bar{\zeta}}\right) + D. \tag{7.4}$$

where $D$ is a constant, In the case of an infinite series of sinks, at distances $2n\ell (n = \pm 1, \pm 2, \ldots)$ from the point $z = \zeta$:

$$F(z) = -\frac{Q}{2\pi}\left(\operatorname{cotg}\frac{\pi(z-\zeta)}{2\ell} + \operatorname{cotg}\frac{\pi(z-\bar{\zeta})}{2\ell}\right) + D \quad .$$

Indeed, since for $z = x$ we have in both cases $\operatorname{Im} F(z) = 0$, or

$$\operatorname{Im} W = -v_y - \frac{k_1}{T_1}\left(H_1 + \frac{\varphi}{k_0}\right) = 0 \quad ,$$

and therefore condition (7.1) appears to be satisfied. Usually the conditions adopted at infinity are also satisfied. The constants $C$ and $D$ are determined from the conditions $\varphi = -k_0H_0$ for $x = \infty$, $\psi = Q$ for $x < 0$. To determine the discharge of the drain $Q$, the condition that $h = h_0$ in the upper point of the drain (for $z = \zeta + \delta i$, where $\delta$, radius of the drain) is used; the y-axis is oriented upward.

## C. Movements in Strata, Adjacent to Weakly Pervious Strata

### §8. Derivation of Equations [7].

In Chapter III, we analyzed a water-bearing stratum, confined by two impervious horizontal planes. We analyzed the confined flow to a "perfect", i.e., fully-penetrating well as a plane flow, and we interpreted the wells by means of sources and sinks.

However, in nature there are no absolutely impervious layers. Usually very pervious layers alternate with weakly pervious layers. The ratio of the seepage coefficients of two adjacent layers may be of the order of $10^{-5}$ to $10^{-10}$, but nevertheless the weakly pervious soil may show influence on the seepage, if this seepage takes place over a large surface. We assume that we have horizontal layers of very pervious soils, alternating with weakly pervious soils (Fig. 278). Let the $n^{th}$ pervious layer have the thickness (depth) $m_n$, and the seepage coefficient $k_n$. The underlying layer of poorly pervious soil may have the depth $\mu_n$ and the seepage coefficient $\lambda_n$. The above

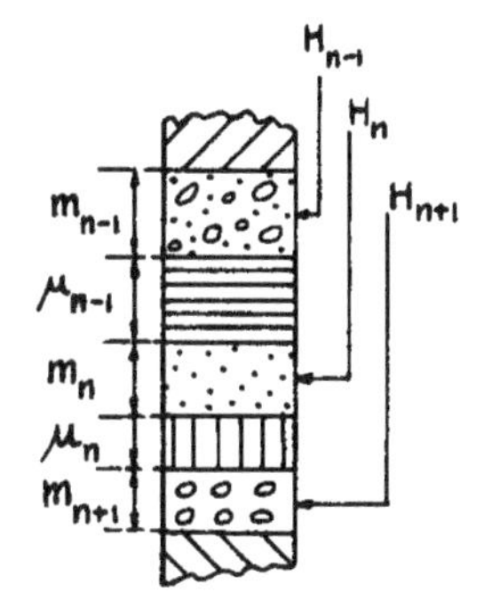


Fig. 278

lying layer of poorly pervious soil may have the depth $\mu_{n-1}$, the seepage coefficient $\lambda_{n-1}$. We consider the quantities $\lambda_n/k_n$ and $\lambda_{n-1}/k_n$ as small. Then any streamline, going from the region $m_n$ to the neighboring region, will be refracted, as we saw in Par. 3 of Chapter II; if the streamline makes the angles $\alpha$ and $\beta$ with the normal to the boundary of the two soils, then

$$\text{tg } \beta : \text{tg } \alpha = \lambda_n : k_n \tag{8.1}$$

For an arbitrary value of $\alpha$, different from $\frac{\pi}{2}$, we find that the angle $\beta$ will be small when the ratio $\lambda_n : k_n$ is small, and, consequently, after refraction the streamline in the poorly pervious soil will be close to the vertical straight line. We will use this fact later.

We designate by $u, v, w$ the projections on the axes $x, y, z$ of the seepage velocity in some point $M(x, y, z)$ of the very pervious stratum.

Considering the fluid to be incompressible, i.e., its density to be constant, we have the equation of continuity

$$\frac{\partial u}{\partial x} + \frac{\partial v}{\partial y} + \frac{\partial w}{\partial z} = 0 \quad . \tag{8.2}$$

We consider the problem to be "hydraulic", (see Chapter X), i.e., we assume that the horizontal velocity components change little with height, and introduce in the analysis the average values of these velocities $U$ and $V$ in height, putting

$$U(x, y) = \frac{1}{m_n} \int_0^{m_n} u(x, y, z)dz \quad ,$$

$$V(x, y) = \frac{1}{m_n} \int_0^{m_n} v(x, y, z)dz \quad .$$

We integrate equation (8.2) along $z$ (as this is done sometimes in the dynamics of eteorology, taking for $m_n$ the thickness of the atmospheric layer). We have the equalities:

$$\int_0^{m_n} \frac{\partial u}{\partial x} dz = \frac{\partial}{\partial x} \int_0^{m_n} u \, dz = m_n \frac{\partial U}{\partial x} \quad ,$$

$$\int_0^{m_n} \frac{\partial v}{\partial y} dz = \frac{\partial}{\partial y} \int v \, dz = m_n \frac{\partial V}{\partial y} \quad ,$$

so that the integration of the third term of equation (8.2) gives:

$$\int_0^{m_n} \frac{\partial w}{\partial z} dz = w_2 - w_1 \quad ,$$

where $w_1$ the value of the vertical velocity at the lower boundary, $w_2$ the value of the vertical velocity at the upper boundary of the stratum.

Now the equation of continuity takes the form

$$m_n\left(\frac{\partial U}{\partial x} + \frac{\partial V}{\partial y}\right) + w_2 - w_1 = 0 \quad . \tag{8.3}$$

Let the velocity potential of the non-averaged movement be $\varphi(x, y, z)$, so that

$$u = \frac{\partial \varphi}{\partial x}, \quad v = \frac{\partial \varphi}{\partial y}, \quad w = \frac{\partial \varphi}{\partial z} .$$

Then, with the z-axis oriented upward, we have:

$$\varphi = -kh(x, y, z) = -k\left(\frac{p}{\rho g} + z\right) .$$

In other words, we may write:

$$u = -k\frac{\partial h}{\partial x}, \quad v = -k\frac{\partial h}{\partial y}, \quad w = -k\frac{\partial h}{\partial z} .$$

We apply average values to the horizontal components of the velocity $u, v$, and do the same to the head $h(x, y, z)$. We designate the average value of this function by $H(x, y)$:

$$H(x, y) = \frac{1}{m_n}\int_0^{m_n} h(x, y, z)dz .$$

Then we may write:

$$U = -k\frac{\partial H}{\partial x}, \quad V = -k_n\frac{\partial H_n}{\partial y} .$$

and applying these equalities to the $n^{th}$ pervious stratum, we obtain

$$U_n = -k_n\frac{\partial H_n}{\partial x}, \quad V_n = -k_n\frac{\partial H_n}{\partial y} . \tag{8.4}$$

Now we must find expressions for the vertical velocities $w_1$ and $w_2$ at the top and bottom of the $n^{th}$ stratum. At the same time we may recall the previously made remark that the seepage in the poorly pervious stratum progresses along the vertical. Designating the functions of the head for the $(n+1)^{th}$ and $(n-1)^{th}$ pervious strata respectively by $H_{n+1}$ and $H_{n-1}$, we will have according to Darcy's law (since the seepage velocity in our assumptions will be proportional to the difference of head at the boundaries of the poorly pervious layers and inversely proportional to the thickness of the strata):

$$w_2 = \lambda_{n-1}\frac{H_n - H_{n-1}}{\mu_{n-1}}, \quad w_1 = \lambda_n\frac{H_{n+1} - H_n}{\mu_n} . \tag{8.5}$$

For $H_{n+1} > H_n$ we have $w_1 > 0$ and vice versa. Substituting the expression (8.4) and (8.5) in equation (8.3) we obtain:

$$-m_n k_n\left(\frac{\partial^2 H_n}{\partial x^2} + \frac{\partial^2 H_n}{\partial y^2}\right) + \lambda_{n-1}\frac{H_n - H_{n-1}}{\mu_{n-1}} - \lambda_n\frac{H_{n+1} - H_n}{\mu_n} = 0 ,$$

or

$$m_n k_n \Delta H_n - \frac{\lambda_n}{\mu_n}\left(H_n - H_{n+1}\right) - \frac{\lambda_{n-1}}{\mu_{n-1}}\left(H_n - H_{n-1}\right) = 0 \quad . \tag{8.6}$$

We introduce the notations:

$$\zeta_n = \frac{\dfrac{\lambda_n}{\mu_n} + \dfrac{\lambda_{n-1}}{\mu_{n-1}}}{m_n k_n} \quad , \tag{8.7}$$

$$\tilde{H}_n = \frac{H_{n-1}\dfrac{\lambda_{n-1}}{\mu_{n-1}} + H_{n+1}\dfrac{\lambda_n}{\mu_n}}{\dfrac{\lambda_{n-1}}{\mu_{n-1}} + \dfrac{\lambda_n}{\mu_n}} \quad . \tag{8.8}$$

Then equation (8.6) can be written in the form

$$\frac{\partial^2 H_n}{\partial x^2} + \frac{\partial^2 H_n}{\partial y^2} - \zeta_n (H_n - \tilde{H}_n) = 0 \quad . \tag{8.9}$$

If the piezometric heads $H_{n-1}$ and $H_{n+1}$ in the neighboring layers change so little that one may consider them to be constant, then also $\tilde{H}_n$ will be constant. If $H_{n-1}$ and $H_{n+1}$ also vary, then one must also set up equations for them, analogous to (8.9). If the stratum $(\mu_n)$ or $(\mu_{n-1})$ is perfectly impervious, then the corresponding values of $\lambda_n$ or $\lambda_{n-1}$ are equal to zero.

§9. Movement in One Pervious Layer [8].

We assume that the movement takes place parallel to the x-axis. We omit the subscript $n$ in equation (8.9) and rewrite the equation in the form

$$\frac{d^2 H}{dx^2} - \zeta (H - \tilde{H}) = 0 \quad . \tag{9.1}$$

The general solution of this equation may be represented in the form

$$H - \tilde{H} = C_1 e^{-x\sqrt{\zeta}} + C_2 e^{x\sqrt{\zeta}} \quad .$$

We must set up two boundary conditions to determine the constants $C_1$ and $C_2$. So, $C_1$ and $C_2$ will be determined if the values of the head $H$ are known in two points of the stratum. If the flow of a fluid from infinity is considered, for example the flow from infinity to a canal, then, requiring the condition of finiteness of the head at $x = +\infty$, $C_2$ must be zero and

$$H = \tilde{H} + C_1 e^{-x\sqrt{\zeta}} \quad .$$

Knowing the magnitude of the head in some point, we determine $C_1$ . For example if $H = H_1$, for $x = 0$, then:

(9.2) $$H = \tilde{H} + (H_1 - \tilde{H})e^{-x\sqrt{\zeta}} .$$

We see that $\tilde{H}$ in this case will be the magnitude of the head at $x = \infty$.

If $H_1 > \tilde{H}$, then curve (9.2) lies above the asymptote $H = \tilde{H}$ and is concave upward; if $H_1 < \tilde{H}$, then the curve lies under its asymptote and is convex upward (Fig. 279). The direction of the convexity of the piezometric curve may be determined directly from equation (9.1), rewritten as:

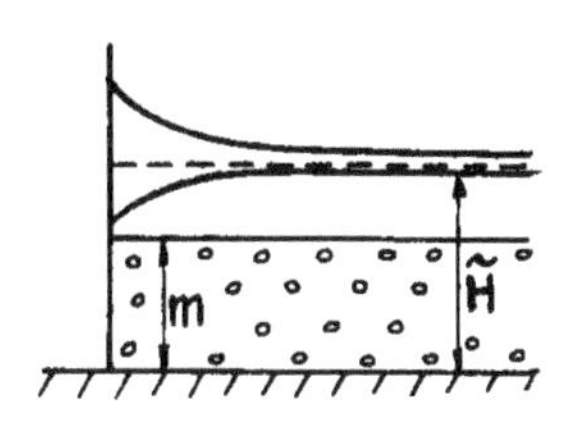


Fig. 279

$$\frac{d^2H}{dx^2} = \zeta(H - \tilde{H}) .$$

Since $\zeta > 0$ for $H > \tilde{H}$ the second derivative is positive, *i.e.*, the curve is concave; for $H < \tilde{H}$ it is, on the contrary, convex.

## §10. Fully Penetrating Wells in a Confined Stratum. [7.8]

We first assume that we have one cylindrical well, with axis coinciding with the z-axis and with identical flow towards it from all sides, so that the flow does not depend upon the polar angle but only upon the radius vector $r$.

We transform the Laplacian operator in the form

$$\Delta H = \frac{\partial^2 H}{\partial x^2} + \frac{\partial^2 H}{\partial y^2} = \frac{d^2H}{dr^2} + \frac{1}{r}\frac{dH}{dr} ,$$

and instead of (8.9) we have:

(10.1) $$\frac{d^2H}{dr^2} + \frac{1}{r}\frac{dH}{dr} - \alpha^2(H - \tilde{H}) = 0 \qquad (\alpha^2 = \zeta) .$$

This is Bessel's equation. Its general solution can be represented in the form

(10.2) $$H - \tilde{H} = C_1 I_0(\alpha r) + C_2 K_0(\alpha r)$$

where $I_0$- the cylindrical function of a complex argument of order zero and of the first kind (modified Bessel function of the first kind), $K_0$ - the corresponding modified Bessel function of the second kind.

If the value of the head is known along two circles of radii $r_1$ and $r_2$, then the constants $C_1$ and $C_2$ are determined.

We consider the movement in an infinite stratum, and require the condition: $H$ must remain finite when $r$ tends to infinity. Since the function $I_0(x)$ becomes infinite for increasing argument, we must put $C_1 = 0$. We obtain:

$$H - \tilde{H} = C_2 K_0(\alpha r) .$$

The function $K_0(x)$ tends to zero for increasing $x$; therefore $H$ tends to $\tilde{H}$ for $r \to \infty$. Consequently, $\tilde{H}$ is the value of the head at rest. We may introduce the quantity

$$S = \tilde{H} - H \quad , \tag{10.3}$$

the so-called "drawdown of head". Then we have for $S$ the expression

$$S = - C_2 K_0(\alpha r) \quad . \tag{10.4}$$

We designate by $\delta$ the radius of the well and compute the discharge $Q_\delta$ through the lateral surface of the cylinder with radius $\delta$ and height $M$

$$Q_\delta = 2\pi\delta M v_\delta \quad , \tag{10.5}$$

where $v_\delta$ the velocity of flow at the wall of the cylinder. We have

$$v_\delta = - k \left.\frac{dH}{dr}\right|_{r=\delta} = - k\alpha K_0'(\alpha\delta) \ , \quad C_2 = C_2 k\alpha K_1(\alpha\delta) \tag{10.6}$$

where $K_1$ is the cylindrical function of a complex argument of the second kind and first order.

Substituting (10.6) in (10.5), we obtain:

$$Q_\delta = 2\pi k\delta C_2 \alpha K_1(\alpha\delta)M \quad . \tag{10.7}$$

From this we find the constant $C_2$

$$C_2 = \frac{Q_\delta}{2\pi k\delta \ \alpha K_1(\alpha\delta)M}$$

and equation (10.3) gives

$$S = \tilde{H} - H = \frac{Q_\delta}{2\pi k\delta\alpha M} \frac{K_0(\alpha r)}{K_1(\alpha r)} \quad . \tag{10.8}$$

Since the seepage of the fluid occurs through horizontal planes that border the very pervious layer, the discharge $Q_r$ through a cylinder of radius $r$ will depend upon $r$. We obtain $Q_r$, by replacing $\delta$ by $r$ in (10.7): $Q_r = 2\pi rkM \ \alpha C_2 K_1(\alpha r)$

Combining this expression with (10.7), we obtain:

$$Q_r = Q_\delta \frac{r}{\delta} \frac{K_1(\alpha r)}{K_1(\alpha\delta)} \quad .$$

If $x < 0.02$, then we may use the approximate formulas

$$K_0(x) \approx - \ln x + \ln 2 - C = 0.1159\ldots - \ln x \tag{10.9}$$

[C = Euler's constant]

$$K_1(x) \approx \frac{1}{x} \tag{10.10}$$

Usually $\alpha\delta < 0.02$, therefore $K_1(\alpha\delta) \approx 1/\alpha\delta$, and the formula (10.8) may be changed into this:

$$H - H \approx - \frac{Q_\delta}{2\pi kM} K_0(\alpha r) \tag{10.11}$$

or

$$S = \frac{Q_\delta}{2\pi kM} K_o(\alpha r) \quad . \tag{10.12}$$

Substituting $r = \delta$ in equation (10.8), we find the drawdown of head at the well

$$\tilde{H} - H(\delta) = S(\delta) = \frac{Q_\delta}{2\pi kM} \frac{K_o(\alpha\delta)}{\alpha\delta K_1(\alpha\delta)} \quad ,$$

where $H(\delta)$ - head at the wall of the well. By means of the approximate expressions for $K_o$ and $K_1$, we rewrite:

$$S(\delta) \approx \frac{Q_\delta}{2\pi kM} \left[ - \ell n \ (\alpha\delta) + 0.1159 \ldots \right] \quad .$$

We find from this for the discharge of the well

$$Q_\delta = \frac{2\pi kM(\tilde{H} - H(\delta))}{0.1159 - \ell n \ (\alpha\delta)} \quad . \tag{10.13}$$

We compare this formula with formula (1.4), obtained for the discharge of a fully-penetrating well, which may be written as:

$$Q_* = \frac{2\pi kM(H(R) - H(\delta))}{\ell n \ R - \ell n \ \delta} \quad .$$

Here $H(R)$ is the head in a point that lies at a distance $R$ (arbitrary) from the center of the well, like $\tilde{H}$ in formula (10.13) is the head at $R = \infty$. Instead of $\ell n \ R$ in the denominator of (10.13), this expression stands

$$\ell n \left(\frac{2}{\alpha}\right) - C = 0.1159 - \ell n \ \alpha \quad , \tag{10.14}$$

depending - through $\alpha$ - upon the intensity of seepage through the poorly pervious soil.

The calculations of A. N Mjatiev [9] have shown that $Q_*$ is closer to $Q$ the closer $R$ is to $\delta$. When water is pumped from a well, observation wells are established at some distances $r_1, r_2, \ldots, r_n$ from this well, and the corresponding drawdowns $S_1, S_2, \ldots S_n$ are observed. If one has at least two pairs of values $(r, S)$, then one may compute the quantities $\alpha$ and kM. If one considers that $\lambda_{n-1} = 0$, then one may compute $\lambda_n$ from formula (8.7) (8.7), knowing $\mu_n$.

As one of his examples, A. N. Mjatiev [9] considers the case that is sketched in Fig. 280.

We notice that the heads $H_o$, $H_1$, $H_2$, $H_3$ in the initial state of the the layers, before pumping started, increase with depth. This shows that in the region where the experiments were run, upward seepage flows must take place.

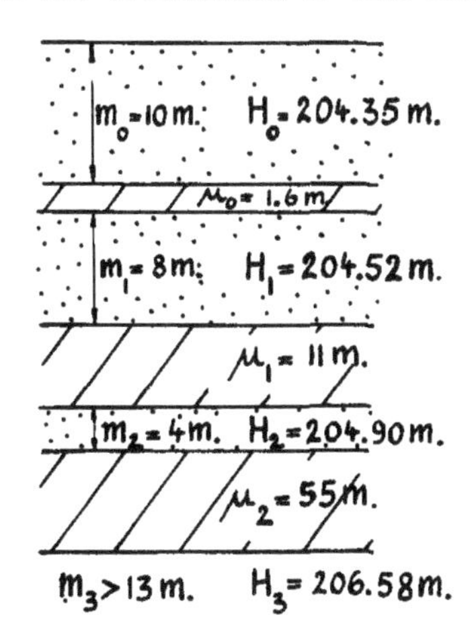


Fig. 280

Waterbearing horizontal layers have been tested by means of successively deeper penetrating wells and in which all except one layer were insulated.

In the stratum which we characterize by the index 1, two pumping tests were made, and the steady state in each of these tests gave results, represented in Table 14.

TABLE 14. RESULTS FROM TWO PUMPING TESTS FROM FOUR WELLS

| No. of well | $H_1$ (m) | r(m) | Test I $Q = 0.01245 m^3/sec$ S(m) | Test II $Q = 0.0178 m^3/sec$ S(m) |
|---|---|---|---|---|
| 1 | 204.53 | 0 | 3.15 | 4.90 |
| 3 | 204.52 | 25.00 | 0.27 | 0.37 |
| 5 | 204.50 | 75.00 | 0.18 | 0.25 |
| 7 | 204.53 | 250.00 | 0.13 | 0.19 |

Well No. 1 is pumped, the remaining three are observation wells. We make two observations, namely in wells No. 3 and No. 5 in the second pumping test. Inserting the corresponding values of $r$ and $S$ in equation (10.8), we obtain:

$$S_3 = \frac{Q}{2\pi kM\delta}\,\frac{K_0(\alpha r_3)}{\alpha K_1(\alpha\delta)}\ , \quad S_5 = \frac{Q}{2\pi kM\delta}\,\frac{K_0(\alpha r_5)}{\alpha K_1(\alpha\delta)}\ , \tag{10.15}$$

and by dividing them term by term, we have:

$$\frac{K_0(25\alpha)}{K_0(75\alpha)} = 1.48\ .$$

We determine the "coefficient of linkage" $\alpha$: $\alpha = 0{,}00116$. From equality (10.15) we may now find the seepage coefficient

$$k = \frac{QK_0(25\alpha)}{S2\pi M} = 0.00356\ .$$

Then the equation of the drawdown surface of well No. 1 will be:

$$S = 5.60QK_0\ (0.00116r)\ .$$

Inserting in this equation the values of $Q$ from the table, and also the values of $r$ for the observation wells, we compute the corresponding drawdowns, which we then may compare with the observed values.

Table 15 was obtained by A. N. Mjatiev.

TABLE 15. RESULTS OF COMPUTED AND OBSERVED DRAWDOWNS IN TWO PUMPING TESTS

| Pump tests | $Q\ \frac{m^3}{sec}$ | Well No. 3 | | | Well No. 5 | | Well No. 7 | | |
|---|---|---|---|---|---|---|---|---|---|
| | | $S_{obs}$ | $S_{comp}$ | $\Delta S = S_{obs} - S_{comp}$ | $S_{obs}$ | $S_{comp}$ | $S_{obs}$ | $S_{comp}$ | $\Delta S$ |
| I | 0.01245 | 0.27 | 0.26 | 0.01 | 0.18 | 0.18 0.00 | 0.13 | 0.10 | 0.03 |
| II | 0.0178 | 0.37 | 0.37 | 0.00 | 0.25 | 0.25 0.00 | 0.19 | 0.14 | 0.05 |

$S_{obs}$ stands for $S_{observed}$ $S_{comp}$ stands for $S_{computed}$

In other examples, when the conditions of the developed theory are satisfied, the results of the computations are close to the results of observation, i.e., they lie within the limits of accuracy of the measurement.

We draw the attention to the fact that in any given observation we always may compute two parameters, occuring in equation (10.12): $A = Q/2\pi kM$ and $\alpha$. These parameters characterize the overall properties of the stratum. If some of the quantities upon which $A$ and $\alpha$ depend are known, then it is also possible to compute some others from the thickness of the layers and the seepage coefficients.

In reality, usually none of the quantities $\lambda_i$, $\mu_i$, $k_i$ are known with sufficient accuracy; therefore one may, if $Q$ is known, look upon the quantities $\alpha$ and $kM$ as upon some constants, characterizing on the average the region of movement of the ground waters. Pumping tests offer the possibility to determine these important constants.

We now turn to the case of $n$ fully penetrating wells [8,9], with centers in the points $(x_1, y_1), \ldots, (x_n, y_n)$. We designate by $r_i$ the distance from an arbitrary point $(x, y)$ to the point $(x_i, r_i)$

$$r_i = \sqrt{(x - x_i)^2 + (y - y_i)^2} \quad .$$

The function $K_0(\alpha r_i)$ for any values $(x_i, y_i)$ is the solution of the equation

$$\frac{\partial^2 H}{\partial x^2} + \frac{\partial^2 H}{\partial y^2} - \alpha^2 (H - \tilde{H}) = 0 \quad .$$

Therefore the drawdown under influence of all the wells will be equal to the sum of the drawdowns, called for by the separate wells. By means of formula (10.8) we may write:

$$S = \sum_{i=1}^{n} \frac{Q_i}{2\pi k \delta_i \alpha M} \frac{K_0(\alpha r_i)}{K_1(\alpha \delta_i)} \quad .$$

We notice that in this theory the concept of "radius of influence" of the well is better explained than in Dupuit's theory. Namely, the radius of influence $R_0$ may be determined as the magnitude of the radius of the cylinder outside which borders the drawdown is practically zero, or, in other words, the magnitude of the drawdown $S$ dows not exceed a given value, for example 1 cm; or the radius for which the ratio $S : S_\delta$ does not exceed a well-defined number, say 0.01 [23]

*§11. Interaction of Pervious Strata, Separated By Poorly Pervious Strata.

There exists an interaction between confined layers of ground waters, so that the discharge from one of the layers must have an influence upon the function of the head in the other layers. In order to elucidate to which degree this influence may exist, we examine here the case of two pervious layers.

We return to equation (8.6). We introduce the notations

$$(11.1)\qquad m_n k_n = \beta_n; \quad \frac{\lambda_n}{\mu_n} = \gamma_n$$

$$\frac{\partial^2 H_n}{\partial x^2} + \frac{\partial^2 H_n}{\partial y^2} = \Delta H_n$$

and give $n$ the values 1, 2. We consider $H_0$ and $H_3$ constant; $H_1$ and $H_2$ the heads of the first and second water-bearing layers, bordered by three poorly pervious strata (Fig. 281). By means of equation (8.6), we obtain a system of two equations

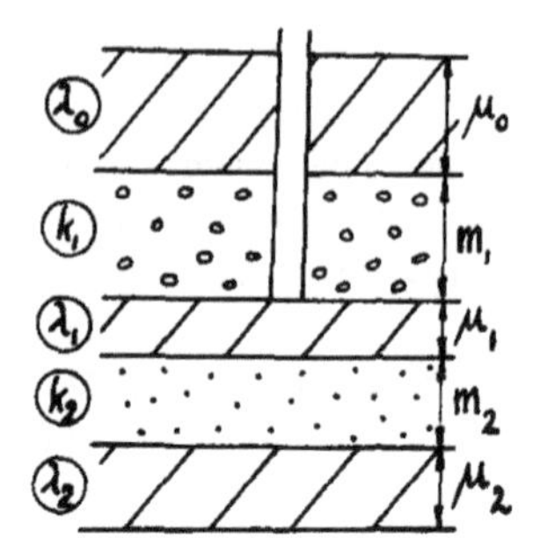


Fig. 281

$$(11.2)\qquad \begin{cases} \beta_1\,\Delta H_1 - \gamma_1(H_1 - H_2) - \gamma_0(H_1 - H_0) = 0 \\ \beta_2\,\Delta H_2 - \gamma_2(H_2 - H_3) - \gamma_1(H_2 - H_1) = 0 \end{cases}$$

or rewritten

$$(11.3)\qquad \begin{cases} \beta_1\,\Delta H_1 - (\gamma_0 + \gamma_1)H_1 + \gamma_1 H_2 = -\gamma_0 H_0 \\ \beta_2\,\Delta H_1 + \gamma_1 H_1 - (\gamma_1 + \gamma_2)H_2 = -\gamma_2 H_3 \end{cases}$$

We observe to start with, that this system has a particualr solution

$$H_1 = H_{10}, \quad H_2 = H_{20},$$

where $H_{10}$, $H_{20}$ are constant. Indeed, in this case $\Delta H_1 = \Delta H_2 = 0$, and the constants are determined by solution of the system of algebraic equations (11.3) in which $\Delta H_1 = \Delta H_2 = 0$

$$-(\gamma_0 + \gamma_1)H_{10} + \gamma_1 H_{20} = -\gamma_0 H_0$$

$$\gamma_1 H_{10} - (\gamma_1 + \gamma_2)H_{20} = -\gamma_2 H_3 .$$

The determinant of this system

$$(11.4)\qquad D = \begin{vmatrix} -\gamma_0 - \gamma_1 & \gamma_1 \\ \gamma_1 & -\gamma_1 - \gamma_2 \end{vmatrix} = \gamma_0\gamma_1 + \gamma_0\gamma_2 + \gamma_1\gamma_2$$

is different from zero, and, consequently, the solution exists always. It is easy to find:

$$H_{10} = \frac{\gamma_0(\gamma_1 + \gamma_2)H_0 + \gamma_1\gamma_2 H_3}{D},$$

$$H_{20} = \frac{\gamma_0\gamma_1 H_0 + \gamma_2(\gamma_0 + \gamma_1)H_3}{D}$$

The obtained values $H_{10}$ and $H_{20}$ give these piezometric water levels that must remain in the strata 1 and 2 for given $H_o$ and $H_3$, if the groundwaters are in constant equilibrium.

To find the general solution of the system (11.3), one must add the general solution of the homogeneous system

$$(11.6) \quad \begin{cases} \beta_1 \Delta H_1 - (\gamma_o + \gamma_1)H_1 + \gamma_1 H_2 = 0 \\ \beta_2 \Delta H_2 + \gamma_1 H_1 - (\gamma_1 + \gamma_2)H_2 = 0 \end{cases}$$

to the particular solution that has been found.

In the consideration of the problem about the circular well, having its axis along the z-axis, as in the previous case, we understand by $\Delta H$ the expression

$$\Delta H = \frac{d^2H}{dr^2} + \frac{1}{r}\frac{dH}{dr} .$$

We look for a solution of system (11.6) in the form

$$H_1 = AK_o(\omega r) + CI_o(\omega r)$$

$$H_2 = BK_o(\omega r) + DI_o(\omega r) .$$

Here $\omega$, A, B, C, D are thus far unknown constants, $K_o$ and $I_o$ are cylindrical functions.

Since we limit the study of the movement to that in an infinite stratum, we put, as in paragraph 10, $C = D = 0$, in order to satisfy the condition of finiteness of the solution at infinity. So, we look for $H_1$, $H_2$ in the form

$$(11.7) \quad H_1 = AK_o(\omega r) , \quad H_2 = BK_o(\omega r) .$$

Remind that

$$\frac{d^2K_o}{dr^2} + \frac{1}{r}\frac{dK_o}{dr} = \omega^2 K_o$$

to obtain

$$\Delta H_1 = A\omega^2 K_o(\omega r) , \quad \Delta H_2 = B\omega^2 K_o(\omega r) .$$

After insertion in (11.6) and elimination of $K_o(\omega r)$, we obtain a system of equations

$$(11.8) \quad \begin{cases} (B_1\omega^2 - \gamma_o - \gamma_1)A + \gamma_2 B = 0 \\ \gamma_1 A + (\beta_2\omega^2 - \gamma_1 - \gamma_2)B = 0 \end{cases}$$

In order to have a solution different from zero for this system of homogeneous equations, it is necessary that the determinant of the system be equal to zero. This gives for $\omega^2$ the equation

$$(11.9) \quad \begin{vmatrix} \beta_1\omega^2 - \gamma_o - \gamma_1 & \gamma_1 \\ \gamma_1 & \beta_2\omega^2 - \gamma_1 - \gamma_2 \end{vmatrix} = 0 .$$

We find two roots of this equation, $\omega_1^2$ and $\omega_2^2$, and obtain two systems of solutions of the form (11.7). And namely, if we insert $\omega = \omega_1$ in (11.8), then we obtain for $A$ and $B$ values, exactly determined up to a constant multiplier, which we designate by $C'$: $A = C'A_1$ $B = C'B_1$. The corresponding particular solution of the system (11.6) will be:

$$H_1^{(1)} = C'A_1K_0(\omega_1 r) \; , \quad H_2^{(1)} = C'B_1K_0(\omega_1 r) \quad . \tag{11.10}$$

In the same way we obtain the solution for the second root,

$$H_1^{(2)} = C''A_2K_0(\omega_2 r) \; , \quad H_2^{(2)} = C''B_2K_0(\omega_2 r) \quad . \tag{11.11}$$

Here $C''$ is an arbitrary constant.

The solution of system (11.3), finite at infinity, is represented by the sum of the solutions (11.5), (11.10), (11.11):

$$H_1 = H_{10} + C'A_1K_0(\omega_1 r) + C''A_2K_0(\omega_2 r) \quad ,$$

$$H_2 = H_{20} + C'B_1K_0(\omega_1 r) + C''B_2K_0(\omega_2 r) \quad .$$

To determine the arbitrary constants one must assign two conditions, for example, give the discharges of the well in the first and second pervious layer or the heads at the walls of the well, and so on.

As an example, we consider the case when the equalities $\beta_1 = \beta_2$, $\gamma_0 = \gamma_1 = \gamma_2$ hold. We put: $\beta_1 = \beta_2 = \beta$, $\gamma_0 = \gamma_1 = \gamma_2 = \gamma$. As is shown by formulas (11.1), the assumed condition means that the pervious layers have a unique value for the product $k_n m_n$, and the poorly pervious layers a unique value for the ratio $\lambda_n/\mu_n$.

In particular, the proposed condition will be satisfied, when both pervious layers have the same thickness $m$ and the same seepage coefficient $k$, and when all the poorly pervious layers are perfectly identical, i.e., have the same thickness and same perviousness. In this case the characteristic equation for $\omega^2$ has the form

$$\begin{vmatrix} \beta\omega^2 - 2\gamma & \gamma \\ \gamma & \beta\omega^2 - 2\gamma \end{vmatrix} = 0 \quad .$$

Its roots are

$$\omega_1^2 = \frac{3\gamma}{\beta} \; , \quad \omega_2^2 = \frac{\gamma}{\beta} \quad .$$

The system (11.8) for the first of them gives: $\gamma A + \gamma B = 0$, i.e., $B = -A$. Taking $A = 1$, we have $B = -1$, so that

$$H_1^{(1)} = C'K_0\left(\sqrt{\frac{3\gamma}{\beta}}r\right) \; , \quad H_2^{(1)} = -C'K_0\left(\sqrt{\frac{3\gamma}{\beta}}r\right) \quad .$$

For the second root we obtain $-A + B = 0$, $A = B$, i.e., we take $A = B = 1$, so that

$$H_1^{(2)} = C''K_0\left(\sqrt{\frac{\gamma}{\beta}}r\right), \quad H_2^{(2)} = C''K_0\left(r\sqrt{\frac{\gamma}{\beta}}\right) \quad .$$

The particular solution of the non-homogeneous equations of (11.5) is

$$H_{10} = \frac{1}{3}(2H_0 + H_3) ,$$

$$H_{20} = \frac{1}{3}(H_0 + 2H_3) .$$

Thus, the most general solution for the case under consideration, for the condition of finiteness of this solution at infinity, will have the form:

$$(11.12) \quad \begin{cases} H_1 = \dfrac{2H_0 + H_3}{3} + C'K_0\left(r\sqrt{\dfrac{3\gamma}{\beta}}\right) + C''K_0\left(r\sqrt{\dfrac{\gamma}{\beta}}\right) , \\ H_2 = \dfrac{H_0 + 2H_3}{3} + C'K_0\left(r\sqrt{\dfrac{3\gamma}{\beta}}\right) + C''K_0\left(r\sqrt{\dfrac{\gamma}{\beta}}\right) . \end{cases}$$

To obtain additional boundary conditions we assume that from the well with radius $\delta$, fluid is withdrawn so that the discharge from the first layer is equal to $Q_1$, from the second $Q_2$. Since

$$Q_1 = -2\pi\delta km \left.\frac{dH_1}{dr}\right|_{r=\delta} ,$$

and putting

$$\omega_1 = \sqrt{\frac{3\gamma}{\beta}} , \quad \omega_2 = \sqrt{\frac{\gamma}{\beta}} .$$

we find by means of (11.12):

$$(11.13) \quad \begin{aligned} Q_1 &= -2\pi\delta km[C'\omega_1K_1(\omega_1\delta) + C''\omega_2K_1(\omega_2\delta)] , \\ Q_2 &= -2\pi\delta km[-C'\omega_1K_1(\omega_1\delta) + C''\omega_2K_1(\omega_2\delta)] . \end{aligned}$$

($K_1$ is the modified cylindrical function of the second kind and first order).

The last equations suffice to determine $C'$ and $C''$. We make additional particular assumptions.

We assume that water is withdrawn only from the upper horizontal layer, so that $Q_2 = 0$. Then from (11.13) we obtain:

$$C'' = C' \frac{\omega_1}{\omega_2} \frac{K_1(\omega_1\delta)}{K_1(\omega_2\delta)} = C' \sqrt{3} \frac{K_1(\omega_1\delta)}{K_2(\omega_2\delta)}$$

and

$$C' = -\frac{Q_1}{2\pi\delta km(\omega_1 + \sqrt{3}\omega_2)K_1(\omega_1\delta)} ,$$

$$C'' = -\frac{Q_1\sqrt{3}}{2\pi\delta km(\omega_1 + \sqrt{3}\omega_2)K_1(\omega_2\delta)} .$$

Now we may rewrite (11.12) as:

$$(11.14) \quad \begin{aligned} H_1 &= \frac{2H_0 + H_3}{3} - \frac{P}{K_1(\omega_1\delta)} K_0(r\omega_1) - \frac{P\sqrt{3}}{K_1(\omega_2\delta)} K_0(r\omega_2) \\ H_2 &= \frac{H_0 + 2H_3}{3} + \frac{P}{K_1(\omega_1\delta)} K_0(r\omega_1) - \frac{P\sqrt{3}}{K_1(\omega_2\delta)} K_0(r\omega_2) \end{aligned}$$

where

$$P = \frac{Q_1}{2\pi\delta km(\omega_1 + \sqrt{3}\,\omega_2)} \quad (11.15)$$

We notice that the limit values of the head in the first and second layer, are respectively equal to:

$$H_{10} = \frac{2H_0 + H_3}{3}, \quad H_{20} = \frac{H_0 + 2H_3}{3} .$$

They represent average weights of the heads $H_0$ and $H_3$, in which the head of the adjacent layer accounts for the larger weight. Designating the drawdowns of head in each layer respectively by $S_1$ and $S_2$: $S_1 = H_{10} - H_1$, $S_2 = H_{20} - H_2$, rewrite the equations (11.14) in the following form:

$$S_1 = \frac{P}{K_1(\omega_1\delta)} K_0(\omega_1 r) + \frac{P\sqrt{3}}{K_1(\omega_2\delta)} K_0(\omega_2 r) ,$$

$$S_2 = -\frac{P}{K_1(\omega_1\delta)} K_0(\omega_1 r) + \frac{P\sqrt{3}}{K_1(\omega_2\delta)} K_0(\omega_2 r) .$$

We see that the drawdown in the first stratum, from which water is withdrawn, is larger than the drawdown in the lower stratum, from which in the assumption no discharge takes place. We compute, in particular, the drawdown for $r = \delta$ in the vicinity of the well itself:

$$S_1(\delta) = \frac{Q_1}{2\pi\delta km(\omega_1 + \sqrt{3}\,\omega_2)} \left[ \frac{K_0(\omega_1\delta)}{K_1(\omega_1\delta)} + \sqrt{3}\,\frac{K_0(\omega_2\delta)}{K_1(\omega_2\delta)} \right] ,$$

$$S_2(\delta) = \frac{Q_1}{2\pi\delta km(\omega_1 + \sqrt{3}\,\omega_2)} \left[ \frac{\sqrt{3}\,K_0(\omega_2\delta)}{K_1(\omega_2\delta)} - \frac{K_0(\omega_1\delta)}{K_1(\omega_1\delta)} \right] .$$

Knowing one of the quantities $S_1(\delta)$ or $S_2(\delta)$, we may compute the discharge $Q_1$. In the more general case, when water is withdrawn from two soil strata, it would be possible to find $Q_1$ and $Q_2$ from the drawdowns $S_1(\delta)$ and $S_2(\delta)$.

## §12. About the Form of the Piezometric Surface.

We rewrite the equation (8.9) for the piezometric head of the $n^{th}$ stratum $H_n(x, y)$ in the form

$$\frac{\partial^2 H_n}{\partial x^2} + \frac{\partial^2 H_n}{\partial y^2} = \alpha^2 (H_n - \tilde{H}) \quad (12.1)$$

where $\tilde{H}$, according to (8.8) is some average of the values of the head in two pervious layers in the neighborhood of the $n^{th}$ stratum:

$$\tilde{H} = \frac{\frac{\lambda_n}{\mu_n} H_{n+1} + \frac{\lambda_{n-1}}{\mu_{n-1}} H_{n-1}}{\frac{\lambda_n}{\mu_n} + \frac{\lambda_{n-1}}{\mu_{n-1}}} . \quad (12.2)$$

This is the magnitude of the head which prevails in the $n^{th}$ layer.

Equation (12.1) has the particular solution $H_n = \tilde{H}$ corresponding to the state at rest.

We assume that, under the influence of some causes (for example pumping or recharge) motion of groundwater occurs.

We consider two cases:

1st case:

$$H_n < \tilde{H} \tag{12.3}$$

In this case, the total vertical velocity (sum of the components) $w = k_m m_n \zeta_n (H_n - \tilde{H})$ will be directed towards the interior of the stratum. Consequently, the $n^{th}$ layer withdraws water from the other horizontal layers. Since

$$w = \frac{\lambda_{n-1}}{\mu_{n-1}} (H_n - H_{n-1}) + \frac{\lambda_n}{\mu_n} (H_n - H_{n+1})$$

consists of two components of the velocities of percolation through each of the two boundaries of the $n^{th}$ stratum, one of these velocities may also be directed outside of the stratum, but the other velocity component, directed inside, must be larger. Therefore, we have three possibilities, sketched in Fig. 282 by means of arrows of different length.

Equation (12.1) shows that in this case $\Delta H_n < 0$, i.e., the piezometric surface is convex (*) (upward). Therefore, when in the observations in the piezometers the piezometric surface is convex, then we must have: $H_n < \tilde{H}$ and the water-bearing layer must withdraw water. A. N. Mjatiev calls the region, in which the inequality (12.3) holds, the "region of intake" of the $n^{th}$ water-bearing layer. He notices that this region is typical for an inter-river space and ridge, i.e., as the geologists express it, is linked with a "positive form of the relief."

2nd case:

$$H_n > \tilde{H} \tag{12.4}$$

In this case we have a general flow out of the stratum (see three possibilities in Fig. 283).

The surface of piezometric head is concave. A. N. Mjatiev calls such a region the "region of outflow" of the $n^{th}$ layer. He notices that usually such regions go with river valleys, foundation areas and depressions, i.e., are linked with "negative forms" of the relief of the country. In Fig. 284 a scheme of the movement of water in a massif between rivers is given by A. N. Mjatiev. A. I. Silin-Bektchourin [11] commenting on Mjatiev's scheme, notices that it only applies to the upper layers.

Now we assume that the ordinate $\tilde{H}$ of the unperturbed free surface

---

(*) As is known, the Laplace operator $\Delta u$ is proportional to the average curvature of the surface $z = u(x, y)$, and the average curvature is positive for a concave and negative for a convex surface.

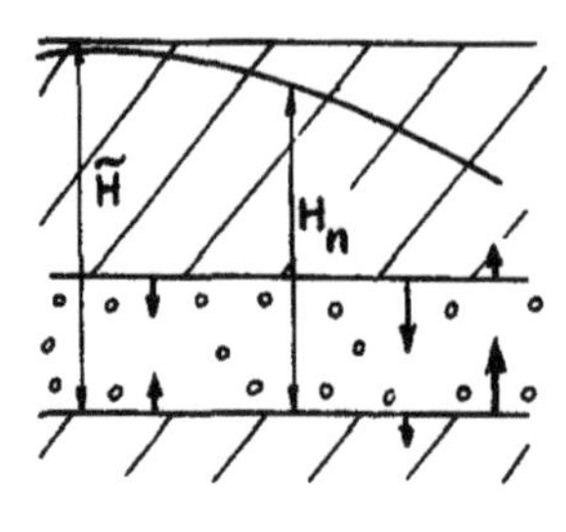


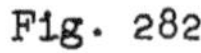
Fig. 282

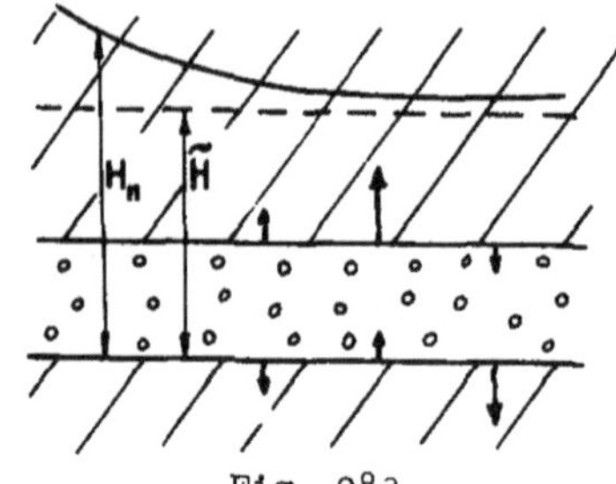


Fig. 283

(i.e., untill the movement starts) is a function of $x, y$

(12.5) $$\tilde{H} = f(x,y)$$

corresponding to some steady movement. Then in the equation

$$\Delta H_n = \frac{\partial^2 H_n}{\partial x^2} + \frac{\partial^2 H_n}{\partial y^2} = \zeta_n\left[H_n - \tilde{H}(x, y)\right]$$

it is convenient to make the substitution

$$H_n - \tilde{H}(x, y) = S \quad .$$

Observing that $\Delta H_n = \Delta\tilde{H} + \Delta S$, we obtain for $S$ :

(12.7) $$\Delta S - \zeta_n S = - \Delta\tilde{H} \quad .$$

This is a nonhomogeneous equation with a known righthand side.

In the particular case, when the initial piezometric surface is a surface of the second order $\tilde{H} = ax^2 + 2bxy + cy^2 + dx + ey + f$, we have $\Delta\tilde{H} = 2(a + c)$, and we obtain a constant for the right side of equation (12.7). We put $-2(a + c) = \zeta$. Then we have instead of (12.7):

(12.8) $$\Delta S - \alpha^2 S = \zeta \qquad (\alpha^2 = \zeta_n) \quad .$$

The substitution $S + \frac{\zeta}{\alpha^2} = \sigma$ reduces equation (12.8) to the homogeneous equation, which we examined previously $\Delta\sigma - \alpha^2\sigma = 0$ .

## § 13. Unconfined Flow to a Well in a Layer With Poorly Pervious Foundation.

Further, in Chapter X, the differential equation for the ordinate $h$ of the free surface (Fig. 285) will be derived:

(13.1) $$\frac{\partial}{\partial h}\left(h \frac{\partial h}{\partial x}\right) + \frac{\partial}{\partial y}\left(h \frac{\partial h}{\partial y}\right) + w + e = 0 \quad .$$

Here $e$ is the amount of precipitation on a unit area per unit time, $w$ the velocity of percolation through the underlying impervious layer:

(13.2) $$w = - \frac{\lambda}{\mu}(h - h_0) \quad ,$$

where $h_0$ is the piezometric level of

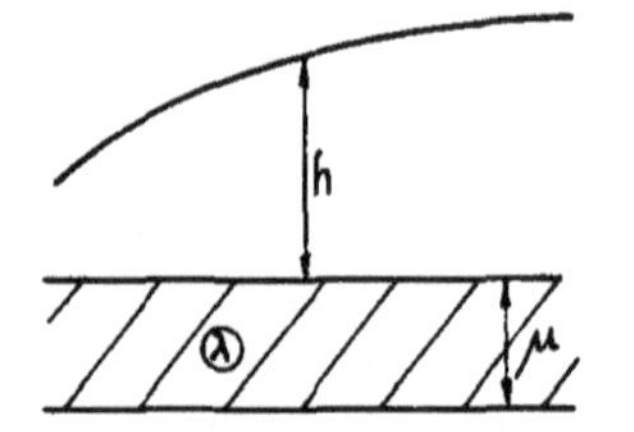


Fig. 285

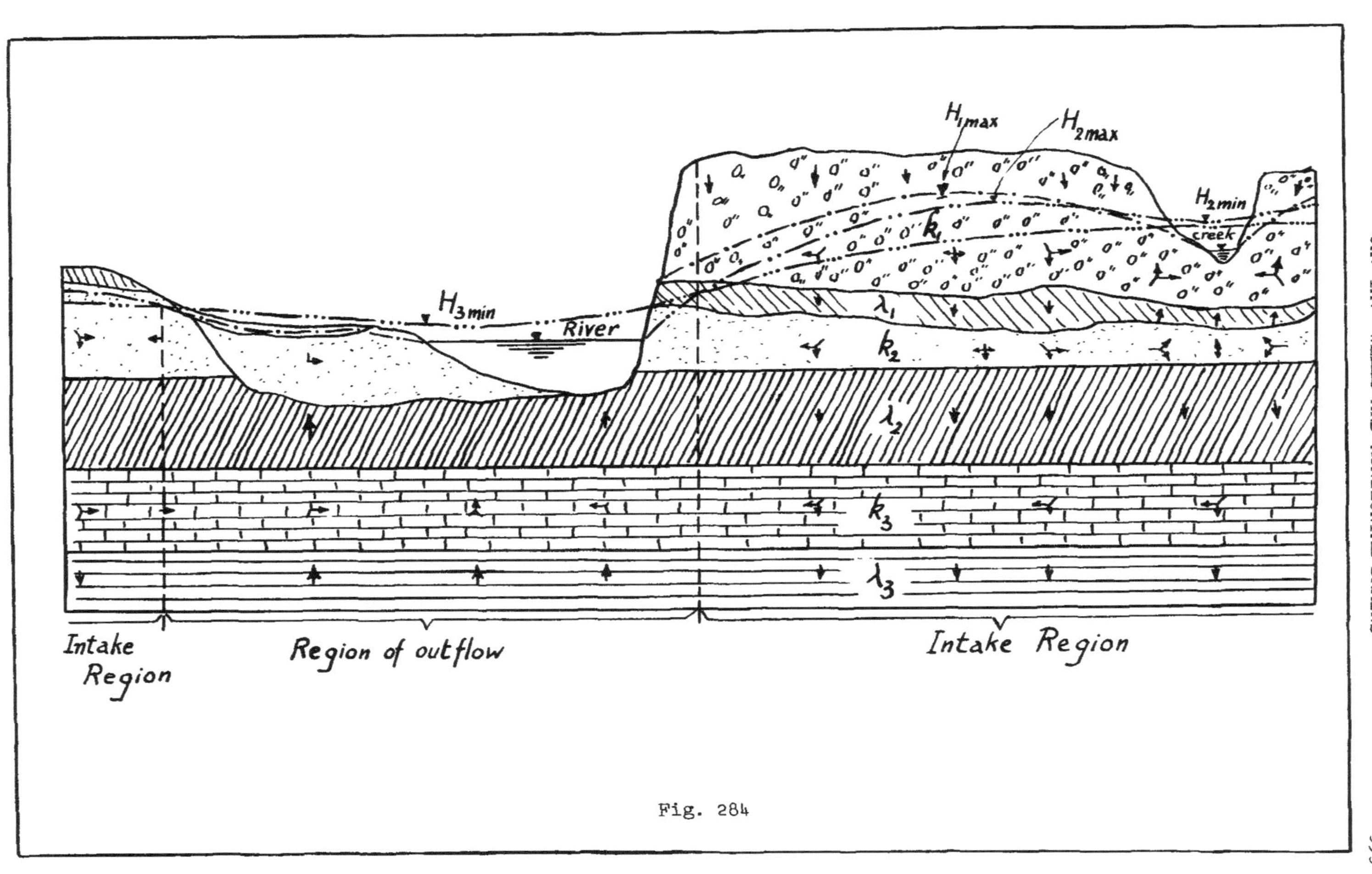
$H_{1max}$
$H_{2max}$
$H_{2min}$
creek
$k_1$
$\lambda_1$
$k_2$
$\lambda_2$
$k_3$
$\lambda_3$
$H_{3min}$
River
Intake Region
Region of outflow
Intake Region

Fig. 284

the lower layer. Equation (13.1) is now rewritten in the form

$$\frac{\partial}{\partial x}\left(h\,\frac{\partial h}{\partial x}\right) + \frac{\partial}{\partial y}\left(h\,\frac{\partial h}{\partial y}\right) - \frac{\lambda}{\mu}\left(h - h_o - \frac{\mu e}{\lambda}\right) = 0 \quad . \tag{13.3}$$

We obtained a nonlinear partial differential equation of the second order. If we make the substitution $h^2 = \mu$, then we obtain a quasilinear equation (i.e., such an equation in which the linear part consists of second derivatives) of the elliptic type [23]

$$\Delta u - \frac{2\lambda}{\mu}\left(\sqrt{u} - h_o - \frac{\mu e}{\lambda}\right) = 0 \quad .$$

In order to obtain a linear equation, we apply a simplification to equation (13.3). Namely, designating by $h_1$ the average depth of the ordinate of the free surface in the region under consideration, we put $h = h_1$ in the multiplier under the differentiation sign. Then we obtain:

$$h_1\left(\frac{\partial^2 h}{\partial x^2} + \frac{\partial^2 h}{\partial y^2}\right) - \frac{\lambda}{\mu}\left(h - h_o - \frac{\mu e}{\lambda}\right) = 0 \quad .$$

This linear equation is of the same form as the corresponding equation in the theory of confined flow that we have analyzed. From this theory we may also borrow the solution of corresponding well problems.

So, by putting

$$H_1 = h_o + \frac{\mu e}{\lambda} \quad , \quad S = h_o + \frac{\mu e}{\lambda} - h = H_1 - h \quad . \tag{13.4}$$

We obtain for the well with radius $\delta$ and water level in the well $h_\delta$:

$$S = \frac{1}{2\pi k\delta\sqrt{\zeta}}\,\frac{Q_\delta}{h_\delta}\,\frac{K_o(r\sqrt{\zeta})}{K_1(\delta\sqrt{\zeta})} \quad . \tag{13.5}$$

The drawdown $S_\delta$ at the well is equal to

$$S_\delta = \frac{1}{2\pi k\delta\sqrt{\zeta}}\,\frac{Q_\delta}{h_\delta}\,\frac{K_o(r\sqrt{\zeta})}{K_1(\delta\sqrt{\zeta})}$$

or

$$S_\delta = \frac{Q_\delta}{h_\delta\varepsilon} \tag{13.6}$$

where by $\varepsilon$ is designated the quantity

$$\varepsilon = 2\pi\delta\sqrt{\zeta}\,\frac{K_1(\delta\sqrt{\zeta})}{K_o(\delta\sqrt{\zeta})} \quad . \tag{13.7}$$

Substituting in (13.6) $h_\delta = H_1 - S_\delta$ we find for $Q_\delta$ the expression $Q_\delta = \varepsilon S_\delta(H_1 - S_\delta)$, representing a polynomial of the second degree depending on the drawdown at the well $S_\delta$.

The function $Q_\delta$ of $S_\delta$ shows that the discharge decreases for $S_\delta > \frac{1}{2}H_1$, which is impossible. From this A. N. Mjatiev [10] drew the conclusion that the drawdown at the well cannot exceed half of the thickness of the water-bearing layer. However, this statement needs verification, since its derivation rests on a series of assumptions, in particular, on the linearization of equation (13.3). If we compute the discharge by Dupuit's formula [formula (8.5) of Chapter X]:

$$Q = A(H_2^2 - H_1^2)$$

$$A = \frac{2\pi k}{\ell n \frac{R}{\delta}} \quad ,$$

Now, considering that $H_2 - H_1 = S_\delta$ where $S_\delta$, the drawdown at the wall of the well, we find: $Q = AS_\delta(2H - S_\delta)$. Thus, in the case of the completely impervious foundation, again a parabolic relation between $Q$ and $S_\delta$ is found, but $S_\delta$ in this relation may be arbitrary, from zero to $H$.

## SOME THREE-DIMENSIONAL PROBLEMS IN THE THEORY OF SEEPAGE

### §14. About Three-dimensional Problems.

At present we do not have any exact solution of three-dimensional problems in the theory of seepage in the presence of a free surface, if we do not consider the single case, described by B. K. Risenkampf and N. K. Kalinin [12] in 1941, but that so far has found no application to whatever practical conditions.

Namely, B. K. Risenkampf and N. K. Kalinin noticed that in the problem of the flow of an incompressible fluid about an ellipsoid with semi-axes a, b, c, in a lateral flow, the velocity potential at the surface of the ellipsoid is a linearfunction of the coordinates. Since at the free surface the ground-water flow must satisfy the condition $\varphi + kz = 0$, for some conditions the surface of a triaxial ellipsoid may be the free surface of a ground-water flow, for which the seepage region must be found outside of the ellipsoid.

Among others, the study of some problems about axisymmetrical flows and three-dimensional flows has an important meaning. We may indicate the wide circle of problems about flow to wells and about interaction of a group of wells, and also the problems about three-dimensional flow about structures, and others.

In the present division we introduce an approximate solution to one axisymmetrical problem and show some formulas for three-dimensional sources, which may be useful in the solution of some three-dimensional problems.

### §15. Partially Penetrating Wells in a Half Space.

A well that does not penetrate to the basis of the pervious layer is called partially penetrating. In a stratum of infinite thickness, a well is always partially penetrating.

Let there be given a well in a stratum, occupying the whole lower half space. We distribute sinks along the axis of the well. The velocity potential for a three-dimensional sink in the point $\zeta$ of the axis of the well is expressed as:

$$\varphi_1 = \frac{q}{4\pi} \frac{1}{\sqrt{(\zeta - z)^2 + r^2}} \tag{15.1}$$

where $r^2 = x^2 + y^2$.

The horizontal plane $z = 0$ is for us a solid wall. Therefore we continue the axis of the well upward and take the integral of expression (15.1) between the limits $-\ell$, $\ell$. We obtain:

$$\varphi = \frac{1}{4\pi} \int_{-\ell}^{\ell} \frac{q(\zeta)\, d\zeta}{\sqrt{(\zeta - z)^2 + r^2}} \,. \tag{15.2}$$

The intensity $q(\zeta)$ of the sinks must be selected so that along the surface of a cylinder with radius $r_o$ the potential has a constant value. Besides this, at the bottom of the cylinder, some given conditions must be satisfied. In such a statement the problems appear to be complicated.

Usually one takes the intensity $q$ constant all along the well [3]. Then the integral (15.2) is easily computed, and we obtain:

$$\varphi(r, z) = \frac{Q}{4\pi\ell} \ln \frac{\sqrt{\left(\frac{\ell - z}{r}\right)^2 + 1} + \frac{\ell - z}{r}}{\sqrt{\left(\frac{\ell + z}{r}\right)^2 + 1} - \frac{\ell + z}{r}} \tag{15.3}$$

where $Q = \ell q$ - discharge of the well.

The surfaces of equipotential here will be ellipsoids of revolution with the foci in the points $(0, 0, \ell)$ and $(0, 0, -\ell)$. Consequently, the surface of the well is substituted for a narrow ellipsoid. We may select the semi-axes of this ellipsoid so that the volume of the ellipsoid is equal to the volume of the cylinder. We obtain

$$Q \approx \frac{2\pi k \ell}{\ln \frac{1.6\ell}{r_o}} S_o \,. \tag{15.4}$$

Here $S_o = -\varphi_o/k$ - drawdown at the well.

Making $z = 0$ in formula (15.3), we obtain the distribution of the potential along the roof of the stratum

$$\varphi(r, 0) = \frac{Q}{2\pi\ell} \ln \frac{\sqrt{\ell^2 + r^2} + \ell}{r} \,. \tag{15.5}$$

The potentials of the linear sources, which we distribute inside of the wells, may be summed up. It is possible, in particular, by this method to compute the distribution of the head along the roof of the stratum for several wells. In Fig. 286, some drawdown curves are constructed for the operation of one and two wells. If we pass on to unconfined flow, considering that the plane $z = 0$ was the surface of the groundwater before the construction of the well, we may assume that after construction of the well the free surface is determined by the equation

$$z = -\frac{\varphi(r, 0)}{k},$$

i.e., assume that the depressed free surface coincides with the surface of piezometric head of the confined flow.

In the case of a pervious layer of finite depth, we substitute $h = -\varphi/k$ of the confined flow by $h^2$ in the transition to unconfined flow. In the case of infinite depth we dispose of a more rational assumption than the one made here. It is justified by the fact that as the pervious layer deepens, the lines of the free surface become more gently sloping.

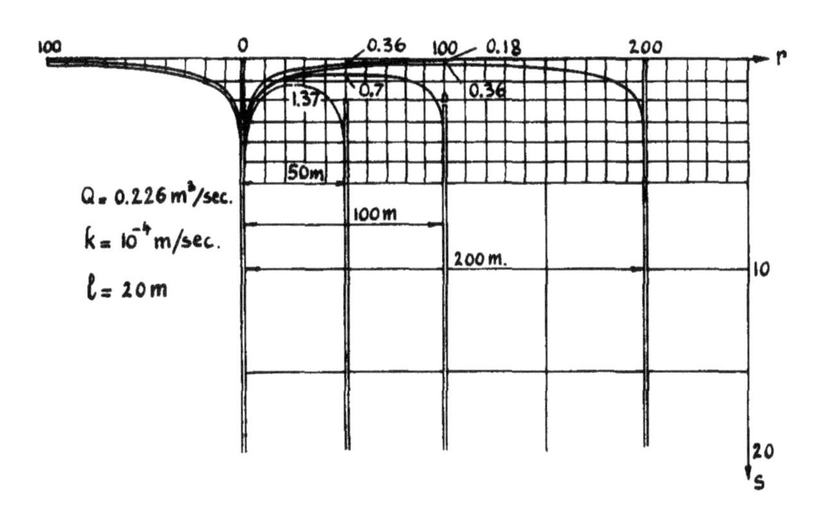


Fig. 286

Returning to the method of construction of the velocity potential by means of sinks, distributed along the axis of the well, we notice that in order to obtain a more accurate solution it would be possible to give a different distribution of the intensity. It is evident that $q(\zeta)$ must increase in the neighborhood of the ends of the segment $(-\ell, \ell)$. In the corresponding plane problem, the distribution of $q(\zeta)$ has the form:

$$q(\zeta) = \frac{q_0}{\sqrt{\ell^2 - \zeta^2}} . \tag{15.6}$$

If we take the same law for the three-dimensional case, then an approximate solution in the form of an elliptical integral is obtained, the more accurate, the smaller the radius $r_0$ of the well. If to expression

(15.6) we add an entire polynomial in $\zeta$, then it is possible to select an even more accurate solution. Before going over to the incompletely penetrating well in a stratum of finite depth, we introduce some formulas for sources in a stratum bordered by horizontal planes.

§16. Source Between Two Horizontal Planes.

Two basic cases offer an interest: when the source is found in a region bordered by two parallel impervious planes, and when one of the boundary planes is impervious but the other is a plane of equipotential.

Let the sink be placed in the point M, which has the coordinates $(0, 0, \zeta)$. For the first problem we may distribute sinks in the points, symmetrical with the given point with respect to both boundary planes, $z = 0$ and $z = c$, and then continue the reflection, going over to the planes $z = nc$, $z = -nc$. Assuming the same intensity for all the sources and writing the series for the velocity potential,

$$\varphi = -\frac{q}{4\pi}\left\{\frac{1}{\sqrt{r^2 + (z-\zeta)^2}} + \frac{1}{\sqrt{r^2 + (z+\zeta)^2}} + \sum_{n=-\infty}^{\infty}\left[\frac{1}{\sqrt{r^2 + (2nc + z - \zeta)^2}} + \frac{1}{\sqrt{r^2 + (2nc + z + \zeta)^2}}\right]\right\}$$

we see that the series diverges. M. Muskat [14] gives expressions applicable for small values of $r$, and others, suitable for large values of $r$.

Here we introduce expressions in the form of definite integrals of two functions: (1) the function $U(r, z, \zeta)$, turning to zero in two planes $z = \pm c$ and having in the point $z = \zeta$ a singularity of the source type, so that

$$U - \frac{1}{\sqrt{r^2 + (z-\zeta)^2}}$$

is a regular harmonic function in the vicinity of the point $(x, y, \zeta)$. (2) the function $V(r, z, \zeta)$, for which the normal derivative is zero at the walls, i.r., $\partial V/\partial z = 0$ for $z = \pm c$. In the point $z = \zeta$, the character of singularity is the same as that of the function.

We have, putting

$$U = \int_0^\infty g(z, \zeta) J_0(\ell r)\, d\ell\,, \quad V = \int_0^\infty \gamma(z, \zeta) J_0(\ell r)\, d\ell$$

where $J_0(x)$, the cylindrical function of zero order:

$$g(z, \zeta) = \begin{cases} \dfrac{2\,\mathrm{sh}\,\ell(c - z)\,\mathrm{sh}\,\ell(c + \zeta)}{\mathrm{sh}\ 2\ell c} & (\zeta \le z \le c)\,, \\[2ex] \dfrac{2\,\mathrm{sh}\,\ell(c - \zeta)\,\mathrm{sh}\,\ell(c + z)}{\mathrm{sh}\ 2\ell c} & (-c \le z \le \zeta)\,, \end{cases}$$

$$\gamma(z, \zeta) = \begin{cases} \dfrac{2\,\mathrm{ch}\,\ell(c - z)\,\mathrm{ch}\,\ell(c + \zeta)}{\mathrm{sh}\ 2\ell c} & (\zeta \le z \le c)\,, \\[2ex] \dfrac{2\,\mathrm{ch}\,\ell(c - \zeta)\,\mathrm{ch}\,\ell(c + z)}{\mathrm{sh}\ 2\ell} & (-c \le z \le \zeta)\,. \end{cases}$$

We may represent U and V in another form, isolating from them the term

$$\int_0^\infty e^{-\ell|z - \zeta|} J_o(r\ell)d\ell = \frac{1}{\sqrt{r^2 + (z - \zeta)^2}} = \frac{1}{R} .$$

In other words, we put:

$$g(z, \zeta) = e^{-\ell|z - \zeta|} + h(z, \zeta) , \quad \gamma(z, \zeta) = e^{-\ell|z - \zeta|} + k(z, \zeta) ,$$

where

$$h(z,\zeta) = \frac{e^{-2\ell c}ch\ell(z-\zeta) - ch\ell(z+\zeta)}{sh\ 2\ell c} , \quad k(z,\zeta) = \frac{e^{-2\ell c}ch\ell(z-\zeta) + ch\ell(z+\zeta)}{sh\ 2\ell c}$$

remain continuous. Now we have

$$U = \frac{1}{R} + \int_0^\infty h(z, \zeta)J_o(r\ell)d\ell , \quad V = \frac{1}{R} + \int_0^\infty k(z, \zeta)J_o(r\ell)d\ell .$$

Then the potentials

$$U_1 = \int_0^\infty h(z, \zeta)J_o(r\ell)d\ell , \ V_1 = \int_0^\infty k(z, \zeta)J_o(r\ell)d\ell$$

remain continuous for $z = \zeta$, and so do the derivatives $\partial U_1/\partial z$, $\partial V_1/\partial z$.

It is possible to check that all the conditions set up for U and V are satisfied and, consequently, U and V are the unknown Green's functions, for a source between solid walls or planes of equal head [24].

If in the points $z = \zeta$ and $z = -\zeta$, symmetrical with respect to the plane $z = 0$, we distribute a source and sink of equal intensity in the case 1 (or sources of the same sign in case 2), then we obtain a source between two planes. One of these planes is a flow plane, the other a plane of equal potential. We may construct a solution for the last problem also in the form of a series, which, being alternating, will converge. This series has the form

$$V = \sum_{n=-\infty}^{\infty} (-1)^n \left[ \frac{1}{\sqrt{r^2 + (z + 2nc - \zeta)^2}} - \frac{1}{\sqrt{r^2 + (z + 2nc + \zeta)^2}} \right]$$

($z = 0$ is an impervious plane; $z = c$ is an equipotential surface). For $r < c$ we may use this series, taking into account some of its first terms. For large r we may take the development into a trigonometric series in $\sin m\pi z/2c$ $(m = 2n + 1)$:

$$V = a_1 \sin \frac{\pi z}{2c} + a_3 \sin \frac{3\pi z}{2c} + a_5 \sin \frac{5\pi z}{2c} + \dots$$

The coefficients of this series are calculated as:

$$a_m = \frac{1}{c} \int_0^{2c} \sin \frac{m\pi z}{2c} \sum (-1)^n \left[ \frac{1}{\sqrt{r^2 + (z+2nc-\zeta)^2}} - \frac{1}{\sqrt{r^2 + (z+2nc+\zeta)^2}} \right] dz .$$

If we put

$$z + 2nc - \zeta = \tau , \quad z + 2nc + \zeta = \tau$$

then, changing in each term the limits of integration, we find:

$$a_m = \frac{1}{c} \int_{-\infty}^{\infty} \left[ \sin \frac{m\pi}{2c}(\tau + \zeta) - \sin \frac{m\pi}{2c}(\tau - \zeta) \right] \frac{d\tau}{\sqrt{r^2 + \tau^2}} =$$

$$= \frac{2}{c} \sin \frac{m\pi\zeta}{2c} \int_{-\infty}^{\infty} \frac{\cos \frac{m\pi\tau}{2c} d\tau}{\sqrt{r^2 + \tau^2}} .$$

But we have the equality

$$K_o(z) = \int_0^{\infty} \frac{\cos u \, du}{(u^2 + z^2)^{\frac{1}{2}}}$$

where $K_o(z)$ - the cylindrical function of zero order and of the second kind. Therefore, we may finally write:

$$V = \frac{4}{c} \sum_{n=o}^{\infty} \sin \frac{(2n + 1)\pi z}{2c} \sin \frac{(2n + 1)\pi\zeta}{2c} K_o \left( \frac{(2n + 1)\pi r}{2c} \right) .$$

We notice that for large $z$, the asymptotic representation of the function holds

$$K_o(z) \approx \sqrt{\frac{\pi}{2}} e^{-z} .$$

The function $V$ may be used [15] in the solution of problems to determine the seepage coefficient by means of pumping tests.

## §17. Partially Penetrating Well in a Stratum of Finite Depth.

For the partially penetrating well in a stratum of finite thickness (Fig. 287, a), one must look for a solution by means of point sinks, distributed along the axis of the well. Their intensity must be so selected that at the surface of the well the velocity potential has a constant value (or a value given after some law). The conditions at the top and bottom of the stratum are usually satisfied by the composition of the velocity potential of each sink. An exact solution of this problem is thus far not available. M. Muskat has given an approximate solution, obtained by means of a convenient selection of the intensities. His formula for the discharge of a well with radius $r$, and length $b$ in a pervious layer of thickness $h$ has the form:

$$(17.1) \qquad Q = \frac{2\pi k b \frac{\Delta p}{\gamma}}{\frac{1}{2}\left\{2 \ln \frac{4h}{r_o} - \ln \frac{\Gamma(0,875 \bar{h}) \Gamma(0,125 \bar{h}}{\Gamma(-0,875 \bar{h}) \Gamma(-0, 25 \bar{h})}\right\} - \bar{h} \ln \frac{4h}{R}} \left(\bar{h} = \frac{b}{h}\right)$$

where $\Gamma(a)$ designates the gamma function, $R$ the distance to the "intake contour", where $p = p_k$. For a group of partially penetrating wells, I.A. Tcharnii gave an approximate method to compute the discharge. He assumed that at a certain distance from the partially penetrating well the flow may be considered to be plane-parallel (this distance is roughly equal to the thickness of the layer). The author chose a cylindrical surface, which we may consider to be a surface of equal potential. Then Muskat's formula may be applied to each well. The potentials at the cylinders chosen around each well are determined by a system of linear equations, in which enter the discharges of fictitious "wells."

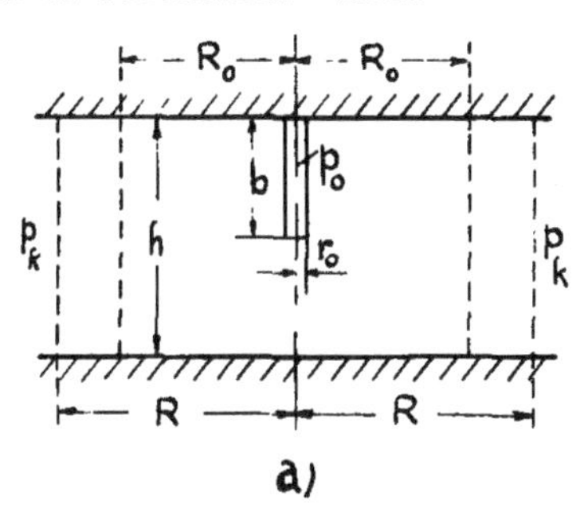


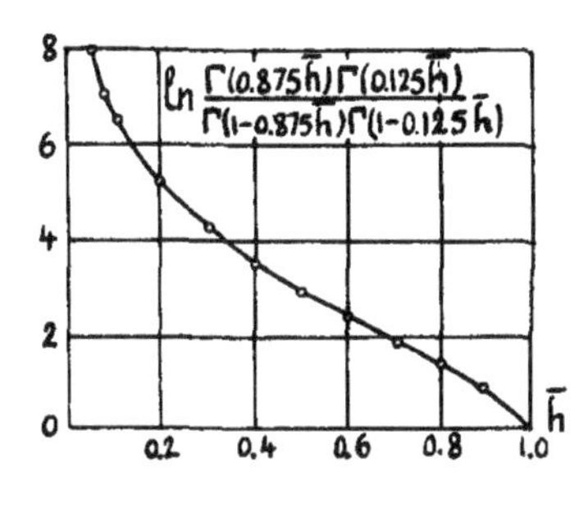


Fig. 287

In Fig. 287, b, a graph is given showing the dependence upon $\bar{h} = \frac{b}{h}$ of an auxiliary quantity, occurring in formula (17.1).

B. I. Segal [17] examines a three-dimensional grid of partially penetrating wells. He distributes along the axes of the wells a line of sinks. Their intensity is selected in the form of a polynomial of the second degree depending upon the ordinates of the places of the sinks. Besides this he distributes a point source at the lower end of the well. The intensity of the latter, and also the coefficients of the polynomial are selected by means of computations.

## §18. The Problem of Recharge of Water in the Soil.

If we put a pipe with impermeable walls and open bottom sufficiently deep in the soil and then fill it with fluid under constant pressure, then a movement with free surface and constant discharge [18] is established after sometime elapses. The velocity potential is found [18] in the form of a sum (the x-axis is oriented vertically downward)

$$\varphi = -\frac{Q}{4\pi\rho} + kx \quad . \tag{18.1}$$

Here

$$\rho = \sqrt{x^2 + y^2 + z^2}$$

is the distance of the point $(x, y, z)$ in the flow region from the point source with intensity $Q$, distributed in the coordinate origin, $k$ the seepage coefficient.

The streamfunction that corresponds to the velocity potential (18.1) is

$$\psi = \frac{kr^2}{2} - \frac{Q}{2\pi}\cos\theta \tag{18.2}$$

$$(r^2 = y^2 + z^2 \ , \quad \theta = \text{arc cos}\ \frac{x}{\rho}) \ .$$

One of the flow surfaces is the so-called "Prandtl's half body" which was taken [18] as a free surface in the study of the flow. However, this surface is not a free surface of the ground water flow, since at the latter the condition of constant pressure must be satisfied

$$\frac{p}{\gamma} = \varphi - kx = \text{const} \ . \tag{18.3}$$

But to satisfy the condition (18.3), we obtain from equation (18.1) spherical surfaces, i.e., the isobar surfaces are spheres $\rho$ = constant.

Instead of this V. M. Nasberg [19] gives a solution, still approximate but more accurate, namely in the right side of (18.1) he adds the term

$$-\frac{Q_1}{\sqrt{(b + x)^2 + r^2}}$$

corresponding to a sink in the point (-b, 0, 0) of intensity $Q_1$. V. M. Nasberg selects $Q_1$ and b so that in the points A and B of Fig. 288 the condition $\varphi - kx = 0$ is exactly satisfied. Therefore these values of $Q_1$ and b depending upon the discharge Q are obtained:

$$Q_1 = 3.50Q \ , \ b = 5.1 \sqrt{\frac{Q}{4\pi k}} \ .$$

The ordinate of the point A and the abscissa of the point B appear to have these values

$$x_A = 0.320 \sqrt{\frac{Q}{k}} , \quad y_B = 0.429 \sqrt{\frac{Q}{k}} \ .$$

Finally we obtain:

$$\varphi = -\frac{Q}{4\pi}\left[\frac{1}{\sqrt{x^2 + r^2}} + \frac{3.50}{\sqrt{(b + x)^2 + r^2}}\right] + kx \ ,$$

$$\psi = \frac{kr^2}{2} - \frac{Q}{4\pi}\left[\frac{x}{\sqrt{x^2 + r^2}} - \frac{3.50(x + b)}{\sqrt{(b + x)^2 + r^2}}\right] \quad (b = 5.1\sqrt{\frac{Q}{4\pi k}}) \ .$$

The equation of the streamline DAC is:

$$r^2 - \frac{Q}{2\pi k}\left[\frac{x}{\sqrt{x^2 + r^2}} - \frac{3.50(x + b)}{\sqrt{(x + b)^2 + r^2}} + 4.50\right] = 0 \ .$$

This confirms the assumption [18], that the upper part of the free surface DAC (see Fig. 288) approaches an ellipsoid of revolution. At infinity we obtain

$$r_\infty = \sqrt{\frac{Q}{\pi k}} .$$

In Fig. 288, the lines of equal head are constructed. The letter "a" designates the quantity

$$a = \sqrt{\frac{Q}{4\pi k}} \approx 0.320 \sqrt{\frac{Q}{k}} .$$

With this we conclude the brief description of the available results of three-dimensional problems.

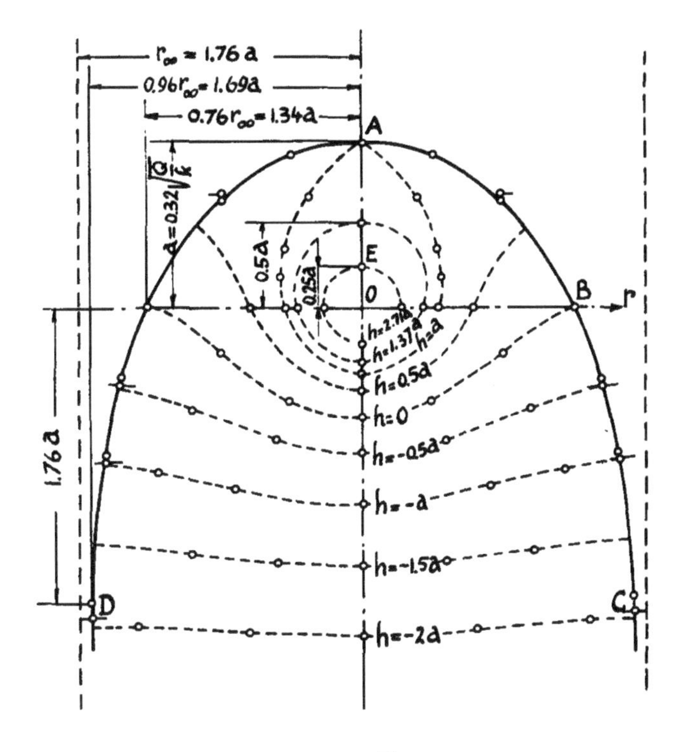


Fig. 288

We notice that we dispose of the solution of a group of spatial problems with two dimensions, which may be applied to the case of wells in an inclined stratum. This represents an interest mainly in the theory of oil flow; these are problems about sinks and sources at the surfaces [20], [21].

CHAPTER X.

# HYDRAULIC THEORY OF STEADY FLOWS

## A. Unconfined Flow in Strata on Bedrock.

### §1. Hydraulic Theory and its Basic Assumptions.

We understand by "hydraulic theory" the theory in which the flow is averaged over the depth. By doing so, the number of dimensions of the investigated flow is decreased: a plane problem is reduced to a one dimensional, i.e., a problem in which the flow elements only depend upon one coordinate $x$ if the flow takes place in the $xz$ plane, and flow in three dimensional space $x,y,z$ is reduced to two-dimensional, in coordinates $x,y$.

In the "hydrodynamic theory" all dimensions – two or three – are considered, according to the problem. Some kind of simplification may be introduced, for example, in the boundary conditions, etc.

In the modern development of the hydraulic theory, one has to make use of a complicated mathematical apparatus (see, for example, Chapter IX, §§8 - 13) to solve some problems; therefore it is not yet possible to oppose the hydraulic theory as a simple theory against the hydrodynamic theory. Comparison of hydraulic and hydrodynamic solutions for a series of problems shows that the hydraulic theory gives good results in many cases.

In computations of seepage through earth dams, methods are applied which could be called semi-hydraulic: the dam is divided into sections for which the average flowrate is computed by means of Dupuit's formula, but the sections adjacent to head – and tailwater reservoirs are computed by various artificial methods.

The free surface of groundwater flow is slightly curved, when flow takes place over a large area, such as in seepage from a canal to a river, in irrigation, etc. Such flow may cover an area, measured in square kilometers, while the depth of flow is measured in meters or decameters and to a large extent – except for the regions immediately adjacent to the canal or river – varies only a fraction of a meter.

At the boundaries of the flow region, i.e., near rivers and canals, the free surface may have a strong curvature, and consequently the flow under a canal has a three-dimensional character. In these parts the flow must be investigated by means of more accurate methods.

The hydraulic theory here exposed applies to these regions of flow, where variations from point to point are slight and where the velocity is almost horizontal. This theory was already used in the study of confined flows (Chapter IX, §8). Here we treat unconfined flows.

We are reminded of the condition that must be satisfied at the free surface: that of constant pressure, i.e., its equality with atmospheric pressure (or capillary).

The relationship between velocity potential and pressure has the form (see Chapter I, §11):

$$\varphi = -k\left(\frac{p}{\rho g} + z\right) \quad . \tag{1.1}$$

The velocity potential is proportional with the head

$$\varphi = -kh \tag{1.2}$$

where

$$h = z + \frac{p}{\rho g} \tag{1.3}$$

in which $h$ is a function of the coordinates $x, y, z$. Considering atmospheric pressure to be zero, we obtain from (1.3) the free boundary condition

$$h(x,y,z) = z \tag{1.4}$$

or from (1.2)

$$\varphi(x,y,z) + kz = 0 \ .$$

If we consider the free surface of the groundwater flow to be slightly curved, it means that $z$ slightly varies, and oscillates about some average value $\bar{z}$. We develop $h$ into a power series of $z - \bar{z}$:

$$h(x,y,z) = h(x,y,\bar{z}) + \left(\frac{\partial h}{\partial z}\right)_{z=\bar{z}} \times (z - \bar{z}) + \ldots$$

We consider the vertical flow velocities to be very small, i.e., we assume that the quantity $\partial h/\partial z$, proportional to the vertical velocity, is small and we neglect the product of this quantity with the small quantity $z - \bar{z}$. In other words, we replace the real head $h(x,y,z)$ by the quantity $h(x,y,\bar{z})$ depending on the seepage depth of the flow $\bar{z}$, but not on the coordinate $z$. We call it $h(x,y)$:

$$h = h(x,y) \ .$$

This means that we consider the head constant along each vertical, or, in other words, we assume that the equipotential surfaces are vertical cylinders [1]. In plane flow, the equipotential lines are vertical lines. (Further we see that we may introduce some corrections to this basic assumption of the hydraulic theory.)

Now we may consider equation (1.4) to be the equation of the free surface:

$$z = h(x,y) \ . \tag{1.5}$$

[We replace $h(x,y,z)$ by the average value over the depth $h(x,y)$].

Thus, in the hydraulic theory, the head in any section is equal to the depth of water in this section.

We derive the equation of continuity for flow with a slightly varying free surface.

The horizontal velocity components $u$ and $v$ are expressed in function of the head $h(x,y)$ as

$$u = -k\frac{\partial h}{\partial x}, \quad v = -k\frac{\partial h}{\partial y}. \tag{1.6}$$

We isolate in the fluid a prism, bounded above by the free surface, below by an horizontal surface $xy$, lying in an impervious plane, and on the sides by vertical planes standing on a rectangle $dx, dy$ in the base plane (Fig. 239).

The flowrate through the left side of this body is equal to area $h\,dy$ times velocity $u$. We designate by $q_x$ the flowrate in the direction of the x-axis (per unit length in the y-axis direction). Then the flowrate through the left side of the prism is

$$q_x dy = -k\left(h\frac{\partial h}{\partial x}\right)_x dy. \tag{1.7}$$

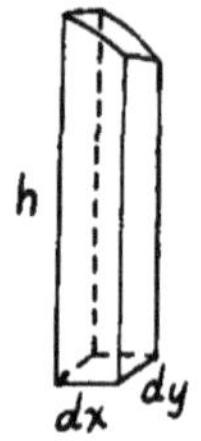

Fig. 289

Here $h$ and $\partial h/\partial x$ are taken in point $(x,y)$. The flowrate through the right side of the prism is equal to the same expression (1.7) in which we only replace $x$ by $x + dx$:

$$q_{x+dx}\, dy = -k\left(h\frac{\partial h}{\partial x}\right)_{x+dx} dy. \tag{1.8}$$

The difference between the amount of fluid flowing through the area at $x + dx$ per unit time and that flowing through the area at $x$, is $(q_{x+dx} - q_x)\,dy$. By Lagrange's formula

$$f(x + dx) - f(x) = f'(\xi)\,dx$$

(where $\xi$ is some number between $x$ and $x + dx$). The change of flowrate in the direction of the s-axis is expressed as

$$(q_{x+dx} - q_x)\,dy = \frac{\partial q_x}{\partial x}dx\,dy = -k\frac{\partial}{\partial x}\left(h\frac{\partial h}{\partial x}\right)dx\,dy. \tag{1.9}$$

In exactly the same way, we find the change of flowrate in the y-axis:

$$\frac{\partial q_y}{\partial y}dx\,dy = -k\frac{\partial}{\partial y}\left(h\frac{\partial h}{\partial y}\right)dx\,dy. \tag{1.10}$$

In expressions (1.9) and (1.10) one must consider the independent variables to be $\xi, \eta$, respectively between $x$ and $x + dx, y$ and $y + dy$. In the passage to the limit, when $dx \to 0$, $dy \to 0$, we obtain the values $(x,y)$ instead of $(\xi, \eta)$.

In the case of steady incompressible flow, the total change of flowrate through the sides of the prism, equal to the sum of expressions (1.9) and (1.10), may be compensated by the inflow or outflow of fluid through the free surface or the base of the stratum.

Rain or irrigation water may hit the free surface, this phenomenon is called "infiltration";"evaporation" causes water to leave the free surface. Slow percolation may take place through the base of the stratum, if the underlying stratum is not completely impervious (this is often the case), but weakly pervious. This case is further examined in this chapter. Let there be a flowrate $e$ on top and a flowrate $w_0$ at the bottom of the prism per unit area (horizontal projection of the free surface).

The respective flowrates through $dx\,dy$ are

$$e\,dx\,dy, \quad w_0\,dx\,dy \;. \tag{1.11}$$

Adding expressions (1.9), (1.10) and equating them with the sum of (1.11), we find after division by $dx\,dy$

$$\frac{\partial q_x}{\partial x} + \frac{\partial q_y}{\partial y} = e + w_0 \,.$$

or

$$k\left[\frac{\partial}{\partial x}\left(h\,\frac{\partial h}{\partial x}\right) + \frac{\partial}{\partial y}\left(h\,\frac{\partial h}{\partial y}\right)\right] + e + w_0 = 0 \quad . \tag{1.12}$$

The quantity $e$ is positive when water reaches the flow region, i.e., in the case of infiltration or when the intensity of infiltration exceeds that of evaporation, and negative in the case of evaporation or evaporation exceeding infiltration. The vertical velocity $w_0$ is positive, when water percolates from lower soil upward, and negative in the opposite case.

If the base of the soil is impervious and in the absence of evaporation and infiltration, $e = w_0 = 0$, and equation (1.12) may be written as

$$\frac{\partial^2(h^2)}{\partial x^2} + \frac{\partial^2(h^2)}{\partial y^2} = 0 \;. \tag{1.13}$$

In this case function $h^2$ satisfies Laplace's equation.

## §2 Plane Flow in Stratum on Horizontal Impervious Base.

We analyze groundwater flow in the $xz$ plane (Fig. 290).

Equation (1.13) becomes in this case:

$$\frac{\partial^2(h^2)}{\partial x^2} = 0 \;. \tag{2.1}$$

Integration gives the parabola ($A$ and $B$ are arbitrary constants)

$$h^2 = Ax + B \;. \tag{2.2}$$

We may also arrive at this parabola, starting from the expression of the flowrate $q$ through section $x$ (per unit width of flow in a direction perpendicular to the plane of the paper , Fig. 290)

$$q = -\,kh\,\frac{dh}{dx} \;. \tag{2.3}$$

Integrating this equation and inserting the coordinates of points $(x_1, h_1)$ and $(x_2, h_2)$ we find

$$q = -\frac{k(h_2^2 - h_1^2)}{2(x_2 - x_1)} \quad . \tag{2.4}$$

Formula (2.4) is called "Dupuit's formula" [1].

We have seen that the same formula is obtained in the exact solution of two-dimensional flow to a horizontal drain (Chapter II, §10). In §9 of Chapter VII, the same formula is derived for an earthdam , where, however, $h_2$ is not the ordinate of the free surface.

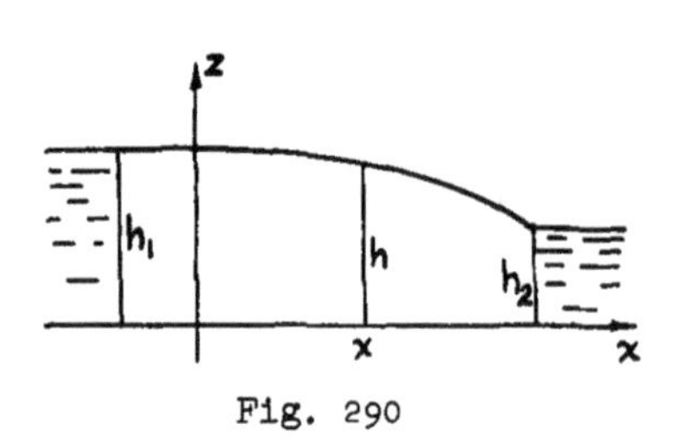


Fig. 290

We already showed that in the hydraulic theory the vertical velocities are usually neglected. However, we may render a method of approximate computation of the vertical flow velocities. We write the equation of continuity

$$\frac{\partial u}{\partial x} + \frac{\partial w}{\partial z} = 0$$

and integrate it along $z$, recalling that $u$ is only a function of $x$, determined by

$$u = -k \frac{dh}{dx} \quad .$$

We obtain:

$$w = -\int_0^z \frac{\partial u}{\partial x} dz = k \int_0^z \frac{d^2h}{dx^2} dz \quad .$$

Since $d^2h/dx^2$ does not depend upon $z$, the integration gives

$$w = kz \frac{d^2h}{dx^2} \quad ,$$

considering the fact that for $z = 0$ (at bedrock) the vertical velocity is zero.

By means of expression (2.3) we obtain:

$$\frac{d^2h}{dx^2} = -\frac{q^2}{k^2h^3} \quad .$$

From this

$$w = -\frac{q^2 z}{kh^3} \quad .$$

At the free surface, where $z = h$, we obtain:

$$w = -\frac{q^2}{kh^2} = -k\left(\frac{q}{kh}\right)^2 \quad .$$

From this it is clear that the vertical velocity is always negative and that its magnitude increases with the depth of flow (Fig. 291).

§3. Free Surface with Infiltration or Evaporation. [9]

We make $w_0 = 0$ in equation (1.12) but leave $e$, indicating infiltration or evaporation. We restrict ourselves to plane flow in the xz plane, and rewrite equation (1.12) as

$$k \frac{d}{dx}\left(h \frac{dh}{dx}\right) + e = 0 . \tag{3.1}$$

Its integration renders

$$kh^2 + ex^2 = C_1 x + C_2 . \tag{3.2}$$

In the case of infiltration, for $e > 0$, we have the equation of an ellips; in case of evaporation, when we may put $e = -\varepsilon$ $(\varepsilon > 0)$, we obtain the equation of an hyperbola

$$kh^2 - \varepsilon x^2 = C_1 x + C_2 . \tag{3.3}$$

To determine the constants $C_1$ and $C_2$, one must have two conditions, such as two points on the free surface. Shifting the origin of

Fig. 291

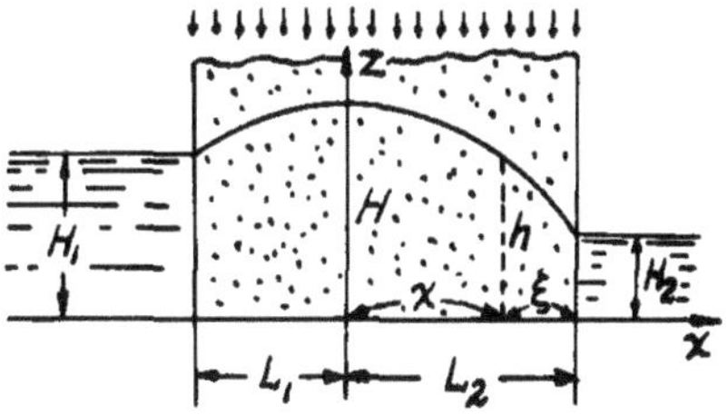

Fig. 292

coordinates along the x-axis, we may achieve that $C_1 = 0$. Let there be infiltration to the free surface in the case of seepage from a reservoir with depth $H_1$ to one with depth $H_2$ (Fig. 292). Calling the maximum ordinate of the free surface $H$, and its distances to the reservoirs respectively $L_1$ and $L_2$, we write equation (3.2) in the form:

$$h^2 + \frac{e}{k} x^2 = H^2 .$$

Therefore, putting $x = -L_1$, $h = H_1$ and $x = L_2$, $h = H_2$, we obtain the relationships

$$H_1^2 + \frac{e}{k} L_1^2 = H^2, \quad H_2^2 + \frac{e}{k} L_2^2 = H^2 . \tag{3.4}$$

If the distance between the two reservoirs is $2L$, then we have

$$L_1 + L_2 = 2L . \tag{3.5}$$

The two equations (3.4) and equation (3.5), in which $L$, $H_1$, $H_2$ are given, are sufficient to determine the unknown $L_1$, $L_2$, $H$.

We may introduce the distance $\xi$ from the reservoir instead of $x$. For the right part, putting

$$x + \xi = L_2$$

we obtain the equation of this part of the free surface

$$h^2 = \frac{e}{k}\,(2L_2\xi - \xi^2) + H_2^2 \quad .$$

For the left part of the free surface, putting

$$x + \xi = L_1$$

we have:

$$h^2 = \frac{e}{k}\,(2L_1\xi - \xi^2) + H_1^2 \quad .$$

In the case of evaporation there are two possibilities depending on the sign of $C_2$ (if we choose $C_1 = 0$) so that we may obtain one of two conjugate hyperbolas.

§4. Seepage in Soils, Slightly Heterogeneous in Vertical Direction.

Already in Chapter VIII we treated the simplest confined flows in heterogeneous — layered — soils. Here we examine flow with free surface, in soil with seepage coefficient that is a function of depth $z$.

We compute the flowrate $q_x$ in the direction of the x-axis, flowing through a rectangle of height $h$, with basis equal to one and lying in a plane perpendicular to the x-axis. Dividing $h$ in elements $dz$, we divide the entire rectangle into elementary rectangles with flowrate

$$-\,k(z)\,\frac{\partial h}{\partial x}\cdot dz \quad .$$

Integrating this expression over $z$ between limits zero and $h$, we find the flowrate $q_x$ through the forementioned rectangle of height $h$:

$$q_x = -\int_0^h k(z)\,\frac{\partial h}{\partial x}\,dz = -\,\frac{\partial h}{\partial x}\int_0^h k(z)\,dz \quad . \tag{4.1}$$

Here $\partial h/\partial x$ may be taken out of the integral sign, because $h$ does not depend upon $z$ (see §1 of this chapter) according to the hydraulic theory.

In exactly the same way, the flowrate $q_y$ through a surface element of height $h$ and unit width, perpendicular to the y-axis, is expressed as:

$$q_x = -\,\frac{\partial h}{\partial y}\int_0^h k(z)\;kz \quad . \tag{4.2}$$

N. K. Girinskii [6], introduced the function

$$\Phi(x,y) = \int_0^h (z - h)k(z)\;dz \quad , \tag{4.3}$$

through which $q_x$ and $q_y$ are simply expressed

$$q_x = \frac{\partial\Phi}{\partial x}\ , \quad q_y = \frac{\partial\Phi}{\partial y}\ . \tag{4.4}$$

We prove the first of equalities (4.4). Therefore we observe that $x$ is a variable in the function $h(x,y)$ and that the latter occurs in the integrand as well as in one of the limits of the integral. Therefore, we apply to (4.3) the formula for the differentiation of an integral with respect to a parameter $\alpha$

$$I = \int_a^{b(\alpha)} f(x,\alpha)\ dx\ ,$$

namely:

$$\frac{dI}{d\alpha} = f(b,\alpha)\frac{db}{d\alpha} + \int_a^b f'_\alpha(x,\alpha)\ dx\ ,$$

We obtain

$$\frac{\partial\Phi}{\partial x} = (h - h)k(h)\frac{\partial h}{\partial x} - \int_0^h \frac{\partial h}{\partial x}k(z)\ dz = -\frac{\partial h}{\partial x}\int_0^h k(z)\ dz\ ,$$

and this proves the first formula of (4.4). The second is proved in the same way.

If the soil is homogeneous, then

$$\Phi(x,y) = k\int_0^h (z - h)dz = k\left.\frac{(z - h)^2}{2}\right|_0^h = -\frac{kh^2}{2}\ . \tag{4.5}$$

Function (4.3) as is easy to check, satisfies Laplace's equation in the variables $x,y$.

Thus function (4.3) is the generalization of function

$$-\frac{kh^2}{2}$$

for the case of soil which is heterogeneous in vertical direction.

Example 1. For continuous variation of $k(z)$ with depth, we take the linear relationship

$$k(z) = k_0(1 + bz)\ .$$

Substitution of this expression in (4.3) and integration give:

$$\Phi(x) = -\frac{k_0h^2}{2} - \frac{k_0bh^3}{6}\ . \tag{4.6}$$

Considering the flowrate $q$ constant for all $x$ and putting

$$\frac{\partial\Phi}{\partial x} = q\ ,$$

we obtain after integration:

$$\Phi = qx + C\ , \tag{4.7}$$

where $C$ is an arbitrary constant. Because of (4.6) we have the equation of a parabola of the third degree

$$-\frac{k_0h^2}{2} - \frac{k_0bh^3}{6} = q_x + C$$

or

$$h^2 + \frac{1}{3}bh^3 = -\frac{2q}{k_0}x + C_1 \quad . \tag{4.8}$$

The knowledge of two points $(x_1,h_1)$ and $(x_2,h_2)$ on the free surface gives for $q$:

$$q = -\frac{k_0(h_2 - h_1)\left[h_2 + h_1 + \frac{b}{3}(h_2^2 + h_1h_2 + h_1^2)\right]}{2(x_2 - x_1)} \quad . \tag{4.9}$$

If $b > 0$, then the seepage coefficient increases with height and the absolute value of $q$ is larger than for $b = 0$; if $b < 0$, then $k$ decreases with height and $q$ is smaller in absolute value than for $b = 0$.

Investigation of the shape of the free surface (we may take $C_1 = 0$) shows that for $b > 0$ a curve is obtained (Fig. 293,a), lying above the parabola

$$h^2 = -\frac{2q}{k_0}x \ .$$

For $b < 0$, only part of the curve that is obtained is possible for the shape of the free surface, namely where $h$ changes from zero to $h_1 = -1/b$, and $x$ from zero to $x_1 = -k_0/3qb^2$, i.e., the part from the top of the parabola (0.0) to the inflection point. This part corresponds to a change of the seepage coefficient from $k_0$ to zero (Fig. 293 b).

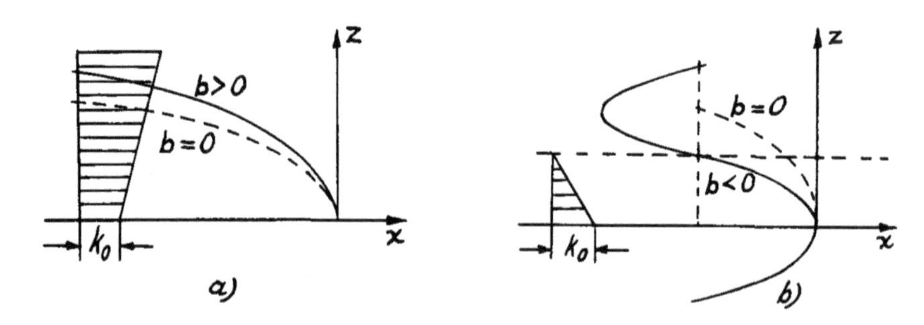


Fig. 293

Example 2. We assume a discontinuous change for $k(z)$

$$k(z) = k_0 \quad \text{for} \quad 0 < z < h_0 \ ,$$

$$k(z) = k_1 \quad \text{for} \quad z > h_0 \ ,$$

in which $k_0$ and $k_1$ are constants. In other words, we examine two layered soil with two seepage coefficients (Fig. 294, 295). Then in the bottom stratum we have the expression for $\Phi$

$$\Phi = k_0 \int_0^h (z - h)dz = -\frac{k_0}{2} h^2 \quad , \tag{4.10}$$

and in the upper stratum

$$\Phi = k_0 \int_0^{h_0} (z - h)dz + k_1 \int_{h_0}^{h} (z - h)dz =$$

$$= -\frac{k_1}{2} \left\{[h + (c - 1)h_0]^2 + c(c - 1)h_0^2\right\} \quad , \tag{4.11}$$

where

$$c = \frac{k_0}{k_1} \quad .$$

We examine the possible flow in plane $xz$. Therefore Laplace's equation, satisfied by function $\Phi$, reduces to

$$\frac{d^2\Phi}{dx^2} = 0 \quad .$$

Integration gives

$$\Phi = C_1 x + C_2 \quad ,$$

where $C_1$ and $C_2$ are arbitrary constants. Recalling the values of $\Phi$ from (4.10) and (4.11), we obtain the equation of the free surface in the bottom stratum:

$$-\frac{k_0}{2} h^2 = C_1 x + C_2$$

or

$$h^2 = Cx + D \quad , \tag{4.12}$$

where $C$ and $D$ are new constants:

$$C = -\frac{2}{k_0} C_1 , \qquad D = -\frac{2}{k_0} C_2 \quad .$$

In the upper stratum

$$[h + (c - 1)h_0]^2 = Ax + B \quad . \tag{4.13}$$

$A, B, C, D$ are constants, subject to determination.

Equation (4.13) represents a parabola with symmetry axis

$$z = -(c - 1)h_0 = b \quad .$$

This axis lies under the x-axis, if $c > 1$ (Fig. 294), i.e., if the lower stratum is more pervious than the upper, and the absolute value of $b$ in this case may be very large (for sufficiently large $c = k_0/k_1$). The larger $b$, the more inclined the free surface will be. For $c < 1$, the axis of parabola (4.13) is above the abscissa axis (Fig. 295), and its distance from the abscissa-axis $b = (1 - c)h_0$ is computed between zero and $h_0$.

We examine the following problem. We assume that the free surface

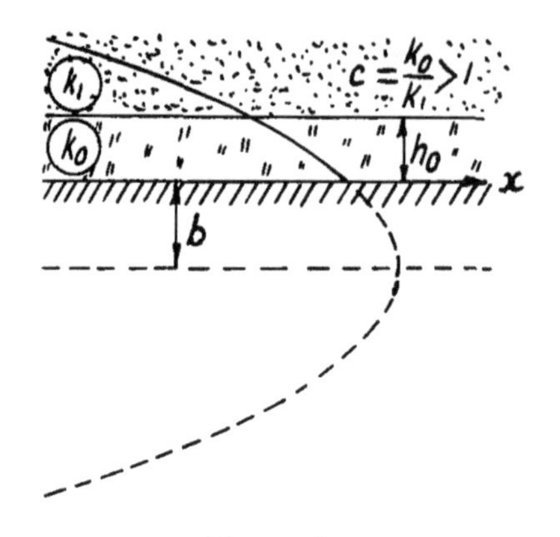


Fig. 294

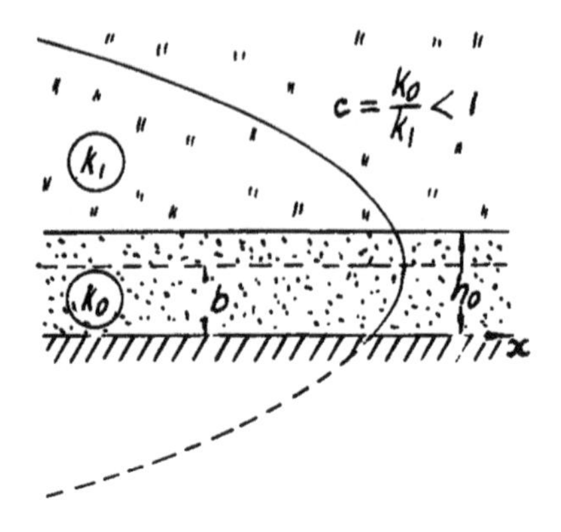


Fig. 295

passes through points $(x_1, h_1)$, $(x_2, h_2)$.. The first lying in the upper stratum, the second in the lower (Fig. 296). We call $x_0$ the unknown abscissa of the point in which $h = h_0$ ($h_0$ is the thickness of the lower stratum). Then the equations of parabolas (4.12) and (4.13) assume the form

$$(h + b)^2 = (h_0 + b)^2 + $$
$$+ (h_1 - h_0)(h_1 + h_0 + 2b) \frac{x_0 - x}{x_0 - x_1} ,$$

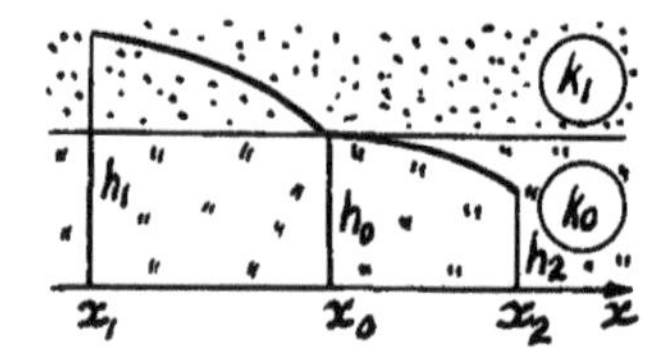


Fig. 296

$$h^2 = h_0^2 -(h_0^2 - h_2^2) \frac{x - x_0}{x_2 - x_Q} ,$$

where

$$b = (c - 1)h_0 .$$

The abscissa $x_0$ is determined by

$$x_0 = \frac{m_1 x_1 + m_2 x_2}{m_1 + m_2} ,$$

where

$$m_1 = c(h_0 + h_2)(h_0 - h_2)$$
$$m_2 = (h_1 - h_0)(h_1 + h_0 + 2b) .$$

For the flowrate, one has

$$q = \frac{k}{2} \frac{(1 - c)(h_1 - h_0)^2 + c(h_1^2 - h_2^2)}{x_2 - x_1} .$$

§5. Seepage for Sloping Bedrock. [2]

Let the angle $\alpha$ of vedrock on horizontal be small. We put

$$i = tg\ \alpha \quad .$$

We call $y$ the ordinate of the free surface above the horizontal x-axis (Fig. 297). To compute the flowrate through section $x$, one must multiply the velocity $u = -k\ (dy/dx)$ with segment AB (Fig. 297) equal to $(y - ix)$. Consequently

$$q = -k(y - ix)\frac{dy}{dx} \quad . \tag{5.1}$$

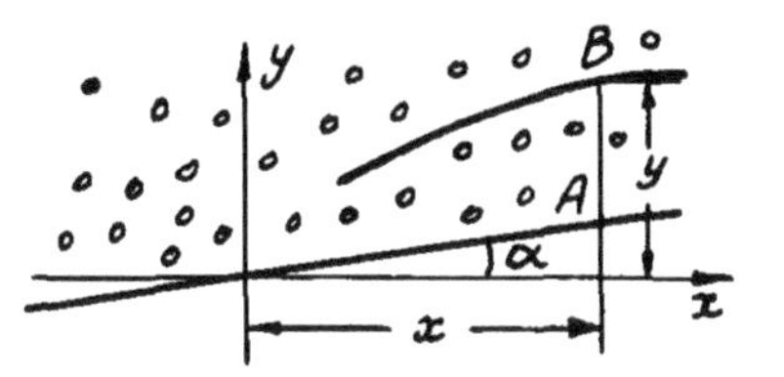


Fig. 297

Flowrate $q$ is constant. To integrate equation (5.1), we rewrite it in this form

$$\frac{dx}{dy} - \frac{ki}{q}x + \frac{k}{q}y = 0 \quad . \tag{5.2}$$

This linear equation has the general solution

$$x = Ae^{\frac{kiy}{q}} + \frac{y}{i} + \frac{q}{ki^2} \quad ,$$

which may be rewritten as

$$y - ix + \frac{q}{ki} = Be^{\frac{kiy}{q}} \quad . \tag{5.3}$$

B is a new arbitrary constant. For $B = 0$ we obtain the particular solution

$$y = ix - \frac{q}{ki} \quad . \tag{5.4}$$

This corresponds to the so called "uniform seepage,"[‡] i.e., seepage with constant velocity

$$u = -k\frac{dy}{dx} = -ki \quad ,$$

parallel to the bedrock (Fig. 298).

The constant depth of flow $h_0$ in case of uniform flow is called "normal depth". The flowrate for uniform seepage is

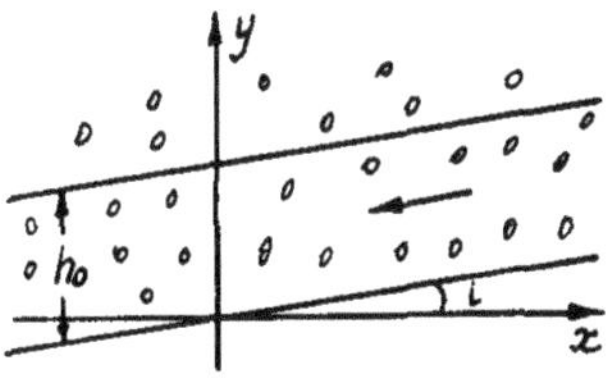

$$q = -kh_0 i \quad . \tag{5.5}$$

Fig. 298

For values of B different from zero in equation (5.3), seepage is called "non-uniform". Curve (5.3) has (5.4) as rectilinear asymptote; by

‡ For uniform seepage, equation (5.4) is hydrodynamically exact. The velocity hodograph degenerates for this case to one point.

means of (5.5), we rewrite equations (5.3) and (5.4) in this form

$$(5.6) \qquad y - ix - h_0 = Be^{-\frac{y}{h_0}}, \qquad y = ix + h_0 .$$

We examine different cases of equation (5.3) which may occur.

1°) $i > 0$, $h_0 > 0$: The asymptotic approximation is reached for $y \to +\infty$. If $B > 0$, the curve lies above the asymptote. It is called a "rising" free surface (curve I, Fig. 299). If $B < 0$, then the entire curve (5.3) lies under the asymptote. It consists of two parts: part II lies above bedrock and therefore corresponds to a possible flow. It is called "falling" curve. Part III is below bedrock and does not correspond to real flow ( it will occur further). Flow takes place down slope.

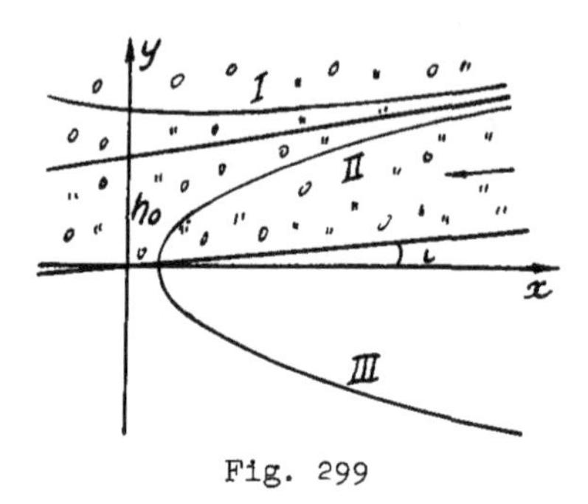

Fig. 299

2°) $i < 0$, $h_0 > 0$. This case is sketched in Fig. 300. The asymptotic approximation is reached for $y \to +\infty$. Again we obtain rising and falling curves and also a branch of a curve in the bedrock region.

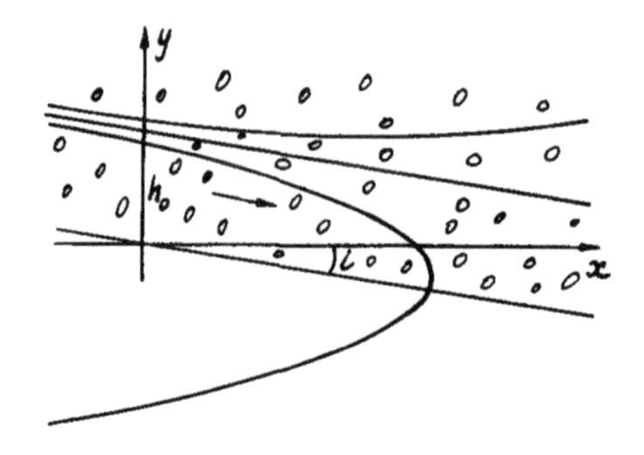

Fig. 300

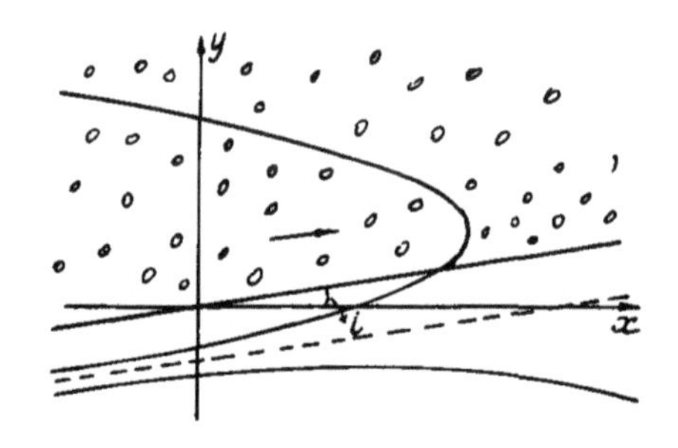

Fig. 301

Case 2° is not essentially different from case 1°; it leads only to another disposition of the figure (further, in the study of two layered soil with inclined boundary line, the case will be essentially different).

We notice that in both cases the flow takes place down slope along to bedrock. This is also apparent from the sign of formula,

$$q = -kh_0 i ,$$

which gives for $q$ a negative value in case 1° and a positive one in case 2°. Formula (5.5) determines the flowrate in an infinitely far cross-section of the flow.

3°) $i > 0$, $h_0 < 0$. We introduce the notation

$$-h_0 = h' \tag{5.7}$$

Then equation (5.6) is rewritten as

$$y - ix + h_0' = Be^{\frac{y}{h_0'}} . \tag{5.8}$$

We see that now curve (5.8) tends asymptotically to

$$y = ix - h_0'$$

for negative values of $y$, in the bedrock region, where we obtain curves which we previously called I and II. In the upper part of Fig. 301, above bedrock, we have the curve which we called III before.

As apparent from formula

$$q = kih_0' \tag{5.9}$$

the flowrate in case 3° is positive, and flow proceeds upwards, against the bedrock slope.

4°) $i < 0$, $h_0 < 0$. This case differs from 3° only by the disposition of the figure. The flowrate according to formula (5.5) is negative (Fig. 302).

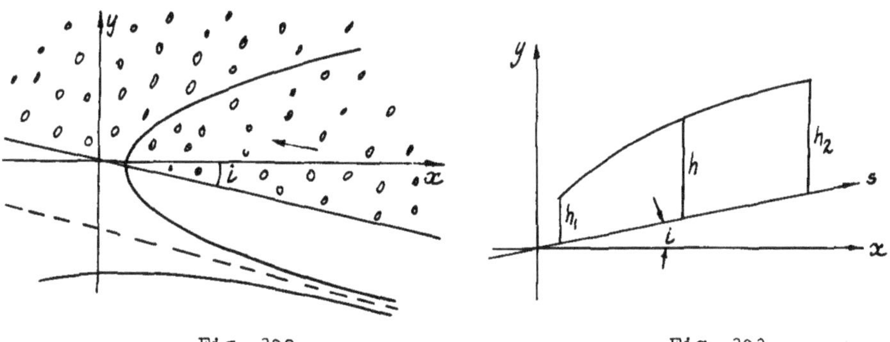


Fig. 302 Fig. 303

We come back to equation (5.3). If we know some point $(x_1, y_1)$ at the free surface, then we may determine the constant B. We substitute $x_1, y_1$ in (5.6) and obtain:

$$y_1 - ix_1 - h_0 = Be^{-\frac{y_1}{h_0}} . \tag{5.10}$$

Dividing (5.3) term by term by (5.10), we eliminate B and obtain:

$$y - ix - h_0 = (y_1 - ix_1 - h_0)e^{-\frac{y - y_1}{h_0}} . \tag{5.11}$$

If we know another point $(x_2, y_2)$ of the free surface, we insert its coordinates in equation (5.11) and obtain, after taking the logarithm:

$$(5.12) \qquad y_2 - y_1 = h_0 \, \ell n \frac{y_1 - ix_1 - h_0}{y_2 - ix_2 - h_0} .$$

We may find $h_0$ from this equation, then formula (5.5) allows us to compute the flowrate.

§6. Change of Variables.

Dupuit [1] and N. N. Pavlovsky [2], who investigated the problem of steady groundwater flow above sloping bedrock, used other variables than $x$ and $y$. To find them, we put

$$(6.1) \qquad h = y - ix$$

and introduce a new abscissa $s$, along the bedrock boundary. (Fig. 303); $s$ and $x$ are related as

$$s = \frac{x}{\cos \alpha} .$$

For sufficiently small angles $\cos \alpha \approx 1$ and we consider

$$s = x$$

(or we measure $x$ along the horizontal axis); in this case equation (5.1) is rewritten as

$$(6.2) \qquad q = - kh \left( \frac{dh}{dx} + i \right) ,$$

and the first equation (5.6) becomes

$$(6.3) \qquad h - h_0 = Be^{-\frac{h+ix}{h_0}} .$$

Instead of (5.11) and (5.12) we have respectively

$$(6.4) \qquad h - h_0 = (h_1 - h_0)e^{-\frac{h-h_1+i(x-x_1)}{h_0}}$$

or after taking the logarithms

$$(6.5) \qquad h - h_1 + i(x - x_1) = h_0 \, \ell n \frac{h_1 - h_0}{h - h_0} ,$$

$$(6.6) \qquad h_2 - h_1 + i(x_2 - x_1) = h_0 \, \ell n \frac{h_1 - h_0}{h_2 - h_0} .$$

The last equation allows us to determine one of the unknowns: $h_1$, $h_2$, $h_0$, $x_2 - x_1$, when three of these quantities are given.

We examine how to construct the free surface. Therefore we write equation (6.5) as follows:

$$(6.7) \qquad i(x - x_1) = h_1 - h - h_0 \, \ell n \left| \frac{h - h_0}{h_0} \right| + h_0 \, \ell n \left| \frac{h_1 - h_0}{h_0} \right| ,$$

We introduce the notations

$$(6.8)\qquad \frac{i(x - x_1)}{h_0} = \xi, \quad \frac{h}{h_0} = \eta, \quad \ell n\left|\frac{h_1 - h_0}{h_0}\right| + \frac{h_1}{h_0} = C$$

and rewrite (6.7) as follows:

$$(6.9)\qquad \xi = -\eta - \ell n \mid \eta - 1 \mid + C .$$

Assigning different values to the constant $C$, we obtain the same curve, only shifted along the $\xi$-axis. Therefore it is sufficient to construct the curve

$$(6.10)\qquad \xi = -\eta - \ell n \mid \eta - 1 \mid ,$$

to have the possibility of constructing the free surface for arbitrary values of $C$.

Curve (6.10) of Fig. 304 was constructed by N. N. Pavlovsky, who considered two cases: $\eta < 1$, when equation (6.10) becomes

$$\xi = -\eta - \ell n(1 - \eta) ,$$

which gives rising and falling curves, and $\eta > 1$ when

$$\xi = -\eta - \ell n(\eta - 1) ,$$

which gives the curve of the image flow.

We stated already that equation (6.6) allows us to determine one of the constants in it, if the remaining ones are known.

If $h_0$, $h_1$ and $h$ are very close to each other so that $(h_1 - h_0)/(h - h_0)$ is close to one, then we may use the development in series of the

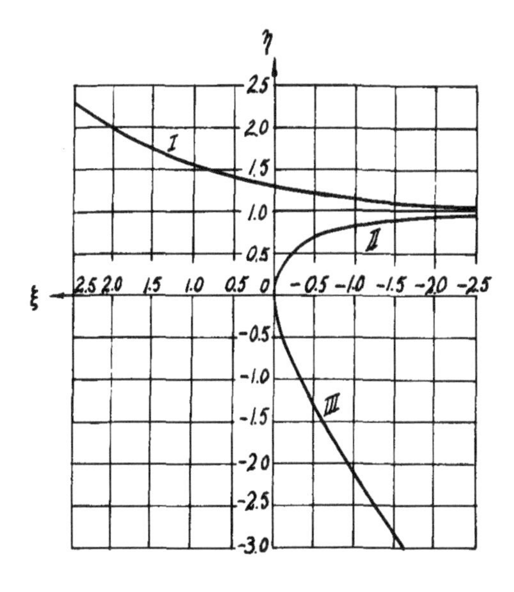


Fig. 304

logarithm in formulas (6.5) and (6.6). We have:

$$\ln u = 2\left[\frac{u-1}{u+1} + \frac{1}{3}\left(\frac{u-1}{u+1}\right)^3 + \ldots\right] \approx 2\,\frac{u-1}{u+1} \quad .$$

The approximate equality holds for $u$ close to one.

We apply it to the expression

$$u = \frac{h_1 - h_0}{h - h_0} \quad .$$

Then equation (6.5) is written in the form:

$$h - h_1 + i(x - x_1) = 2h_0 \; \frac{h_1 - h}{h + h_1 - 2h_0} \quad .$$

Solving this equation for $h_0$ and replacing $h_0$ by its value in (5.5), we find:

$$q = -k\left[\frac{h^2 - h_1^2}{2(x - x_1)} + i\,\frac{h + h_1}{2}\right] = -k\,\frac{h + h_1}{2}\left[\frac{h - h_1}{x - x_1} + i\right] \quad . \tag{6.11}$$

Equation (6.11) is the generalization of Dupuit's equation and is reduced to it for $i = 0$.

The flowrate may be computed, knowing one more point $(x_2, h_2)$:

$$q = -k\left[\frac{h_2^2 - h_1^3}{2(x_2 - x_1)} + i\,\frac{h_2 + h_1}{2}\right] = -k\,\frac{h_1 + h_2}{2}\left[\frac{h_2 - h_1}{x_2 - x_1} + i\right] \quad . \tag{6.12}$$

This formula may also be obtained from (6.2) if we replace $h$ by the average value $(h_1 + h_2)/2$ and $dh/dx$ by $(h_1 - h_2)/(x_1 - x_2)$ .

We notice that formula (6.12) is only useful when $h_1$ is near $h_2$. In this case $x_2 - x_1$ may not be large.

## §7. Seepage in Two-layered Soil with Inclined Boundary Line. [12]

Let the equation of the boundary between two soils be

$$y = ix \quad . \tag{7.1}$$

Below this line there is soil with seepage coefficient $k_0$, above it, soil with seepage coefficient $k$.

The flowrate through a section with abscissa $x$ and ordinate $h$ of the free surface, is composed of two parts: that through segment AB and that through segment BC (Fig. 305). We assume that the piezometric slope is constant and equal to $dh/dx$. We have then:

$$q = -k_0 y\,\frac{dh}{dx} - k(h - y)\,\frac{dh}{dx} = -\,[k_0 ix + k(h - ix)]\,\frac{dh}{dx} \quad .$$

We integrate this equation, considering $q$ constant. Therefore, we rewrite the equation as follows:

$$\frac{dx}{dh} - \frac{(k - k_0)i}{q}\,x + \frac{k}{q}\,h = 0 \quad . \tag{7.2}$$

This equation is linear and of the first order, of the same form as equation (5.2) for seepage in soil resting on bedrock. It reduces to the latter

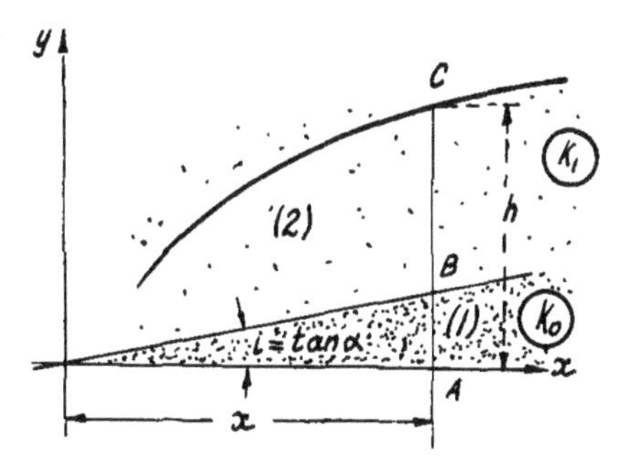


Fig. 305

for $k_0 = 0$. If $k_0$ is negligibly small compared with $k$ within the limits of accuracy of the computations, then (7.2) is also reduced to (5.2); consequently, if $k_0/k$ is small (practically $\leq 0.001$), the boundary plays the roll of impervious bedrock.

If we introduce the notations

$$\left(1 - \frac{k_0}{k}\right) i = J, \quad H_0 = -\frac{q}{kJ} = -\frac{q}{(k - k_0)i},$$

then equation (7.2) becomes:

$$\frac{dx}{dh} - \frac{kJ}{q} x + \frac{k}{q} h = 0 ,$$

consistent with (7.2), and its solution, passing through point $(x_1, y_1)$, is

$$h - J(x - x_1) = Ce^{-\frac{h-h_1}{H_0}} .$$

Putting $h - Jx = H$, $h_1 - Jx_1 = H_1$, we obtain an equation similar to equation (5.3). However, here the four cases, examined in §5, are essentially different from each other, so that the asymptotes of the integral

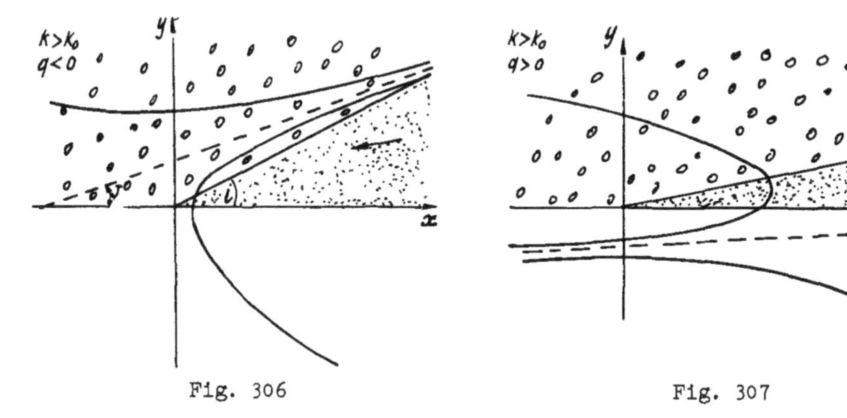


Fig. 306 Fig. 307

curves are not parallel to the boundary between the two soils. These four cases are represented in Fig. 306-309.

To derive the two last cases, when $k < k_0$, i.e., $J < 0$, we observe that if the ratio of the seepage coefficients $k_0/k$ is large, the asymptote

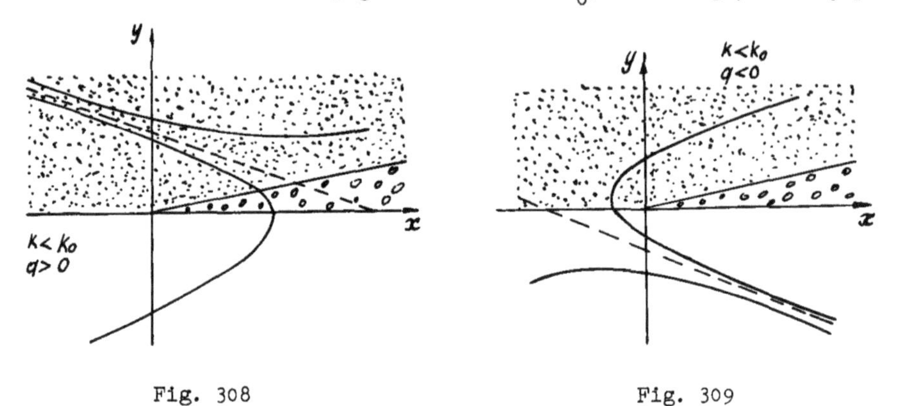


Fig. 308 Fig. 309

becomes steep, the free surface is strongly curved, and the hydraulic theory becomes inapplicable. The case of large $k_0/k$ represents considerable interest, but must be investigated by other methods.

### B. Unconfined and Semi-Confined Three-Dimensional Flows.

#### §8. Connection Between Three-dimensional Unconfined and Two-dimensional Confined Flows.

In §1 we showed that the problem of three-dimensional groundwater flow with slightly varying free surface, in soil on horizontal bedrock, leads to Laplace's equation for function $h^2(x,y)$. Here $h(x,y)$ is the average of the head over the height and at the same time the ordinate of the surface. On the other side, in two-dimensional confined flow, in a stratum bounded by two horizontal planes, the function of the head, which we call $h_H(x,y)$, satisfies Laplace's equation

$$\Delta h_H = 0 \quad .$$

If the boundary conditions are the same for the two problems, then a linear relationship [3] must exist between $h^2$ and $h_H$ [3].

It is convenient to use the concept introduced by N. N. Pavlovsky about the "reduced head $h_r$," i.e., the head divided by the effective head H.

Pavlovsky assumed that $h_r = 1$ along the boundary of the headwater reservoir, and $h_r = 0$ along the boundary of the tailwater reservoir.

We assume that the effective head h along the boundary of the headwater reservoir is $h_1$, along the lower reservoir $h_2$. Then the relationship between h and $h_r$ may be expressed as

$$\frac{h - h_2}{h_1 - h_2} = \frac{h_r}{1} \quad . \tag{8.1}$$

If we examine the more general case, when the correspondence between the heads $h$ and $h_r$ is given in two arbitrary points, for which $h_r = h_{r_1}$ corresponds to $h = h_1$, and $h_r = h_{r_2}$ corresponds to $h = h_2$, then we have:

$$\frac{h - h_2}{h_1 - h_2} = \frac{h_r - h_{r_2}}{h_{r_1} - h_{r_2}} \quad . \tag{8.2}$$

Now we assume that we introduce two flows in the relationship: one unconfined, the other confined and "reduced". Thus we must replace $h, h_1, h_2$, in the left part of equation (8.2) respectively by $h^2$, $h_1^2$ $h_2^2$ and we obtain:

$$\frac{h^2 - h_2^2}{h_1^2 - h_2^2} = \frac{h_r - h_{r_2}}{h_{r_1} - h_{r_2}} \quad . \tag{8.3}$$

We study for example the flow to a foundation pit, circular in plane cross-section. Such a pit is none other than a well with large radius. We call this radius $r_1$ and $h_1$ the corresponding value of $h$. Let $h = h_2$ for $r = r_2$.

The corresponding confined flow is determined by a function $h_r$ of the form

$$h_r = A \ln r + B \quad . \tag{8.4}$$

For $r = r_1$, we have $h_r = h_{r_1}$, for $r = r_2$, $h_r = h_{r_2}$, where $h_{r_1}$, $h_{r_2}$ are arbitrary numbers. From (8.2), by means of (8.4) we find:

$$\frac{h_r - h_{r_2}}{h_{r_1} - h_{r_2}} = \frac{\ln \frac{r}{r_2}}{\ln \frac{r_1}{r_2}} \quad .$$

Then (8.3) gives:

$$h^2 = h_2^2 + (h_1^2 - h_2^2) \frac{\ln \frac{r}{r_2}}{\ln \frac{r_1}{r_2}} \quad .$$

From this and by means of

$$Q = - k \left. \frac{dh}{dr} \right|_{r=r_1} \cdot 2\pi r_1 h_1 \ ,$$

where $r_1$ is the radius of the well, we obtain Dupuit's formula for the flowrate of a well $Q$ (see also Chapter VII, §9):

$$Q = \frac{\pi k (h_1^2 - h_2^2)}{\ln \frac{r_2}{r_1}} \quad . \tag{8.5}$$

§9. About Flow To a Pit with Polygonal Plane Cross-section.

V. I. Aravin [4] gives an approximate method to compute groundwater flow to pits with polygonal plane cross-sections. Here we might apply Christoffel's formula to map the exterior of a polygon onto a half-plane. V. I. Aravin however, proceeds as follows.

He considers the equation for three-dimensional flow towards a slit

$$h^2 - h_*^2 = \frac{4q}{\pi} \sinh^{-1} \frac{1}{\ell} \sqrt{\frac{x^2 + y^2 - \ell^2}{2} + \sqrt{\left(\frac{x^2 + y^2 - \ell^2}{2}\right)^2 + \ell^2 y^2}} =$$

$$= f(q, \ell, x, y) \; . \tag{9.1}$$

Here $h_*$ is an unknown flow constant, $q$ the flowrate through half of the segment.

Formula (9.1) may be obtained from the solution of the plane problem about the flow of a fluid to a rectilinear segment, completely similar to the problem of flow about a cut-off in the half-plane . (Chapter II, §13).

For $n$ segments, bounding our region (Fig. 310), we have the sum

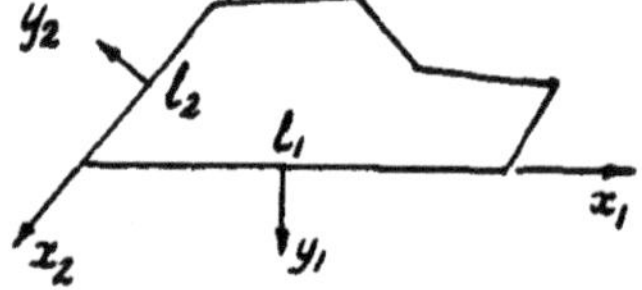


Fig. 310

$$h^2 - h_*^2 = \sum_{k=1}^{n} f(q_k, \ell_k, x_k, y_k) \; , \tag{9.2}$$

where index $k$ indicates the number of the segment of length $\ell_k$, the axis $x_k$ is directed along this segment and the axis $y_k$ passes through its midpoint.

We apply (9.2) to the verticals, passing through the center of each of the $n$ sides of the perimeter of the pit. Since the depth of all these points is the same, we call it $h_0$ and obtain:

$$\left\{ \begin{aligned} h_0^2 - h_*^2 &= \sum_{k=1}^{n} f_1(q_k, \ell_k, x_k, y_k) , \; \ldots , \\ h_0^2 - h_*^2 &= \sum_{k=1}^{n} f_n(q_k, \ell_k, x_k, y_k) \end{aligned} \right. \tag{9.3}$$

This equation gives us n-1 equations to determine

$$q_1, \, q_2, \, \ldots, \, q_{n-1} \; .$$

Considering the contour as a circle, at a far distance $R_k$ from the midpoint of segment $\ell_k$, we have a flowrate computed by the formula for a circular well:

$$f(q_k, \ell_k, x_k, y_k) = \frac{4q_k}{\pi} \ln \frac{2R_k}{\ell_k} \quad ,$$

and therefore, assuming $h = H$ in points $R_k$, we obtain:

$$H^2 - h_*^2 = \sum_{k=1}^{n} \frac{4q_k}{\pi} \ln \frac{2R_k}{\ell_k} .$$

Assuming approximately $R_1 = R_2 = \dots = R_n$, we write

$$H^2 - h_*^2 = \frac{4(q_1 + \dots + q_n)}{\pi} \ln R -$$

$$- \frac{4}{\pi} \left(q_1 \ln \frac{\ell_1}{2} + q_2 \ln \frac{\ell_2}{2} + \dots + q_n \ln \frac{\ell_n}{2}\right) - \sum_{k=1}^{n} f_1(q_k, \ell_k, x_k, y_k) . \tag{9.4}$$

When the polygon is regular, then $\ell_1 = \ell_2 = \dots = \ell_n = \ell$ and

$$H^2 - h_*^2 = \frac{4(q_1 + \dots + q_n)}{\pi} \ln \frac{2R}{\ell} .$$

One of equations (9.3) may serve to determine $h_*$.

Expressing the flowrates of all segments in function of that of the first

$$q_2 = a_2 q_1 , \dots , q_n = a_n q_1$$

and substituting these equations in (9.4), we obtain:

$$H^2 - h_0^2 = \frac{4q_1}{\pi} (1 + a_2 + \dots + a_n) \ln R -$$

$$- \frac{4q_1}{\pi} \left(\ln \frac{\ell_1}{2} + \dots + a_n \ln \frac{\ell_n}{2}\right) - \sum_{k=1}^{n} f_1(a_k q_1, \ell_k, x_k, y_k) .$$

From this we find $q_1$ and the total flowrate $Q$ to the pit:

$$Q = 4kq_1(1 + a_1 + \dots + a_n) .$$

For a regular polygon we obtain:

$$H^2 - h_0^2 = \frac{4nq}{\pi} \ln \frac{2R}{\ell} - \frac{4Mnq}{\pi} ,$$

$$M = \frac{1}{n} \sum_{k=1}^{n} \sinh^{-1} \frac{1}{\ell} \sqrt{\frac{x_k^2 + y_k^2 + \ell^2}{2} + \sqrt{\left(\frac{x_k^2 + y_k^2 + \ell^2}{2}\right) + \ell^2 y_k^2}} .$$

Comparing the flowrates obtained for a triangular, square and hexagonal cross-section of the well, with the flowrate of a circular well, the author comes to the conclusion that in the case of a polygonal well the flowrate may be computed with sufficient accuracy by the formula for a circular well, if one takes the average radius of the circles inscribing and circumscribing the polygon.

## §10 Semi-confined Flows. [9]

In §9 of Chapter II, we recalled the existence of semi-confined flows, i.e., flows for which the flow region is divided in a series of regions with unconfined and confined flows.

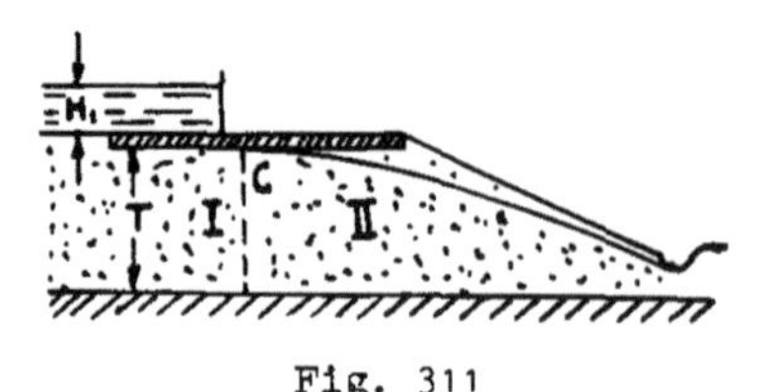

Fig. 311

Thus, one may consider that the confined flow under a weir separates and further becomes free surface flow (Fig. 311). We have confined flow in region I, unconfined flow in region II. At the boundary between them we have $h = h_{II}$ (head of the boundary cross-section ). In this case $h_{II} = T$. For region I formula (8.2) is applied, which may be rewritten, recalling the $h_2 = T$, $h_1 = H_1$, $h_{r_2} = 1$, in this way

$$\frac{h - T}{H_1 - T} = \frac{h_r - h_{rII}}{1 - h_{rII}} \quad . \tag{10.1}$$

$h_{rII}$ designates the unknown head at the boundary line. For region II, we must use formula (8.3) which gives, if we consider that now $h_1 = T$, $h_2 = H_2$, $h_{r1} = T$, $h_{r2} = 0$:

$$\frac{h^2 - H_2^2}{T^2 - H_2^2} = \frac{h_r}{h_{rII}} \quad . \tag{10.2}$$

To determine the unknown head $h_{rII}$ in the boundary section, we may use the condition of equal flowrate through this section in regions I and II.

We examine the case of Fig. 311. The groundwater flow separates from the weir foundation in point C, forming a free surface and flowing further into a river or drain. Separating the flow in two parts, we apply to the left part, the formula for the flow about a two-dimensional flat bottom foundation of length $2\ell$ (Chapter III):

$$Q = \kappa \frac{H_1 - T}{K} K', \quad k = \operatorname{th} \frac{\pi i}{2T} .$$

For the right part we compute the flowrate by Dupuit's formula

$$Q = \kappa \frac{T^2 - H_2^2}{2\ell_2} = \kappa \frac{T^2 - H_2^2}{2(\ell_1 - \ell)} \quad .$$

Comparison of these formulas allows us to find the location of the separation point of the flow, i.e., segment $\ell$ .

## §11. Seepage Around Structures.

We examine groundwater flow in riverbanks, due to the difference in water level in head — and tailwater reservoirs, when a screen protecting against seepage is available. The screen may be a vertical wall, disposed along the riverbank or cut in it, and has the purpose of lengthening the seepage path from the reservoir (headwater) to the lower water body.

We assume that the groundwater flow before the construction of the structure was directed from infinity normal to the rectilinear segment of the river and had a constant velocity $v = kI$, where $I$ is the initial gradient of the flow. We assume that the riverbanks are vertical and extend till the horizontal bedrock, which serves as a base for the construction.

Two types of screens are sketched in Fig. 312, 313. The riverbank coincides with the x-axis, so that the flow takes place in the upper half-plane. The construction of the structure disturbs the initial flow and adds a supplementary flow from upper to lower reservoir.

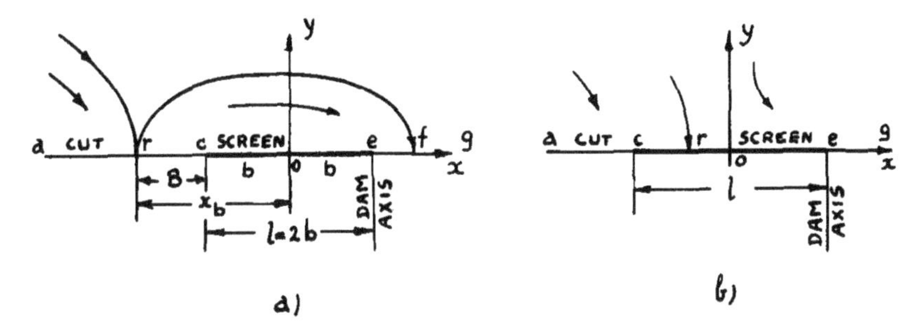


Fig. 312

The flow may branch in two different cases; when point $r$ of zero velocity is outside the screen (Fig. 312 a, and 313 a) and when that point is on the screen (fig 312 b and 313 b).

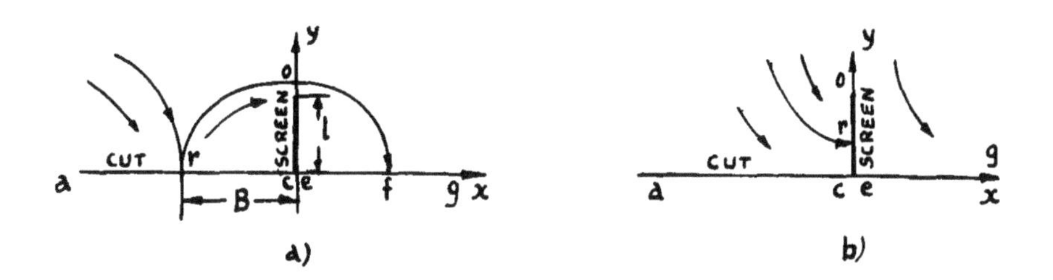


Fig. 313

First we find a solution for the confined flow corresponding to the sketches of Figs. 312-313. The region of the complex potential (reduced)

$$\omega = - H + \frac{i\psi}{k}, \quad H = - \frac{\varphi}{k}$$

is sketched in Figs. 314 and 315. Mapping of one of them onto the half-plane $\zeta$ gives:

$$\omega = Ai \int_0^\zeta \frac{(\alpha_1 + \zeta)d\zeta}{\sqrt{1 - \zeta^2}} + C.$$

For segment $-1 < \zeta < 1$

$$\omega = \frac{H_0}{\pi} \text{ arc sin } \zeta + \frac{H_0}{\pi\alpha} \sqrt{1 - \zeta^2} - \frac{H_1 + H_2}{2} ,$$

where $H = H_1 - H_2$, and $H_1$, $H_2$ are respectively the heads in upper and lower reservoirs.

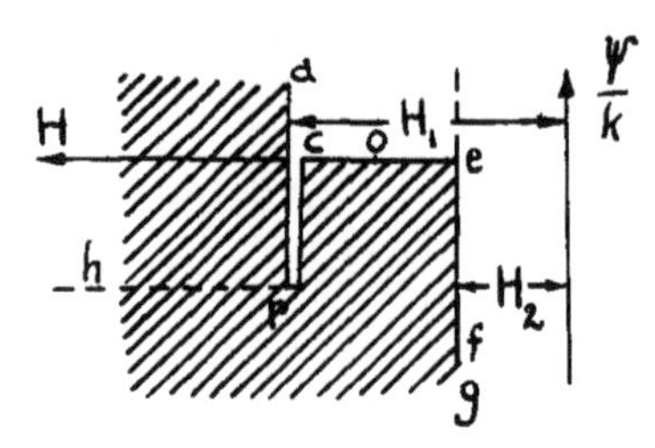


Fig. 314

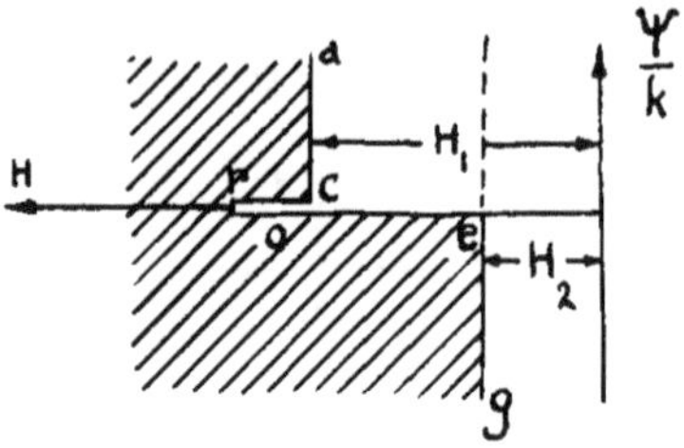

Fig. 315

For the segment $-\infty < \zeta < -1$ we have:

$$\omega = \frac{iH_0}{\pi} \text{ arch } \zeta + \frac{iH_0}{\pi\alpha} \sqrt{\zeta^2 - 1} - H_1 .$$

Mapping the flow region onto a half-plane, we obtain respectively: for the screen in prolongation of the riverbank

$$z = b\zeta , \tag{11.1}$$

where $b$ is the half-length of the screen, and for the screen normal to the riverbank

$$z = \ell \sqrt{\zeta^2 - 1} . \tag{11.2}$$

The coordinates of the point where the flow branches are found from the condition

$$v = kI, \quad u = 0$$

at infinity. From this we find for the prolongated screen

$$\alpha = - \frac{H_0}{\pi I b} ,$$

and for the perpendicular screen (called horizontal cut-off)

$$\alpha = - \frac{H_0}{\pi I \ell} .$$

The reduced potential is expressed in $z$ as follows:

for the prolongated screen: on segment $coe$

$$\omega = \frac{H_0}{\pi} \text{ arc sin } \frac{z}{b} - Ib \sqrt{1 - \frac{z^2}{b^2}} - \frac{H_1 + H_2}{2} ,$$

on segment $ca$

$$\omega = \frac{iH_0}{\pi} \text{ arch } \frac{z}{b} - iIb \sqrt{\frac{z^2}{b^2} - 1} - H_1 ;$$

For the horizontal cut-off: on segment coe

$$\omega = \pm \frac{H_0}{\pi} \operatorname{arch} \frac{z}{b} - iIz - \frac{H_1 + H_2}{2}$$

(plus for $-1 < \zeta < 0$, minus for $0 < \zeta < 1$), on segment ca

$$\omega = -\frac{iH_0}{\pi} \operatorname{arsh} \frac{z}{\ell} + iIz - H_1 .$$

For $\alpha = -1$, point b coincides with point c, consequently there is no seepage from upper to lower reservoir. In general the width of the zone of seepage around a structure in the region of the lower reservoir will be

$$B = x_b - b ,$$

for the prolongated screen, and

$$B = x_b$$

for the horizontal cut-off.

The abscissa $x_b$ is determined from equations (11.1) and (11.2) for $\zeta = -\alpha$.

For the prolongated screen, we have the head on the screen expressed as

$$H = Ib \sqrt{1 - \frac{x^2}{b^2}} - \frac{H_0}{\pi} \arcsin \frac{x}{b} + \frac{H_1 + H_2}{2}$$

and the flowrate

$$(11.3) \qquad Q = kH_0 A, \quad A = \frac{1}{\pi}\left(\operatorname{arch} \frac{1}{\beta} - \sqrt{1 - \beta^2}\right), \quad B = \frac{H_0}{\pi I} - b ,$$

where

$$(11.4) \qquad \beta = \frac{\pi I b}{H_0} .$$

For the flow with branchpoint on the screen, we have

$$Q = 0, \quad B = 0 .$$

The limit length of the screen $\ell_n = 2b_m$, for which the round about seepage from upper to lower reservoir does not exist, is expressed as

$$(11.5) \qquad \ell_m = \frac{2H_0}{\pi I} \approx \frac{2}{3} L .$$

Here $L = H_0/I$, distance from the line of the riverbank to the line of intersection of the plane of the upper reservoir with the initial plane of the piezometric heads.

In the case of a horizontal cut-off, we obtain for the head:

$$H = \frac{H_1 + H_2}{2} \pm \frac{H_0}{\pi} \arccos \frac{y}{\ell} + Iy .$$

Here $y$ is the distance from the point where the head is calculated to the bank line, $\ell$ the total length of the screen.

For flows of the first type, the flowrate is determined by means of formulas (11.3) and (11.4), where $b$ is replaced by the full length $L$ of the horizontal cut-off. The width of the zone of roundabout seepage in $B$ is

$$B = \ell \sqrt{\left(\frac{H_0}{\pi I \ell}\right)^2 - 1} \quad .$$

For flows of the second type $Q = 0$, $B = 0$. The limit length of the screen follows from the condition that there be no roundabout seepage:

$$\ell_m = \frac{H_0}{\pi I} \approx \frac{1}{3} L \quad .$$

From this it is evident that to avoid roundabout seepage, the prolongated screen must be twice as long as the horizontal cut-off.

Passing on to unconfined flow, we must replace $H$ by $h^2/2$. We obtain for the prolongated screen:

$$h^2 = \frac{h_1^2 + h_2^2}{2} - \frac{h_1^2 - h_2^2}{2} \text{arc sin} \frac{x}{b} + \frac{2Qb}{k} \sqrt{1 - \frac{x^2}{b^2}} \quad .$$

In case of flow of the first type,

$$Q = \frac{k(h_1^2 - h_2^2)}{2} A \quad ,$$

where $h_1$, $h_2$ are heads in upper and lower reservoirs.

$$A = \frac{1}{\pi}\left(\text{arch} \frac{1}{\beta_0} - \sqrt{1 - \beta_0^2}\right), \quad \beta_0 = \frac{2\pi Q b}{k(h_1^2 - h_2^2)} = \frac{\pi b}{L} \quad .$$

Here $L$ has the same value as in (11.5). For the value of $B$ we had:

$$B = \frac{k}{Q} \frac{h_1^2 - h_2^2}{2\pi} - b = \frac{L_c}{\pi} - b \quad .$$

For flow of the second type $Q = 0$, $B = 0$

$$\ell_m = 2b_m = \frac{k}{Q} \frac{h_1^2 - h_2^2}{2\pi} = \frac{2}{\pi} \frac{h_1 + h_2}{h_b + h_2} \cdot \frac{h_0}{I_{ave}} \approx \frac{2}{3} L \; .$$

Here $h_b$ is the depth of groundwater at the waterdivide, $I_{ave}$ is its average gradient on the segment from bedrock to river, $h_0 = h_1 - h_2$.

For the horizontal cut-off, we have the equalities

$$h^2 = \frac{h_1^2 + h_2^2}{2} - \frac{h_1^2 - h_2^2}{\pi} \text{arc cos} \frac{y}{\ell} + \frac{Q}{k} y \quad ,$$

$$B = \ell \sqrt{\frac{k}{Q}\left(\frac{h_1^2 - h_2^2}{2\pi \ell}\right)^2 - 1}$$

In the case of flow of the second type

$$\ell_m = \frac{k}{Q} \frac{h_1^2 - h_2^2}{2\pi} = \frac{1}{\pi} \frac{(h_1 + h_2) h_0}{\pi (h_b + h_2) I_{ave}} \approx \frac{1}{3} L \; .$$

The present theory is explained in reference [11]. Reference [10] treats these problems in a somewhat different way; in this book solutions are also given for problems where horizontal cut-offs and abutments are combined.

CHAPTER XI

# GRAPHICAL, NUMERICAL AND EXPERIMENTAL METHODS IN THE STUDY OF GROUNDWATER FLOW.

## A. Graphical and Numerical Methods.

### §1. Flownet.

The construction of a family of streamlines and equipotential lines (see Chapter II), sometimes called "hydrodynamical net" or "flownet" has a significant value in the study of two-dimensional steady flows.

It is convenient to replace the stream function and the velocity potential by proportional functions — the reduced flowrate $q$ and the head $h$:

$$(1.1) \qquad q = -\frac{\psi}{k}, \quad h = -\frac{\varphi}{k}.$$

The complex potential $\omega = \varphi + i\psi$ is replaced by

$$(1.2) \qquad \tilde{\omega} = h + iq .$$

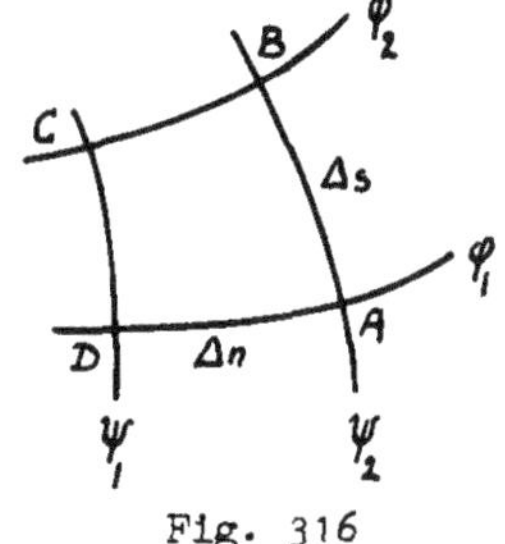

Fig. 316

Function $\tilde{\omega} = h + iq$, as well as $\omega$, is a function of the complex variable $z$, its real and imaginary parts are harmonic functions. The net $h$ = constant, $q$ = constant may be chosen equilateral, i.e., representing a curvilinear square. Indeed, let in point A (Fig. 316) $\Delta n$ be the distance between adjacent streamlines, $\Delta s$ the distance between adjacent equipotential lines. The velocity in point A is approximately

$$v_A \approx \frac{\Delta \varphi}{\Delta s} \approx \frac{\Delta\psi}{\Delta n} .$$

If we choose $\Delta\varphi = \Delta\psi$, then $\Delta s = \Delta n$, i.e., two adjacent sides of the rectangle ABCD are equal. This is called an equilateral net.

One may also construct a net composed of rectangular elements having a constant length/width ratio [2]. Putting $\Delta\varphi = m\Delta\psi$, where $m$ is a given number, obtains $\Delta s = m\Delta n$ .

We had examples of flownets in Chapter III for simple flows, when flowlines and equipotential lines were simple curves, with easy equations. Approximate methods are used in more complicated cases. We show the most important ones.

§2. Graphical Method for Flownet Construction [2].

We examine the confined flow under an hydraulic structure of arbitrary form (Fig. 317). Two streamlines are given, the contour of the base

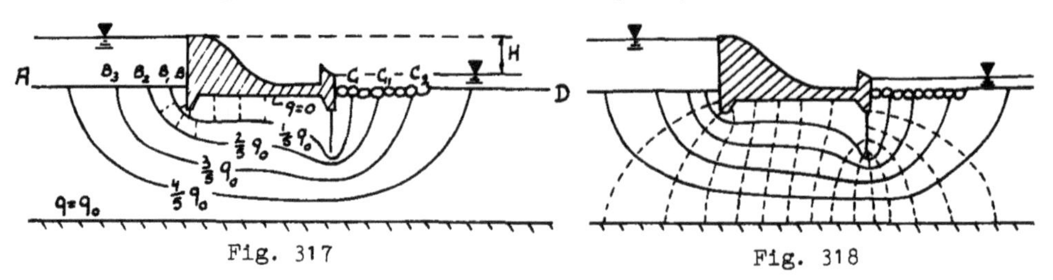


Fig. 317 Fig. 318

of the hydraulic structure and the boundary of the bedrock, and also two lines of equipotential, the boundaries of head – and tailwater reservoirs. We may assume $h = H$ and $h = 0$ for the latter lines.

We draw by eye streamlines $q = q_1, q_2, \ldots, q_5 = q_0$ (in Fig. 317 this means $q = 0, \frac{1}{5} q_0, \frac{2}{5} q_0, \frac{3}{5} q_0, \frac{4}{5} q_0$ and $q_0$). The value of $q_0$ is still unknown.

Now we construct equipotential lines. We start, for example with a line close to AB. By means of segments $BB_1, B_1B_2, \ldots$ we construct curvilinear squares. We notice that a curvilinear rectangle may be considered to be square if the lengths of its median lines are equal.

The squares, adjacent to line AB of Fig. 317, are such that their sides (in dotted line) do not lie in each other's prolongation, i.e., they do not form a constant potential line, as they should. Consequently, we must correct the streamlines, by shortening first the segment $BB_1$. Smoothing out the first series of squares we pass on to the second, etc. The last series, adjacent to line CD, must not consist of squares, in general, but will consist of rectangles. Finally we obtain the net of Fig. 318.

To check the correctness of the net, we may draw diagonal squares which also must form an orthogonal net.

We obtain other figures than squares in the vicinity of singular points. So, equipotential lines ending in corners of an impervious contour, divide the angles in two. At the tip of a cutoff, the tangent to an equipotential line is in the prolongation of the cutoff.

We assume now that we have constructed the entire net and that contour BC is divided in n parts (n may be a fractional number, when the last series of elements of the net consists of rectangles). The equipotential lines may now be labeled as

$$h = 0\,, \quad \frac{H}{n}\,, \quad \frac{2H}{n}\,, \ldots, \quad H\,.$$

We constructed the net equilateral; consequently, $q_0/s$ (where $s$ is the number of segments into which AB and CD are divided) must be equal to $H/n$:

$$\frac{q_0}{s} = \frac{H}{n} \ .$$

From this we find an approximate value for the reduced flowrate $q_0$ and the real flowrate $Q$

$$q_0 = \frac{s}{n} H , \quad Q = \frac{ks}{n} H \ . \tag{2.1}$$

It would have been possible to construct a net of rectangles instead of squares, subject to the condition of a constant length-width ratio of all rectangles.

If the pervious soil reaches till infinity, then one may take for streamline $q = q_0$, a half circle with sufficiently large radius, for example three times the length of the wetted contour (Chapter II, Par. 12).

In unconfined flow, one must satisfy the condition $h = y$ at the free surface, or $\Delta h = \Delta y$, where $\Delta h$, the head loss between two adjacent equipotential lines, $\Delta y$ the distance along the vertical of the intersections of adjacent equipotential lines with the free surface.

## §3. Method of Arithmetical Averages. Finite Differences.

This method is based on a property of the harmonic function: the value of the function in some point is equal to the average of its values along a circle centered around the point:

$$\varphi_0 = \frac{1}{2\pi} \int_0^{2\pi} \varphi_M \, d\theta \ .$$

Here $\varphi_0$ is the value of the harmonic function in point 0, $\varphi_M$ its value in point M of the perimeter. From this we obtain the approximate formula

$$\varphi_0 = \frac{\varphi_1 + \varphi_2 + \dots + \varphi_n}{n} \ ,$$

where $\varphi_1, \varphi_2, \dots$ — values of $\varphi$ in equidistant points along the perimeter. In particular, if we take only four points,

$$\varphi_0 = \frac{\varphi_1 + \varphi_2 + \varphi_3 + \varphi_4}{4} \ , \tag{3.1}$$

the four points are in the vertices of a square, $\varphi_0$ is the value of the function in the center of the square (Fig. 319).

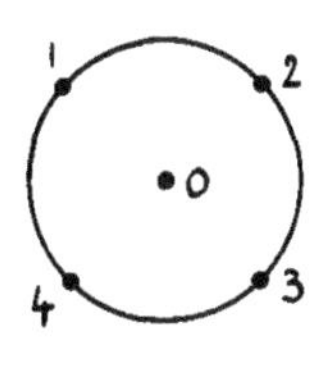

Fig. 319

We assume now that we have roughly sketched a system of iso-lines on graph paper. By interpolation between adjacent iso-lines we find approximate values of the function in the vertices of square ABCD (Fig. 320). We take their arithmetic average for the value of the function in point E, center of the square.

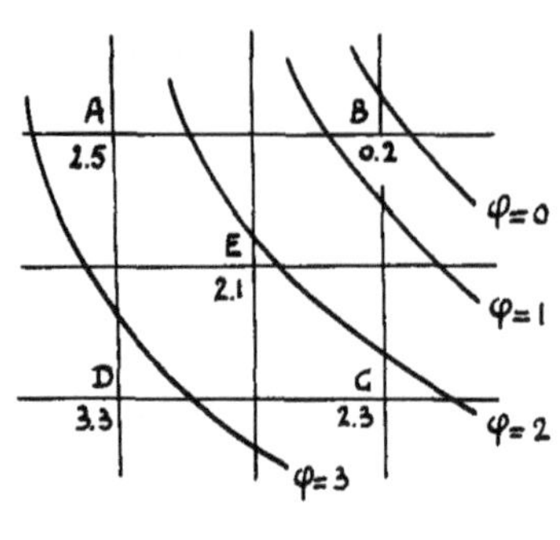


Fig. 320

Filling in this way the centers of all possible squares, we form the arithmetic average of the new values. They give us more precise values of our function in the vertices A, B, C, D, .... We may erase the old values and write the new ones instead. If we continue this process, we must obtain a system of numbers that tends to a well-defined limit — the exact values of the harmonic function in the vertices of the square [4]. Practically, after 3 or 4 trials, the values of the constants do not vary any more. Having written the values of $\varphi$ at the vertices of the squares, we shift the system of iso-lines accordingly.

By this method, we may construct one family of lines, for example, those of equal flowrate. The other family — the equipotentials — may be constructed orthogonally to the first or even independently from the first by the method of the arithmetical average.

The method of finite differences [4], applied to Laplace's equation, replaces the equation

$$\Delta u = 0$$

by its corresponding difference equation

$$u_{xx} + u_{yy} = 0 \quad , \tag{3.2}$$

where

$$u_{xx} = \frac{1}{h^2}\,[u(x+h,\,y) - 2u(x,\,y) + u(x-h,\,y)] \quad ,$$

$$u_{yy} = \frac{1}{h^2}\,[u(x,\,y+h) - 2u(x,\,y) + u(x,\,y-h)] \quad ,$$

or in other notations

$$u_{xx} = \frac{1}{h^2}\,(u_{k+1,i} - 2u_{k,i} + u_{k-1,i}) \quad ,$$

$$u_{yy} = \frac{1}{h^2}\,(u_{k,i+1} - 2u_{k,i} + u_{k,i-1}) \quad .$$

Consequently, equation (3.2) becomes

$$u_{k+1,i} + u_{k,i+1} - 4u_{k,i} + u_{k-1,i} + u_{k,i-1} = 0 \; . \tag{3.3}$$

From (3.3) one may obtain formula (3.1) for $u_{k,i}$

Given the value of the function or of its normal derivative (which we must express through its difference), along the contour of the region, or given the value of the function along some parts of the contour, and the value of the normal derivative along other parts, we may find a system of equations for $u_{k,i}$. Usually the number of equations and the number of unknowns is very

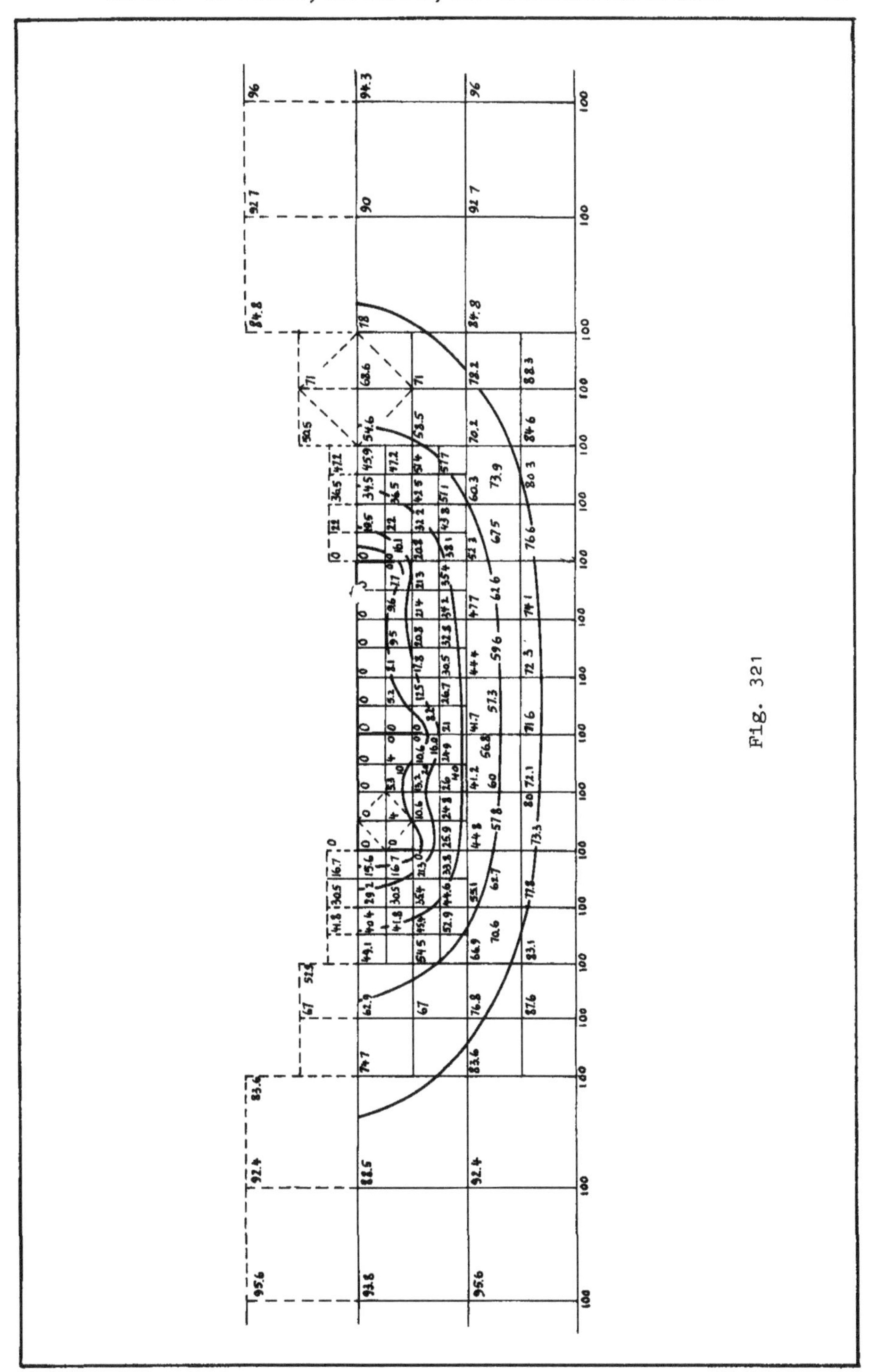

Fig. 321

large.

Methods of approximate solution of these systems have been examined by different authors [6]. Southwell [12] and his co-authors have applied these methods to a number of problems in various fields, including that of seepage.

The construction of the streamlines for a flat bottom wetted contour with three cut-offs on a stratum of finite depth, is given in Fig. 321 [7].

For non-homogeneous soil, one must consider the law of refraction of the streamlines at the boundary of the two soils. Examples of construction of flownets in such conditions are given in works [3, 8].

Examples of the construction of flownets for two different soils are given in Fig. 322 and 323.

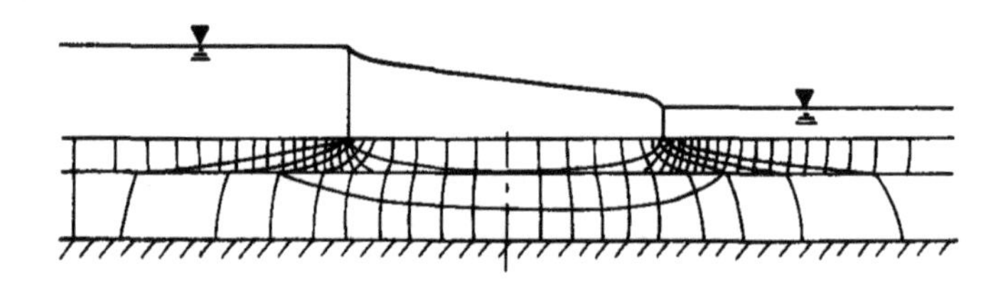

Fig. 322

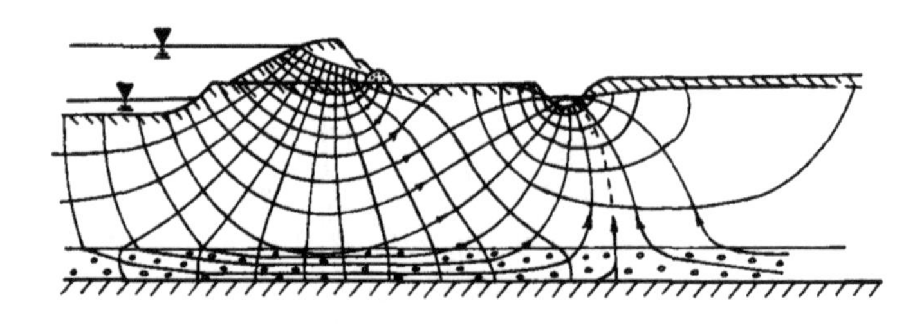

Fig. 323

§4. Axisymmetrical Flownet.

In the case of flow of an incompressible fluid with axial symmetry, the velocity components are expressed in terms of the stream function $\psi$:

$$(4.1)\qquad \begin{cases} v_r = \dfrac{1}{r}\dfrac{\partial\psi}{\partial z} , \\ v_z = -\dfrac{1}{r}\dfrac{\partial\psi}{\partial r} \end{cases}$$

The z-axis is the axis of revolution, $r$ the polar radius vector.

Function $\psi$ satisfies the equation

$$(4.2)\qquad \frac{\partial^2\psi}{\partial r^2} + \frac{\partial^2\psi}{\partial z^2} - \frac{1}{r}\frac{\partial\psi}{\partial r} = 0 .$$

When the flow is derived from a potential $\varphi$, the latter satisfies Laplace's equation in cylindrical coordinates

$$\frac{\partial^2\varphi}{\partial r^2} + \frac{\partial^2\varphi}{\partial z^2} + \frac{1}{r}\frac{\partial\varphi}{\partial r} = 0 \ . \tag{4.3}$$

We assume that we have constructed a net of lines $\psi$ = constant and $\varphi$ = constant. These lines are orthogonal, but it is impossible to represent them as an equilateral net, i.e., a net of squares. Indeed from equation (4.1) and from the property of function $\varphi$ we have:

$$v_r = \frac{1}{r}\frac{\partial\psi}{\partial z} = \frac{\partial\varphi}{\partial r} \ , \qquad v_z = \frac{\partial\varphi}{\partial z} = -\frac{1}{r}\frac{\partial\psi}{\partial r} \ .$$

If we assign to $\varphi$ and $\psi$ a series of equally spaced values, putting $\Delta\psi = \Delta\varphi$, then we must satisfy conditions

$$r\,\Delta z_\psi = \Delta r_\varphi \ , \qquad \Delta z_\varphi = r\,\Delta r_\psi \ . \tag{4.4}$$

Indices $\varphi$ and $\psi$ indicate between which lines the increment is taken. In the lower part of Fig. 325 $\Delta z_\varphi$ is almost constant, and therefore $\Delta r_\psi$ decreases with increasing $r$.

We derive a formula [5] to compute the values of the stream function $\psi_m$ in the center of a square with sides $2n$, when the values $\psi_a$, $\psi_b$, $\psi_c$, $\psi_d$ of this function in the vertices of the square are known (fig. 324). We draw the lines $\psi = \psi_a$, ... $\psi = \psi_d$ through the vertices of the square and compute the flowrate between the surface of revolution $\psi = \psi_a$ and $\psi = \psi_m$. The differential flowrate $dQ$ between the surfaces $\psi$ and $\psi + d\psi$ is

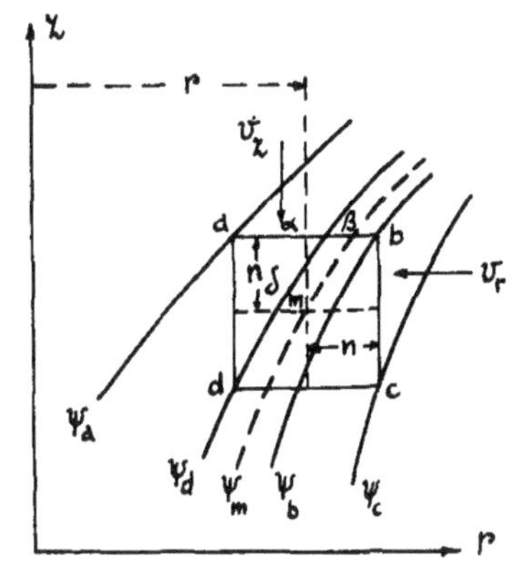

Fig. 324

$$dQ = 2\pi r v_z \, dr = 2\pi \frac{\partial\psi}{\partial r} dr \ .$$

By an appropriate choice of the integration constant, we obtain

$$Q = 2\pi\psi \ . \tag{4.5}$$

For the flowrate $q_{am}$ between surfaces $\psi_a$ and $\psi_m$ we find, considering that the flowrate between the sides $\alpha\beta$ and $m\alpha$ is the same:

$$q_{am} = q_{a\alpha} + q_{\alpha\beta} = q_{a\alpha} + q_{\alpha m} = 2\pi\left(r - \frac{n}{2}\right)nv_z + 2\pi r \cdot nv_r \ .$$

From this, expressing the flowrates in terms of the stream function by means of formula (4.5) and reducing by $2\pi$, we find

$$\psi_m = \psi_a + \left(r - \frac{n}{2}\right)nv_z + rnv_r \ .$$

In the same way, computing the flowrate between surfaces $\psi_m$ and $\psi_b$, we have

$$\psi_m = \psi_b - \left(r + \frac{n}{2}\right)nv_z + rnv_r \quad .$$

Similarly we find

$$\psi_m = \psi_c - \left(r + \frac{n}{2}\right) nv_z - (r + n)nv_r \quad ,$$

$$\psi_m = \psi_d + \left(r - \frac{n}{2}\right) nv_z - (r - n)nv_r \quad .$$

Adding the four expressions for $\psi_m$, and dividing by 4, we find:

$$\psi_m = \frac{\psi_a + \psi_b + \psi_c + \psi_d}{4} - \frac{n^2}{2} v_z \; .$$

Therefore we may take for $v_z$ one of the approximate expressions

$$v_z = \frac{1}{r}\frac{\partial\psi}{\partial r} \approx \frac{1}{r}\frac{\psi_b - \psi_a}{2n} \approx \frac{1}{r}\frac{\psi_c - \psi_d}{2n} \approx \frac{1}{r}\frac{\psi_b + \psi_c - \psi_a - \psi_d}{4n} \quad .$$

In the following paragraph, we examine an application of the graphical method.

## §5. Net for Steady Well Recharge (Axial Symmetry).

Permeability of soils is determined by two field test methods: pumping and recharge methods.

We assume that we have steady conditions in a recharge test. Therefore we must carry out the test until the flowrate to the well is practically constant. V. M. Nasberg [10] analyzed such a case; the ratio

$$\frac{h}{d} = 50 \quad , \tag{5.1}$$

where $h$ is the height of the water column in the borehole, above its bottom, $d$ the diameter of the borehole. The acting head above some plane $a_1c_1$ (Fig. 325) is designated by $H$. Equipotential surfaces are constructed through equal intervals of head $\Delta H$, so that

$$\Delta H = \frac{H}{n} \quad , \tag{5.2}$$

where $n$ is the number of head zones. The walls of the streamtubes are constructed through equal flowrate intervals, i.e., the flowrates $\Delta Q$ of each streamtube made by two arbitrary adjacent walls, must be unique:

$$\Delta Q = \frac{Q}{t} \quad , \tag{5.3}$$

where $Q$ is the total flowrate to the bore hole and $t$ the number of streamtubes.

The condition $h = z$ at the free surface must be satisfied and the vertical distance between horizontal planes through the intersection of the equipotential surfaces with the free surface must be a constant $\Delta H$.

In the construction of the net, we make use of some methods applied in the representation of electrostatic fields with axial symmetry [11].

We call the quantity

$$\Delta R = k\,\frac{\Delta H}{\Delta Q} = \text{constant} \quad , \tag{5.4}$$

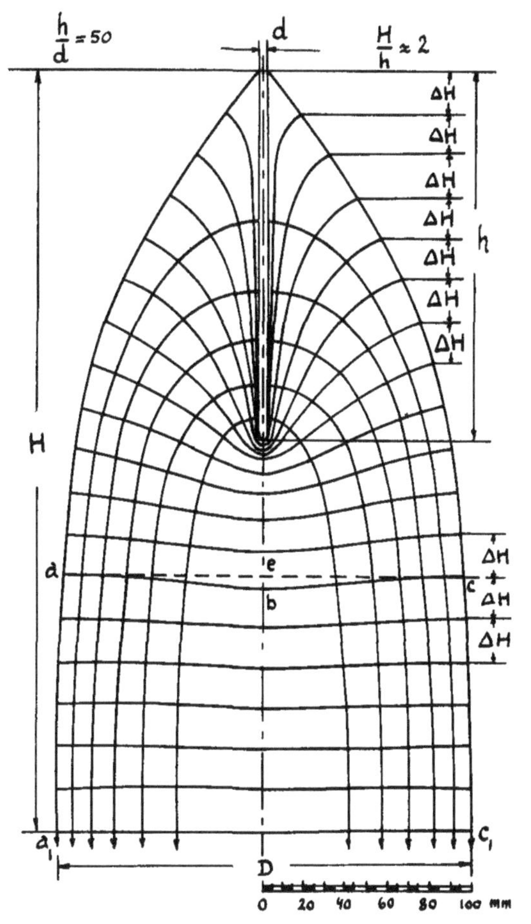


Fig. 325

the "reduced resistance" of the streamtube segment between two adjacent equipotential surfaces. From (4.1) we have the approximate equalities:

$$\frac{\Delta\varphi}{\Delta s} = \frac{1}{r}\frac{\Delta\psi}{\Delta n}, \quad \frac{\Delta\varphi}{\Delta\psi} = \frac{\Delta s}{r\,\Delta n},$$

where $\Delta s$, $\Delta n$ are two mutually orthogonal directions. Putting $\Delta\varphi = k\Delta H$, $2\pi\Delta\psi = \Delta Q$, $\Delta s = \ell$, $\Delta n = b$, we obtain:

$$\Delta R = \frac{\ell}{2\pi r b} \quad . \tag{5.5}$$

For $r$ small (close to the well in Fig. 325), the stream surfaces are almost horizontal planes and the equipotential surfaces are close to vertical cylinders and we may use formula (1.4) of Chapter IX for the flow-rate $\Delta Q$. Therefore

$$\Delta R = \frac{\ln r_1/r_2}{2\pi b} \quad . \tag{5.6}$$

The following notations are used in these formulas (Fig. 326):

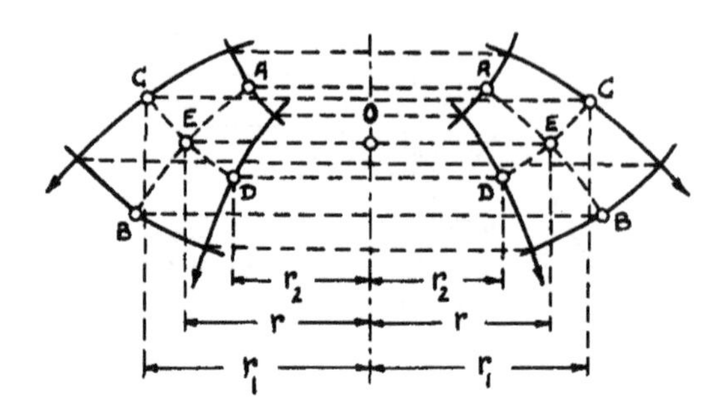


Fig. 326

$\ell$, average length of a streamtube segment between adjacent equipotential surfaces ($\ell$ = AB in Fig. 326); b, average width of a diagonal cross-section of a streamtube segment (b = CD); $2\pi rb$, area of the average cross-section of a streamtube segment; r, distance from the center of the streamtube cross-section to the axis of the bore hole (r = OE); $r_1$, distance from point C to that axis; $r_2$, distance from point D to that axis.

From equations (5.2) (5.3) (5.4), we obtain a formula to determine the seepage coefficient k:

$$k = \frac{\Delta R}{H} \cdot \frac{n}{t} \cdot Q = \frac{\Delta R}{\Delta H} \cdot \frac{1}{t} \cdot Q = CQ . \tag{5.7}$$

It is apparent from this formula that, knowing the constant C and having determined Q by experiment, we may compute the unknown seepage coefficient.

Constant C has dimensions: $L^{-2}$. To find the natural value of C, we must divide the constant $C_m$, found on the sketch of the flownet (on scale 1:m) by $m^2$. We find

$$C = \frac{C_m}{m^2} . \tag{5.8}$$

In the net of Fig. 325,

$$\frac{h}{d} = 50, \quad h = 175\text{mm}, \quad \Delta H = 20\text{mm}, \quad t = 6, \quad \Delta R = 3.82\times10^{-3} \frac{1}{\text{mm}} .$$

By means of (5.7) we have

$$C_m = \frac{\Delta R}{\Delta H} \times \frac{1}{t} = 31.8 \times 10^{-6} \frac{1}{\text{mm}^2} . \tag{5.9}$$

As a check, we may take several segments of Fig. 326 and compare the results with $C_m$. If in nature we have a bore hole, drilled as in Fig. 325, with water depth h', then m = h': h.

Let us take h' = 5 m. Then the diameter of the bore hole must be d' = 5/50 = 0.1 m.

For the value of C we find by formula (5.8):

$$C = 31.8 \times 10^{-6} : (28.6)^2 = 0.0389 \frac{1}{\text{m}^2} .$$

If Q is expressed in $m^3$/sec and k in m/sec, then we have for k

$$k = 0.0389Q$$

for $h' = 5$ m, $d' = 0.1$ m, and Q.

If for a given well the ratio $h' : d' \neq 50$, then it would be necessary to construct another flownet. However we may use an approximate method to compute the flowrate for a well diameter $d''$, if the flowrate is known for a diameter $d'$, all other conditions being the same for both wells.

Let us consider two partially penetrating wells in a finite confined layer, both of length $h'$, head H, but with different diameters: $d'$ and $d''$. Assigning the values $C'$ and $C''$ respectively for C, V. M. Nasberg derives the formula

$$\frac{C'}{C''} = \frac{\ell n \frac{4h'}{d'}}{\ell n \frac{4h'}{d''}} .$$

This formula is not exact in the case of recharge that we considered, but the author proves that it may be used as an approximation when $25 < \frac{h'}{d'} < 100$.

In §15 of Chapter IX, we treat the problem of a well in an infinite stratum and introduce N. K. Girinsky's function. Among some other formulas, we arrive at an expression that differs little from the one above:

$$\frac{C'}{C''} = \frac{\ell n \frac{3.2h'}{d'}}{\ell n \frac{3.2h'}{d''}} .$$

## §6. Construction of an Isobar.

Having families of curves $\varphi$ = constant and $\psi$ = constant, by graphical addition we may construct the lines $\chi = \varphi + \psi$ = constant. Fig. 327 gives an example of such a construction.

In Chapter II we derived the equation

$$p = h + y , \tag{6.1}$$

where p is expressed in meters of water. If we construct the lines $h = C_1$ and $y = C_2$, then in the points of intersection of these lines we have

$$p = C_1 + C_2 .$$

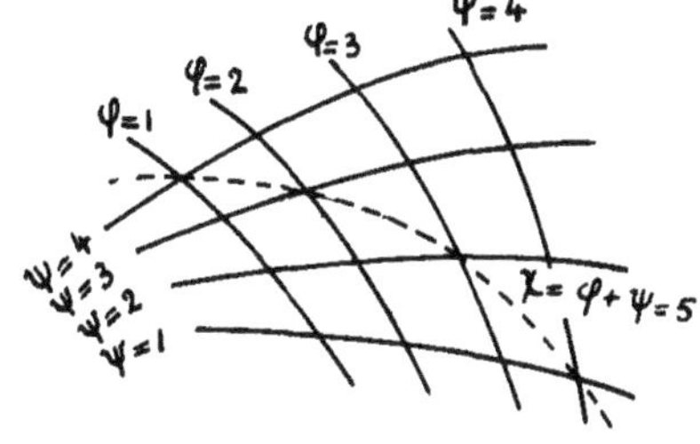


Fig. 327

Choosing values of $C_1$ and $C_2$ such that their sum is equal to a given number $C_1 + C_2 = C$, and drawing a smooth line (the diagonal of the rectangles made by $h = C_1$ and $y = C_2$), we have the isobar

$$p(x, y) = C.$$

A family of isobars for an earthdam on bedrock is given in Fig. 328;

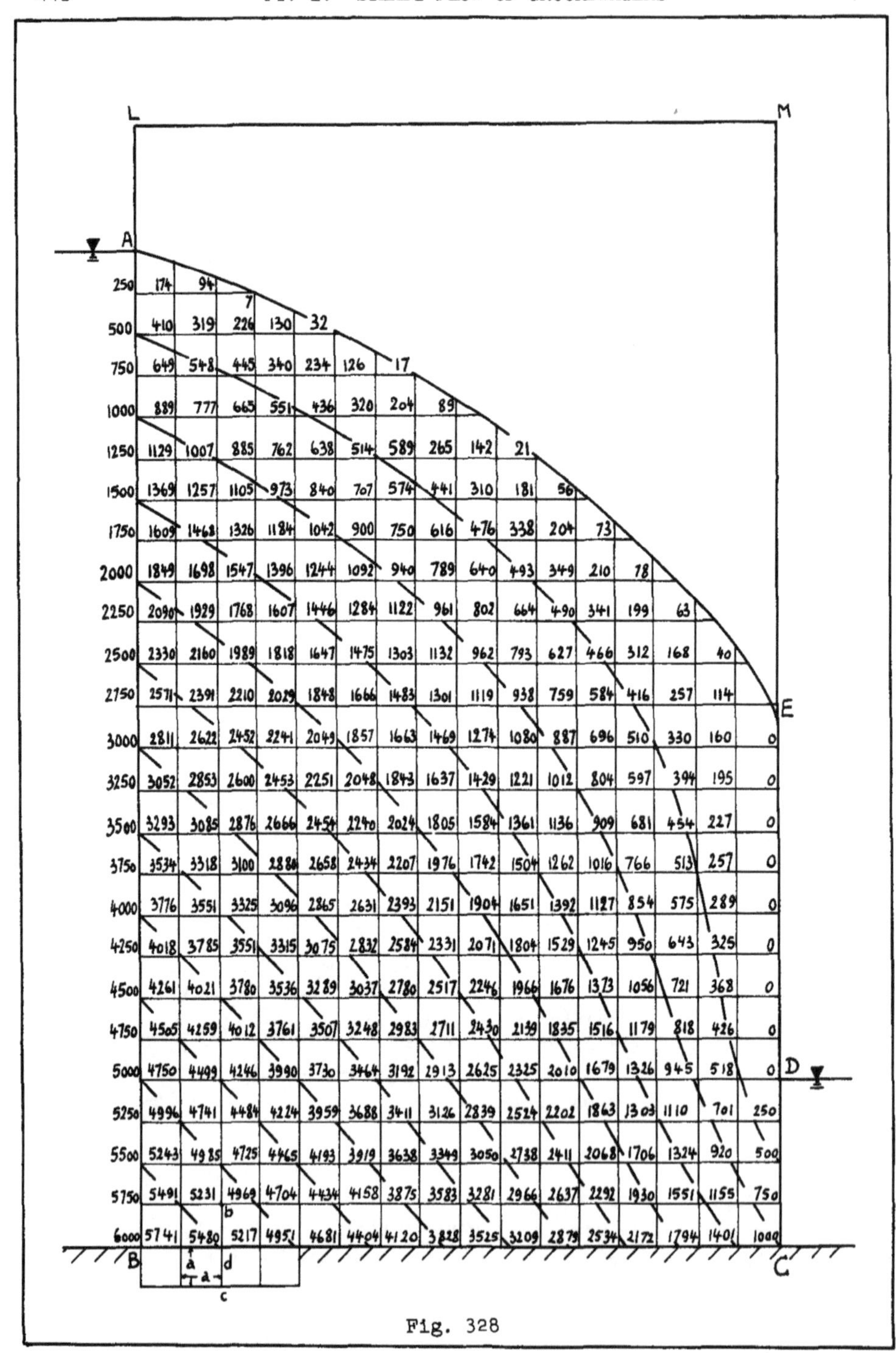
L
M
A
E
D
B
C
a
b
c
d
250
500
750
1000
1250
1500
1750
2000
2250
2500
2750
3000
3250
3500
3750
4000
4250
4500
4750
5000
5250
5500
5750
6000

Fig. 328

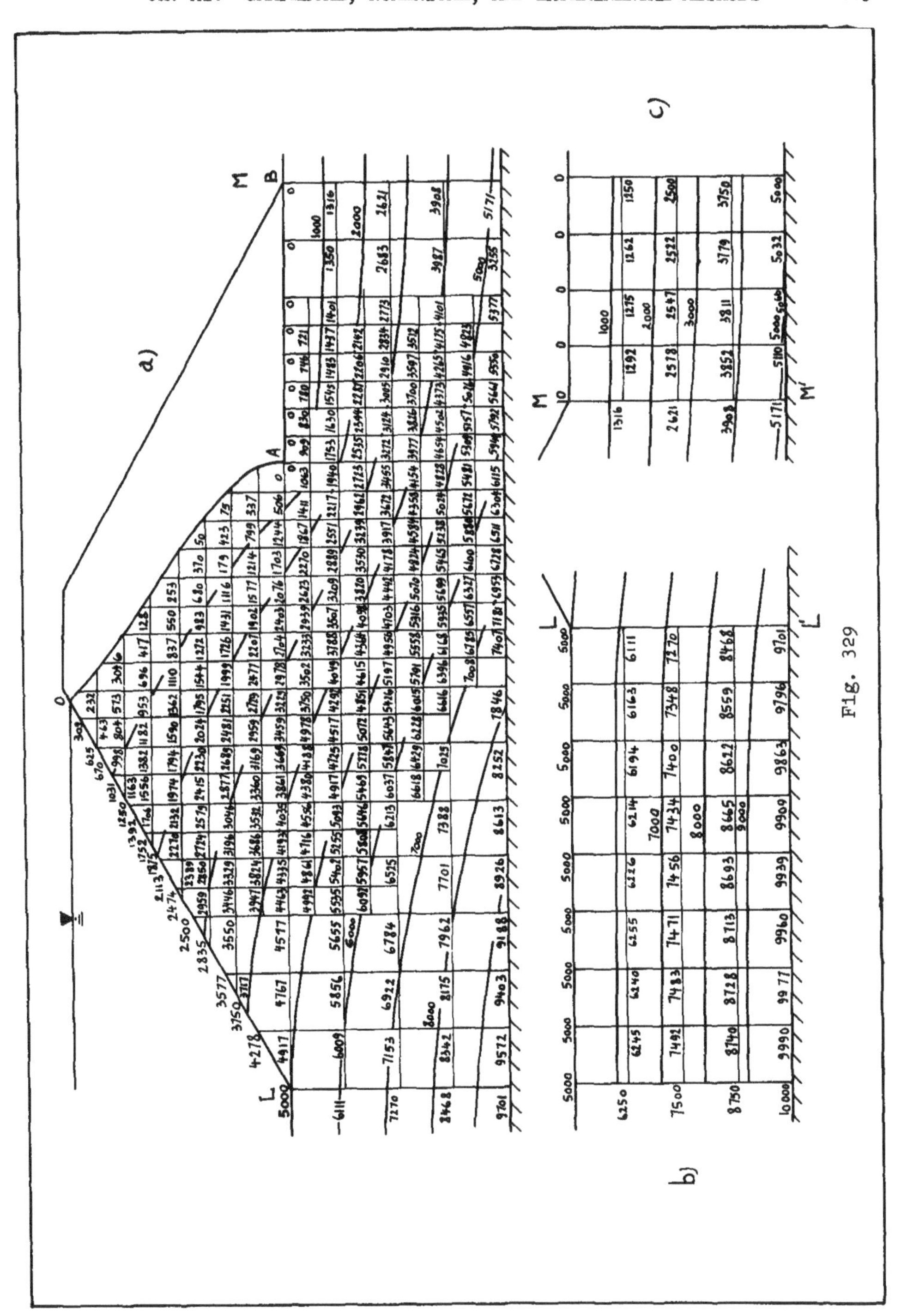


Fig. 329

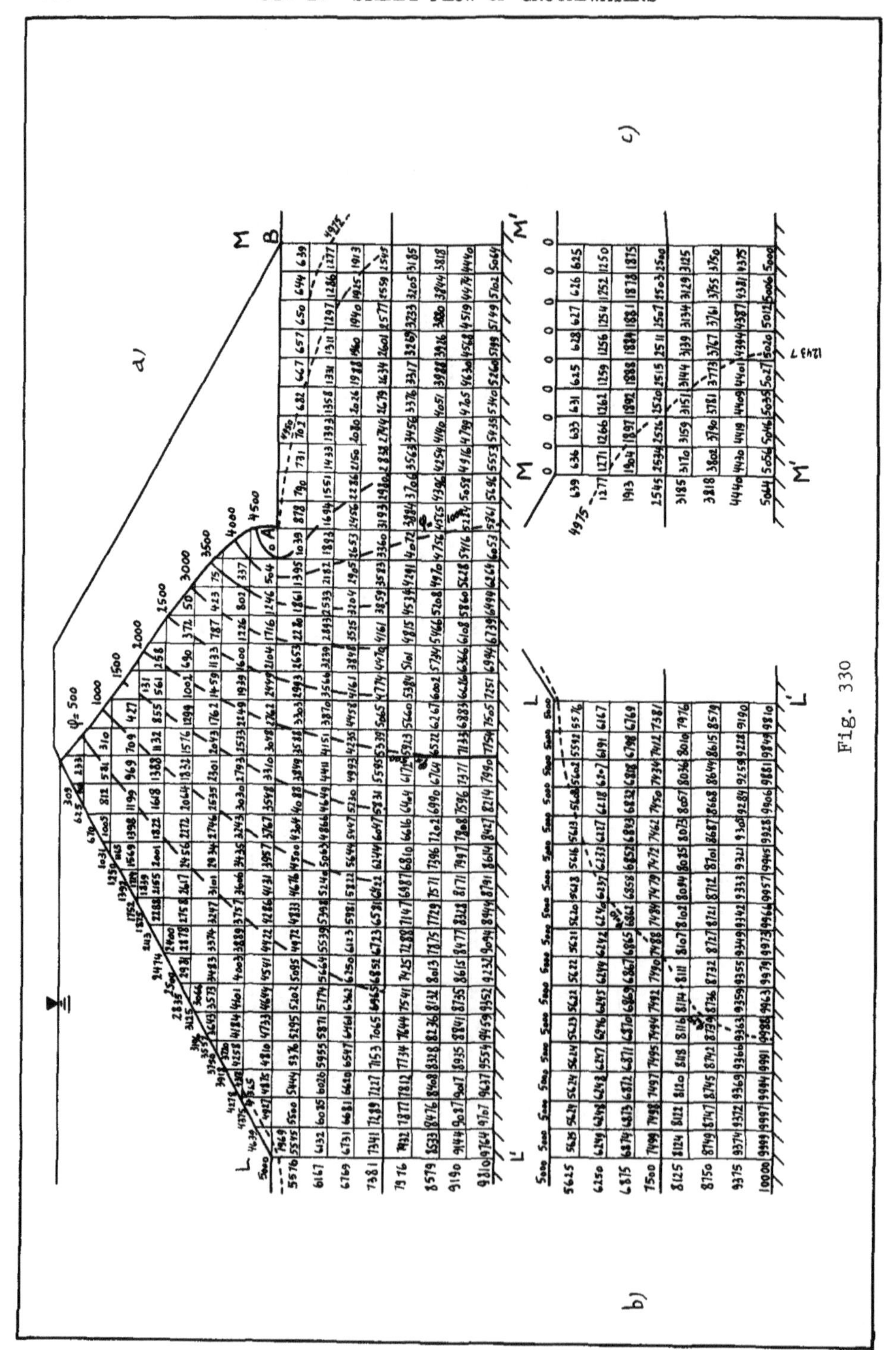


Fig. 330

for a dam on a pervious stratum of finite depth, see Fig. 329. In Fig. 330, to construct the net of pressure lines, equipotential lines are plotted for an earthdam on a non-homogeneous stratum (the seepage coefficients of upper and lower parts have the ratio 4 : 1) [12]. By graphical addition one may derive the equipotential lines from the isobars.

We observe that in the construction of isobars for the seepage through an earthdam, we encounter the difficulty that the free surface is not known in advance, and it must be determined as a streamline along which $h = y$.

Using the fact that the free surface here is an isobar, we may find along the free surface

$$\frac{\partial p}{\partial n} = - \cos(n, y) \, . \tag{6.2}$$

where $n$ the normal to the free surface outward from the flowregion.

Equation (6.2) may be derived from the condition that at the free surface

$$\frac{\partial \varphi}{\partial n} = 0 \, . \tag{6.3}$$

Since $\varphi = - k(p + y)$, for $\rho g = 1$,

$$\frac{\partial \varphi}{\partial n} = - k\left(\frac{\partial p}{\partial n} + \frac{\partial y}{dn}\right) = 0 \, . \tag{6.4}$$

But $dy/dn$, as apparent from Fig. 331, is equal to the cosine of the angle between $n$ and $y$:

$$\frac{dy}{dn} = \sin \alpha = \cos(n, y) \, .$$

Inserting this value of $dy/dn$ in equation (6.4) we obtain (6.2). The latter may serve to check the construction of the free surface or the construction of the isobar $p = p_1$, when either one is known. In the latter case, we have approximately

$$p_1 - p_0 \approx \Delta n \cos(n, y) \, ,$$

thus allowing us to find the length of the normal segments $\Delta n$ between $p_0$ and $p_1$, after construction of the normal in a series of points of the free surface and computation of $\cos(n, y)$.

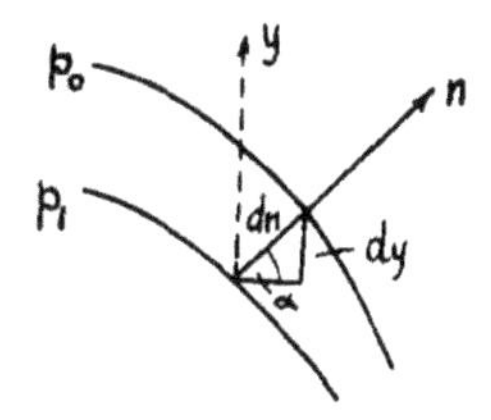

Fig. 331

§7. Fictitious Flow in the Hodograph Plane.

We assume that we mapped the complex potential plane onto the $w = u - iv$ plane. Then $\omega(w)$ represents a flow in the hodograph plane, i.e., a fictitious flow that must satisfy the same boundary conditions for $\varphi$ and $\psi$, as the flow in the real z-plane [13].

To illustrate this, we treat the flow from infinity towards a vertical seepage surface of a dam on impervious foundation, when there is no tailwater [14] (Fig. 332). This is a particular case of the vertical cofferdam treated in Chapter VII.

The hodograph of the cofferdam is sketched in Fig. 333. The free surface corresponds to an arc of the circle $u^2 + v^2 + kv = 0$, the impervious foundation to the u-axis, the seepage surface to a straight line parallel to the abscissa axis and passing through the point $u = 0$, $v = -k$. Lines AB and AC represent streamlines of the fictitious flow in the hodograph plane. We label them respectively $\psi = 0$ and $\psi = 1$. (When we find Q, then we know the value of the streamfunction.)

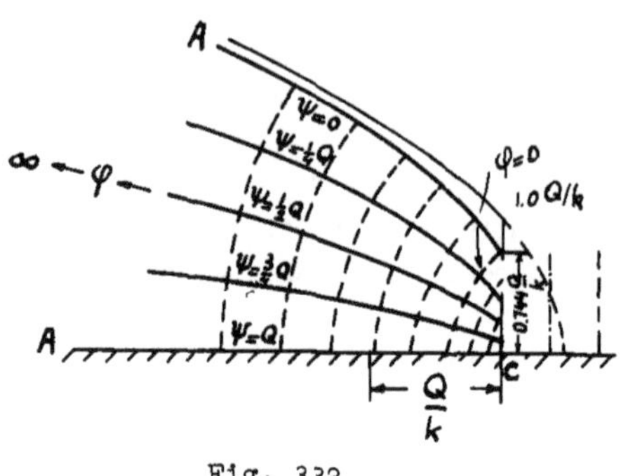

Fig. 332

Line BC represents a seepage surface. To investigate the behaviourof the streamline along this straight line, we may construct a family of lines that form a constant angle with the velocity (u, v). It is a bundle of straight lines passing through the origin and not sketched in Fig. 333. The angle $\theta$ of each of them with the abscissa axis varies from zero in point C till 90° in point B.

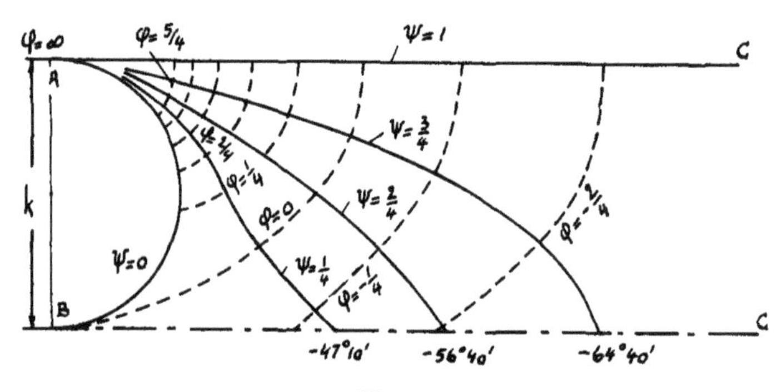


Fig. 333

Since $z$ is mapped conformally onto $u + iv$, with preservation of angles (but with change of orientation of angles, conformal mapping of the second kind), streamlines must intersect the segment BC under the same angle $\alpha$, as do streamlines with the seepage surface in the z-plane. Therefore $\alpha$ is the complement of $\theta$ to 90°.

Indicating the values of $\alpha$ along segment BC in Fig. 333, we construct a family of flowlines, starting from A, and also equipotential lines. We assume zero for the potential line passing through B. In A, the equipotentials tend to $\varphi = \infty$ and the velocity of the fictitious flow will be infinite there.

## §8. Construction of the Net in the Flowregion.

Streamlines of the actual flow in the z-plane correspond to

streamlines of the fictitious flow. The same holds for the equipotential lines. We show how these isolines are transformed from the $u + iv$ plane onto the $x + iy$ plane.

From the equation

$$\frac{d\omega}{dz} = w$$

we obtain

$$z = \int \frac{d\omega}{w} = \int \frac{d\omega}{u - iv} \quad . \tag{8.1}$$

Introducing the modulus $V$ of the velocity and the argument $\theta$ of the velocity vector with the x-axis,

$$w = Ve^{-i\theta}, \quad \frac{1}{u - iv} = \frac{1}{V} e^{i\theta} \quad .$$

The expression

$$\frac{d\omega}{w} = \frac{1}{V} e^{i\theta} d(\varphi + i\psi)$$

may be developed into its real and imaginary parts:

$$z = x + iy = \int \frac{1}{V}(\cos\theta \, d\varphi - \sin\theta \, d\psi) + i \int \frac{1}{V} (\sin\theta \, d\varphi + \cos\theta \, d\psi) \, . \tag{8.2}$$

Along a streamline $d\psi = 0$, and therefore streamlines in the z-plane are determined by the equations

$$x = \int_{\varphi_1}^{\varphi} \frac{\cos\theta}{V} d\varphi, \quad y = \int_{\varphi_1}^{\varphi} \frac{\sin\theta}{V} d\varphi \tag{8.3}$$

Along equipotential lines $d\varphi = 0$, and therefore equipotential lines in the plane of flow are

$$x = -\int_{\psi_1}^{\psi} \frac{\sin\theta}{V} d\psi, \quad y = \int_{\psi_1}^{\psi} \frac{\cos\theta}{V} d\psi \, . \tag{8.4}$$

Here $\varphi_1$, $\psi_1$ are some arbitrary values of $\varphi$ and $\psi$.

If we have the lines $\varphi$ = constant and $\psi$ = constant in the hodograph plane, then, by graphical or numerical integration of formulas (8.3) and (8.4), we may construct the flownet in the z-plane.

We take for example the equipotential $\varphi = 0$ in the hodograph plane (Fig. 333). We locate on this line points with values $\psi = 0, 1/4, 2/4, 3/4, 1$, take in these points values of $V$ and $\theta$ and construct the graphs $(\cos\theta)/V$ and $(\sin\theta)/V$ as functions of $\psi$ (Fig. 334). Carrying out the numerical integration, we find values of $x$ and $y$ for the equipotential $\varphi = 0$. The integral for $y$ between limits 0 and 1 gives the reduced length of the seepage surface segment:

$$\bar{h}_0 = \int_0^1 \frac{\cos\theta}{V} d\psi \, .$$

Actual computations [12] gave for $\bar{h}_0$ the value 0.744. As will be shown in §10 of Chapter VII, one must have $\bar{h}_0 = 0.7425$. The graphoanalytical result 0.744 differs only from the real one by 0.2%. If we carry out the integration along other isolines of $\varphi$ and $\psi$ in the hodograph plane, then we obtain the flownet of the z-plane (Fig. 332). In particular, integration along the half circle of the hodograph plane gives the coordinates of points of the free surface. The condition $\varphi + k(y - h_0) = 0$ at the free surface may serve to check the accuracy of the construction of this curve.

§9. Isoclines and Isotaches of the Fictitious Flow [13].

Considering the complex potential $\omega$ as a function of $w$, we may take its derivative with respect to $w$:

$$\frac{d\omega}{dw} = \tilde{W} = \tilde{V}e^{-i\tilde{\theta}} , \tag{9.1}$$

This may represent the complex velocity of the fictitious flow. $\tilde{V}$ is the modulus of the velocity, $\tilde{\theta}$ the angle made by the velocity with the u-axis. Taking $\ln \tilde{W}$ and separating its real and imaginary parts, we obtain:

$$\ln \tilde{W} = \ln \tilde{V} - i\tilde{\theta} .$$

The lines $\tilde{V}$ = constant and $\theta$ = constant represent an orthogonal net of isotaches and isoclines of the fictitious flow. By means of this net, one may more accurately construct the streamlines and equipotential lines of the fictitious flow.

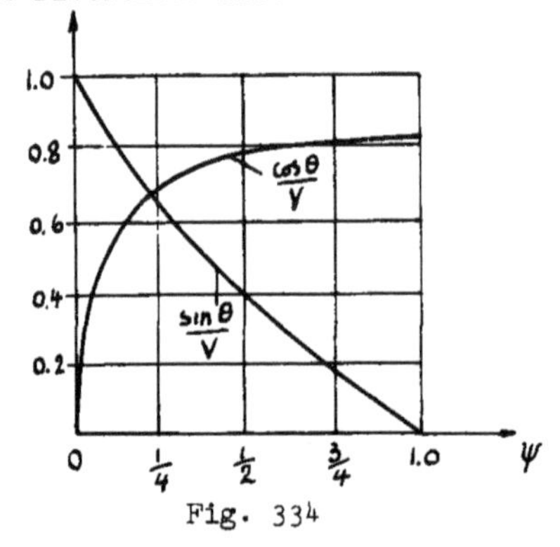


Fig. 334

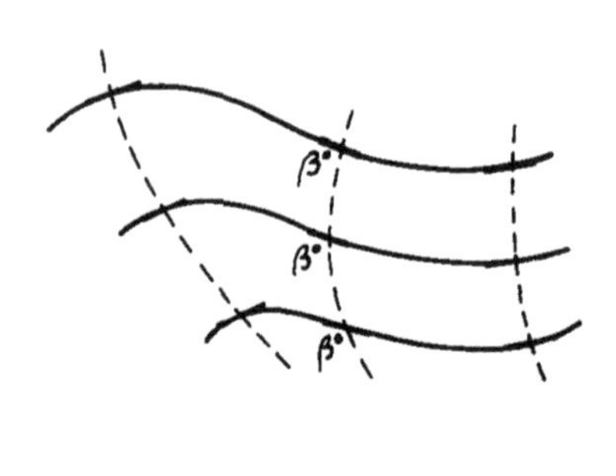


Fig. 335

Let us assume that we have a rough sketch of the fictitious flow (it is desired to construct a dense net). We draw tangents to the streamlines making a constant angle $\beta$ with the abscissa axis and join the corresponding points by a smooth line. This is the isocline (Fig. 335). After construction of the isoclines corresponding for example to $\beta = 0°, 5°, 10°...$, we construct the lines $\tilde{V}$ = constant, orthogonal to the isoclines. If we obtain an equilateral net, then the values of $\tilde{V}$ along adjacent isotaches, $n$ and $n + 1$, will satisfy

$$\ln \frac{V_{n+1}}{\tilde{V}_n} = \text{arc } 5° = \frac{\pi}{36}$$

or

$$\tilde{V}_{n+1} = e^{\frac{\pi}{36}} \tilde{V}_n = 1.091\tilde{V}_n .$$

Therefore, $\tilde{V}_1$, initially arbitrary, may be determined at the end of all constructions.

The net obtained in this way must adjusted as long as the correct system of squares is not obtained.

Inside of the hodograph region, one finds points in which isoclines with the same value intersect each other, usually under 90°.

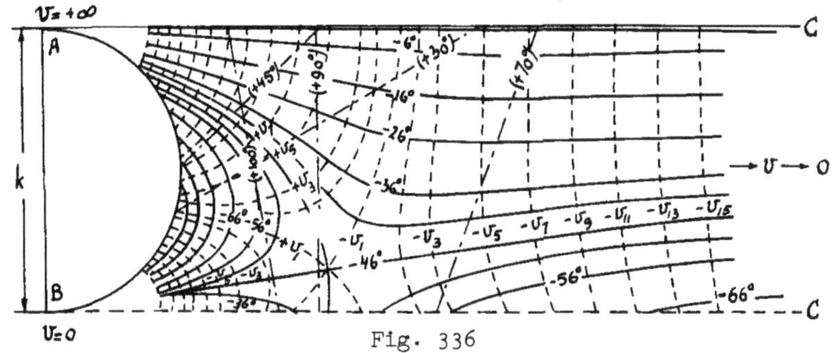


Fig. 336

The isotaches intersect each other also at the same point, also under a 90° angle, making a 45° angle with the isoclines. Such a singular point is a branch point. Isoclines and isotaches of the fictitious flow that we analyzed in the previous chapter, are sketched in Fig. 336 [14]. They have one branch point. Drawing the tangent in some point of the half circle of the velocity hodograph, we find the angle that corresponds to the isocline that passes through this point. The isoclines leave the half circle and in part tend to infinity, in part converge to the lower part of the half circle.

From the system of isoclines and isotaches of the fictitious flow, we may pass to the system of lines $\varphi$ = constant and $\psi$ = constant in the hodograph plane, if we integrate equation (9.1) along an arbitrary path in the w plane. We put

$$\omega = \int \tilde{V} e^{-i\tilde{\theta}}\, dw \tag{9.2}$$

and let

$$dw = e^{i\theta_s}\, ds ,$$

where ds is an element of arc length of the integration path, $\theta_s$ is the angle of the tangent to this arc with the abscissa axis. We have:

$$\omega = \int_{s_0}^{s} \tilde{V} e^{i(\theta_s - \tilde{\theta})}\, ds ,$$

or

$$\varphi = \int_{s_0}^{s} \tilde{V}\cos(\theta_s - \tilde{\theta})ds, \quad \psi = \int_{s_0}^{s} \tilde{V}\sin(\theta_s - \theta)ds . \tag{9.3}$$

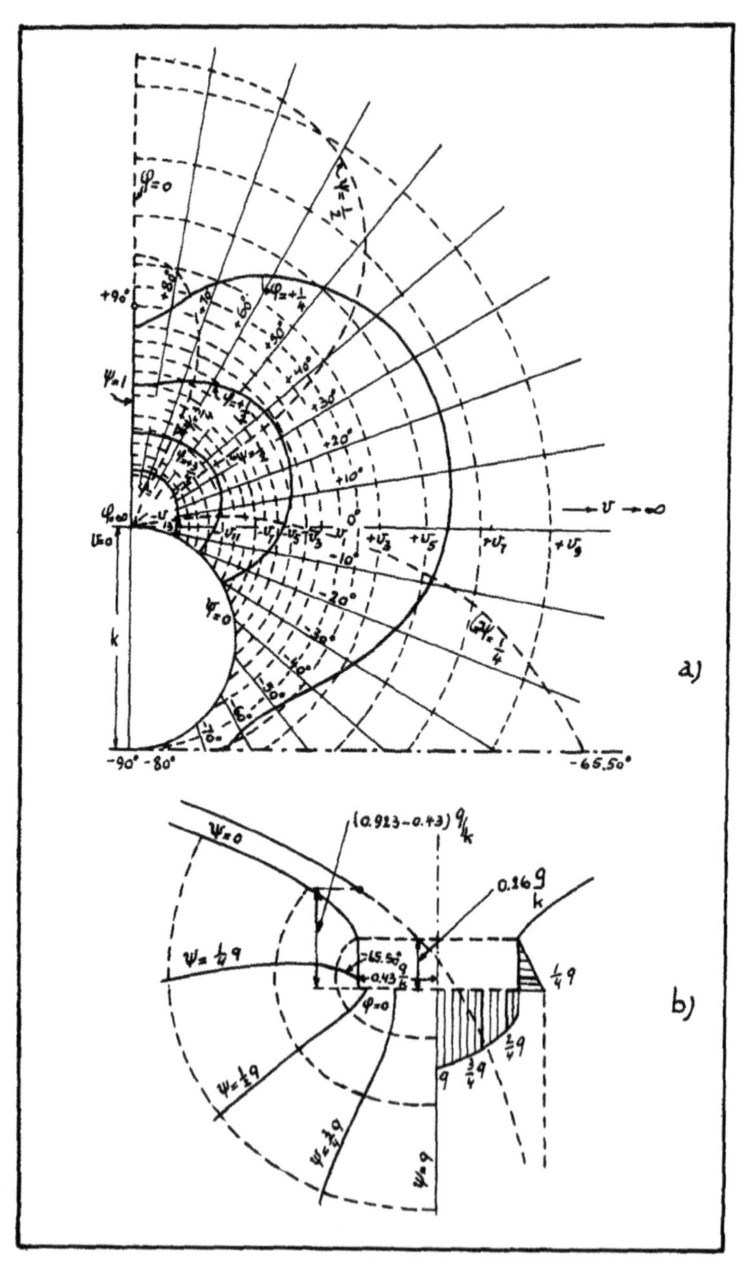

Fig. 337

Here $s_0$ is the beginning and $s$ the final value of the arc length of the integration path. The values $\varphi$ and $\psi$ in any point, obtained along different integration paths, must be the same.

By means of the integrals (9.3) one may find the values of the velocity potential and the stream function in different points of the hodograph region and so introduce clarity in the construction of the net.

Some examples were examined by means of such a method by Weinig and Childs [13], and later by Breitenoder [14]. Fig. 337 a and b represent the construction of the flownet in the case of flow of groundwaters towards a ditch of rectangular cross-section.

One may show [15] the method to determine the radii of curvature of the lines $\varphi$ = constant and $\psi$ = constant, which offers the possibility of a more accurate construction of these lines, as composed of arcs of circular curvature. The method is based on this theorem:

Consider lines $\varphi = C_1$ and $\psi = C_2$ and lines diagonal to them. The centers of curvature of the diagonal lines are found on the same straight line with the centers of curvature of the streamlines and equipotential lines (Fig. 338). The direction of this line coincides with the direction of the tangent to the contour of the "modified" velocity hodograph — under this is understood the line, traced by the end of a vector, leaving an arbitrary pole $\Omega$, with length equal to the magnitude of the velocity vector, but perpendicular to the velocity direction.

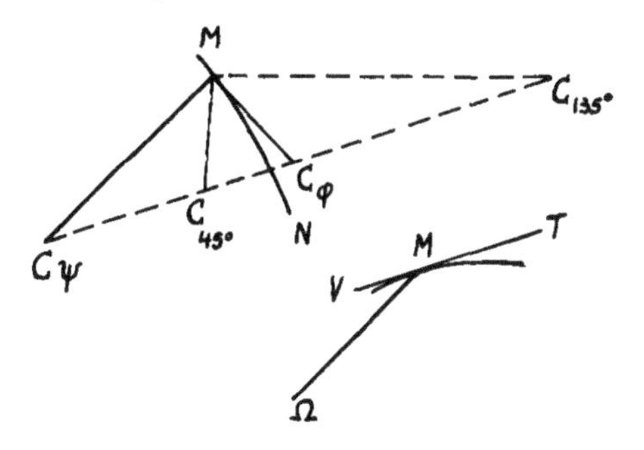


Fig. 338

Thus in Fig. 338 the line MN is a streamline, $\Omega$ the pole of the modified hodograph. If the center of curvature $C_\psi$ is known and if we know the direction of the tangent to the modified hodograph MT, then if we draw a straight line through $C_\psi$, parallel to MT, we have found the axis on which $C_\varphi$, $C_{45°}$, $C_{135°}$ are located.

## §10. Graphoanalytical Method To Compute Three Dimensional Flows.

Under natural conditions, plane groundwater flows are found very seldom, and actually the flow is usually three dimensional.

Such flows are sketched in Fig. 339. Fig. a) shows the groundwater flow resulting from infiltration losses of an irrigation system; b) represents a typical case of groundwater flow in the flood planes of a river; c) gives a groundwater flow, resulting from infiltration of precipitation and percolating down slope; d) sketches the seepage from a reservoir and the flow about a dam.

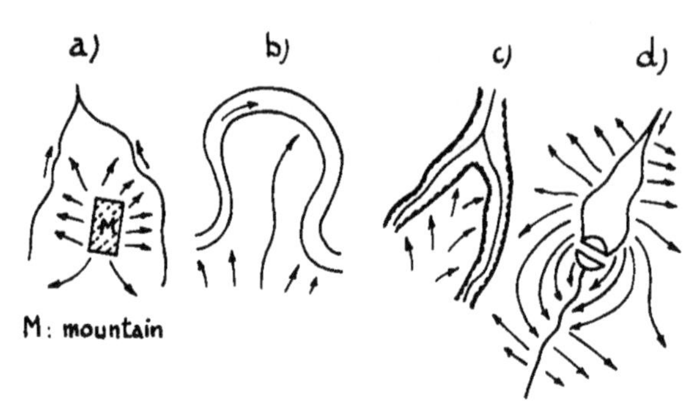


Fig. 339

S. F. Averjanov [16] introduced these examples and showed which problems arise in their analysis. In Fig. a) the question of the rise of groundwater that leaves agricultural soils saline, crops up. Indeed, when the water table is close to the earth's surface, intensive evaporation of groundwater may take place and also capillary rise; in this way salts are brought to the soil surface (see Literature to Chapter XIV [11]).

In case b) the question arises about the possibility of improving agricultural soils by using the flood plain of a river and draining it.

In case c) it is important to locate the spot of the most intense outflow of groundwater and to determine the necessary measures in the prevention of landslides.

In case d) it is important to be able to determine the seepage losses from the reservoir and to find measures to fight the rise of groundwater in overlying agricultural soils.

S. F. Averjanov proposed the following method for numerical computations of three dimensional steady flows.

In three dimensions, one must consider a flow surface instead of a streamline, represented by a streamtube, and instead of a line of constant head an equipotential surface. Both surfaces are orthogonal to each other.

Along each streamtube, the flowrate is constant. We may choose the equipotentials so that the head loss between adjacent equipotential surfaces is constant, say $\Delta h$. If $n$ is the number of these surfaces, then $n\Delta h = H$, where $H$ is the difference in head between the beginning and final equipotential surface (Fig. 340). Let us consider an element of streamtube with flowrate $\Delta Q$, expressed as

$$\Delta Q = \frac{k\Delta h}{\Delta \ell} \times \Delta t \times \Delta b \; . \tag{10.1}$$

Here $k$ is the seepage coefficient, $\Delta \ell$ the length of the element of streamtube, $\Delta b$ the width, $\Delta t$ the depth of the element.

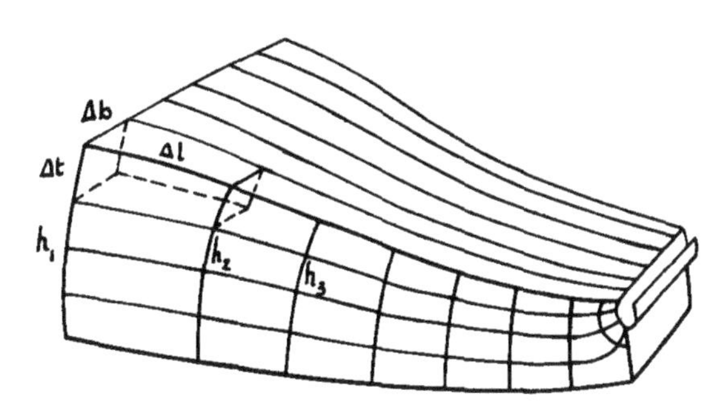


Fig. 340

Inserting the value $\frac{H}{n}$ for $\Delta h$, we obtain

$$\Delta Q = \frac{kH}{n} \frac{\Delta t \times \Delta b}{\Delta \ell} = \text{constant.} \tag{10.2}$$

We divide the flowregion in streamtubes for which the flowrate $\Delta Q$ = constant. Since $\frac{kH}{n}$ is constant, we must have

$$\frac{\Delta t \times \Delta b}{\Delta \ell} = \text{constant.} \tag{10.3}$$

This condition together with the condition of orthogonality of the streamlines and the equipotential surfaces gives a criterion to check the construction of three dimensional flownets.

For flows over a large area and with slightly changing free surface, as shown in Chapter X, we may assume that the equipotential surfaces are cylindrical surfaces with vertical axis. Adding in this case the elements of streamtubes along a vertical, we find for the flowrate $\Delta Q$ through a bundle of elements

$$\Delta Q = k\Delta h \frac{t \times \Delta b}{\Delta \ell} . \tag{10.4}$$

Here $t$ is the average depth of flow for a bundle of elements, distributed along one vertical.

Therefore one must observe the condition for one bundle

$$\frac{\Delta Q}{k\Delta h} = t \frac{\Delta b}{\Delta \ell} = \tau = \text{constant} . \tag{10.5}$$

We may call $\tau$ the "form coefficient" of the three dimensional net.

The full flowrate for $\tau$ = constant in all the nets is

$$Q = k\Delta h \times m \times \tau = \frac{kH}{n} m\tau , \tag{10.6}$$

where $m$ is the number of bundles for the width of the flow, $n$ the number of equipotential surfaces.

From this it follows that if we construct a two dimensional net with constant form coefficient $\tau$, we may obtain all the flow characteristics. The two dimensional equipotential surfaces are represented by lines of constant head for confined flow and by iso-hypsometric curves, i.e., lines of constant free surface elevation, in unconfined flow. Along these lines one may determine the flowrates and velocities of the different flow elements. The elements of the net are curvilinear rectangles, with varying ratio of the sides, since for three dimensional flow condition (10.3) must be satisfied, but not

$$\frac{\Delta b}{\Delta \ell} = \text{constant}$$

which holds for two dimensional flow. Only for constant depth of flow, the three dimensional net becomes a two-dimensional one.

For large values of $B$ and $L$ — total width and length of the flow, — compared with $t$, in first approximation we may take the net of two dimensional flow with constant depth of flow $t_{ave}$, equal to the average value of the depth $t$ of the flow. Then this net may be corrected by moving the equipotentials and streamlines, preserving their orthogonality and observing condition (10.5) for each parallelepiped of the net.

In the general case where the equipotential surfaces are not cylindrical surfaces with vertical axis, the full flowrate is obtained by summing expressions (10.2)

$$Q = \frac{kH}{n} \sum_1^m \sum_1^r \frac{\Delta t \times \Delta b}{\Delta \ell}, \tag{10.7}$$

where $m$ and $r$ are the number of elements of streamtube for the width and depth of flow.

In case the flow is only caused by infiltration of constant intensity $\varepsilon$, we have for an arbitrary parallelepiped of the net:

$$\Delta Q = \varepsilon S = k\Delta h \frac{t\Delta b}{\Delta \ell},$$

where $S$ is the region of intake, equal to the area of all the squares of the net from the plane under consideration till the water divide (source of the flow); $\Delta Q$ is no longer constant over the length of the element of streamtube. The criterion to check the construction of the three dimensional net now becomes

$$\frac{\varepsilon}{k} \frac{1}{\Delta h} = \frac{t \times \Delta b}{S\Delta \ell} = \text{constant}.$$

If the soil is non-homogeneous (i.e., the seepage coefficient is not constant), then

$$h = -\frac{\varphi}{k} = \frac{p}{\rho g} + z$$

does not satisfy Laplace's equation. But if Darcy's equations

$$u = -k\frac{\partial h}{\partial x}, \quad v = -k\frac{\partial h}{\partial y}, \quad w = -k\frac{\partial h}{\partial z}$$

hold, then they show that the flowrate $\Delta Q$ is expressed as before by formula (10.4). However condition (10.5) is not satisfied; instead we have

$$\frac{\Delta Q}{\Delta h} = k\frac{t \times \Delta b}{\Delta \ell} = \text{constant},$$

where $k$ is the average seepage coefficient along the vertical.

§11. Application of the Method of Flownets to Anisotropic Soils.

We examine the following case [7]. Given a flat bottom base for which [Fig. 341] $\ell_1 = 6$ m, $\ell_2 = 10$ m, $s = 3.5$ m.

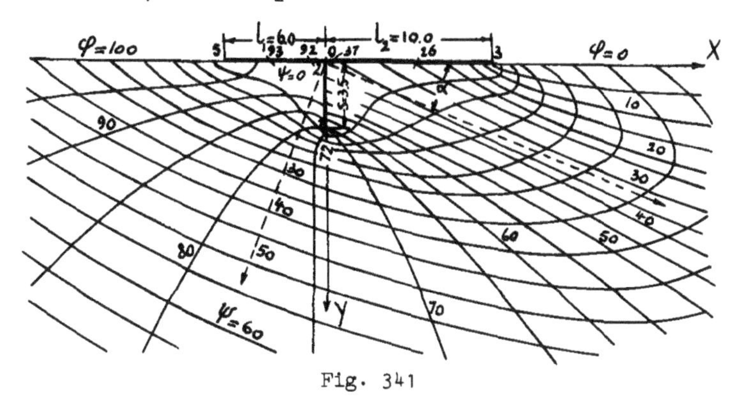


Fig. 341

The seepage coefficients in the directions of the principal seepage axes have the following values:

$$k_1 = 0.000049 \text{ m/sec}, \quad k_2 = 0.000004 \text{ m/sec}$$

so that the "coefficient of anisotropy"

$$\lambda = \frac{k_2}{k_1} = 0.0816.$$

The angle $\alpha$ between the x-axis and the principal seepage axis $\mu$, designating the slope of the strata, be 19°7', the difference in the water-levels of the reservoirs be $H = 3$ m.

The flownet must be constructed. According to the results of Par. 6, Chapter VIII, we pass from the coordinates $X$, $Y$ to the coordinates $\mu$, $\nu$ (here we change the notations, assuming $\lambda = 1/\nu^2$ and introducing $\mu$, $\nu$ instead of $x_1$, $y_1$), by means of the equations

$$(11.1) \qquad \mu = X \cos\alpha + Y \sin\alpha, \qquad \nu = \frac{1}{\sqrt{\lambda}}(-X \sin\alpha + Y \cos\alpha).$$

The straight line $Y = 0$ becomes the straight line

$$\frac{\nu}{\mu} = -\frac{\operatorname{tg}\alpha}{\sqrt{\lambda}}.$$

If we designate by $\delta$ the angle of the new straight line with the $\mu$-axis, we obtain

$$(11.2) \qquad \operatorname{tg}\delta = -\frac{\operatorname{tg}\alpha}{\sqrt{\lambda}},$$

and in our example we find: $\operatorname{tg}\delta = -1$, $\delta = -45°$.

In the same way we find that the old ordinate axis $X = 0$ transforms into the straight line

$$\frac{\nu}{\mu} = \frac{\operatorname{ctg}\alpha}{\sqrt{\lambda}}.$$

Its angle $\beta$ with the $\mu$-axis is determined by the equation

$$\text{(11.3)} \qquad \operatorname{tg} \beta = \frac{\operatorname{ctg} \alpha}{\sqrt{\lambda}} .$$

In our example we find $\beta = 84°20'$.

Inserting in (11.1) the given values of $\alpha$ and $\lambda$, we find

$$\mu = 0.9453X + 0.3279Y ,$$

$$\nu = - 1.148X + 3.309Y .$$

From this we may find the lengths of the segments of the flat bottom dam of Fig. 342: 8.92 m and 14.87 m and that of the cut-off: 11.64 m.

With respect to the coordinate axes $\mu$ and $\nu$, an orthogonal net of streamlines and equipotential lines is constructed.

Solving (11.1) for X and Y, we obtain

$$\text{(11.4)} \qquad X = \mu \cos \alpha - \nu \sqrt{\lambda} \sin \alpha, \quad Y = \mu \sin \alpha + \nu \sqrt{\lambda} \cos \alpha .$$

These equations allow to transform the obtained flownet into the X, Y plane. The new, no longer orthogonal, flownet is represented in Fig. 341. This is the picture of the flow about a flat bottom base with cut-off in the case under consideration.

We notice that the net of Fig. 342 may be constructed by graphical methods explained further, and also in another way. V. S. Kozlov constructed it by means of an electrodynamical analog method, treated in paragraphs 12-14.

As second example we take a flat bottom base with three cut-offs, sketched in Fig. 343. In this case $\ell_1 = 6$ m, $\ell_2 = 15$ m, $s_1 = 3$ m, $s_2 = 4$ m, $s_3 = 2$ m. The base overlies anisotropic soil with depth $T = 24.5$ m.

The seepage coefficients in the directions of the principal axes are chosen as: $k_1 = 0.00001$ m/sec, $k_2 = 0.0000019$ m/sec, so that $\lambda = k_2/k_1 = 0.19$. The angle $\alpha$ between the axes X and $\mu$ is equal to 15°. The head $H = 5$ m.

We obtain the equations of the coordinate axes:

$$\mu = 0.9659X + 0.2588Y, \quad \nu = - 0.5942X + 2.222Y .$$

With these we determine the coordinates of the points corresponding to the vertices of the flowregion.

| X, Y | -5; 3 | -6; 0 | 0; 0 | 0; 4 | 15; 0 | 15; 2 | 0; 24.5 |
|---|---|---|---|---|---|---|---|
| $\mu$ | -5.019 | -5.795 | 0 | 1.035 | 14.49 | 15.01 | 6.341 |
| $\nu$ | 10.23 | 3.565 | 0 | 8.886 | -8.913 | -13.36 | 54.43 |

The last point of the table is arbitrarily taken on the real axis.

The flowregion in the $\mu$, $\nu$ plane develops into a parallelogram, with sides parallel to the base and the direction of the cut-off (Fig. 344).

The streamfunction $\psi$ has prescribed values: zero along the contour of the structure and 100 along the impervious base. If we draw by eye

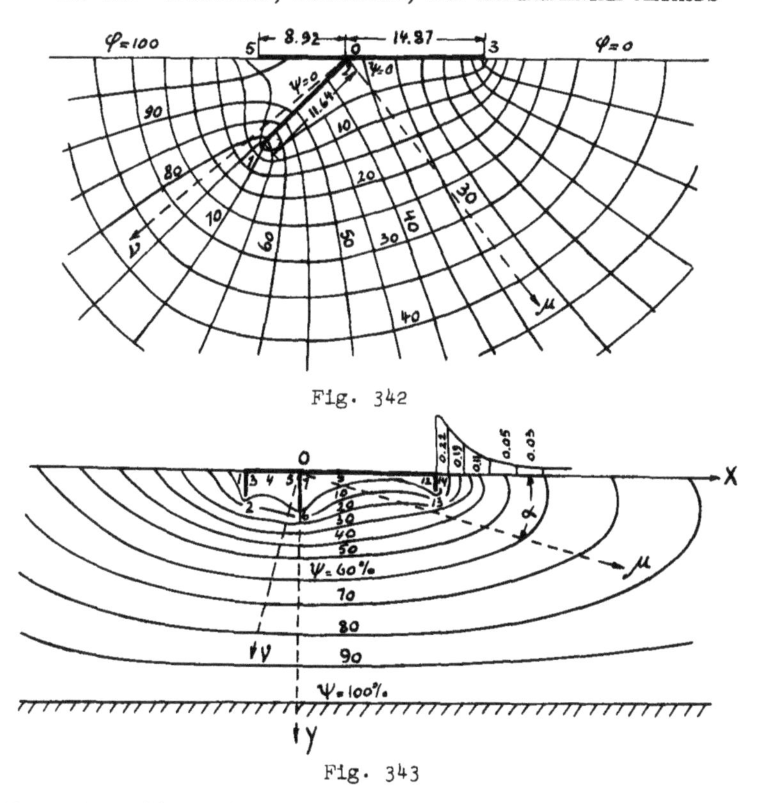


Fig. 342

Fig. 343

intermediate streamlines, then we may arrive at approximate values of function $\psi$ in the vertices of the parallelograms.

The values of the harmonic function $\psi$, in the center of the parallelogram, may be determined by a formula given by D. J. Panov [4], (Fig. 345):

(11.5)
$$\psi = \frac{\sigma_1\lambda_1 + \sigma_2\lambda_2 + \sigma_3\lambda_3}{\lambda_1 + \lambda_2 + \lambda_3} .$$

Here

$$\sigma_1 = \frac{1}{2}(\psi_5 + \psi_7), \quad \sigma_2 = \frac{1}{2}(\psi_6 + \psi_8), \quad \sigma_3 = \frac{1}{2}(\psi_2 + \psi_4),$$

$$\lambda_1 = \frac{1}{c^2} - \frac{\cos \delta_0}{ac}, \quad \lambda_2 = \frac{1}{a^2} - \frac{\cos \delta_0}{ac}, \quad \lambda_3 = \frac{\cos \delta_0}{ac},$$

in which a and c are the sides of the element of the net, $\delta_0$ is the angle between the sides of the parallelogram. This angle is determined as the difference between the angles $\delta$ and $\beta$, for which formulas (11.2) and (11.3) hold. Therefore we obtain:

$$\operatorname{tg} \delta_0 = \frac{\sqrt{\lambda}}{(1 - \lambda) \sin \alpha \cos \alpha} .$$

In our example $\delta_0 = 65°$.

In first approximation, instead of using formula (11.5), we may take

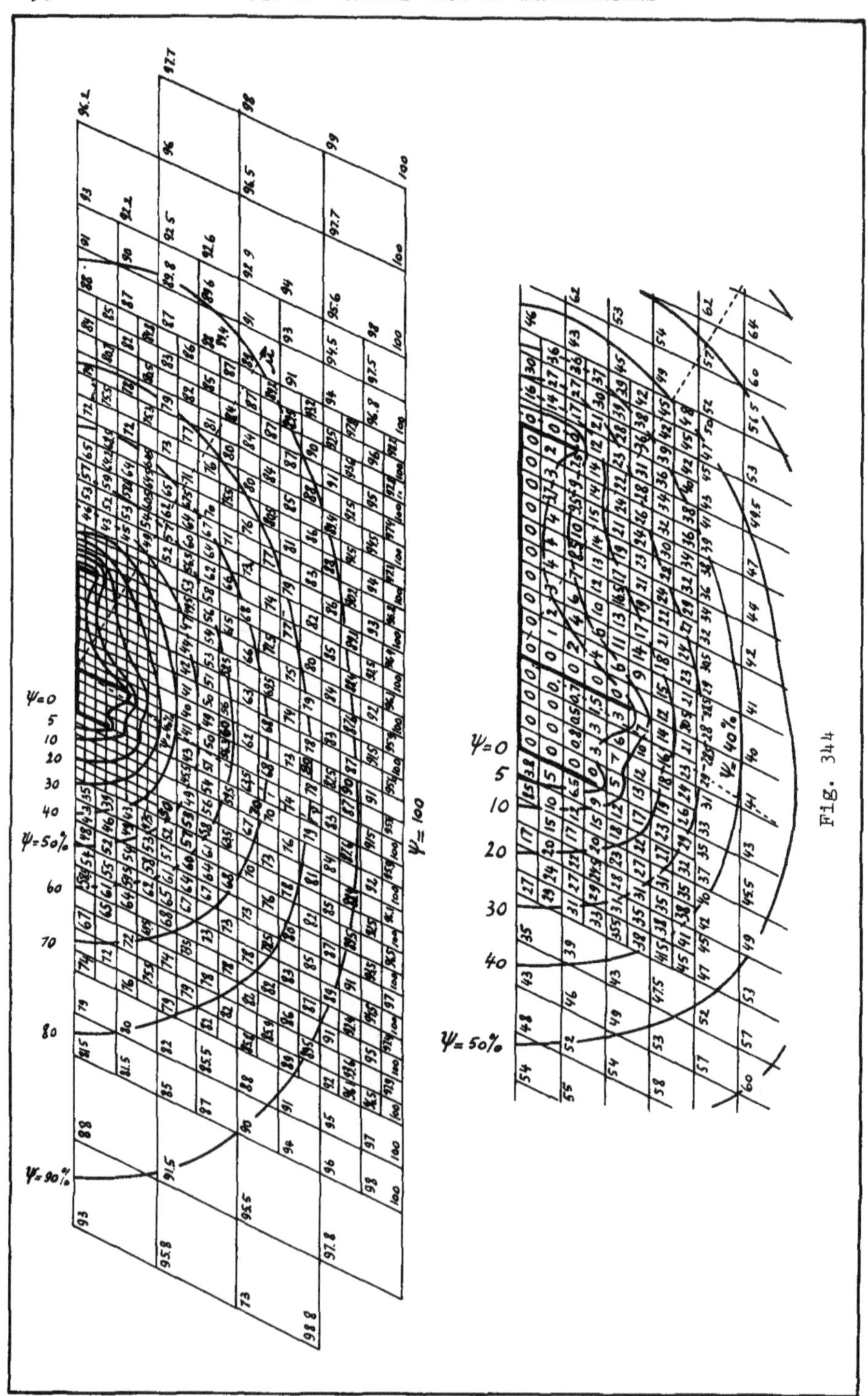


Fig. 344

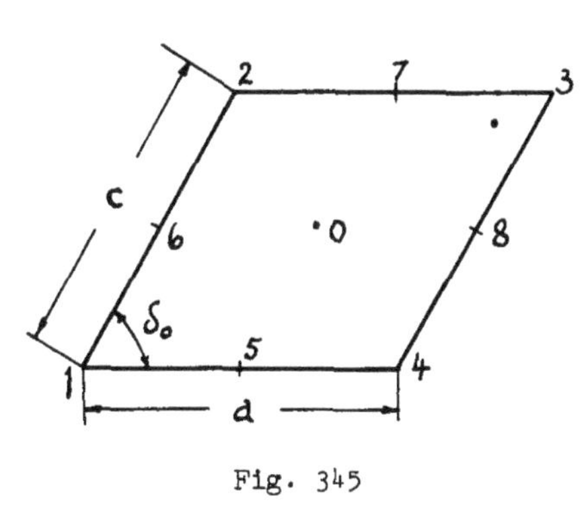


Fig. 345

the arithmetic average:

$$\psi = \frac{\psi_1 + \psi_2 + \psi_3 + \psi_4}{4},$$

and for further approximations we use formula (11.5).

The streamlines are then carried over from the $\mu$, $\nu$ plane to the X, Y plane (Fig. 343).

From Figs. 341 and 343 it is clear that the streamlines are not orthogonal to the boundaries of the reservoirs.

Fig. 343 gives the construction of the graph of the velocity along the boundary of the tailwater reservoir, where the velocity is devided by the seepage coefficient.

## B. METHOD OF ELECTRO-DYNAMICAL ANALOG [EGDA]

### §12. Basic Concepts.

Between steady groundwater flow and electric current flow, there exists an analogy because both phenomena are described by differential equations of the same form with identical boundary conditions. This analogy was used by N. N. Pavlovsky, who already in 1918 applied it to the experimental solution of a series of groundwater problems.

Suppose we have a thin layer of a fluid or solid conductor, with thickness $z$ that may be variable. The differential equations of the steady electric current flow arc

$$i_x = -c\frac{\partial V}{\partial x}, \quad i_y = -c\frac{\partial V}{\partial y}, \tag{12.1}$$

$$\frac{\partial(zi_x)}{\partial x} + \frac{\partial(zi_y)}{\partial y} = 0, \tag{12.2}$$

where $i_x$, $i_y$ are the components of the current density, $V$ the electric potential or tension function, $c$ the coefficient of electric conductivity, $x$, $y$ the coordinates of points of the electric model.

We introduce the streamfunction $\Psi$ by means of the equalities:

$$\frac{\partial\Psi}{\partial x} = -zi_y, \quad \frac{\partial\Psi}{\partial y} = zi_x. \tag{12.3}$$

Equating these expressions with (12.1) we obtain

$$\frac{\partial V}{\partial x} = -\frac{1}{cz}\frac{\partial\Psi}{\partial y}, \quad \frac{\partial V}{\partial y} = \frac{1}{cz}\frac{\partial\Psi}{\partial x}. \tag{12.4}$$

We now return to the equations of the seepage theory

$$u = -k\frac{\partial h}{\partial x}, \quad v = -k\frac{\partial h}{\partial y}, \tag{12.5}$$

$$\frac{\partial u}{\partial x} + \frac{\partial v}{\partial y} = 0, \tag{12.6}$$

where $h$ is the head. Introducing the streamfunction $\psi$, we find the dependence between $h$ and $\psi$:

$$\frac{\partial h}{\partial x} = -\frac{1}{k}\frac{\partial \psi}{\partial y}\,, \quad \frac{\partial h}{\partial y} = \frac{1}{k}\frac{\partial \psi}{\partial x}\,. \tag{12.7}$$

Comparison of (12.7) and (12.4) shows that the electric current lines correspond to the lines of groundwater flow, and the lines of equal electric potential correspond to the lines of constant head. If we model both through the condition

$$k = cz$$

and construct a region filled with conducting material and similar to the plane region of the groundwater flow (Fig. 346), then, if we measure the electric potential in various points, we may know the value of the head proportional to this potential. Similarly, the measurement of the current density in various points gives the magnitude of the velocity, and the magnitude of the force field is proportional to the flowrate.

S. A. Khristianovich [17] demonstrated the application of the EGDA method in cases where the groundwater flow does not satisfy the linear Darcy law, i.e., when $k$ is a function of the velocity.

In case of flow in homogeneous medium, when $k$ = constant and $c$ = constant, the equations for $V$ and $h$ are Laplace's equation

$$\frac{\partial^2 V}{\partial x^2} + \frac{\partial^2 V}{\partial y^2} = 0\,, \tag{12.8}$$

$$\frac{\partial^2 h}{\partial x^2} + \frac{\partial^2 h}{\partial y^2} = 0\,. \tag{12.9}$$

In the analysis of basic seepage problems, we had boundary conditions of this kind: along boundaries of reservoirs, $h$ = constant; along impervious walls and the free surface, $\partial h/\partial n = 0$. In the electric model, it is easy to satisfy at a given line conditions of the form

$$V = \text{constant} \quad \text{or} \quad \frac{\partial V}{\partial n} = 0\,.$$

Models for seepage regions are made of conductors with high resistance: tin foil plates with a thickness of 0.01 – 0.02 mm, electrolytes – fluids or dissolved salts, graphite or a mixture of graphite with marble, a jelly-like mass of gelatine or agar, agar dissolved in common salt, electroconducting paper etc. To obtain a contour of equal tension, brass bus bars of insignificant small resistivity (compared with that of the conductor of the flowregion) are used. In Fig. 346 such bars are indicated by the number 1.

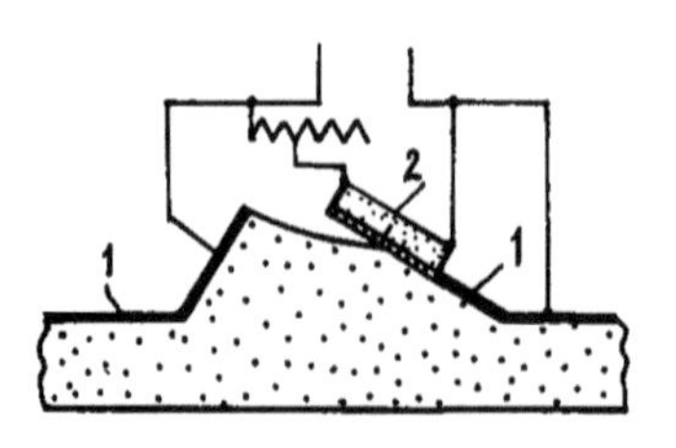


Fig. 346

To obtain, on the model, lines that corespond to streamlines of a given flow, one must put insulation along the corresponding segment of the contour, such as parafine, wax, glass, etc.

Air is an insulator for conducting paper and therefore the streamlines follow simply the edge cut in the paper figure.

Along the seepage surface, as we know, the velocity potential and head are linear functions of the ordinate:

$$h = y + \text{constant}.$$

One uses a collector bar (2 in Fig. 346) for a given tension that varies linearly. In its simplest form, this bar may consist of a series of alternating brass and ebonite segments.

At the free surface, as we know, we also must satisfy the condition $h = y + \text{constant}$. Moreover, the free surface is a streamline, except when infiltration and evaporation are accounted for. However, its location is not known in advance and one must choose it so as to satisfy two conditions

$$V = ay + b, \quad \frac{\partial V}{\partial n} = 0 .$$

Usually one does this: the upper part of the model is cut along the assumed shape of the free surface, but with some extra material left. One obtains for this region a line of constant potential and finds the points of intersection with the lines $y$ = constant and joins them by a smooth line. The model is cut along the new line, again leaving some margin, etc.

When the model is ready, in the electric circuit it is wired in parallel with a potentiometer, i.e., a conductor with known tension in every point, if the magnitude of the electric current intensity along it does not change. Model and potentiometer are joined in a Wheatstone bridge. When no current goes through the circuit, tension across potentiometer and model are the same.

The EGDA apparatus works for steady and unsteady flow. A telephone (or some other indicator – recently a light spot) is used in the EGDA method; the absence of or minimum in sound shows that in the point under investigation the tension across the model is the same as established across the potentiometer. This may simply be shown by a galvanometer dial.

One determines a series of points with the same potential on the model, joins them with a smooth line and obtains a line of constant potential. Streamlines may be constructed orthogonal to the equipotential lines. They may however be constructed independently, by interchanging the roles of equipotential and streamlines.

To determine the seepage rate $Q$ along an arc length of constant potential, one may compare

$$Q = - k \int_L \frac{\partial h}{\partial s} \, d\ell ,$$

where $L$ is the length of the arc of constant potential, with the formula for the current intensity $I$, passing through an arc of constant potential:

$$I = - c \int_L \frac{\partial V}{\partial s} \, d\ell .$$

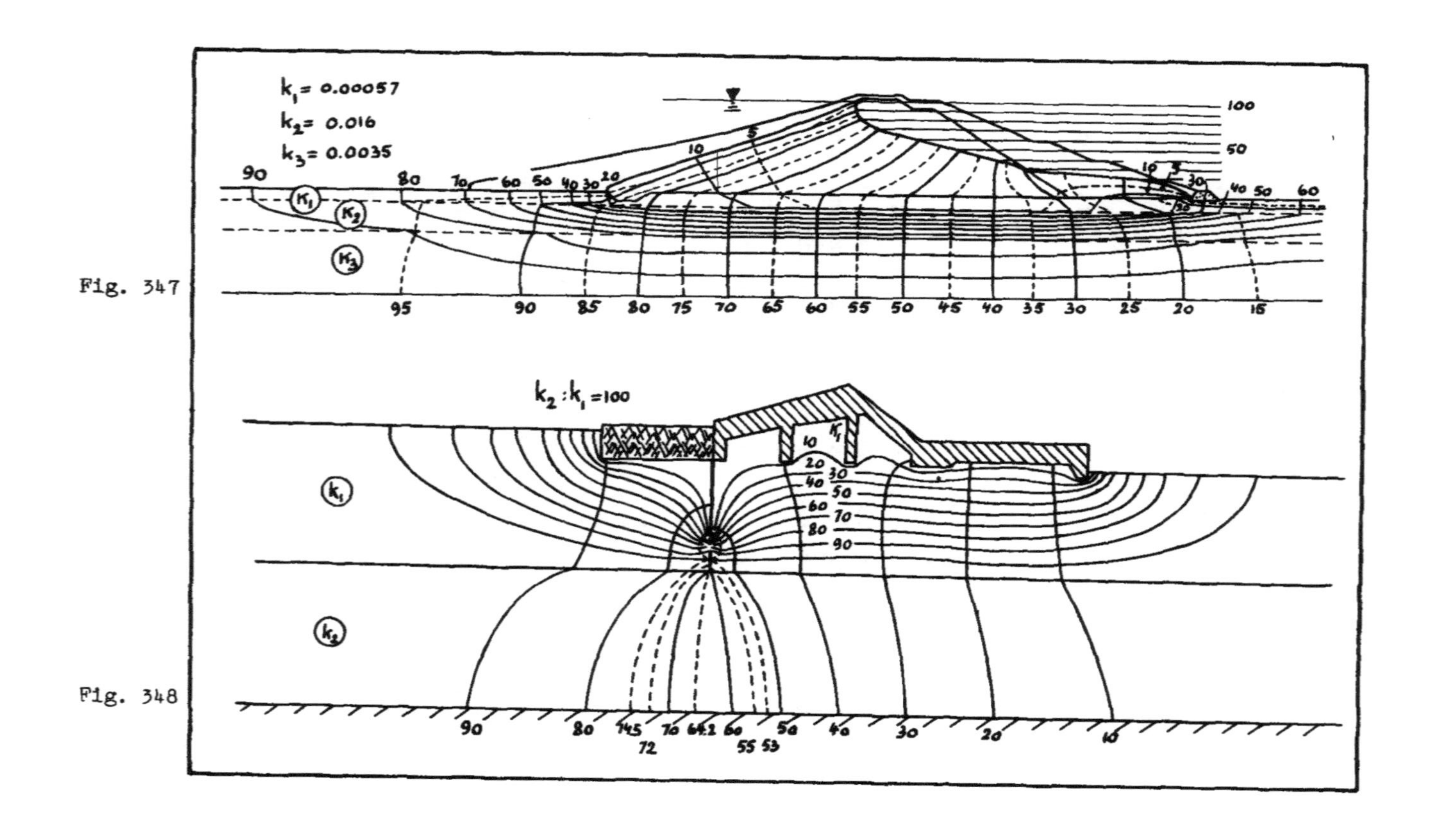


Fig. 347

Fig. 348

Some authors [17] recommend methods based on the measurement of the current density, rather than of the potential in different points of the model.

When it becomes necessary to model a region, composed of some parts of different hydraulic conductivity, it suffices to take different electrolytes or, as we showed before, to use the same electrolyte of variable thickness. It is also convenient to use paper of different electric conductivity.

At present, there is also a widespread application of the EGDA method method to three-dimensional problems [18, 19]. We do not treat this subject in detail, and limit ourselves to one application (Par. 14).

Recently an application of a so-called "electro-integrator" [20,21] to an oilbearing stratum became known.

## §13. Application of the EGDA Method to Two-dimensional Problems.

The EGDA method received wide-spread application in various scientific research institutes and projects. By means of it, flownets for models of real dams have been constructed. (Figs. 347 and 348 give examples of this kind [22].) It is apparent, from Fig. 348, how important the influence of the cracks is that may be formed at the end of the cut-off.

Applications of the EGDA method to simplified seepage schemes are important. S. M. Proskurnikov [23] communicated results of seepage tests through rectangular and trapezoidal dams; they gave the dependence of the flowrate and length of seepage surface on the angle of the inclined slope (the other face of the dam is vertical). He also gave results about seepage through dams with two sloping faces.

P. F. Filchakov [24] analyzed models of flow through rectangular and other dams in non-homogeneous soil.

## §14. Application of the EGDA Method to Determine the Seepage Coefficient.

It is very important for the construction of hydraulic structures to know the seepage coefficient of soils underlying a riverbed, therefore pumping tests are carried out.

N. I Druzjinin [25] analyzed 120 schemes of water flowing to a perforated pipe, put in the soil, by means of a three-dimensional EGDA model. Thus it bacame possible to construct graphs relating the reduced flowrate to the place and length of the filter (Fig. 349 and 350).

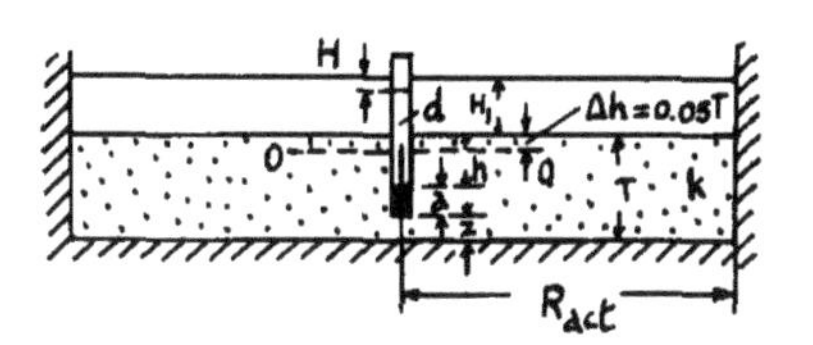

Fig. 349

We give a short description of the tests. We assume that there is an active zone of seepage, characterized by the magnitude of the radius $R_{act}$, which is determined for a given depth of flow and does not exceed three times the depth of flow. If we consider a cylinder of radius $R_{act}$ around the well and obtain some flowrate Q through it, Q will only slightly vary when we increase the radius of the region and we may neglect this variation.

The groundwater flow model is constructed in this way. The active zone of depth $T$ and radius $R_{act}$ is modelled at a scale 1/100 by means of an electrolyte $\rho$ (Fig. 350) in a bath $nnmm$, made of paraffine. A copper sheet L, representing the bottom of the waterbearing body, covers the bottom of the bath. Electric current is fed to this sheet by means of a conductor $n_d$. In the center of the bath, a mobile brass pivot $\Phi$ is established on the frame P, and is insulated over its entire perimeter with lacquer. Therefore, the lacquer is stripped off segment $a$ of the pivot, representing the length of the filter. The upper part of the pivot is wired to the conductor $n_\Phi$, along which electric current from the model

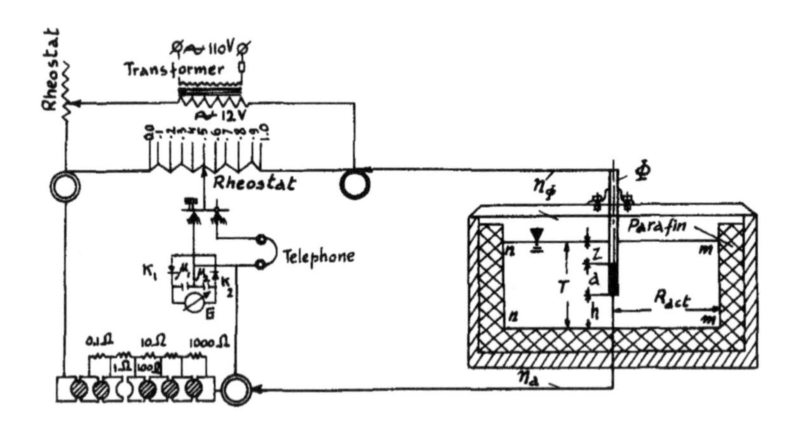


Fig. 350

is withdrawn. The pivot is moved vertically; changing $a$, one obtains different measures of the filter for different locations. The model is connected with a EGDA apparatus and its electric resistance is determined (Fig. 350). The flowrate follows from

$$Q = \frac{\rho}{R_0 M} kH \quad . \tag{14.1}$$

Here $\rho$ is the specific electric resistivity, i.e., the resistance of 1 $cm^3$ of model material (ohm cm), $R_0$ is the total electric resistance of the model (ohms), $M$ is the scale of the model, $k$ is the seepage coefficient and $H$ is the head.

For the so-called reduced flowrate $q_r$, (value of the flowrate for $k = 1$, $H = 1$, $T = 1$), we rewrite (13.1) as

$$Q = \frac{\rho}{R_0 MT} kHT = kHTq_r \quad .$$

The seepage coefficient is

$$k = \frac{Q}{HTq_r} = \frac{\rho}{R_0 T_m}$$

where $T_m = MT$ is the depth of the electro - conducting layer of the model in cm.

The results of the tests are given in Fig. 351 and 352.

In the second graph, $1/q_r$ is constructed in order to be able to determine $q_r$ for small values of $h/T$. When the upper limit of the filter becomes an horizontal line (OO in Fig. 349), at $\Delta h = 0.05T$ from the bottom, $q_r$ increases sharply.

The graphs are constructed for $R_{act}/T = 3.25$ and $d/T = 0.025$. In case $d \neq 0.025T$, V. I. Aravin and S. N. Numerov proposed the following formula for k:

$$k = \frac{Q}{q_r HT\left[1 + 0.37\,\frac{Tq_r}{a}\,\lg_{10}\left(\frac{40d}{T}\right)\right]} .$$

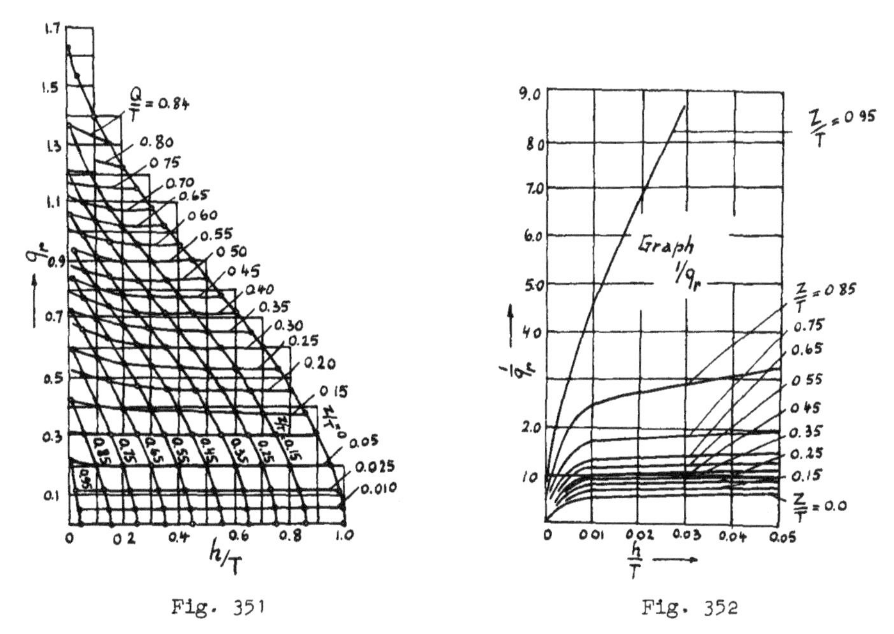


Fig. 351

Fig. 352

Thus we saw that the EGDA method not only may be used to investigate seepage in cases where the flowregion actually has a complicated form, but also to model some individual flows.

## C. EXPERIMENTS WITH PARALLEL PLATE MODELS *

### §15. The Theory of the Parallel Plate Model.

It is well.known that plane irrotational flow is well reproduced by means of the flow of a fluid between two parallel plates, sufficiently close to each other.

* The literal translation of the Russian term is "slit-shaped gutters."

The analogy between laminar flow of a fluid in a slit between parallel plates and the movement of water in soil with a constant coefficient of seepage has been used by E. A. Zamarin. The apparatus, called "slit-shaped gutter" (Fig. 353), has at present a widespread application in scientific research and educational laboratories. The parallel plate model is convenient, because it gives a visual picture of the flow, and easily allows reproduction of unsteady movement. Contrary to the EGDA apparatus, the free surface is generated automatically.

The theory of the parallel plate model is explained in the literature (see [26]). In courses of hydrodynamics, the problem of the flow of a viscous incompressible fluid between two parallel plane walls is treated (Fig. 354). Let the flow be "one-dimensional," occurring parallel to the x-axis, so that $v = w = 0$.

As is well-known [32], in this case we have

$$\frac{\partial p}{\partial x} = \mu \frac{\partial^2 u}{\partial z^2} , \tag{15.1}$$

where $\mu$ is the viscosity of the fluid; $\frac{\partial p}{\partial x}$ has here a constant magnitude.

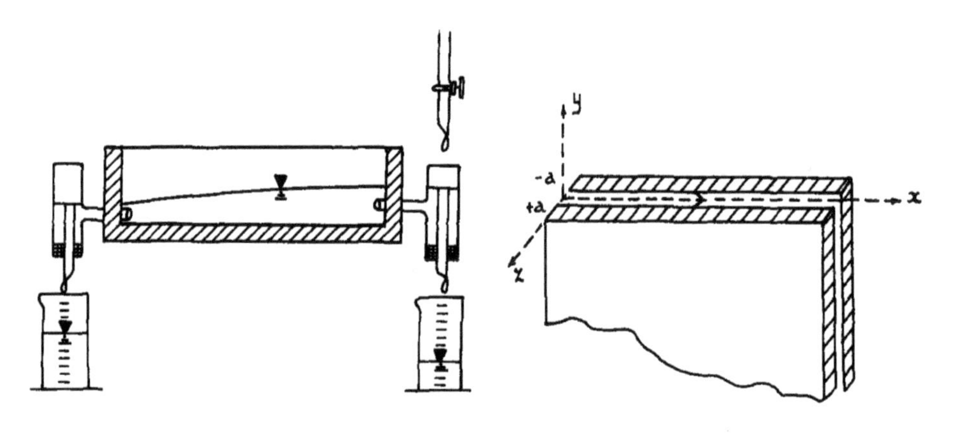


Fig. 353 Fig. 354

Integration of equation (15.1) along $z$ between limits from $a$ to $z$, taking into account the conditions of adhesion of the fluid to the walls, i.e., the conditions $u = 0$ for $z = \pm a$, gives:

$$u(z) = - \frac{1}{2\mu}(a^2 - z^2) \frac{\partial p}{\partial x} .$$

Integrating the obtained expression of the velocity along $z$ between the limits $-a$ and $+a$, we find the discharge of fluid through a cross section of the slit (considering the thickness of the layer, in the direction parallel to the y-axis of Fig. 354, to be unit):

$$q = - \frac{2a^3}{3\mu} \frac{\partial p}{\partial x} .$$

From this equation, the average velocity $\bar{u}$ of the flow parallel to the x-axis is obtained

$$\bar{u} = -\frac{a^2}{3\mu}\frac{\partial p}{\partial x} . \tag{15.2}$$

If we take account of the action of the gravity force, then in formula (15.2) instead of the pressure we have to take the quantity

$$P = p + \rho g y$$

or the quantity proportional to it — the head

$$h = y + \frac{p}{\rho g} .$$

In the latter case, equation (15.2) becomes

$$\bar{u} = -\frac{a^2 g}{3\nu}\frac{\partial h}{\partial x} . \tag{15.3}$$

We return now to two-dimensional flow between parallel walls, i.e., such a flow in which the component of the velocity along the z-axis, (perpendicular to the parallel walls) is zero, but both $u$ and $v$ are different from zero. Then we have the system of equations of the flow of a viscous fluid

$$\begin{cases} \frac{\partial u}{\partial t} + u\frac{\partial u}{\partial x} + v\frac{\partial u}{\partial y} = -\frac{1}{\rho}\frac{\partial p}{\partial x} + \frac{\mu}{\rho}\left(\frac{\partial^2 u}{\partial x^2} + \frac{\partial^2 u}{\partial y^2} + \frac{\partial^2 u}{\partial z^2}\right) , \\ \frac{\partial v}{\partial t} + u\frac{\partial v}{\partial x} + v\frac{\partial v}{\partial y} = -\frac{1}{\rho}\frac{\partial p}{\partial y} + \frac{\mu}{\rho}\left(\frac{\partial^2 v}{\partial x^2} + \frac{\partial^2 v}{\partial y^2} + \frac{\partial^2 v}{\partial z^2}\right) - g , \end{cases} \tag{15.4}$$

$$0 = \frac{\partial p}{\partial z} .$$

The flow of a viscous fluid in a narrow slit, in the same way as the movement of a fluid in porous medium, belongs to the group of "creeping" flows. These flows are characterized by small velocities, so that one may neglect in the left side of equations (15.4) the inertia terms, and in the right sides of these equations — the values of the derivatives along $x$ and $y$ in comparison with the derivatives along $z$. (*) Then we obtain the equations:

$$\begin{aligned} 0 &= -g\frac{\partial h}{\partial x} + \frac{\mu}{\rho}\frac{\partial^2 u}{\partial z^2} , \\ 0 &= -g\frac{\partial h}{\partial y} + \frac{\mu}{\rho}\frac{\partial^2 v}{\partial z^2} , \\ \frac{\partial p}{\partial z} &= 0 \left(h = y + \frac{p}{\rho g}\right) . \end{aligned} \tag{15.5}$$

As in the case of one-dimensional flow, we will have for the average values of the velocity components $\bar{u}$ and $\bar{v}$, the expressions

$$\bar{u} = -\frac{a^2 g}{3\nu}\frac{\partial h}{\partial x} , \quad \bar{v} = -\frac{a^2 g}{3\nu}\frac{\partial h}{\partial y} \quad \left(\nu = \frac{\mu}{\rho}\right) \tag{15.6}$$

From this it is clear that the average velocities of the flow of a

---

* In some cases, namely in the consideration of similarity ratios, we will maintain in equations (15.4) the terms $\partial u/\partial t$, $\partial v/\partial t$ .

fluid in a narrow slit when satisfying our proposed assumptions, are indeed derived from a potential.

Putting

$$m = \frac{ga^2}{3\nu} \tag{15.7}$$

and calling $m$ "the coefficient of permeability of the slit," we may rewrite the equations for the average velocities of the steady flow of a fluid in a parallel plate model

$$\bar{u} = -m\frac{\partial h}{\partial x}, \quad \bar{v} = -m\frac{\partial h}{\partial y}, \tag{15.8}$$

we observe a complete analogy between them. The boundary conditions in the parallel plate model are the same as for groundwater flow, and namely: along impervious boundaries, $\psi$ = constant, along boundaries of water bodies $h$ = constant, at free surfaces and seepage surfaces, the pressure $p$ = constant. One ought to notice that in the laws of flow of a viscous fluid, this fluid sticks to the walls, where the velocities are zero, and forms a boundary layer. However, this layer is extremely thin, and as experiments of flow about a profile have shown, the streamline picture in the parallel plate model coincides with the one that is obtained in the theoretical solution of the problem of flow about a two-dimensional profile.

We also observe that the derivation of the formula for the flow in a parallel plate model assumes the laminar character of the flow, and not the turbulent type, and therefore the Reynolds number for the model

$$R_e = \frac{Va}{\nu}$$

where $V$ — the characteristic velocity of flow, should not exceed the critical value, for which V. I. Aravin [26] proposes to take

$$R_{e,crit} = 500 .$$

Furthermore, for very narrow spacing of the plates, law (15.8) for the distribution of the velocities becomes inexact.

However, for the usual experiments with glycerine and dimensions of the slit of the order of 1 mm. or parts of a mm., this danger does not appear.

Finally, we notice that a steady picture of laminar flow is observed at some distance $\ell$ from the entrance to the model. Indication is available that $\ell = 0.1aR_e$ .

## §16. Determination of the Seepage Picture.

In order to construct a model of the seepage flow in a parallel plate apparatus, the body of the earthdam is made out of paraffine, the head - and tailwater basins are modelled by means of pans, much wider than the slit.

The seepage through the body of the dam is modelled in this way: in a wide collector filled with glycerine, a model of the dam in paraffine is standing (abcdefgh in Fig. 355). The thickness of the paraffine coat must be so that the spacing between the glass walls and the paraffine is sufficiently narrow. In Fig. 355, the form of a parallel plate model is given. On each side of the model, glass cylinders serve to maintain desired fluid levels at the boundaries of the slit. The formula

(16.1) $$m = \frac{ga^2}{3\nu}$$

shows that if we wish to obtain a well-determined value of the permeability, we have to take the width $2a$ of the slit smaller if the viscosity of the fluid in the experiment is smaller. If water is used, one has to take a slit of the order of 0.1 to 0.3 mm., for glycerine it is possible to take 1 to 2mm. We now suppose that we have a gutter filled with soil (for example with sand) and a parallel plate model (with glycerine or water) of identical dimensions. The flowrate of the groundwater flow in the gutter per unit width (perpendicular to the plane of the drawing, i.e., along the z-axis) is designated by $q_p$. The flowrate in the parallel plate model – through the full width $2a$ of the slit – is designated by $q_m$. We have

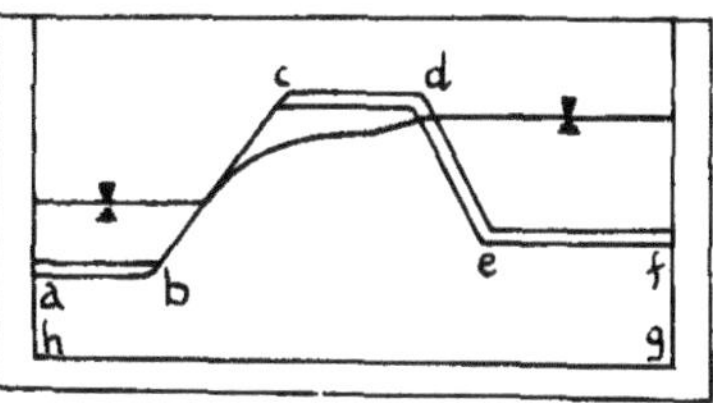


Fig. 355

(16.2) $$q_p = - k \frac{dh}{ds} ,$$

where $s$ is the direction of the streamlines.

In the soil gutter and in the parallel plate model, we establish identically operating heads and we make all the geometrical dimensions identical. Therefore we may write for the flowrate $q_m$ through the slit of width $2a$

(16.3) $$q_m = - m \frac{dh}{ds} 2a .$$

Comparing the formulas for $q_p$ and $q_m$, we obtain

(16.4) $$\frac{q_p}{q_m} = \frac{k}{2am} .$$

We suppose now that in the parallel plate model, the groundwater flow is reproduced at the scale $\lambda$. This means that all the lengths of the groundwater flow in nature are $\lambda$ times larger than the lengths in the plate model. Then, since the flow velocities in the soil gutter and in nature are identical, the flowrate in nature $q_n$ will be $\lambda$ times larger than the flowrate $q_p$ in the soil gutter. Consequently we obtain

$$\frac{q_n}{q_m} = \frac{\lambda k}{2am} .$$

Taking into account equation (16.1) we obtain the formula for the transition from the flowrate $q_m$ in the parallel plate model to the flowrate $q_n$ in nature:

(16.5) $$q_n = \frac{3\lambda k\nu}{2ga^3} q_m .$$

## §17. Modelling of Heterogeneous Soils.

Let us consider a model, composed of two parts with different permeabilities, i.e., with different widths of slit, for which AB (Fig. 356) is the border line. To elucidate the seepage laws, it is convenient not to consider the average velocities $\bar{u}$ and $\bar{v}$ here, but the flowrates through the width of each slit.

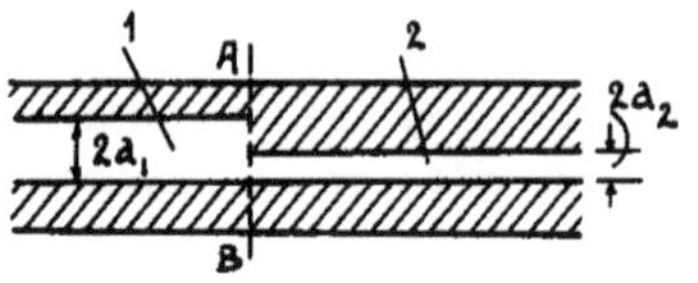


Fig. 356

Let, in the first region, the half width of the slit be $a_1$, in the second part $a_2$. We introduce the flowrate vector $\vec{W}$ with components $W_x$ and $W_y$:

$$W_x = q_x = 2a\bar{u} = -n\frac{\partial h}{\partial x}, \quad W_y = q_y = 2a\bar{v} = -n\frac{\partial h}{\partial y} .$$

Here, by $n$ is designated the quantity

$$n = 2am$$

the so-called "second permeability of the slit." We have respectively for the regions 1 and 2, designating by $t$ and $n$ tangential and normal components:

$$W_{t1} = -n_1\frac{\partial h_1}{\partial t}, \quad W_{n1} = -n_1\frac{\partial h_1}{\partial n},$$

$$W_{t2} = -n_2\frac{\partial h_2}{\partial t}, \quad W_{n2} = -n_2\frac{\partial h_2}{\partial n} .$$

Along the boundary AB, the head remains continuous, so that $\partial h_1/\partial t = \partial h_2/\partial t$. Therefore we obtain

$$\frac{W_{t1}}{n_1} = \frac{W_{t2}}{n_2} .$$

Due to the continuity of the flow, the normal components of the flowrate must be equal, i.e.,

$$W_{n1} = W_{n2} .$$

It is easy to see, as in Par. 3 of Chapter II, that

$$\frac{W_{t1}}{W_{t2}} = \frac{\operatorname{tg}\theta_1}{\operatorname{tg}\theta_2} = \frac{n_1}{n_2},$$

where $\theta_1$, $\theta_2$ are the angles formed by the vectors $\vec{W}_1$ and $\vec{W}_2$ with the normal to the border line. This formula expresses the law of refraction of the flowlines.

For heterogeneous soils, we had in Chapter II the refraction law in the form

$$\frac{\operatorname{tg}\theta_1}{\operatorname{tg}\theta_2} = \frac{k_1}{k_2} .$$

Thus we obtain

$$\frac{k_1}{k_2} = \frac{n_1}{n_2}$$

or, considering equation (16.1) and $n = 2am$

$$\frac{k_1}{k_2} = \frac{a_1^3}{a_1^3} = \frac{(2a_1)^3}{(2a_2)^3} \quad . \tag{17.1}$$

We observe that the ratio of the seepage coefficients is proportional to the ratio of the cubes of the plate spacings. This offers a possibility to model soils which differ rather strongly in permeability; so, when $a_1 : a_2 = 2$, $k_1 : k_2 = 8$. If we have several different soils, then if we designate by $k_i$, $a_i$ respectively the seepage coefficient of the $i^{th}$ soil and half-width of slit in the $i^{th}$ region of the model, we have for the flowrate $q$ of the groundwater flow the same equation

$$q = \frac{3\lambda k_i \nu}{2ga_i^3} q_m \quad . \tag{17.2}$$

We notice that even in the case of homogeneous soil, we may limit our considerations to only the second permeability of the model, if instead of the velocities $\overline{u}$ and $\overline{v}$, we take the flowrates $W_x$ and $W_y$ and compare them with the natural seepage velocities $u$ and $v$.

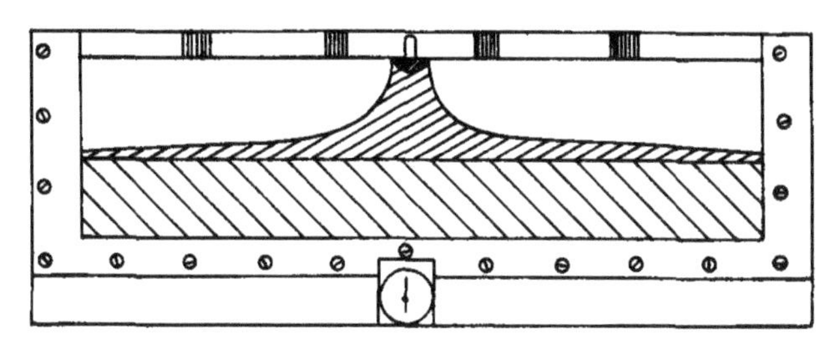

Fig. 357

Fig. 357 gives the picture of the steady seepage from a canal; the permeability of the upper soil stratum is smaller than the permeability of the lower stratum.

*§18. Unsteady Flow.

Neglecting the quadratic part of the inertia terms in the equations of motion, we may write the differential equations of two-dimensional groundwater flow:

$$\left\{ \begin{aligned} \frac{1}{\sigma}\frac{\partial u}{\partial t} &= -\frac{1}{\rho}\frac{\partial p}{\partial x} - \frac{g}{k}u \ , \\ \frac{1}{\sigma}\frac{\partial v}{\partial t} &= -\frac{1}{\rho}\frac{\partial p}{\partial y} - \frac{g}{k}v - g \quad . \end{aligned} \right. \tag{18.1}$$

We assume that we go from the natural flow to the groundwater model, by introducing the scale coefficients in this way:

$$\left\{ \begin{aligned} u &= \lambda_V u_p \ , \quad v = \lambda_V v_p \ , \\ t &= \lambda_t t_p \ , \quad \sigma = \lambda_\sigma \sigma_p \ , \end{aligned} \right. \qquad k = \lambda_h k_p \ . \tag{18.2}$$

We designate by $\lambda$ the scale coefficient of the piezometric head, i.e., of the quantity $p/\rho g$, and also of the length.

$$\frac{p}{\rho g} = \lambda \frac{p_p}{\rho g}, \quad x = \lambda x_p, \quad y = \lambda y_p \tag{18.3}$$

Inserting in the second equation of (18.1) the expressions (18.2) and (18.3), we obtain

$$\frac{\lambda_V}{\lambda_\sigma \lambda_t \sigma_p} \frac{\partial v_p}{\partial t_p} = -\frac{1}{\rho_p} \frac{\partial p_p}{\partial y_p} - \frac{g}{\lambda_k k_p} \lambda_V u_p - g \ . \tag{18.4*}$$

In order to have the same form for equations (18.4) * and (18.1), it is necessary to satisfy the equality

$$\frac{\lambda_V}{\lambda_\sigma \lambda_t} = \frac{\lambda_V}{\lambda_k} = 1 \ .$$

From this it follows that

$$\lambda_k = \lambda_V, \quad \lambda_t = \frac{\lambda_V}{\lambda_\sigma} \ . \tag{18.5}$$

Instead of (18.4) we have:

$$\frac{1}{\sigma_p} \frac{\partial v_p}{\partial t_p} = -\frac{1}{\rho_p} \frac{\partial p_p}{\partial y_p} - \frac{g}{k_p} u_p - g \ . \tag{18.6*}$$

These ratios show that the larger we take $\lambda_k$, i.e., the smaller grained we take the soil in the model, the smaller will be the velocities in the model and the smaller will be the time period obtained in the model (if $\sigma_p = \sigma$, i.e., if the porosity of the soil and in the model are the same).

We now write the equations of unsteady flow for the parallel plate model, indicating all corresponding quantities by the subscript "m":

$$\frac{\partial u_m}{\partial t_m} = -\frac{1}{\rho_m} \frac{\partial p_m}{\partial x} - \frac{g}{m} u_m \ ,$$

$$\frac{\partial v_m}{\partial t_m} = -\frac{1}{\rho_m} \frac{\partial p_m}{\partial y} - \frac{g}{m} v_m - g \ .$$

Let us assume that the soil model and the parallel plate model are constructed at the same scale.

Let us compare the groundwater flow in nature with the flow in the parallel plate model; therefore, in the formulas (18.2), we replace the index "p" by the index "m", taking into account that

$$\sigma_m = 1 \ .$$

We obtain:

$$u = \lambda_V u_m, \quad v = \lambda_V v_m \ ,$$

$$t = \lambda_t t_m, \quad \sigma = \lambda_\sigma \ ,$$

$$k = \lambda_k m \ .$$

We observe that $\lambda_\sigma$ cannot be taken arbitrarily; this scale is equal to the

---

* Equations (18.4) and (18.6) have been corrected for printing errors.

magnitude of the soil porosity.

The width of the slit can be computed after one of the scales, for example $\lambda_t$, has been selected. We obtain

$$m = \frac{k}{\lambda_k} .$$

But, according to (18.5)

$$\lambda_k = \lambda_V = \lambda_t \lambda_\sigma = \lambda_t \sigma .$$

Consequently

$$m = \frac{k}{\sigma \lambda_t} , \quad \lambda_t = \frac{k}{m\sigma} .$$

Taking into account formula (16.1), we find for the time scale the expression

$$\lambda_t = \frac{3k\nu}{ga^2\sigma} , \tag{18.7}$$

and from there the half width of the slit:

$$a = \sqrt{\frac{3\nu k}{\sigma \lambda_t}} .$$

This formula allows us to find the width of the slit, if we know $k$ and $\sigma$ for a given soil and $\nu$ for the parallel plate model, and if we assume a definite time scale $\lambda_t$, wishing to obtain in the model a quicker movement than in nature.

Formula (18.7) allows us to find the time scale, if we have a model with half-width $a$, glycerine of viscosity $\nu$ and given soil constants $k$ and $\sigma$. Then, if we obtained a duration $t_m$ of some process in the parallel plate model, the corresponding lapse of time in nature will be

$$t = \lambda_t t_m = \frac{3k\nu}{ga^2\sigma} t_m .$$

For the relationship between the flow rate in nature (perunit width of flow) and the flowrate through the slit we have formula (16.5) as before.

Applications of the parallel plate model to unsteady flows are given in Chapters XIII and XIV.

## §19. Calculation of the Capillarity of the Soil.

We notice that we may take into account the capillarity of the soil in the parallel plate model.

If we know the height of capillary rise $h_\kappa$ of the soil, then we must have the height of capillary rise $h_\kappa/\lambda$ in the model, where $\lambda$ is the linear scale of the model. Let $h_{\kappa,m}$ be the height of capillary rise in the slit. According to Laplace,

$$h_{\kappa,m} = \frac{c}{\gamma}\left(\frac{1}{r_1} + \frac{1}{r_2}\right) ,$$

where $c$ — capillary constant, depending upon the fluid and the solid walls, between which the fluid flows, $\gamma$ — specific weight of the fluid $r_1$ and $r_2$ radii of curvature of the menisci. In the slit, one of the radii of curvature is infinite; let $r_2 = \infty$. One may consider approximately $r_1 = a$, then

$$h_{\kappa,m} = \frac{c}{\gamma a} .$$

For water and glass, we have $c = 0.075$ gram/cm. for a temperature of 15°C. Therefore, for water flowing between glass plates

$$h_{\kappa,m} = \frac{0.075}{a} \text{ (cm)} .$$

If in the parallel plate model one wall is of glass and the other metallic or made of paraffine wax, then the menisc may have a more complicated form (Fig. 358). The dependence between $h_{\kappa,m}$ and $a$ is determined experimentally for a given model.

Fig. 358

§20. Modelling of Anisotropic Soil. Modelling at Distorted Scale.

Anisotropic soil can be modelled by making excavations in the walls so that we have two values $2a_1$ and $2a_2$ for the width of the slit, to which correspond the second coefficients of permeability $n_1$ and $n_2$.

Let us assume that the anisotropy of the soil is caused by the alternation of strata of the same thickness, having seepage coefficients $k_1$ and $k_2$. Then the steady flow parallel along these strata, as we saw in Chapter VIII, will be as if we had a seepage coefficient

$$k_x = \frac{k_1 + k_2}{2} .$$

The transversal flow, perpendicular to these layers, will have (see Chapter VIII) the average seepage coefficient

$$k_y = \frac{2}{\frac{1}{k_1} + \frac{1}{k_2}} .$$

As is well-known (*), the arithmetical average is larger than the harmonic average

$$\frac{k_1 + k_2}{2} \geq \frac{2}{\frac{1}{k_1} + \frac{1}{k_2}} .$$

Considering $k_1$ and $k_2$ different, we will always have

$$k_x > k_y .$$

Therefore let us adopt the notation

$$k_x = k_{max}, \quad k_y = k_{min} .$$

---

* In the same way, from this inequality results the following:

$$(k_1 + k_2)^2 \geq 4k_1k_2$$

which is true, since it can be rewritten in this way:

$$(k_1 - k_2)^2 \geq 0$$

If one puts

$$\frac{k_1}{k_2} = t, \quad \frac{k_{max}}{k_{min}} = d,$$

then it is easy to find

$$d = \frac{1}{4}\left(2 + t + \frac{1}{t}\right),$$

from which we obtain for $t$ as function of $d$

$$t = 2d - 1 + 2\sqrt{d(d-1)} \quad . \tag{20.1}$$

On the other side, the ratio of the seepage coefficients of the alternating strata is equal to the ratio of the permeabilities of the slits

$$\frac{k_1}{k_2} = \frac{n_1}{n_2} = \frac{a_1^3}{a_2^3} = t \quad .$$

Having chosen the value of $d$, we find $t$ by formula (20.1). If we fix, according to the conditions of the experiment, one of the half-widths $a_2$, then we obtain the other from the equality

$$a_1 = a_2 \sqrt[3]{t} \quad .$$

In anisotropic soil, as we saw, the Laplace equation for the head $h$ is replaced by:

$$k_x \frac{\partial^2 h}{\partial x^2} + k_y \frac{\partial^2 h}{\partial y^2} = 0 \quad , \tag{20.2}$$

where the $x$ and $y$ axes are taken in the direction of the maximum and minimum seepage coefficient $k_x$ and $k_y$.

Equation (20.2) allows us to consider the problem of modelling at a distorted scale. It is a fact that in nature usually there are regions where the groundwater flow in the vertical direction is significantly smaller than in the horizontal, and therefore modelling of such flows was difficult. We introduce the scales of the seepage coefficients $K_x$, $K_y$, the scale of the heads $\lambda_h$ and the length scales $\lambda_x$, $\lambda_y$, different for the two coordinate axes.

We shall consider that equation (20.2) corresponds to the flow in the parallel plate model, by means of which we like to reproduce the flow in nature, with seepage coefficient $k$, head $h_n$ coordinates $X$ and $Y$. We have:

$$k = K_x k_x = K_y k_y, \quad h_n = \lambda_h h, \quad X = \lambda_x x, \quad Y = \lambda_y y \; .$$

Inserting these expressions into (20.2), we obtain:

$$\frac{K_x}{k}\frac{\lambda_h}{\lambda_x^2}\frac{\partial^2 h_n}{\partial x^2} + \frac{K_y}{k}\frac{\lambda_h}{\lambda_y^2}\frac{\partial^2 h_n}{\partial y^2} = 0 \quad .$$

From this we obtain, since $K_x : K_y = k_x : k_y$

$$\frac{K_x}{\lambda_x^2} = \frac{K_y}{\lambda_y^2}\,{}^{*}, \quad \frac{k_y}{k_x} = \frac{\lambda_x^2}{\lambda_y^2} \quad .$$

---

* This condition implies $\nabla^2 h_n = 0$, equation for isotropic soil. (Translator's remark)

We put

$$\frac{\lambda_x}{\lambda_y} = \Delta;$$

then

$$\frac{k_y}{k_x} = \Delta^2 ,$$

and for

$$t = \frac{k_1}{k_2} = \frac{n_1}{n_2}$$

we obtain, substituting $\Delta^2$ for $d$ in formula (20.1)

$$t = 2\Delta^2 - 1 + 2\Delta\sqrt{\Delta^2 - 1} .$$

After choosing the quantity $\Delta$, we find $t$, and according to formula $a_1 = a_2\sqrt[3]{t}$ we find also the ratio of the widths of the slits for an anisotropic soil, representing the model of an isotropic soil at a distorted scale. To increase the vertical scale one has to arrange protuberances and fissures in a vertical way in the parallel plate model.

*§21. Remark Concerning Experiments with Parallel Plate Models.

V.I. Aravin [26] shows other applications of the parallel plate model. Thus radial flow can be modelled, by giving the split a changing width, in the form of a sector, made by two plates with a small angle $\theta$ between them.

It is possible to carry out the study of three-dimensional flows with slightly curved free surface if one models a confined flow, corresponding to the given unconfined flow (Fig. 359). Thus, for the flow of water towards a foundation pit (Fig. 359a), an auxiliary artesian flow in a stratum of thickness $t$ is introduced, for which the form of the well in the plane is the same as that of the pit. At different points of the stratum the head $h_m$ is measured. The ordinates of the free surface of the unconfined flow are designated by $h$.

Then

$$\frac{1}{2}(h^2 - h_0^2) = \frac{nt}{k}(h_m - h_{m0}) ;$$

$h_0$ — height of the water in the foundation pit, $h_{m0}$ — piezometric height in the artesian aquifer.

At some distance from the center of the foundation pit — already at

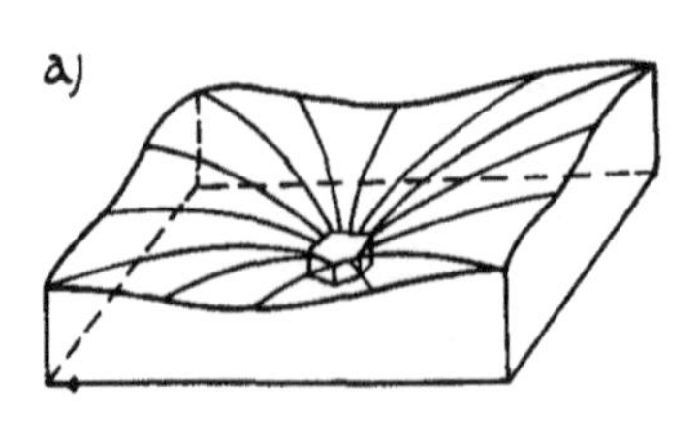


Fig. 359a

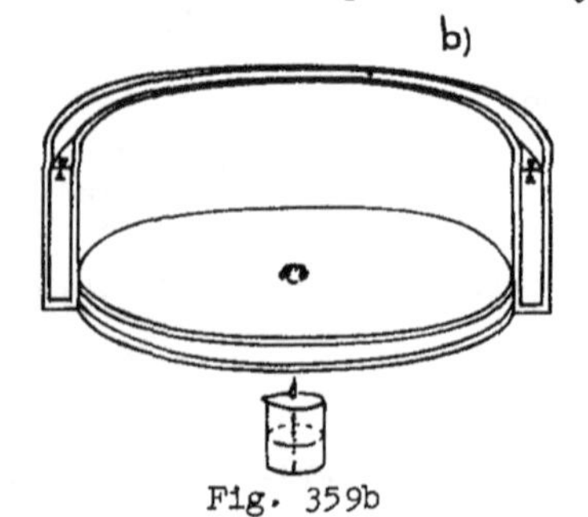


Fig. 359b

1.5 times the maximum dimension of the pit in the plane – the "hydroisohypses" i.e., the lines of equal height or the corresponding lines of equal head $h_m$ may be substituted by circles; therefore one may construct the parallel plate model with two circular horizontal plates. In Fig. 359b, a section of such a kind of model is represented.

## D. OBSERVATIONS OF SEEPAGE OF WATER IN THE SOIL.

### §22. Experiments in Soil Boxes.

For the study of special movements of groundwaters in different conditions, soil gutters have been constructed – i.e., boxes filled with soil. In the walls of the gutter, apertures are established which are joined by means of rubber tubes to glass pipes – "piezometers." The latter show which head is available in that point of the soil in which the end of the tube is introduced.

In Fig. 360, a soil box is represented in which water flows under unfluence of a head differential. The piezometers show the values of the head in the points 1, 2, ..., located at the bottom of the gutter. For a slightly changing free surface, these values would be close to the ordinates of the free surface flow, if there would be no capillary rise in the soil. But at the upper part of the soil, a capillary fringe of larger or smaller width is always observed. If particles of potassium manganese are placed in the soil, they trace the movement of the water in the soil as painted stripes, exhibiting the stream lines. Such streamlines can be observed even in the capillary zone (Fig. 360). However, if we put transparent, perforated (i.e., with holes in the walls) pipes in the soil, then the water levels in these pipes will not reach till the end of the moistened soil, but will perfectly correspond to the boundary of the wetted region in the absence of capillarity. We remark that the introduction itself of such pipes disturbs somewhat the original movement.

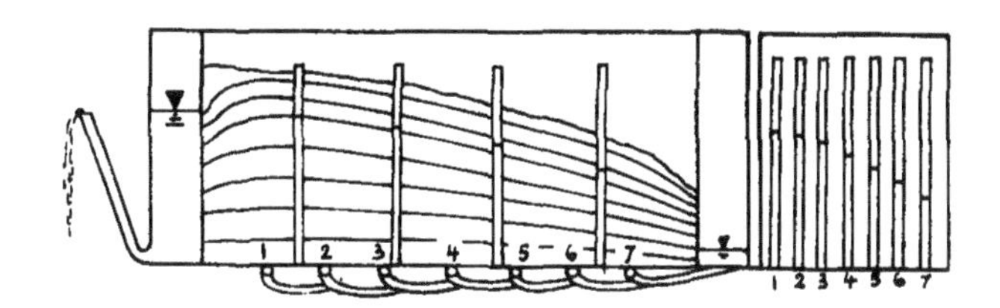


Fig. 360

In Fig. 361, are represented the free surface lines for the flow of water from a half "canal" into sand, for successive times, expressed in seconds and minutes. The length of the gutter is 1 m.

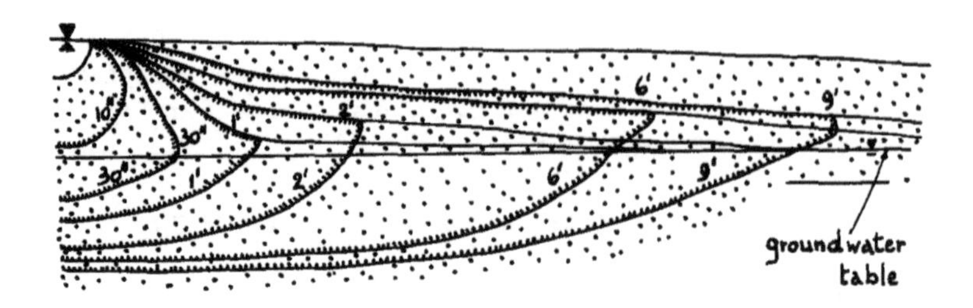


Fig. 361

§23. Observations in Nature.

Performing observations in nature of the groundwater level in the regions of projected and already constructed structures has an important value. The works [28, 29] are devoted to observations in nature.

In book [29] one may find the explanation of the basic methods of observation of groundwater flow in nature. In order to determine groundwater levels, bore holes are drilled in a series of points. In these holes are lowered a rod, that allows us to observe the moment of contact with the water: a flapper, whistle, contact cylinder, closing a contact, that send a light or sound signal, etc.

1

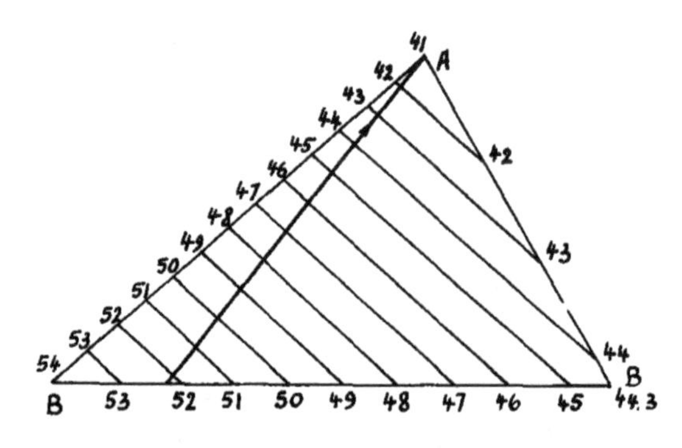


Fig. 362

Let us assume that the location of the free surface of the groundwater flow has been obtained in three points: A, B, C, (the distance between them will be of the order of 100 m.). We construct the triangle ABC, divide its sides in segments proportional to the corresponding level marks and through the points with the same marks we draw the hydroisohypses, i.e., lines of equal height of water. The lines, which are orthogonal to the hydroisohypses, oriented along the drop in value of the hydroisohypses, determine the direction of the ground water flow. (Fig. 362).

To find the average velocity of groundwater flow, one may lower a tracer in a test hole (dye matter, slat etc. ), which then attracts the eye in an observation point, after water has been sampled in well-determined time intervals. The distance between the test holes, divided by the elapsed time, gives the average velocity of flow.

In books [29, 30] various methods to determine the seepage coefficient of soils are shown.

---

PART II.

## Unsteady Flows of Groundwaters

# CHAPTER XII.

## ABOUT INERTIA TERMS IN UNSTEADY FLOWS. CONFINED FLOWS.

### §1. About Confined Flows With Operating Heads Depending On Time.

In Chapter I, Par. 11, we already gave an estimate of the linear part of the inertia terms and showed that for values of the seepage coefficient as they are present in real conditions, the terms

$$\frac{1}{m}\frac{\partial u}{\partial t}, \quad \frac{1}{m}\frac{\partial v}{\partial t} \tag{1.1}$$

may be neglected indeed.

In connection with this, a simple solution is given to the problem of confined flow in unsteady conditions, arising on account of the fact that the water levels of reservoirs change in time. Let us namely assume that we have $n$ reservoirs, respectively with water levels

$$H_1(t),\ H_2(t),\ \ldots,\ H_n(t) \quad ,$$

where $H_1(t), H_2(t), \ldots, H_n(t)$ – given functions of time. Between reservoirs are located bases of hydraulic structures (Fig. 363). If we have a solution of some problem with constant values $H_1, H_2, \ldots, H_n$, the same solution is valid for a changing $H_m(t)$, in which the time $t$ must be considered as a parameter.

Let us take the simplest case of flow about a two-dimensional wetted contour of width $2b$ on a pervious layer of infinite depth, for which we may consider that the head reservoir has a constant level $H_1 = h_1$, but that

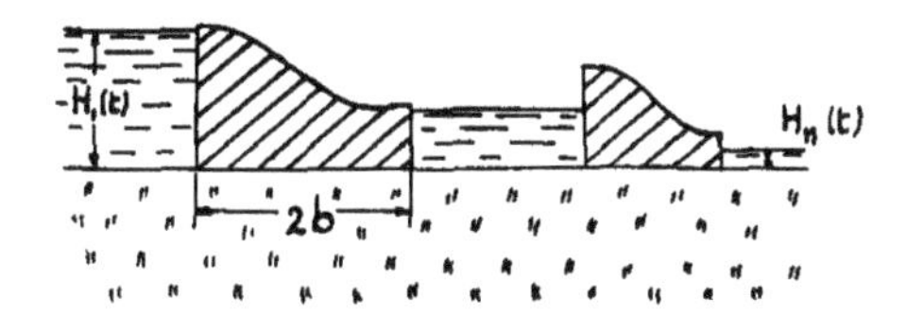


Fig. 363

tailwater level changes according to the sinus law

$$H_2(t) = h_2 + a \sin \lambda t \quad ,$$

under conditions $h_2 > a$, $h_1 > h_2 + a$.

Then it is sufficient to substitute the given values $H_1(t)$ and $H_2(t)$ in the formulas which we obtained in Chapter III for the complex potential and velocity

$$\omega = \varphi + i\psi = \frac{k(H_1 - H_2)}{\pi} \sin^{-1} \frac{z}{b}, \quad w = u - iv = \frac{k(H_1 - H_2)}{\pi \sqrt{b^2 - z^2}}$$

We obtain:

$$w = u - iv = \frac{k(h_1 - h_2 - a \sin \lambda t)}{\pi \sqrt{b^2 - z^2}} \quad . \tag{1.2}$$

We now return to another solution of the same problem [1], [2], when terms of form (1.1) are taken into account in the equations of motion, and show the illusory character of this calculation.

If we write the equations of two dimensional flow with terms (1.1) but without the quadratic terms in the expression of the acceleration, then we obtain the following equations of motion:

$$\left\{ \begin{aligned} \frac{1}{m} \frac{\partial u}{\partial t} &= - \frac{1}{\rho} \frac{\partial p}{\partial x} - \frac{gu}{k} , \\ \frac{1}{m} \frac{\partial v}{\partial t} &= - \frac{1}{\rho} \frac{\partial p}{\partial y} - \frac{gv}{k} - g \quad , \end{aligned} \right. \tag{1.3}$$

and the equation of continuity

$$\frac{\partial u}{\partial x} + \frac{\partial v}{\partial y} = 0 \quad . \tag{1.4}$$

If we introduce a vector $\vec{W}(U, V)$, where

$$U = u + \frac{k}{mg} \frac{\partial u}{\partial t} , \quad V = v + \frac{k}{mg} \frac{\partial v}{\partial t} \quad , \tag{1.5}$$

then equations (1.3) and (1.4) can be written as:

$$U = - k \frac{\partial}{\partial x} \left(\frac{p}{\rho g} + y\right), \quad V = - k \frac{\partial}{\partial y}\left(\frac{p}{\rho g} + y\right) \quad , \tag{1.6}$$

$$\frac{\partial U}{\partial x} + \frac{\partial V}{\partial y} = 0 \quad . \tag{1.7}$$

We introduce the notation

$$\varphi = - k\left(\frac{p}{\rho g} + y\right) + \text{constant} \quad . \tag{1.8}$$

We have

$$U = \frac{\partial \varphi}{\partial x} , \quad V = \frac{\partial \varphi}{\partial y} , \tag{1.9}$$

and the function $\varphi$ satisfies Laplace's equation.

If we know U, V, then the components of the seepage velocity u, v are determined by the solution of linear equations (1.5) in the following form $(\alpha = \frac{mg}{k})$:

$$(1.10)\quad \begin{cases} u = u_0 e^{-\alpha(t-t_0)} + \alpha e^{-\alpha t}\int_{t_0}^{t} e^{\alpha t}\, U(t)\, dt , \\ v = v_0 e^{-\alpha(t-t_0)} + \alpha e^{-\alpha t}\int_{t_0}^{t} e^{\alpha t}\, V(t)\, dt . \end{cases}$$

The expression of the complex velocity may be written as:

$$(1.11)\qquad w = u - iv = w_0 e^{-\alpha(t-t_0)} + \alpha e^{-\alpha t}\int_{t_0}^{t} e^{\alpha t}\, W(t)\, dt ,$$

where $W = U - iV$, $w_0$ — the value of $w$ for $t = t_0$.

Then, for the magnitude $W$ we have such an expression as is obtained in the theory of the steady movements, i.e., in our example

$$W = U - iV = \frac{k(H_1 - H_2)}{\pi\sqrt{b^2 - z^2}} .$$

Therefore, we find for the velocity $u - iv$

$$w = w_0 e^{-\alpha(t-t_0)} + \frac{\alpha k}{\pi}\frac{e^{-\alpha t}}{\sqrt{b^2 - z^2}}\int_{t_0}^{t} e^{\alpha t}\,(h_1 - h_2 - a\sin\lambda t)\, dt ,$$

where

$$(1.12)\qquad \alpha = \frac{mg}{k} .$$

If we carry out the integration, then we obtain Davison's solution:

$$(1.13)\qquad w = w_0 e^{-\alpha(t-t_0)} + \frac{k}{\pi\sqrt{b^2 - z^2}}\left[(h_1 - h_2)\left(1 - e^{-\alpha(t-t_0)}\right) - a\left(1 - e^{-\alpha(t-t_0)}\right)\frac{\sin\lambda t - \frac{\lambda}{\alpha}\cos\lambda t}{1 + \frac{\lambda^2}{\alpha^2}}\right] .$$

Let us estimate the terms that occur in this expression. In order to have the maximum possible influence of the terms that contain the parameter $\alpha$, we must take such values of $k$ and $m$ for which this parameter is as small as possible. Therefore we take

$$k = 100\,\frac{\text{meter}}{\text{day}} = \frac{10000}{86400}\frac{\text{cm}}{\text{sec}} = \frac{100}{864}\frac{\text{cm}}{\text{sec}} ; \quad m = \frac{1}{10} , \quad g = 1000\,\frac{\text{cm}}{\text{sec}^2} .$$

Then we have $\alpha = \frac{mg}{k} = 800\,\frac{1}{\text{sec}}$ and the term $e^{-800(t-t_0')}$ will be of the order of 0.001, already for $t - t_0 = 0.01$ sec.

Consequently, we may eliminate from the solution the terms which contain the multiplier $e^{-\alpha(t-t_0)}$. We obtain the expression

$$w = \frac{k}{\pi\sqrt{b^2 - z^2}}\left[h_1 - h_2 - a\,\frac{\sin\lambda t - \frac{\lambda}{\alpha}\cos\lambda t}{1 + \frac{\lambda^2}{\alpha^2}}\right]$$

which Davison considered to be corresponding to the value of $t_0 = -\infty$, i.e., to the corresponding "steady" process after elapse of a sufficiently long time. We notice that $t_o$ in essence is of the order of small fractions of a second. But even in the now remaining expression, the terms that contain $\alpha$ are extraordinarily small. Indeed, $\lambda$ is the frequency of oscillation of the water in the lower reservoir; it corresponds to the period of oscillation $T$:

$$\lambda = \frac{2\pi}{T} .$$

If we take the period of oscillation of the order of fractions of a minute, for example,

$$T = \frac{2\pi}{\lambda} = 10 \text{ sec}, \quad \lambda = \frac{2\pi}{10}\,\frac{1}{\text{sec}} ,$$

then we have

$$\frac{\lambda}{\alpha} = \frac{2\pi}{10 \times 800} \approx \frac{1}{1000} .$$

Consequently, the term $\frac{\lambda}{\alpha}\cos\lambda t$ may be appreciable only for very small periods of oscillation, of the order of one second for strongly pervious soil and still smaller for weakly permeable soil. For ordinary conditions the terms $c\,\frac{\lambda}{\alpha}$ may be neglected, and we come back to the formula which was obtained without consideration of the inertia terms.

## §2. About the Influence of Waviness On Seepage Under Hydraulic Structures.

In problems about steady ground water movements under hydraulic structures, we have assumed that the water in the reservoir, bordering the seepage region, was motionless, i.e., that the water surface of the reservoir was lying in a horizontal plane. However, in reality, waviness of the free surface of the reservoir is often observed. This circumstance changes the picture of seepage under the dam, in a way that the influence of waviness is felt especially in the tailwater reservoir.

Let us consider the simplest case of a two-dimensional flat bottom dam on soil of infinite depth. We assume that the water levels of head and tail-reservoirs are given functions of $x$ and $t$ (Fig. 364), which we designate by $H_1(x, t)$ and $H_2(x, t)$.

We choose $\psi = 0$ along the base of the dam. Then we obtain the problem of determining the complex potential $\omega(z, t)$ the imaginary part of which is known along the segment $(-b, b)$ and the real part along the segments $(-\infty, -b)$ and $(b, \infty)$: on the first of these

$$\varphi = -kH_1(x, t) ,$$

and on the second

$$\varphi = -kH_2(x, t) .$$

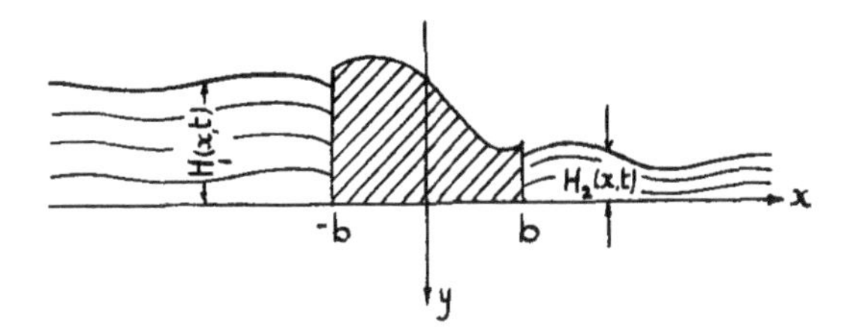


Fig. 364

We may take advantage of the method which is exposed in the beginning of Chapter VI, namely to apply a formula analogous to formula (1.6) of Chapter VI, to the function

$$\frac{\omega(z)}{\sqrt{z^2 - b^2}} \ .$$

We obtain, considering that $\sigma(\zeta) = 0$ for $-b < \zeta < b$, $\sigma(\zeta) = kH_1$ for $-\infty < \zeta < -b$, $\sigma(\zeta) = -kH_2$ for $b < \zeta < \infty$:

$$\frac{\omega(z)}{\sqrt{z^2 - b^2}} = -\frac{1}{\pi i}\int_{-\infty}^{\infty}\frac{\sigma(\zeta)\,d\zeta}{\sqrt{\zeta^2 - b^2}(\zeta - z)} \tag{2.1}$$

(the minus sign in front of the integral because we evaluate the value of the function in the lower, and not in the upper half-plane).

The constant on the right side of this equality, is zero in the present case, since in the neighborhood of $z = \infty$

$$\omega(z) = O(\ln z), \qquad \lim_{z \to \infty} \frac{\omega(z)}{\sqrt{z^2 - b^2}} = 0 \ .$$

Inserting the values of $\sigma(\zeta)$ at the corresponding segments, we obtain:

$$\omega(z) = -\frac{\sqrt{z^2 - b^2}}{\pi i}\int_{-\infty}^{-b}\frac{kH_1(\zeta,\ t)\,d\zeta}{\sqrt{\zeta^2 - b^2}\,(\zeta - z)} + \frac{\sqrt{z^2 - b^2}}{\pi i}\int_{b}^{\infty}\frac{kH_2(\zeta,\ t)\,d\zeta}{\sqrt{\zeta^2 - b^2}\,(\zeta - z)} \ . \tag{2.2}$$

We consider for example the case when

$$H_1(\zeta,\ t) = H_1, \quad H_2(\zeta,\ t) = H_2 + a \sin \omega t \sin \frac{\pi\zeta}{c},$$

which corresponds to a standing wave in the tailwater reservoir. We obtain

$$\omega(z) = -\frac{k\sqrt{z^2 - b^2}}{\pi i}\left\{H_1\int_{-\infty}^{-b}\frac{d\zeta}{\sqrt{\zeta^2 - b^2}(\zeta - z)} + \right.$$

$$\left. + a \sin \omega t \int_{+b}^{\infty}\frac{\sin\frac{\pi\zeta}{c}\,d\zeta}{\sqrt{\zeta^2 - b^2}(\zeta - z)} + H_2\int_{b}^{\infty}\frac{d\zeta}{\sqrt{\zeta^2 - b^2}(\zeta - z)}\right\} .$$

The first and third terms, after integration and reduction must give such an expression as is obtained for the complex potential in the flow about a two-dimensional wetted contour in a half plane with water levels $H_1$ and $H_2$ in upper and lower reservoirs, i.e.,

$$- k \frac{H_1 - H_2}{\pi} \sin^{-1} \frac{z}{b} + k \frac{H_1 + H_2}{2}$$

Therefore we obtain

$$\omega(z, t) = - k \frac{H_1 - H_2}{\pi} \sin^{-1} \frac{z}{b} + k \frac{H_1 + H_2}{2} +$$

$$+ \frac{ka \sqrt{z^2 - b^2}}{\pi i} \sin \omega t \int_b^\infty \frac{\sin \frac{\pi \zeta}{c} \, d\zeta}{\sqrt{\zeta^2 - b^2}(\zeta - z)} .$$

V. I. Aravin examined this problem in simplified form [3], considering that along the whole real x-axis the head changes according to a sinus law (the multiplier $\sin \omega t$ is dropped in all formulas)

$$(2.3) \quad h = \frac{h_1 - h_2}{2} \sin \frac{\pi x}{c} + \frac{h_1 + h_2}{2}$$

where $h_1$ and $h_2$ correspond to the largest and smallest ordinate of the half wave sinusoid. (Fig. 365). For the velocity potential, one may write an analogous equality

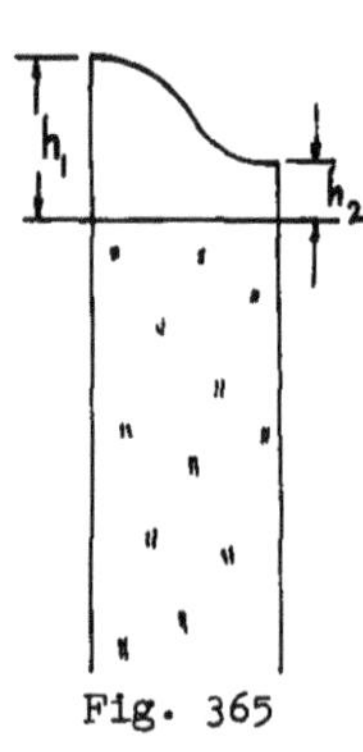

Fig. 365

$$(2.4) \quad \varphi = - \frac{\varphi_2 - \varphi_1}{2} \sin \frac{\pi x}{c} + \frac{\varphi_1 + \varphi_2}{2} ,$$

where

$$\varphi_1 = - kh_1, \quad \varphi_2 = - kh_2 ,$$

But then one may take the available solution from hydrodynamics, where the problem of wavy motion with velocity potential and stream function is treated [5]:

$$\varphi = - \frac{\varphi_2 - \varphi_1}{2} e^{-\frac{\pi y}{c}} \sin \frac{\pi x}{c} + \frac{\varphi_1 + \varphi_2}{2}$$

$$\psi = \frac{\varphi_2 - \varphi_1}{2} e^{-\frac{\pi y}{c}} \cos \frac{\pi x}{c} .$$

The complex potential will have the form:

$$\omega = + i \frac{\varphi_2 - \varphi_1}{2} e^{\frac{\pi z i}{c}} + \frac{\varphi_1 + \varphi_2}{2} .$$

The seepage velocity will be:

$$u = - \frac{\pi(\varphi_2 - \varphi_1)}{2c} e^{-\frac{\pi y}{c}} \cos \frac{\pi x}{c}, \qquad v = \frac{\pi(\varphi_2 - \varphi_1)}{2c} e^{-\frac{\pi y}{c}} \sin \frac{\pi x}{c} ,$$

Along the boundary of the reservoir, for $y = 0$, we obtain:

$$u_0 = -\frac{\pi(\varphi_2 - \varphi_1)}{2c} \cos \frac{\pi x}{c}, \quad v_0 = \frac{\pi(\varphi_2 - \varphi_1)}{2c} \sin \frac{\pi x}{c}.$$

As is evident from the formula for the velocity components, the magnitude of the velocity is equal to

$$V = \frac{\pi(\varphi_2 - \varphi_1)}{2c} e^{-\frac{\pi y}{c}}.$$

The lines $y$ = constant are isotaches, i.e., lines of equal velocity, the lines $x$ = constant are isoclines, i.e., lines of equal velocity direction. For $y = 0$, the velocity is maximum, equal to $\frac{\pi(\varphi_2 - \varphi_1)}{2c}$.

The formula for $\psi$ shows that for a distance of four sinus waves (Fig. 366), the discharge of the flow from the right side of the fragment to the left is equal to

$$q = -\frac{\varphi_1 - \varphi_2}{2} = k\frac{h_1 - h_2}{2}.$$

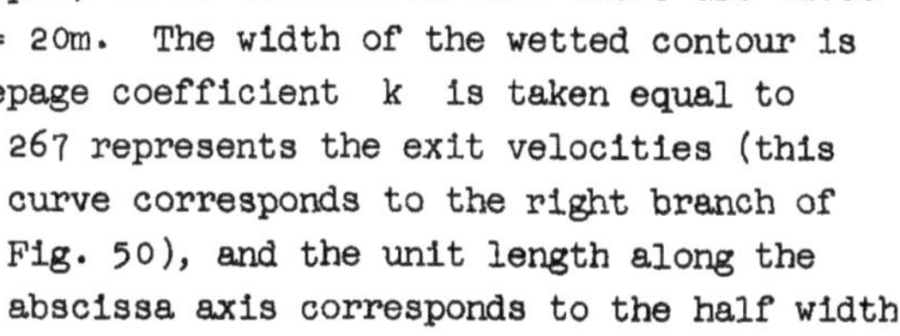

Fig. 366

To illustrate the influence of waviness in the lower basin upon the exit velocities, i.e., the velocities along the border of the lower basin, V. I. Aravin analyzes the following example. Consider a dam, with headwater at the 15m. mark, tailwater at 4m. Let the height of the wave in the tailwater reservoir be 2m. (in principle, it is considered that there are waves everywhere), the wavelength 2C = 20m. The width of the wetted contour is 2b (an arbitrary number), the seepage coefficient $k$ is taken equal to unity. Then, curve I of Fig. 267 represents the exit velocities (this curve corresponds to the right branch of Fig. 50), and the unit length along the abscissa axis corresponds to the half width b of the wetted contour.

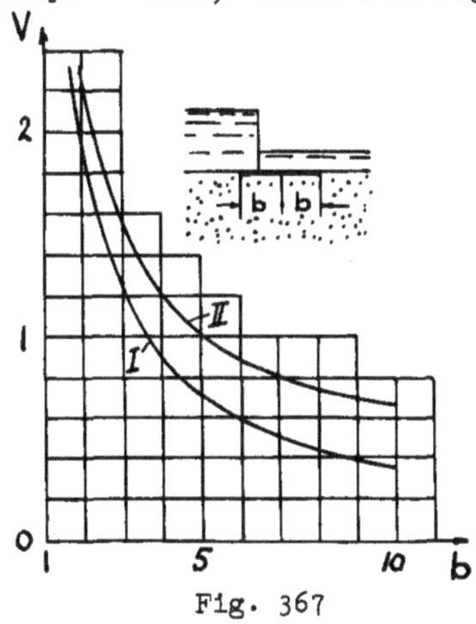


Fig. 367

Furthermore, with formula

$$V_{max} = \frac{\pi(\varphi_2 - \varphi_1)}{2c}$$

the maximum seepage velocity is computed, arising through waviness in the tailwater reservoir. In each point of the line for the original seepage velocity, the velocity arising only from the waviness is added. The result is the curve of maximum possible velocities (Curve II on Fig. 367). It shows that significant differences in the velocity are possible under influence of waviness. In Fig. 366 are given the lines of equal head and equal (reduced) flow rate.

## §3. One-dimensional Vertical Flow for Constant Operating Head.

We examine the flow of water along a vertical line in the soil, considering the seepage coefficient $k$ and the porosity $m$ to be constant quantities.

We take the general equation of motion with the inertia terms (the y-axis is directed downward)

$$\frac{\partial v_y}{\partial t} + \frac{\partial v_y}{\partial y} v_y = -\frac{1}{\rho}\frac{\partial p}{\partial y} + g - \frac{mg}{k} v_y . \tag{3.1}$$

We drop the rest of the equation, since we consider one-dimensional flow, parallel to the y-axis. Therefore, the equation of continuity takes on the form:

$$\frac{\partial v_y}{\partial y} = 0 .$$

This shows that $v_y$ only depends upon time (which was noticed by N. N. Pavlovsky). Therefore (3.1) may be rewritten, introducing instead of $v_y$ the seepage velocity $v$ through

$$v_y = \frac{1}{m} v ,$$

in the following form

$$\frac{1}{m}\frac{\partial v}{\partial t} = -\frac{1}{\rho}\frac{\partial p}{\partial y} + g - \frac{g}{k} v . \tag{3.2}$$

But we have shown in Chapter I (and also in Chapter II), that the term $\frac{1}{m}\frac{\partial v}{\partial t}$ may be neglected in all practically interesting cases. If we introduce the quantity

$$h = \frac{p}{\rho g} - y , \tag{3.3}$$

we come back to Darcy's law

$$v = -k\frac{dh}{dy} , \tag{3.4}$$

valid, consequently, even in the case of unsteady flows. We assume now that water percolates in the soil under a constant head $H$ (Fig. 368), and that at the moment $t$ it has seeped to the depth $y_0$ from the boundary of the reservoir. The y-axis will be oriented vertically downward. Let $y_0 = 0$ at the origin of time $t = 0$.

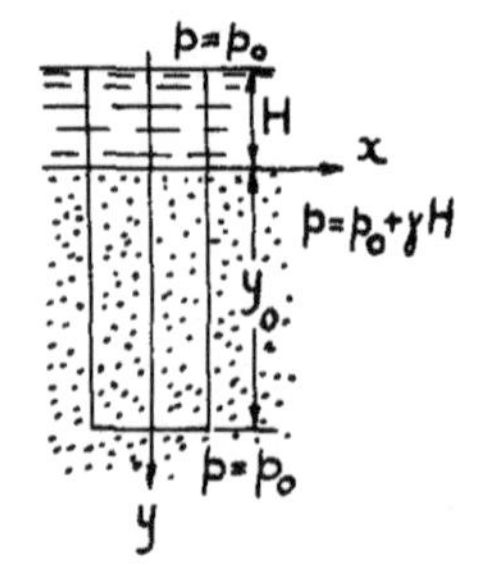

Fig. 368

Since $v = v(t)$ depends only upon time and not upon $y$, $h$ is a linear function of $y$:

$$h = a(t)y + b(t) .$$

For $y = 0$, the head is equal to $H$.

$$h(0) = b(t) = H . \tag{3.5}$$

For $y = y_0$, considering the atmospheric pressure to be zero, we have after (3.3):

$$h(y_0) = -h_k - y_0 = ay_0 + b\ , \tag{3.6}$$

where by $h_k$ the height of capillary rise is designated.

$$h_k = -\frac{p_k}{\rho g}\ .$$

Therefore for "a" we may write this expression [after (3.5) and (3.6)]:

$$a \doteq \frac{dh}{dy} = -\frac{h_k + y_0 + H}{y_0}\ . \tag{3.7}$$

Seepage velocity $v$ and derivative $\frac{dy_0}{dt}$ are related as

$$v = m\frac{dy_0}{dt}\ . \tag{3.8}$$

Comparison of (3.7) and (3.8) leads to the equation for $y_0$:

$$m\frac{dy_0}{dt} = k\frac{H + h_k + y_0}{y_0}\ . \tag{3.9}$$

We notice that according to the obtained equation, the capillary height is added to the acting head, as if instead of the head $H$ we had the head $H + h_k$.

To integrate equation (3.9) it suffices to write it in the form

$$\frac{y_0 dy_0}{y_0 + H + h_k} = \frac{k}{m}dt$$

or

$$dy_0 - \frac{H + h_k}{y_0 + H + h_k}dy_0 = \frac{k}{m}dt\ ,$$

after which we find, considering that $y_0 = 0$ for $t = 0$:

$$\frac{y_0}{H + h_k} - \ell n\left(1 + \frac{y_0}{H + h_k}\right) =$$

$$= \frac{kt}{m(H + h_k)}\ . \tag{3.10}$$

Introducing the dimensionless quantities:

$$\left\{\begin{aligned} \frac{y_0}{H + h_k} &= \eta\ , \\ \frac{kt}{m(H + h_k)} &= \tau\ , \end{aligned}\right. \tag{3.11}$$

we rewrite (3.10) in the following form:

$$\tau = \eta - \ell n(1 + \eta)\ . \tag{3.12}$$

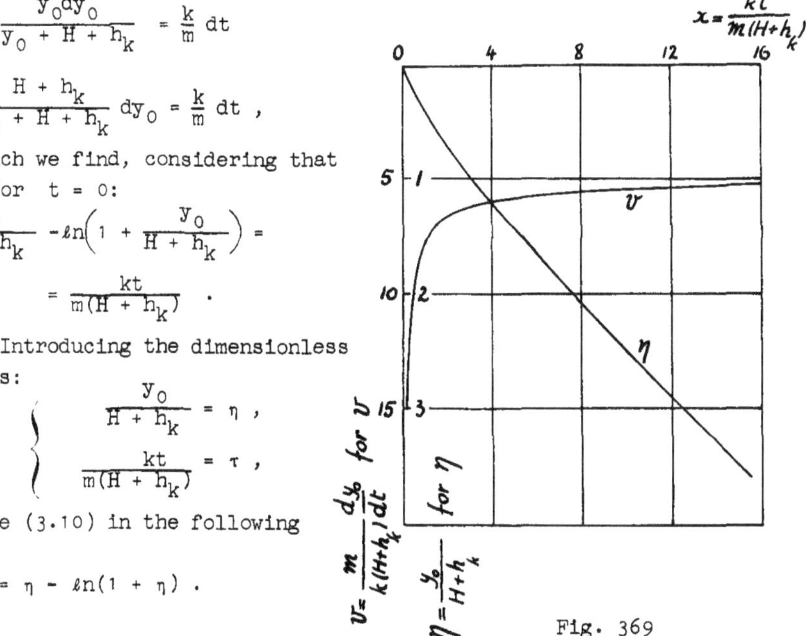

Fig. 369

The graph of the dependence of $\eta$ on $\tau$ is given in Fig. 369. Also, the dependence of $v$ upon $\tau$ is given there.

For small values of $\eta$ it is possible to carry out calculation of equation (3.12), if we develop $\ln(1 + \eta)$ into a power series. We obtain:

$$\tau = \frac{\eta^2}{2} - \frac{\eta^3}{3} - \frac{\eta^4}{4} - \ldots = \frac{\eta^2}{2}\left(1 - \frac{2}{3}\eta + \frac{1}{2}\eta^2 + \ldots\right), \tag{3.13}$$

from where we find, extracting the quadratic root:

$$\sqrt{2\tau} = \eta\left(1 - \frac{2}{3}\eta + \frac{1}{2}\eta^2 + \ldots\right)^{\frac{1}{2}} = \eta - \frac{1}{3}\eta^2 + \frac{7}{36}\eta^8 + \ldots \tag{3.14}$$

We put

$$\sqrt{2\tau} = \mu \tag{3.15}$$

and express $\eta$ in the form of a power series of $\mu$:

$$\eta = \mu + A\mu^2 + B\mu^3 + \ldots, \tag{3.16}$$

where A, B, — unknown coefficients of the flow. Inserting the expression (3.16) into the series (3.14) and equating the coefficients for the same powers of $\mu$, we find

$$\eta = \mu + \frac{1}{3}\mu^2 + \frac{1}{36}\mu^3 + \ldots = \sqrt{2\tau} + \frac{2}{3}\tau + \frac{\sqrt{2}}{18}\tau^{3/2} + \ldots \tag{3.17}$$

## §4. Seepage for a Given Constant Flow Rate.

Let us assume that we took a certain quantity of water Q and quickly poured it in a pipe (of unit cross-section) and further added no more water. Then at the moment t, the quantity Q is used up in the head H(t) and the moistening of the soil over a depth $y_0$ so that we will have (Fig. 370):

$$Q = H(t) + my_0 .$$

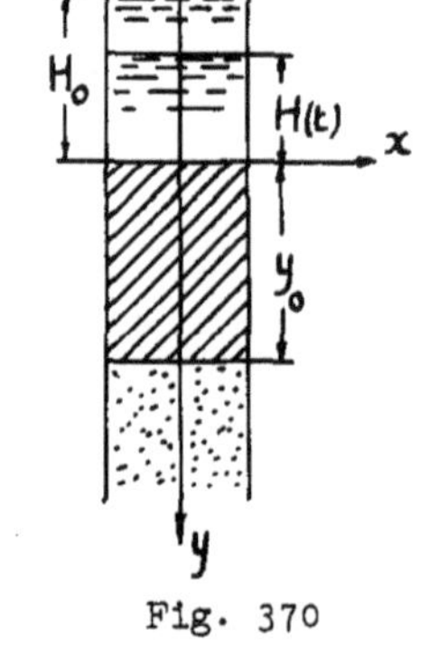

Fig. 370

If we now replace H in equation (3.9) by its value

$$H = Q - my_0,$$

then we obtain the equation:

$$m\frac{dy_0}{dt} = k\frac{Q + h_k + (1 - m)y_0}{y_0} =$$

$$= k(1 - m)\frac{y_0 + \frac{Q + h_k}{1 - m}}{y_0} . \tag{4.1}$$

The solution of this equation has the form (3.12) if we put

$$\eta = \frac{1 - m}{Q + h_k} y_0 , \quad \tau = \frac{k(1 - m)^2}{m(Q + h_k)} t \tag{4.2}$$

(considering $y_0 = 0$ for $t = 0$).

A flow of such nature may take place in the following way. We drive a vertical pipe in the soil. In the upper part of this pipe, emerging above the surface of the soil, we quickly pour some water. If we now watch the change in time of the water level H(t) in the upper part of the pipe, then we also will know the variation of $y_0$ in time. Comparing the curve H(t) or $y_0(t)$, obtained in nature, with the theoretical one, it is possible to determine some parameters, characteristic of soil, for example the parameters

$$\frac{1 - m}{Q + h_k}, \quad \frac{k(1 - m)^2}{m(Q + h_k)} .$$

From these, knowing $Q$ and $h_k$, we may find $k$ and $m$.

Such kinds of experiments were carried out by a series of investigators, amongst whom M. M. Protodjakonov.

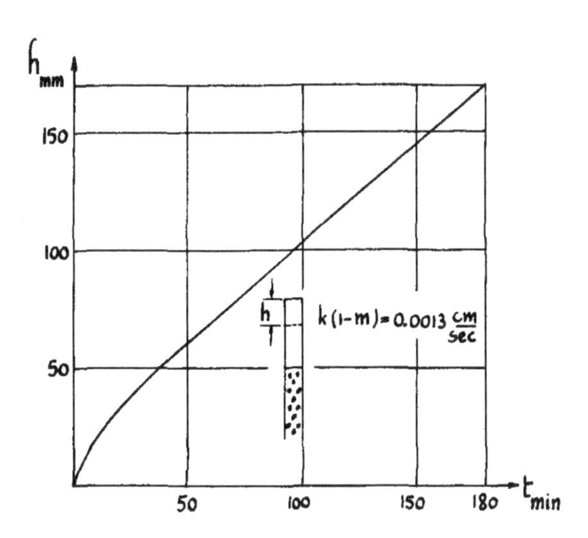


Fig. 371

However, he drove the cylinder only at a slight depth in the soil, so that the seepage that resulted was not one-dimensional, but three-dimensional. In order to obtain a movement much closer to the vertical, M. M. Protodjakonov took a second cylinder, concentric with the first one, of large diameter, and poured also water between the cylinders. In Fig. 371 is given one of the curves, resulting from such an experiment. Putting $h = H(0) - H(t)$, we find from (4.1) for large $t$: $\frac{dh}{dt} \approx k(1 - m)$. Therefore for large values of $t$ the direction coefficient of the line of Fig. 371 gives the value of $k(1 - m)$.

## §5. Gradual Filling With Water.

Actually it is impossible to carry out with precision the sudden filling of the upper part of the pipe up to a given height. And even if this were possible, then we would have a discontinuity in the pressure at the initial moment of time, when $y_0 = 0$. Indeed, at that time, the water column of height $H$ exerts on the horizontal area in the plane $y = 0$ a pressure, equal to the weight of this column per unit area. On the other side the pressure is zero (or corresponds to the capillary vacuum when capillarity is accounted for) in the soil at the boundary with the moistened region, i.e., for $y = 0$. This discontinuity in the pressure is also explained by the fact that in the problems examined by us, an infinite velocity is obtained at the origin of time.

In connection with the permanent practice of gradually adding water in the pipe (the same takes place in the filling of a canal), we examine the following problem. We assume that in the pipe at all times the flowrate $q \frac{m^3}{m^2 \sec}$ (cubic meters per second per unit area) is constant, so that at the moment $t$ the quantity of added fluid is:

$$Q = qt \tag{5.1}$$

This amount of fluid accounts for the buildup of the height of water $H(t)$ and for the moistening of the soil over a depth $y_0$, i.e., we have:

$$qt = H(t) + my_0 \ .$$

Inserting $H(t)$ from this expression into equation (3.9), we obtain:

$$m \frac{dy_0}{dt} = k \frac{qt + h_k + (1 - m)\, y_0}{y_0} \ . \tag{5.2}$$

Let us examine at first the case $h_k = 0$, i.e., we neglect the influence of capillarity. The equation

$$my_0 \frac{dy_0}{dt} = k[qt + (1 - m)\, y_0] \tag{5.3}$$

has a particular solution of the form

$$y_0 = ct \ . \tag{5.4}$$

To determine the value of the constant $c$, we insert (5.4) in (5.3) and obtain a quadratic equation for $c$. .

$$mc^2 = k[q + (1 - m)\, c] \ .$$

We take the positive root of this equation, since we are interested in the downward motion:

$$c = \frac{k(1 - m)}{2m} + \sqrt{\frac{k^2(1 - m)^2}{4m^2} + \frac{kq}{m}} \ .$$

Since the obtained solution satisfies the initial condition $y_0(0) = 0$, this solution is the one we need.

From this it is clear that if we add water in the pipe to have a constant flow rate $q$, it will percolate into the soil with constant velocity (we recall that we did not account for the capillarity of the soil). The water level above the soil surface $H(t)$ will be a linear function of time

$$H(t) = (q - mc)\, t \ . \tag{5.5}$$

It is evident that in order to have movement, the inequality

$$q \geq mc$$

must be satisfied.

In the case of the equality we have percolation in the absence of a water layer on the soil surface. This percolation can take place with velocity $c = k/m$ .

If we return to the general case of equation (5.2), when the capillarity of the soil is taken into account, then we see that its solution has a more complicated form. But through the substitution $t + h_k/q = \tau$ , equation (5.2) is reduced to the homogeneous equation

$$m \frac{dy_0}{dt} = k \frac{q\tau + (1 - m) y_0}{y_0} .$$

Its general integral can be written in the form

$$C\tau = (A - u)^{-\frac{1}{2} - \alpha} (B + u)^{-\frac{1}{2} + \alpha} ,$$

(5.6)

$$y_0 = u\tau ,$$

where

$$A = \frac{k}{2m} + \sqrt{\frac{k^2}{4m^2} + \frac{kq}{m}} , \quad B = -\frac{k}{2m} + \sqrt{\frac{k^2}{4m^2} + \frac{kq}{m}} ,$$

(5.7)

$$\alpha = \frac{1}{2\sqrt{1 + \frac{4mq}{k}}} .$$

Equations (5.6) present $\tau$ and $y_0$ in parameter form, in function of a parameter $u$. By elimination of $u$, it is possible to write the general integral (5.6) in the form

$$(A\tau - y_0)^{\frac{1}{2} + \alpha} (B\tau + y_0)^{\frac{1}{2} - \alpha} = C_1 .$$

The condition $y_0 = 0$ for $t = 0$ gives the initial value $\tau = \tau_0$:

$$\tau_0 = \frac{h_k}{q} .$$

Therefore we find for the arbitrary constant $C_1$, the following expression:

$$C_1 = A^{\frac{1}{2} + \alpha} B^{\frac{1}{2} - \alpha} \frac{h_k}{q} .$$

## §6. Seepage In Two-layered Soil.

We consider the problem of seepage of water in two-layered soil, when in the upper layer seepage coefficient and porosity have respectively the values $k_1$ and $m_1$, in the lower layer — the values $k_2$ and $m_2$, (Fig. 372). Above the soil, there is a water layer of thickness $H$. The depth of water, percolated in the soil at time $t$, is usually designated by $y_0$ . From the conditions of continuity of flow, we obtain that the seepage velocities $v_1$ and $v_2$ in the layers are equal and depend only upon time

(6.1)

$$v_1 = v_2 = v(t) .$$

We assume that

$$v_1 = \frac{\partial \varphi_1}{\partial y}, \quad v_2 = \frac{\partial \varphi_2}{\partial y},$$

and

$$\varphi_1 = -k_1\left(\frac{p_1}{\rho g} - y\right), \quad \varphi_2 = -k_2\left(\frac{p_2}{\rho g} - y\right). \tag{6.2}$$

Since $\varphi_1$ and $\varphi_2$ must satisfy the Laplace equation in $y$, we have, just as for $h$ in the case of homogeneous soil, for $\varphi_1$ and $\varphi_2$ linear functions of $y$:

$$\varphi_1 = v(t)y + b_1(t), \quad \varphi_2 = v(t)y + b_2(t).$$

At the border of the two soils for $y = h$ we have from the condition of equality of pressure

$$\frac{\varphi_1}{k_1} = \frac{\varphi_2}{k_2},$$

This can be brought in the form

$$\frac{vh + b_1}{k_1} = \frac{vh + b_2}{k_2}.$$

Further, for $y = 0$, we have,

$$\varphi_1 = b_1 = -k_1 H,$$

and for $y = y_0$

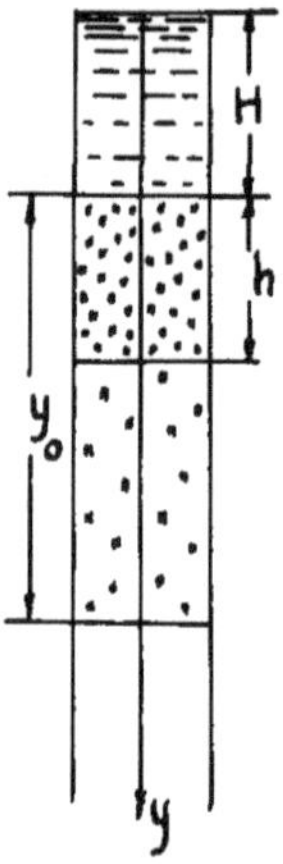

$$\varphi_2 = vy_0 + b_2 = k(h_k + y_0).$$

Fig. 372

Eliminating $b_1$ and $b_2$ from the last equations, we obtain for the percolation in the lower layer of soil:

$$v = m_2 \frac{dy_0}{dt} = \frac{H + h_k + y_0}{\frac{h}{k_1} + \frac{y_0 - h}{k_2}} \tag{6.3}$$

or

$$\frac{dy_0}{dt} = \frac{k_2}{m_2} \frac{H + h_k + y_0}{y_0 + \left(\frac{k_2}{k_1} - 1\right)h}. \tag{6.4}$$

Integration of the last equation gives

$$\frac{y}{H + h_k} - \left(1 - A\right) \ln \frac{y + H + h_k + h}{H + h_k + h} = \frac{k_2 t}{m_2 (H + h_k)}. \tag{6.5}$$

In this equation are designated

$$y = y_0 - h, \quad A = \left(\frac{k_2}{k_1} - 1\right)\frac{h}{H + h_k}.$$

The time origin is chosen as the moment at which percolation starts in the lower layer. It is evident, that percolation in the upper layer in our assumptions is calculated with the formulas covered in the previous paragraphs.

For $k_1 = k_2$ we have $A = 0$.

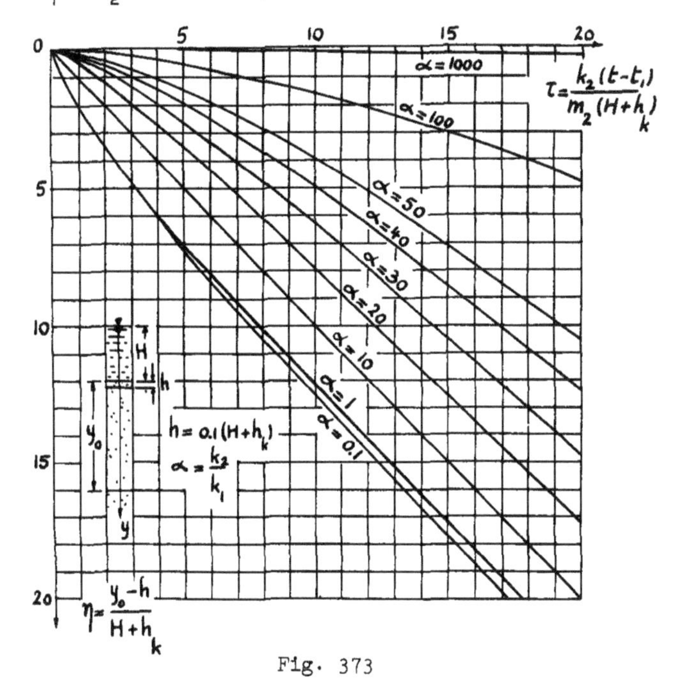


Fig. 373

We have examined the case where the upper layer consists of fines. This happens in the case of "clogging" (Chapter IV, Par. 3).

Often artificial conditions are responsible for such clogging of the soil at the bottom of a canal, by the introduction in the canal of turbid water that contains small clayey particles. The clogged layer may have a thickness of the order of 10 - 20 cm.

We assume that $h$ is small in comparison with the magnitude of $H + h_k$.

Then, ignoring $h$ under the logarithm sign (but preserving it in the expression for $A$), we introduce the notations:

$$\frac{y}{H + h_k} = \eta , \quad \frac{k_2}{m_2} \frac{t}{H + h_k} = \tau .$$

We obtain the equation

$$\tau = \eta - (1 - A) \ell n (1 + \eta) .$$

Assigning various values to $A$, we may construct the graphs for the dependence of $\eta$ on $\tau$. In Fig. 373 are given such curves for $\frac{h}{H + h_k} = 0.1$.

We see that clogging may to a significant degree decrease the velocity of percolation of water in the soil.

One ought to notice that if water would pass through a layer of weakly pervious soil and then reach sufficiently dry soil, percolation in the latter would occur under the form of isolated streams, instead of continuous ground water flow, and the exposed theory would be inapplicable. The latter is only applicable in the case of sufficiently moist soil, and in this case under $m$ we must understand "effective porosity" or deficit moisture content.

We remark that we may apply formula (6.3) to the case of the velocity of steady flow in two-layered medium, if we consider $y_0$ as a constant.

We put

$$y_0 - h = h_2 , \quad h = h_1$$

and write (6.3) in the form

$$v = \frac{H + h_k + h_1 + h_2}{\frac{h_1}{k_1} + \frac{h_2}{k_2}} .$$

If we introduce, as we did in Chapter VIII, the "equivalent" coefficient $k$ by means of the equality

$$\frac{h_1 + h_2}{k} = \frac{h_1}{k_1} + \frac{h_2}{k_2} ,$$

then the equation of the seepage velocity may be written in the form

$$v = k \frac{H + h_k + h_1 + h_2}{h_1 + h_2} .$$

This expression assumes a form as if seepage took place in a pipe of length $h_1 + h_2$, with seepage coefficient $k$. We notice that in the exposed theory, no account has been made of the change in saturation of the soil.

## CHAPTER XIII

## NON-LINEAR EQUATIONS OF UNSTEADY FLOWS WITH A FREE SURFACE

### §1. Derivation of Basic Relations.

In the general case, when inertia terms are not accounted for in the equations of motion, the flow of ground water is determined by the equations (Chapter I, Par. 11)

$$u = \frac{\partial \varphi}{\partial x}, \quad v = \frac{\partial \varphi}{\partial y}, \quad w = \frac{\partial \varphi}{\partial z}, \tag{1.1}$$

$$\frac{\partial u}{\partial x} + \frac{\partial v}{\partial y} + \frac{\partial w}{\partial z} = 0, \tag{1.2}$$

where

$$\varphi = -k\left(\frac{p}{\rho g} + z\right). \tag{1.3}$$

Here $u, v, w$ are the projections of the seepage velocity, so that

$$u = m\frac{dx}{dt}, \quad v = m\frac{dy}{dt}, \quad w = m\frac{dz}{dt}, \tag{1.4}$$

where $dx/dt, dy/dt, dz/dt$ are the projections of the velocity of a fluid particle, $m$ the soil porosity. The quantity $k$ is the seepage coefficient, $p$ — the pressure, $\rho$ — the density, $g$ — the acceleration of gravity. The z-axis is oriented upward.

We consider the motion with free surface.

If the equation of the latter is written in the form $F(x, y, z, t) = 0$, then, after differentiation with respect to time and insertion of the equalities (1.4), we obtain:

$$m\frac{\partial F}{\partial t} + \frac{\partial F}{\partial x}u + \frac{\partial F}{\partial y}v + \frac{\partial F}{\partial z}w = 0. \tag{1.5}$$

When the equation of the free surface is written in the form

$$z = \delta(x, y, t) \quad \text{or} \quad \delta(x, y, t) - z = 0,$$

we have instead of (1.5)

$$m\frac{\partial \delta}{\partial t} + \frac{\partial \delta}{\partial x}u + \frac{\partial \delta}{\partial y}v - w = 0. \tag{1.6}$$

At the surface, where pressure $p$ is constant, this equation follows from (1.3):

$$\varphi(x, y, z, t) + kz = \text{constant}. \tag{1.7}$$

First Assumption:

We assume that the free surface is slightly curved (Fig. 374) and oscillates about an average height $h$. Developing the function $\varphi$ into a power series of $(z - h)$ and neglecting the terms of smaller order, we have at the free surface:

$$\varphi(x, y, z, t) = \varphi(x, y, h, t) + \frac{\partial \varphi}{\partial z}\Big|_{z=h}(z - h) + \dots \approx \varphi(x, y, h, t),$$

and equation (1.7) may be rewritten as

$$x = \delta(x, y, t) = -\frac{\varphi(x, y, h, t)}{k}. \tag{1.8}$$

The constant that occurs in equation (1.7) may be considered as included in the function $\varphi$. Thus, if we would have found the velocity potential, then equation (1.8) would be the equation of the free surface.

Fig. 374

Second Assumption:

We assume that the horizontal velocities do not depend upon the height $z$. Then the equation of continuity (1.2) may be integrated with respect to $z$. We obtain:

$$w - w_0 = -\int_0^z \left[\frac{\partial u}{\partial x} + \frac{\partial v}{\partial y}\right] dz$$

or

$$w(x, y, z, t) = -z\left(\frac{\partial u}{\partial x} + \frac{\partial v}{\partial y}\right) + w_0(x, y, t). \tag{1.9}$$

Here $w_0$ is the vertical velocity at the lower basis layer for $z = 0$. But since we have essentially averaged the flow for its depth, we may also include in the expression of $w_0$ other factors creating a vertical velocity, for example, to take account of the infiltration or evaporation from the free surface.

Equation (1.9) gives a linear dependence of the vertical velocity $w$ upon $z$.

## §2. Derivation of Non-linear Equation.

We show that from equation (1.6), under the forementioned two assumptions, we may obtain the non-linear equation of Boussinesq [1]. We introduce in the analysis the function $H(x, y, t)$, related to the velocity potential $\varphi(x, y, h, t)$ by

$$H = -\frac{\varphi}{k}. \tag{2.1}$$

The equation of the free surface takes the form

$$z = H(x, y, t), \tag{2.2}$$

and for the horizontal velocities we have the equations

$$u = -k\frac{\partial H}{\partial x}, \quad v = -k\frac{\partial H}{\partial y}. \tag{2.3}$$

Equation (1.6), in which we put $H$ instead of $\delta$, gives:

$$m \frac{\partial H}{\partial t} - k\left[\left(\frac{\partial H}{\partial x}\right)^2 + \left(\frac{\partial H}{\partial y}\right)\right] - w = 0 . \tag{2.4}$$

For $w$ we take (1.9), with $z = H$. Then (2.4), after application of (2.3) gives:

$$m \frac{\partial H}{\partial t} - k\left[\left(\frac{\partial H}{\partial x}\right)^2 + \left(\frac{\partial H}{\partial y}\right)^2\right] - kH\left(\frac{\partial^2 H}{\partial x^2} + \frac{\partial^2 H}{\partial y^2}\right) - w_0 = 0 ,$$

which may also be rewritten in the form

$$\frac{\partial H}{\partial t} = \frac{k}{m}\left[\frac{\partial}{\partial x}\left(H \frac{\partial H}{\partial x}\right) + \frac{\partial}{\partial y}\left(H \frac{\partial H}{\partial y}\right)\right] + \frac{w_0}{m} . \tag{2.5}$$

## §3. The Method of the Small Parameter.

We consider the equation

$$\frac{\partial H}{\partial t} = \frac{k}{m} \frac{\partial}{\partial x}\left(H \frac{\partial H}{\partial x}\right) + \lambda f(x, t) , \tag{3.1}$$

in which we consider $\lambda$ as some small parameter; $f(x, t)$, a function that accounts for an external influence on the flow: infiltration, evaporation and so on.

When we put

$$H(x, t) = H_0 + \lambda H_*(x, t) + \lambda^2 H_{**}(x, t) + \ldots, \tag{3.2}$$

where $H_0$ — some constant, and insert series (3.2) in equation (3.1) then we obtain a system of equations to determine $H_*$, $H_{**}$...after we equate the coefficients of the same powers of $\lambda$:

$$\left\{\begin{aligned} \frac{\partial H_*}{\partial t} &= a^2 \frac{\partial^2 H_*}{\partial x^2} + f(x, t) , \\ \frac{\partial H_{**}}{\partial t} &= a^2 \frac{\partial^2 H_{**}}{\partial x^2} + a^2 H_0 \frac{\partial}{\partial x}\left(H_* \frac{\partial H_*}{\partial x}\right) , \\ &\cdots\cdots\cdots\cdots\cdots \\ &\left(a^2 = \frac{kH_0}{m}\right) . \end{aligned}\right. \tag{3.3}$$

Let the initial and boundary conditions of equation (3.1) be given:

$$H(x, 0) = \varphi(x) , \tag{3.4}$$

$$H(0, t) = F_1(t), \quad H(\ell, t) = F_2(t) . \tag{3.5}$$

We apply the exposed method to the case where one of the expressions (3.4) or (3.5) reduces to a constant, i.e., when one of the functions $\varphi(x)$, $F_1(t)$, $F_2(t)$ is a constant. Let, for example, $H(x, 0) = C$. Then we take $H_0 = C$. Furthermore, we may assume that the functions $H_{**}$, $H_{***}$ and so on, satisfy the zero boundary conditions (+), so that condition (3.5)

(+) i.e., are zero at $x = 0$ and $x = \ell$ at all times.

is satisfied for the first two terms of series (3.2):

$$
(3.6) \quad \begin{cases} H_*(x,\ 0) = H_{**}(x,\ 0) = \ldots = 0\ , \\ H_0 + \lambda H_*(0,\ t) = F_1(t),\quad H_0 + \lambda H_*(\ell,\ t) = F_2(t)\ , \\ H_{**}(0,\ t) = H_{***}(0,\ t) = \ldots = 0\ , \\ H_{**}(\ell,\ t) = H_{***}(\ell,\ t) = \ldots = 0\ . \end{cases}
$$

In this way, in some cases the problem of the integration of the non-linear equation (3.1) may be reduced to the integration of a sequence of equations of the heat conduction type. Practical computations of a large number of terms of series (3.2) are impossible, but the second approximation is sometimes obtained without difficulty. The case of the seepage from a reservoir, examined later on, is an example where the terms of the series are computed till the third order included.

## §4. Seepage By Variable Water Level In a Reservoir.

We examine here in detail the simplest problem, namely that of the integration of the non-linear equation

$$
(4.1) \qquad \frac{\partial H}{\partial t} = \frac{k}{m} \frac{\partial}{\partial x}\left(H \frac{\partial H}{\partial x}\right)
$$

for the initial and boundary conditions:

$$
(4.2) \qquad H(x,\ 0) = H_2\ ,\quad H(0,\ t) = H_1\ ,
$$

where $H_1$, $H_2$ – constants which correspond to the two cases of Fig. 375 and 376.

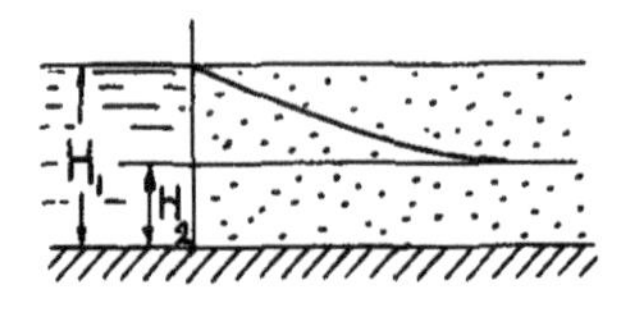

Fig. 375

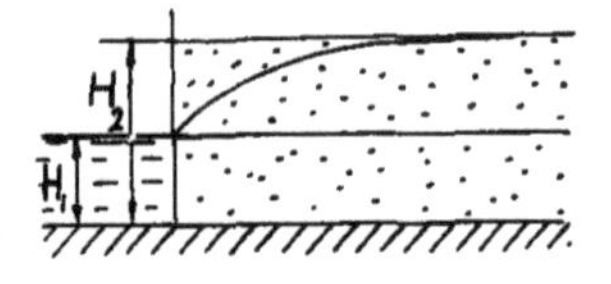

Fig. 376

A detailed analysis of this problem is interesting because it offers the possibility of comparing the solutions obtained after linearization of equation (4.1) with the exact solution of equation (4.1). There exist two methods of linearization of this equation.

The first method consists in replacing the multiplier H in the brackets of equation (4.1) by some constant value $\bar{h}$. A linear equation is obtained

$$
(4.3) \qquad \frac{\partial H}{\partial t} = a^2 \frac{\partial^2 H}{\partial x^2}\ ,
$$

where

$$
a^2 = \frac{k\bar{h}}{m}\ .
$$

In the second method, applied by a series of authors (see, for example, [13]), not H but $H^2$ is adopted as the unknown function. Assuming

(4.4) $$H^2 = u ,$$

this equation is obtained

$$\frac{\partial u}{\partial t} = \frac{k}{m} \sqrt{u} \frac{\partial^2 u}{\partial x^2} .$$

Now we replace in this equation $\sqrt{u}$ by some constant value $\tilde{h}$. Then we have the linear equation

(4.5) $$\frac{\partial u}{\partial t} = a^2 \frac{\partial^2 u}{\partial x^2} ,$$

where

(4.6) $$a^2 = \frac{k\tilde{h}}{m} .$$

In favor of the reduction of the second form of linearization is the fact that in the transition from the non-steady motion to the steady one, in the first method of linearization we obtain for $H$ the equation

$$\frac{\partial^2 H}{\partial x^2} = 0 ,$$

giving for solution the linear function

$$H = C_1 x + C_2 .$$

In the second way of linearization we have the equation

$$\frac{\partial^2 u}{\partial x^2} = \frac{\partial^2 H^2}{\partial x^2} = 0 ,$$

with general solution:

$$H^2 = C_1 x + C_2 .$$

In other words, in the first method the equation of the free surface of the steady flow is obtained in the form of the straight line $y = C_1 x + C_2$, which is only suited for $C_1 = 0$ (and for larger values of $H_2$ compared with $H_1 - H_2$), whereas in the second method we obtain for the free surface the parabola $y^2 = C_1 x + C_2$ (Dupuit's parabola). Later on we will see that in some cases results can be obtained by the first method of linearization which are closer to the exact solution; in other cases this is obtained by the second method.

In our problem, $x$ changes between limits from $0$ to infinity, and the boundary and initial conditions are such that it is possible to make a well-known substitution, applied in the equation of heat conduction, in the theory of the boundary layer, and so on, and namely:

(4.7) $$\eta = \frac{x}{\alpha \sqrt{t}} ,$$

where $\alpha$ — a constant which we will determine further. This substitution was indicated by Wiener and Boltzmann in the equation of diffusion with variable coefficient of diffusivity, applied by Boussinesq [1] in the theory of seepage and by L. S. Leibenzon [4] to the equation of motion of a gas in porous medium, but further computations with the equations obtained in this way were not carried out. (The results of the computations for a case, somewhat different from ours, are available [2]). With this substitution, equation (4.1) is transformed into the non-linear ordinary differential equation

$$\frac{d^2H^2}{d\eta^2} + \frac{m\alpha^2}{k}\,\eta\,\frac{dH}{d\eta} = 0\ .$$

Introducing further these substitutions:

$$H = H_0 u\ ,\quad \eta = \frac{x\sqrt{m}}{2\sqrt{kH_0 t}}\ ,\quad \alpha = 2\sqrt{\frac{kH_0}{m}}\ ,$$

we obtain the equation [6]:

$$\frac{d^2u^2}{d\eta^2} + 4\eta\,\frac{du}{d\eta} = 0\ . \tag{4.8}$$

Parallel with this equation we examine the equation that is obtained from (4.8) by the substitution [5], $u^2 = v$ i.e., the equation

$$\frac{d^2v}{d\eta^2} + \frac{2\eta}{\sqrt{v}}\,\frac{dv}{d\eta} = 0\ . \tag{4.9}$$

We look for a solution of each of the equations (4.8), and (4.9) in the form of a power series of the parameters $\ell$ and $\lambda$ respectively:

$$u = 1 + \ell u_1 + \ell^2 u_2 + \ell^3 u_3 + \dots, \tag{4.10}$$

$$v = 1 + \lambda v_1 + \lambda^2 v_2 + \lambda^3 v_3 + \dots . \tag{4.11}$$

$H_0$ may be taken arbitrarily to be equal to $H_1$ as well as to $H_2$. If we want to obtain a family of curves passing through the origin, then we put $H_0 = H_1$; if we want to obtain curves, corresponding to the lowering of the water level in a canal, then we take $H_0 = H_2$.

In both cases, to determine the coefficients of the series (4.10) we obtain such a system of equations for $u$:

$$\begin{cases} u_1'' + 2\eta u_1' = 0\ , \\ u_2'' + 2\eta u_2' = -\frac{1}{2}\,(u_1^2)''\ , \\ u_3'' + 2\eta u_3' = -\,(u_1u_2)''\ , \\ \dots\dots\dots\dots \end{cases} \tag{4.12}$$

For the coefficients of the series (4.11), simpler equations with simpler right sides are obtained

$$\begin{cases} v_1'' + 2\eta v_1' = 0\ , \\ v_2'' + 2\eta v_2' = \eta v_1 v_1'\ , \\ v_3'' + 2\eta v_3' = \eta(v_1v_2)' - \frac{3}{4}\,\eta v_1^2\, v_1'\ , \\ \dots\dots\dots\dots \end{cases} \tag{4.13}$$

We return to the first case, where we take $H_0 = H_1$. Here we make the the functions $u_n$ and $v_n$ subject to the conditions

$$u_1(0) = u_2(0) = \dots = 0\ ,\quad v_1(0) = v_2(0) = \dots = 0\ ,$$

$$u_2(\infty) = u_3(\infty) = \dots \quad 0 \,, \quad v_2(\infty) = v_3(\infty) = \dots = 0 \,.$$

In order to satisfy the conditions at infinity

$$u(\infty) = \frac{H_2}{H_1} \,, \qquad v(\infty) = \left( \frac{H_2}{H_1} \right)^2,$$

we must take respectively

$$1 + \ell u_1(\infty) = \frac{H_2}{H_1} \,, \; 1 + \lambda v_1(\infty) = \left( \frac{H_2}{H_1} \right)^2 \,,$$

from which we obtain for the parameters $\ell$ and $\lambda$ the values

$$\ell = \frac{H_2 - H_1}{H_1} \,, \qquad \lambda = \frac{H_2^2 - H_1^2}{H_1^2} \tag{4.14}$$

(we will see further that $u_1(\infty) = v_1(\infty) = 1$)

The first approximation gives both for $u_1$ and $v_1$ the probability function, i.e.,

$$u_1(\eta) = v_1(\eta) = \Phi(\eta) = \frac{2}{\sqrt{\pi}} \int_0^{\eta} e^{-\tau^2} \, d\tau \,. \tag{4.$\phi$5}$$

For the second approximations we obtain respectively

$$u_2(\eta) = \frac{1}{\pi} \left(1 - e^{-2\eta^2}\right) - \frac{1}{\sqrt{\pi}} \eta e^{-\eta^2} u_1 - \frac{1}{2} u_1^3 + \left( \frac{1}{2} - \frac{1}{\pi} \right) u_1 \,, \tag{4.16}$$

$$v_2(\eta) = \frac{1}{2\pi} \left(1 - e^{-2\eta^2}\right) - \frac{1}{2\sqrt{\pi}} \eta e^{-\eta^2} v_1 - \frac{1}{2\pi} v_1 \,. \tag{4.17}$$

Finally, for the third approximations we find these expressions:

$$\begin{aligned} u_3(\eta) = {} & \frac{1}{2} u_1^3 + \frac{9}{4\sqrt{\pi}} \eta e^{-\eta^2} u_1^2 - \frac{1}{2\sqrt{\pi}} \eta^3 e^{-\eta^2} u_1^2 + \frac{3}{\pi} e^{-\eta^2} u_1 - \frac{1}{\pi} \eta^2 e^{-2\eta^2} u_1 - \\ & - \frac{1}{\pi\sqrt{\pi}} \eta e^{-\eta^2} - \frac{1}{2\pi\sqrt{\pi}} \eta e^{-3\eta^2} - \frac{3\sqrt{3}}{4\pi} \Phi(\eta\sqrt{3}) + \left(1 - \frac{2}{\pi}\right) u_2 + \left( \frac{3\sqrt{3}}{4\pi} - \frac{1}{2} \right) u_1, \end{aligned} \tag{4.18}$$

$$\begin{aligned} v_3(\eta) = {} & \left( \frac{5\eta}{16\sqrt{\pi}} - \frac{\eta^3}{8\sqrt{\pi}} \right) e^{-\eta^2} v_1^2 + \left( \frac{1}{2\pi^2} + \frac{3\sqrt{3}}{16\pi} + \frac{e^{-2\eta^2}}{2\pi} - \frac{\eta^2 e^{-2\eta^2}}{4\pi} \right) v_1 - \\ & - \frac{1}{\pi} v_2 - \frac{\eta e^{-\eta^2}}{4\pi\sqrt{\pi}} - \frac{\eta e^{-3\eta^2}}{8\pi\sqrt{\pi}} - \frac{3\sqrt{3}}{16\pi} \Phi(\eta\sqrt{3}) \,. \end{aligned} \tag{4.19}$$

We show for example, how the expression for $v_2$ may be found. The second of equations (4.13) may be written as:

$$e^{-\eta^2} \left(v_2' e^{\eta^2}\right)' = \eta v_1 v_1' = \frac{2}{\sqrt{\pi}} \eta e^{-\eta^2} v_1 \,.$$

From this, we find after integration by parts:

$$v_2' e^{\eta^2} = \frac{2}{\sqrt{\pi}} \int \eta v_1 \, d\eta = \frac{1}{\sqrt{\pi}} \eta^2 v_1 - \frac{1}{2\sqrt{\pi}} v_1 + \frac{1}{\pi} \eta e^{-\eta^2} \,.$$

Finally for $v_2$ we obtain

$$v_2 = \frac{1}{\sqrt{\pi}} \int \eta^2 e^{-\eta^2} v_1 \, d\eta - \frac{1}{2\sqrt{\pi}} \int v_1 e^{-\eta^2} \, d\eta + \frac{1}{\pi} \int \eta e^{-2\eta^2} \, d\eta =$$

$$= - \frac{1}{2\sqrt{\pi}} \eta e^{-\eta^2} v_1 - \frac{1}{2\pi} e^{-2\eta^2} + c_1 v_1 + c_2 .$$

The arbitrary constants are determined by the boundary conditions.

When we come back to the second case — the family of lines that tend to infinity, when we take $H_0 = H_2$, it will be convenient to introduce other designations, to find a solution of equation (4.9) in the form of a series

$$v = 1 + \mu w_1 + \mu^2 w_2 + \dots \tag{4.20}$$

The functions $w_1$, $w_2$ and so on satisfy the equations (4.13), but for other boundary conditions. We write these equations:

$$\left\{ \begin{aligned} & w_1'' + 2\eta w_1' = 0 , \\ & w_2'' + 2\eta w_2' = \eta w_1 w_1' , \\ & w_3'' + 2\eta w_3' = \eta (w_1 w_2)' - \tfrac{3}{4} \eta w_1^2 w_1' , \\ & \dots\dots\dots\dots\dots\dots \end{aligned} \right. \tag{4.21}$$

The boundary conditions for $w_1$, $w_2$ and so on are chosen in this way:

$$w_1(\infty) = w_2(\infty) = \dots = 0 ,$$

$$w_2(0) = w_3(0) = \dots = 0 .$$

Then we obtain

$$w_1(\eta) = 1 - \Phi(\eta) = 1 - v_1 ,$$

$$w_2(\eta) = \frac{1}{2\pi} (1 - e^{-2\eta^2}) - \frac{1}{2\sqrt{\pi}} \eta e^{-\eta^2} v_1 - \frac{1}{2\pi} v_1 + \frac{1}{2\sqrt{\pi}} \eta e^{-\eta^2} = v_2(\eta) + \frac{1}{2\sqrt{\pi}} \eta e^{-\eta^2}$$

The value of the parameter $\mu$ is determined from the condition

$$1 + \mu w_1(0) = \frac{H_1^2}{H_2^2} ,$$

so that

$$\mu = \frac{H_1^2 - H_2^2}{H_2^2} .$$

With the help of the obtained formulas, the computations of Table 16 were carried out. It turned out that in the extreme cases — the flow into the soil for zero ground water level and the seepage into an empty reservoir — the degree of approximation, adopted in the restriction of the terms up to the third degree included, was insufficient. These cases were investigated separately by means of numerical integration.

TABLE 16 COEFFICIENTS OF SERIES (4.10) AND (4.11) .

| $\eta$ | $u_1 = v_1$ | $u_2$ | $u_3$ | $v_2$ | $v_3$ |
|---|---|---|---|---|---|
| 0 | 0 | 0 | 0 | 0 | 0 |
| 0.1 | 0.1125 | +0.0141 | -0.0039 | -0.0179 | +0.0084 |
| 0.2 | 0.2227 | +0.0160 | -0.0081 | -0.0353 | +0.0165 |
| 0.3 | 0.3286 | +0.0073 | -0.0090 | -0.0515 | +0.0241 |
| 0.4 | 0.4284 | -0.0092 | -0.0049 | -0.0658 | +0.0307 |
| 0.5 | 0.5205 | -0.0300 | +0.0039 | -0.0774 | +0.0358 |
| 0.6 | 0.6039 | -0.0519 | +0.0159 | -0.0858 | +0.0390 |
| 0.7 | 0.6778 | -0.0718 | +0.0280 | -0.0905 | +0.0401 |
| 0.8 | 0.7421 | -0.0874 | +0.0373 | -0.0915 | +0.0389 |
| 0.9 | 0.7969 | -0.0975 | +0.0422 | -0.0892 | +0.0357 |
| 1.0 | 0.8427 | -0.1017 | +0.0418 | -0.0840 | +0.0310 |
| 1.1 | 0.8802 | -0.1004 | +0.0368 | -0.0766 | +0.0254 |
| 1.2 | 0.9103 | -0.0946 | +0.0281 | -0.0677 | +0.0194 |
| 1.3 | 0.9340 | -0.0855 | +0.0194 | -0.0581 | +0.0136 |
| 1.4 | 0.9523 | -0.0744 | +0.0078 | -0.0486 | +0.0085 |
| 1.5 | 0.9661 | -0.0626 | -0.0011 | -0.0395 | +0.0043 |
| 1.6 | 0.9764 | -0.0510 | -0.0079 | -0.0303 | +0.0012 |
| 1.7 | 0.9838 | -0.0394 | -0.0125 | -0.0241 | -0.0010 |
| 1.8 | 0.9891 | -0.0310 | -0.0147 | -0.0182 | -0.0022 |
| 1.9 | 0.9928 | -0.0232 | -0.0151 | -0.0133 | -0.0029 |
| 2.0 | 0.9953 | -0.0169 | -0.0141 | -0.0096 | -0.0031 |
| 2.5 | 0.9996 | -0.0024 | -0.0047 | -0.0013 | -0.0011 |
| 3.0 | 0.9999 | -0.0002 | -0.0006 | -0.0001 | -0.0002 |
| 3.5 | 1 | -0.0000 | -0.0001 | 0 | 0 |
| 4.0 | 1 | -0.0000 | -0.0001 | 0 | 0 |

§5. Numerical Integration.

By the substitution

$$\eta\sqrt{2} = \xi \,, \tag{5.1}$$

we transform equation (4.8) into the following

$$\frac{d^2u^2}{d\xi^2} + 2\xi\frac{du}{d\xi} = 0 \,,$$

or in developed form as

$$uu'' + u'^2 + \xi u' = 0 \,. \tag{5.2}$$

We will look for a solution of (5.2) for small values of $\xi$ in the form of a series

$$u = 1 + u_0'\xi + \frac{u_0''}{1.2}\xi^2 + \dots \tag{5.3}$$

We put

$$u_0' = \alpha$$

where $\alpha$ will be chosen arbitrarily. Then we find:

$$u_0'' = -\alpha^2, \quad u_0''' = 3\alpha^3 - \alpha, \quad u_0^{IV} = -15\alpha^4 + 6\alpha^2 \,,$$

$$u_0^{V} = 105\alpha^5 - 49\alpha^3 + 3\alpha, \quad u_0^{VI} = -945\alpha^6 + 504\alpha^4 - 52\alpha^2 \,,$$

$$u_0^{VII} = 10{,}395\alpha^7 - 6237\alpha^5 + 882\alpha^3 - 15\alpha \,.$$

Assigning to $\alpha$ a series of values, one may construct the family of integral curves of equation (5.2). If one starts with arbitrary values of $\xi$, then the series (5.3) ceases to be fit for calculations and one has to turn to numerical integration of equation (5.2). By this method, the family of integral curves of equation (5.2) has been constructed for $\alpha = \pm .1; \pm .2; \ldots; \pm .6$. All these curves have straight asymptotes, parallel to the $\xi$-axis (see further Fig. 381). By means of interpolation, the curves were so evaluated that the asymptotes bacame equidistant straight lines: $u = \pm .1;\ \pm .2 \ldots.$ The value of $\alpha = -0{,}628$ corresponds to the curve that progresses up to the abscissa-axis.

## §6. Seepage Into an Empty Reservoir.

We link this particular case of equation (4.8) to such a system of two equations that makes it possible to take benefit of some of the results of the calculations that occur in the theory of the boundary layer of a plate. We therefore obtain a solution of our problem in an obvious form for small values of the independent variable $\eta$.

We eliminate from equation (4.8) the differentiation along the variable $\eta$. Therefore we rewrite the equation in the form

$$\frac{d(uu')}{du} + 2\eta = 0 .$$

Now we introduce a new auxiliary variable $\xi$, and consider $u$ and $\eta$ as functions of $\xi$, through definition of the variable $\xi$ by the equality

$$\frac{d\eta}{d\xi} = u .$$

Consequently, if we have found $u$ as a function of $\xi$, then we also obtained $\eta$ in the form

$$\eta = \int_0^\xi u\, d\xi .$$

Now we have:

$$u = \frac{d\eta}{d\xi}, \quad u' = \frac{du}{d\eta} = \frac{1}{u}\frac{du}{d\xi} = \frac{1}{u}\frac{d^2\eta}{d\xi^2},$$

from where

$$\frac{d}{d\xi}\left(\frac{d^2\eta}{d\xi^2}\right)\frac{d\xi}{du} + 2\eta = 0$$

and, finally,

$$\frac{d^3\eta}{d\xi^3} + 2\eta\frac{d^2\eta}{d\xi^2} = 0 . \tag{6.1}$$

The same equation is found in the theory of the boundary layer of a plate [14]. We may represent its solution in the form of a power series:

$$\eta = a_0 + a_1\xi + a_2\xi^2 + \ldots$$

with boundary conditions

$$\eta(0) = 0, \quad \eta'(0) = 0, \quad \eta'(\infty) = 1 .$$

We may immediately borrow this solution from the theory of the boundary layer. We obtain for $\eta$ as a function of $\xi$ the series

$$(6.2) \qquad \eta = \alpha\xi^2\left(1 - \frac{\alpha}{120}\xi^3 + \frac{11\alpha^2}{2.8!}\xi^6 - \frac{375\alpha^3}{4.11!}\xi^9 + \dots\right) \qquad (\alpha = 0.33206) .$$

Taking from this the derivative along $\xi$, we obtain $u$ as a function of $\xi$.

$$(6.3) \qquad u = 2\alpha\xi - \frac{1}{4!}\alpha^2\xi^4 + \frac{11}{2.7!}\alpha^3\xi^7 - \frac{375}{4.10!}\alpha^4\xi^{10} + \dots$$

Let

$$Y = \sqrt{\eta\alpha^{-1/3}(5!)^{-2/3}} \qquad X = \alpha^{\frac{1}{3}}(5!)^{-\frac{1}{3}}\xi \quad ,$$

Then we obtain instead of (6.2):

$$Y = X(1 - X^3 + \dots)^{\frac{1}{2}} = X - \frac{1}{2}X^4 + \dots$$

The inversion of this series gives

$$X = Y + \frac{1}{2}Y^4 + \frac{1}{7}Y^7 - \frac{4}{11}Y^{10} + \dots$$

Inserting this expression of $X$ in the series for $u$, we find finally:

$$(6.4) \qquad u = B\left(Y - 2Y^4 + 3Y^7 - \frac{4}{11}Y^{10} - 4.77Y^{13} + \dots\right) .$$

In this series the following abbreviations are introduced

$$B = \alpha^{\frac{2}{3}}(5!)^{\frac{1}{3}} = 2.3652, \quad \alpha = 0.33206 \dots, \quad Y = \frac{\sqrt{\eta}}{\alpha^{1/3}(5!)^{1/3}} = 0.4873\sqrt{\eta} .$$

By means of series (6.4), the calculations are carried out up to $\eta = 0.6$. More points are further found by means of the numerical integration:

$$\eta = \int_0^\xi u(\xi)\, d\xi$$

and the graph (Fig. 377) is constructed.

Furthermore, it is interesting to compare the exact solution, with the approximate solutions, obtained by means of the first and second methods of linearization (see Par. 4). In both cases we take

$$\bar{h} = \tilde{h} = H_2 .$$

In the first method of linearization, the solution has the form

$$u = \tilde{u} = \Phi(\eta) = \frac{2}{\sqrt{\pi}}\int_0^\eta e^{-\tau^2} d\tau .$$

In the second method of linearization, we have:

$$u = \tilde{\tilde{u}} = \sqrt{\Phi(\eta)} .$$

In Fig. 377, the two approximate solutions are plotted side by side with the exact solution. We see that the first method of linearization gives a very poor result, but that the second method gives a solution, sufficiently close to the exact solution.

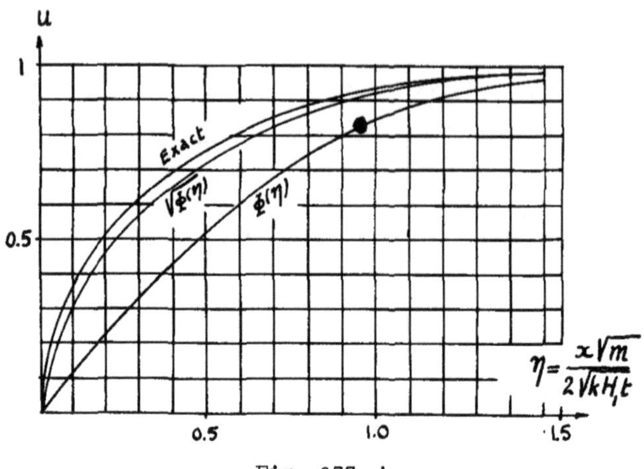


Fig. 377

§7. Flow In the Soil With Zero Ground Water Level [7, 14].

The other extreme case, for which the adopted degree of precision – up to the third degree of the parameter included – is insufficient, is the case of flow into the soil in which before, that is, prior to the filling of the reservoir with water, there was no ground water (but which may, finally, be wet; in this case porosity will be replaced by "deficiency of saturation" or "active" porosity – see Chapter I, Par. 2). The insufficient precision in the limitation of the number of terms till the third order included is revealed both by the behaviour of the terms containing the third power of $\ell$, and by the graph of $u(\eta)$, in which at some place small negative ordinates are obtained. (Fig. 378). Therefore, in this case, the integral curve has been constructed by means of the following development in series.

If, in equation (5.2) we put $u = 0$, then, since $u'' \neq \infty$ we have:

$$u'(u' + \xi) = 0 .$$

From this, if $u' \neq 0$, we obtain:

$$u' + \xi = 0 . \tag{7.1}$$

Assuming that in the case under consideration the point of intersection of the integral curve with the abscissa axis exists, we designate its abscissa by the letter $c$. In this point we have:

$$u = 0, \quad u' = -c . \tag{7.2}$$

After differentiation of (5.2) and making use of (7.2), we find the series development in powers of the difference $\xi - c$:

$$u = -c(\xi - c) - \frac{1}{4}(\xi - c)^2 - \frac{1}{72c}(\xi - c)^3 + \frac{(\xi - c)^4}{576\, c^2} + \dots \tag{7.3}$$

We put $\xi = 0$ in the obtained equality, We have:

$$1 = c^2\left(1 - \frac{1}{4} + \frac{1}{72} - \dots\right).$$

Neglecting the rapidly vanishing terms of the series between brackets, we find $c^2$ and then $c$ :

$$c = 1.14277 \dots \tag{7.4}$$

By means of series (7.3), Table 17 giving the dependence of $u$ upon $\xi$, is obtained.

TABLE 17

Dependence of $u$ upon $\xi$ in the case of the flow of water from a canal into soil with zero ground water level.

| $\xi$ | 0 | 0.1 | 0.2 | 0.3 | 0.4 | 0.5 | 0.6 | 0.7 | 0.8 | 0.9 | 1.0 | 1.1 | 1.143 |
|---|---|---|---|---|---|---|---|---|---|---|---|---|---|
| u | 1 | 0.936 | 0.867 | 0.794 | 0.716 | 0.635 | 0.549 | 0.458 | 0.363 | 0.263 | 0.058 | 0.0487 | 0 |

It is not difficult to re-evaluate this table, passing from the independent variable $\xi$ to the variable $\eta$, where $\eta = \frac{\xi}{\sqrt{2}}$ . This computation is not carried out, but we give a graph (Fig. 378) in which the dependence of $u$ upon $\eta$ is shown, and also the graphs $(1 - \Phi(\eta))$ and $\sqrt{1 - \Phi(\eta)}$. corresponding to solutions of the equation of heat conduction, linearized after the first and second methods. We see that the second method of linearization, when for the unknown function in the equation of heat $H^2$ is adop ted, gives significantly the worst result. The curve $\ell = -1$ is obtained as the result of computation with formula (4.10).

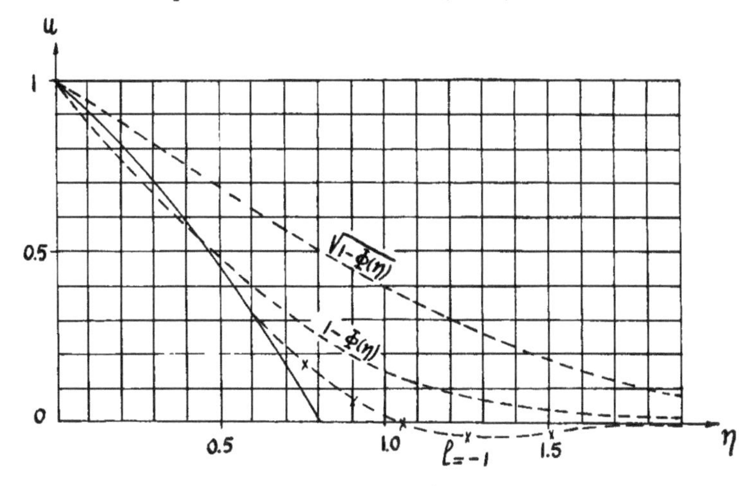


Fig. 378

Returning to the exact equation (7.3) of the free surface line, we notice that we only took part of this line, for $0 \leq \xi \leq c$ , since negative values of $u$ have no meaning for us. For $\xi > c$ we must take $u = 0$. Therefore $u'$ must undergo a discontinuity, but the quantity $uu'$, proportional to the flow rate in the various cross sections of the flow, remains continuous. It is interesting that in the extreme case under consideration the tongue of ground water moves with a finite velocity (for $t > 0$). From the equation $\xi = c$ we find, considering that $\xi = \eta\sqrt{2}$ $\frac{x\sqrt{m}}{\sqrt{2kH_1 c}}$, for the abscissa of the front of the tongue: $x = c\sqrt{\frac{2kH_1 t}{m}}$. Consequently, the diffusion velocity of the front is: $\frac{dx}{dt} = \frac{c\sqrt{kH_1}}{\sqrt{2mt}}$.

We must bear in mind that the integration of the non-linear equation gives us a more precise form of the free surface, than the integration of a linearized equation, though we only obtain an "hydraulic" solution, i.e., averaged in height. An exact solution of two-dimensional problems is not known to us. G. K. Mihailov [8] took notice of the fact that the equation of Boussinesq for the moving tongue of ground water (case of this paragraph) exactly corresponds to anistropic soil with $k_y = \infty$.

In the case $k_y = 0$, all the streamlines are horizontal and therefore the equation of the free surface takes on the form

$$\frac{dx}{dt} = \frac{k}{m}\frac{H_1 - y}{x} \quad (7.5)$$

Its integration for the condition $x = 0$, for $t = 0$, gives for the free surface:

$$y = H_1 - \frac{mx^2}{2kt} \quad (7.6)$$

or, with the previous notations, $u = 1 - \xi^2$.

Experiments [9] have shown that the form of the free surface in a two-dimensional problem lies between curves (7.3) and (7.6). In Fig. 379, both theoretical curves, the lower ends of which converge into one point, are represented.

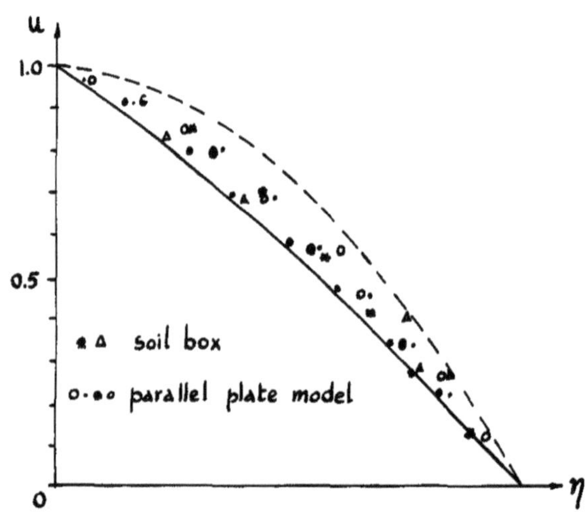


Fig. 379

Points that are obtained as the result of a series of experiments [9], fall in the region bordered by the curves (7.3) and (7.6). (Fig. 379).

§8. Construction of Other Curves.

For the construction of other curves, at first the method of numerical integration of equation (5.3) was applied for given values of $\alpha = \pm 0.1;\ \pm 0.2;\ \ldots;\ -0.6$. Then one may find the values of $y_\infty$, i.e., the ordinates of the straight lines to which the integral lines tend asymptotically. A table giving the dependence of $\alpha$ on $y_\infty$ for equidistant values of $y_\infty$ was found by interpolation and computations for new values of $\alpha$ were carried out. The results are represented in Fig. 380, where along the abscissa axis, values of the quantity $\eta = \frac{\xi}{\sqrt{2}} = \frac{x\sqrt{m}}{2\sqrt{kH_1 t}}$ are laid out.

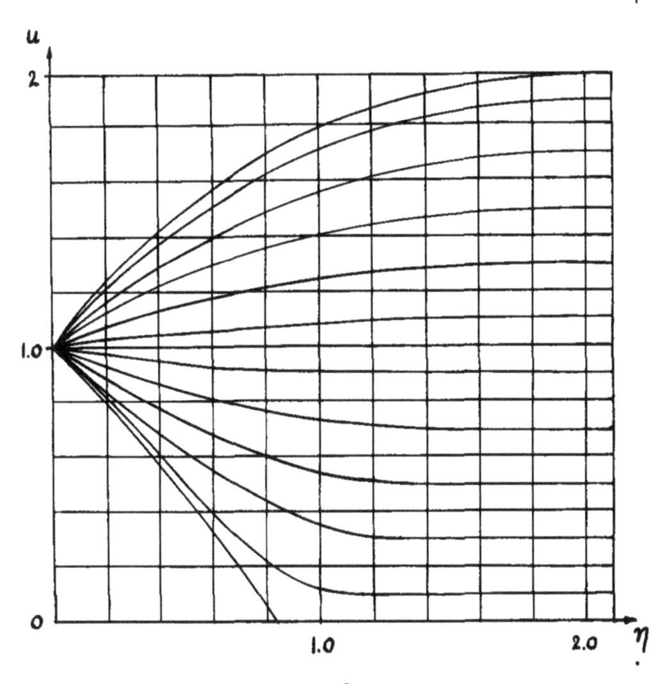


Fig. 380

Let us assume that we have constructed the free surface $u = u(\eta)$ for some time $t_1$. We take on the free surface the point $(\eta_1, u_1)$. To this point corresponds the abscissa $x_1 = 2\sqrt{\frac{kH_1}{m}}\,\eta_1\sqrt{t_1}$. The same ordinate $u_1$ will be reached at the moment $t_2$ for the value $x = x_2$ for which $x_2 = 2\sqrt{\frac{kH_1}{m}}\,\eta_1\sqrt{t_2}$. Then we have the correlation

$$x_1 : x_2 = \sqrt{t_1} : \sqrt{t_2},$$

which shows that, to obtain the free surface at different times, one needs to take the ordinates of the graph of Fig. 380 and change their abscissas in a given ratio.

It is possible to reconstruct the graph of Fig. 380 for this problem. We designate by $H_0$ the depth of ground water flow above an impervious foundation, by $H_1$ the depth of water in the canal above the ground water flow. After designating the ratio $H_1/H_0$ by the letter $\beta$, we may construct the graph for the quantity $u = (H - H_0)/H_1$, where $H$ depth of flow above impervious foundation. Several cases are given in Fig. 381. If we had linearized equation (4.1) then we would have found a single curve for all $\beta$. (See Chapter XIV.)

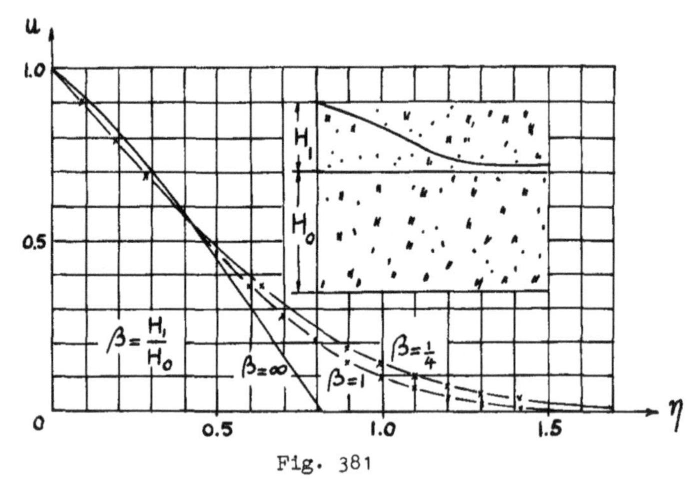


Fig. 381

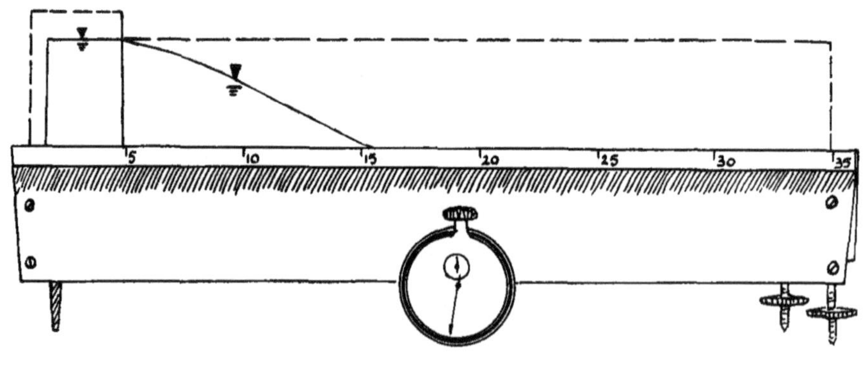


Fig. 382

In Fig. 382 and 383, pictures are given of a parallel plate model with glycerine, exhibiting the seepage from a canal with vertical wall. (At times $t = 92$ sec. and $t = 322$ sec.).

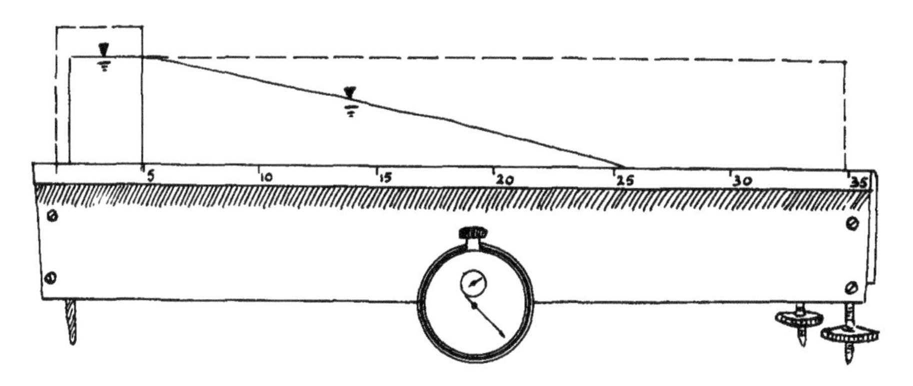


Fig. 383

§9. Uniform Rise of Water Level In a Canal. Solutions of the Non-linear Equation of the Source Type.

It is possible to generalize the problem that we have considered in the previous paragraphs of this chapter [10]. It has been shown that the non-linear differential equation

$$\frac{\partial h}{\partial t} = a^2 \frac{\partial^2 h^k}{\partial x^2} \qquad \left(a^2 = \frac{\kappa}{2m}\right) \tag{9.1}$$

in the case where for its solution the conditions must be satisfied

$$h(x,\ 0) = 0, \quad h(0,\ t) = \sigma t^p \quad (p \geq 0)\ , \tag{9.2}$$

can be reduced to an ordinary differential equation (9.4) by means of the substitution

$$h = \sigma t^p f(\xi), \quad \xi = \frac{x}{a\sqrt{\sigma^{k-1} t^{1+p(k-1)}}} \tag{9.3}$$

$$\frac{d^2 f^k}{d\xi^2} + \frac{1 + p(k-1)}{2}\, \xi \frac{df}{d\xi} - pf = 0\ . \tag{9.4}$$

Therefore $f$ must satisfy the conditions

$$f(0) = 1, \quad f(\infty) = 0\ . \tag{9.5}$$

If the solution of equation (9.1) must satisfy the conditions

$$h(x,\ 0) = 0, \quad \frac{\partial h^k(0,\ t)}{\partial x} = -\ \tau t^q \quad (q \geq Q)\ , \tag{9.6}$$

then the substitution

$$h = (a\tau)^{\frac{2}{k+1}}\, t^{\frac{2q+1}{k+1}}\, f_1(\xi), \quad \xi = x\tau^{\frac{1-k}{1+k}}\, a^{\frac{2(k+1)}{k}}\, t^{\frac{q(1-k)-k}{k+1}} \tag{9.7}$$

leads to the equation

$$\frac{d^2 f_1^k}{d\xi^2} + \frac{1}{2}\left[1 + \frac{(2q+1)(k-1)}{k+1}\right] \xi \frac{df_1}{d\xi} - \frac{2q+1}{k+1} f_1 = 0, \tag{9.8}$$

for which

$$\frac{df_1^k(0)}{d\xi} = -\ 1\ .$$

Of special interest to us is the case $k = 2$, $p = 1$ corresponding to the uniform rise of water in a canal. In this case equation (9.4) reduces to the following:

$$\frac{d^2f^2}{d\xi^2} + \xi \frac{df}{d\xi} - f = 0, \quad \xi = \frac{x}{a\sqrt{\sigma t}} \quad . \tag{9.9}$$

It is easy to show that this equation has a solution of the form

$$f = 1 + b\xi \ .$$

Insertion in equation (9.9) gives for the constant $b$ the value $\pm \frac{1}{\sqrt{2}}$, from which we choose $-\frac{1}{\sqrt{2}}$. If we consider the function of the following shape:

$$\begin{cases} f = 1 - \frac{\xi}{\sqrt{2}} & \text{for} \quad 0 \leq \xi \leq \sqrt{2} , \\ f = 0 & \text{for} \quad \sqrt{2} < \xi < \infty \quad , \end{cases} \tag{9.10}$$

then this represents a continuous solution of equation (9.9) with continuous values of the quantity $ff'$, proportional to the flowrate. This solution corresponds to the posed problem. Returning to the function $h$, we obtain the solution which corresponds to the conditions

$$h(x, 0) = 0, \quad h(0, t) = \sigma t$$

for the equation

$$\frac{\partial h}{\partial t} = \frac{k}{2m} \frac{\partial^2 h^2}{\partial x^2} \tag{9.11}$$

in the following form:

$$\begin{cases} h = \sigma t - x\sqrt{\frac{m\sigma}{k}} & \text{for} \quad 0 \leq x \leq \sqrt{\frac{k\sigma}{m}}\, t , \\ h = 0 & \text{for} \quad t\sqrt{\frac{k\sigma}{m}} < x . \end{cases} \tag{9.12}$$

Consequently, the free surface is represented by a straight line that advances parallel to itself (Fig. 384) with a constant velocity of diffusion of its front

$$v_0 = \sqrt{\frac{k\sigma}{m}} \ .$$

It is possible to obtain the same solution of the source type for equation (9.11) in the case of symmetry relative to the ordinate axis [10]. We have

$$h = \frac{m}{6kt}(\tau_0^2 - x^2) , \quad \tau_0 = \left( \frac{9kt}{m} \right)^{\frac{1}{2}} \quad . \tag{9.13}$$

It is possible to consider this also as the scheme of spreading of a ground water mound along the surface of an impervious foundation. (Fig. 385; see the problem of the spreading of a ground water mound in Paragraph 5 of Chapter XIV). In the case of cylindrical symmetry we have instead of (9.11) the equation

$$\frac{\partial h}{\partial t} = \frac{a^2}{r} \frac{\partial}{\partial r}\left( r \frac{\partial h^2}{\partial r} \right) \quad \left( a^2 = \frac{k}{2m} \right) . \tag{9.14}$$

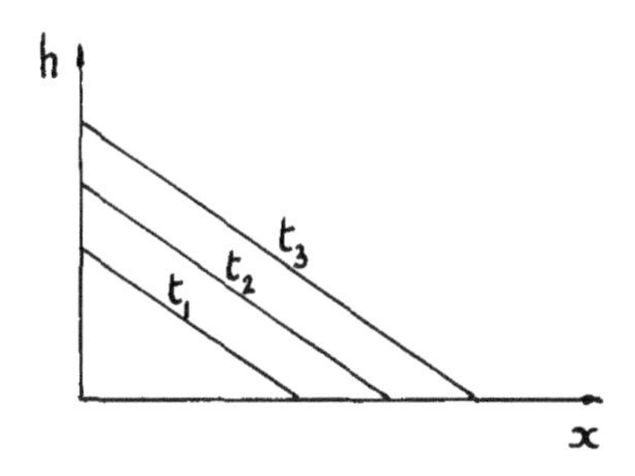

Fig. 384

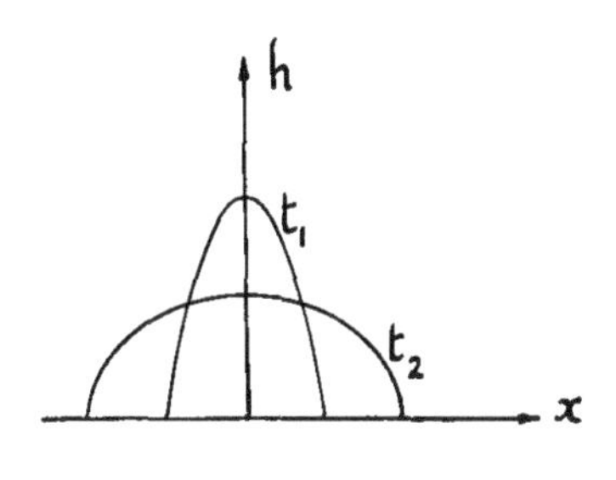


Fig. 385

The solution of the instantaneous source type has here for form

$$h = \frac{\sqrt{\pi}}{4}\left(\frac{m}{kt}\right)^{\frac{3}{2}}\left(\frac{2k}{m\pi}t - r^2\right). \tag{9.15}$$

It represents the picture of the spreading of an axisymmetrical mound of ground water under concentrated flow along an impervious foundation. In both cases the free surface has in cross-section the form of a parabola that widens in time. However, the largest abscissa increases according to different laws: in the plane problem — proportional to $t^{1/3}$, in the axismmetrical one — proportional to $t^{1/2}$.

§10. Boussinesq's Problem.

Boussinesq [1] looked for a solution of the equation

$$\frac{\partial h}{\partial t} = \frac{k}{m}\frac{\partial}{\partial x}\left(h\frac{\partial h}{\partial x}\right) \tag{10.1}$$

in the form of a product of two functions

$$h = T(t)\,X(x)\,, \tag{10.2}$$

one of which depends only on the time, the other on the coordinate. Insertion of (10.2) in (10.1) and separation of the variables gives us:

$$\frac{T'}{T^2} = \frac{k}{m}\frac{(XX')'}{X} = -A\,, \tag{10.3}$$

where A — an arbitrary constant. Equation (10.3) separates into two equations, one in T and one in X. Integration of both of them gives:

$$T = \frac{1}{At + C}\,, \tag{10.4}$$

$$x = \int \frac{XdX}{\sqrt{C_1 - \frac{2}{3}\frac{Am}{k}X^3}} + C_2\,. \tag{10.5}$$

In the expression for X, we replace the arbitrary constants by new constants, by putting

$$x = D\int_0^X \frac{XdX}{\sqrt{H^3 - X^3}}\,, \quad D = \sqrt{\frac{3k}{2mA}}\,.$$

Considering $X$ as varying from zero to $H$, we may carry out the integration between limits from zero to $H$, by choosing the constant $D$ so that $x = L$ for $X = H$. We have

$$x = \frac{3L}{B\left(\frac{2}{3}, \frac{1}{2}\right)\sqrt{H}} \int_0^X \frac{X\,dX}{\sqrt{H^3 - X^3}}, \tag{10.6}$$

where $B\left(\frac{2}{3}, \frac{1}{2}\right)$ designates the beta-function. By means of the substitution $X^3 = H^3\tau$, one may transform the integral of equality (10.6):

$$\int_0^H \frac{X\,dX}{\sqrt{H^3 - X^3}} = \frac{\sqrt{H}}{3} \int_0^1 \tau^{-\frac{1}{3}} (1 - \tau)^{-\frac{1}{2}} d\tau = \frac{\sqrt{H}}{3} B\left(\frac{2}{3}, \frac{1}{2}\right).$$

It is evident then that

$$D = \frac{3L}{B\left(\frac{2}{3}, \frac{1}{2}\right)\sqrt{H}} = \frac{L}{0.86236\sqrt{H}}.$$

In the expression for $T$, we choose the constant $C$ equal to one, so that $T = 1$ for $t = 0$.

Designating by $f\left(\frac{x}{L}\right) = X$ the result of equation (10.6) and taking into account the dependence of $A$ upon $D$, we may rewrite (10.2) in the form:

$$h(x, t) = \frac{Hf\left(\frac{x}{L}\right)}{1 + 1.115\frac{kHt}{mL^2}}. \tag{10.7}$$

The graph of the function $y = f(\frac{x}{L})$ is represented in Fig. 386.

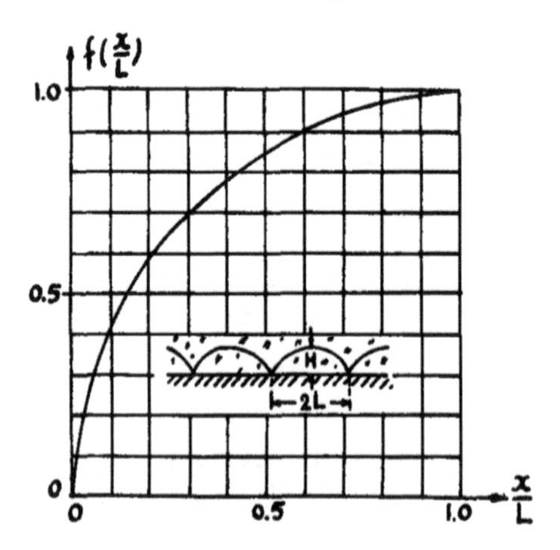


Fig. 386

One may consider this curve as the initial form of the free surface, which later on drops according to (10.7). This scheme corresponds to the problem of an infinite series of drains, lying on an impervious basis, and for which the free surface initially has the equation

$$y = Hf\left(\frac{x}{L}\right) .$$

We find the expression for the discharge in each drain after making the product

$$2kh \frac{\partial h}{\partial x}$$

and going over to the limit for $h \to 0$. We have

$$q = \frac{1.725kH^2}{L\left(1 + 1.115 \frac{kH}{mL^2} t\right)^2} . \tag{10.8}$$

Boussinesq tried to prove the steadiness of the movement, the so-called "regulated regimen." However, he only showed the existence of a category of perturbations that tend quicker to zero than $h$ for $t \to \infty$, which is insufficient for a complete proof of steadiness.

Having finished herewith the analysis of the non-linear equation (4.1), we observe that non-linear equations in partial derivatives also occur in other important divisions of the seepage theory.

Here, we do not touch upon problems of mechanical and chemical nature, problems of consolidation of soil media [11]. These problems have a very important practical meaning, but the mathematical part of their theory is still little developed. Similarly, we leave on the side the theory of the unsteady flows with full account of the inertai terms [12]. Here one obtains a system of non-linear equations in partial derivatives, resembling the one that is obtained in the unsteady flow of water in a prismatic channel. Maybe it is possible to integrate this system by means of the method of characteristics.

CHAPTER XIV.

LINEAR EQUATION OF UNSTEADY GROUNDWATER FLOW.

A. Seepage From a Canal On Sloping Bedrock.

§1. Seepage from a Canal on Horizontal Bedrock and without Infiltration.

Before investigating the influence of bedrock slope, we examine the problem of water seeping from a canal on horizontal bedrock.

In this chapter we consider a linearized equation of unsteady flow, which is permissible if we want to obtain crude results as a guidance. Let $H_0$ be the original groundwater level at $t = 0$. Furthermore, a water level $H_0 + H_1$ is maintained in the canal. Assuming that the ordinate of the free surface satisfies the equation

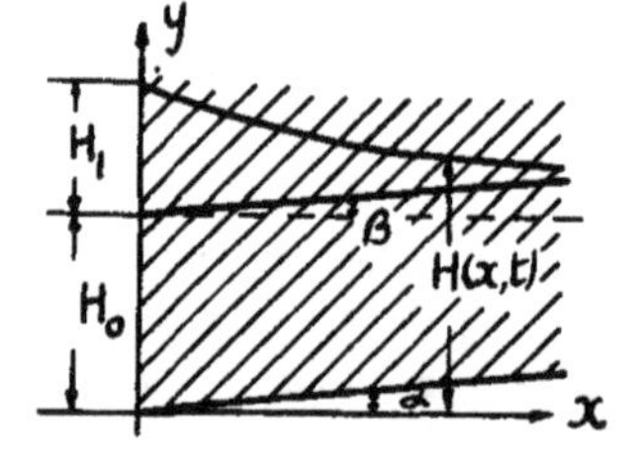


Fig. 387

$$\frac{\partial H}{\partial t} = a \frac{\partial^2 H}{\partial x^2} \tag{1.1}$$

where

$$a = \frac{k\bar{h}}{m}$$

(here $k$ is the seepage coefficient, $m$ the porosity, $\bar{h}$ the average depth of the seeping water), we obtain a known solution from the theory of heat conduction

$$H(x, t) = H_0 + H_1\left[1 - \Phi\left(\frac{x}{2\sqrt{at}}\right)\right] . \tag{1.2}$$

Here $\Phi(\xi)$ is the probability integral

$$\Phi(\xi) = \frac{2}{\sqrt{\pi}} \int_0^{\xi} e^{-\xi_1^2}\, d\xi_1 . \tag{1.3}$$

In practical problems it is usually of particular interest to know the interval of time in which the ordinate of the free surface attains some given value. We call $y$ the value of the ordinate of the free surface, measured from $H_0$ (Fig. 388).

$$y = H(x,t) - H_0 . \tag{1.4}$$

Then from (1.2) we obtain

$$\frac{H_1 - y}{H_1} = \Phi\left(\frac{x}{2\sqrt{at}}\right) . \tag{1.5}$$

For a given $H_1$ and $y$, one may find the argument $\xi = \frac{x}{2\sqrt{at}}$, and hence $t$, in tables of the function $\Phi(\xi)$.

The following problem was treated by S. F. Averjanov. At what time does the groundwater level reach a height which is equal to a given fraction of the height $H_1$ (i.e., additional to the initial head $H_0$)?

We call $z$ the drop of ground water level (compared with $H_0 + H_1$), i.e., the height measured from the final groundwater level $H_0 + H_1$ (Fig. 388):

$$z = H_1 - y.$$

By means of (1.5) we obtain

$$\frac{z}{H_1} = \Phi\left(\frac{x}{2\sqrt{at}}\right). \tag{1.6}$$

For given $z$ we obtain from this an expression for $t$

$$t = \mu_1 \frac{x^2}{4a} = \mu \frac{mx^2}{k\bar{h}}. \tag{1.7}$$

Coefficient $\mu$, depending on $\frac{z}{H_1}$, may be called the "coefficient to reach the value $\frac{z}{H_1}$." Thus, if we put $\frac{z}{H_1} = 0.05$, i.e., if we consider that $z$ attains 5% of $H_1$, we may write

$$\mu_{5\%} = 125$$
$$t_{5\%} = \frac{125mx^2}{k\bar{h}}. \tag{1.8}$$

Formula (1.8), at any distance from the canal, for given $k$, $\bar{h}$, and $m$, allows us to find the moment at which the rising groundwater level reaches a value 5% smaller than the maximum possible (i.e., horizontal level). We communicate table 18 with values of $\mu$.

TABLE 18

Values of the coefficient $\mu$ to reach the value of $\frac{z}{H_1}$.

| $\frac{z}{H_1}$ | 0.01 | 0.03 | 0.05 | 0.10 | 0.20 | 0.30 | 0.40 | 0.50 |
|---|---|---|---|---|---|---|---|---|
| $\mu$ | 3190 | 353 | 127 | 31.5 | 7.69 | 3.26 | 1.81 | 1.10 |
| $\lg \mu$ | 3.504 | 2.548 | 2.104 | 1.50 | 0.886 | 0.526 | 0.258 | 0.041 |

| $\frac{z}{H_1}$ | 0.60 | 0.70 | 0.80 | 0.90 | 0.95 | 0.99 |
|---|---|---|---|---|---|---|
| $\mu$ | 0.704 | 0.463 | 0.305 | 0.184 | 0.130 | 0.0752 |
| $\lg \mu$ | -0.141 | -0.334 | -0.516 | -0.735 | -0.886 | -1.124 |

We still examine the same problem in a somewhat different form: find the moment of time, at which the height of the free surface, measured from the initial groundwater level, reaches some given percent of the height $H_1$. We put

$$t = \nu \frac{mx^2}{k\bar{h}}.$$

Computation of $\nu$ is similar to computation of $\mu$.

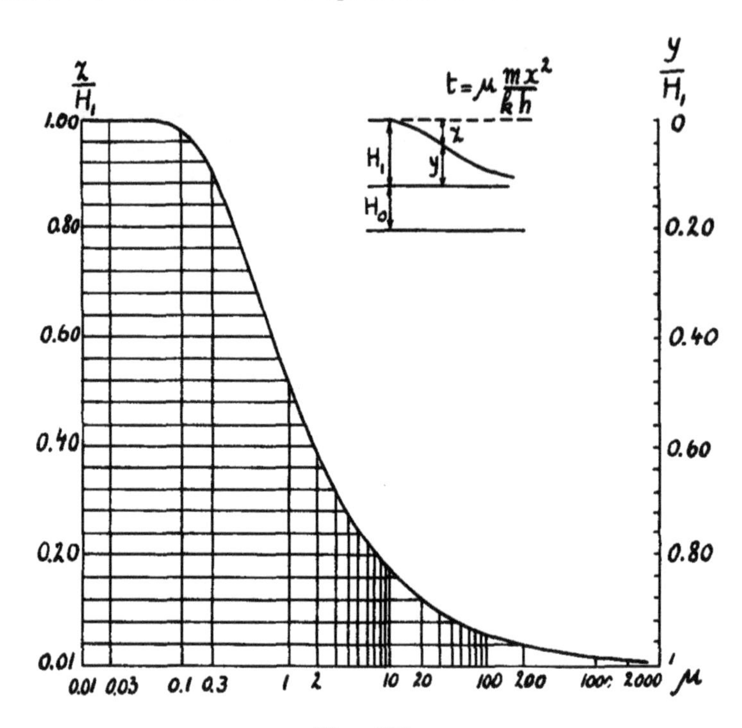


Fig. 388

The results are rendered by Fig. 388.

These problems are very important in cases of soil salinization, which may take place for a large rise of the groundwater level. Indeed, if groundwater is found at a depth of 2 to 3 m. below the earth's surface, it will rise under intensive evaporation and effectuate dissolution of the salts it contains. A high groundwater level may also cause soil to become a bog.

As is evident from formula (1.7), knowledge of the quantity $m$ is very important to determine the time to reach a given rise of the water level.

## §2. Seepage from a Canal When the Underlying Bedrock has a Slight Slope.

Let

$$i_1 = \operatorname{tg} \alpha \approx \sin \alpha \approx \alpha \tag{2.1}$$

be the bedrock slope, $\alpha$ the angle between bedrock and an horizontal line.

Let the unknown equation of the free surface be

$$y = H(x, t) \; . \tag{2.2}$$

We consider the variation of $H$ with $x$ so small that the flow velocities are almost horizontal. The horizontal velocity is expressed as

(2.3) $$u = -k \frac{\partial H}{\partial x},$$

where $k$ is the seepage coefficient.

The flowrate through a vertical section, at the abscissa coordinate $x$, is equal to the velocity $u$ times the depth $z$ of the water, i.e.,

(2.4) $$q = -kz \frac{\partial H}{\partial x}, \qquad z = H - i_1 x .$$

The change of the flowrate along the x-axis is compensated by the variation in time of the product of the depth $H$ and the porosity $m$ and by the possible addition of water along the vertical from above or below:

(2.5) $$-\frac{\partial q}{\partial x} = \frac{\partial (Hm)}{\partial t} - w,$$

where $w$ is the vertical velocity.

Inserting (2.4) in equation (2.5), we find:

(2.6) $$m \frac{\partial H}{\partial t} = k \frac{\partial}{\partial x}\left[(H - i_1 x \frac{\partial H}{\partial x}\right] + w$$

In the case of an upward slope from left to right, as in Fig. 387, the number $i_1$ is positive; in the opposite case it is negative. To simplify equation (2.6), we carry out differentiation along $x$ and replace the variable quantity

$$z = H + H_0 - i_1 x$$

by the constant number

(2.7) $$h = H_1 + H_0.$$

The error generated by this substitution must not be important, if $x$ is not very large. Thus, if $i_1 = 0.0003$, at a distance of 30 km, we find $i_1 x = 0.009$ km = 9 m, at a distance of 15 km, $i_1 x = 4.5$ m and so on.

Now equation (2.6) may be written as:

(2.8) $$\frac{\partial H}{\partial t} = \frac{kh}{m} \frac{\partial^2 H}{\partial x^2} - \frac{ki_1}{m} \frac{\partial H}{\partial x} .$$

We introduce the variable $\tau$ instead of $t$, putting

(2.9) $$\frac{m}{kh} t = \tau$$

Then the coefficient of $\frac{\partial^2 H}{\partial x^2}$ is equal to one, and we may write

(2.10) $$\frac{\partial H}{\partial \tau} = \frac{\partial^2 H}{\partial x^2} - 2\alpha \frac{\partial H}{\partial x},$$

where we used the notation

(2.11) $$2\alpha = \frac{i_1}{h} .$$

Now we must solve equation (2.10) for the boundary condition

(2.12) $$H(0, \tau) = H_1 + H_0 .$$

We assume that when flow started, we had a constant slope of the free surface, equal to $i_2$. Then we take as initial condition:

$$H(x, 0) = i_2 x . \tag{2.13}$$

We look for a solution of equation (2.8) with conditions (2.12) and (2.13) by means of operational calculus. Therefore we multiply (2.8) term by term with $e^{-p\tau}d\tau$ and integrate along $\tau$ between limits $0$ to $\infty$. Integrating by parts, we obtain

$$\begin{aligned} \int_0^\infty \frac{\partial H}{\partial \tau} e^{-p\tau}\, d\tau &= He^{-p\tau}\Big|_0^\infty + p\int_0^\infty He^{-p\tau} d\tau = \\ &= - H(x, 0) + p\int_0^\infty He^{-p\tau}\, d\tau . \end{aligned} \tag{2.14}$$

We introduce the notation

$$\bar{H}(x,p) = \int_0^\infty He^{-p\tau}\, d\tau \tag{2.15}$$

and call $\bar{H}(x,p)$ the Laplace transform of function $H(x,t)$. We write (2.14) in this form

$$\int_0^\infty \frac{\partial H}{\partial \tau} e^{-p\tau}\, d\tau = - H(x, 0) + p\bar{H}(x, p). \tag{2.16}$$

The transform of $\frac{\partial H}{\partial x}$ and $\frac{\partial^2 H}{\partial x^2}$ gives:

$$\int_0^\infty \frac{\partial H}{\partial x} e^{-p\tau}\, d\tau = \frac{\partial}{\partial x}\int_0^\infty He^{-p\tau}\, d\tau = \frac{\partial \bar{H}}{\partial x} ,$$

$$\int_0^\infty \frac{\partial^2 H}{\partial x^2} e^{-p\tau}\, d\tau = \frac{\partial^2}{\partial x^2}\int_0^\infty He^{-p\tau}\, d\tau = \frac{\partial^2 \bar{H}}{\partial x^2} .$$

Since the derivative of $\bar{H}(x,p)$ with respect to $\tau$ is zero, only the $x$-derivative remains and we may write ordinary derivatives instead of partial ones, i.e., replace $\frac{\partial \bar{H}}{\partial x}$ by $\frac{d\bar{H}}{dx}$ and $\frac{\partial^2 \bar{H}}{dx^2}$ by $\frac{d^2\bar{H}}{dx^2}$.

Thus, the transform of equation (2.10) gives us this equation for $\bar{H}$:

$$- H(x, 0) + p\bar{H} = \frac{d^2\bar{H}}{dx^2} - 2\alpha \frac{d\bar{H}}{dx} .$$

Using formula (2.13) we obtain

$$\frac{d^2\bar{H}}{dx^2} - 2\alpha \frac{d\bar{H}}{dx} - p\bar{H} = - i_2 x . \tag{2.17}$$

This is a linear equation with constant coefficients. Its characteristic equation is

$$s^2 - 2\alpha s - p = 0 .$$

Its roots are $\alpha \pm \sqrt{\alpha^2 + p}$. A particular solution of equation (2.17) may be written as

$$H^* = Ax + B .$$

Determining the constants A and B, we find

$$H^* = \frac{i_2 x}{p} - \frac{2\alpha i_2}{p^2} .$$

The general solution of (2.17) has the form

$$\bar{H} = (C_1 e^{x\sqrt{\alpha^2+p}} + C_2 e^{-x\sqrt{\alpha^2+p}}) e^{\alpha x} + \frac{i_2 x}{p} - \frac{2\alpha i_2}{p^2} . \tag{2.18}$$

We make $C_1 = 0$ because otherwise $\bar{H}$ would increase indefinitely with $x$. Now we find the value of $\bar{H}$ for $x = 0$. According to formula (2.15) we have

$$\bar{H}(0, p) = \int_0^\infty H(0, \tau) e^{-p\tau} \, d\tau .$$

Because of (2.12) we obtain further

$$\bar{H}(0, p) = (H_0 + H_1) \int_0^\infty e^{-p\tau} \, d\tau = \frac{H_0 + H_1}{p} . \tag{2.19}$$

We now make $x = 0$ in (2.18). We have

$$\frac{H_0 + H_1}{p} = C_2 - \frac{2\alpha i_2}{p^2} ,$$

and

$$C_2 = \frac{H_0 + H_1}{p} + \frac{2\alpha i_2}{p^2} . \tag{2.20}$$

Consequently, we finally obtain for $\bar{H}(x, p)$:

$$\bar{H}(x, p) = \left(\frac{H_0 + H_1}{p} - \frac{2\alpha i_2}{p^2}\right) e^{\alpha x} e^{-x\sqrt{\alpha^2+p}} + \frac{i_2 x}{p} - \frac{2\alpha i_2}{p^2} . \tag{2.21}$$

Now we must return to the function $H(x, t)$. We find in a table of Laplace transforms that the function

$$\frac{e^{-x\sqrt{\alpha^2+p}}}{p}$$

corresponds to the function [1]

$$F(x, \tau) = \frac{1}{2} \left\{ e^{\alpha x}\left[1 - \Phi\left(\frac{x}{2\sqrt{\tau}} + \alpha\sqrt{\tau}\right)\right] + \right.$$

$$\left. + e^{-\alpha x}\left[1 - \Phi\left(\frac{x}{2\sqrt{\tau}} - \alpha\sqrt{\tau}\right)\right]\right\} , \tag{2.22}$$

where $\Phi(x)$ is the probability function, determined by (1.3).

Then function

$$\frac{e^{-x\sqrt{\alpha^2+p}}}{p^2}$$

corresponds to the integral

$$\int_0^\tau F(x, \tau)\, d\tau = \tau \operatorname{ch} \alpha x - \frac{1}{2} e^{\alpha x} \int_0^\tau \Phi\left(\frac{x}{2\sqrt{\tau}} + \alpha\sqrt{\tau}\right) d\tau - \tag{2.23}$$

$$- \frac{1}{2} e^{-\alpha x} \int_0^\tau \Phi\left(\frac{x}{2\sqrt{\tau}} - \alpha\sqrt{\tau}\right) d\tau .$$

We introduce the notations

$$\xi = \frac{x}{2\sqrt{\tau}} + \alpha\sqrt{\tau} , \tag{2.24}$$

$$\eta = \frac{x}{2\sqrt{\tau}} - \alpha\sqrt{\tau} . \tag{2.25}$$

From these equalities we find $\sqrt{\tau}$ in function of $\xi$ and $\eta$

$$\sqrt{\tau} = \frac{\xi}{2\alpha} \pm \sqrt{\frac{\xi^2}{4\alpha^2} - \frac{x}{2\alpha}} , \tag{2.26}$$

$$\sqrt{\tau} = \frac{\eta}{2\alpha} \pm \sqrt{\frac{\eta^2}{4\alpha^2} + \frac{x}{2\alpha}} . \tag{2.27}$$

When we construct $\xi$ as a function of $\tau$, we have two branches. For

$$\tau \leq \tau_1 \tag{2.28}$$

where $\tau_1 = \frac{x}{2\alpha}$, we must take the left branch of the $\xi$ function; for $\tau > \tau_1$ the right branch. The result is the same for both branches.

For the time being we assume that $\tau < \tau_1$. Then one must take the minus sign in front of the root of formula (2.26) and the plus sign in (2.27), i.e.,

$$\sqrt{\tau} = \frac{\xi}{2\alpha} - \sqrt{\frac{\xi^2}{4\alpha^2} - \frac{x}{2\alpha}} ,$$

$$\sqrt{\tau} = -\frac{\eta}{2\alpha} + \sqrt{\frac{\eta^2}{4\alpha^2} + \frac{x}{2\alpha}} .$$

From this

$$\tau = \frac{\xi^2}{2\alpha^2} - \frac{x}{2\alpha} - \frac{\xi}{\alpha}\sqrt{\frac{\xi^2}{4\alpha^2} - \frac{x}{2\alpha}} , \tag{2.29}$$

$$\tau = \frac{\eta^2}{2\alpha^2} + \frac{x}{2\alpha} - \frac{\eta}{\alpha}\sqrt{\frac{\eta^2}{4\alpha^2} - \frac{x}{2\alpha}} . \tag{2.30}$$

We now compute the integrals

$$J_1 = \int_0^\tau \Phi\left(\frac{x}{2\sqrt{\tau}} + \alpha\sqrt{\tau}\right) d\tau, \qquad J_2 = \int_0^\tau \Phi\left(\frac{x}{2\sqrt{\tau}} - \alpha\sqrt{\tau}\right) d\tau . \tag{2.31}$$

We have

$$(2.32)\quad \begin{cases} J_1 = \displaystyle\int_0^\tau \Phi(\xi)\, d\tau = \tau\Phi(\xi)\Big|_0^\tau - \int_\infty^\xi \tau\Phi'(\xi)\, d\xi = \\ \quad = \tau\Phi(\xi) + \dfrac{2}{\sqrt{\pi}} \displaystyle\int_\xi^\infty e^{-\xi^2}\left(\frac{\xi^2}{2\alpha^2} - \frac{x}{2\alpha} - \frac{\xi}{2\alpha^2}\sqrt{\xi^2 - 2\alpha x}\right) d\xi , \\ J_2 = \displaystyle\int_0^\tau \Phi(\eta)\, d\tau = \tau\Phi(\eta)\Big|_0^\tau - \int_\infty^\eta \tau\Phi'(\eta)\, d\eta = \\ \quad = \tau\Phi(\eta) + \dfrac{2}{\sqrt{\pi}} \displaystyle\int_\eta^\infty e^{-\eta^2}\left(\frac{\eta^2}{2\alpha^2} + \frac{x}{2\alpha} - \frac{\eta}{2\alpha^2}\sqrt{\eta^2 + 2\alpha x}\right) d\eta . \end{cases}$$

We use the formulas

$$(2.33)\quad \begin{cases} \dfrac{2}{\sqrt{\pi}} \displaystyle\int_x^\infty e^{-\xi^2}\, d\xi = 1 - \Phi(x) , \\ \displaystyle\int_x^\infty x^2 e^{-x^2} dx = \frac{1}{2} x e^{-x^2} + \frac{\sqrt{\pi}}{4}[1 - \Phi(x)] , \\ \displaystyle\int_x^\infty x\sqrt{x^2 + m}\, e^{-x^2} dx = \frac{e^{-x^2}}{2}\sqrt{x^2 + m} + \frac{\sqrt{\pi}}{4} e^m[1 - \Phi(\sqrt{x^2 + m})] . \end{cases}$$

Then we obtain

$$J_1 = \tau\Phi(\xi) + \frac{1}{2\alpha^2\sqrt{\pi}}\left[\xi e^{-\xi^2} + \frac{\sqrt{\pi}}{2}(1 - \Phi(\xi))\right] - \frac{x}{2\alpha}(1 - \Phi(\xi)) -$$

$$- \frac{1}{2\alpha^2\sqrt{\pi}}\left\{e^{-\xi^2}\sqrt{\xi^2 - 2\alpha x} + \frac{\sqrt{\pi}}{2} e^{-2\alpha x}[1 - \Phi(\sqrt{\xi^2 - 2\alpha x})]\right\} .$$

Observing that $\xi^2 - 2\alpha x = \eta^2$, we find:

$$J_1 = \tau\Phi(\xi) + \frac{\xi - \eta}{2\alpha^2\sqrt{\pi}} e^{-\xi^2} + \frac{1 - 2\alpha x}{4\alpha^2}(1 - \Phi(\xi)) - \frac{e^{-2\alpha x}}{4\alpha^2}(1 - \Phi(\eta)) .$$

In the same way we find for $J_2$:

$$J_2 = \tau\Phi(\eta) + \frac{1}{2\alpha^2\sqrt{\pi}}\left[\eta e^{-\eta^2} + \frac{\sqrt{\pi}}{2}(1 - \Phi(\eta))\right] + \frac{x}{2\alpha}(1 - \Phi(\eta)) -$$

$$- \frac{1}{2\alpha^2\sqrt{\pi}}\left[e^{-\eta^2}\sqrt{\eta^2 + 2\alpha x} + \frac{\sqrt{\pi}}{2} e^{2\alpha x}(1 - \Phi(\sqrt{\eta^2 + 2\alpha x}))\right] =$$

$$= \tau\Phi(\eta) + \frac{\eta - \xi}{2\alpha^2\sqrt{\pi}} e^{-\eta^2} + \frac{1 + 2\alpha x}{4\alpha^2}(1 - \Phi(\eta)) - \frac{e^{2\alpha x}}{4\alpha^2}(1 - \Phi(\xi)) .$$

Now, considering formula (2.23) we find:

$$\int_0^\tau F(x,\tau)d\tau = \tau \operatorname{ch} \alpha x + \frac{x}{2\alpha} \operatorname{sh} \alpha x - \frac{1}{4\alpha} e^{\alpha x}(x + 2\alpha\tau)\Phi(\xi) + \frac{1}{4\alpha} e^{-\alpha x}(x - 2\alpha\tau)\Phi(\eta)$$

$$= \frac{x + 2\alpha\tau}{4\alpha} e^{\alpha x}[1 - \Phi(\xi)] - \frac{x - 2\alpha\tau}{4\alpha} e^{-\alpha x}[1 - \Phi(\eta)]$$

(2.34)

$$= \frac{1}{2}\left\{\left(\tau + \frac{x}{2\alpha}\right) e^{\alpha x}[1 - \Phi(\xi)] + \left(\tau - \frac{x}{2\alpha}\right) e^{-\alpha x}[1 - \Phi(\eta)]\right\} .$$

Returning to formula (2.21), we notice that one corresponds to the transform $1/p$ and $\tau$ to $1/p^2$. Therefore we obtain for $H(x, t)$:

$$H(x, t) = \frac{(H_1 + H_0)e^{\alpha x}}{2} \{e^{\alpha x}(1 - \Phi(\xi)) + e^{-\alpha x}(1 - \Phi(\eta))\}$$

$$- 2\alpha i_2 e^{\alpha x}\left\{\frac{x + 2\alpha\tau}{4\alpha} e^{\alpha x}(1 - \Phi(\xi)) - \frac{x - 2\alpha\tau}{4\alpha} e^{-\alpha x}(1 - \Phi(\eta))\right\} -$$

$$- i_2 x - 2\alpha i_2 \tau$$

or

$$H(x, t) = \frac{e^{2\alpha x}}{2} [H_1 + H_0 - i_2(x + 2\alpha\tau)](1 - \Phi(\xi)) +$$

(2.35)

$$+ \frac{1}{2} [H_1 + H_0 - i_2(x - 2\alpha\tau)](1 - \Phi(\eta)) - i_2(2\alpha\tau - x).$$

## §3. Seepage from One Canal to Another on Sloping Bedrock.

Up to now we examined such problems where groundwater, seeping from a canal, may percolate as far as one wants from the canal without encountering any obstacle on its path. But in practicality, groundwater sooner or later reaches another canal or river. Therefore we study here this problem, which is a generalization of the one treated in §2. Seepage takes place from a canal in which at all times the level $H_1$ is kept constant, towards another canal with constant level $H_2$. (Fig. 389) Bedrock has a slope $i_1$, sufficiently small so that we may consider $i_1$ to be equal to the tangent of the angle bedrock x-axis and also to the angle itself. The initial groundwater level is at a height $H_0$ near the first canal and has a small angle $i_2$.

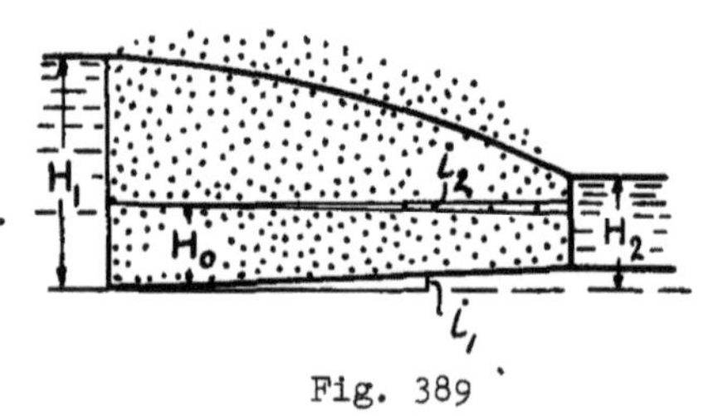

Fig. 389

We take equation (2.10) previously derived:

(3.1)

$$\frac{\partial H}{\partial \tau} = \frac{\partial^2 H}{\partial x^2} - 2\alpha \frac{\partial H}{\partial x} ,$$

where

$$\tau = \frac{k\bar{h}}{m}\, t, \qquad \alpha = \frac{i_1}{2\bar{h}} \tag{3.2}$$

and find its solution for conditions on $H(x, t)$:

$$H(0, t) = H_1, \qquad H(L, t) = H_2, \tag{3.3}$$

$$H(x, 0) = H_0 + i_2 x . \tag{3.4}$$

Designating, as before, the Laplace transform of $H(x, t)$ by $\bar{H}(x, p)$:

$$\bar{H}(x, p) = \int_0^\infty e^{-p\tau} H(x, \tau)\, d\tau , \tag{3.5}$$

we obtain the equation for $\bar{H}$:

$$\frac{d^2\bar{H}}{dx^2} - 2\alpha \frac{d\bar{H}}{dx} - p\bar{H} = - H(0, x) = - H_0 - i_2 x . \tag{3.6}$$

For $\bar{H}$ given by (3.5), we have the following boundary conditions:

$$\bar{H}(0,p) = H_1 \int_0^\infty e^{-p\tau}\, d\tau = \frac{H_1}{p} , \tag{3.7}$$

$$\bar{H}(L,p) = H_2 \int_0^\infty e^{-p\tau}\, d\tau = \frac{H_2}{p} . \tag{3.8}$$

The general solution of equation (3.6) has the form ($A$, $B$ are arbitrary constants)

$$\bar{H} = [A \operatorname{ch}(x\sqrt{\alpha^2 + p}) + B \operatorname{sh}(x\sqrt{\alpha^2 + p})]e^{\alpha x} + \frac{i_2 x}{p} - \frac{2\alpha i_2}{p^2} + \frac{H_0}{p} . \tag{3.9}$$

Putting here $x = 0$, $H = \frac{H_1}{p}$ and $x = L$, $H = \frac{H_2}{p}$, we obtain:

$$\frac{H_1}{p} = A - \frac{2\alpha i_2}{p^2} + \frac{H_0}{p} ,$$

$$\frac{H_2}{p} = (A \operatorname{ch} L\sqrt{\alpha^2 + p} + B \operatorname{sh} L\sqrt{\alpha^2 + p})e^{\alpha L} + \frac{i_2 L}{p} + \frac{H_0}{p} - \frac{2\alpha i_2}{p^2} .$$

From this we find:

$$A = \frac{H_1 - H_0}{p} + \frac{2\alpha i_2}{p^2} ,$$

$$B = \frac{\left(\frac{H_2 - H_0 - i_2 L}{p} + \frac{2\alpha i_2}{p^2}\right) e^{-\alpha L} - \left(\frac{H_1 - H_0}{p} + \frac{2\alpha i_2}{p^2}\right) \operatorname{ch} L\sqrt{\alpha^2 + p}}{\operatorname{sh} L\sqrt{\alpha^2 + p}} .$$

Inserting these values of $A$ and $B$ in (3.9), we obtain after some transformations:

$$\bar{H}(x,\ p) = e^{\alpha x}\left(\frac{H_1 - H_0}{p} + \frac{2\alpha i_2}{p^2}\right)\frac{\operatorname{sh}(L-x)\sqrt{\alpha^2+p}}{\operatorname{sh} L\sqrt{\alpha^2+p}} +$$

(3.10)

$$+ e^{-\alpha(L-x)}\left(\frac{H_2 - H_0 - i_2 L}{p} + \frac{2\alpha i_2}{p^2}\right)\frac{\operatorname{sh} x\sqrt{\alpha^2+p}}{\operatorname{sh} L\sqrt{\alpha^2+p}} + \frac{H_0 + i_2 x}{p} - \frac{2\alpha i_2}{p^2} .$$

In order to construct the solution $H(x, t)$, we first find the roots of equation

$$\operatorname{sh} L\sqrt{\alpha^2+p} = i \sin iL\sqrt{\alpha^2+p} = 0 . \tag{3.11}$$

We find:

$$iL\sqrt{\alpha^2+p} = k\pi \qquad (k \text{ integer}) \tag{3.12}$$

from which we obtain the roots $p_k$ of this equation:

$$p_k = -\alpha^2 - \frac{k^2\pi^2}{L^2} . \tag{3.13}$$

Now we use the formula for the inverse transform of the ratio of two entire transcendental functions [1]:

$$L^{-1}\left[\frac{\Phi(p)}{\Psi(p)}\right] = \sum_{n=0}^{\infty}\frac{\Phi(p_n)}{\Psi'(p_n)}e^{p_n\tau} , \tag{3.14}$$

where $\Phi(p)$ has a non-zero term, in its series development of powers of $p$, and $\Psi(p)$ starts with the first power $p$. Here one assumes that the roots of the denominator are simple.

We apply (3.14) to function $\dfrac{\operatorname{sh} x\sqrt{\alpha^2+p}}{p \operatorname{sh} L\sqrt{\alpha^2+p}} = \dfrac{\sin(ix\sqrt{\alpha^2+p})}{p\sin(iL\sqrt{\alpha^2+p})}$ .

Besides roots (3. 3), the denominator has also the root $p = 0$. We put

$$\begin{cases} \Phi(p) = \sin(ix\sqrt{\alpha^2+p}) \\ \Psi(p) = p\sin(Li\sqrt{\alpha^2+p}) , \end{cases} \tag{3.15}$$

and find

$$\Psi'(p) = \sin(Li\sqrt{\alpha^2+p}) + \frac{Lpi}{2\sqrt{\alpha^2+p}}\cos(Li\sqrt{\alpha^2+p}) .$$

From this

$$\begin{cases} \Psi'(0) = i \operatorname{sh} \alpha L, \\ \Psi'(p_n) = \dfrac{Lp_n i}{2\sqrt{\alpha^2+p_n}}\cos L\sqrt{\alpha^2+p_n} = \dfrac{(-1)^n(L^2\alpha^2 + n\pi^2)}{2n\pi} \end{cases} \tag{3.16}$$

Moreover, because of (3.12):

$$\begin{aligned} \Phi(p_n) &= \sin(ix\sqrt{\alpha^2+p}) = \sin\frac{n\pi x}{L} , \\ \Phi(0) &= \sin(xi\alpha) = i \operatorname{sh} \alpha x . \end{aligned} \tag{3.17}$$

According to (3.14) one may write:

$$L^{-1}\left[\frac{\operatorname{sh} x\sqrt{\alpha^2+p}}{p\,\operatorname{sh} L\sqrt{\alpha^2+p}}\right] = \frac{\operatorname{sh}\alpha x}{\operatorname{sh}\alpha L} + 2\pi\sum_{n=1}^{\infty}\frac{(-1)^n\, n\sin\frac{n\pi x}{L}}{L^2\alpha^2+n^2\pi^2}\, e^{-\left(\alpha^2+\frac{n^2\pi^2}{L^2}\right)\tau} \tag{3.18}$$

Since $1/p$ corresponds to integration over $\tau$ between limits from $o$ to $\tau$, we find further for the inverse transform

$$L^{-1}\left[\frac{\operatorname{sh} x\sqrt{\alpha^2+p}}{p^2\,\operatorname{sh} L\sqrt{\alpha^2+p}}\right] = \tau\frac{\operatorname{sh}\alpha x}{\operatorname{sh}\alpha L} + 2\pi L^2\sum_{n=1}^{\infty}\frac{(-1)^n\, n\sin\frac{n\pi x}{L}\left[1-e^{-\left(\alpha^2+\frac{n^2\pi^2}{L^2}\right)\tau}\right]}{(L^2\alpha^2+n^2\pi^2)^2} \tag{3.19}$$

In the same way we obtain

$$L^{-1}\left[\frac{\operatorname{sh}\alpha(L-x)\sqrt{\alpha^2+p}}{p\,\operatorname{sh} L\sqrt{\alpha^2+p}}\right] = \frac{\operatorname{sh}\alpha(L-x)}{\operatorname{sh}\alpha L} - 2\pi\sum_{n=1}^{\infty}\frac{n\sin\frac{\pi n x}{L}}{L^2\alpha^2+n^2\pi^2}\, e^{-\left(\alpha^2+\frac{n^2\pi^2}{L^2}\right)\tau} \tag{3.20}$$

$$L^{-1}\left[\frac{\operatorname{sh}\alpha(L-x)\sqrt{\alpha^2+p}}{p^2\,\operatorname{sh} L\sqrt{\alpha^2+p}}\right] = \tau\frac{\operatorname{sh}\alpha(L-x)}{\operatorname{sh}\alpha L} - 2\pi L^2\sum_{n=1}^{\infty}\frac{n\sin\frac{n\pi x}{L}\left[1-e^{-\left(\alpha^2+\frac{n^2\pi^2}{L^2}\right)\tau}\right]}{(L^2\alpha^2+n^2\pi^2)^2}\,. \tag{3.21}$$

Recalling also that

$$L^{-1}\left(\frac{1}{p}\right) = 1, \qquad L^{-1}\left(\frac{1}{p^2}\right) = \tau \tag{3.22}$$

and collecting expressions (3.18) — (3.21), we obtain, taking account of equation (3.10):

$$\begin{aligned} H(x,t) = {} & (H_1 - H_0)e^{\alpha x}\left[\frac{\operatorname{sh}\alpha x}{\operatorname{sh}\alpha L} + 2\pi\sum_{n=1}^{\infty}\frac{(-1)^n\, n\sin\frac{n\pi x}{L}}{L^2\alpha^2+n^2\pi^2}\, e^{-\left(\alpha^2+\frac{n^2\pi^2}{L^2}\right)\tau}\right] + \\ & + (H_2-H_0-i_2L)e^{-\alpha(L-x)}\left[\frac{\operatorname{sh}\alpha(L-x)}{\operatorname{sh}\alpha L} - 2\pi\sum_{n=1}^{\infty}\frac{n\sin\frac{n\pi x}{L}}{L^2\alpha^2+n^2\pi^2}\, e^{-\left(\alpha^2+\frac{n^2\pi^2}{L^2}\right)\tau}\right] - \\ & - 2\alpha i_2 e^{\alpha x}\left[\tau\frac{\operatorname{sh}\alpha x}{\operatorname{sh}\alpha L} + 2\pi L^2\sum_{n=1}^{\infty}\frac{(-1)^n\, n\sin\frac{n\pi x}{L}\left[1-e^{-\left(\alpha^2+\frac{n^2\pi^2}{L^2}\right)\tau}\right]}{(L^2\alpha^2+n^2\pi^2)^2}\right] - \\ & - 2\alpha i_2 e^{-\alpha(L-x)}\left[\tau\frac{\operatorname{sh}\alpha(L-x)}{\operatorname{sh}\alpha L} - 2\pi L^2\sum_{n=1}^{\infty}\frac{n\sin\frac{\pi n x}{L}\left[1-e^{-\left(\alpha^2+\frac{n^2\pi^2}{L^2}\right)\tau}\right]}{(L^2\alpha^2+n^2\pi^2)^2}\right] + \\ & + H_0 + i_2 x - 2\alpha i_2\tau\,. \end{aligned} \tag{3.23}$$

We may somewhat transform this expression, representing it in the form

$$H(x, t) = \left(H_1 - H_0 - \frac{i_1 i_2 \tau}{h}\right) \frac{e^{-\frac{i_1 x}{h}} - 1}{2 \operatorname{sh} \frac{i_1 L}{2h}} +$$

$$+ \left(H_2 - H_0 - i_2 L - \frac{i_1 i_2 \tau}{h}\right) \frac{1 - e^{\frac{i_1 (L-x)}{h}}}{2 \operatorname{sh} \frac{i_1 L}{2h}} +$$

$$+ 2\pi (H_1 - H_0) e^{-\frac{i_1 x}{2h}} \sum_{n=1}^{\infty} \frac{(-1)^n n \sin \frac{n\pi x}{L}}{\frac{L^2 i_1^2}{4h^2} + n^2\pi^2} e^{-\left(\frac{i_1^2}{4h^2} + \frac{n^2\pi^2}{L^2}\right)\tau} -$$

$$- 2\pi (H_2 - H_0 - i_2 L) e^{\frac{i_2 (L-x)}{2h}} \sum_{n=1}^{\infty} \frac{n \sin \frac{n\pi x}{L}}{\frac{L^2 i_1^2}{4h^2} + n^2\pi^2} e^{-\left(\frac{i_1^2}{4h^2} + \frac{n^2\pi^2}{L^2}\right)\tau} -$$

$$- \frac{2\pi L^2 i_1 i_2}{h} e^{-\frac{i_1 x}{2h}} \sum_{n=1}^{\infty} \frac{(-1)^n n \sin \frac{n\pi x}{L}}{\left(\frac{L^2 i_1^2}{4h^2} + n^2\pi^2\right)^2} \left[1 - e^{-\left(\frac{i_1^2}{4h^2} + \frac{n^2\pi^2}{L^2}\right)\tau}\right] -$$

$$- \frac{2\pi L^2 i_1 i_2}{h} e^{-\frac{i_1 (L-x)}{2h}} \sum_{n=1}^{\infty} \frac{n \sin \frac{n\pi x}{L} \left[1 - e^{-\left(\frac{i_1^2}{4h^2} + \frac{n^2\pi^2}{L^2}\right)\tau}\right]}{\left(\frac{L^2 i_1^2}{4h^2} + n^2\pi^2\right)^2} +$$

$$+ H_0 + i_2 x - 2\alpha i_2 \tau . \tag{3.24}$$

We introduce some simplifications in the formula here obtained for small values of $i_1$ and $i_2$ and a large value of $L$. For small values of time $\tau$, we may give another form of solution to our problem. Therefore we return to expression (3.10) for function $\bar{H}$ and develop the functions occurring. In it

$$\frac{\operatorname{sh} (L - x) \sqrt{\alpha^2 + p}}{\operatorname{sh} L \sqrt{\alpha^2 + p}}, \quad \frac{\operatorname{sh} x \sqrt{\alpha^2 + p}}{\operatorname{sh} L \sqrt{\alpha^2 + p}}$$

in the following series:

$$\frac{\operatorname{sh} x\sqrt{\alpha^2+p}}{\operatorname{sh} L\sqrt{\alpha^2+p}} = \frac{e^{x\sqrt{\alpha^2+p}}(1-e^{-2x\sqrt{\alpha^2+p}})}{e^{L\sqrt{\alpha^2+p}}(1-e^{-2L\sqrt{\alpha^2+p}})} =$$

$$= e^{-(L-x)\sqrt{\alpha^2+p}}[1+e^{-2L\sqrt{\alpha^2+p}}+\ldots-e^{2x\sqrt{\alpha^2+p}}-e^{-2(x+L)\sqrt{\alpha^2+p}}+\ldots]$$

$$= e^{-(L-x)\sqrt{\alpha^2+p}}-e^{-(L+x)\sqrt{\alpha^2+p}}+e^{-(3L-x)\sqrt{\alpha^2+p}}-e^{-(3L+x)\sqrt{\alpha^2+p}}+\ldots$$

$$\frac{\operatorname{sh}(L-x)\sqrt{\alpha^2+p}}{\operatorname{sh} L\sqrt{\alpha^2+p}} = e^{-x\sqrt{\alpha^2+p}}\,\frac{1-e^{-2(L-x)\sqrt{\alpha^2+p}}}{1-e^{-2L\sqrt{\alpha^2+p}}} =$$

$$= e^{-x\sqrt{\alpha^2+p}}(1+e^{-2L\sqrt{\alpha^2+p}}+\ldots-e^{-2(L-x)\sqrt{\alpha^2+p}}-$$

$$-e^{-2(2L-x)\sqrt{\alpha^2+p}}-\ldots) = e^{-x\sqrt{\alpha^2+p}}+e^{-(2L+x)\sqrt{\alpha^2+p}}-e^{-2L\sqrt{\alpha^2+p}}-\ldots$$

Keeping only the first terms of these series, we write equation (3.10) in the form

$$\bar{H}(x, p) \approx e^{\alpha x}\left(\frac{H_1-H_0}{p}+\frac{2\alpha i_2}{p^2}\right)e^{-x\sqrt{\alpha^2+p}}+$$

$$+e^{-\alpha(L-x)}\left(\frac{H_2-H_0-i_2L}{p}+\frac{2\alpha i_2}{p^2}\right)e^{-(L-x)\sqrt{\alpha^2+p}}+\frac{H_0+i_2x}{p}-\frac{2\alpha i_2}{p^2}.$$

Finally we obtain for small $\tau$:

$$H(x, t) \approx \frac{e^{2\alpha x}}{2}\left\{\left(H_1-H_0-\frac{i_1i_2}{h}\tau+xi_2\right)(1-\Phi(\xi))\right\}+$$

$$+\frac{1}{2}\left(H_1-H_0-\frac{i_1i_2}{h}\tau-i_2x\right)(1-\Phi(\eta))+$$

$$+\frac{1}{2}\left[H_2-H_0-i_2L-\frac{i_1i_2}{h}\tau+i_2(L-x)\right](1-\Phi(\tilde{\xi}))+$$

$$+\frac{1}{2}\left[H_2-H_0-i_2L-\frac{i_1i_2}{h}\tau-i_2(L-x)\right](1-\Phi(\tilde{\eta}))\,e^{-2\alpha(L-x)}+$$

$$+H_0+i_2x+\frac{i_1i_2}{h}\tau,$$

where

$$\xi=\frac{x}{2\sqrt{\tau}}+\alpha\sqrt{\tau}, \qquad \eta=\frac{x}{2\sqrt{\tau}}-\alpha\sqrt{\tau},$$

$$\tilde{\xi}=\frac{L-x}{2\sqrt{\tau}}+\alpha\sqrt{\tau}, \qquad \tilde{\eta}=\frac{L-x}{2\sqrt{\tau}}-\alpha\sqrt{\tau}.$$

## B. Dynamics of Groundwater Spreading.

### §4. Spreading Strip – Stratum Overlying Impervious Bedrock.

The groundwater surface in a region subject to artificial irrigation has a complicated form, composed of mounds and depressions, deforming in time.

In this chapter we examine some problems about unsteady groundwater flow, due to well-planned spreading. Among the elementary basic schemes we analyze the case of spreading on a strip in its last stage, the damping of the groundwater mound.

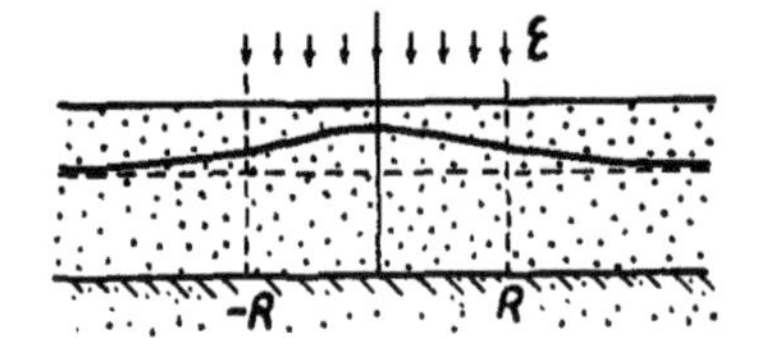


Fig. 390

Assume that, at the initial moment $t = 0$, the groundwater surface is a horizontal plane. Further, a constant flowrate $\varepsilon (m^3/m^2 \text{ day})$ hits the strip $-R < x < R$. It comes from uniform spreading or rain on a strip-shaped region (Fig. 390). The question arises as to know how the groundwater level rises in time in different points of the region [2].

We assume that the problem is "hydraulic," i.e., we consider the flow to be averaged in depth.

Our problem is then reduced to the solution of the differential equation

$$\frac{\partial H}{\partial t} = a^2 \frac{\partial^2 H}{\partial x^2} + \frac{w}{m}, \tag{4.1}$$

where

$$a^2 = \frac{kh}{m}.$$

Here $k$ is the seepage coefficient, $h$ the average depth of flow (we may take the initial depth for it), $m$ the active soil porosity. $H$ is the head, related to the velocity potential $\varphi$ as

$$\varphi = -kH = -k\left(\frac{p}{\rho g} + y\right).$$

We write the initial condition as

$$H(x, 0) = H_0,$$

where $H_0$ is the constant initial groundwater level.

We put $w = \varepsilon$ for $|x| < R$, $w = 0$ for $|x| > R$. We introduce a new variable

$$H - H_0 = u. \tag{4.2}$$

Then for $u$ we have the initial zero condition

$$u(x, 0) = 0. \tag{4.3}$$

The solution of

$$\frac{\partial u}{\partial t} = a^2 \frac{\partial^2 u}{\partial x^2} + f(x, t) \tag{4.4}$$

for the initial condition $u = 0$ has the form [14]

$$(4.5) \qquad u(x, t) = \frac{1}{2a\sqrt{\pi}} \int_0^t \frac{d\tau}{\sqrt{t-\tau}} \int_{-\infty}^{\infty} e^{-\frac{(x-\zeta)^2}{4a^2 t}} f(\zeta, \tau)\, d\zeta$$

Inserting in (4.5) instead of $f(\zeta, \tau)$ the expression $\frac{\varepsilon}{m}$ for $|\zeta| < R$ and zero for $|\zeta| > R$, we obtain:

$$H(x, t) = \frac{\varepsilon}{2ma\sqrt{\pi}} \int_0^t \frac{d\tau}{\sqrt{t-\tau}} \int_{-R}^{R} e^{-\frac{(x-\zeta)^2}{4a^2(t-\tau)}} d\zeta + H_0 .$$

The substitution

$$\frac{\zeta - x}{2a\sqrt{t-\tau}} = \lambda$$

gives

$$(4.6) \qquad u = \frac{\varepsilon}{m\sqrt{\tau}} \int_0^t d\tau \int_{r_1}^{r_2} e^{-\lambda^2} d\lambda \qquad \left(r_1 = -\frac{R+x}{2a\sqrt{t-\tau}}, \quad r_2 = \frac{R-x}{2a\sqrt{t-\tau}}\right)$$

Using the notation of the probability function

$$(4.7) \qquad \Phi(x) = \frac{2}{\sqrt{\pi}} \int_0^x e^{-\lambda^2} d\lambda ,$$

we write the expression for $u$ as:

$$(4.8) \qquad u(x, t) = \frac{\varepsilon}{2m} \int_0^t \left[\Phi\left(\frac{R-x}{2a\sqrt{t-\tau}}\right) + \Phi\left(\frac{R+x}{2a\sqrt{t-\tau}}\right)\right] d\tau .$$

We integrate each term of formula (4.8) by parts, and recalling that function $\Phi(x)$ is odd:

$$(4.9) \qquad \Phi(-x) = -\Phi(x) .$$

We obtain different expressions for $u$ in the intervals $|x| < R$ and $|x| > R$. We communicate the final result in dimensionless quantities,

(4.10)

$$U = \frac{hk}{R^2\varepsilon} u = \frac{hk}{R^2\varepsilon}(H - H_0), \qquad \xi = \frac{x}{R}, \quad \tau = \frac{a^2 t}{R^2} = \frac{hk}{mR^2} t .$$

For $|\xi| < 1$, we have:

(4.11)

$$U = \frac{1}{\sqrt{\pi}} \left\{\tau\Phi\left(\frac{1-\xi}{2\sqrt{\tau}}\right) + \frac{(1-\xi\sqrt{\tau}}{\sqrt{\tau}} e^{-\frac{(1-\xi)^2}{4\tau}} - \frac{(1-\xi)^2}{2}\left[1 - \Phi\left(\frac{1-\xi}{2\sqrt{\tau}}\right)\right] + \right.$$

$$\left. + \tau\Phi\left(\frac{1+\xi}{2\sqrt{\tau}}\right) + \frac{(1+\xi)\sqrt{\tau}}{\sqrt{\pi}} e^{-\frac{(1+\xi)^2}{4\tau}} - \frac{(1+\xi)^2}{2}\left[1 - \Phi\left(\frac{1+\xi}{2\sqrt{\tau}}\right)\right]\right\} ;$$

For $|\xi| > 1$

(4.12)

$$U = \frac{1}{\sqrt{\pi}} \left\{ \tau\Phi\left(\frac{1+\xi}{2\sqrt{\tau}}\right) + \frac{(1+\xi)\sqrt{\tau}}{\sqrt{\pi}} e^{-\frac{(1+\xi)^2}{4\tau}} - \frac{(1+\xi)^2}{2}\left[1 - \Phi\left(\frac{1+\xi}{2\sqrt{\tau}}\right)\right] - \right.$$

$$\left. - \tau\Phi\left(\frac{\xi - 1}{2\sqrt{\tau}}\right) - \frac{(\xi+1)\sqrt{\tau}}{\sqrt{\pi}} e^{-\frac{(1-\xi)^2}{4\tau}} + \frac{(\xi-1)^2}{2}\left[1 - \Phi\left(\frac{\xi-1}{2\sqrt{\tau}}\right)\right] \right\} .$$

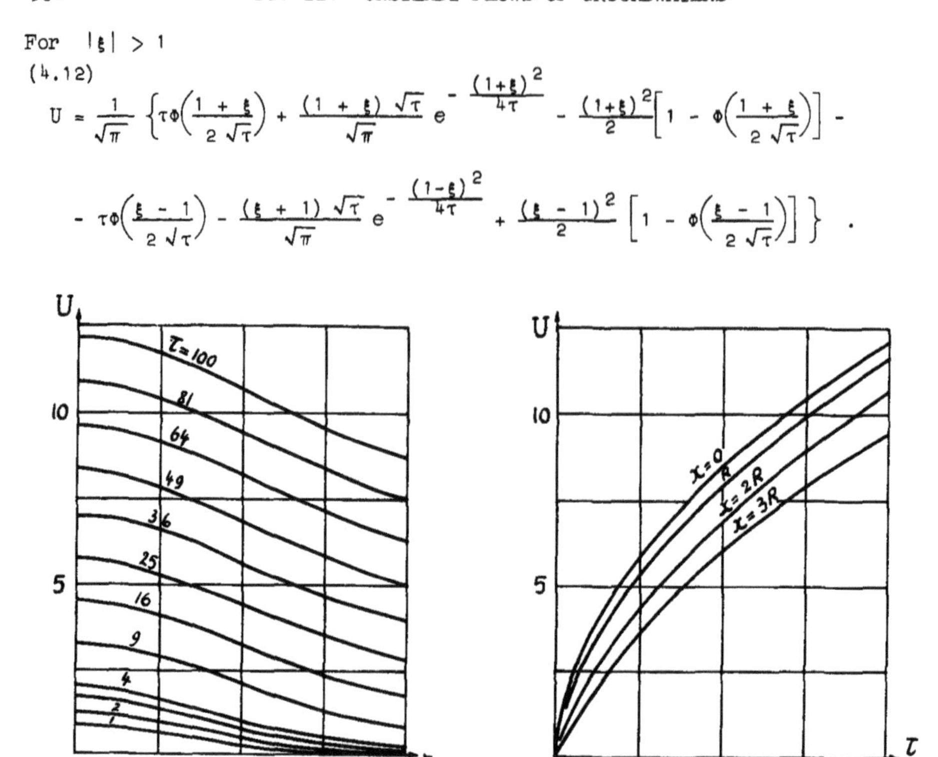


Fig. 391 Fig. 392

Graphs of $U$ in function of $\xi$ for various values of $\tau$ are given in Fig. 391. Fig. 392 renders $U$ in function of $\tau$ for $\xi = 0, 1, 2, 3$. The case [11] has been examined, where spreading takes place during a finite time interval $T$, then is discontinued and after a time $T_1$ is resumed, and so on.

## §5. Damping of a Groundwater Mound.

We now assume that spreading is discontinued; the mound of groundwater that is created must spread out. The question is to know at which velocity this phenomenon will proceed. To solve this problem, one must know the initial form of the mound.

First we examine the case of a rectangular mound. Such a mound may be formed in case of complete replenishment of the soil in the strip subject to spreading (Fig. 393). We must find the solution of equation (4.1) for the original condition

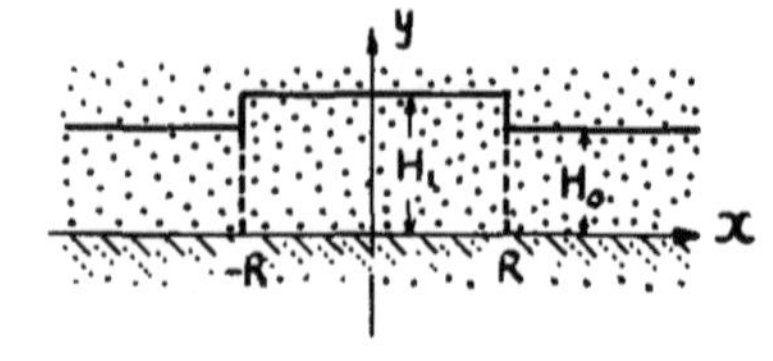


Fig. 393

(5.1) $$H(x, 0) = H_0 \text{ for } |x| > R, \quad H(x, 0) = H_1 \text{ for } |x| < R.$$

The boundary condition is that of finiteness of $H$ at $\infty$.

As known, the solution of the homogeneous equation of (4.4) (for $f(x, t) = 0$), satisfying the initial condition

$$u(x, 0) = f(x) ,$$

has the form

$$u(x, t) = \frac{1}{2a\sqrt{\pi t}} \int_{-\infty}^{\infty} e^{-\frac{(x-\zeta)^2}{4a^2 t}} f(\zeta)\, d\zeta . \tag{5.2}$$

In our case, calling $u = H - H_0$, we have $u = H_1 - H_0$ for $|x| < R$, $u = 0$ for $|x| > R$. Therefore

$$H(x, t) = H_0 + \frac{H_1 - H_0}{2a\sqrt{\pi t}} \int_{-R}^{R} e^{-\frac{(x-\zeta)^2}{4a^2 t}} d\zeta .$$

By means of probability function (4.7), this solution is presented as

$$H(x, t) = H_0 + \frac{H_1 - H_0}{2} \left[ \left( \frac{R - x}{2a\sqrt{t}} \right) + \Phi \left( \frac{R + x}{2a\sqrt{t}} \right) \right] . \tag{5.3}$$

Fig. 394 gives graphs of the groundwater level in function of time for the center of the mound and for values of $x = R, 2R$. We see that the groundwater level in point $x = 0$ drops steeply in the beginning and then tends to an unperturbed level, at an extremely slow rate.

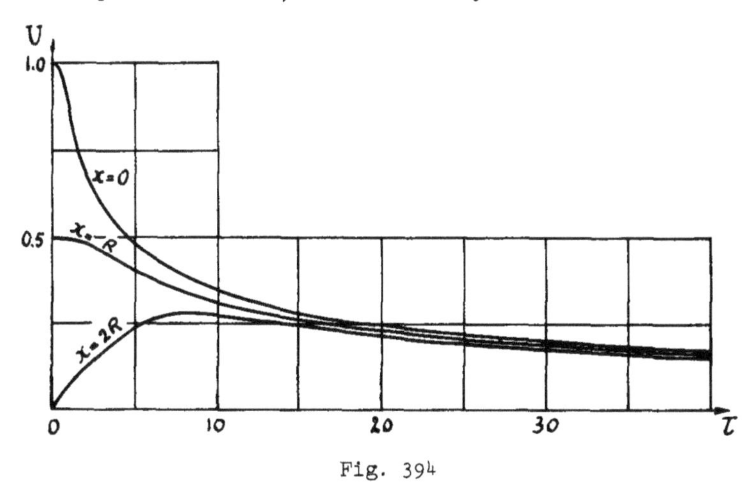


Fig. 394

Fig. 395 renders curves of the rise of groundwaters and their subsequent subsidence as observed in one of the strips [3] (depths in centimeters, measured from the soil surface, are laid out along the ordinate axis). Comparing them with the theoretical curves of Fig. 394, we see that their general character is the same.

We notice that observation of groundwater levels during spreading and in the period following, may make it possible to compute an important numerical soil characteristic, the quantity

$$a^2 = \frac{kh}{m} .$$

This quantity accounts for the average soil permeability and the depth of groundwater flow.

We still mention that it might be possible to consider other initial forms of groundwater mounds. For example, taking $H(x, 0) = Ae^{-\alpha x^2}$, we find the solution

$$H(x, t) = \frac{A}{(1 + 4\alpha a^2 t)^{3/2}} e^{-\frac{\alpha x^2}{1+4\alpha a^2 t}} .$$

But in first approximation we may use a rectangular mound, equivalent (as regards to the region) with the given mound.

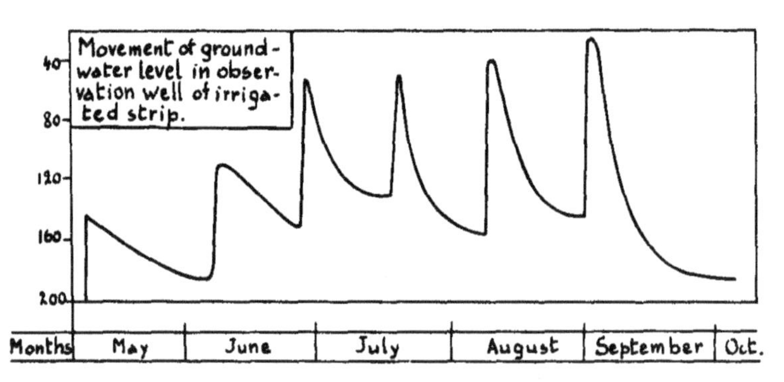


Fig. 395

§6. Account for Evaporation and Transpiration from Plants.

Recently an important roll in groundwater flow has been attributed to evaporation, either directly from the groundwater surface or indirectly from plants, which absorb soil moisture and evaporate it then from the surface of their leaves (transpiration).

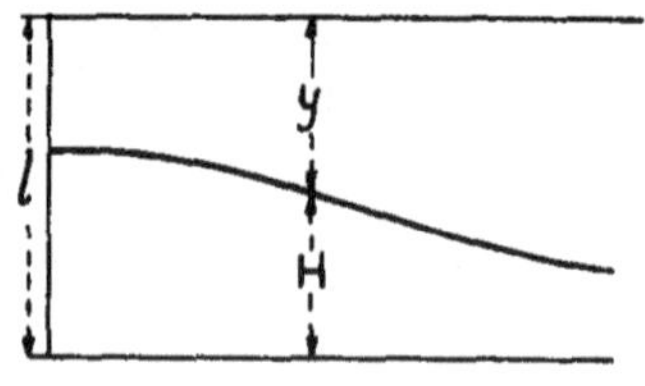


Fig. 396

Based on a series of observations of irrigated strips, M. M. Krilov [3] found the dependence of evaporation and transpiration intensity on the depth of the free groundwater surface below the earth surface. This dependence appeared to be close to linear.

Designating the intensity of evaporation by $c\left(\frac{m^3}{m^2\ day}\right)$, we may write this relationship in the form

$$c = \beta - \alpha y,$$

where $y$ is the ordinate, measured from the earth surface (Fig. 396). Putting

$$y + H = \ell ,$$

we obtain:

$$c = \alpha H + \beta - \alpha \ell .$$

Returning now to equation (4.1), we notice that in the case of evaporation one must consider vertical flow to be *negative*, i.e., introduce in equation (4.1) the quantity ($\varepsilon$ = intensity of infiltration)

$$\frac{w}{m} = -\frac{c}{m} + \frac{\varepsilon}{m} = -b^2(H - H^*) + \frac{\varepsilon}{m} . \tag{6.1}$$

Here $b^2$ is a new constant introduced instead of $\frac{\alpha}{m}$, $H^* = \ell - \frac{\beta}{\alpha}$. Instead of (4.1) we now obtain the equation

$$\frac{\partial H}{\partial t} = a^2 \frac{\partial^2 H}{\partial x^2} - b^2(H - H^*) + \frac{\varepsilon}{m} . \tag{6.2}$$

The same equation, only with other values of the constants $b$ and $H^*$, is obtained when one considers the vertical velocity, which takes place at the boundary with poorly pervious bedrock. We communicate the solution of equation (6.2) for the case of rain or spreading on a strip [4].

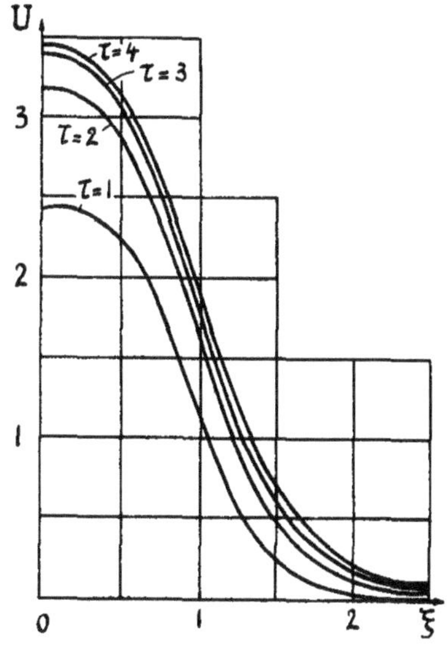


Fig. 397

We reproduce here, without derivation, the formulas, introducing the dimensionless quantities

$$U = \frac{4b^2m}{\varepsilon}(H - H^*), \quad \xi = \frac{x}{R}, \qquad \tau^2 = b^2 t, \quad \alpha = \frac{bR}{2a} . \tag{6.3}$$

For $U(\xi, \tau)$ we have for spreading on a strip, accounting for evaporation (or slight percolation into the underlying soil) for $|\xi| < 1$:

$$\begin{aligned} U = 4 &- 2e^{-\tau^2}\left[\Phi\left(\frac{\alpha(1+\xi)}{\tau}\right) + \Phi\left(\frac{\alpha(1-\xi)}{\tau}\right)\right] - \\ &- e^{-2\alpha(1+\xi)}\left[1 - \Phi\left(\frac{\alpha(1+\xi)}{\tau} - \tau\right)\right] - e^{-2\alpha(1-\xi)}\left[1 - \Phi\left(\frac{\alpha(1-\xi)}{\tau} - \tau\right)\right] - \\ &- e^{2\alpha(1+\xi)}\left[1 - \Phi\left(\frac{\alpha(1+\xi)}{\tau} + \tau\right)\right] - e^{2\alpha(1-\xi)}\left[1 - \Phi\left(\frac{\alpha(1-\xi)}{\tau} + \tau\right)\right] . \end{aligned} \tag{6.4}$$

For $|\xi| > 1$

$$\begin{aligned} U = &- 2e^{-\tau^2}\left[\Phi\left(\frac{\alpha(\xi+1)}{\tau}\right) - \Phi\left(\frac{\alpha(\xi-1)}{\tau}\right)\right] - \\ &- e^{-2\alpha(\xi+1)}\left[1 - \Phi\left(\frac{\alpha(\xi+1)}{\tau} - \tau\right)\right] - e^{2\alpha(\xi+1)}\left[1 - \Phi\left(\frac{\alpha(\xi+1)}{\tau} + \tau\right)\right] + \end{aligned}$$

$$+ e^{-2\alpha(\xi-1)}\left[1 - \Phi\left(\frac{\alpha(\xi - 1)}{\tau} - \tau\right)\right] + e^{2\alpha(\xi-1)}\left[1 - \Phi\left(\frac{\alpha(\xi - 1)}{\tau} + \tau\right)\right] . \quad (6.5)$$

In the case under consideration, the free surface, does not rise indefinitely, as in the absence of evaporation; the free surface lines tend to a limit position. For $t \to \infty$ function $U$ inside and outside the interval $|\xi| < 1$ tends respectively to

$$4[1 - e^{-2\alpha} \operatorname{ch} 2\alpha\xi] \quad \text{and} \quad 4 \operatorname{sh} 2\alpha e^{-2\alpha\xi} . \quad (6.6)$$

Fig. 397 gives graphs for the dependence of $U$ on $\xi$ for $\alpha = 1$ for some moments of time: $\tau = 1,2,3,4$. Fig. 398 shows the limit positions of the free surface for a series of values of parameter $\alpha$.

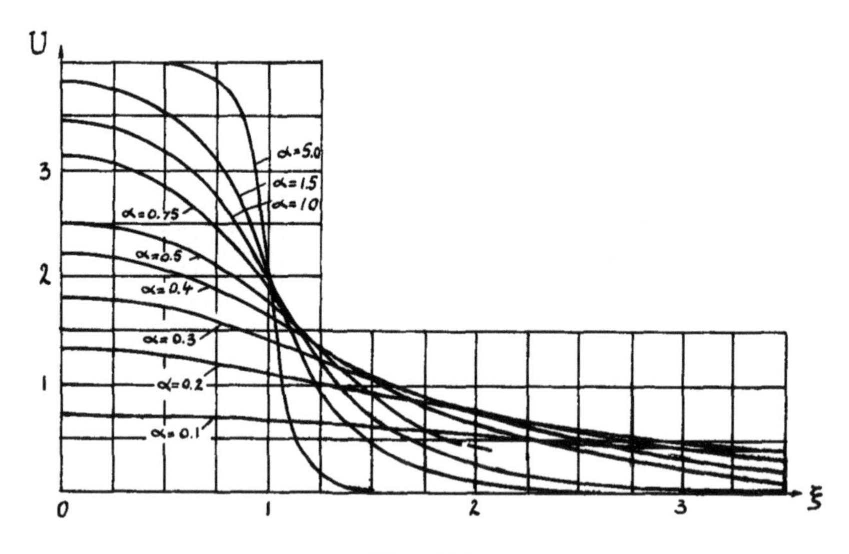


Fig. 398

§7. <u>Damping of Groundwater Mound in Case of Evaporation [2].</u>

In this case the problem reduces to the integration of equation (6.2) with initial condition (5.1) in which we consider $\varepsilon = 0$. Substitution

$$H - H^* = e^{-b^2 t} U \quad (7.1)$$

reduces equation (6.2) to the simplest equation of heat conduction:

$$\frac{\partial U}{\partial t} = a^2 \frac{\partial^2 U}{\partial x^2} .$$

For function $U$, the initial conditions are

$$\begin{cases} U(x, 0) = H_1 - H^* = U_1 & \text{for } |x| < R, \\ U(x, 0) = H_0 - H^* = U_0 & \text{for } |x| > R . \end{cases} \quad (7.2)$$

To determine $U(x, t)$ we may use formula (5.3), replacing in it $H_0$, $H_1$ respectively by $U_0$, $U_1$. We obtain:

$$U(x,t) = U_0 + \frac{U_1 - U_0}{2}\left[\Phi\left(\frac{R - x}{2a\sqrt{t}}\right) + \Phi\left(\frac{R + x}{2a\sqrt{t}}\right)\right] .$$

Returning to $H(x, t)$, we have:

$$H(x, t) = H^* + e^{-b^2 t}\left\{H_0 - H^* + \frac{H_1 - H_0}{2}\left[\Phi\left(\frac{R - x}{2a\sqrt{t}}\right) + \Phi\left(\frac{R + x}{2a\sqrt{t}}\right)\right]\right\} . \tag{7.3}$$

We see that for $t \to \infty$, the groundwater level tends to a constant value $H^*$.

## C. Unsteady Flow in Multilayered Medium. Flow of Two Fluids.

### §8. Unsteady Flow for a Seepage Coefficient, Slightly Changing with Depth of Flow.

We analyzed some problems of steady flow in heterogeneous soils in Chapter X. There we already introduced a function, initially given by N. K. Girinsky [5].

$$\Phi(x, y) = \int_0^h (z - h)\, k(z)\, dz . \tag{8.1}$$

Here $h$ is the head, which we consider as a function of $x$, $y$ and $t$; $k$ seepage coefficient, depending on the depth of flow. We assume that $k$ varies slightly. For unconfined flow, $h$ is taken for the ordinate of the free surface.

In the case of unsteady flow, we may construct the equation of motion in the following way.

We call $m$ the soil porosity, which we assume to be constant in time. Then, expressing the condition that the change of flowrate along the coordinates, is compensated by a change in time of the quantity $mh$, we may write:

$$\frac{\partial^2\Phi}{\partial x^2} + \frac{\partial^2\Phi}{\partial y^2} = - \frac{\partial(mh)}{\partial t} . \tag{8.2}$$

Further we have:

$$\frac{\partial\Phi}{\partial t} = \Phi'(h)\,\frac{\partial h}{\partial t} . \tag{8.3}$$

Eliminating $\frac{\partial h}{\partial t}$ from (8.2) and (8.3), we obtain:

$$\frac{\partial\Phi}{\partial t} = - \frac{\Phi'(h)}{m}\left[\frac{\partial^2\Phi}{\partial x^2} + \frac{\partial^2\Phi}{\partial y^2}\right] . \tag{8.4}$$

If we consider that the quantity

$$\frac{1}{m}\,\Phi'(h) = - \frac{1}{m}\int_0^h k(z)\, dz$$

changes slightly and replace it by a constant value, putting

$$- \frac{1}{m} \Phi'(h) \approx \frac{1}{m} \int_0^h k(z)\, dz = a^2 \tag{8.5}$$

then we obtain the equation of heat conduction

$$\frac{\partial \Phi}{\partial t} = a^2 \left( \frac{\partial^2 \Phi}{\partial x^2} + \frac{\partial^2 \Phi}{\partial y^2} \right) . \tag{8.6}$$

For example we consider soil with two strata, the lower one having a thickness $h_0$, a seepage coefficient $k_0$, an active porosity $m_0$, the upper one having characteristics $k_1$, $m_1$ (Fig. 399). The free surface is assumed to be in the upper soil stratum.

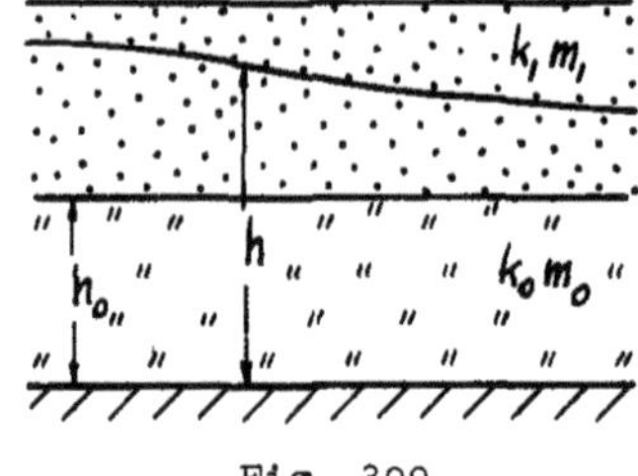

Fig. 399

We have then, designating by $c$ the ratio of the seepage coefficients

$$\frac{k_0}{k_1} = c ,$$

the following expression for $\Phi$ in the upper stratum:

$$\Phi = k_0 \int_0^{h_0} (z - h)\, dz + k_1 \int_{h_0}^{h} (z - h)\, dz =$$

$$= - \frac{k_1}{2} \{ [h + (c - 1) h_0]^2 + c(c - 1) h_0^2 \} . \tag{8.7}$$

For $a^2$ we take the quantity

$$a^2 = \frac{k_0 h_0 + k_1 \bar{h}_1}{m_1} , \tag{8.8}$$

where $\bar{h}_1$ is the average value of $h - h_0$.

In the case of seepage from a canal with vertical wall, assuming

$$h(x, 0) = H_2, \quad h(0, t) = H_1$$

and considering $h$ finite for $x = \infty$, we write the initial and boundary conditions for $\Phi$:

$$\begin{cases} \Phi(x, 0) = - \frac{1}{2} k_1 \{ [H_2 + (c - 1) h_0]^2 + c(c - 1) h_0^2 \} = \Phi_2 , \\ \Phi(0, t) = - \frac{1}{2} k_1 \{ [H_1 + (c - 1) h_0]^2 + c(c - 1) h_0^2 \} = \Phi_1 . \end{cases} \tag{8.9}$$

For $\Phi(x, t)$ we obtain the known solution

$$\Phi(x, t) = (\Phi_2 - \Phi_1)\, \Phi\left( \frac{x}{2a\sqrt{t}} \right) + \Phi_1 \tag{8.10}$$

where

$$\Phi(x) = \frac{2}{\sqrt{\pi}} \int_0^x e^{-\lambda^2}\, d\lambda . \tag{8.11}$$

Further

$$\Phi_2 - \Phi_1 = -\frac{1}{2} k_1\left\{ \frac{1}{2} (H_1 + H_2) + (c - 1)h_0\right\} (H_2 - H_1) . \tag{8.12}$$

The expression of the flowrate in section $x$ has the form

$$q(x, t) = \frac{\Phi_2 - \Phi_1}{a\sqrt{\pi t}} e^{-\frac{x^2}{4a^2t}} .$$

For $x = 0$, we obtain:

$$q(0, t) = \frac{\Phi_2 - \Phi_1}{a\sqrt{\pi t}} = -\frac{k_1(H_2 - H_1)}{a\sqrt{\pi t}}\left[\frac{H_1 + H_2}{2} + (c - 1)h_0\right] . \tag{8.13}$$

For homogeneous soil, putting $k_0 = k_1$, $h_0 + \bar{h}_1 = \bar{h}$, we have for the flowrate, referred to as $q_0(0, t)$, the expression

$$q_0(0, t) = -\frac{k_1(H_2 - H_1)(H_2 + H_1)}{2a_0\sqrt{\pi t}} , \tag{8.14}$$

where

$$a_0^2 = \frac{k_1\bar{h}}{m_1} .$$

Computing the ratio of the flowrates, we find

$$\frac{q}{q_0} = \left[1 + (c - 1)\frac{2h_0}{H_1 + H_2}\right]\sqrt{\frac{h_0 + \bar{h}_1}{ch_0 + \bar{h}_1}} \quad \left(c = \frac{k_0}{k_1}\right) . \tag{8.15}$$

We observe that in the equation of heat conduction, and consequently, also in the solution of the present equation, time is multiplied by $a^2$. For homogeneous soil, we have called this factor $a_0^2$. Computing the ratio

$$\frac{a^2}{a_0^2} = 1 + (c - 1)\frac{h_0}{\bar{h}_1} ,$$

we see that for $c < 1$, the inequality holds

$$a^2 < a_0^2 ,$$

and for $c > 1$, the opposite inequality

$$a^2 > a_0^2 .$$

This shows that in the first case the duration of the process (for otherwise equal conditions) is much shorter for the heterogeneous soil than for the homogeneous ( the upper stratum kind); in the second case we have the opposite phenomenon.

To find the form of the free surface at different times, we write the expression for $\Phi$, inserting in it the right part of (8.7).

§9. Variable Interface Between Two Liquids of Different Density.

We already talked about the occurrence in nature of flow of two liquids of different densities, in particular fresh and salt water (Chapter VIII). Here we analyze the problem of determining equations for the location of the interface of such fluids in time, under the usual condition, which we always assume, of small variation of the interface.

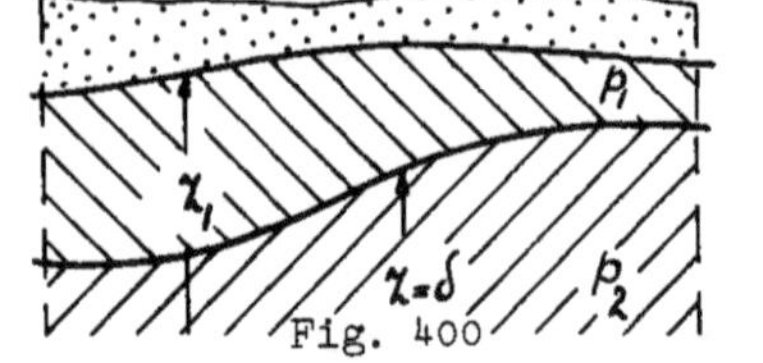
Fig. 400

Let there be two fluids: the upper, for which all quantities carry the subscript 1, and the lower, with larger density, in which we use the subscript 2 (Fig. 400). We have respectively for the velocity potentials:

$$\varphi_1 = -k_1\left(\frac{p_1}{\rho_1 g} + z\right), \tag{9.1}$$

$$\varphi_2 = -k_2\left(\frac{p_1}{\rho_2 g} + z\right). \tag{9.2}$$

The condition of continuity of pressure at the interface $(p_1 = p_2)$ gives:

$$\rho_1\left(\frac{\varphi_1}{k_1} + z\right) = \rho_2\left(\frac{\varphi_2}{k_2} + z\right),$$

from which we obtain an approximate equation of the interface

$$z = \alpha_1\varphi_1(x, y, h, t) - \alpha_2\varphi_2(x, y, h, t) = \delta(x, y, t), \tag{9.3}$$

where

$$\alpha_1 = \frac{\rho_1}{k_1(\rho_2 - \rho_1)}, \quad \alpha_2 = \frac{\rho_2}{k_2(\rho_2 - \rho_1)}.$$

Differentiating this equation along the interface and inserting in it once the velocities $u_1, v_1, w_1$ and once $u_2, v_2, w_2$, we obtain two equations at the interface

$$\begin{cases} m\dfrac{\partial\delta}{\partial t} + \dfrac{\partial\delta}{\partial x}u_1 + \dfrac{\partial\delta}{\partial y}v_1 - w_1 = 0, \\ m\dfrac{\partial\delta}{\partial t} + \dfrac{\partial\delta}{\partial x}u_2 + \dfrac{\partial\delta}{\partial y}v_2 - w_2 = 0. \end{cases} \tag{9.4}$$

We assume as before that the horizontal velocities do not depend upon $z$. Then from the equation of continuity, we obtain for the lower soil:

$$w_2 = -z\left(\frac{\partial u_2}{\partial x} + \frac{\partial v_2}{\partial y}\right) + w_{20} = -\delta\left(\frac{\partial^2\varphi_2}{\partial x^2} + \frac{\partial^2\varphi_2}{\partial y^2}\right) + w_{20}, \tag{9.5}$$

where $w_{20}$ is the velocity for $z = 0$, and for the upper stratum

$$w_1 = -(z - \delta)\left(\frac{\partial u_1}{\partial x} + \frac{\partial v_1}{\partial y}\right) + w_{10}. \tag{9.6}$$

Here $z$ is the level of the free surface, at which we take

$$z = -\frac{\varphi_1}{k}.$$

Therefore, for $w_1$, we may write:

$$w_1 = \left(\frac{\varphi_1}{k} + \delta\right)\left(\frac{\partial^2\varphi_1}{\partial x^2} + \frac{\partial^2\varphi_1}{\partial y^2}\right) + w_{10} \quad (9.7)$$

The quantity $w_{10}$ is the vertical velocity at the free surface, as a result of precipitation, evaporation, etc.

Inserting the obtained equations for $w_1$ and $w_2$ in equation (9.4), we obtain a system of two non-linear equations in partial derivatives of functions $\varphi_1$ and $\varphi_2$ (the quantity $\delta$ may be eliminated from these equations by means of $\delta = \alpha_1\varphi_1 - \alpha_2\varphi_2$).

Linearization of these equations is possible, but we do not dwell on this subject. A non-linear homogeneous system may be reduced to a system of ordinary differential equations, for the case of the displacement of a tongue of the lower liquid in a confined stratum [6].

The problem of the displacement of the interface between fresh and salt water under a dam foundation has been examined [7]. The inclusion of linear inertia terms lead to the Telegraph equation. Because of what was said in Chapter XII, it makes no sense to include these terms, and then instead of the telegraph equation, the much simpler equation of heat conduction is obtained. The problem of a two fluid system was also examined by N. K. Girinsky [9].

## D. Some Three-dimensional Problems.

### §10. Example of Three-dimensional Problem. [4].

Consider a canal with vertical walls and impervious bottom. At the initial moment of time, the ground water level is at a constant $H_0$. Suddenly the canal is filled so that in one of its parts, for $x < 0$, the water level is $H_1$, and in the other part, for $x > 0$, the water level is $H_2$. These levels are then kept constant (Fig. 401 gives a view from above). We must find the equation of the free surface for the upper half space

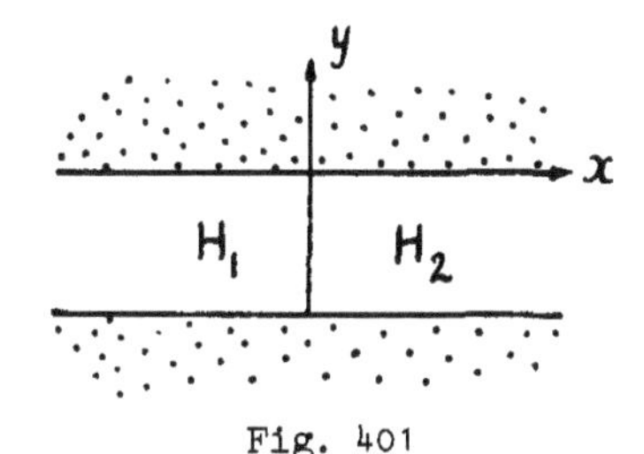


Fig. 401

$$z = H(x, y, t) .$$

The unknown function $H$ satisfies the following equation:

$$\frac{\partial H}{\partial t} = a^2\left(\frac{\partial^2 H}{\partial x^2} + \frac{\partial^2 H}{\partial y^2}\right) \quad \left(a^2 = \frac{k\bar{h}}{m}\right) . \quad (10.1)$$

The solution of the equation of heat conduction (10.1) for the initial condition

$$H(x, y, 0) = f(x, y) \quad (10.2)$$

and the boundary condition

$$H(x,\ 0,\ t) = F(x,\ t) \tag{10.3}$$

has the form [16]

$$H(x,y,t) = \frac{1}{4\pi a^2 t}\int_{-\infty}^{\infty} d\xi \int_0^{\infty} f(\xi,\eta)\left[e^{-\frac{(x-\xi)^2+(y-\eta)^2}{4a^2t}} - e^{-\frac{(x-\xi)^2+(y+\eta)^2}{4a^2t}}\right] d\eta\ +$$

$$+ \frac{y}{4a^2\pi}\int_0^t d\tau \int_{-\infty}^{\infty} \frac{F(\xi,\tau)}{(t-\tau)^2}\, e^{-\frac{(x-\xi)^2+y^2}{4a^2(t-\tau)}}\, d\xi\ . \tag{10.4}$$

In our case

$$f(x,\ y) = H_0, \quad F(x,\ t) = \begin{cases} H_1 & \text{for } x < 0\ , \\ H_2 & \text{for } x > 0\ . \end{cases} \tag{10.5}$$

Since here the boundary function $F$ does not depend upon time, we transform the second integral beforehand, performing in it the integration over $\tau$. We obtain:

$$H(x,y,t) = \frac{1}{4\pi a^2 t}\int_{-\infty}^{\infty} d\xi \int_0^{\infty} f(\xi,\eta)\left[e^{-\frac{(x-\xi)^2+(y-\eta)^2}{4a^2t}} - e^{-\frac{(x-\xi)^2+(y-\eta)^2}{4a^2t}}\right] d\eta\ +$$

$$+ \frac{y}{\pi}\int_0^{\infty} \frac{F(\xi)}{(x-\xi)^2+y^2}\, e^{-\frac{(x-\xi)^2+y^2}{4a^2t}}\, d\xi\ . \tag{10.6}$$

We insert in this equation the expressions of (10.5) and recall the equality

$$\frac{1}{4a^2\pi t}\int_{-\infty}^{\infty} e^{-\frac{(x-\xi)^2}{4a^2t}}\, d\xi \int_0^{\infty}\left[e^{-\frac{(y-\eta)^2}{4a^2t}} - e^{-\frac{(y+\eta)^2}{4a^2t}}\right] d\eta = \Phi\left(\frac{y}{2a\sqrt{t}}\right).$$

We obtain:

$$H(x,y,t) = H_0\Phi\left(\frac{y}{2a\sqrt{t}}\right) + \frac{H_1 y}{\pi}\int_{-\infty}^{0} \frac{e^{-\frac{(x-\xi)^2+y^2}{4a^2t}}}{(x-\xi)^2+y^2}\, d\xi + \frac{H_2 y}{\pi}\int_0^{\infty} \frac{e^{-\frac{(x-\xi)^2+y^2}{4a^2t}}}{(x-\xi)^2+y^2}\, d\xi$$

Substitution $\xi - x = \eta$ gives:

$$H(x,y,t) = H_0\Phi\left(\frac{y}{2a\sqrt{t}}\right) + \frac{H_1 y}{\pi}\int_x^{\infty} e^{-\frac{\eta^2+y^2}{4a^2t}}\, \frac{d\eta}{\eta^2+y^2}\ +$$

$$+ \frac{H_2 y}{\pi}\int_0^{x} e^{-\frac{\eta^2+y^2}{4a^2t}}\, \frac{d\eta}{\eta^2+y^2}\ , \tag{10.7}$$

We may rewrite this:

$$H(x, y, t) = H_0\Phi\left(\frac{y}{2a\sqrt{t}}\right) + \frac{(H_1 + H_2)y}{\pi}\int_0^\infty e^{-\frac{\eta^2+y^2}{4a^2t}}\frac{d\eta}{\eta^2 + y^2} - \frac{(H_1 - H_2)y}{\pi}\int_0^x e^{-\frac{\eta^2+y^2}{4a^2t}}\frac{d\eta}{\eta^2 + y^2} . \tag{10.8}$$

The integral with constant limits, occurring in the last formula, is reduced to the error function:

$$\frac{2y}{\pi}\int_0^\infty e^{-\frac{\eta^2+y^2}{4a^2t}}\frac{d\eta}{\eta^2 + y^2} = 1 - \Phi\left(\frac{y}{2a\sqrt{t}}\right) . \tag{10.9}$$

To derive equation (10.9) we notice that

$$\frac{h}{u^2 + h^2} = \int_0^\infty e^{-h\xi}\cos u\xi\, d\xi , \qquad \int_0^\infty e^{-mu^2}\cos u\xi\, du = \frac{\sqrt{\pi}}{2\sqrt{m}}e^{-\frac{\xi^2}{4m}} .$$

We may carry out the following transformations of the integral:

$$J = \frac{2h}{\pi}\int_0^\infty \frac{e^{-m(u^2+h^2)}}{u^2 + h^2}du = \frac{2}{\pi}\int_0^\infty d\xi\int_0^\infty e^{-m^2h^2-h\xi-mu^2}\cos u\xi\, du =$$

$$= \frac{1}{\sqrt{m\pi}}\int_0^\infty e^{-\left(\frac{\xi}{2\sqrt{m}}+h\sqrt{m}\right)}d\xi .$$

The substitution $\xi/2\sqrt{m} + h\sqrt{m} = \lambda$ gives $J = 1 - \Phi(h\sqrt{m})$ .

Using (10.9) we reduce expression (10.8) to

$$H(x,y,t) = H_0\Phi\left(\frac{y}{2a\sqrt{t}}\right) + \frac{H_1 + H_2}{2}\left[1 - \Phi\left(\frac{y}{2a\sqrt{t}}\right)\right] - \frac{(H_1 - H_2)y}{\pi}\int_0^x e^{-\frac{\eta^2+y^2}{4a^2t}}\frac{d\eta}{\eta^2 + y^2} . \tag{10.10}$$

It follows from (10.10) that function $H$ is composed of three terms: the first is derived from the initial groundwater level, the second corresponds to the homogeneous problem with constant head $\frac{1}{2}(H_1 + H_2)$, and the third, depending upon $x$, determines the asymmetrical flow due to the difference in head between head-and tailwater.

For $t = \infty$, we obtain steady flow, for which the free surface has the equation:

$$z = \frac{H_1 + H_2}{2} - \frac{H_1 - H_2}{\pi}\operatorname{arc\,tg}\frac{x}{y} = H_2 + \frac{H_1 - H_2}{\pi}\operatorname{arc\,tg}\frac{y}{x} . \tag{10.11}$$

This is the equation of an helicoidal surface: it is made of straight lines issuing from various points of the z-axis and lying in horizontal planes, and has the form of a fan, unfolded over 180°.

Along the y-axis (for $x = 0$), we obtain $z = \frac{1}{2}(H_1 + H_2)$; for $y = 0$, $x > 0$, we obtain $z = H_2$; for $y = 0$, $x < 0$, we have $z = H_1$.

We now examine the same problem assuming that the flowregion underlies slightly pervious soil with thickness $h_0$ and seepage coefficient $k_0$ (see §8, Chapter IX). Then we may take the equation

$$\frac{\partial H}{\partial t} = a^2\left(\frac{\partial^2 H}{\partial x^2} + \frac{\partial^2 H}{\partial y^2}\right) - b^2(H - H_0), \quad b^2 = \frac{k_0}{mh_0} \quad . \tag{10.12}$$

We put

$$H - H_0 = e^{-b^2 t}\, u. \tag{10.13}$$

where $H_0$ initial water level in canal and groundwater.

One must determine the flow in the half-plane $y > 0$, if given $H = H_0$ for $t = 0$, and for $t > 0$

$$\begin{cases} H = H_1 & \text{for } x < 0, \ y = 0 , \\ H = H_2 & \text{for } x > 0, \ y = 0 . \end{cases} \tag{10.14}$$

For function $u$ these conditions become

$$\begin{cases} u = 0 & \text{for } t = 0 , \\ u = (H_1 - H_0)e^{b^2 t} & \text{for } x < 0, \ y = 0 , \\ u = (H_2 - H_0)e^{b^2 t} & \text{for } x > 0, \ y = 0 . \end{cases} \tag{10.15}$$

In formula (10.4) one has to put

$$f(\xi, \eta) = 0$$

and

$$F(\xi, \tau) = \begin{cases} (H_1 - H_0)e^{b^2\tau} & (\xi < 0) , \\ (H_2 - H_0)e^{b^2\tau} & (\xi > 0) . \end{cases}$$

Then we obtain:

$$u = (H - H_0)e^{b^2 t} = \frac{y(H_1 - H_0)}{4a^2\pi}\int_0^t\int_{-\infty}^0 \frac{e^{b^2\tau - \frac{(x-\xi)^2+y^2}{4a^2(t-\tau)}}}{(t-\tau)^2}\, d\tau\, d\xi +$$

$$+ \frac{(H_2 - H_0)y}{4a^2\pi}\int_0^t\int_0^\infty \frac{e^{b^2\tau - \frac{(x-\xi)^2+y^2}{4a^2(t-\tau)}}}{(t-\tau)^2}\, d\tau\, d\xi \ . \tag{10.16}$$

From this

$$H = H_0 + \frac{(H_1 - H_0)y}{4a^2\pi}\int_0^t\int_0^\infty \frac{e^{-b^2(t-\tau) - \frac{(x+\xi)^2+y^2}{4a^2(t-\tau)}}}{(t-\tau)^2}\, d\tau\, d\xi +$$

$$+ \frac{(H_2 - H_0)y}{4a^2\pi}\int_0^t\int_0^\infty \frac{e^{-b^2(t-\tau) - \frac{(x-\xi)^2+y^2}{4a^2(t-\tau)}}}{(t-\tau)^2}\, d\tau\, d\xi \ . \tag{10.17}$$

We introduce the substitutions

$$\frac{(x + \xi)^2 + y^2}{4a^2(t - \tau)} = v , \quad \frac{(x - \xi)^2 + y^2}{4a^2(t - \tau)} = w \tag{10.18}$$

respectively for the first and second integrals of formula (10.17). We obtain, rearranging the limits:

$$H = H_0 + \frac{b^2(H_1 - H_0)y}{4a^2\pi} \int_0^\infty \int_{v_1}^\infty e^{-v - \frac{\alpha}{v}} \frac{dv\, d\xi}{\alpha} +$$

$$+ \frac{b^2(H_2 - H_0)y}{4a^2\pi} \int_0^\infty \int_{w_1}^\infty e^{-w - \frac{\beta}{w}} \frac{dw\, d\xi}{\beta} . \tag{10.19}$$

Here we introduce the notations:

$$\begin{cases} v_1 = \dfrac{(\xi + x)^2 + y^2}{4a^2 t} & w_1 = \dfrac{(\xi - x)^2 + y^2}{4a^2 t} , \\ \alpha = \dfrac{b^2}{4a^2} [(x + \xi)^2 + y^2], & \beta = \dfrac{b^2}{4a^2} [(x - \xi)^2 + y^2] . \end{cases} \tag{10.20}$$

We find the form of the free surface for $t = \infty$, corresponding to steady motion. We have to compute the integral

$$J_1 = \int_0^\infty e^{-v - \frac{\alpha}{v}} dv . \tag{10.21}$$

Substitution

$$v = \sqrt{\alpha}\, e^{-t}$$

reduces the integral to the following

$$J_1 = \sqrt{\alpha} \int_{-\infty}^\infty e^{-2\sqrt{\alpha}\,(\mathrm{ch}\, t) - t}\, dt. \tag{10.22}$$

There exists the following formula for the modified Bessel function of the second kind:

$$K_p(x) = \frac{\pi}{2} \frac{I_{-p}(x) - I_p(x)}{\sin p\pi} = \frac{1}{2} \int_{-\infty}^{+\infty} e^{-x(\mathrm{ch}\, t) - pt} dt . \tag{10.23}$$

By means of (10.23), integral (10.22) is expressed in function of $K_1(2\sqrt{\alpha})$:

$$J_1 = 2\sqrt{\alpha}\, K_1(2\sqrt{\alpha}) = \frac{b}{a} \sqrt{(x + \xi)^2 + y^2}\; K_1\left( \frac{b}{a} \sqrt{(x + \xi)^2 + y^2} \right) .$$

In the same way we find:

$$J_2 = \int_0^\infty e^{-w - \frac{\beta}{w}} dw = \frac{b}{a} \sqrt{(x - \xi)^2 + y^2}\; K_1\left( \frac{b}{a} \sqrt{(x - \xi)^2 + y^2} \right) .$$

Inserting these expressions in formula (10.19) we obtain:

$$H(x,y,\infty) = H_0 + \frac{(H_1 - H_0)y}{\pi} \int_0^\infty \frac{K_1\left(\frac{b}{a}\sqrt{(x+\xi)^2 + y^2}\right)}{\sqrt{(x+\xi)^2+y^2}}\, d\xi + $$

$$+ \frac{(H_2 - H_0)y}{\pi} \int_0^\infty \frac{K_1\left(\frac{b}{a}\sqrt{(x-\xi)^2 + y^2}\right)}{\sqrt{(x-\xi)^2+y^2}}\, d\xi \,.$$

We put in the first of the obtained integrals

$$\xi + x = \zeta \,,$$

and in the second

$$\xi - x = \zeta \,.$$

We rewrite the last equation as

$$H(x,y,\infty) = H_0 + \frac{(H_1 - H_0)y}{\pi} \int_x^\infty \frac{K_1\left(\frac{b}{a}\sqrt{\zeta^2 + y^2}\right)}{\sqrt{\zeta^2+y^2}}\, d\zeta + $$

$$+ \frac{(H_2 - H_0)y}{\pi} \int_0^x \frac{K_1\left(\frac{b}{a}\sqrt{\zeta^2 + y^2}\right)}{\sqrt{\zeta^2+y^2}}\, d\zeta + \frac{(H_2 - H_0)y}{\pi} \int_0^\infty \frac{K_1\left(\frac{b}{a}\sqrt{\zeta^2 + y^2}\right)}{\sqrt{\zeta^2+y^2}}\, d\zeta \tag{10.24}$$

In the particular case $H_1 = H_2$, we obtain a one-dimensional problem of seepage from a canal. The solution depends only on the $y$ coordinate and we have instead of (10.12):

$$\frac{\partial H}{\partial t} = a^2 \frac{\partial^2 H}{\partial y^2} - b^2(H - H_0) \,.$$

Substitution (10.13) reduces this equation to the ordinary heat conduction equation. The solution of the latter, satisfying the conditions $u(y,0) = f(y)$, $u(0,t) = F(t)$, has the form $u(y,t)$

$$u(y,t) =$$

$$= \frac{1}{2a\sqrt{\pi t}} \int_0^\infty f(\lambda)\left[e^{-\frac{(y-\lambda)^2}{4a^2t}} - e^{-\frac{(y+\lambda)^2}{4a^2t}}\right] d\lambda + \frac{2}{\sqrt{\pi}} \int_X^\infty F\left(t - \frac{y^2}{4a^2\lambda^2}\right) e^{-\lambda^2}\, d\lambda \tag{10.25}$$

$$\left(X = \frac{y}{2a\sqrt{t}}\right) .$$

In our case $f(y) = H_0$, $F(t) = e^{b^2t}(H_1 - H_0)$ and (10.25) gives:

$$H(y,t) = H_0 + (H_2 - H_0)\, e^{-b^2 t}\, \Phi\left(\frac{y}{2a\sqrt{t}}\right) +$$

$$+ (H_1 - H_0)\left[\operatorname{ch}\frac{by}{a} - \frac{1}{2}\, e^{\frac{by}{a}}\, \Phi(\xi) + \frac{1}{2}\, e^{-\frac{by}{a}}\, \Phi(\eta)\right]$$

(10.26)

$$\left(\xi = b\sqrt{t} + \frac{y}{2a\sqrt{t}}, \qquad \eta = b\sqrt{t} - \frac{y}{2a\sqrt{t}}\right).$$

To derive formula (10.26), use was made of the integral

$$\frac{2}{\sqrt{\pi}} \int_m^\infty e^{-\lambda^2 - \frac{\alpha^2}{\lambda^2}}\, d\lambda = \frac{1}{2}\, e^{-2\alpha}\left[1 - \Phi\left(m - \frac{\alpha}{m}\right)\right] + \frac{1}{2}\, e^{2\alpha}\left[1 - \Phi\left(m + \frac{\alpha}{m}\right)\right].$$

For $t = \infty$, steady flow, we obtain from (10.26) the simple form of the free surface: $H(y,\infty) = H_0 + (H_1 - H_0)\, e^{-by/a}$.

It may also be obtained from equation (10.24) for $H_1 = H_2$.

## §11. Unsteady Flow to a Well in Confined Stratum.

We write equation (10.1) in polar coordinates $r$, $\theta$, where

$$x = r\cos\theta, \quad y = r\sin\theta.$$

We have:

$$\frac{\partial H}{\partial t} = a^2\left(\frac{\partial^2 H}{\partial r^2} + \frac{1}{r}\frac{\partial H}{\partial r} + \frac{1}{r^2 \sin^2\theta}\frac{\partial^2 H}{\partial \theta^2}\right). \tag{11.1}$$

For axisymmetrical flow, this equation becomes

$$\frac{\partial H}{\partial t} = a^2\left(\frac{\partial^2 H}{\partial r^2} + \frac{1}{r}\frac{\partial H}{\partial r}\right). \tag{11.2}$$

We may look for a solution of this equation, depending only on a combination of the variables $r$ and $t$:

$$\eta = \frac{r^2}{4a^2 t}.$$

Passing to the variable $\eta$, instead of (11.2) we obtain

$$\eta\frac{d^2 H}{d\eta^2} + (1 + \eta)\frac{dH}{d\eta} = 0. \tag{11.3}$$

The general integral of this equation has the form

$$H = C_1 \int \frac{e^{-\eta}}{\eta}\, d\eta + C_2.$$

Introducing the notation of the exponential integral

$$\operatorname{Ei}\left(-\frac{r^2}{4a^2 t}\right) = -\int_{\frac{r^2}{4a^2 t}}^{\infty} \frac{e^{-\eta}}{\eta}\, d\eta, \tag{11.4}$$

we may write the obtained solution as:

$$H = C \, \mathrm{Ei}\left(-\frac{r^2}{4a^2t}\right) + C_2 \quad .$$

We consider a well with radius $r_0$ and compute the flowrate $Q$ per unit length of this well. It is equal to the velocity, multiplied by the perimeter of radius $r_0$:

$$Q = -2\pi r_0 k \left.\frac{\partial H}{\partial r}\right|_{r=r_0} = 4\pi k C e^{-\frac{r_0^2}{4a^2t}} \quad .$$

Putting

$$4\pi k C = Q_0 \; ,$$

we obtain:

$$Q = Q_0 e^{-\frac{r_0^2}{4a^2t}} \tag{11.5}$$

and

$$H(r,t) = -\frac{Q_0}{4\pi} \int_{\frac{r^2}{4a^2t}}^{\infty} \frac{e^{-\eta}}{\eta} \, d\eta + C \; . \tag{11.6}$$

We see that initially the well discharge is zero, that it increases extremely rapidly and tends to a constant value $Q_0$. In numerous cases of unsteady well flow in oilbearing strata [12], it was found practical to assume that the well discharge, determined by (11.5) is constant. The head, determined by (11.6) tends to infinity for $t \to \infty$.

V. N. Tsjelkachev applied the above formulas to the following problem. An oilwell in operation was stopped, so that the pressure in it started to rise. The results of the observations were compared with the results of the computations. Values of the pressure rise $S$ in centimeters of water, computed by means of (11.6) are laid out along the ordinate axis of Fig. 402. Here coefficient $a^2$ has not the same value as the one we considered usually (we have $a^2 = kh/m$ in confined flow) — this quantity takes into account the compressibility of the fluid and the elasticity of the stratum. Its value is determined from observations.

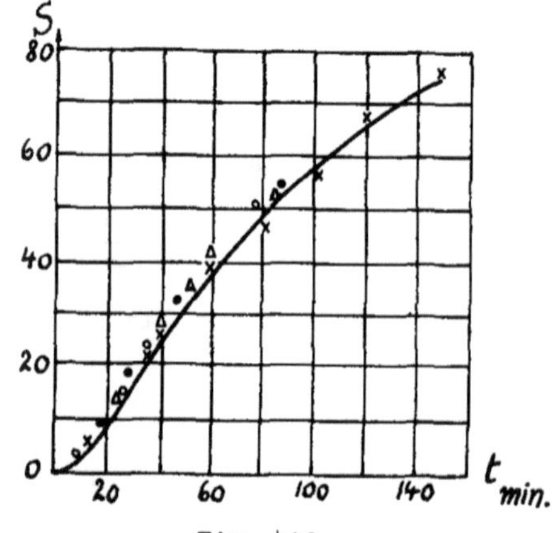


Fig. 402

In the case of a series of wells, we may add the velocity potentials of the individual wells. In the analysis of recharge wells and discharging wells, potentials containing

$+Q_0$ and $-Q_0$ are added. If we pass from confined to unconfined flow, then we may similarly obtain an approximate solution for the problem of a well near a river.

## §12. Irrigation of a Circular Area.

We examined before the problem of spreading on a strip shaped area. We communicate here formulas which allow us to solve a similar problem, when infiltration occurs in a circular area of radius $R$. The solution of the non-homogeneous equation of heat conduction

$$\frac{\partial H}{\partial t} = a^2\left(\frac{\partial^2 H}{\partial r^2} + \frac{1}{r}\frac{\partial H}{\partial r}\right) - b^2 H + f(r,t) \tag{12.1}$$

for the initial condition $H(r,0) = 0$ is given as [13]

$$H(r,t) = \int_0^t \frac{d\tau e^{-b^2(t-\tau)}}{2a^2(t-\tau)} \int_0^\infty f(\rho,\tau) e^{-\frac{r^2+\rho^2}{4a^2(t-\tau)}} I_0\left(\frac{r\rho}{2a^2(t-\tau)}\right)\rho\, d\rho \tag{12.2}$$

($I_0(x)$ is the cylindric function with complex argument). To derive the last formula we may start from the solution of the equation

$$\frac{\partial u}{\partial t} = a^2\left(\frac{\partial^2 u}{\partial x^2} + \frac{\partial^2 u}{\partial y^2}\right) + f(x, y, t) ,$$

corresponding to the initial condition $u(x, y, 0) = 0$:

$$u(x,y,t) = \frac{1}{4\pi a^2} \int_0^t d\tau \int_{-\infty}^{\infty}\int_{-\infty}^{\infty} f(\xi,\eta,\tau) e^{-\frac{(x-\xi)^2+(y-\eta)^2}{4a^2(t-\tau)}} d\xi\, d\eta . \tag{12.3}$$

Putting $x = r\cos\theta$, $y = r\sin\theta$, $\xi = \rho\sin\theta_0$ and considering that

$$\int_0^{2\pi} e^{\frac{r\rho\cos(\theta-\theta_0)}{2a^2(t-\tau)}} d\theta = 2\pi I_0\left(\frac{r\rho}{2a^2(t-\tau)}\right) , \quad .$$

we obtain formula (12.2).

If $f(r,t) = w/m$ for $r \leq R$ and $f(r,t) = 0$ for $r > R$, then, expressing the multiplier of $f(\rho,\tau)$ in (12.2) by means of the Weber-Sonine formula and completing two integrations in the obtained triple integral, we find:

$$H(r,t) = \frac{w}{mR}\left[\int_0^\infty \frac{I_0(\lambda r)\, I_1(\lambda R)}{b^2 + a^2\lambda^2} d\lambda - e^{-b^2 t}\int_0^\infty \frac{I_1(\lambda R)\, I_0(\lambda r)\, e^{-b^2 t\lambda}}{b^2 + a^2\lambda^2} d\lambda\right] . \tag{12.4}$$

## CHAPTER XV.

## TWO-DIMENSIONAL UNSTEADY GROUNDWATER FLOW.

### §1. Introduction.

Up till now we studied unsteady groundwater flow with slightly varying free surface, taking place in a stratum of finite thickness. Such flows were considered to be "hydraulic", i.e., the horizontal velocities did not vary with depth (vertical velocities were considered to be small) and therefor the number of independent variables was reduced to one.

In cases where one has to take into account the form of the boundaries of the flow region, when there are singularities in the flow region – such as drainage pipes, – or when the flow proceeds to a great depth, etc., the above method to solve the problem is not sufficient.

Here we present an account of the basic results available at present in the study of two-dimensional unsteady groundwater flow in a vertical plane. Investigations of this kind started with a problem that was called "the problem of the migration of oil in hydrostatic environment. Paragraph 7 of this chapter is devoted to it.

### §2. Conditions at the Free Surface.

In Chapter XII, we came to the conclusion that we may neglect the inertia terms in unsteady groundwater flow and assume that the seepage velocity vector is derived from a potential $\varphi(x,y,t)$.

$$\vec{v} = \operatorname{grad} \varphi, \tag{2.1}$$

satisfying Laplace's equation

$$\frac{\partial^2\varphi}{\partial x^2} + \frac{\partial^2\varphi}{\partial y^2} = 0 . \tag{2.2}$$

Function $\varphi$ is related to the pressure by means of

$$\varphi = -k\left(\frac{p}{\rho g} + y\right) + C \tag{2.3}$$

The y-axis is oriented upward.

We see that $\varphi(x,y,t)$ contains time $t$ only has a parameter. In unsteady flow with free surface, the dependence of $\varphi$ on time comes from the condition at the free surface, which we now derive.

Assuming the pressure at the free surface to be constant, from (2.3) we obtain the following conditions at the free surface (take $C = 0$):

$$\varphi + ky = 0 . \tag{2.4}$$

We assume that at the free surface $L$ of the groundwater flow, some function remains zero at all times:

$$F(x, y, t) = 0 . \tag{2.5}$$

If at the moment $t + \Delta t$ line $L$ moves on to position $L'$, then some point $M(x, y)$ of line $L$ moves into the position of point $M'(x + \Delta x, y + \Delta y)$ of line $L'$. (We may consider $L$ as a boundary line, in this case a "stationary" one [1].) We obtain:

$$F(x + \Delta x, y + \Delta y, t + \Delta t) = 0 . \tag{2.6}$$

Subtracting (2.5) from (2.6), dividing by $\Delta t$ and passing to the limit for $\Delta t = 0$, we obtain the equation

$$\frac{\partial F}{\partial t} + \frac{\partial F}{\partial x}\frac{dx}{dt} + \frac{\partial F}{\partial y}\frac{dy}{dt} = 0 . \tag{2.7}$$

By means of (2.1) we have

$$m\frac{dx}{dt} = \frac{\partial \varphi}{\partial x}, \quad m\frac{dy}{dt} = \frac{\partial \varphi}{\partial y} \tag{2.8}$$

($m$ is the active porosity of the soil). Inserting these expressions in (2.7), we find the condition at the free surface

$$m\frac{\partial F}{\partial t} + \frac{\partial F}{\partial x}\frac{\partial \varphi}{\partial x} + \frac{\partial F}{\partial y}\frac{\partial \varphi}{\partial y} = 0 . \tag{2.9}$$

In particular, assuming

$$F = \varphi + ky$$

and considering that

$$\frac{\partial F}{\partial t} = \frac{\partial \varphi}{\partial t}, \quad \frac{\partial F}{\partial x} = \frac{\partial \varphi}{\partial x}, \quad \frac{\partial F}{\partial y} = \frac{\partial \varphi}{\partial y} + k ,$$

we find the equation for $\varphi$

$$m\frac{\partial \varphi}{\partial t} + \left(\frac{\partial \varphi}{\partial x}\right)^2 + \left(\frac{\partial \varphi}{\partial y}\right)^2 + k\frac{\partial \varphi}{\partial y} = 0 , \tag{2.10}$$

which must be satisfied at the free surface [2].

We transform this condition into complex variables [2]. We consider simultaneously the complex potential $\omega = \varphi + i\psi$ and Zhoukovsky's function

$$\theta = \omega - ikz = \varphi + ky + i(\psi - kx) . \tag{2.11}$$

Rewriting equation (2.10) in the form

$$\left(\frac{\partial \varphi}{\partial x}\right)^2 + \frac{\partial \varphi}{\partial y}\left(k + \frac{\partial \varphi}{\partial y}\right) + m\frac{\partial \varphi}{\partial t} = 0 ,$$

it is easy to show that it may be replaced by the following equation:

$$\mathrm{Re}\left(\frac{\partial \omega}{\partial z}\overline{\frac{\partial \theta}{\partial z}}\right) + m\,\mathrm{Re}\,\frac{\partial \theta(z,t)}{\partial t} = 0 ,$$

which may also be written as:

$$\frac{\partial \omega}{\partial z}\overline{\frac{\partial \theta}{\partial z}} + \overline{\frac{\partial \omega}{\partial z}}\frac{\partial \theta}{\partial z} + m\left[\frac{\partial \theta(z,t)}{\partial t} + \frac{\partial \overline{\theta(z,t)}}{\partial t}\right] = 0 . \tag{2.12}$$

Here the dash above the quantities indicates complex conjugates.

We assume that the region occupied by the flowing fluid is mapped conformally onto the half-plane of the auxiliary complex variable $\zeta = \xi + i\eta$ such that the free surface always corresponds to the real axis. Designating the mapping function respectively by $z = z(\zeta, t)$ and $\zeta = \zeta(z, t)$ and passing to the variables $\zeta$ and $t$, we have:

$$\frac{\partial\theta(z,t)}{\partial t} = \frac{\partial\theta(\zeta,t)}{\partial t} + \frac{\partial\theta}{\partial\zeta}\dot{\zeta} \qquad \left(\dot{\zeta} = \frac{\partial\zeta(z,t)}{\partial t}\right). \tag{2.13}$$

Observing then that at the free surface for $\zeta = \xi$

$$\theta(\zeta,t) + \overline{\theta(\zeta,t)} = 0, \qquad \frac{\partial\theta(\zeta,t)}{\partial t} + \frac{\overline{\partial\theta(\zeta,t)}}{\partial t} = 0\ , \tag{2.14}$$

we write condition (2.12) in the form

$$\frac{\partial\omega}{\partial\zeta}\,\frac{\overline{\partial\theta}}{\partial\zeta}\,\frac{\partial\zeta}{\partial z}\,\frac{\overline{\partial\zeta}}{\partial z} + \frac{\overline{\partial\omega}}{\partial\zeta}\,\frac{\partial\theta}{\partial\zeta}\,\frac{\partial\zeta}{\partial z}\,\frac{\overline{\partial\zeta}}{\partial z} + m\left(\frac{\partial\theta}{\partial\zeta}\dot{\zeta} + \frac{\overline{\partial\theta}}{\partial\zeta}\bar{\dot{\zeta}}\right) = 0\ .$$

Because for differentiation along the real variable $\zeta = \xi$ the equality holds

$$\frac{\partial\theta(\zeta,t)}{\partial\zeta} + \frac{\overline{\partial\theta(\zeta,t)}}{\partial\zeta} = 0$$

and since

$$\frac{\partial\zeta}{\partial z} = 1 : \frac{\partial z}{\partial\zeta}\ , \qquad \frac{\overline{\partial\zeta}}{\partial z} = 1 : \frac{\overline{\partial z}}{\partial\zeta}\ ,$$

we obtain

$$\frac{\overline{\partial\omega}}{\partial\zeta} - \frac{\partial\omega}{\partial\zeta} + m\,\frac{\partial z}{\partial\zeta}\,\frac{\overline{\partial z}}{\partial\zeta}\,(\dot{\zeta} - \bar{\dot{\zeta}}) = 0\ . \tag{2.15}$$

Further, if in equation $z = f(\zeta, t)$ we insert $\zeta = \zeta(z, t)$ then we obtain the identity $z = f[\zeta(z,t), t]$. Partial differentiation of this identity with respect to $t$ gives:

$$\frac{\partial z}{\partial t} + \frac{\partial z}{\partial\zeta}\dot{\zeta} = 0\ .$$

From which

$$\dot{\zeta} = -\frac{\partial z}{\partial t} : \frac{\partial z}{\partial\zeta}\ , \tag{2.16}$$

and this allows to write (2.15) in the form

$$\frac{\overline{\partial\omega}}{\partial\zeta} - \frac{\partial\omega}{\partial\zeta} - m\left(\frac{\partial z}{\partial t}\,\frac{\overline{\partial z}}{\partial\zeta} - \frac{\overline{\partial z}}{\partial t}\,\frac{\partial z}{\partial\zeta}\right) = 0\ , \tag{2.17}$$

or in other notations:

$$m\ \mathrm{Im}\left(\frac{\partial z}{\partial t}\,\frac{\partial\bar{z}}{\partial\zeta}\right) = -\ \mathrm{Im}\left(\frac{\partial\omega}{\partial\zeta}\right)\ . \tag{2.18}$$

If we go back to Zhoukovsky's function, then the equations at the free surface will be

$$\mathrm{Im}\left(i\frac{\partial\theta}{\partial\zeta}\right) = 0, \quad \mathrm{Im}\left(\frac{\partial z}{\partial t}\,\frac{\partial\bar{z}}{\partial\zeta}\right) = -\frac{1}{m}\,\mathrm{Im}\left(\frac{\partial\theta}{\partial\zeta}\right) + \frac{k}{m}\,\mathrm{Im}\left(i\,\frac{\partial z}{\partial\zeta}\right)\ . \tag{2.19}$$

The expression of functions $\theta(\zeta)$ or $\omega(\zeta)$ is known for various groups of problems, and this allows us to add equations (2.18) and (2.19) to the condition for function $z(\zeta,t)$ at the free surface.

§3. Problem of the Damping of a Groundwater Mound in the Half-plane.

After rainfall, after cessation of spreading, etc., the groundwater surface assumes the shape of a mound, which later on spreads out. A problem of this kind, about the deformation of the free surface, with equation given at the initial moment, and for a flowregion extending down to infinity is the simplest problem with boundary condition (2.19).

Here Zhoukovsky's function is simply proportional to $\zeta$, namely $\theta = -ki\zeta$. Indeed we recall that $\theta$ is proportional to the complex pressure:

$$\theta = -\frac{k}{\rho g}(p + ip') \ . \tag{3.1}$$

For $\eta = 0$, we have $p = 0$, $\mathrm{Re}\,\theta = 0$; at infinity the pressure behaves as in the case of an unperturbed surface, i.e., as $\rho g \eta$. Therefore

$$\theta = -ki\zeta. \tag{3.2}$$

Therefore we consider that the function, mapping the flow region onto the half-plane $\zeta$, has the form

$$z = \zeta + \Phi(\zeta,t) \tag{3.3}$$

where $\Phi(\zeta,t)$ is holomorphic in the lower half plane.

Inserting (3.2) in the second equation (2.19), we obtain the condition at the free surface for $z(\zeta,t)$:

$$\mathrm{Im}\left(\frac{\partial z}{\partial t}\frac{\partial \bar z}{\partial \zeta} - \frac{ki}{m}\frac{\partial z}{\partial \zeta}\right) = \frac{k}{m} \quad \text{for} \quad \zeta = \xi \ . \tag{3.4}$$

Condition (3.4) may be written somewhat differently:

$$\mathrm{Re}\left(\frac{k}{m}\frac{\partial z}{\partial \zeta} + i\,\frac{\partial z}{\partial \zeta}\frac{\partial \bar z}{\partial t}\right) = \frac{k}{m} \quad \text{for} \quad \zeta = \xi \ . \tag{3.5}$$

Now it is easy to obtain a condition for the complex potential $\omega$ at the free surface. Because of (2.11) and (3.2) we have:

$$z(\zeta,t) = \zeta - \frac{1}{k}\,\omega(\zeta,t) \ , \tag{3.6}$$

so that

$$\frac{\partial z}{\partial \zeta} = 1 - \frac{1}{k}\frac{\partial \omega}{\partial \zeta} \ , \quad \frac{\partial \bar z}{\partial t} = \frac{1}{k}\frac{\partial \bar\omega}{\partial t} \ .$$

Inserting these expressions in (3.5), after some simplifications, we find

$$\mathrm{Re}\left(\frac{\partial \bar\omega}{\partial t} + \frac{ki}{m}\frac{\partial \omega}{\partial \zeta}\right) = \mathrm{Re}\left(\frac{i}{k}\frac{\partial \bar\omega}{\partial t}\frac{\partial \omega}{\partial \zeta}\right) \quad \text{for} \quad \zeta = \xi \ . \tag{3.7}$$

On the left side we have a linear expression in derivatives, on the right side a product of derivatives. If we discard the right part, then condition (3.7) is significantly simplified. To estimate the possibility of such simplification and to obtain more accurate solutions,

L. A. Galin [4] proposed to apply the method of successive approximations in the following form.

We take the first approximation $\omega^{(1)}(\zeta,t)$ satisfying the linear condition

$$\operatorname{Re}\left(\frac{\partial\omega^{(1)}}{\partial t}+\frac{ki}{m}\frac{\partial\omega^{(1)}}{\partial\zeta}\right)_{\zeta=\xi}=0 \tag{3.8}$$

and the initial condition ($\omega_0(\zeta)$ is a given function)

$$\omega^{(1)}(\zeta,0)=\omega_0(\zeta)\ . \tag{3.9}$$

Having found $\omega^{(1)}(\zeta,t)$ we insert it in the right part of (3.7) and consider that $\omega^{(2)}(\zeta,t)$ satisfies the boundary condition

$$\operatorname{Re}\left(\frac{\partial\omega^{(2)}}{\partial t}+\frac{ki}{m}\frac{\partial\omega^{(2)}}{\partial\zeta}\right)=\operatorname{Re}\left(\frac{i}{k}\frac{\partial\bar{\omega}^{(1)}}{\partial t}\frac{\partial\omega^{(1)}}{\partial\zeta}\right)\quad\text{for}\quad\zeta=\xi \tag{3.10}$$

and the initial condition

$$\omega^{(2)}(\zeta,0)=\omega_0(\zeta)\ .$$

This process may be continued.

We consider an example. Let at $t = 0$ groundwater occupy a region which is mapped onto a half-plane by means of the function

$$z_0(\zeta)=\zeta+\frac{\alpha}{\zeta-i}\ , \tag{3.11}$$

so that for $\zeta = \xi$ we have the equations of the free surface (in Fig. 403 this is the curve $\tau = 0$)

$$x_0=\xi+\frac{\alpha\xi}{\xi^2+1}\ ,\quad y_0=\frac{\alpha}{\xi^2+1}\ .$$

Then, because of (3.6), we have for the initial value $\omega(\zeta,0)=\omega_0$

$$\omega_0(\zeta)=\frac{ki\alpha}{\zeta-i}\ . \tag{3.12}$$

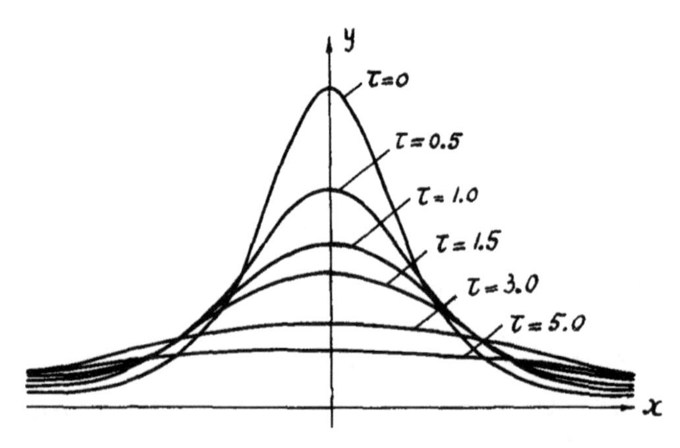


Fig. 403

Having found the first approximation $\omega^{(1)}$ by means of condition

(3.8), we notice that if we have a function $F(\zeta)$ regular in the lower half-plane, for which the real part is zero on the real axis: $\operatorname{Re} F(\zeta) = 0$ for $\zeta = \xi$, then this function has a constant value in the entire half-plane.

It follows from (3.8) and the above observation, that

$$\frac{\partial \omega^{(1)}}{\partial t} + \frac{ki}{m}\frac{\partial \omega^{(1)}}{\partial \zeta} = 0$$

in the entire lower half-plane. From this it follows that $\omega^{(1)}$ is a function of one argument $\zeta - \frac{ki}{m}t$:

$$\omega^{(1)} = \Phi\left(\zeta - \frac{kit}{m}\right) .$$

And since for $t = 0$

$$\omega^{(1)}(\zeta, 0) = \Phi(\zeta) = \omega_0(\zeta) ,$$

we find at once

$$\omega^{(1)}(\zeta, t) = \omega_0\left(\zeta - \frac{kit}{m}\right) .$$

In our case

$$\omega^{(1)}(\zeta, t) = \frac{ki\alpha}{\zeta - i\left(1 + \frac{kt}{m}\right)} \tag{3.13}$$

and consequently, because of (3.6)

$$z^{(1)}(\zeta, t) = \zeta + \frac{\alpha}{\zeta - i\left(1 + \frac{kt}{m}\right)} \tag{3.14}$$

A family of curves is constructed in Fig. 403, rendering the location of the groundwater mound at different times. Here $\alpha = \frac{1}{2}$, $\tau = \frac{kt}{m}$ were assumed.

To pass to the second approximation, we compose the expression

$$\operatorname{Re}\left[\frac{i}{k}\frac{\partial \bar{\omega}^{(1)}}{\partial t}\frac{\partial \omega^{(1)}}{\partial \zeta}\right]_{\zeta=\xi} = \operatorname{Re}\left[\frac{i}{k}\,\omega_0'\left(\xi + \frac{kit}{m}\right)\frac{ki}{m}\omega_0'\left(\xi - \frac{kit}{m}\right)\right]$$

$$= -\frac{1}{m}\left|\omega_0'\left(\xi - \frac{kit}{m}\right)\right|^2 , \tag{3.15}$$

and condition (3.10) for $\omega^{(2)}$ becomes:

$$\operatorname{Re}\left[\frac{\partial \omega^{(2)}}{\partial t} + \frac{ki}{m}\frac{\partial \omega^{(2)}}{\partial \zeta}\right]_{\zeta=\xi} = -\frac{1}{m}\frac{k^2\alpha^2}{\left[\xi^2 + \left(1 + \frac{kt}{m}\right)^2\right]^2} .$$

Applying formula (1.6) of Chapter VI to function

$$F(\zeta, t) = \frac{\partial \omega^{(2)}}{\partial t} + \frac{ki}{m}\frac{\partial \omega^{(2)}}{\partial \zeta}$$

and considering that $F(\zeta,t) = 0$ for $\zeta = \infty$, we find

$$F(\zeta,t) = -\frac{1}{m\pi i}\int_{-\infty}^{\infty}\left|\omega_0'\left(\xi - \frac{kit}{m}\right)\right|^2 \frac{d\xi}{\xi - \zeta} ,$$

and this gives:

$$\frac{\partial\omega^{(2)}}{\partial t} + \frac{ki}{m}\frac{\partial\omega^{(2)}}{\partial\zeta} =$$

$$= \frac{k^2\alpha^2}{2m\left(1 + \frac{kt}{m}\right)^2\left(\zeta - i - \frac{ki}{m}t\right)^2} + \frac{k^2\alpha^2 i}{2m\left(1 + \frac{kt}{m}\right)^3\left(\zeta - i - \frac{kit}{m}\right)} . \tag{3.16}$$

We notice that we may find expression (3.16) not by means of integration, but by means of a method which we shall apply further (§5) in the solution of a problem about a drainage pipe. In order to find a solution of equation (3.16), L. A. Galin examined the equation of the form

$$\frac{\partial U}{\partial t} + \beta\frac{\partial U}{\partial\zeta} = f(t)\Phi(\beta t - \zeta) ,$$

for which he indicated a particular solution of the form

$$U(t,\zeta) = \Phi(\beta t - \zeta)\int_0^t f(\tau)\, d\tau .$$

This expression becomes zero for $t = 0$. Adding to it the solution of the homogeneous equation

$$\frac{\partial\omega^{(2)}}{\partial t} + \frac{ki}{m}\frac{\partial\omega^{(2)}}{\partial\zeta} = 0 ,$$

satisfying the condition $\omega^{(2)}(\zeta,0) = \omega_0(\zeta)$ and, consequently, coinciding with the one we already found for $\omega^{(1)}(\zeta,t)$, we obtain the second approxi-

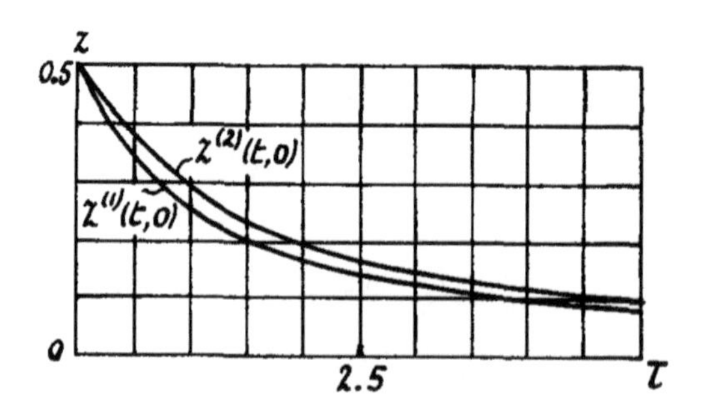


Fig. 404

mation for $\omega$:

$$\omega^{(2)}(\zeta,t) = \frac{ki\alpha}{\zeta - i\left(1 + \frac{kt}{m}\right)} + \alpha^2 k\left\{\frac{1}{2}\left(1 - \frac{1}{1 + \frac{kt}{m}}\right)\frac{1}{\left[\zeta - \left(1 + \frac{kt}{m}\right)i\right]^2} + \right.$$

$$\left. + \frac{1}{4}\left(1 - \frac{1}{\left(1 + \frac{kt}{m}\right)^2}\right)\frac{i}{\zeta - \left(1 + \frac{kt}{m}\right)i}\right\} . \tag{3.17}$$

The second approximation for $z$ is:

$$z^{(2)}(\zeta,t) = \zeta - \frac{i}{k}\,\omega^{(2)}(\zeta,t) = \zeta + \left[\alpha + \frac{\alpha^2}{4}\left(1 - \frac{1}{\left(1 + \frac{kt}{m}\right)^2}\right]\times\right.$$

$$\times\frac{1}{\zeta - \left(1 + \frac{kt}{m}\right)i} - \frac{\alpha^2}{2}\left(1 - \frac{1}{1 + \frac{kt}{m}}\right)\frac{i}{\left[\zeta - \left(1 + \frac{kt}{m}\right)i\right]^2} .$$

We may construct graphs for the dependence on time of the maximum ordinates of the free surface, computed by the first and second approximations:

$$z^{(1)}(0,t) = \frac{\alpha i}{1 + \frac{kt}{m}} ,$$

$$z^{(2)}(0,t) = \frac{\alpha i}{1 + \frac{kt}{m}} + \alpha^2 i\left[\frac{1}{4\left(1 + \frac{kt}{m}\right)} + \frac{1}{2\left(1 + \frac{kt}{m}\right)^2} - \frac{3}{4\left(1 + \frac{kt}{m}\right)^3}\right] .$$

They are represented in Fig. 404, for which $\alpha = \frac{1}{2}$, $\tau = \frac{kt}{m}$ .

§4. Same Problem for Linearized Condition.

If we neglect terms $(\partial\varphi/\partial x)^2$, $(\partial\varphi/\partial y)^2$ in equation (2.10), we obtain a linear equation holding for the free surface

$$(4.1) \qquad \frac{\partial\varphi}{\partial t} + c\,\frac{\partial\varphi}{\partial y} = 0 \quad \left(c = \frac{k}{m}\right) .$$

Assuming that the free surface is slightly curved, we transfer the boundary condition to the abscissa-axis. Therefore we determine the equation of the free surface from condition (2.4); we rewrite it in the form

$$y = -\frac{\varphi(x,y,t)}{k}$$

and make $y = 0$ in its right side, to have the equation of the free surface finally as

$$(4.2) \qquad y = -\frac{\varphi(x,0,t)}{k} .$$

The assumption here is similar to that made in the hydrodynamics of shallow waves.

We analyze the problem: groundwater occupies the lower half-plane $y < 0$. Initially the free surface has the form of a curve with given equation

$$(4.3) \qquad y = f(x)$$

Comparing (4.2) and (4.3), we find the boundary condition for $\varphi(x,y,t)$ for $t = 0$

$$(4.4) \qquad \varphi(x,0,0) = -kf(x) .$$

Knowing the value of an harmonic function on the real axis and given the condition that $\varphi$ is zero at infinity, we may construct the initial value of the function $\omega = \varphi + i\psi$, by means of the integral

$$\omega(z,0) = -\frac{k}{\pi i}\int_{-\infty}^{\infty}\frac{f(\xi)\,d\xi}{\xi - z} \quad . \tag{4.5}$$

We make a remark concerning equality (4.1). Since function $\varphi$ is harmonic, so is the function

$$\Phi(x,y) = \frac{\partial\varphi}{\partial t} + c\,\frac{\partial\varphi}{\partial y}$$

in the lower half-plane. From the condition that the value of $\Phi(x,y)$ is zero at the x-axis and $\Phi(x,y)$ finite at infinity, follows that $\Phi(x,y)$ is zero in the entire half-plane, i.e., equation (4.1) holds in the entire half-plane. From this follows that $\varphi(x,y,t)$ depends on the linear combination of $y$ and $t$

$$\varphi(x,y,t) = \varphi(x,y - ct) \ .$$

The same applies to the complex potential (see a similar remark in §3.). Therefore, to find $\omega(z,t)$, it is sufficient to replace $z = x + iy$ by $x + i(y - ct)$ in the expression obtained for $\omega(z)$. We obtain:

$$\omega(x,y,t) = \varphi + i\psi = -\frac{k}{\pi i}\int_{-\infty}^{\infty}\frac{f(\xi)\,d\xi}{\xi - x - i(y - ct)} \quad . \tag{4.6}$$

Separating real and imaginary parts, we find $\varphi(x,y,t)$. Making $y = 0$ in it, we find by means of (4.2) the equation of the free surface

$$y(x,t) = \frac{1}{\pi}\int_{-\infty}^{\infty}\frac{ctf(\xi)\,d\xi}{(\xi - x)^2 + c^2t^2} \quad . \tag{4.7}$$

The same results may be obtained in another way [5], giving at once a visual geometric flow net.

First we examine the case where an initial perturbation is concentrated in an infinitely small circle around point $x = 0$, $y = 0$. Let $Q$ be the magnitude of the area, comprised between the profile of the initial disturbance of the free surface and the x-axis.

Following N. E. Kochin [6], we compose the dimensionless expression

$$\frac{\varphi(0,y,t)}{Q}\cdot\frac{y}{c} = U(u) \ ,$$

which must be a function $U$ of the dimensionless variable

$$u = \frac{ct}{y} \ .$$

Condition (4.1) must not only be satisfied at the boundary, but in the entire flowregion, because the function

$$\frac{\partial\varphi}{\partial t} + c\,\frac{\partial\varphi}{\partial y}$$

satisfies Laplace's equation (4.1). Inserting equation

$$\varphi(0,y,t) = \frac{cQ}{y}\,U(u)$$

in equation (4.1), we obtain the equation for $U$,

$$U'(1 - u) = U .$$

The general solution of this equation is

$$U = \frac{A}{1 - u} = \frac{Ay}{y - ct} .$$

where $A$ is an arbitrary constant.

Consequently, for $\varphi(0,y,t)$ we obtain the expression

$$\varphi(0,y,t) = \frac{cQA}{y - ct} .$$

Replacing in it $y$ by $y - ix$, we obtain the complex potential $\omega(z,t)$ of the non-steady flow under consideration

$$\omega(z,t) = \varphi(x,y,t) + i\psi(x,y,t) = \frac{cQAi}{z - ict} .$$

Separating real and imaginary part, we find

$$\varphi(x,y,t) = \frac{cQA(y - ct)}{(y - ct)^2 + x^2} , \quad \psi = \frac{cQAx}{(y - ct)^2 + x^2} .$$

To find the free surface equation, we must make $y = 0$ in the expression for $\varphi(x,y,t)$. We obtain:

$$y = \frac{Qc^2At}{k(c^2t^2 + x^2)} .$$

We determine the arbitrary constant $A$ so that the area between the abscissa axis and the free surface line is equal to $Q$. We find:

$$A = \frac{k}{c\pi} ,$$

and consequently

$$\omega(z,t) = \frac{Qki}{\pi(z - ict)} , \quad \varphi = \frac{kQ(y - ct)}{\pi[(y - ct)^2 + x^2]} . \tag{4.8}$$

The equation of the free surface finally is written as:

$$y = \frac{Qct}{\pi(x^2 + c^2t^2)} .$$

As is apparent from the expression for the complex potential, we have flow with a dipole in point $z = ict$. Initially, the dipole is in the origin of coordinates, next it moves up along the y-axis with velocity $c = k/m$. Thus, at a time close to the initial moment, the dipole hits the flowregion. We find the moment of time when the dipole is at the maximum of the free surface. Therefore one equates the maximum ordinate of the free surface $Q/\pi ct$ with the quantity $ct$. We obtain:

$$t = t_1 = \frac{1}{c}\sqrt{\frac{Q}{\pi}}$$

For $t > t_1$, the flowregion does not contain the singularity any more.

If we take for the initial moment a value of $t_0 > t_1$, then we obtain flow without singularity and with initial form of the free surface

$$y = \frac{Qct_0}{\pi(x^2 + c^2t_0^2)} .$$

Let the initial flowlines be circles (Fig. 405), passing through point $x = 0$, $y = ct$, and the initial equipotential lines be orthogonal to these circles.

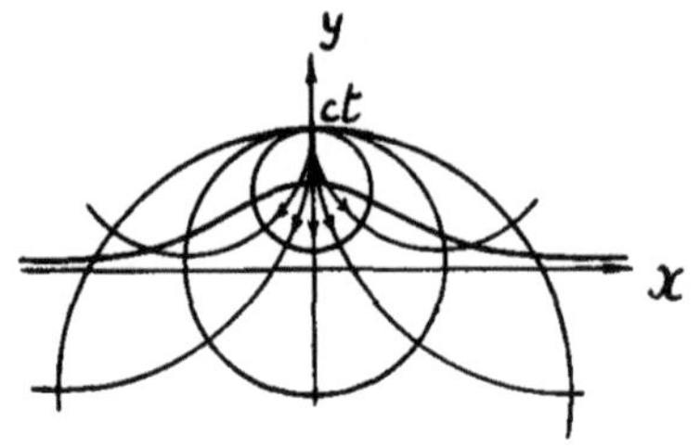

Fig. 405

We now pass to the general case, where the initial form of the free surface is given by the equation

$$y = F(x)$$

We show that we obtain the solution of the problem, i.e., the value of the velocity potential by replacing in (4.8) $Q$ by $F(\zeta)$, $x$ by $x-\zeta$ and by integration along $\zeta$ between limits from $-\infty$ to $\infty$. We have

$$\varphi(x,y,t) = \frac{k}{\pi} \int_{-\infty}^{\infty} \frac{F(\zeta)(y - ct)\, d\zeta}{(y - ct)^2 + (x - \zeta)^2} .$$

Putting $y = 0$ and using the equation of the free surface

$$y = -\frac{\varphi(x,0,t)}{k} ,$$

we find for the form of the free surface at time $t$

$$y(x,t) = \frac{ct}{\pi} \int_{-\infty}^{\infty} \frac{F(\zeta)\, d\zeta}{c^2t^2 + (x - \zeta)^2} .$$

If we make the substitution

$$\zeta - x = ct(\mathrm{tg}\ \alpha) ,$$

then we obtain another expression for $y(x,t)$:

$$y(x,t) = \frac{1}{\pi} \int_{-\pi/2}^{\pi/2} F\Big[x + ct(\mathrm{tg}\ \alpha)\Big]\, d\alpha ,$$

From this expression it is apparent that at $t = 0$, the form of the free surface is reduced to the given initial condition, i.e., the equation

$$y(x,0) = F(x) .$$

We consider for example the case of a mound, rectangular initially, when the initial form of the free surface has the equation (Fig. 406)

$$F(x) = y(x,0) = \begin{cases} \varepsilon & \text{for } |x| < R , \\ 0 & \text{for } |x| > R . \end{cases}$$

The form of the free surface in the following moments of time is determined by the equation

$$y = \frac{\varepsilon}{\pi}\left[\text{arc tg}\,\frac{x + R}{ct} - \text{arc tg}\,\frac{x - R}{ct}\right] .$$

The complex potential of the flow

$$\omega(z,t) = \frac{ik\varepsilon}{\pi}\ \ell n\,\frac{z + R - ict}{z - R - ict}$$

represents plane parallel flow with a pair of vertices in points $(\pm R, ct)$. These point vertices move up parallel to the ordinate axis with speed

$$c = \frac{k}{m} .$$

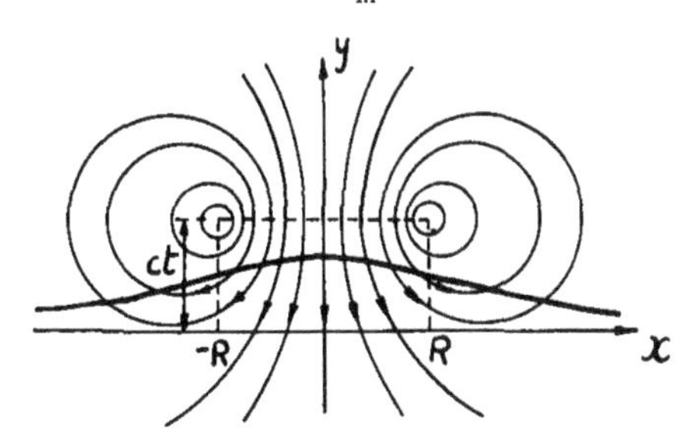


Fig. 406

Initially the singular points $(\pm R,0)$ are at the free surface. They leave it at time $t_1$, determined by

$$\frac{\varepsilon}{\pi}\,\text{arc tg}\,\frac{2R}{ct_1} = ct_1 .$$

If we consider the equation of the free surface at $t > t_1$, then the singular points of the flow are already found outside the region occupied by the flowing fluid.

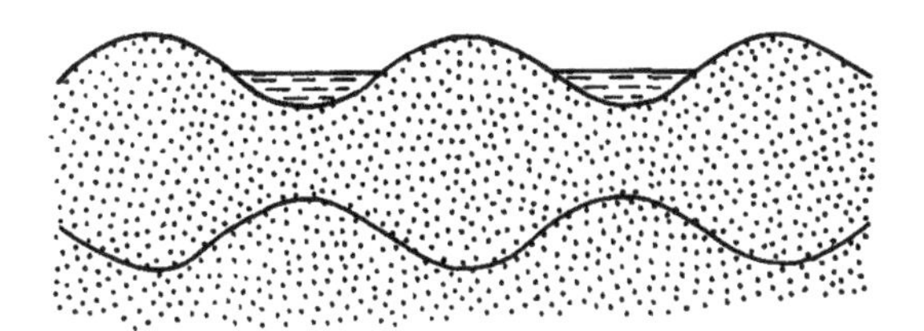

Fig. 407

The initial flowlines are here circles of Apollonius, the initial equipotential lines are orthogonal to them.

Where irrigation is done by means of furrows, the free groundwater surface has a wavy nature. A groundwater mound prevails directly under the furrow (Fig. 407). Therefore we assume as initial form of the free surface.

$$y = \alpha\ .\ \sin\omega x .$$

Then it is easy to obtain the equation of the free surface in time

$$y = ae^{-\omega ct} \sin \omega x \ .$$

The streamlines have the form, shown in Fig. 408.

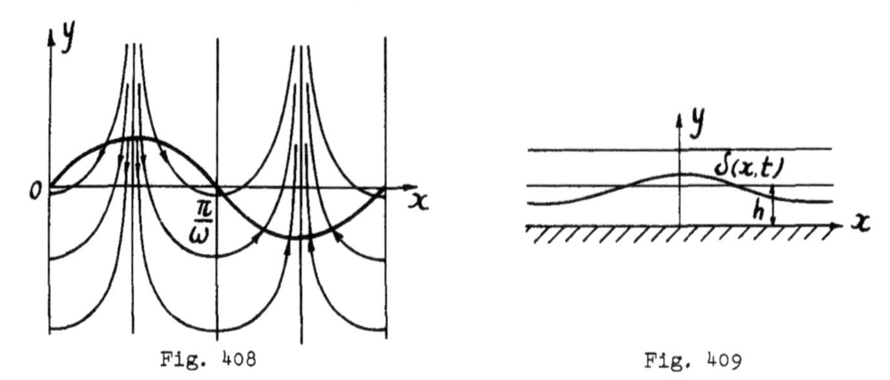


Fig. 408

Fig. 409

§5. Problem of Spreading of a Groundwater Mound in a Layer of Finite Depth.

For the case of Fig. 409, we have the following conditions (in linearized form):

$$(5.1) \qquad \begin{cases} \dfrac{\partial\varphi}{\partial t} + \dfrac{k}{m}\dfrac{\partial\varphi}{\partial y} = 0 & \text{for } y = h \ , \\ \dfrac{\partial\varphi}{\partial y} = 0 & \text{for } y = 0 \ . \end{cases}$$

Moreover, one has to satisfy the initial condition

$$(5.2) \qquad t = 0, \quad y = h, \quad \varphi = -k\delta_0(x) \ .$$

Here $\delta_0(x)$ is the initial groundwater level [4].

We look for an harmonic function $\varphi(x,y,t)$ in the case of a symmetric problem in the form of a Fourier integral

$$(5.3) \qquad \varphi(x,y,t) = \int_0^\infty A(\alpha,t) \operatorname{ch} \alpha y \cos \alpha x \, d\alpha \ .$$

The constructed function satisfies, as is easy to see, the second condition (5.1). To satisfy the first condition (5.1) it is necessary to satisfy the equation

$$(5.4) \qquad \left(\frac{\partial\varphi}{\partial t} + \frac{k}{m}\frac{\partial\varphi}{\partial y}\right)_{y=h} = \int_0^\infty \left[\operatorname{ch} \alpha h \frac{\partial A}{\partial t} + \frac{k\alpha}{m} A \operatorname{sh} \alpha h\right] \cos \alpha x \, d\alpha = 0 \ .$$

From this it follows that function $A(\alpha,t)$ must satisfy the equation

$$(5.5) \qquad \operatorname{ch} \alpha h \frac{\partial A}{\partial t} + \frac{k}{m} A \alpha \operatorname{sh} \alpha h = 0 \ .$$

Since $\varphi(x,h,0) = -k\delta_0(x)$ for $t = 0$, we obtain from (5.3):

$$(5.6) \qquad -k\delta_0(x) = \int_0^\infty A(\alpha,0) \operatorname{ch} \alpha h \cos \alpha x \, d\alpha \ .$$

The solution of (5.5) may be written as

$$A(\alpha,t) = A(\alpha,0) \exp\left\{- \frac{k\alpha \operatorname{th} \alpha h}{m} t\right\}$$

and consequently, for $\varphi(x,y,t)$, we have:

$$\varphi(x,y,t) = \int_0^\infty A(\alpha,0) \operatorname{ch} \alpha y \cos \alpha x \exp\left\{- \frac{k\alpha \operatorname{th} \alpha h}{m}\right\} d\alpha . \tag{5.7}$$

Function $A(\alpha,0)$ is found by taking the inverse of equation (5.6):

$$A(\alpha,0) = - \frac{2k}{\pi \operatorname{ch} \alpha h} \int_0^\infty \delta_0(\beta) \cos \alpha\beta \, d\beta . \tag{5.8}$$

In particular, when $\delta_0(x)$ is equal to the constant $\delta_0$ in the interval $-a < x < a$ and zero outside this interval, we find for $A(\alpha,0)$:

$$A(\alpha,0) = - \frac{2k\delta_0}{\pi \operatorname{ch} \alpha h} \int_0^\infty \cos \alpha\beta \, d\beta = - \frac{2k\delta_0}{\pi \operatorname{ch} \alpha h} \frac{\sin a\alpha}{\alpha} .$$

From this we find the value of the velocity potential

$$\varphi(x,y,t) = - \frac{2k\delta_0}{\pi} \int_0^\infty \frac{\sin \alpha a \operatorname{ch} \alpha y \cos \alpha x}{\alpha \operatorname{ch} \alpha h} \exp\left\{- \frac{k\alpha \operatorname{th} \alpha h}{m} t\right\} d\alpha$$

and of the groundwater level

$$\delta(x,t) = \frac{2\delta_0}{\pi} \int_0^\infty \frac{\sin a\alpha \cos \alpha x}{\alpha} \exp\left\{- \frac{k\alpha \operatorname{th} \alpha h}{m} t\right\} d\alpha .$$

## §6. Groundwater Flow to a Drain in a Stratum of Infinite Thickness.

We analyze the behaviour of the groundwater level under influence of a drainage pipe. Let for $t = 0$ the groundwater level be horizontal and coincide with the x-axis (Fig. 410). Groundwater occupies the entire lower half-plane. We put the center of the drain on the y-axis in the point with ordinate $h_0 i$. After elapse of some time $t$, the groundwater level drops and is represented by the curve BAC (Fig. 410). We map the region below BAC onto the lower half-plane $\zeta$. Point D of the z-plane transforms into a point of the $\zeta$-plane with abscissa $\zeta = hi$, where $h$ is a quantity, changing (decreasing) in time, which we try to find in the form of a series

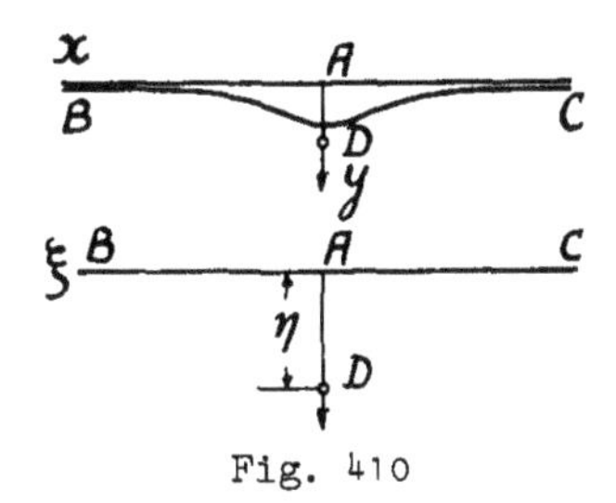

Fig. 410

$$h = h_0 + h_1 t + h_2 t^2 + \dots , \tag{6.1}$$

where $h_1, h_2, \dots$ are constants, subject to determination.

We look for a mapping function in the form of a series in powers of $t$

$$z = \zeta + z_1(\zeta) t + z_2(\zeta) t^2 + \dots , \tag{6.2}$$

where $z_1, z_2, \ldots$ — are unknown functions, regular in the half-plane $\eta > 0$, vanishing at infinity and satisfying boundary conditions, obtained from (2.19). Thus we consider a non-linear condition at the free surface.

We construct the function occurring in (2.19). Based on the equalities

$$\left(\frac{\partial z}{\partial \zeta}\right)_{\zeta=\infty} = 1, \quad \left(\frac{\partial \omega}{\partial \zeta}\right)_{\zeta=\infty} = 0, \quad \frac{\partial \theta}{\partial \zeta} = \frac{\partial \omega}{\partial \zeta} + ki\,\frac{\partial z}{\partial \zeta} \tag{6.3}$$

we find

$$\left(\frac{\partial \theta}{\partial \zeta}\right)_{\zeta=\infty} = ki\ .$$

Moreover, in the point $\zeta = hi$ this function has a pole of the first order of the form $-q/2\pi(\zeta - ih)$, where $q$ is the flowrate of the drain, i.e., the amount of fluid absorbed by the drain per unit length. This quantity depends on the difference in pressure at the free surface and at the circumference of the drain and on the radius of the drain. It is a function of time which we may try to find in the form

$$q = q_0 + q_1 t + q_2 t^2 + \ldots\ , \tag{6.4}$$

where $q_0, q_1, \ldots$ are constants subject to determination. Taking into account the condition for $\partial\theta/\partial\zeta$, we find

$$\frac{\partial \theta}{\partial \zeta} = ki - \frac{q}{2\pi}\left(\frac{1}{\zeta - ih} - \frac{1}{\zeta + ih}\right)\ . \tag{6.5}$$

We now come to boundary condition (2.19). We bring its right part in a form which contains only regular functions in the half-plane $\eta > 0$. Observing that

$$\frac{\partial \bar{\omega}}{\partial \zeta} - \frac{\partial \omega}{\partial \zeta} + m\,\frac{\partial z}{\partial \zeta}\cdot\overline{\frac{\partial z}{\partial \zeta}}\,(\dot{\xi} - \bar{\dot{\xi}}) = 0\ ,$$

$$\operatorname{Im}\frac{1}{\zeta - ih} = -\operatorname{Im}\frac{1}{\zeta + ih} \qquad \text{for } \zeta = \xi\ ,$$

we have for $\zeta = \xi$

$$\operatorname{Im}\frac{\partial \theta}{\partial \zeta} = k + \operatorname{Im} W, \qquad W = \frac{q}{\pi(\zeta + ih)}\ . \tag{6.6}$$

Now condition (2.19) may be written as

$$\operatorname{Im}\left(\frac{\partial z}{\partial t}\cdot\overline{\frac{\partial z}{\partial \zeta}}\right) = -\frac{k}{m} - \frac{1}{m}\operatorname{Im} W + \frac{k}{m}\operatorname{Im}\left(i\,\frac{\partial z}{\partial \zeta}\right)\ . \tag{6.7}$$

Based on (6.2), (6.4), (6.5), (6.6) we have the series

$$\begin{gathered}\frac{\partial z}{\partial t} = z_1 + 2z_2 t + 3z_3 t^2 + \ldots, \quad \frac{\partial z}{\partial \zeta} = 1 + \frac{\partial z_1}{\partial \zeta}t + \frac{\partial z_2}{\partial \zeta}t^2 + \ldots, \\ W = W_0 + W_1 t + W_2 t^2 + \ldots,\end{gathered} \tag{6.8}$$

where

$$W_0 = \frac{q_0}{\pi(\zeta + ih_0)}\ , \quad W_1 = \frac{1}{\pi}\left[\frac{q_1}{\zeta + ih_0} - \frac{iq_0 h_1}{(\zeta + ih_0)^2}\right]\ ,$$

$$W_2 = \frac{1}{\pi}\left[\frac{q_2}{\zeta + ih_0} - \frac{iq_1 h_1 + iq_0 h_2}{(\zeta + ih_0)^2} + \frac{q_0 h_1^2}{(\zeta + ih_0)^3}\right]\ .$$

Inserting these series in (6.7) and equating the coefficients of the same powers in $t$, we obtain a system of boundary conditions to determine functions $z_1(\zeta)$, $z_2(\zeta)$. Thus, for $z_1$ the condition is

$$\operatorname{Im} z_1 = -\frac{1}{m}\operatorname{Im} W_0 \quad \text{for} \quad \zeta = \xi . \tag{6.9}$$

Because $z_1$ and $W_0$ are regular in the half-plane $\eta > 0$ and vanish for $\zeta \to \infty$, we find at once from (6.9)

$$z_1 = -\frac{1}{m} W_0 = -\frac{q_0}{\pi m}\cdot\frac{1}{\zeta_0 + h_0 i} . \tag{6.10}$$

To determine $z_2$ we have the condition: for $\zeta = \xi$

$$\operatorname{Im}(2z_2) + \operatorname{Im}\left(z_1\frac{\partial \bar{z}_1}{\partial\zeta}\right) = -\frac{1}{m}\operatorname{Im} W_1 + \frac{k}{m}\operatorname{Im}\left(i\frac{\partial z_1}{\partial\zeta}\right) . \tag{6.11}$$

Putting $z_2 = z_{21} + z_{22}$, we split it in two conditions

$$\begin{cases} \operatorname{Im}(2z_{21}) = -\dfrac{1}{m}\operatorname{Im} W_1 + \dfrac{k}{m}\operatorname{Im}\left(i\dfrac{\partial z_1}{\partial\zeta}\right), \\ \operatorname{Im}(2z_{22}) = -\operatorname{Im}\left(z_1\dfrac{\partial\bar{z}_1}{\partial\zeta}\right) = \operatorname{Im}\left(\bar{z}_1\dfrac{\partial z_1}{\partial\zeta}\right) . \end{cases} \tag{6.12}$$

From (6.11) we find at once

$$z_{21} = -\frac{1}{2m}W_1 + \frac{ki}{2m}\frac{\partial z_1}{\partial\zeta} = -\frac{q_1}{2m\pi}\frac{1}{\zeta + h_0 i} + \frac{iq_0(mh_1 + k)}{2m^2\pi(\zeta + ih_0)^2} . \tag{6.13}$$

Here $q_0$, $q_1$ and $h_1$, are constants subject to determination. Therefore we use the condition that for $\zeta = ih$, $z = h_0 i$:

$$h_0 i = z(ih) = ih + z_1(ih)t + z_2(ih)t^2 + \ldots \tag{6.14}$$

But

$$z_1(ih) = -\frac{q_0}{m\pi}\cdot\frac{1}{ih + ih_0} = \frac{iq_0}{(m\pi h_0)\left[2 + \dfrac{h_1}{h_0}t + \dfrac{h_2}{h_0}t^2 + \ldots\right]} =$$

$$= \frac{iq_0}{2m\pi h_0} - \frac{iq_0 h_1}{4m\pi h_0^2}t + \ldots$$

Therefore (6.14) gives

$$h_0 i = i(h_0 + h_1 t + \ldots) + \frac{iq_0}{2m\pi h_0}t - \frac{iq_0 h_1}{4m\pi h_0^2}t^2 + \ldots$$

Making the coefficient for $t_1$ zero, we find

$$h_1 = -\frac{q_0}{2m\pi h_0} . \tag{6.15}$$

We determine the quantities $q_0$ and $q_1$. From (6.5), by integration, we find

$$\theta = P_\kappa + ki\zeta + \frac{q}{2\pi}\ln\frac{\zeta + ih}{\zeta - ih} , \tag{6.16}$$

where $P_\kappa = -\dfrac{p_\kappa k}{\gamma}$, and $p_\kappa$ is the pressure at the boundary of the region, i.e., at the free surface. In points of the surface of the pipe, where we may put $z = ih_0 + \delta e^{i\theta}$, the pressure is equal to a constant quantity $p_{pi}$.

In particular, $p = p_{pi}$ in the point $ih_0 + \delta$, i.e., for $\theta = 0$. Because

$$\zeta(ih_0 + \delta) = \zeta(ih_0) + \frac{\partial\zeta}{\partial z} \cdot \delta + \dots,$$

in the point $z = ih_0 + \delta$ of the pipe

$$\theta_{pi} \approx p_\kappa + ki\left[ih_0 + \frac{\delta}{\left(\frac{\partial z}{\partial\zeta}\right)_{ih}}\right] + \frac{q_0}{2\pi}\,\ell n\,\frac{2ih_0}{\left(\frac{\partial\zeta}{\partial z}\right)_{ih}\cdot\delta} .$$

From this, recalling that

$$\left(\frac{\partial\zeta}{\partial z}\right)_{ih} = \frac{1}{\left(\frac{\partial z}{\partial\zeta}\right)_{ih}} ,$$

we obtain

$$p_{pi} = \mathrm{Re}(\theta_{pi}) = p_\kappa - kh_0 + \frac{q_0}{2\pi}\,\ell n\,\frac{2h_0}{\delta} . \tag{6.17}$$

Thus we find the initial flowrate of the pipe

$$q_0 = \frac{2\pi\left(kh_0 + \frac{k\Delta p}{\gamma}\right)}{\ell n\left(\frac{2h_0}{\delta}\right)} , \tag{6.18}$$

where

$$\Delta p = p_\kappa - p_{pi} . \tag{6.19}$$

If we collect the terms of the first power in $t$ in the equation for $\theta_{pi}$, then we find

$$q_1 = \frac{q_0(3q_0 - 4k\,\pi h_0)}{4m\pi h_0^2\,\ell n\,\frac{2h_0}{\delta}} . \tag{6.20}$$

Now equation (6.13) for $z_{21}$ is completely determined. Next we find $z_{22}$ from condition (6.12). We bring this equation in a form to have a regular function in the half-plane $\eta > 0$.

$$z_1\,\overline{\frac{\partial z_1}{\partial\zeta}} = \frac{q_0^2}{m^2\pi^2}\cdot\frac{1}{\zeta + ih_0}\cdot\frac{1}{(\zeta - ih_0)^2} =$$

$$= \frac{q_0^2}{h_0^2m^2\pi^2}\left[-\frac{1}{4(\zeta + ih_0)} + \frac{h_0}{2i(\zeta - ih_0)^2} + \frac{1}{4(\zeta - ih_0)}\right] .$$

Therefore we obtain from (6.12)

$$\mathrm{Im}\,2z_{22} = \frac{q_0^2}{2m^2\pi^2h_0^2}\,\mathrm{Im}\left[\frac{1}{2(\zeta + ih_0)} - \frac{h_0}{i(\zeta + ih_0)^2} - \frac{1}{2(\zeta + ih_0)}\right] .$$

Using the fact that $\mathrm{Im}\,[f(\zeta)] = -\,\mathrm{Im}\,[\overline{f(\zeta)}]$ for $\zeta = \xi$, we write the latter equality as:

$$\mathrm{Im}(2z_{22}) = \frac{q_0^2}{2m^2\pi^2h_0^2}\ \mathrm{Im}\left[\frac{1}{2(\zeta + ih_0)} - \frac{h_0}{i(\zeta + ih_0)^2} + \frac{1}{2(\zeta + ih_0)}\right] =$$

$$= \frac{q_0^2}{2m^2\pi^2h_0^2}\ \mathrm{Im}\left[\frac{1}{\zeta + ih_0} - \frac{h_0}{i(\zeta + ih_0)^2}\right] .$$

From this we find $z_{22}$:

$$z_{22} = \frac{q_0^2}{4m^2\pi^2h_0^2} \cdot \left[\frac{1}{\zeta + ih_0} - \frac{h_0}{i(\zeta + ih_0)^2}\right] . \tag{6.21}$$

Adding expressions (6.13) and (6.21) we obtain:

$$z_2 = -\frac{1}{(2m\pi h_0)^2}\left[-\frac{q_0^2 + 2m\pi q_1 h_0^2}{\zeta + ih_0} + \frac{iq_0h_0(-q_0 - 2kh_0\pi - 2m\pi h_1 h_0)}{(\zeta + ih_0)^2}\right] . \tag{6.22}$$

We may calculate the following coefficients of the z series in powers of t in a similar way. However, with each following term the difficulty of computation increases strongly.

To track the displacement in time of point A on the free surface, we put $\zeta = 0$ in formulas (6.10) and (6.22) for $z_1$, $z_2$. Then we find for $y_A$ the approximate expression

$$y_A = \frac{q_0}{m\pi h_0}\, t + \frac{q_0^2 + m\pi(q_1h_0 - q_0h_1)h_0 - 2k\pi q_0 h_0}{2m^2\pi^2h_0^3}\, t^2 + \ldots \tag{6.23}$$

We now examine the expression for the flowrate of the pipe (6.23). For $q_0$ and $q_1$ we found respectively (6.18) and (6.20). We also communicate the expression

$$q_2 \ln\frac{2h_0}{\delta} = \frac{q_1(9q_0 - 4k\pi h_0)}{8m\pi h_0^2} + \frac{q_0^2(27q_0 - 28k\pi h_0)}{8(2m\pi)^2h_0^4} + \frac{(k\pi)^2 q_0}{(2m\pi)^2h_0^2} . \tag{6.24}$$

From formula (6.18) for $q_0$ it is apparent, that if $\Delta p = 0$, $q_0$ is still different from zero, as the flow proceeds under influence of gravity. Further, expression (6.20) for $q_1$ may be positive as well as negative. This means that for moments in time close to the initial, flowrate q may be increasing or decreasing. We compose the quantity:

$$F = \frac{1}{2\pi k}\sqrt{\ln\frac{2h_0}{\delta}}\,(3q_0 - 4k\pi h_0)_{h_0=1} = 3 + 3\,\frac{\Delta p}{\gamma} - 2\ln\frac{2}{\delta} .$$

Putting $\gamma \approx 0.001$ kilogram / cm$^3$, we find the following table for values of $\delta$, the radius of the pipe, for which $F = 0$:

| $\Delta p$ | 0 | 0.001 | 0.002. | 0.003 |
|---|---|---|---|---|
| $\delta_1$ | 0.45 | 0.11 | 0.022 | 0.005 |

$q_1 > 0$ for $\delta > \delta_1$.

In a similar way, one may study the case of a drainage pipe in a stratum of finite thickness [7]. One must notice that the series obtained in the method explained above, converge only for small values of $t$.

## *§7. Problem About the Migration of Oil in Hydrostatic Environment.

Somewhat simpler and more investigated is the following problem. Given a region, occupied with fluid (oil) in a horizontal plane at the origin of time. Inside the contour $L$, bounding this region, wells are given as point sources. The question arises as to how the contour $L$ will change in time, if one considers that along the contour, the pressure remains constant at all times.

This is a simplified scheme of the problem of the displacement of oil, surrounded by water in a confined stratum [8]. Indeed, for the equations of plane flow of water and oil, we may take the following equations

$$\vec{v} = -\frac{k_0}{\mu}\,\text{grad}\,p, \quad \vec{v}_1 = -\frac{k_{01}}{\mu_1}\,\text{grad}\,p_1 \quad . \tag{7.1}$$

Here the symbols without subscript 1 refer to the region filled with oil, while those with subscript 1 refer to the region filled with water surrounding oil. If we assume that the viscosity of water is very small in comparison with that of oil, so that coefficient $\mu_1$ may be made zero, then one has to take $p_1$ = constant (otherwise $\vec{v}_1$ becomes infinite). From this it follows that the pressure on contour $L$ no longer depends on the coordinates and may be a function only of time, and in particular, a constant. Then also the velocity potential, proportional in this case to the pressure, must preserve a constant value at the contour. Differentiating for $t$ the equation

$$\varphi(x, y, t) = \text{constant on } L$$

as we did in §2, we obtain the equation

$$m\frac{\partial\varphi}{\partial t} + \left(\frac{\partial\varphi}{\partial x}\right)^2 + \left(\frac{\partial\varphi}{\partial y}\right)^2 = 0 \quad , \tag{7.2}$$

which must be satisfied on $L$ at all times during the flow.

In the works of P. P. Kyfarev [11-14], solutions in closed form are obtained for problems of one well and a series of equidistant wells in a half-plane, strips and circles being taken for contour $L$. The simplest problem is that of the cardioid [9], the equation of which at $t = 0$ is given as

$$z = \zeta + a\zeta^2, \quad \zeta = e^{i\varphi} \ ,$$

where one assumes that $0 < a < \frac{1}{2}$, i.e., that the cardioid has no corner point. Then, if the center of the well is in point $z = 0$, contour $L$ has the following equation in subsequent times

$$z = A_1(t)\zeta + A_2(t)\zeta^2$$

where $A_1$, $A_2$ are functions of time.

When $A_1$ becomes equal to $2A_2 = (2a)^{1/2}$, the cardioid receives a turning point, and the solution of the problem becomes multivalued and does not suit. The same circumstance, namely the appearance of corner points on the contour and simultaneous multivaluedness of the solution before points of the contour reach the contour of the well (the latter is assumed to be close to a circle, with radius small compared to the dimensions of the region), takes place in all problems shown above [10-14]. The existence of a solution for small $t$ is shown for rather general assumptions about the initial form of the contour [14]. The property mentioned — loss of single valuedness — restricts the interest of the theory. It is possible that one has to consider the inertia terms, which we rejected in the equations of motion in the analysis of problems of §§6 and 7.

## CHAPTER XVI

## APPROXIMATE NUMERICAL AND GRAPHICAL METHODS IN THE STUDY OF UNSTEADY GROUNDWATER FLOW.

### §1. Method of Successive Changes of Steady State Values.

In some cases it is possible to apply an approximate method, in which the flow at each moment of time is assumed to be steady, but the boundary of the flow region in regard to which some simplifying assumption is made, is changing. Therefore the flowrate for an elementary lapse of time is equal to the change of area, filled with fluid, divided by this lapse of time.

Apparently, K. E. Lembke [1] was the first to apply an approximation of this kind (1886). He examined the problem of the drainage of a waterbearing stratum in the case of an horizontal drain and in the case of a vertical well. We spend some time on the first of these problems.

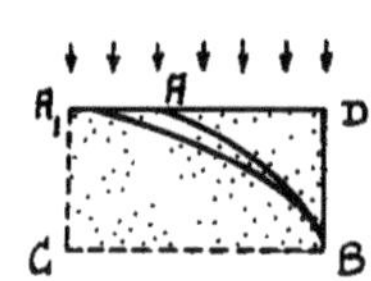

Fig. 411

We assume that seepage with constant intensity $\varepsilon$ takes place in the area under consideration. In point B (Fig. 411), there is a drain. At times $t$ and $t + dt$ the drainage area will respectively be ABD, $A_1BD$. We assume that the line AB is a parabola $y^2 = (H^2/L)x$, where $L$ is a length, changing in time, and $H$ the thickness of the waterbearing stratum (BD). Then the flowrate per unit time will be $q = ky(dy/dk) = k(H^2/2L)$. During the interval $dt$, the amount of water, withdrawn by drainage from part of the stratum is equal to $q\ dt$. On the other side it is also equal to the product of the change of area $A_1BD$ (equal to $HL/3$) with the porosity, to which is added the amount of fluid, derived from infiltration through the surface $A_1B$.

As a result we have the equation

$$q\ dt = k\frac{H^2}{2L}\,dt = \frac{m}{3}H\,dL + L\varepsilon\,dt$$

or, separating the variables,

$$\frac{L\,dt}{\frac{H^2k}{2} - L^2\varepsilon} = \frac{3}{Hm}\,dt\ . \tag{1.1}$$

Integrating this equation and recalling the initial condition $L(0) = 0$, we obtain:

$$L(t) = \frac{H\sqrt{k}}{\sqrt{2\varepsilon}}\sqrt{1 - e^{-\alpha t}} \quad \left(\alpha = \frac{6\varepsilon}{H}\right)\ . \tag{1.2}$$

For $t \to \infty$, we obtain the limit value of the length of the water-collector, corresponding to steady flow

$$L_\infty = H\sqrt{\frac{k}{2\varepsilon}} \quad (1.3)$$

Lembke called $L_\infty$ the limit of the "radius" of the water collecting structure. The change of flowrate in time is expressed by the formula

$$q = \frac{H\sqrt{k\varepsilon}}{\sqrt{2(1 - e^{-\alpha t})}} \quad (1.4)$$

For $t = \infty$ we obtain the value of the flowrate in steady flow, when it is entirely derived from infiltration

$$q = H\sqrt{\frac{k\varepsilon}{2}} = L_\infty \varepsilon . \quad (1.5)$$

Lembke examined also the case where the dimensions of the waterbearing stratum do not allow the free surface lines to reach a length corresponding to the normal action of the drain, and he distinguished two stages of flow. In the second of these stages, the free surface drops with preservation of its length.

In particular, when there is no infiltration, equation (1.1) gives after integration:

$$L = \sqrt{\frac{3kH}{m} t} = 1.732\sqrt{\frac{kH}{m} t} . \quad (1.6)$$

By means of the same method we may obtain an approximate solution of the problem about the advancement of a groundwater tongue seeping from a canal with vertical wall, when evaporation from the free surface at constant intensity $c$ takes place. Since now we must consider the wetted area $A_1BC$, with magnitude $2/3HL$, we must replace $1/3$ by $2/3$ in the solutions of problems (1.1) and (1.6). We obtain:

$$L = \frac{H\sqrt{k}}{\sqrt{2\varepsilon}}\sqrt{1 - e^{-\alpha t}} \quad \left(\alpha = -\frac{3c}{H}\right) , \quad (1.7)$$

and when there is no evaporation

$$L = \sqrt{\frac{3}{2}\frac{kH}{m} t} = 1.225\sqrt{\frac{kH}{m} t} . \quad (1.8)$$

The problem about the movement of a groundwater tongue when there is evaporation, has been examined by V. M. Makkaveev [2]. He assumed that the equation of the free surface is written as

$$h = H\left(\frac{L - x}{L}\right)^n \quad (1.9)$$

where $n$ is an exponent that may be chosen.

We may assume that $n$ varies in time, but sufficiently slow, so that we may neglect the value of the derivative $dn/dt$. To determine this exponent one has to substitute equation (1.9) in Bousinesq's differential equation.

When there is no evaporation, the case $n = 1/2$ as examined before prevails. For the steady flow, when the total flowrate, passing through $A_1C$ must evaporate from $A_1B$, V. M. Makkaveev considers that one must take $A_1B$ as a straight line, in order to have a constant seepage velocity u. The equation of the straight line is

$$h = \frac{H}{L_0}(L_0 - x) \quad , \tag{1.10}$$

where $L_0$ is the limit value of L. We have for the seepage velocity

$$u = \frac{kH}{L_0} \quad .$$

Therefore, at any x, the equality of flowrate through the ordinate h and of flowrate (as evaporation) from the segment $L_0 - x$ (Fig. 411) gives

$$\frac{kH}{L_0} h = c(L_0 - x) \quad ,$$

From this we obtain the value of $L_0$:

$$L = H\sqrt{\frac{k}{c}} \quad . \tag{1.11}$$

We notice that formulas (1.10) and (1.11) give an exact hydrodynamic solution of the problem about the steady seepage from a canal with vertical wall in case of evaporation (Chapter VII, §11; in Fig. 229 the hodograph degenerates into point C). Concerning formula (1.8), the coefficient 1.225 in it differs significantly from the coefficient 1.616, which may be obtained from the solution of the non-linear equation of Bousinesq (Chapter XIII, §7). Therefore it is hard to say beforehand, which degree of accuracy is attained by the application of the method of successive changes of steady state values. This method is applied in the theory of oil seepage.

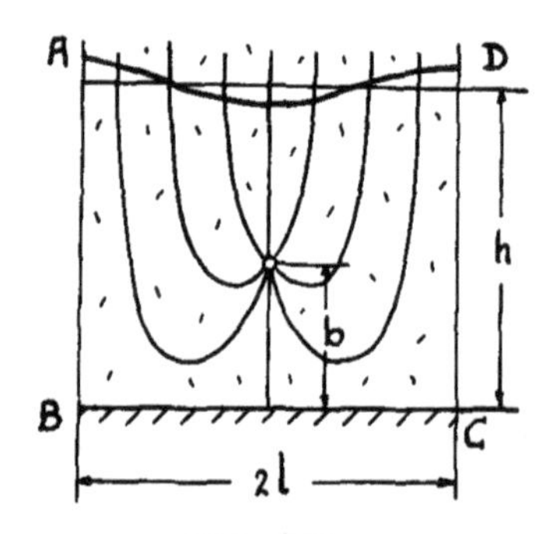

Fig. 412

We show here the possibility of application of this method in the solution of some problems, considered as two-dimensional. Let there be a system of horizontal drains at equal distance. Line BC is the impervious boundary, the centers of the drains are at a distance b from the impervious boundary (Fig. 412). We place sinks with intensity Q in the centers of the drains and consider the infinite chain of sinks in the points

$$z = bi, \quad \pm 2\ell + bi, \quad \pm 4\ell + bi, \; \ldots$$

We place the same chain of sinks in points with symmetrical abscissas

$$z = -bi, \quad \pm 2\ell - bi, \quad \pm 4\ell - bi, \; \ldots$$

The complex potential of this flow, as in known from hydrodynamics [5], has the form

$$\omega = -\frac{Q}{2\pi}\, \ell n \left[\sin\frac{\pi}{2\ell}(z - bi)\,\sin\frac{\pi}{2\ell}(z + bi)\right] + C =$$

$$\omega = -\frac{Q}{2\pi}\, \ell n \left(\cos\frac{\pi z}{\ell} - \mathrm{ch}\,\frac{\pi b}{\ell}\right) + C.$$

Separating real and imaginary part, we obtain the following expression for the velocity potential

$$(1.12)\quad \varphi = -\frac{Q}{4\pi}\, \ell n \left\{\left[\mathrm{ch}\,\frac{\pi(y - b)}{\ell} - \cos\frac{\pi x}{\ell}\right]\left[\mathrm{ch}\,\frac{\pi(y + b)}{\ell} - \cos\frac{\pi x}{\ell}\right]\right\} + C.$$

We assume that formula (1.12) also holds for unsteady flow, in which the flowrate is a function of time, subject to determination. Let $y = h_0$ be the equation of the free surface at $t = 0$. Later the free surface will change its form. But we first assume that at all times it remains an horizontal line $y = h(t)$, dropping as time goes on. In other words, we take the straight line $y = h(t)$ for some average line, about which the real free surface oscillates.

The flowrate in the drain per element of time $dt$ is equal to the change during this time of the area, filled with water, between the limits of the region of influence of one drain. This gives the equation

$$(1.13)\qquad Q\, dt = -2\ell\, dh\ .$$

To find the equation for $Q$, we proceed in this way. First we take the point $x = 0$ at the free surface. For it $y = h - \eta_1$. Considering that at the free surface there is a capillary rise of height $h_k$, we have condition $\varphi = -k(y - h_k)$, which gives $\varphi = -k(h - \eta_1 - h_k)$.

We insert these equations in (1.12), and neglect $\eta_1$ as compared with $h$ in the right part of the equation; we obtain:

$$-k(h - \eta_1 - h_k) = -\frac{Q}{4\pi}\, \ell n \left[\left(\mathrm{ch}\,\frac{\pi(h - b)}{\ell} - 1\right)\left(\mathrm{ch}\,\frac{\pi(h + b)}{\ell} - 1\right)\right] + C\ .$$

Exactly in the same way, putting $x = \ell$, $y = h + \eta_2$, $\varphi = -k(h + \eta_2 - h_k)$, we obtain the equation

$$-k(h + \eta_2 - h_k) = -\frac{Q}{4\pi}\, \ell n \left[\left(\mathrm{ch}\,\frac{\pi(h - b)}{\ell} + 1\right)\left(\mathrm{ch}\,\frac{\pi(h + b)}{\ell} + 1\right)\right] + C\ .$$

We take the half sum of the left and right parts of the latter equations and neglect in the left part the quantity $\eta_1 - \eta_2$ in comparison with $h$. After some transformations we find:

$$(1.14)\qquad -k\,(h - h_k) = -\frac{Q}{4\pi}\, \ell n \left[\mathrm{sh}\,\frac{\pi(h - b)}{\ell}\,\mathrm{sh}\,\frac{\pi(h + b)}{\ell}\right] + C\ .$$

Now we assume that on the contour of the drain, which we assume to be a circle of radius $\delta$ (this radius is small in comparison with the basic dimensions of the flow region: $\ell$, $h$), the head is equal to the quantity $h_c$ and, consequently, the velocity potential is $-kh_c$. The circle of radius $\delta$ with center in the point $x = 0$, $y = b$, generally speaking, is not an equipotential, but we assume the opposite in first approximation. In particular, we may consider that in point $x = 0$, $y = b + \delta$, the velocity potential is $-kh_c$. Substitution in equation (1.12) gives

$$(1.15) \qquad -kh_c = -\frac{Q}{4\pi} \ln \left\{ \left( \operatorname{ch} \frac{\pi\delta}{\ell} - 1 \right) \left[ \operatorname{ch} \frac{\pi(\delta + 2b)}{\ell} - 1 \right] \right\} + C .$$

Subtracting (1.15) from (1.14), we eliminate the constant $C$ and obtain the expression of the flowrate of the drainage pipe (per unit length)

$$(1.16) \qquad Q = \frac{4k\pi(h - h_k - h_c)}{\ln \left[ \operatorname{sh} \frac{\pi(h - b)}{\ell} \operatorname{sh} \frac{\pi(h + b)}{\ell} \right] - \gamma} .$$

Here $\gamma$ stands for the expression

$$(1.17) \qquad \gamma = \ln \left( 4 \operatorname{sh} \frac{\pi\delta}{2\ell} \operatorname{sh} \frac{\pi b}{\ell} \right) \approx \ln \left( \frac{2\pi\delta}{\ell} \operatorname{sh} \frac{\pi b}{\ell} \right) .$$

We insert the expression for the flowrate in formula (1.13) and separate the variables in the equation thus obtained. After integration and some simplifications we have:

$$(1.18) \qquad t = -\frac{\ell}{2\pi k} \int_{h_c}^{h} \frac{\ln \left[ \frac{1}{2} \left( \operatorname{ch} \frac{2\pi h}{\ell} - \operatorname{ch} \frac{2\pi b}{\ell} \right) \right] - \gamma}{h - h_k - h_c} \, dh .$$

The expression under the log sign of the numerator, may be transformed as

$$\ln \left( \frac{e^{\frac{2\pi h}{\ell}} + e^{-\frac{2\pi h}{\ell}}}{4} - \frac{1}{2} \operatorname{ch} \frac{2\pi b}{\ell} \right) =$$

$$= \frac{2\pi h}{\ell} - \ln 4 + \ln \left( 1 + e^{-\frac{4\pi h}{\ell}} - e^{-\frac{2\pi(h-b)}{\ell}} - e^{-\frac{2\pi(h+b)}{\ell}} \right) .$$

During the flow the following inequality holds at all times

$$h \geq b + h_k + h_c ;$$

Therefore, if $h_k + h_c > 0$, all three powers of $e$ in the logarithmic expression are negative, and we may develop the logarithm in a power series. In the initial stage of flow, when $h$ is still large, the logarithmic term is small, and we have the following approximate expression for $t$

$$t \approx -\frac{\ell}{2\pi k}\int_{h_c}^{h}\frac{\frac{2\pi h}{\ell} - \ell n\, 4 - \gamma}{h - h_k - h_c}\, dh =$$

$$(1.19)\qquad = \frac{h_c - h}{k} - \frac{h_k + h_c - \ell n\, 4 - \gamma}{k}\,\ell n(h - h_k - h_c)\ .$$

After we found the dependence of $h$ on $t$, we may compute $Q$ as a function of $t$ by means of equality (1.16).

In order to obtain an approximate equation of the free surface, we use equation (1.12) for the velocity potential, from which we eliminate the arbitrary constant by means of (1.14). Recalling that at the free surface $\varphi = -k(y - h_k)$, we obtain:

$$(1.20)\qquad y = h + \frac{Q}{4\pi k}\,\ell n\,\frac{\left(\mathrm{ch}\,\frac{\pi(h - b)}{\ell} - \cos\frac{\pi x}{\ell}\right)\left(\mathrm{ch}\,\frac{\pi(y + b)}{\ell} - \cos\frac{\pi x}{\ell}\right)}{\mathrm{sh}\,\frac{\pi(h - b)}{\ell}\ \mathrm{sh}\,\frac{\pi(h + b)}{\ell}}\ .$$

Putting $b = 0$, we obtain the particular case of a drain on an impervious foundation. Fig. 413 renders two positions of the free surface in a parallel plate model (with glycerine), corresponding to a drain on impervious foundation.

For this case a comparison was made between the results of the computations and those of the experiments with the parallel plate model [6].

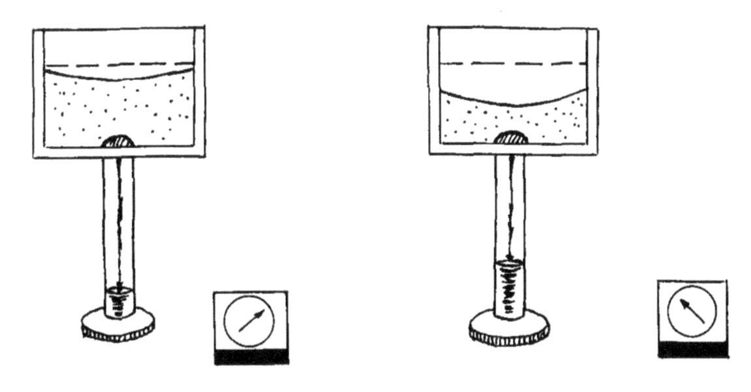

Fig. 413

§2. Method of Finite Differences.

In Chapter XIII we had the differential equation of groundwater flow with free surface

$$(2.1)\qquad \frac{\partial h}{\partial t} = \frac{\kappa}{2m}\frac{\partial^2 h^2}{\partial x^2} + \frac{w}{m}\ .$$

Here $w$ is the seepage from above, $m$ the active porosity. The ordinate $h$ of the free surface is measured from bedrock.

Equation (2.1) may be changed into an equation of finite differences

$$\frac{h_{i,k+1} - h_{i,k}}{\Delta t} = \frac{\kappa}{2m} \frac{h^2_{i+1,k} - 2h^2_{i,k} + h^2_{i-1,k}}{\Delta x^2} + \frac{w}{m} , \tag{2.2}$$

where the indices $i-1, i, i+1$ designate successive sections of the groundwater flow, $\Delta x$ is the distance between sections (Fig. 414), the numbers $k, k+1$ are ordinal indices for the moments in time, $\Delta t$ is the interval for the computations.

The quantities $\kappa, m$ and $w$ may here be considered as variables. G. N. Kamensky [7] communicates the results of computations for a reach of the Moscow-Volga Canal and a comparison with observed levels. It is of interest that intervals $\Delta x$ and $\Delta t$ may be taken rather large. Thus G. N. Kaminsky took $\Delta x = 100$ m, $\Delta t = 5$ days.

For computations, formula (1.2) is rewritten in a form where it is solved for $h_{i,k+1}$ :

$$h_{i,k+1} = h_{i,k} + \frac{\kappa \Delta t}{2m\Delta x^2} (h^2_{i+1,k} - 2h^2_{i,k} + h^2_{i-1,k}) + \frac{w\Delta t}{m} , \tag{2.3}$$

in which one must start with given initial and boundary conditions.

By means of finite differences, one may also carry out computations for the case where bedrock is not an horizontal plane but some curved surface. Let the equation of this surface be

$$z = h_0(x) .$$

Measuring the ordinates of the free surface above some horizontal line,

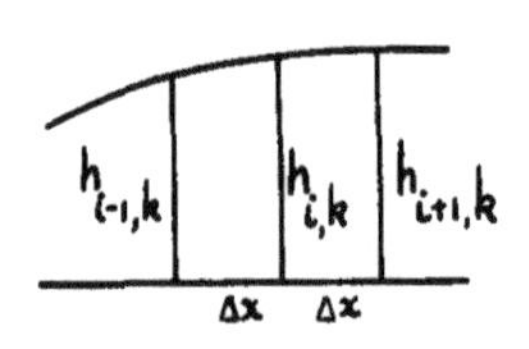


Fig. 414

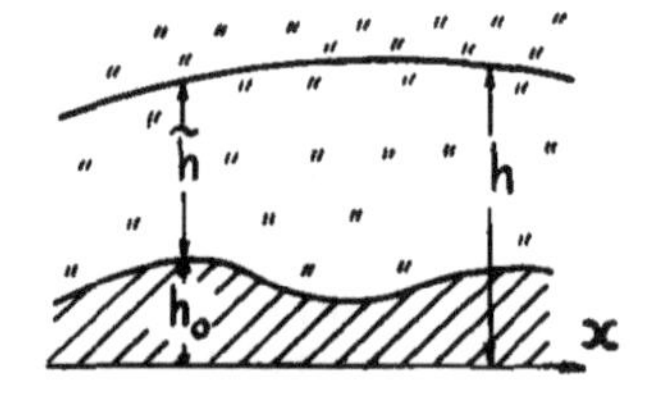


Fig. 415

we call the difference $h - h_0 = \tilde{h}$, i.e., the depth of flow above bedrock. Then equation (2.1) must be changed in this way (Fig. 415):

$$\frac{\partial h}{\partial t} = \frac{\kappa}{m} \frac{\partial}{\partial x} \left( \tilde{h} \frac{\partial h}{\partial x} + \frac{w}{m} \right) , \tag{2.4}$$

and equation (2.3) is replaced by

$$h_{i,k+1} = h_{i,k} + \frac{\kappa\Delta t}{m\Delta x^2} \tilde{h}_{i,k}(h_{i+1,k} - 2h_{i,k} + h_{i-1,k}) + \frac{w}{m} \Delta t . \tag{2.5}$$

§3. Graphical Method to Integrate the Equation of Heat Conduction.

We may show [8] a graphical method to integrate the heat equation

$$\frac{\partial u}{\partial t} = a^2 \frac{\partial^2 u}{\partial x^2} \quad . \tag{3.1}$$

The formula corresponding to (2.3) has the following form for the equation of heat conduction:

$$u_{i,k+1} = \left(1 - \frac{2a^2\Delta t}{\Delta x^2}\right) u_{i,k} + \frac{a^2\Delta t}{\Delta x^2}(u_{i-1,k} + u_{i+1,k}) \quad . \tag{3.2}$$

We may chose various relationships between $\Delta t$ and $\Delta x$ [9]. For a special choice of $\Delta t$ and $\Delta x$, namely for

$$\Delta t = \frac{\Delta x^2}{2a^2} \tag{3.3}$$

the equation simplifies and becomes

$$u_{i,k+1} = \frac{u_{i-1,k} + u_{i+1,k}}{2} \quad . \tag{3.4}$$

We introduce a system of coordinates $(x,t,u)$. We assume that we must find a solution of equation (3.1) for the initial condition

$$u(x,0) = f(x) \tag{3.5}$$

and the boundary conditions

$$u(0,t) = \varphi_1(t), \quad u(\ell,t) = \varphi_2(t) \quad . \tag{3.6}$$

We layout a length $\ell$ along the x-axis and divide it into $n$ parts, each of them being equal to $\Delta x = \ell/n$. In the plane $t = 0$, we construct a line $u = f(x)$, and in the planes $x = 0$ and $x = \ell$, we construct respectively the lines $u = \varphi_1(t)$ and $u = \varphi_2(t)$. Along the t-axis, we lay out segments $\Delta t$, computed by formula (3.3).

Now, considering the ordinates of the initial curve $u_{00}$ and $u_{20}$, we make their half sum corresponding to formula (3.4). We obtain the ordinate $u_{11}$. It is possible to obtain the half sum of the ordinates by joining the points $u_{00}$ and $u_{20}$ by a line and by taking the average ordinate.

Joining the ends of the ordinates $u_{10}$ and $u_{30}$ by a straight line, we mark the point of intersection of this straight line with the ordinate $u_{20}$, and the distance of this point of intersection to the x-axis gives the ordinate $u_{21}$, etc.

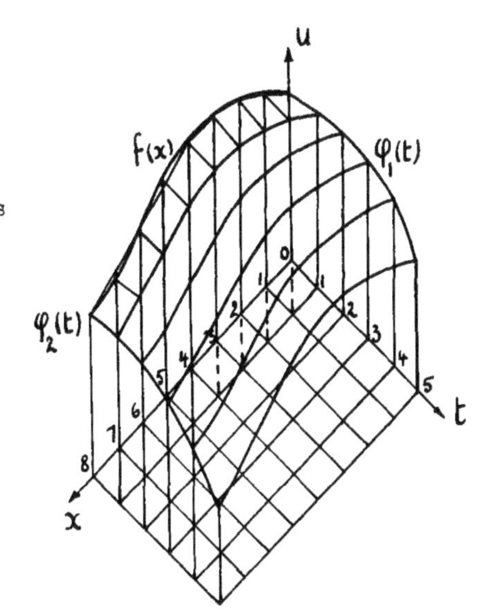


Fig. 416

We carry the obtained ordinates parallel to the u-axis in points (1,1), (2,1),...,(n-1, 1) of the first layer. Joining the ends of ordinates $u_{01}, u_{11}, \dots, u_{n1}$ by a smooth line, we obtain a curve for the moment of time $t_1 = \Delta t$.

Constructing the average arithmetic ordinates of this curve and shifting them to the points (1,2),(2,2),...,(n-1,2), we know the magnitude of the ordinates $u_{02}, u_{12}, \dots, u_{n2}$, by means of which we construct a curve for the following moment of time $t_2 = 2\Delta t$, etc.

For the construction, it is recommended to sketch the (x,t) region as a rectangle (Fig. 416), and the u ordinates as vertical segments. In the work mentioned [8], the equation of heat conduction for three independent variables (x,y,t) and also for four (x,y,z,t) is treated.

## §4. A Graphical Method to Compute Plane Unsteady Groundwater Flow.

As we already saw, the pressure is constant at the free surface of groundwater flow, equal to the atmospheric pressure or to the atmospheric pressure minus the pressure corresponding to the capillary rise.

The line of the free surface is thus a line of constant pressure or isobar, with equation

$$p(x,y,t) = p_0 \,. \qquad (p_0 = \text{constant}) \tag{4.1}$$

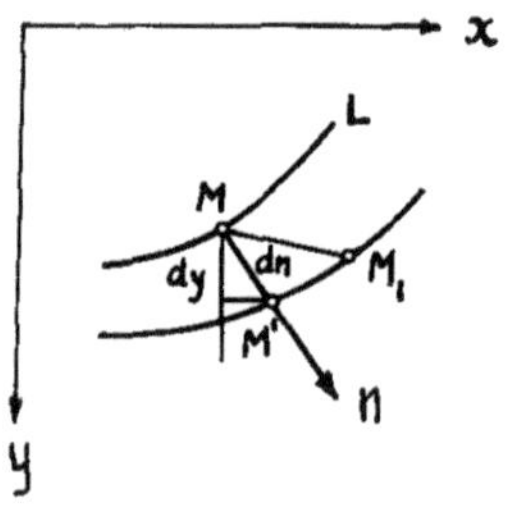


Fig. 417

We take point $M(x,y)$ on the free surface at time $t$. At time $t + \Delta t$, it assumes the position $M_1$ (Fig. 417). We erect the normal $n$ to $L$ in point $M$, so that $\Delta n = MM'$. The limit of the ratio $\Delta n/\Delta t$ for $\Delta t = 0$

$$N = \lim_{\Delta t \to 0} \frac{\Delta n}{\Delta t} = \frac{dn}{dt} \tag{4.2}$$

is called the displacement velocity of line $L$. This velocity is connected with the normal component $v_n$ of the seepage velocity in point $M$ by the equality

$$\frac{dn}{dt} = \frac{v_n}{m} = \frac{1}{m}\frac{\partial\varphi}{\partial n} \,. \tag{4.3}$$

By means of the relationship

$$\varphi = -k\left(\frac{p}{\gamma} - y\right) \tag{4.4}$$

($\gamma = \rho g$ - unit weight of the fluid), for the y-axis oriented downward, we have

$$N = \frac{k}{m}\left(\frac{dy}{dn} - \frac{1}{\gamma}\frac{\partial p}{\partial n}\right) \,. \tag{4.5}$$

As apparent from Fig. 417,

$$\frac{dy}{dn} = \cos(n,y) \,,$$

where $n$ is the normal, oriented in the direction of motion of line $L$. Therefore we obtain:

$$\frac{dn}{dt} = N = \frac{k}{m}\left(\cos(n,y) - \frac{1}{\gamma}\frac{\partial p}{\partial n}\right) \tag{4.6}$$

We notice that formula (4.6) also holds in three-dimensional space.

In case of unsteady flow, $N = 0$, and therefore

$$\cos(n,y) = \frac{1}{\gamma}\frac{\partial p}{\partial n} \quad . \tag{4.7}$$

We examine the particular case when water seeps down vertically under head $H$ (Fig. 368). Because the pressure $p$ is an harmonic function, depending only on the coordinate $y$, it is a linear function of $y$, and consequently $\Delta p$ of formula (4.6) is a constant:

$$\Delta p = (\gamma H + p_0) - p_0 = \gamma H \ .$$

For $\Delta n$ we must take the entire distance $y$, i.e.,

$$-\frac{1}{\gamma}\frac{\partial p}{\partial n} = \frac{H}{y} \quad .$$

Since the direction of the normal coincides with that of the $y$-axis,

$$\frac{dy}{dt} = \frac{k}{m}\left(1 + \frac{H}{y}\right) .$$

We obtained equation (3.12) of Chapter XII; we gave a graph and table for its solution.

In our case we have

$$\tau = \eta - \ln(1 + \eta), \quad \tau = \frac{kt}{mH}, \quad \eta = \frac{y}{H} \ . \tag{4.8}$$

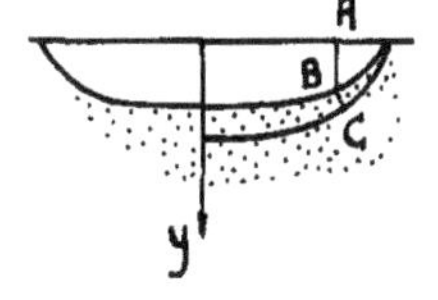

Fig. 418

We may compute the initial stage of flow by means of formula (4.8) or the approximate formulas of §3 of Chapter XII, taking segment $BA$ (Fig. 418) for the depth $H$ of the canal, varying from one section to the other.

## §5. Examples of Graphical Computation.

We assume that the form of the free surface is known at some time $t_1$. One has to determine the form of this surface at time $t_1 + \Delta t$, where $\Delta t$ is sufficiently small so that the displacement velocity $dn/dt$ of the free surface may be considered constant during $\Delta t$. Therefore, it is sufficient by means of formula (4.6) to compute the cosine of the angle between the normal to the surface and the y-axis and to find $\Delta p/\Delta n$ at $t_1$. If the field of the isobar is known at $t_1$, then $\Delta p$ may be computed along isobar $p = p_0$ and along isobar $p = p_1$, close to $p_0$. Thus the problem is reduced to that of the construction of a system of isobars at a given moment.

Since the pressure satisfies Laplace's equation, $p(x,y)$ is harmonic. We call its complex conjugate $p'$. Then the quantity

$$P = p + p'$$

differs only by a multiplier from Zhoukovsky's $\theta$ function.

Between the stream function and velocity potential on one side and the quantities $p$ and $p'$ on the other side, we have

$$\varphi = -k\left(\frac{p}{\gamma} - y\right), \quad \psi = -k\left(\frac{p'}{\gamma} + x\right) .$$

It is convenient to introduce two other functions between $\varphi$ and $\psi$, namely the head

$$h = \frac{p}{\gamma} - y = -\frac{\varphi}{k} \tag{4.9}$$

and a function, determined by

$$\varphi = -\frac{\psi}{k} = \frac{p'}{\gamma} + x \tag{4.10}$$

This function may be called "derived flowrate."

Assuming $\gamma = 1$ for seepage of water, we write the latter equation in the form ($p$ is expressed in meters)

$$h = p - y, \quad q + p' + x , \tag{4.11}$$

We have three families of orthogonal lines

$p$ = constant, $p'$ = constant; $h$ = constant, $q$ = constant; $x$ = constant, $y$ = constant.

We need the lines $p$ = constant, but as a check the properties of the other lines are useful.

We give some directions in the solution of various cases.

Case 1. Given the cross section of a canal with arbitrary form (Fig. 419), assume the pressure is zero at the free surface. We may determine the value of the pressure along the canal perimeter, considering that in the canal the pressure is distributed according to the hydrostatic law.

Let line ABC be the line of $p = 0$ at $t$. We may, by eye, draw the lines $p = 1$, $p = 0.25$, $p = 0.50, \ldots$ and construct the lines $p'$ = constant, orthogonal to them. The resulting net must be straightened out using methods described in Chapter XI. From (4.4), along the isobars we have

$$h = -y + p_1 ,$$

i.e., along the isobars $h$ is a linear function of $y$. Exactly in the same way, along the streamlines $q = q_1$ we have $p'$ as a linear function of $x$

$$p = -x + q_1 .$$

We construct the surfaces $h$ = constant and $q$ = constant.

Therefore we lay out values of $h$ along the contour $p = 0$ (Fig. 420) and we draw the family of lines $y$ = constant. Constructing the diagonals of this net, we find the family of lines $h$ = constant.

Exactly in the same way, constructing the diagonals of the net composed of lines $p'$ = constant, $x$ = constant (Fig. 421), we find the family of lines $q$ = constant.

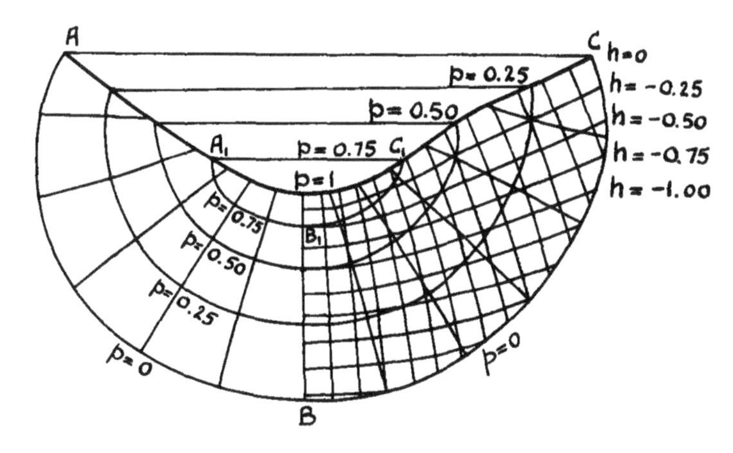


Fig. 419

Lines $h$ = constant and $q$ = constant must be mutually orthogonal, if the original net $p$ = constant, $p'$ = constant was correct. The net, represented in Fig. 419, is obtained by superposition of figures 420 and 421.

Inversely, having the net of lines $h$ = constant, $q$ = constant, and drawing the families of lines $y$ = constant, $x$ = constant, by graphical addition one may construct the families of lines $p$ = constant, $p'$ = constant (the latter play an auxiliary roll).

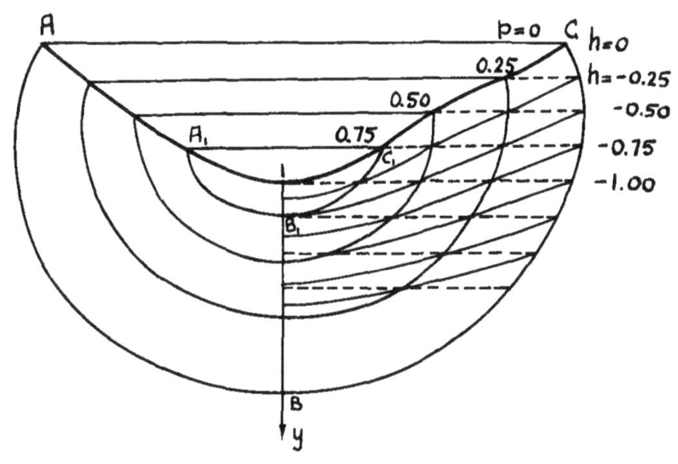


Fig. 420

When the position of the isobar $p = p_1$ is fixed, close to the isobar $p = p_0$, by means of formula (4.6) we may compute $N = \Delta n/\Delta t$ 'in different points of the surface $p = 0$ and then along the normal to this surface we may lay out segments $\Delta n = N\Delta t$ for a given value of $\Delta t$.

To construct the free surface at $t_2 = t_1 + \Delta t$, one must continue the process of construction of the isobars for this time interval, etc. The flowrate may be determined by the number of divisions along the canal for $q$ or by measuring the wetted area.

We notice that soil capillarity may be taken into account by measuring the level in the canal at a height increased by the effective capillary

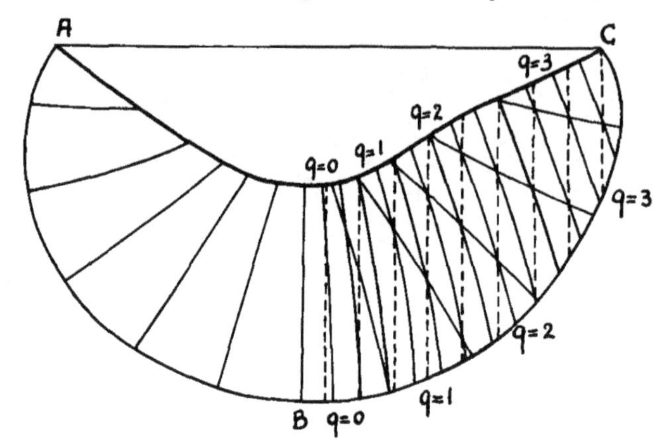


Fig. 421

rise (Fig. 422), equal to some fraction of the rise $h_k$ in a capillary pipe (see Chapter V).

Case 2. When water, seeping from a canal, reaches bedrock, which is an impervious wall, the construction of the isolines is complicated. Near the bedrock, the streamlines change their direction.

Lines of constant head $h$ are orthogonal to streamlines and consequently some of the lines $h$ = constant must be normal to a horizontal streamline, others to a vertical line (Fig. 423). Lines of constant head are drawn on Fig. 424 for the case where water seeps from a strip at the earth's surface, on which rests an infinitely thin layer of water

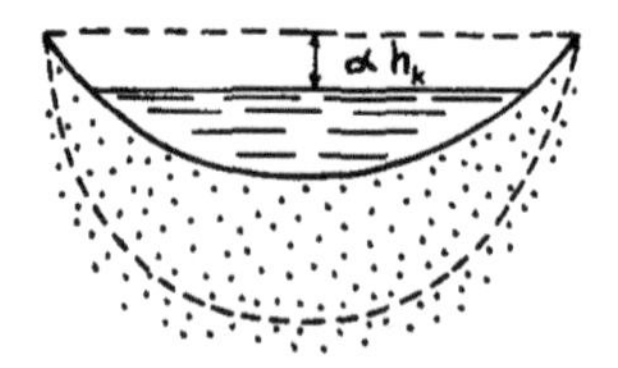


Fig. 422

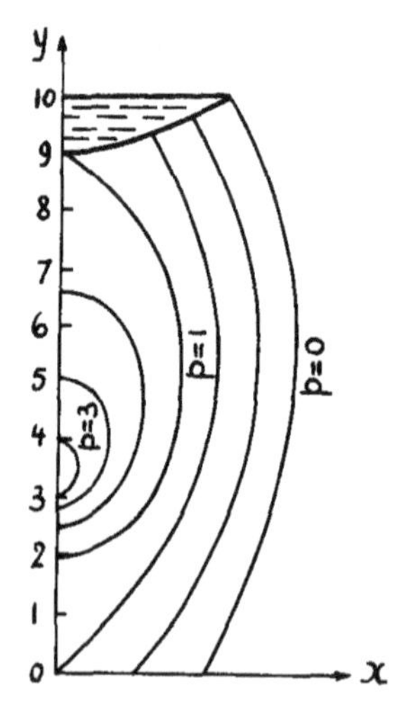


Fig. 423

when water reaches bedrock. The lines $h(x,y)$ = constant are constructed by using arithmetic averages of the values of function $h(x,y)$. The lines of equal flowrate are sketched by eye in this figure.

In the same figure 424, isolines are constructed for the moment of contact of the seeping water with the bedrock. In the construction, use is made of the relationship $p = h - y$. Lines $t = t_2$ and $t = t_3$ are locations of the groundwater tongue at two successive moments of time.

Fig. 425 renders the sketch of a case when water, seeping from a canal, comes in contact with the surface of stagnant groundwater. We see that close to the zone of contact a region of increased pressure exists.

Observations of the flow of glycerine in a parallel plate model have shown that when the fluid flowing from above reaches the underlying fluid, the free surface of the "groundwater" rises at once in the vicinity of the point of contact (see Fig. 426). This is due to the fact that the isobars, originally horizontal in the region of stagnant groundwater, now rise (in the left part of Fig. 426) and come closer together under maximum pressure, causing here the steepest pressure gradients.

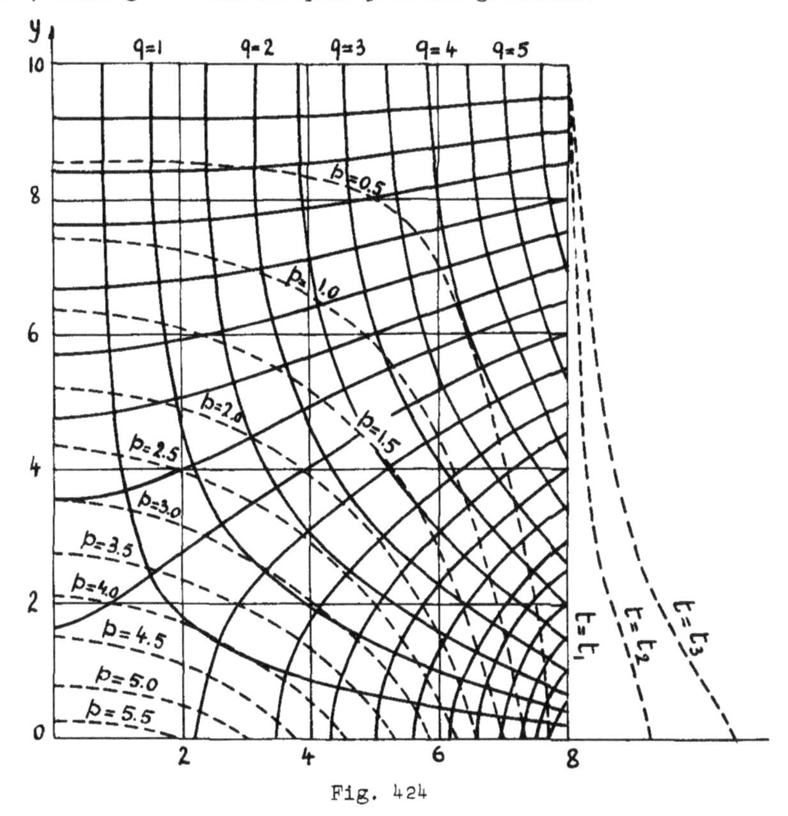


Fig. 424

The circumstance that the isobar of the free surface initially horizontal must instantaneously rise near the zone of contact with groundwater flowing from the canal, follows from the fact that otherwise the equality $p = h + y$ would lead to a contradiction: if we had $y$ = constant, $p$ = constant, then we would have $h$ = constant, i.e., an horizontal line would be a line of constant head at all times of the groundwater flow.

Attention should be paid to the fact that the isobars in Fig.425 are strongly condensed in the vicinity of the region of maximum pressure, and

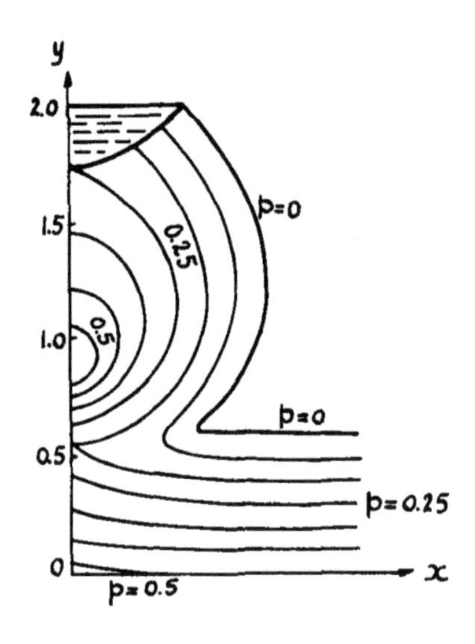


Fig. 425

moreover in the low part of the figure.

This means that in the region below the maximum pressure the pressure gradients are significant, and consequently large velocities to the right and downward prevail. Therefore one may expect that water seeping from the canal will reach the underlying groundwater and squeeze it out in such a way that at some distance from the canal we have a rise of groundwater. Such a picture of the seepage phenomenon was developed by hydrogeologists on the basis of their observations of the varying groundwater level in seepage from a canal and of the decreasing salt content of water. Indeed, groundwater contains more minerals than water seeping from a canal, as may also be concluded from the flow picture. Experiments with the parallel plate model and with samdboxes confirm also the hypothesis of the hydrogeologists.

In Fig. 427 a picture is given of the following experiment with a parallel plate model. A coat of paraffin is placed in the upper strip,

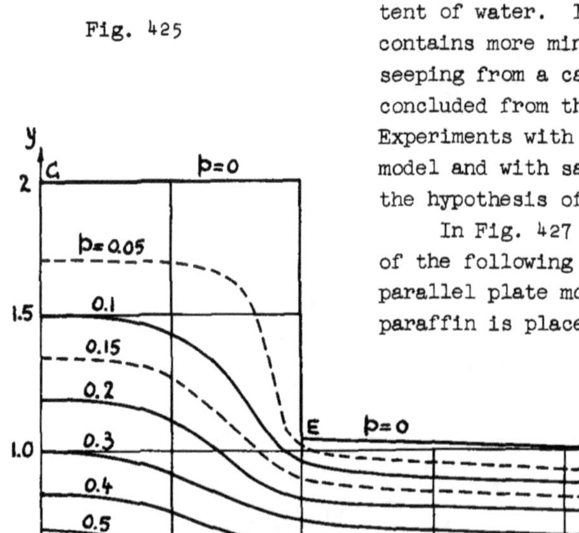


Fig. 426

to decrease the width of the spacing of the plates. Thus the two parts of the model correspond to two soil strata with different permeability. In the lower part glycerine is poured, corresponding to the initial position of the groundwater table. In the middle of the upper part of the model, an excavation is made to represent the cross section of the canal. It is filled at all times with glycerine of a different color than that of the lower part; therefore, it is easy to observe the boundary between the two fluids: that flowing from the canal and that in the lower region of the model. It is apparent that the upper fluid squeezes the lower one out, causing it to rise at the zone of contact between the two fluids (the fluid in the lower part flows then through lateral openings in the walls of the model).

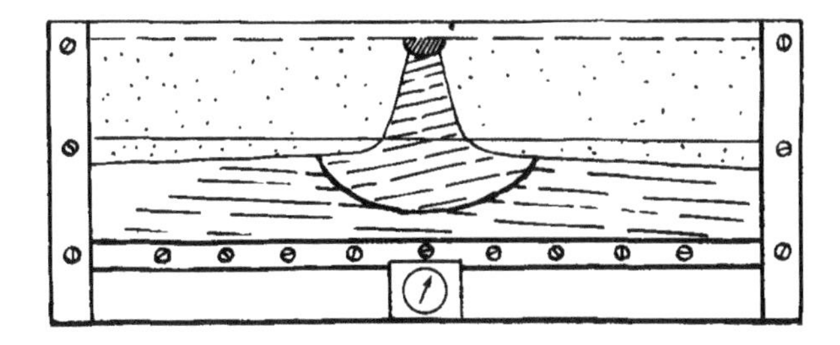

Fig. 427

§6. Problem with Axial Symmetry.

This case may occur when one studies the flow of water under irrigation of a small strip of land, seepage from a well, etc. We introduce cylindrical coordinates $r, z$, taking the z-axis as axis of rotation; we may write the equation of the free surface as

$$p(z,r,t) = p_0 \quad . \tag{6.1}$$

We examine flow with a velocity potential $\varphi(r,z,t)$. As is known, in the case of axial symmetry, there exists a streamfunction $\psi(r,z,t)$, determined by means of the equations

$$\frac{\partial\varphi}{\partial r} = \frac{1}{r}\frac{\partial\psi}{\partial z} \quad , \qquad \frac{\partial\varphi}{\partial z} = -\frac{1}{r}\frac{\partial\psi}{\partial r} \quad . \tag{6.2}$$

$\varphi$ and $\psi$ satisfy the differential equations

$$\frac{\partial^2\varphi}{\partial r^2} + \frac{\partial^2\varphi}{\partial z^2} + \frac{1}{r}\frac{\partial\varphi}{\partial r} = 0 \quad ,$$

$$\frac{\partial^2\psi}{\partial r^2} + \frac{\partial^2\psi}{\partial z^2} - \frac{1}{r}\frac{\partial\psi}{\partial r} = 0 \quad .$$

Velocity potential and pressure are connected as

$$\varphi = -k\left(\frac{p}{\gamma} - z\right) \tag{6.3}$$

(the z-axis is oriented downward).

Differentiating (6.1) for $t$ and using (6.3), and also that

$$m \frac{dz}{dt} = \frac{\partial \varphi}{\partial z} ,$$

$$m \frac{dr}{dt} = \frac{\partial \varphi}{\partial r} ,$$

we obtain the condition for the pressure at the free surface in the following form

$$\frac{\partial p}{\partial t} = \frac{k}{\gamma m} \left[ \left( \frac{\partial p}{\partial z} \right)^2 + \left( \frac{\partial p}{\partial r} \right)^2 \right] - \frac{k}{m} \frac{\partial p}{\partial z} .$$

This equation, as also holds for the two.dimensional case, may be written in the form

$$N = \frac{dn}{dt} = \frac{k}{m} \left[ \cos(n, z) - \frac{1}{\gamma} \frac{\partial p}{\partial n} \right] .$$

The pressure satisfies the same equation as function $\varphi$ does:

$$\frac{\partial^2 p}{\partial r^2} + \frac{\partial^2 p}{\partial z^2} + \frac{1}{r} \frac{\partial p}{\partial r} = 0 .$$

If we consider functions $\varphi$ and $\psi$, defined by (6.2) as "conjugate," then we may choose function $p'$, conjugate to function $p$, determined as

$$\frac{k}{\gamma} p' = -\psi - \frac{kr^2}{2}$$

or

$$\psi = -k \left( \frac{p'}{\gamma} + \frac{r^2}{2} \right) .$$

Functions $p$ and $p'$ satisfy the equations

$$\frac{\partial p}{\partial r} = \frac{1}{r} \frac{\partial p'}{\partial z} , \quad \frac{\partial p}{\partial z} = - \frac{1}{r} \frac{\partial p'}{\partial r}$$

and $p'$ satisfies the same equation as $\psi$ does:

$$\frac{\partial^2 p'}{\partial z^2} + \frac{\partial^2 p'}{\partial r^2} - \frac{1}{r} \frac{\partial p'}{\partial r} = 0 .$$

The families of lines $\varphi$ = constant and $\psi$ = constant are mutually orthogonal, as are the families of lines $p$ = constant and $p'$ = constant. But these systems, as we saw in Chapter XI, §4, are not "isothermic" i.e., do not form a quadratic net.

LITERATURE

Abbreviations

| | |
|---|---|
| AN [CCCP]: | Akademiya Nauk, Academy of Sciences, U.S.S.R. |
| Izv. NIIG: | Izvestiya, Nauchno-issledovatelskovo Instituta Gidrotekhniki<br>Proceedings of the Scientific Research Institute of Hydromechanics. |
| Inzh. Zh: | Inzhenerny Zhurnal<br>Engineering Journal |
| Mosc. gos. un-tet:<br>(M G U) | Moskovskaya gosudarstvennaya universitet<br>Moscow State University |
| D A N: | Doklady Akademii Nauk<br>Proceedings of the Academy of Sciences |
| P M M: | Prikladnaya Matematika i Makhanika<br>Applied Mathematics and Mechanics. |
| Gostekhizdat: | State Press for Technology. |
| Inzh. sb: | Inzhenerny sbornik<br>Engineering Collection |
| Z A M M: | Zeitschrift für angewandte Mathematik und Mechanik<br>Periodical of Applied Mathematics and Mechanics (Germany). |

## LITERATURE

INTRODUCTION

[1]. Danilevsky, V. V., Russian Technology, Leningrad, 1947. History of Hydropower Plants in Russia up to the 19th Century. Moscow, 1940.

[2]. Gordeev, D. I., "Subterranean water from the viewpoint of M. V. Lomonosov." Works in Hydrogeology, F. P. Savarensky Laboratory. AN [CCCP], Vol. 1, 1948.

[3]. Lembke, K. E., (1) The Engineer, Journal of the Ministry of Communications, 1886 no. 2. (2) Journal of the Ministry of Communications, 1887, nos. 17-19; 1888, no. 5.

[4]. Evnevich, I. A., Inzh. Zh. 1890, no. 7 and 8. "A few words about the laws of motion of underground water."

[5]. Krasnopolsky, A. A., "Unconfined and artesian wells. Mining Journal, nos. 3-6, 1912.

CHAPTER I.

[1]. Zjoukovsky (Zhukovsky), N. E., "Theoretical investigations about the movement of subterranean waters" (1889), Collected Works, Vol. III, Gostekhizdat (State Press): 1949.

[2]. Zjoukovsky (Zhukovsky), N. E., "About the influence of pressure on saturation of sands with water" (1890), Collected Works, Vol. III, Gostekhizdat: 1949.

[3]. Zhoukovsky (Zhukovsky), N. E., "Percolation of water through dams" (1920), Collected Works, Vol. VII, Gostekhizdat: 1950.

[4]. Pavlovsky, N. N., The Theory of Groundwater Flow under Hydraulic Structures, Petrograd: 1922 (lithographic).

[5]. Lebedev, A. F., Subterranean and Ground Waters, edited by AN [CCCP]: 1936.

[6]. Leibenzon, L. S., The Flow of Natural Fluids and Gases in Porous Medium, Gostekhizdat, Moscow: 1947.

[7]. Puzirevskaya, T. N., "The percolation of water through sandy soils," Izv. NIIG, Vol. 1: 1931.

[8]. Sergeev, E. M., Selected Chapters of General Soil Science, edited by Mosc. gos. un-ta., Moscow: 1946. General Soil Behaviour, edited by M G U: 1952.

[9]. Vasileev, A. M., "Nature and Forms of Moisture in Subsoils and Soils," Soviet Geology, Collection no. 35, Gosgeolizdat (State Geology Press) 1948.

[10]. Ivanova, M. V., "The dependence of some physical properties of soils upon the amount of colloids," Works of the Institute of Construction, Vol. 1, Moscow: 1928.

[11]. Priklonsky, V. A., General Soil Science, Part I, Moscow-Leningrad: 1948.

[12]. Kachinsky, N. A., The Soil, Its Properties and Life, edited by A N [CCCP]: 1951.

[13]. Williams, V. R., Soil Science, Parts I and II (no date given).

[14]. Tzitovich, N. A., Soil Mechanics, 3rd edition, State Press of Literature for Construction and Architecture, Moscow-Leningrad: 1951.

[15]. Cherkasov, A. A., Reclamation and Argicultural Watersupply, Agricultural Press, Moscow: 1950.

[16]. Zamarin, E. A., Calculation of Groundwater Flow, Tashkent: 1928.

[17]. Vedernikov, V. V., "Results of experiments with free surface flow," Izv. A N [CCCP], Division of the Technological Sciences, no. 8, 1947; see also D A N [CCCP], Vol. 55 no. 3 (1947).

[18]. Aravin, V. I., and Numerov, S. N., Seepage Calculation for Hydraulic Structures, Gostroiizdat, Moscow: 1958.

[19]. Lomize, G. M., Seepage in Fissured Rocks, State Press, Moscow-Leningrad: 1951.

[20]. Averjanov, S. F., (Aver'yanov), "About permeability of subsurface soils in case of incomplete saturation," Engineering Collection, Vol. vii, (1950); see also "Dependence of permeability of subsurface soils on their air content," D A N [CCCP], Vol. 69, no 2. (1949).

[21]. Gorbunov, B. P., "About permeability of Loess soils with fractured structure," Works of the Institute of Construction, A N, Vol. 2, Tashkent: 1951.

[22]. Darcy, H., Les Fontaines Publiques de la Ville de Dijon, Paris: 1856.

[23]. Dupuit, J., Etudes Théoriques et pratiques sur le Mouvement des Eaux dans les Canaux Découverts et à Travers les Terrains Perméables, 2nd edition, Paris: 1863.

[24]. Kamensky, G. N., Principles of Groundwater Dynamics, 2nd edition, State Geology Press, Moscow: 1943.

[25]. Khristianovich, S. A., "Flow of groundwater, not obeying Darcy's Law," P M M, Vol. IV, no. 1.

[26]. Risenkampf, B. K., Hydraulics of Groundwater, State University of Saratovsky, Series F M N, Vol. 1 (XIV), no. 1, (1938).

[27]. Mikhailov, G. K., "On the geometry of a fictitious soil," P M M, Vol. XVI, no. 4, (1952).

[28]. Gilbert, D., and Confossen, S., Descriptive Geometry, 2nd edition, State Press: 1951.

[29]. Girinsky, N. K., "Some problems in groundwater dynamics," Hydrogeology and Engineering Geology, Collection of articles, no. 9; Moscow-Leningrad: 1941.

[30]. "Definitions of geotechnical characteristics of soil," All- union Scientific Research Institute for Hydromechanics, Gosstroiizdat, Leningrad: 1941.

[31]. Technological Conditions and Norms for Projects of Earth Dams, State Press: 1941.

[32]. Kochin, N. E., Kibel, I. A., Rose, N. V., Theoretical Hydrodynamics, Parts I and II, State Press: 1948.

[33]. Prandtl, L., Hydro – and Aeromechanics, 2nd edition, 1934; Dover Publications, New York.

[34]. Sokolovsky, V. V., "About non-linear seepage of groundwater," P M M, Vol. XIII, no. 5, (1949).

[35]. Rode, A. A., Subsoil Moisture, edited by A N [CCCP]: 1952.

[36]. Grishin, M. M., Hydraulic Structures, Part I, State Press, Moscow: 1947.

CHAPTER II.

[1]. Zhukovsky, N. E., Theoretical Investigations about the Movement of Subterranean Water (1889), Collected Works, Gostekhizdat: 1949.

[2]. Pavlovsky, N. N., "About the flow of water to an horizontal drain," Izv. NIIG, Vol. 21, 1937.

[3]. Vedernikov, V. V., Seepage Theory and its Application in the Field of Irrigation and Drainage, State Press: 1939.

[4]. Risenkampf, B. K., Hydraulics of Groundwater, State University of Saratovsky, Part I, Vol. 1, no. 1, 1938; Part II, Vol. 1, no. 2, 1938.

[5]. Polubarinova-Kochina, P. Ya., "About a continuous change in the velocity hodograph in steady two-dimensional groundwater flow," D A N [CCCP], Vol. 24, no. 4, (1939).

[6]. Polubarinova-Kochina, P. Ya., "Some problems of two-dimensional groundwater flow," edited by A N [CCCP], Moscow-Leningrad: 1942.

[7]. Davison, B. B., "Flow of groundwater," in the book by Khristianovich, S. A., Mikhlin, S. G., Davison, B. B., Some new problems in the mechanics of continuous media.

[8]. Grishin, M. M., Hydraulic Structures, Part I, State Press, Moscow: 1947.

[9]. Zamarin, E. A., Popov, A. V., Gininsky, N. K., Kobek, D. N., Course of Hydraulic Structures, Part I, State Press, Moscow: 1940.

[10]. Melechenko, N. J., Movement of Groundwater under Hydraulic Structures, (Methods of Computation), Moscow-Leningrad: 1937.

[11]. Polubarinova-Kochina, P. Ya., "On the problem of the flow of groundwater in a drained dam," M G U, Scientific Proceedings, no. 39, 1940.

[12]. "Discussions about problems of uplift pressures," Proceedings NIIG, Vol. 41, 1949.

[13]. Danilevsky, V. V., History of Hydropower Plants in Russia up to the 19th Century, Moscow: 1940. "Kozma Dimitrevich Frolov" in the book Russian Men of Science, Vol. 2, State Press.

[14]. Smirnov, V. I., Course of Advanced Mathematics, Vol. III, Part II, State Press: 1951.

[15]. Forchheimer, F., Hydraulics, Leipzig: 1933.

CHAPTER III

[1]. Pavlovsky, N. N., Theory of Groundwater Flow under Hydraulic Structures and its Basic Applications, Petrograd: 1922.

[2]. Zamarin, E. A., Computations of Groundwater Flow, Tashkent: 1928.

[3]. Zamarin, E. A., Shipenko, P. I., Homographic Computations of Hydraulic Structures, Moscow: 1934.

[4]. Kozlov, V. S., Hydromechanic Design of a Dam Foundation, State Press for Energy, Moscow-Leningrad: 1941.

[5]. Kozlov, V. S. "Hydromechanic design of dams with cut-offs," Izv.A N, Division of Technological Sciences, no. 6, 1939.

[6]. Aravin, V. I., and Numerov, S. N., Seepage Computations for Hydraulic Structures, State Press, Moscow: 1948.

[7]. Bazanov, M. I., "About seepage under overflow weirs in two-dimensional conditions," Izv. NIIG, Vol. 28, 1940.

[8]. Melechenko, N. T., "Approximate method to compute seepage under structures resting on pervious strata, unlimited in depth," Izv. NIIG, Vol. 28, 1940.

[9]. Filchakov, P. F., "Mechanic principles of the hydromechanical design of dams," Dissertation, Institute of Mathematics, A N YCCP, 1951.

[10]. Filchakov, P. F., Approximate Method of Hydromechanical design of Dams, edited, A N YCCP: 1951.

[11]. Harza, L. F., "Uplift and seepage under dams on sand," Proceedings Am. Soc. of Civ. Engrs., Vol. 60, no. 7, 1934.

[12]. Hoffman, R., "Grundwasser strömung unter Wehren," Wasserwirtschaft, no. 18-19, 20-21, 1934.

[13]. Muskat, M., Flow of Homogeneous Fluids in Porous Medium, McGraw Hill, New York, 1937.

[14]. Girinsky, N. K., "Basic theory of groundwater flow under hydraulic structures," Hydraulic Construction (Journal of), no. 6, 1936.

[15]. Verigin, N. N., "Seepage under dam foundations with inclined screens and cut-offs," Journal of Hydraulic Construction, 1940, no. 2.

[16]. Melechenko, N. T., "Computation of groundwater flow under hydraulic structures having drainage pipes," Izv. NIIG, Vol. XIX, 1936.

[17]. Mkhitarian, A. M., "On the computations of seepage through an earth-dam," Inzh. Zh., Vol. XIV, 1952.

[18]. Davison, B. B., "On a simplified method of computation of groundwater flow," Works of the State Institute of Hydraulics, no. 5, 1937.

[19]. Melechenko, N. T. Groundwater Flow under Hydraulic Structures, (Method of Computation), Moscow:1937.

[20]. Grishin, M. M., Hydraulic Structures, Part I, State Press: 1947.

[21]. "Proceedings of the All-union Scientific Research Institute for Hydromechanics," Vol. 41, 1949. Works of Reltov, B.F., Chugaev, R. R., Byazemsky, O. V., about water pressure in soils.

[22]. Lavrentiev, M. A., Conformal Mapping, Gostekhizdat, Moscow: 1946.

[23]. Lavrentiev, M. A., and Shabat, B. V., Methods in the Theory of Functions of a Complex Variable, Gostekhizdat, Moscow: 1951.

[24]. Zhuravsky, A. M., Reference Book on Elliptic Functions, edited, A N [CCCP], Moscow-Leningrad: 1941.

[25]. Sikorsky, Ju. S., Elements of the Theory of Elliptic Functions with Application to Mechanics, Moscow-Leningrad: 1936.

[26]. Privalov, I. I., Introduction to the Theory of Functions of a Complex Variable, Gostekhizdat: 1948.

[27]. Polozhy (Polozjü), G. N., Ukranian Mathematical Journal, Vol. 6, no. 1, 1952.

[28]. Kochin, N. E., Kibel, I. A., Roze, N. V., Theoretical Hydromechanics, Part I, Moscow: 1948.

[29]. Pavlovsky, N. N., Hydromechanic Computation of Dams of the Senkov Type, State Press: 1937.

CHAPTER IV.

[1]. Zhukovsky, N. E., "Seepage through dams (1920)," Collected Works, Vol. VII, Gostekhizdat: 1950.

[2]. Pavlovsky, N. N., "Free surface seepage to infinity from open channel with curvilinear perimeter," Izv. NIIG, Vol. XIV, 1936.

[3]. Pavlovsky, N. N., "Basic methods of the hydromechanical solution of the problem of free surface seepage from open channels," Izv. NIIG, Vol. XIX, 1936.

[4]. Vedernikov, V. V., Seepage Theory and its Applications in the Fields of Irrigation and Drainage, State Press: 1939.

[5]. Risenkampf, B. K. "Hydraulics of groundwater," Proceedings, State University of Saratovsky, Vol. XV, no. 5, 1940.

[6]. Aravin, V. I., "Groundwater flow to a collector," Izv. NIIG, Vol. 18, 1936.

[7]. Numerov, S. N., "About a method to solve seepage problems when infiltration or evaporation from the free surface takes place," Izv. NIIG, Vol. 38, 1948.

[8]. Votshinin, A. P., "Seepage in soil and in homogeneous earthdams with horizontal filter on a pervious stratum of finite depth," D A N [CCCP], Vol. 25, no. 9, (1939).

[9]. Mkhitarian, A. M. "On the computation of seepage through an earthdam with cut-off and drain," Inzh. sb., Vol. XV.

[10]. Nelson-Skornyakov, F. B., Seepage in Homogeneous Medium, 2nd edition, . 1949, Moscow: 1947.

[11]. Bazanov, M. I., "Investigations of seepage in the case of flow of water to a drainage canal," P M M, Vol. 2, no. 2, 1938.

[12]. Vedernikov, V. V., "Seepage through an earthdam on pervious stratum," D A N [CCCP], Vol. 50, (1945). Hydrotechnical Construction (Journal of), no. 1, 1947.

[13]. Gersevanov, N. M., "Application of functional analysis in the solution of seepage problems," Izv. A N [CCCP], Division of Technological Sciences, no. 7, 1943. Book: Calculus of Iterations and its Applications, State Press, Moscow: 1950.

[14]. Aravin V. I., "Flow of groundwater to a foundation area bounded by cut-offs," Izv. NIIG, Vol XX, 1937.

[15]. Aptekar, L. D., "Questions of seepage computations for horizontal drains of navigable locks and dry docks, Izv. NIIG, Vol. 44, and 46 (1951).

[16]. Polubarinova-Kochina, P. Ya., and Falkovich, S. V., "Theory of seepage in porous media," P M M, Vol. XI, no. 6, 1947. Also: Advances in Applied Mechanics, Part II, pp. 153-225, edited, R. van Mises and Th. von Karman, Acad. Press, New York, 1951.

[17]. Numerov, S. N., "About seepage from canals and irrigation systems," Izv. NIIG, Vol. 34 (1947).

[18]. Nelson-Skornyakov, F. B., "Hydromechanical solution for a dam with cut-off on a pervious stratum of finite thickness," Izv. A N [CCCP], Division of Technological Sciences, no. 9-10, 1943.

[19]. Aravin V. I., "On seepage from reservoirs," Izv. NIIG, Vol. 16, 1935.

CHAPTER V.

[1]. Zhukovsky, N. E., "Seepage through dams," (1923), Collected Works, Vol. VII, 1949.

[2]. Vedernikov, V. V., "Hydromechanical methods to compute free surface seepage flows," Scientific notes of the Hydrometric Institute of of Moscow, vol. IV, 1937.

[3]. Vedernikov, V. V., Seepage Theory and its Applications in the field of Irrigation and Drainage, State Press: 1939.

[4]. Sokolov, Ju. D., "On seepage to a drainage ditch with trapezoidal cross section," P M M, Vol. XV, no. 6, (1951).

[5]. Sokolov, Ju. D., "On the computation of seepage from a canal with trapezoidal cross section," D A N [CCCP], Vol. 79, no. 5, 1951.

[6]. Risenkampf, B. K., "Hydraulics of groundwater," 3rd part, Proceedings, State University Saratovsky, Vol. XV, no. 25, 1940.

[7]. Vedernikov, V. V., "Account of soil capillarity on seepage from a canal," D A N [CCCP], Vol. 28, no. 5, 1940.

[8]. Averjanov, S. F., "Approximate appraisal of the role of seepage in the zone of capillary fringe," D A N [CCCP], Vol. 69, no. 3, 1949.

[9]. Verigin, N. N., "Seepage from drainage and irrigation systems," D A N, Vol. 66, no. 4, 1949.

[10]. Vedernikov, V. V., "On the theory of drainage," D A N, Vol. 59, no. 6, 1948.

[11]. Vedernikov, V. V., "Seepage in case of draining or waterbearing stratum," D A N, Vol. 69, no. 5, 1949.

[12]. Numerov, S. N., "Approximate methods of computations of seepage through earth dams on pervious strata," Works of the Leningrad Polytechnic Institute, named after M. I. Kalinin, no. 4, 1947.

[13]. Khomovskaya, E. D., "Hydromechanical solution of the problem of groundwater flow to a collector pipe in a highly pervious stratum," Works of the State Hydrologic Institute, Chapter 5, 1937.

[14]. Votshinin, A. P., "On the application of streamlined and cut-off subterranean contours in the construction of hydraulic structures on pervious strata," Inzh. sb. Vol. VII, 1950.

[15]. Kochina, I. N., and Polubarinova-Kochina, P. Ya., "On the application of smooth contours as bases for hydraulic structures," P M M, Vol. XVI, no. 1, 1952.

CHAPTER VI.

[1]. Numerov, S. N., "On seepage in earthdams with drain on impervious base," Izv. NIIG, Vol. 25, 1939.

[2]. Numerov, S. N., "On the influence of infiltration or evaporation from the free surface in seepage calculations," P M M, Vol. IV, nos. 5-6, 1940.

[3]. Numerov, S. N., "On seepage in earthdams with drain on permeable base," Izv. NIIG, Vol. 27, 1940.

[4]. Numerov, S. N., "On the problem of seepage through earthdams," Izv. NIIG, Vol. 28, 1940.

[5]. Numerov, S. N., "Hydromechanical method to compute bank drainage," Izv. NIIG, Vol. 28, 1940.

[6]. Numerov, S. N., "Practical method to compute seepage through earthdams with drain on impervious bases." Izv. NIIG, Vol. 28, 1940.

[7]. Numerov, S. N., "Solution of the seepage problem without seepage surface and infiltration or evaporation from the free surface," P M M, Vol. VI, bulletin 1, 1942.

[8]. Numerov, S. N., "Practical method of computation of seepage through earthdams with drain on impervious stratum when tailwater exists," Izv. NIIG, Vol. 31, 1946.

[9]. Numerov, S. N., "Seepage computations of horizontal drain of hydropower plants, for permeable foundation of finite thickness," Izv. NIIG, Vol. 34, (1947).

[10]. Numerov, S. N., "On seepage towards horizontal drain above inclined bedrock," Izv. NIIG, Vol. 46, 1951.

[11]. Numerov, S. N., "On seepage to a ditch with rectilinear cross-section," Izv. NIIG, Vol. 46, 1951.

[12]. Aravin V. I., and Numerov, S. N., Seepage Computations for Hydraulic Structures, State Press, Moscow, 1948.

[13]. Shankin, P. A., Seepage Computations for Earthdams, Moscow-Leningrad, 1947.

[14]. Verigin, N. N., "Drainage computations taking into account the existence of a seepage surface," D A N, Vol. LXX, no. 4, 1950.

[15]. Smirnov, V. I., Course of Higher Mathematics, Vol. III, part 2, Moscow-Leningrad: 1951.

[16]. Muskhelishvili, N. I., Singular Integral Equations, Gostekhizdat, Moscow-Leningrad: 1946.

[17]. Sedov, L. I., Two-dimensional Problems of Hydrodynamics and Aerodynamics, Gostekhizdat: 1950

[18]. Polozhy, G. N., Ukranian Mathematical Journal, Vol. 6, no. 1, 1952.

CHAPTER VII.

[1]. Polubarinova-Kochina, P. Ya., "Application of the theory of linear differential equations to some problems of groundwater flow," Izv. A N [CCCP], math. series., no. 3, 1938.

[2]. Polubarinova-Kochina, P. Ya., "Application of the theory of linear differential equations to some problems of groundwater flow (case of three singular points), Izv. A N, no. 3, 1939.

[3]. Polubarinova-Kochina, P. Ya., "Same title as in 1 and 2 above, Case where the number of singular points exceeds three," Izv. A N, no. 5-6, 1939.

[4]. Polubarinova-Kochina, P. Ya., "Example of groundwater flow through an earthdam when evaporation takes place," Izv. A N [CCCP], Division of Technological Sciences, no. 7, 1939.

[5]. Polubarinova-Kochina, P. Ya., "Computations of seepage through an earthdam," P M M, Vol. IV, no. 1, 1940.

[6]. Polubarinova-Kochina, P. Ya., "Some problems of two-dimensional groundwater flow," edited by A N [CCCP], 1942.

[7]. Polubarinova-Kochina, P. Ya., and Falkovich, S. V., Theory of Seepage in Porous Media, P M M, Vol. XI, no. 6, 1947. Also: Advances in Applied Mechanics, Part II, pp. 153-225, ed. R. von Mises and Th. von Karman, Acad. Press, New York, 1951.

[8]. Risenkampf, B. K. "Hydraulics of groundwater flow, Part 3, Chapter IV, Proceedings, State University of Saratovsky, Vol. XV, 1940, no. 5.

[9]. Kalinin, N. K. "On the method by P. Ya. Polubarinova-Kochina to solve seepage problems," D A N, Vol. 45, no. 3, 1944.

[10]. Mikhailov, G. K., "On the problem of seepage in anisotropic earthdams with trapezoidal profile on horizontal bedrock," D A N, Vol. 80, no. 4, 1951.

[11]. Mikhailov, G. K., "On seepage in trapezoidal dams on horizontal bedrock," Hydromechanics and Reclamation, no. 1, 1952.

[12]. Charny, I. A., "A rigorous derivation of Dupuit's formula for unconfined seepage with seepage surface," D A N, Vol. 79, no. 6, 1951.

[13]. Davison, B. B., "On unsteady groundwater flow through earthdams," Proceedings, State Hydrologic Institute, Vol. VI, 1932. See also:

Works of the Hydrologic Conference of the Baltic States, 1933.

[14]. Hamel, G., Ueber Grundwasserströmung," Z A M M, Vol. 14, no. 3, 1936.

[15]. Proskurnikov, S. M., "New computational data on seepage through earth dams," Works of the State Hydrologic Institute, no. 5, 1937.

[16]. Melechenko, N. T., "About computations of seepage through earthdams by the method of Prof. N. N. Pavlovsky, Journal of Hydraulic Construction, no. 2-3, 1932.

[17]. Pavlovsky, N. N., On Seepage through Earthdams, Leningrad: 1931.

[18]. Mkhitarian, A. M., "On computation of seepage through an earthdam," Inzh. sb. Vol. 14.

[19]. Nelson-Skornyakov, F. B., Seepage in Homogeneous Medium, 2nd edition, Moscow:1949.

[20]. Smirnov, V. I., Course of Higher Mathematics, Vol. III, Gostekhizdat: 1951.

[22]. Whittaker and Watson, A Course of Modern Analysis, 1947, The Macmillan Co., New York.

[23]. Golubev, V. V., Lectures on the Analytic Theory of Differential Equations, 2nd edition, Gostekhizdat: 1950.

[24]. Goursat, E., Propriétés genérales des équations d'Euler et de Gauss, Paris: 1936.

[25]. Gauss, K., "Allgemeine Untersuchungen über die unendliche Reihe $1 + \frac{\alpha\beta}{1.\gamma} + \dots$," Berlin: 1888.

[26]. Gakhov, F. D., "Regional problems of analytic functions and singular integral equations," Proceedings, Phys. Math, State Univ. Kazan, Vol. XIV, series 3, 1949.

CHAPTER VIII.

[1]. Polubarinova-Kochina, P. Ya., "Simplest case of groundwater flow in two strata with different seepage coefficient," Izv. A N, Div. Techn. Sci. no.6, 1939.

[2]. Polubarinova-Kochina, P. Ya., "Some two-dimensional seepage problems," ed. A N [CCCP], Chapters II, VI, VIII, May, 1942.

[3]. Risenkampf, B. K., "On a case of seepage in multilayered soil," Proceedings, State Univ. of Saratovsky, Vol. XV, no. 5, 1940.

[4]. Kalinin, N. K., "Some approximate methods to solve seepage problems in two.layered medium," D A N, Vol. 30, no. 7, 1941.

[5]. Kalinin, N. K., "Reduction of some seepage problems in a medium with two strata to seepage problems in homogeneous soil," D A N, Vol. 10, no. 5, 1943.

[6]. Kalinin, N. K. "Seepage through a wedge composed of two strata," P M M, Vol. XVI, no. 2, 1952.

[7]. Aravin, V. I., "On the problem of seepage in anisotropic pervious soils," Works of the Industrial University, Leningrad, no. 9, 1937

[8]. Polubarinova-Kochina, P. Ya., "On seepage in anisotropic soil," P M M, Vol. IV, no. 2, 1940.

[9]. Kozlov, V. S., "On the problem of computing groundwater flow in anisotropic pervious soils," Izv. A N, Div. Techn. Sci. no. 3, 1940.

[10]. Poluvarinova-Kochina, P. Ya., "About unsteady groundwater flow in two strata of different compactness," Izv. A N, Div. Techn. Sci. no. 6, 1940.

[11]. Braginskaya, V. A., "Some seepage problems in anisotropic soil," P M M, Vol. VI, nos. 2-3, 1942.

[12]. Pavlov, A. T., "Steady groundwater flow for two layers of fluids of different densities," P M M, Vol. VI, no. 2-3, 1942.

[13]. Girinsky, N. K., "Some problems of dynamics of underground waters," Hydrology and Engineering Geology, (Journal of), no. 9, 1947.

[14]. Girinsky, N. K., "Complex potential of flow of fresh water in contact with brackish water," D A N, Vol. 58, no. 4, 1947.

[15]. Silin-Bekchurin, A. I., "On the influence of the kinematic density, reduced pressure and rock permeability on the seepage velocity of brines in oilbearing strata of the Ural-Volga region," D A N, [CCCP], Vol. 58, no. 6, 1947.

[16]. Silin-Bekchurin, A. I., "Method of approximate calculation of seepage velocity and subterranean flow of brines," Works of the State Hydrogeologic Laboratory, Vol. II, 1949, Moscow.

[17]. Patrashev, A. N., and Arutyunyan, N. Kh., "Diffusion of salts in homogeneous seepage," Izv. NIIG, Vol. 30, 1941.

[18]. Numerov, S. N., and Patrashev, A. N., "Diffusion of dissoluble matter in bases of hydraulic structures," Works of the Polytechnic Institute, (named after M. I. Kalinin), 1947, no. 4, Leningrad.

[19]. Kamensky, G. N., Korchebokov, N. A., Razin, K. I., Flow of Subterranean water in Heterogeneous Strata, State Press, 1935.

[20]. Frank, Ph., and R. von Mises, Differential and Integral Equations of Mathematical Physics, Vieweg-Braunschweig: 1935.

[21]. Kochin, N. E., Kibel, I. A., Rose, N.V., Theoretical Hydromechanics, Vol. II, Gostekhizdat: 1948.

CHAPTER IX.

[1]. Tschelkachev, V. N., and Pikhachev, G. B., Interference of Wells and Theory of Confined Systems, State Press: 1939.

[2]. Polubarinova-Kochina, P. Ya., "On the direct and inverse problem of hydraulics in oilbearing strata," P M M, Vol. VII, 1943, no. 5.

[3]. Tschelkachev, V. N., Principles of Subterranean Hydraulics Applied to Oil, State Press: 1945.

[4]. Polubarinova-Kochina, P. Ya., "On the flow to a well in heterogeneous medium," D A N, Vol. 34, no. 2, 1942.

[5]. Vedernikov, V. V., Seepage Theory and its Applications in the Field of Irrigation and Drainage, State Press, Moscow: 1939.

[6]. Verigin, N. N., "On the problem of the computation of subterranean water divides in conditions of two-dimensional groundwater flow," D A N, Vol. LXIV, no. 2. 1949.

[7]. Mjatiev, A. N., "Effect of a well on a confined groundwater basin," Turkmenian Journal of the A N [CCCP], No. 3-4, 1946.

[8]. Girinsky, N. K., "Some problems in the dynamics of groundwater," Journal of Hydrogeology and Engineering Geology, no.9, 1947, Moscow.

[9]. Mjatiev, A. N., "Pressure complex of underground water and wells," Izv. A N, [CCCP] Div. Techn. Sci. no.9, 1947.

[10]. Mjatiev, A. N., "Problem about wells in a stratum of groundwater," Izv. A N, [CCCP], Div. Techn. Sci. no.3, 1948.

[11]. Silin-Bekchurin, A. I., Special Hydrogeology, State Press, Moscow: 1951.

[12]. Risenkampf, B. K., and Kalinin, N. K., "Flow of groundwater in three-dimensions with ellipsoidal free surface," P M M, Vol. V, 1941, no.2.

[13]. Girinsky, N. K., Determination of the Seepage Coefficient, State Geology Press: 1950.

[14]. Muskat, M., Flow of Homogeneous Fluids in Porus Medium, McGraw-Hill New York, 1937.

[15]. Povalo-Shveikovsky, N. T. "Determination of the seepage coefficient by means of pumping tests," Works of the Central Institute of Experimental Hydrology and Meteorology, bulletin 11(44), Moscow: 1935.

[16]. Charny, I. A., Subterranean Hydromechanics, Gostekhizdat: 1948.

[17]. Segal, B. I., "Some three-dimensional problems in potential theory and their applications," Izv. AN [CCCP], Math. Series, Vol. 10, no. 4, 1946.

[18]. Ivakin, V. V., "On seepage during recharge of water in unsaturated soil," Izv. A N [CCCP], Div. Tech. Sci. no. 6, 1947.

[19]. Nasberg, V. M., "On recharge of water in unsaturated soil," Izv. A N [CCCP], Div. Tech. Sci. no.9, 1951.

[20]. Polubarinova-Kochina, P. Ya., "About sources and sinks at the free surface," P M M, Vol. XIV, bulletin 1, 1950.

[21]. Golubeva, O. V., "Equations of two-dimensional flow of an ideal fluid with curvilinear surface and their application in seepage theory," P M M, Vol. XIV, bulletin 3, 1950.

[22]. Poluvarinova-Kochina, P. Ya., "On the hydraulic theory of wells in multilayered medium," P M M, Vol. XI, bulletin 3, 1947.

[23]. Matveenko, T. J., "Problem of seepage to a well from one and two strata," Inzh. sb. Vol. XIV.

CHAPTER X.

[1]. Dupuit, J., Etudes Théoriques et Pratiques sur le Mouvement des Eaux, 2nd edition, Paris: 1863.

[2]. Pavlovsky, N. N., "Non-uniform flow of groundwater," Leningrad: 1930. "Non-uniform flow of groundwater. Further development of the problem," Leningrad: 1932.

[3]. Aravin, V. I., "Approximate methods for computation of seepage around dam foundations in three-dimensional flow," Works of the Leningrad Polytechnic Institute, no.4, 1947.

[4]. Aravin, V. I., "Flow of groundwater to foundation pits, resting on horizontal impervious strata," Izv. NIIG, Vol. XXI, 1937.

[5]. Aravin, V. I., "Computation of seepage about hydraulic structures," Izv. NIIG, Vol. 27, 1940.

[6]. Girinsky, N. K., "Complex potential of flow with free surface in a stratum of relatively small thickness and $k = f(z)$," D A N, Vol. 51, no.5, pp. 337-338, 1946.

[7]. Girinsky, N. K., "Some problems in the dynamics of groundwater," Hydrogeology and Engineering Geology, (Journal of), no.9, 1947.

[8]. Kamensky, G. N., Korchelokov, N. A., Razin, K. I., Flow of Groundwater in Heterogeneous Strata, State Press, 1935.

[9]. Aravin, V. I., and Numerov, S. N., Seepage Computations for Hydraulic Structures, State Press, Moscow: 1948.

[10]. Nedriga, V. P., "Computation of seepage about hydraulic structures," Journal of Hydraulic Construction, no.5, 1947.

[11]. Verigin, N. N., "Seepage about dams and efficiency of cut-offs against uplift pressures," Journal of Hydraulic Construction, no.5, 1947.

[12]. Polubarinova-Kochina, P. Ya., "About seepage in two strata with inclined boundary," Izv. A N [CCCP], Div. Tech. Sci. no.1, 1952.

CHAPTER XI.

[1]. Zamarin, E. A., "Hydrodynamic flownets," Scientific Notes of Math. and Hydromechanic Institute, M G U, bulletin IV, 1937.

[2]. Girinsky, N. K., "Graphical Construction of hydrodynamic flownets," Scientific Notes, M G U, Vol. VII, bulletin 11, 1939.

[3]. Girinsky, N. K., "Basic theory of groundwater flow under hydraulic structures in heterogeneous soils," Scientific Notes, Math. and Hydromech. Inst., M G U, bulletin 5, 1938.

[4]. Panov, D. Ju., Reference book on Numerical Integration of Partial Differential Equations, 5th edition, Gostekhizdat, 1951.

[5]. Proskura, G. F., Hydrodynamics of Turbomachines, State Press, 1934.

[6]. Nikolaeva, M. V., "On Southwell's relaxation method," Works of the Mathematical Institute, named after V. A. Steklov, AN [CCCP], Vol. XXVIII, 1949.

[7]. Kozlov, V. S., Hydromechanic Computation of Damfoundations, State Press for Energy, Moscow-Leningrad: 1941.

[8]. Ayvazyan, V., "Graphical method to compute earth dams," Hydraulic Construction, Journal of, no.7, 1933.

[9]. Melentev, P. V., "Method to compute blades of hydroturbomachines," edited by, A N [CCCP], Moscow: 1939.

[10]. Nasberg,V. M., "Hydrodynamic flownet for seepage from a well in a stratum of large thickness and application of the net to determine soil permeability," Proceedings of the Institute of Hydraulic Energy at Tbilisk, Vol. III, 1950.

[11]. Korsuntsev, V. A., "Auxiliary nomograph to construct electric fields with axial symmetry," Elektrichestvo, no.14, 1937.

[12]. Southwell, R. V., "Relaxation methods in theoretical physics," Oxford: 1946.

[13]. Weinig, F., and Shields A., "Graphisches Verfahren zur Ermittlung der Sickerströmung durch Staudämme," Wasserkraft und Wasserwirtschaft, 1936, no. 18.

[14]. Breitenöder, M., "Ebene Grundwasserströmungen mit freier Oberfläche," Berlin: 1942.

[15]. Barillon, E. G., "Note sur les rayons de courbure intervenant dans la construction des réseaux hydrodynamiques," Revue générale de l'Hydraulique, no. 8, 1936.

[16]. Averjanov, S. F., "On the study of steady groundwater flow by means of flownets," D A N, bulletin 4, 1949.

[17]. Khristianovich, S. A., "Flow of groundwater, not obeying Darcy's law," P M M, Vol. IV, 1940, series 1.

[18]. Reltov, B. F., "Investigation of seepage in three-dimensional conditions by the electric analog method devised by academician, N. N. Pavlovsky," Izv. NIIG, Vol. XV, 1935.

[19]. Aravin, V. I., and Druzhinin, N. I., "Some questions about the experimental method of investigating three-dimensional seepage problems by electric analog," Izv. NIIG, Vol. 40, 1949.

[20]. Tolstov, Ju. G., "Application of electric models of physical phenomena to the solution of some problems of groundwater hydraulics," Journal of Technological physics, Vol. XII, no.10, 1942.

[21]. Hutenmacher, L. I., Electric Models, edited by A N [CCCP], 1949.

[22]. Filchakov, P. F., Mechanical Principles of Computations of Dams, Kiev, 1952.

[23]. Proskurnikov, S. M., "Application of electric analogs in the computation of seepage through earth embankments and dams," Works of the State Hydrologic Institute, bulletin 8, (62), 1948; see also bulletin 5, 1937.

[24]. Filchakov, P. F., "Electric model for seepage problem in heterogeneous soil," D A N [CCCP], Vol. 66, no.4, 1949.

[25]. Druzhinin, N. I., "About the accuracy of determining the flowrate in the investigation of three-dimensional seepage by the electric analog method," Izv. NIIG, Vol. 44 (1951), [see also Izv. NIIG, Vol. 37, 1948, Vol. 38, 1948, Vol. 40, 1949, Vol. 42, 1950.

[26]. Aravin, V. I. "Basic problems in the experimental investigation of the flow of groundwater by means of a parallel plate model," Izv. NIIG, Vol. 23, 1938.

[27]. Prandtl, L., Hydro-and Aeromechanics, Vieweg und Son-Brunswick, 1931; see also: Essentials of Fluid Mechanics, Hafner, New York, 1952.

[28]. Shankin, P. A., Seepage Investigations in the Construction of Hydraulic Structures, State Press, Moscow: 1947.

[29]. Silin-Bekchurin, A. I., Special Hydrogeology, State Geology Press, Moscow: 1951.

[30]. Girinsky, N. K., Determination of the Seepage Coefficient, State Geology Press, 1950.

[31]. Kochin, N. E., Kibel, I. A., Rose, N. V., Theoretical Hydromechanics, Vol. I, and II, Gostekhizdat: 1948.

[32]. Lukyanov, A. V., "On electrolytic models of three-dimensional problems," D A N, Vol. LXXV, no. 5, 1950.

CHAPTER XII.

[1]. Khristianovich, S. A., Mikhlin, S. G., Davison, B. B., "Some new problems in the mechanics of continuous media," Part 3, Groundwater Flow, edited by A N [CCCP], 1938.

[2]. Numerov, S. N., "Application of the method of phragments in the computation of two-dimensional unsteady seepage," Works of the Polytechnic Institute (named after M. I. Kalinin ), 1948, no.5.

[3]. Aravin, V. I., "On the influence of waviness on seepage under hydraulic structures," Izv. NIIG, Vol. 28 (1940).

[4]. Protodyakonov, M. M., Theory of Surface Water Flow, State Press: 1932.

[5]. Kochin, N. E., Kibel, I. A., Rose, N. V., Theoretical Hydromechanics, Vol. I, Ch. VIII, Gostekhizdat: 1948.

CHAPTER XIII.

[1]. Boussinesq, J., "Recherches théoriques sur l'écoulement des nappes d'eau infiltreés dans le sol," Journal de Math. Pures et Appl., Series 5, Tome X, fascicule 1, 1904.

[2]. Tikhonov, A. I. and Samarsky, A. A., Equations of Mathematical Physics, Gostekhizdat: 1951.

[3]. Boltzmann, Annalen der Physik, 53, 959, 1894.

[4]. Leibenzon, L. S., "Flow of gaseous fluids in porous medium," Izv. A N [CCCP], Geophysical Series, 1941, no.4-5.

[5]. Polubarinova-Kochina, P. Ya., "On unsteady flow of groundwater seeping from reservoirs," P M M, Vol. XIII, no.2, 1949.

[6]. Polubarinova-Kochina, P. Ya., "On a non-linear partial differential equation, occurring in seepage theory," D A N, Vol. 36, no.6, 1948.

[7]. Polubarinova-Kochina, P. Ya., "On the displacement of the groundwater tongue in seepage from a canal," D A N, Vol. 82, no.6, 1952.

[8]. Mikhailov, G. K., "Application of a model of anisotropic soil to estimate the solution of some regional problems of groundwater flow," Inzh. sb., Vol. 16.

[9]. Semchinova, M. M., "Conparison of experimental data with theory for the case of unsteady seepage from a canal, resting on horizontal bedrock," Inzh. sb., Vol. 15.

[10]. Barenblatt, G. I., "On some unsteady flows of fluids and gases in porous medium," P M M, Vol. XVI, no.1, 1952.

[11]. Florin, V. A., "Infiltration of soil and seepage when porosity varies, considering the phenomenon of bound water," Izv. A N, Div. Tech. Sci., no.11, 1951.

[12]. Numerov, S. N., "Slowly varying unsteady seepage," Izv. NIIG, Vol. 37, 1948.

[13]. Verigin, N. N., "On unsteady flow of groundwater near reservoirs," D A N, Vol. 66, no.6, 1949.

[14]. Verigin, N. N., "On seepage from canals into dry soil," D A N, Vol. 79, no.4, 1951.

[15]. Kochin, N. E., Kibel, I. A., Roze, N. N., Theoretical Hydrodynamics, Vol. II, p. 29, Gostekhizdat: 1948.

CHAPTER XIV.

[1]. Likov, A. V., Heat Conduction of Unsteady Processes, Moscow: 1948.

[2]. Polubarinova-Kochina, P. Ya., "On the dynamics of groundwater under spreading," P M M, Vol. XV, no.6, 1951.

[3]. Krilov, M. M., "On the study of the dynamic equilibrium of groundwater for hydrogeologic forecasting," Izv. A N, no.2, pp. 106-117, 1947.

[4]. Polubarinova-Kochina, P. Ya., "On unsteady groundwater flow in seepage from reservoirs," P M M, Vol. XIII, no.2, 1949.

[5]. Girinsky, N. K., "Some problems of the dynamics of groundwater," Journal of Hydrogeology and Engineering Geology, no.9, 1947.

[6]. Polubarinova-Kochina, P. Ya., "On unsteady seepage with interphase surface," D A N, Vol. 66, no.2, 1949.

[7]. Polubarinova-Kochina, P. Ya., "On unsteady groundwater flow in two layers of different density," Izv. A N, Div. Tech. Sci.,no.6, 1940.

[8]. Polubarinova-Kochina, P. Ya., "Some problems of two-dimensional groundwater flow," edited by A N [CCCP], Moscow: 1942.

[9]. Girinsky, N. K., "Complex potential of fresh groundwater flow in contact with brackish water," D A N, Vol. LVIII, no.4, 1947.

[10]. Silin-Bekchurin, A. I., "Method of approximate calculation of seepage velocity and subterranean flow of brines," Works of the Laboratory of Hydrogeology, A N [CCCP], Vol. II, 1949.

[11]. Kovda, V. A., Origin and Regimen of Saltridden Lands, Chapters I and II, Moscow: 1946.

[12]. Tschelkachev, V. N., "Some applications of the theory of the elastic regimen of water-oil bearing strata," D A N, Vol. 52, no.5, 1946; see also D A N, Vol. 79, no.4, 1951.

[13]. Amosov, S. I., Works of the Industrial Institute of Leningrad, 1937, Bulletin II, no.4, (phys. Math. sciences).

[14]. Sobolev, S. L., Equations of Mathematical Physics, Gostekhizdat, 1947.

[15]. Polubarinova-Kochina, P. Ya., "On the influence of bedrock slope and infiltration on unsteady seepage," D A N, Vol. 75, 1950, no.4,

[16]. Carslaw, H. S., and Jaeger, J. C., Conduction of Heat in Solids, Oxford University Press, London 1947 (2nd edition, 1959).

CHAPTER XV.

[1]. Kochin, K. E., Kibel, I. A., Roze, N. V., Theoretical Hydromechanics, Part II, Gostekhizdat: 1948.

[2]. Polubarinova-Kochina, P. Ya., "On unsteady flows in the theory of seepage," P M M, Vol. IX, no.1, 1945.

[3]. Kalinin, N. K., and Polubarinova-Kochina, P. Ya, "On unsteady groundwater flow with free surface," P M M, Vol. XI, 1957, no.2.

[4]. Galin, L. A., "Some problems of unsteady groundwater flow," P M M, Vol. XV, 1951, no.6.

[5]. Kochina, N. N., "Two-dimensional problem on the spreading of a groundwater mound in a stratum of infinite thickness," P M M, Vol. XV, 1951, no.6.

[6]. Kochin, N. E., "On the theory of the Cauchy-Poisson Law," Collected Works, Vol. II, edited by A N, 1949.

[7]. Kalinin, N. K., "On unsteady seepage in the case of drain in permeable stratum of finite thickness," P M M, Vol. XII, 1948, no.2.

[8]. Leibenzon, L. S., Mechanics of Oil Production, Part II, 1934.

[9]. Polubarinova-Kochina, P. Ya., "On the problem of the displacement of the oil-water interphase," D A N, Vol. 47, 1945, no.4.

[10]. Galin, L. A., "Unsteady seepage with free surface," D A N, Vol. 47, 1945, no.4.

[11]. Kufarev, P. P., and Vinogradov, Ju. P., "On some partial solutions of seepage problems," D A N, Vol. 57, 1947, no.4.

[12]. Kufarev, P. P., "Solution of the problem of the contour of the oil-bearing region in a strip with a series of wells," D A N, Vol. 75, 1950, no.3.

[13]. Kufarev, P. P., "Problem of contour of oil-bearing region for a circle with arbitrary number of wells," D A N, Vol. 75, 1950, no.4.

[14]. Vinogradov, Ju. P., and Kufarev, P. P., "On a seepage problem," P M M, Vol. XII, 1948, bulletin 2.

CHAPTER XVI.

[1]. Lembke, K. E., "Groundwater flow and the theory of water collectors," The Engineer, Journal of the Ministry of Communications, 1886, no.2; Journal of the Ministry of Communications, 1887, no.17-19.

[2]. Makkaveev, V. M., "Particular case of unsteady groundwater flow with free surface (On the question of bogging of floodlands)," Works of the State Hydrologic Institute, no.5, 1937.

[3]. Charny, I. A., Subterranean Hydromechanics, Gostekhizdat: 1948.

[4]. Polubarinova-Kochina, P. Ya., "On some unsteady groundwater laws," Izv. A N [CCCP], Div. Tech. Sci. no.6, 1949.

[5]. Kamensky, G. N., "Equations of unsteady groundwater flow in finite differences and their application to an investigation of the phenomenon of head," Izv. A N, Div. Tech. Sci. no.4, 1940.

[6]. Kharkeevich, Ju. F., "Graphical solution of the partial differential equation of the parabolic type," P M M, Vol. XIV, 1950, no.3.

[7]. Panov, D. Ju., Reference Book for Numerical Integration of Partial Differential Equations, 5th edition, Gostekhizdat, 1951.

[8]. Lukyanov, V. S., "Hydraulic devices for technological computations," Izv. A N [CCCP], Div. Tech. Sci. 1939, no.2.

[9]. Polubarinova-Kochina, P. Ya., "Graphical method to compute unsteady groundwater flow," Inzh. sb., Vol. IX, 1951.

[10]. Konkov, B. S., and Petrov, G. N., Steady State Groundwater Flow in the Conditions Prevailing in the Golodnoi steppe, Tashkent, 1929.

[11]. Kochin, N. E., Kibel, I. A., Roze, N. V., Theoretical Hydrodynamics, Part I, Gostekhizdat: 1948.

INDEX

Lightning Source UK Ltd.
Milton Keynes UK
UKHW022002060123
414950UK00009B/617